INDUSTRIAL ELECTRONICS

James Maas

Prentice Hall
Englewood Cliffs, New Jersey   Columbus, Ohio

Library of Congress Cataloging-in-Publication Data

Maas, James W.
    Industrial electronics / James Maas.
      p. cm.
    Includes index.
    ISBN 0-02-373023-4
    1. Industrial electronics. I. Title.
  TK7815.M15 1995
  629.8'9—dc20
                                      94-34333
                                          CIP

Cover Photo Copyright Superstock, Inc.
Editor: Dave Garza
Production Editor: Rex Davidson
Production Coordination: Spectrum Publisher Services
Cover Designer: Gryphon III
Production Buyer: Deidra M. Schwartz
Illustrations: North Market Street Graphics

This book was set in Times Roman by Bi-Comp, Inc. and was printed and bound by R. R. Donnelley & Sons. The cover was printed by Phoenix Color Corp.

© 1995 by Prentice-Hall, Inc.
A Simon & Schuster Company
Englewood Cliffs, New Jersey 07632

Printed in the United States of America

10 9 8 7 6 5 4 3 2 1

ISBN 0-02-373023-4

Prentice-Hall International (UK) Limited, *London*
Prentice-Hall of Australia Pty. Limited, *Sydney*
Prentice-Hall Canada Inc., *Toronto*
Prentice-Hall Hispanoamericana, S. A., *Mexico*
Prentice-Hall of India Private Limited, *New Delhi*
Prentice-Hall of Japan, Inc., *Tokyo*
Simon & Schuster Asia Pte. Ltd., *Singapore*
Editora Prentice-Hall do Brasil, Ltda., *Rio de Janeiro*

PREFACE

This book is intended to serve as the primary theoretical text within the industrial electronics portion of a 2- or 3-year electronic engineering technology program. It could also be used in a 4-year electrical engineering program, with the understanding that a supplemental text on control theory, written at a higher level of mathematics, would be required. The target audience of this text is students preparing for careers as maintenance technicians and maintenance supervisors, supporting the operation of complex industrial control systems. For this reason, emphasis is placed on devices and circuits that are used widely in various forms of industrial control machinery.

Prerequisites to the use of this text include a fundamental understanding of diodes and a basic knowledge of both bipolar-junction and field-effect transistors. A fundamental knowledge of digital logic gates and logic circuits is also required. While a background in op amps and other linear ICs would be beneficial, it is not an immediate requirement for the use of this text.

The target audience of this book is electronics technicians rather than systems engineers; therefore the mathematics used in developing control concepts and explaining device and circuit operation is limited to linear algebra. Since this text is aimed at the individual who performs component-level troubleshooting and repair (rather than system design), the theory of closed-loop control systems is given only introductory coverage. While acknowledging that a basic understanding of the industrial control modes is essential for the prospective industrial electronics technician, this author believes that the student is best served by a text that provides detailed information on input transducers, signal-conditioning circuitry, output control devices, and microprocessor-based control sources. For each of these topics, several actual industrial circuit examples are presented. Also, for each major subject presented in this text, examples, problems, and solutions, along with end-of-section and end-of-chapter review questions, are amply provided.

Within this text, the internal structure and operation of the microprocessor, microcomputer, microcontroller, and programmable logic controller are examined in detail. Emphasis is placed on applications of these devices as embedded controllers that govern the operation of subsystems within larger industrial machines. Personal computers are also briefly mentioned, since they often serve as host computers in modern industrial control systems, monitoring and governing overall system operation. However, internal structure and operation of personal computers and their peripheral devices are not covered. These topics, along with local area network operation, are assumed to be addressed as part of a comprehensive course in microcomputer technology, which would preferably (but not necessarily) be presented before the industrial electronics course.

Industrial applications of lasers and robots are given only cursory coverage. The text is designed with the assumption that either of these subjects, especially robotics, would become a course of study in its own right. While not addressing robotics in detail, this book does provide a thorough foundation for future study of that subject. Many of the input transducers and position controllers found within mechanical manipulators, including strain gauges, high-resolution rotary shaft encoders, resolvers, stepper motors, and brushless dc servomotors, are covered in detail.

This book begins with a survey of on/off control devices, including mechanical, electromechanical, and solid-state switches. The text continues with op amps. After the internal structure and operation of these devices are briefly investigated, op amp and comparator circuits commonly used within industrial signal conditioning circuits are surveyed.

Input transducers commonly found in industrial control systems are then investigated. Those devices used to sense temperature, position, level, flow rate, pressure, and weight, along with their associated signal-conditioning circuits, are examined in detail. Light sources and light-sensing and amplifying devices are also introduced. Applications of these devices within conveyance and materials handling systems are examined in detail.

Industrial pulse circuits are then introduced. Typical applications of monostable, astable, and bistable multivibrators within industrial timing-control circuits are identified. Integrated circuits commonly used in industrial timing applications, including TTL one-shots, the 555 timer, and TTL counters are also introduced. Advanced forms of industrial pulse circuits, including pulse-width modulators, voltage-controlled oscillators, and phase-locked loops are then briefly surveyed.

Unijunction transistors and thyristors are covered in detail. After the structure and operation of silicon-controlled rectifiers and TRIACs are investigated, common forms of unilateral and bilateral breakover devices are examined. Examples of control circuits utilizing these various forms of thyristor are then analyzed.

Industrial power sources are then surveyed. After single-phase and three-phase power distribution systems are introduced, safety and noise elimination devices used within these systems are identified. Methods by which system control electronics can sense the status of AC branch circuits are also investigated. Linear and switching power supplies are then analyzed, with the advantages of off-of-the-line, high-frequency switching power supplies being stressed.

Motors most commonly used in modern industrial control systems are then surveyed. This study begins with a detailed examination of the structure and operation of single- and three-phase ac motors and conventional dc motors. After the control circuitry for conventional dc motors is introduced, synchros and angular position-sensing transducers are surveyed. Next, the internal structure and operation of the brushless dc servomotor is investigated and the use of a resolver as the feedback device within such a motor is examined in detail. The study ends with a detailed analysis of stepper motors.

After digital logic families are briefly surveyed, applications of Boolean algebra and DeMorgan's theorems in combinatorial logic circuit design are investigated. Applications of sequential logic within industrial control circuits are also presented. Examples of such circuits include two basic forms of stepper motor controller. Digital-to-analog and analog-to-digital conversion methods are then introduced.

The text continues with the study of microprocessor and microcomputer architecture. After the internal structure of a typical 8-bit microprocessor is examined, techniques of programming this device in machine language are introduced. Methods by which the microprocessor interfaces with memory devices and ports are then examined. Next, the internal structure and operation of a typical industrial microcontroller are investigated.

Programming techniques for this device and methods of connecting it to external circuitry are briefly surveyed.

Programmable logic controllers are then introduced, with advantages of these devices over embedded microcomputers being stressed. After the various forms of input and output modules used in conjunction with the programmable logic controller are introduced, programming languages and techniques for this device are surveyed.

Next, the basic modes of closed-loop control are presented. After on/off control is investigated, the various forms of continuous-state control are analyzed. Methods by which microcomputers and programmable logic controllers may exercise sequential on/off control and continuous-state control are briefly presented.

The text concludes with an introduction to automated machine control and industrial robotics. After the concept of the servo loop is presented, the internal structure and operation of an actual microcomputer-based servomotor controller is analyzed in detail. Methods of programming this device are also briefly presented. Finally, the structure and operation of the mechanical manipulator, the most common form of industrial robot, are covered at an introductory level.

The author would like to thank the following for reviewing the manuscript: Richard Ackerman, ITT Technical Institute; Robert Allen, Columbus Technical Institute; Michael Charek, Ph.D., Southeast Missouri State University; James Davis, Muskingum Area Technical College; Leon E. Drouin, Memphis State University; Kenneth B. Ferguson, Midlands Technical College; Gregory Marinakis, State University of New York–Farmingdale; James Marks, IVY Technical Institute; David L. Newton, ITT Technical Institute; Dr. Lee Rosenthal, Fairleigh Dickinson University; Dr. Norman Sprankle, Humboldt State University; and Mark Williams, Memphis State University.

CONTENTS

1 BASIC ON AND OFF CONTROL DEVICES

CHAPTER OUTLINE

LEARNING OBJECTIVES

On completion of this chapter, the student should be able to:

- Identify the various forms of mechanical switches.
- Describe the structure and operation of a microswitch.
- Describe the structure and operation of an electromechanical relay.
- Describe the structure and function of a solenoid.
- Describe the function of a bipolar junction transistor as a switch.
- Describe the function of a JFET as a switch.
- Describe the function of a MOSFET as a switch.
- List the advantages of solid-state devices over mechanical and electromechanical switches.
- Explain the Hall effect and describe the operation of a Hall effect switch.

INTRODUCTION

In the first portion of this chapter, the various forms of mechanical and electromechanical switching devices are surveyed. The study begins with the simplest forms of on/off switches and extends through the operation of relays and solenoids.

Next, simple electronic switching devices are introduced, beginning with bipolar junction transistors. JFET and MOSFET switches are also surveyed. The study concludes with an introduction to Hall effect devices, which are used prevalently as proximity sensors.

During this chapter, comparisons are made between electromechanical and solid-state switches, with the advantages of the latter being stressed. Simple switching systems are introduced, incorporating both forms of devices.

1-1   MECHANICAL SWITCHES

The operation of mechanical switches is often neglected in the general study of electronics. However, in the study of industrial electronics, the operation of mechanical switches must be covered, even though many of the tasks they once performed are now assigned to solid-state devices. Those mechanical switches that serve as actuators within electrical control systems require routine adjustment and eventual replacement. Often, what may appear to be an electrical problem within a system may actually be the result of a faulty switch. For this reason, the technician should not only know the electronic operation of a given system, but must also be thoroughly familiar with the mechanical switching methods being utilized.

The simplest form of mechanical switch is the **single-pole, single-throw** (SPST) switch. This type of switch is often placed in series with a load and is used to make or break the path for current from a power source. The schematic symbols for this type of switch are shown in Figure 1–1. Figure 1–1(a) represents a switch that would remain open until being either manually or mechanically actuated. For this type of switch, actuation would result in closure, thus allowing the flow of current to a load. (Note that the **normally open** condition is abbreviated as NO.) The schematic symbol for a **normally closed** SPST switch is shown in Figure 1–1(b). (Note that the normally closed state is indicated as NC.) For this type of switch, the contacts would remain engaged until actuation, which would result in an open condition. Thus, actuation of this form of switch would interrupt the flow of current to a load. Figure 1–1(c) represents a SPST switch with a spring return. For this device, since it is normally open, the contacts would be engaged only when force is applied to the wiper. This type of switching action is referred to as **momentary.**

Figure 1–2 contains two simple circuit applications of the SPST switch. In Figure 1–2(a), power will be transferred from the dc source to the load until the actuation occurs. The flow of load current is interrupted only while force is being applied to the wiper. The circuit in Figure 1–2(b) performs a complementary action to that in Figure 1–2(a). Here, power is transferred to the load only while actuating force is being applied to the wiper of the switch.

A single-pole, double-throw switch (SPDT) is represented in Figure 1–3(a). This device has a common terminal, to which the wiper is attached, and two additional terminals designated as NO and NC. With this form of switch, prior to actuation, the wiper rests against the contact at the NC terminal. When actuation occurs, the wiper pivots, breaking contact with the NC terminal and making contact with the NO terminal. Such a switching

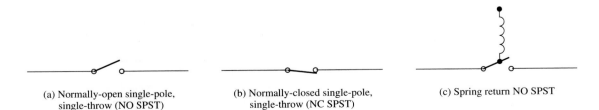

(a) Normally-open single-pole,
single-throw (NO SPST)

(b) Normally-closed single-pole,
single-throw (NC SPST)

(c) Spring return NO SPST

Figure 1–1
Schematic symbols for SPST switches

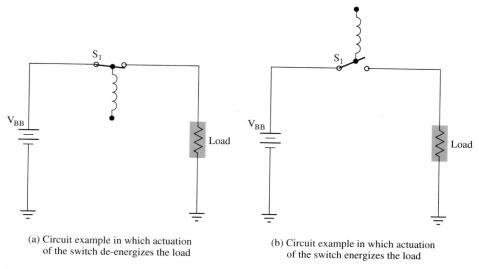

(a) Circuit example in which actuation
of the switch de-energizes the load

(b) Circuit example in which actuation
of the switch energizes the load

Figure 1–2
Two simple circuit examples that make use of the SPST switch

action is required in many industrial systems in situations where one load is energized
and another load is de-energized through a single actuating motion. This switching method
is represented in Figure 1–3(b). The common terminal of the SPDT would be connected
to a common line within the control circuit. (In this example, the common line extends
to circuit ground.)

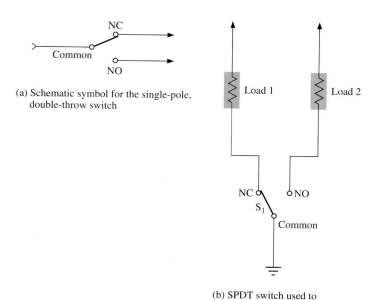

(a) Schematic symbol for the single-pole,
double-throw switch

(b) SPDT switch used to
control two loads

Figure 1–3
The single-pole, double-throw switch

Figure 1–4(a) contains the schematic symbol for a SPDT switch that has a center-off position for the wiper. With the wiper in the off position, both of the possible conduction paths through the switch would be interrupted. Thus, if this form of switch were to replace the basic SPDT used in Figure 1–3(b), both loads could be held in the off condition at the same time. A very common application of the center-off SPDT switch is shown in Figure 1–4(b). Here, with the wiper in the off position, the path for current to the lamp would be broken. If the wiper were moved upward, due to the rectifying action of D_1, only the positive alternation of the source waveform would be developed by the lamp. Thus, because it is receiving only half of the available source energy, the lamp would appear relatively dim. If the wiper were moved downward from the off position, a direct path for current would be created between the source and the lamp, which would then be fully illuminated. The lamp waveforms for dim and full illumination are shown in Figure 1–4(c).

Figure 1–5(a) shows the schematic symbol for a double-pole, single-throw switch (DPST). This four-terminal device may be considered simply as two SPST switches that are controlled simultaneously by the same actuator. (The dashed line extending between the two wipers indicates that they are ganged together.) An application of the DPST

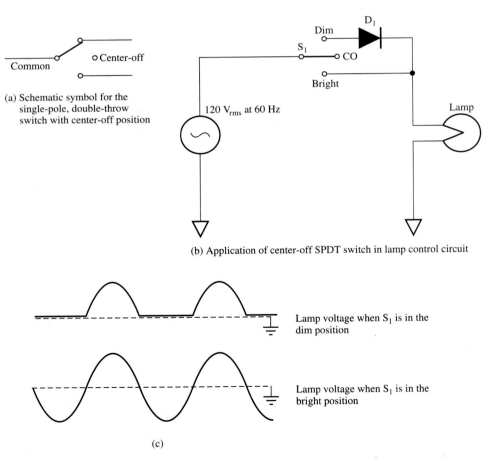

(a) Schematic symbol for the single-pole, double-throw switch with center-off position

(b) Application of center-off SPDT switch in lamp control circuit

Lamp voltage when S_1 is in the dim position

Lamp voltage when S_1 is in the bright position

(c)

Figure 1–4

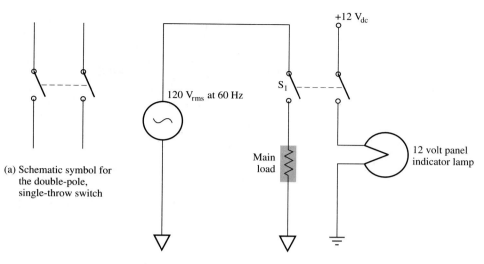

(a) Schematic symbol for
the double-pole,
single-throw switch

120 V$_{rms}$ at 60 Hz

+12 V$_{dc}$

S$_1$

Main
load

12 volt panel
indicator lamp

(b) Application of the DPST switch, simultaneously controlling
the operation of a load and an indicator lamp

Figure 1–5
The DPST switch

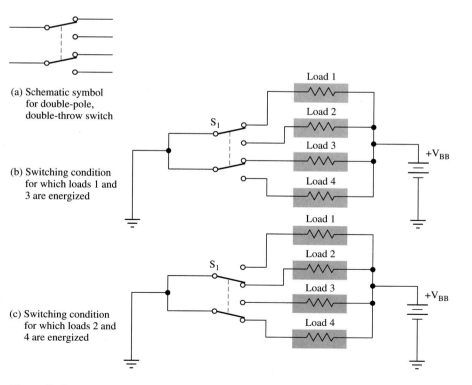

(a) Schematic symbol
for double-pole,
double-throw switch

(b) Switching condition
for which loads 1 and
3 are energized

(c) Switching condition
for which loads 2 and
4 are energized

Load 1

Load 2

Load 3

Load 4

S$_1$

+V$_{BB}$

Load 1

Load 2

Load 3

Load 4

S$_1$

+V$_{BB}$

Figure 1–6
The DPDT switch

switch is illustrated in Figure 1–5(b). Here, one wiper of the switch is used to make or break the path between a load and its ac power source. A small panel indicator lamp, which shows the condition of the ac load, is controlled by the second wiper. Note that the lamp is powered by a 12-V_{dc} source. This is possible due to the fact that, although the wipers are ganged, they may operate within different circuits.

The schematic symbol for a double-pole, double-throw switch (DPDT) is contained in Figure 1–6(a). This six-terminal device consists of a pair of SPDT switches with ganged wipers. As with the DPST switch, the wipers are controlled simultaneously by the same actuator. The DPDT switch may control as many as four separate loads, as illustrated in Figures 1–6(b) and (c). In Figure 1–6(b), loads 1 and 3 are receiving power from the common dc source. Actuation of the switch, as represented in Figure 1–6(c), would break the circuit paths to loads 1 and 3 while allowing loads 2 and 4 to be energized.

For wiper-action switches that are actuated manually, common forms of operation include the toggle, rocker, and slide mechanisms, as illustrated in Figures 1–7(a) to (c).

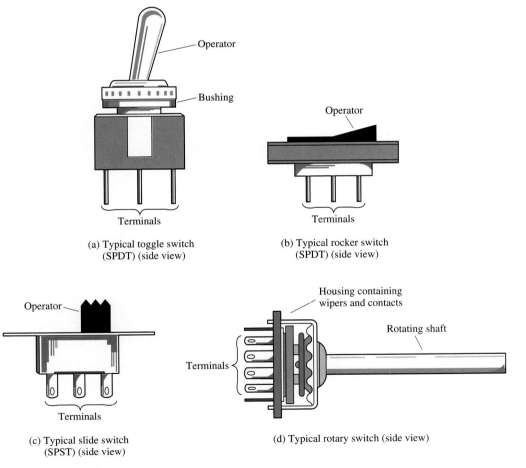

(a) Typical toggle switch
(SPDT) (side view)

(b) Typical rocker switch
(SPDT) (side view)

(c) Typical slide switch
(SPST) (side view)

(d) Typical rotary switch (side view)

Figure 1–7
Four common forms of manually operated switches

Figure 1–7(d) represents a rotary switch, which is often used to satisfy the more complex switching requirements of electronic instrumentation and test equipment.

Figure 1–8(a) is a cutaway view of a SPDT toggle switch, representing its internal structure and operation. With the operator positioned as shown in Figure 1–8(a), continuity would exist between the common terminal and terminal 2. At the same time, the circuit path between the common terminal and terminal 1 would be broken. As illustrated in Figure 1–8(b), movement of the operator to the opposite position would break continuity between the common terminal and terminal 2 while creating a circuit path between the common terminal and terminal 1.

A toggle switch must be able to withstand prolonged use. For this reason, the toggle mechanism must be housed within a highly torque-resistant bushing. Toggle switches, especially the miniature type, are particularly susceptible to mechanical failure—primarily because the operator and its bushing must absorb a high degree of physical force. To expedite replacement, most large- and medium-size mechanical switches are fitted with standard-size terminals, to which wires with solderless connectors may easily be fastened.

Rocker switches are frequently used as the on/off controls for various forms of electronic equipment. For this reason, they are readily available in SPST and DPDT packages. A comparison of Figures 1–7(a) and (b) would indicate that an advantage of the rocker switch is its relatively low profile. This physical attribute makes it ideally suited for mounting on a control panel or on a piece of portable equipment where a toggle switch is likely to be damaged.

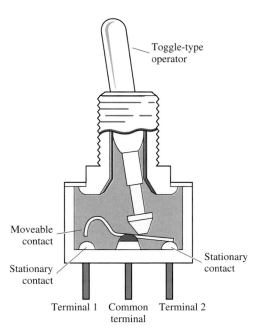

(a) Structure of a typical SPDT
toggle switch

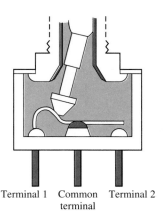

(b) SPDT switch with operator
in opposite position

Figure 1–8
Structure and operation of a SPDT toggle switch

Slide switches are available in a wide variety of sizes and are used primarily in medium- to low-power circuit applications. [The typical package design for a slide switch is illustrated in Figure 1–7(c).] Slide switches are readily found in SPST, SPDT, and DPDT packages. Subminiature slide switches are manufactured in sizes that fit conveniently onto printed circuit boards.

As illustrated in Figure 1–9(a), SPST slide switches are often fabricated in an eight-position package referred to as the **dual in-line package** (DIP) switch. As shown in Figure 1–9(b), the eight-position DIP switch has 16 terminal pins and is designed to fit into a standard-size 16-pin DIP socket.

DIP switches are used within digital circuitry for the purpose of manually encoding input data. Because they fit easily onto breadboards, they are often used in the process of system prototyping. As represented in Figure 1–9(c), the eight SPST switches within the DIP package may be individually actuated. This would allow a **byte** (eight bits) of binary data to be encoded by the device, as represented in Figure 1–9(d). In this example, the eight terminals on the upper side of the switch are connected to a common 5-V source, while the lower eight terminals are connected to load resistors. Closing one of the switches would allow the supply voltage to be developed over its load resistor. This 5-V level could be detected as a logic high by a digital circuit. An open switch would inhibit the flow of current to a load resistor, resulting in a logic low at its output point. Thus, with

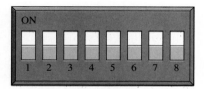

(a) Top view: eight individually
activated slide switches

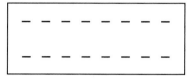

(b) Bottom view: sixteen terminal pins
which will fit into a standard-size
sixteen pin DIP socket

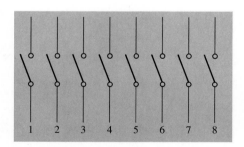

(c) Schematic representation of sixteen pin DIP switch

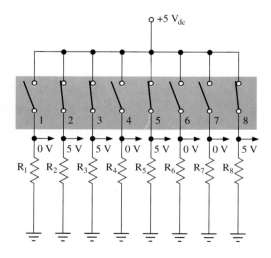

(d) Application of a DIP switch as an encoder

Figure 1–9
Structure and application of the DIP switch

the eight switches positioned as shown in Figure 1–9(d), the binary number developed over the resistors would equal 01101001 (read over R_1 through R_8, respectively).

The most complex form of manually operated switch is the rotary dial, the basic structure of which is illustrated in Figure 1–7(d). A common application of such a device would be the TIME/DIV or VOLTS/DIV controls of an oscilloscope. Rotary switches are readily available in one-, two-, or three-pole design formats. If a single-pole rotary dial switch has 12 positions, it would have an angular distance of 30° between its contact points (assuming the device is capable of a full 360° of rotation). For a two-pole rotary switch with the same contact spacing, six switching positions are possible. If there were three poles, with contact spacing remaining the same, the number of switching positions would be limited to four. This last switching format is illustrated in Figure 1–10.

Figure 1–10(a) contains the schematic diagram for a three-pole, four-position rotary switch, while the electromechanical wafer diagram for the device is shown in Figure 1–10(b). The dashed line in Figure 1–10(a) indicates that the three wipers, which maintain

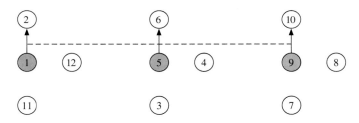

(a) Schematic representation of a three-pole, four-position rotary switch

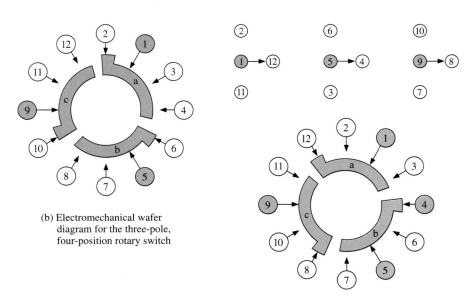

(b) Electromechanical wafer diagram for the three-pole, four-position rotary switch

(c) Switching condition after rotation of the shaft

Figure 1–10
Structure and operation of the rotary switch

permanent contact with terminal points 1, 5, and 9, are ganged and thus move simultaneously. With the switching condition represented in Figure 1–10(a), point 1 is making contact with point 2, point 5 is making contact with point 6, and point 9 is making contact with point 10. The corresponding mechanical condition for this switching arrangement is illustrated in Figure 1–10(b). The electromechanical operation of the rotary switch may be better understood through comparison of the switching condition in Figure 1–10(c) to that of Figures 1–10(a) and (b). In Figure 1–10(c), 30° of angular rotation has occurred. Terminal point 1 is now making contact with point 12, terminal point 5 is making contact with point 4, and terminal point 9 is making contact with point 8. Carefully note the corresponding change in the electromechanical wafer diagram.

Rotary switches are specified as either **shorting** or **nonshorting.** These terms describe the electrical condition that occurs between the contacts of the switch as it is being actuated. The electromechanical wafer diagrams in Figure 1–10 represent a nonshorting switching action. Referring now to Figure 1–11(a), as contact is broken between points 1 and 2 and then made between points 1 and 12, a transitional stage occurs during which neither point 2 nor point 12 is making contact with point 1. This nonshorting switching action is appropriately referred to as **break-before-make** (BBM).

If the contact points on the wipers of the rotary switch were wider, as illustrated in Figure 1–11(b), a shorting condition would result. Note that during the switching process, contact points 12 and 2 are temporarily shorted before contact is established between just points 1 and 12. Thus, this shorting action may be described as **make-before-break** (MBB).

Push-button switches represent another major type of manually operated device. Rather than having a wiper, the push-button form of switch has at least one movable contact. For the SPST NO push-button switch, as represented in Figure 1–12(a), actuation results in the making of a circuit path. For the NC SPST push-button switch seen in Figure 1–12(b), the opposite action would occur. Here, the circuit path would be broken by the actuation of the switch. Push-button switches are likely to be found in the more

Figure 1–11
Nonshorting and shorting action for the rotary switch

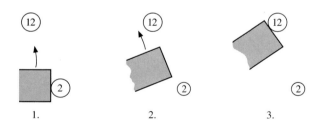

(a) Action of wiper in non-shorting (BBM) switching

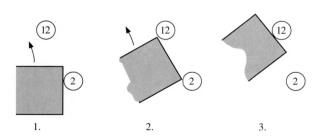

(b) Action of wiper in shorting (MBB) switching

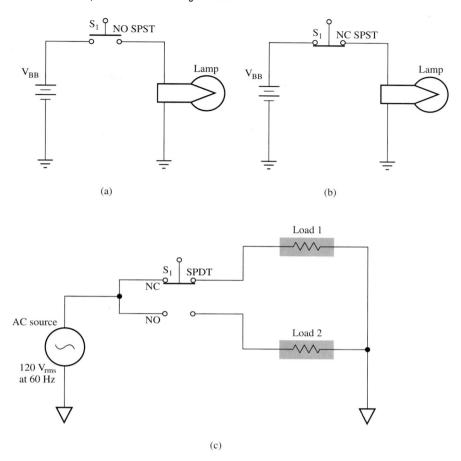

Figure 1–12
Operation of the push-button switch

versatile SPDT configuration illustrated in Figure 1–12(c). Here, actuation would break
the circuit path for load 1 while making the circuit path for load 2.

Nearly all push-button switches are spring-return devices. *Momentary* push-button
switches remain actuated only as long as pressure is applied to the operator. (A common
application of such a push-button device would be the RESET switch on a personal
computer.) Referring again to Figure 1–12(a), if the action of that switch were momentary,
the lamp would be illuminated only while the operator was being pressed. In Figure
1–12(b), if the action of the switch were momentary, the lamp would remain illuminated
as long as pressure was *not* applied to the operator.

For a **maintained-action** push-button switch, the actuated condition would remain
in effect even after pressure is removed from the operator. (Referring again to the personal
computer, an example of such a push-button switch would be the TURBO control.) In
Figure 1–12(c), if the push-button switch were a maintained-action switch rather than
momentary, pressing the operator once would turn load 2 on while turning load 1 off as
before. However, the two loads would remain latched in these conditions after pressure
was removed from the operator. Pressing the operator a second time would return the

operator to its original position, with load 1 being latched on and load 2 being latched off. The type of operator action just described is often referred to as **push-push.**

Manually operated switches represent only one form of mechanical switch used in industrial electronics. A mechanical device used prevalently within moving machinery is the **microswitch.** Microswitches are utilized to detect the location of moving parts within industrial machinery and, for this reason, are often referred to as **position-sensing switches.** Position-sensing switches are typically **snap-action switches,** which means they may be actuated by a very low operating force. The internal structure and operation of a typical microswitch are illustrated in Figure 1–13.

Like a manually operated SPDT switch, the microswitch contains a common terminal to which the wiper is connected. The leaf spring wiper maintains contact with the NC terminal until the plunger is actuated. This causes the wiper to move downward, breaking contact with the NC terminal and making contact with the NO terminal. Thus, two separate

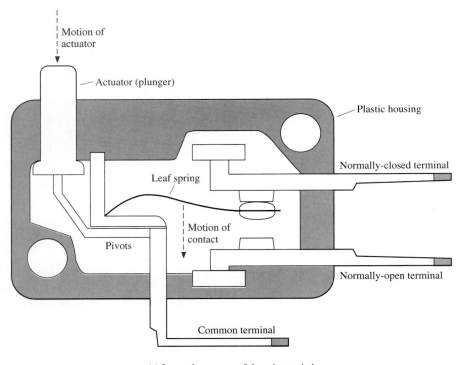

(a) Internal structure of the microswitch

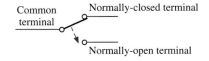

(b) Schematic representation of the microswitch

Figure 1–13
Structure and operation of the microswitch

circuits could be affected by the actuation of the microswitch. A circuit path involving the NC contact would be broken while, at the same instant, a circuit path involving the NO contact would be closed. (The common terminal could, of course, be tied to either a common power source or a circuit ground.)

Figure 1–14 illustrates a common application of the microswitch. Note that here a roller lever has been connected to the housing of the device, which allows it to function as a **limit switch.** The cam would be attached to the bottom of a moving part of a piece of machinery, an example of which might be the oscillating table of a plane. As the cam makes contact with the roller lever of the microswitch, the plunger is pressed downward, breaking the circuit path involving the common and NO terminals of the switch. This action would break the circuit controlling the left-hand (LH) motion of the table, preventing any further movement in that direction. The same switching action that stopped the LH motion of the table may also initiate its right-hand (RH) movement. Continuity for the circuit controlling this reverse motion would be provided via the common and NO contacts of the microswitch. The standard industrial symbols for the limit switch are shown in Figure 1–15. In the simple switching example just covered, the common and NC contacts of the microswitch correspond to the symbol in Figure 1–15(b), while the common and NO contacts correspond with Figure 1–15(a).

Figure 1–16 illustrates the structure and operation of a **mercury bulb switch.** This device is used as a position sensor, specifically in applications where we want to detect the position of an object with respect to the horizontal plane. Mercury is an element that remains in a liquid form over an extremely wide range of temperatures. Thus, the mercury switch functions reliably in virtually all industrial environments. In constructing a mercury

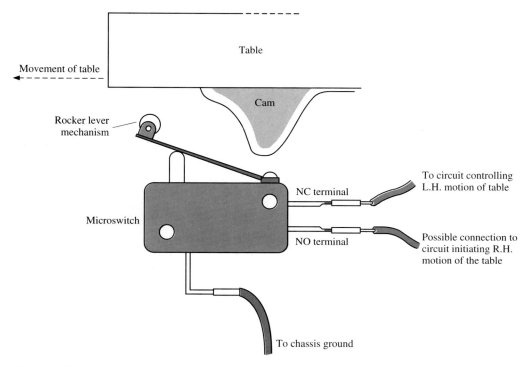

Figure 1–14
Microswitch with roller lever mechanism functioning as a limit switch

Figure 1–15

(a) Schematic symbol for the normally-open limit switch

(b) Schematic symbol for the normally-closed limit switch

bulb switch, a small drop of mercury is hermetically sealed within a glass envelope, which also contains the three contact points for the switch.

The operation of the mercury bulb switch in Figure 1–16 is equivalent to that of a SPDT mechanical switch. Its design allows the liquid mercury to maintain continuous contact with the common terminal. When the switch is tilted as shown in Figure 1–16(a), the position of the mercury allows continuity between the common terminal and terminal 1. If the switch is tilted as shown in Figure 1–16(b), the new position of the mercury allows continuity between the common terminal and terminal 2. Mercury switches are often mounted on the bimetallic temperature sensors commonly used within thermostats. As the metal expands and contracts with changes in temperature, the tilt of the mercury switch will change, either making or breaking the path for the circuit controlling a heating or air conditioning unit.

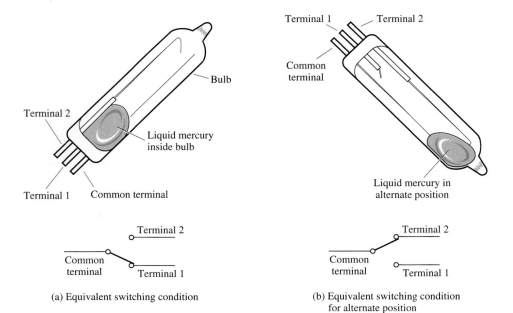

Figure 1–16
Structure and operation of the mercury bulb switch

SECTION REVIEW 1-1

1. What type of switch is represented in Figure 1–17(a)?
2. True or false?: The switch represented in Figure 1–17(a) could control as many as three separate loads.
3. What type of switch is represented in Figure 1–17(b)?
4. How many loads could be controlled by the switch represented in Figure 1–17(b)?
5. What type of switch is represented in Figure 1–17(c)?
6. True or false?: The switch represented in Figure 1–17(c) could control as many as four separate loads.
7. What type of switch is represented in Figure 1–17(d)?
8. True or false?: During the operation of the switch shown in Figure 1–17(d), all the contacts would remain stationary.
9. What kind of switch is represented in Figure 1–17(e)?

Figure 1–17

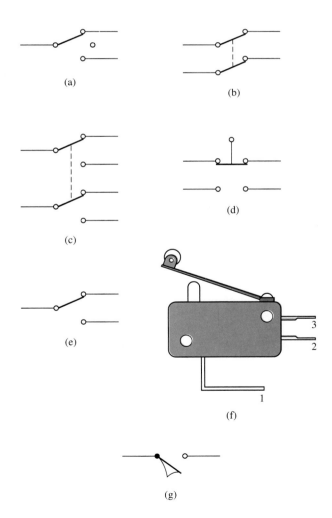

10. The device represented in Figure 1–17(f) would be described correctly as
    a. a DPDT position-sensing switch
    b. a SPST microswitch
    c. a DPST roller lever switch
    d. a SPDT limit switch
11. True or false?: The switching symbol in Figure 1–17(e) could represent the device in Figure 1–17(f).
12. What type of switch is represented in Figure 1–17(g)?
13. The device represented in Figure 1–17(f) could perform the switching function symbolized in Figure 1–17(g) if
    a. terminals 2 and 3 are utilized
    b. terminals 1 and 3 are utilized
    c. terminals 1 and 2 are utilized
    d. terminals 1, 2, and 3 are utilized

1–2   ELECTROMECHANICAL ACTUATING DEVICES

Electromechanical actuating devices, such as relays and solenoids, rely on the forces of attraction and repulsion exhibited by electromagnets in order to produce a motion as a result of an electrical current. The main advantage of the **electromechanical relay** is the fact that it allows several high-power ac loads to be controlled by a low-power dc or ac control circuit. The structure of the relay also allows the dc control source to be located at a safe distance from such ac loads. The term **solenoid** may be used in a general sense to describe any coil of wire used as an electromagnet. With reference to industrial electronics, the term solenoid applies to an electromechanical device containing an armature that is used as an actuator. Solenoids have a wide variety of applications; they are used to open and close valves and to change the position of gears. In many industrial systems, solenoids serve as the ac loads for relays.

Figure 1–18(a) shows the structural diagram for an electromechanical relay (EMR). The coil contained within the relay is electrically isolated from the contacts, as indicated by the schematic representation of the device in Figure 1–18(b). Within the general-purpose relay represented here, the six contacts are arranged in a DPDT format. The two main contacts are connected to the armature of the relay, and the other four contacts are held in fixed positions. The action of the six relay contacts is that of two SPDT switches operating simultaneously. Thus, for the sake of simplicity, in Figure 1–18(a), only one set of contacts is illustrated.

When current flows through the coil of the relay, it produces an electromagnetic field. Magnetic flux lines would, therefore, cut across the armature, which consists of a ferrous material. As a result of this action, the armature, now magnetized, is attracted toward the coil. For the relay to operate reliably, the current flow through the coil must be of sufficient strength to overcome the force of the spring that holds the main contacts against the NC contacts. Once this level of current is attained, the armature can move the main contacts to the alternative switching position. The main contacts would be held against the NO contacts only as long as the electromagnetic field of the winding is of sufficient strength to overcome the force of the spring. When coil current is interrupted, the electromagnetic field of the coil collapses, allowing the force of the spring to return the armature to its original position. Thus, the operation of the relay in Figure 1–18 is simply that of a spring-return DPDT switch, with actuation being accomplished as the result of electromechanical rather than manual force.

To use a relay properly, essential data about the device must first be obtained. With regard to the coil, this information would include its voltage and current ratings as well

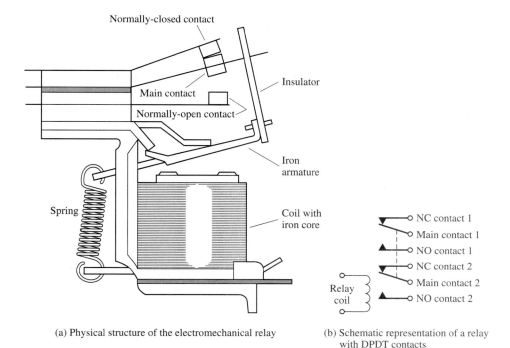

(a) Physical structure of the electromechanical relay

(b) Schematic representation of a relay with DPDT contacts

Figure 1–18
Structure and operation of an electromechanical relay

as its dc resistance. Assume for the relay in Figure 1–18 that the coil is rated at 12 V_{dc}, 75 mA, and 150 Ω. This would indicate that the coil must have 12 V_{dc} developed over it, with 75 mA of current flowing through it, for actuation of the armature to occur. Ohm's law may be applied to the relationship of these three variables. Therefore:

$$V_C = I_C \times r_C \tag{1–1}$$

Note that the product of the specified values of current and coil resistance is slightly lower than the designated voltage:

$$V_C = I_C \times r_C = 75 \text{ mA} \times 150 \ \Omega = 11.25 \text{ V}$$

Knowledge of this Ohm's law relationship is extremely beneficial in the design process. For instance, assume that only the voltage rating of a relay is known. Even though the coil resistance is not specified, it may be easily measured with an ohmmeter. Once this is accomplished, the required level of actuating current may be approximated using a derivation of Ohm's law.

Example 1–1

Assume that the relay coil in Figure 1–18 is rated at 12 V_{dc} and has a measured resistance of 274 Ω. How much current must be flowing through the coil for actuation to occur?

Solution:
$$I_C = \frac{V_C}{r_C} = \frac{12 \text{ V}}{274 \ \Omega} = 43.8 \text{ mA} \cong 44 \text{ mA}$$

Figure 1–19 illustrates a simple experimental procedure involving the dc relay. Assume that the relay under test has a coil resistance of 320 Ω and is rated at 12 V. With potentiometer R_1 set at its maximum ohmic value, the voltage developed over the relay winding would be well below the level required for actuation of the armature. Thus, the condition of the relay contacts would be as shown in Figure 1–19(a). This minimum value of coil voltage would be determined as follows:

$$V_C = \frac{320\ \Omega}{1\ k\Omega + 320\ \Omega} \times 25\ V = 6.06\ V$$

With this condition, series current would then be determined as

$$I_C = \frac{6.06\ V}{320\ \Omega} \cong 19\ mA$$

As the resistance of the potentiometer is slowly decreased, the series current and coil voltage gradually increase. Finally, when nearly 12 V is attained over the relay winding, as shown in Figure 1–19(b), actuation of the armature will occur. The value of the current needed to cause actuation is estimated as follows:

$$I_C = \frac{12\ V}{320\ \Omega} = 37.5\ mA$$

At this point, the ohmic value of R_1 would have been reduced to the following value:

$$R_1 = \frac{25\ V - 12\ V}{37.5\ mA} = \frac{13\ V}{37.5\ mA} = 347\ \Omega$$

Now assume that, with the armature actuated, the ohmic value of the potentiometer is slowly increased. Ideally, the armature would immediately return to its original position at the instant series current falls below 37.5 mA. In reality, the armature is likely to remain actuated until the coil voltage falls well below the 12-V level. Thus, the armature

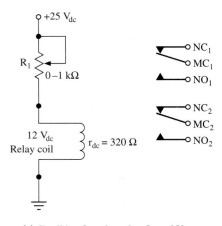

(a) Condition for relay when I_L and V_L are below the minimum pick-up level

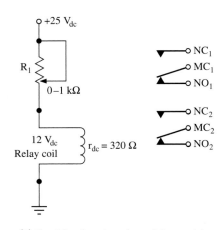

(b) Condition for relay when minimum pick-up levels of I_L and V_L have been attained

Figure 1–19

would return to its original position at a much higher setting of R_1 than the setting that allowed actuation.

The fact that the movement of the armature occurs at different levels of coil current, depending on the direction in which the wiper of R1 is being turned, is an example of a phenomenon called **hysteresis.** With an EMR, hysteresis may be desirable because it dampens the action of the armature. Also, it results in more efficient operation of the device because it allows actuation of the armature to be maintained at a lower level of current than that required to initiate the motion. That value of coil current required to actuate the armature is referred to as **pick-up current.** (The voltage at which actuation occurs may be referred to as the pick-up voltage.) As has been explained, a much lower value of current is adequate for holding the relay in the actuated condition. Below the minimum value of **holding current,** the **drop-out current** would occur. When coil current has been reduced to the drop-out point, the armature will return to its normal position.

The schematic representation of the relay used in Figures 1–18 and 1–19 is not practical for most industrial circuit diagrams. The symbols contained in Figure 1–20 allow the windings and contacts of a given relay to be shown at various points within a schematic or control ladder diagram. As shown in Figure 1–20(a), a circle may be used to symbolize the relay coil. The symbol in Figure 1–20(b) represents a NO contact. As shown in Figure 1–20(c), a diagonal line placed across this symbol would indicate a NC contact.

Figure 1–21 illustrates how the relay symbols just described might be used and identified within a circuit. As illustrated here, four different loads could be controlled by a single relay containing a DPDT contact configuration. If S_2 is open, none of the four loads is energized. However, if S_2 is closed while S_1 remains open, loads 1 and 3 will be energized since they are being controlled by the relay's NC contacts. If S_1 is then closed while S_2 remains closed, the relay coil would be energized, causing the armature to be actuated. As a result, loads 1 and 3 are turned off while loads 2 and 4 are switched on. A possible wiring diagram for the relay control circuit in Figure 1–21 is shown in Figure 1–22. Careful comparison of these two figures should help clarify how the switching of the loads actually takes place.

While the required value of dc current for the coil of a general-purpose relay could be less than 100 mA, the contacts of the device are likely to be rated at ac current values ranging between 3 and 10 A_{rms}. The ac voltage rating for the contacts is typically 125 V_{rms}. At this juncture, the primary advantage of the EMR should become quite clear. The coil of the DPDT relay, which consumes a very small amount of power, is able to control as many as four high-power ac loads.

Figure 1–20
Standard relay symbols

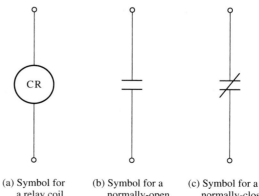

(a) Symbol for a relay coil

(b) Symbol for a normally-open contact

(c) Symbol for a normally-closed contact

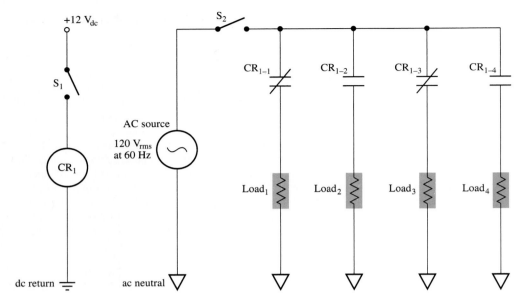

Figure 1–21
Symbolic representation of a relay controlling four loads

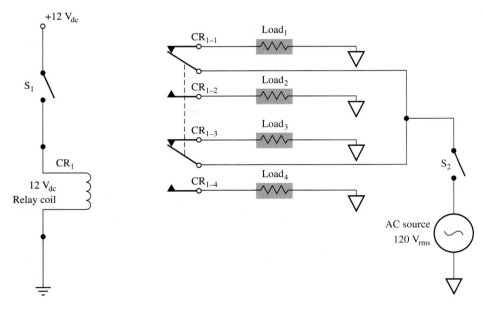

Figure 1–22
Possible wiring configuration for the relay circuit in Figure 1–21

For the dc coil circuit, power consumption may be estimated simply as the product of the rated values of voltage and current:

$$P_C = V_C \times I_C \tag{1-2}$$

If the dc coil resistance is known:

$$P_C = I_C^2 \times r_C \tag{1-3}$$

or

$$P_C = \frac{V_C^2}{r_C} \tag{1-4}$$

For each of the ac loads, maximum power consumption may be determined by the same basic formula:

$$P(\max) = V(\max) \times I(\max) \tag{1-5}$$

Substituting the maximum rated values of contact voltage and current into this formula would yield the *maximum* value of power that could be consumed by a load placed in series with one of the contacts. This result, in turn, would allow the designer to determine the *minimum* value of load resistance.

Example 1–2

Assume the dc resistance of a relay coil is 177 Ω, and its voltage rating is 12 V_{dc}. What would be the required current flow through the coil for actuating the armature? What would be the power consumed by the winding when this current is flowing?

Solution:
$$I_C = \frac{V_C}{r_C} = \frac{12\ V_{dc}}{177\ \Omega} = 67.8\ \text{mA}$$
$$P_C = V_C \times I_C = 12\ V_{dc} \times 67.8\ \text{mA} = 814\ \text{mW}$$

Assuming that the contacts of the same relay are rated at 125 V_{rms} with 5 A_{rms}, solve for the maximum power dissipation of the ac load.

Solution:
$$P(\max) = V(\max) \times I(\max)$$
$$= 125\ V_{rms} \times 5\ A_{rms} = 625\ \text{W}$$

What would be the minimum safe value of resistance for an ac load?

Solution:
$$P(\max) = \frac{V(\max)}{I(\max)} = \frac{125\ \text{W}}{5\ \text{A}} = 25\ \Omega$$

As seen in Figure 1–23, relays may be used to control dc loads. Here, a relay with DPDT contacts is used to control the direction of rotation for a dc motor. For a permanent magnet dc motor, the direction of current flow through its armature windings will determine the direction of armature rotation. The function of the relay in controlling the motor operation is better understood by first analyzing the switching diagram in Figure 1–23(a). When S_2 is open, dc power is removed from the motor, which would then remain idle, regardless of the position of S_1. If S_2 is closed, power is transferred to the motor. Direction of armature rotation then depends on the position of S_1. First we assume that, with S_1 open, the direction of armature rotation is counterclockwise. This would occur with the wipers in their NC positions. In this switching condition, point A would be common to

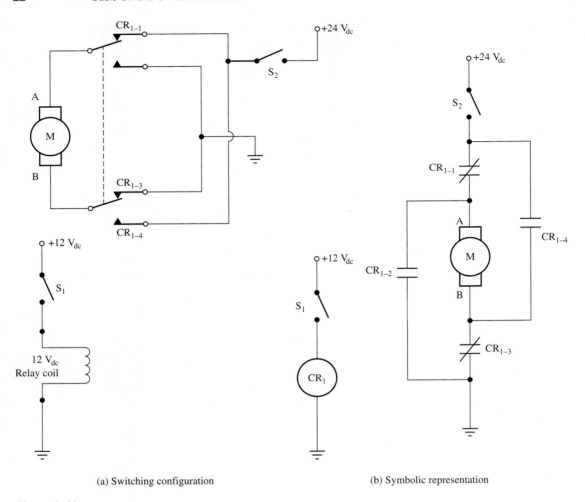

(a) Switching configuration   (b) Symbolic representation

Figure 1–23
Bidirectional motor-control circuit that makes use of a DPDT relay

a 24-V_{dc} source, while point B would be connected to dc ground. If S_1 is closed, actuation of the relay armature would occur. As a result, point B would then be connected to the 24-V_{dc} source, while point A would be common to dc ground. This reversal of polarity over the motor armature windings would cause the motor to rotate in the clockwise direction.

Figure 1–24 illustrates how a relay may be latched on by one of its own NO contacts. The major advantage of this switching method is the fact that the push button needs to be pressed only long enough for the armature to be actuated. Once the NO contacts are closed, an alternative path for coil current is provided, in parallel with S_1. This allows the armature to remain actuated after the push button is released. A disadvantage of this latching technique is, of course, the fact that one of the NO contacts must be committed to the process of maintaining the flow of current through the relay winding. With the relay now latched on, the only way the load may be turned off is through the actuation

Figure 1–24
Relay being latched on by one of its own NO contacts

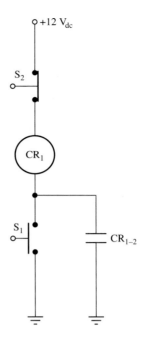

of the NC push-button switch S_2, which is placed in series with the relay coil. This action would interrupt the flow of coil current, causing the armature to return to its normal position. Since the path for coil current in parallel with S_1 would no longer exist, S_2 returns to its NC position but the coil remains de-energized until S_1 is again actuated.

Figure 1–25 shows a motor control circuit that demonstrates the interaction of two relays. Assume that relay 1 is used to control the clockwise rotation of the dc motor, which in turn causes a table to move in the LH direction. Relay 2 would then be used to control the counterclockwise rotation of the motor, causing the RH motion of the table. As shown in Figure 1–26, two cams attached to the bottom of the table are used to actuate limit switches S_2 and S_3.

In Figure 1–26(a), the table is shown in its home or starting position. With this condition, S_3 is open and neither relay coil is energized. As a result, the motor is idle. Careful examination of the contact configuration around the motor would indicate that no path for current through its windings as yet exists.

Now assume that S_1 is pressed. This action completes the current path for the coil of relay 1, allowing contacts CR_{1-2} and CR_{1-4} to close. With CR_{1-4} closed, relay is latched on, allowing the operator to remove pressure from S_1 once the armature of relay 1 has been actuated. A path for motor current is now completed, via CR_{2-1} and CR_{1-2}. As the LH motion of the table begins, S_3 closes. However, because S_2 remains open, no current path as yet exists through the coil of relay 2. Thus, the table continues to move in the LH direction. Once S_2 is actuated by the LH cam, the coil of relay 2 is energized. When the armature of relay 2 is actuated, CR_{2-3} opens, causing relay 1 to drop out. With CR_{1-1} and CR_{2-2} now providing the path for motor current, motor rotation is reversed and the table begins to move in the RH direction. Since relay 2 is now latched on by its own NO contact CR_{2-4}, the RH motion of the table is allowed to continue until limit switch S_3 is opened by the RH cam. With the table again resting in the home position, both relays

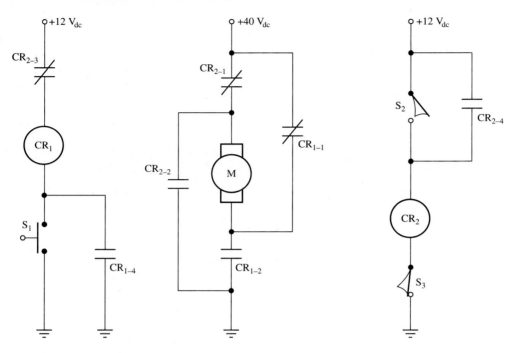

Figure 1–25
Bidirectional motor-control circuit involving two relays

and the motor remain de-energized until S_1 is again pressed, causing the switching cycle to be repeated.

Note at this point that additional control circuitry may be required in order to control both the speed of rotation and the breaking action of the motor represented in Figure 1–25. Such circuitry is studied in detail later in this text.

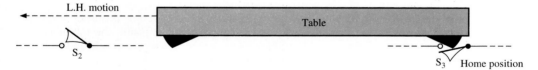

(a) Starting position for the table (Limit switches S_2 and S_3 are both open)

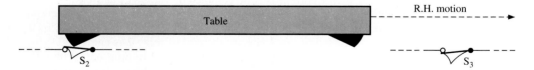

(b) Table at left hand limit (Limit switches S_2 and S_3 are both closed)

Figure 1–26
Motion of the table controlled by the relay circuit in Figure 1–25

Relay logic, as demonstrated in the previous circuit example, involves the interaction of two or more relays. The basic digital logic functions, including the OR, AND, NOR, and NAND gates, actually had their first electrical applications with relay circuits. In the following five examples of basic relay logic, in addition to the contact configurations, the equivalent digital logic functions and corresponding truth tables are shown. For the truth tables, *logic 1* represents an energized relay coil, whereas *logic 0* represents an off condition for the coil. For the Y output conditions, logic 1 indicates that the load is receiving power, whereas a 0 indicates that the load is de-energized.

As shown in Figure 1–27, an OR function is created when a load is given two or more possible sources of current. The possible paths for load current are through NO contacts A and B, which belong to two different relays. If either or both of the coils controlling these contacts are energized, a path for current flow will be provided to the load. Only when both coils are de-energized at the same time will the flow of load current be inhibited.

The basic AND function may be created as shown in Figure 1–28(a) by placing the normally open contacts of two relays in series. With this contact configuration, both coils must be energized at the same time in order for the load to receive power.

To understand how relay contacts may be arranged to create NAND and NOR functions, DeMorgan's theorems should first be reviewed. According to these fundamental

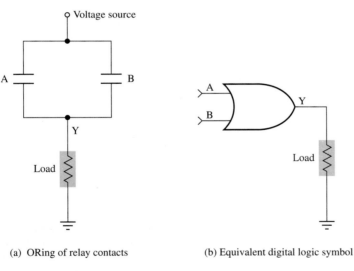

(a) ORing of relay contacts   (b) Equivalent digital logic symbol

$$Y = A + B$$

A	B	Y
0	0	0
0	1	1
1	0	1
1	1	1

(c) Boolean expression and truth table for the OR function

Figure 1–27
The OR function

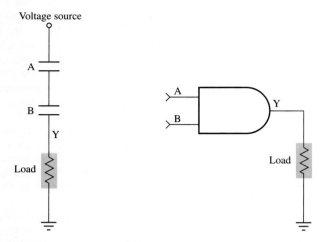

(a) ANDing of relay contacts   (b) Equivalent digital logic symbol

A	B	Y
0	0	0
0	1	0
1	0	0
1	1	1

$$Y = A \cdot B$$

(c) Boolean expression
and truth table for
the AND function

Figure 1–28
The AND function

Boolean principles:

$$\overline{A + B} = \overline{A} \cdot \overline{B} \quad \text{and} \quad \overline{A \cdot B} = \overline{A} + \overline{B}$$

Thus, the contact arrangement in Figure 1–29(a) would satisfy the Boolean expression for the NOR function. Since the NC contacts of the two relays are now placed in series, the load will receive current only when both relay coils are de-energized. Therefore, as shown by the truth table in Figure 1–29(c), Y would represent an on condition for the load only when A and B represent off conditions. When either coil is energized, its NC contacts open, blocking the flow of current to the load.

As shown in Figure 1–30(a), placement of the NC contacts of the two relays in an OR configuration results in a NAND function. As is evident in the truth table of Figure 1–30(c), when one or both of the coils is de-energized, a current path is provided for the load. Only when both relay coils are energized will the load be in the off condition.

As shown in Figure 1–31, more complex relay logic circuits may be developed using three or more relays. For the contact configuration shown here, providing that relay C has dropped out, the load will be allowed to turn on when either relay A has dropped out or relay B has picked up, or when both of these actions have occurred at the same time. However, when the coil of relay C is energized, the load will remain off regardless of the switching conditions for relays A and B.

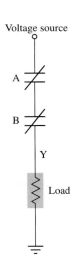

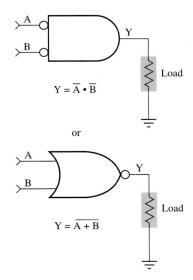

(a) NORing of relay contacts

(b) Equivalent digital logic symbols and Boolean expressions

A	B	Y
0	0	1
0	1	0
1	0	0
1	1	0

(c) Truth table for NOR function

Figure 1–29
The NOR function

To troubleshoot efficiently a control circuit containing several relays, the technician must understand the overall sequential logic of the system. The most challenging aspect of troubleshooting relay logic circuits is identifying the true reason for a system failure among all of the possible causes of a malfunction. For instance, the failure of a relay to pick up within the operating sequence of a given system might cause a mechanical device such as a valve or a clutch to malfunction. An inexperienced technician might immediately replace the relay, or even an expensive mechanical part, only to discover later that the actual cause of failure was something as simple (and inexpensive) as a faulty microswitch. If this part controlled the coil of the relay, its failure due to breakage or improper adjustment could cause both the relay and its load to malfunction.

Some relays are manufactured to operate with ac coil voltage. The only significant structural feature of this type of relay is the presence of a shaded pole at the end of the magnetic core material. Its purpose is to create a magnetic field that lags behind the magnetic field produced by the actual relay winding. As a result of this effect, magnetic flux is present constantly within the air gap that exists between the electromagnet and the armature. This allows the armature to remain in the actuated position while ac current is flowing within the relay winding, thus inhibiting 60-cycle hum and contact chatter.

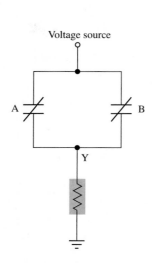

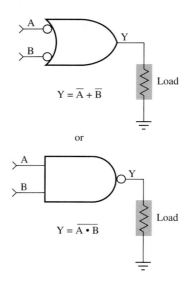

$$Y = \overline{A} + \overline{B}$$

or

$$Y = \overline{A \cdot B}$$

Load

Load

(a) NANDing of relay contacts

(b) Equivalent digital logic symbols and Boolean expressions

A	B	Y
0	0	1
0	1	1
1	0	1
1	1	0

(c) Truth table for NAND function

Figure 1–30
The NAND function

Due to their dynamic impedance, which consists of both series resistance and inductive reactance, the coils of ac relays may operate directly from a 120-V_{rms} source while drawing only a few milliamperes. A typical coil impedance value is 4.5 kΩ, whereas the required values of pick-up current are as low as 15 mA. Although the contacts of ac relays tend to respond more slowly than those of dc relays, they do offer distinct advantages, the most obvious of which is the fact that they allow both the relay coils and the loads within an electrical control system to be operated from a single ac source. Thus, the need for a separate dc voltage source within a given system might be alleviated.

Problems associated with EMRs include contact oxidation and arcing. Surface oxidation greatly increases the resistivity of relay contacts. However, this problem may be minimized through the use of noble metal alloys as contact materials. (Here the term *noble* is applied to materials that are chemically inert, particularly with regard to the process of oxidation.) The resistance of relay contacts may also be reduced significantly through the optimization of their physical design.

Various types of contact design are illustrated in Figure 1–32. For relays with contact ratings as high as 10 A_{rms}, the single-button silver-cadmium oxide contact is most often used, as shown in Figure 1–32(a). Contact resistance is reduced significantly through the use of a **bifurcated** design, which, as illustrated in Figure 1–32(b), allows two paths for

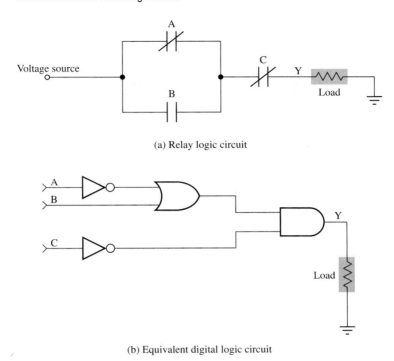

(a) Relay logic circuit

(b) Equivalent digital logic circuit

Figure 1–31

contact current. While the contacts illustrated in Figures 1–32(a) and (b) are likely to be used for currents ranging from 2 through 10 A, the button-type contacts represented in Figures 1–32(c) and (d) are designed for lower current applications, ranging from 1 mA to 2 A. These smaller, rounded contacts are often diffused with fine silver and gold. The **bifurcated crossbar** form of relay contact shown in Figure 1–32(e) is reserved for **dry circuit** (extremely low current) applications. Crossbar contacts are made from gold, silver, and platinum alloys.

When utilized in dc circuits, relay contacts are particularly susceptible to damage from arcing. In ac circuits, however, the life expectancy of the contacts is much greater, with damage usually being limited to only a slight surface roughening. In low-voltage dc circuits, the problem of arcing is actually more acute because needle-type metal transfer is likely to occur. The less damaging dome-type transfer may occur at higher dc voltages. These forms of contact damage are illustrated in Figure 1–33.

Contact arcing and bouncing are a major source of electrical noise. The harmonic content of the voltage spikes produced during arcing is within the RF range. Thus, the resultant noise travels easily throughout an industrial environment and, if not suppressed, can cause sensitive electronic equipment to malfunction.

Due to the reasons just mentioned, EMRs have been replaced to a great extent by solid-state devices. This is especially true for dc control applications. At the same time, however, improved forms of EMRs are being manufactured. An example of such a device is the **DIP relay** designed for mounting on printed circuit boards. Such relays are used for interfacing low-power dc control circuitry directly with high-power ac and dc loads.

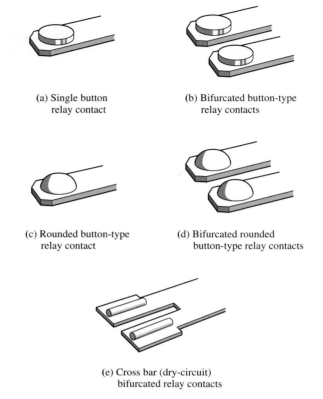

Figure 1–32
Typical forms of relay contacts

(a) Single button
relay contact

(b) Bifurcated button-type
relay contacts

(c) Rounded button-type
relay contact

(d) Bifurcated rounded
button-type relay contacts

(e) Cross bar (dry-circuit)
bifurcated relay contacts

A typical DIP relay could have coil ratings of 5 V and 72 mA, which would allow it to operate from the same power source as most digital logic circuits. The SPDT contacts of the device could be rated at 125 V_{rms} with 2 A, thus allowing the logic circuit to control both a normally on and a normally off high-power load.

An alternative form of electromechanical switching device, designed for high-speed operation, is the **reed relay.** The fast switching speeds of the reed relay are attributed to its unique structure, which is illustrated in Figure 1–34. Rather than a heavy armature, the reed relay contains two lightweight flattened reed contacts, which are comprised of ferromagnetic material. These contacts, which may either be dry or **mercury wetted,** are

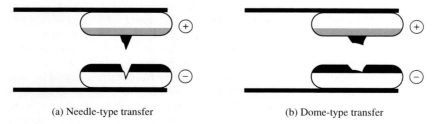

(a) Needle-type transfer

(b) Dome-type transfer

Figure 1–33
Damage incurred by relay contacts in dc circuit applications

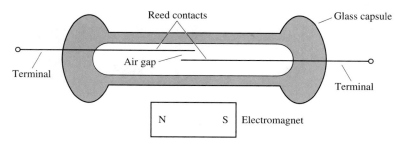

(a) Reed relay in normally-open condition

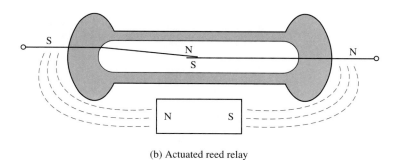

(b) Actuated reed relay

Figure 1–34
Structure and operation of the reed relay

hermetically sealed within a glass capsule. For high-voltage switching, the glass container is often kept in a tight-vacuum condition. Otherwise, it could contain an inert gas such as nitrogen.

As seen in Figure 1–34(a), the contacts of the reed relay must overlap slightly in order to allow a narrow air gap to be formed when the electromagnet is energized. Although not shown in the figure, the coil of the electromagnet may actually encompass both the electromagnet and the capsule. When current flows through the coil of the electromagnet, the reed contacts, now polarized, are pulled together.

Reed relays are readily available in designs suitable for printed circuit board mounting. The coil of such a device is likely to require only 5 V and 20 mA for operation, which would make it directly compatible with most digital control circuits. The SPST contacts of this same relay could easily operate at 125 V_{rms} with 1 A.

Solenoids are similar to relays in that they rely on electromagnetism to produce mechanical force. The **linear solenoid** contains a **plunger** (or armature), which is usually spring loaded. Because the plunger is likely to consist of a ferrous material, the magnetic flux produced by the coil will form closed loops, as illustrated in Figure 1–35. When the magnetic flux moves, as shown, through the plunger and frame, the plunger is magnetized and pulled downward toward the stop. Considering the direction of the flux shown, the bottom of the plunger represents a north magnetic pole, whereas the stop, being at the opposite side of the air gap, represents a south magnetic pole. If the flow of coil current is of sufficient magnitude, the magnetic attraction between the stop and the plunger becomes strong enough to overcome the force of the spring, thus resulting in downward motion of the plunger.

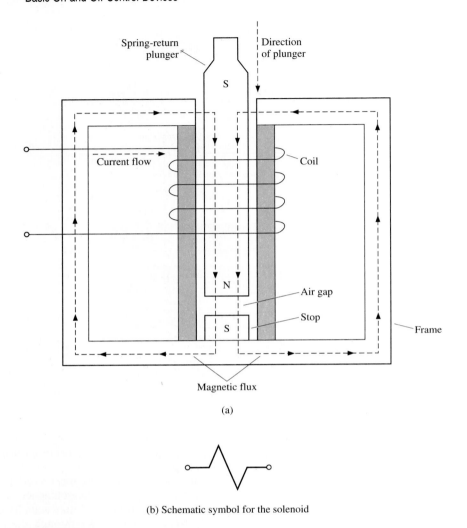

(a)

(b) Schematic symbol for the solenoid

Figure 1–35
Structure and operation for a basic solenoid

Figure 1–36 illustrates how a solenoid may be used to control the operation of a valve. In this example, the plunger of the solenoid is connected directly to the valve stem. When no current is flowing through the coil, the spring is able to press the stem firmly against the valve seat. As a result, the flow of liquid is inhibited. If current is then allowed to flow through the coil, the resultant motion of the plunger causes the valve stem to be pulled upward, temporarily overcoming the force of the spring. This action allows the flow of liquid through the valve.

We should stress that solenoids, when used to control valves, provide only basic on/ off control. Thus, if the solenoid is functioning properly, it will cause the valve to be either completely opened or completely seated. The ac solenoids are used more commonly than dc, although ac solenoids do exhibit an inherent design weakness, especially when used as valve actuators. If the valve stem becomes stuck in or near the closed position,

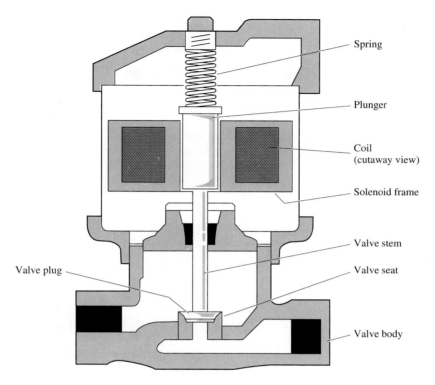

Figure 1–36
Solenoid-operated valve

the plunger of the solenoid may not be able to fully enter the core of the coil. As a result, the coil effectively becomes an air-core rather than an iron-core inductor, which greatly reduces its value of inductance. Because the value of inductive reactance offered to the 60-cycle source is substantially reduced, the flow of current through the windings increases to a level that could eventually overheat and destroy the coil.

The size of mechanical load that a solenoid is capable of actuating depends on the amount of magnetomotive force produced by the coil. Industrial solenoids are available with load capabilities ranging from a fraction of an ounce to nearly 100 lb. To overcome the inertia of a mechanical load, the actuating capability of the solenoid should be at least 25% greater than the weight of the given load.

Example 1–3

Assume the load being actuated by a solenoid weighs 14 lb. What would be the minimal actuating force required for the solenoid?

Solution: Actuating force = 1.25 × 14 lb = 17.5 lb

Another important consideration in the selection of a solenoid is its **in-rush current,** which is likely to be several times greater than its **holding current.** As explained earlier with regard to the ac solenoid, until the plunger is fully actuated, the coil draws a large amount of current. As an example, an ac solenoid operating with 125 V_{rms} at a frequency of 60 Hz could have a coil resistance of 90 Ω. Before the plunger becomes fully actuated, the inductive reactance of the coil represents a very small ohmic value. Thus, a typical in-rush current value for this solenoid could be as high as 1.2 A_{rms}. Once the plunger has been fully actuated, the current being drawn by the coil could be reduced to around 250 mA. This indicates that the impedance now being offered by the coil has increased to nearly 480 Ω. The amount of time during which the in-rush current will flow is a function of the **stroke** of the solenoid, which is the distance the plunger must travel prior to being fully seated.

The **duty cycle** of a solenoid is simply the ratio of the time the device is actuated (T_1) to the total time period of its operating cycle ($T_1 + T_2$). The duty cycle of a solenoid is usually expressed as a percentage, as calculated here:

$$\text{Duty cycle} = \frac{T_1}{T_1 + T_2} \times 100\% \qquad (1-6)$$

(we assume here that $T_1 + T_2$ is a predictable unit of time.)

Example 1–4

Assume that a solenoid normally remains actuated for 12 sec and is de-energized for 55 sec. What is the duty cycle of the solenoid?

$$\text{Duty cycle} = \frac{T_1}{T_1 + T_2} \times 100\% = \frac{12 \text{ sec}}{12 \text{ sec} + 55 \text{ sec}} \times 100\% = 17.9\%$$

Some solenoids, such as the small dc devices used to operate fuel injectors in automobiles, can be operated several times within a second. Larger ac solenoids, however, such as those used to control valves, can only be operated once every several minutes. The approximate duty cycle for a solenoid that is to be actuated for no more than 30 sec can be as high as 50%. If the actuation time for the solenoid is more than 30 sec, but less than 3 min, the duty cycle should be limited to 25%. For a solenoid that must be actuated for 3 to 5 min, the duty cycle should be limited to 10%.

A major advantage of a **continuous-duty** dc solenoid is its ability to withstand prolonged operation without overheating. The major limitation of the ac solenoid is its susceptibility to thermal damage caused by the repeated flow of in-rush current. As an ac solenoid overheats, the insulating material surrounding its windings breaks down and the device is destroyed. For this important reason, the duty cycles specified earlier should be strictly adhered to when operating an ac solenoid.

Figure 1–37 illustrates how a solenoid-operated valve might be controlled by a float switch and an ac relay. While the NO float switch detects the level of fluid within the tank, the solenoid-controlled valve attempts to hold the fluid at a constant level. As the fluid begins to fall below the desired level, the float switch falls open, causing the coil of the relay to be de-energized. As this occurs, the NC contacts of the relay close, causing the coil of the solenoid to be energized. As the valve stem is pulled up, fluid is allowed to enter the tank. As the fluid attains the desired level, the float switch closes, causing the relay coil to be energized. The NC contact then opens, de-energizing the solenoid winding and causing the valve stem to return to the seated position.

The representation of the control circuitry in Figure 1–37(b) is in the form of a **control ladder.** Placing the float switch and relay coil on the same rung of the ladder represents their series relationship. A series relationship is also shown for the solenoid and the relay contact. From this simple form of circuit diagram, a technician could quickly

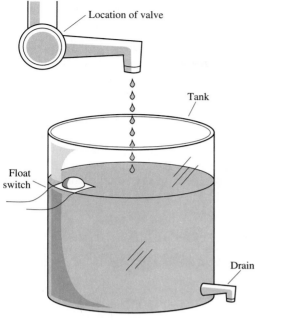

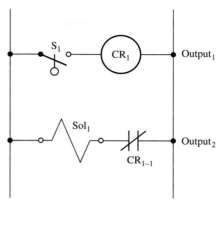

(a) Operation of the solenoid-controlled valve in an automatic tank leveling application

(b) Control ladder diagram

Figure 1–37
Application of solenoid-controlled valve

deduce that the coil of relay 1 is controlled by the float switch, while the solenoid, in turn, is controlled by the NC contact of relay 1.

Example 1–5

Assume the ac solenoid that controls the valve in Figure 1–37 remains actuated for up to 1.5 min. What duty cycle should be maintained for the device? How long should the minimum off time be?

Solution:
The duty cycle of the solenoid should be limited to 25%, since its on time falls between 30 sec and 3 min. Thus, 1.5 min would represent 25% of the total cycle time, which is determined as follows:

$$T = 4 \times 1.5 \text{ min} = 6 \text{ min}$$

Thus, the minimum off time is determined as follows:

$$T(\text{off}) = 0.75 \times 6 \text{ min} = 4.5 \text{ min}$$

From Example 1–5, it becomes evident that additional control circuitry might be required to prevent the solenoid from being damaged due to a lack of cooling time. One practical solution to this problem would be the placement of a thermal (temperature-

Figure 1–38
Modified section of control ladder containing
thermal switch

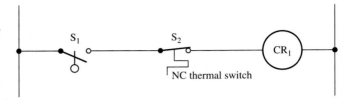

sensitive) switch close to the solenoid coil. This device would detect an overheated condition for the solenoid and cause the flow of coil current to be interrupted until the temperature of the solenoid is reduced to a safe level. Electrically, the thermal switch could be placed in series with the relay coil as shown in the modified portion of the control ladder of Figure 1–38.

SECTION REVIEW 1–2

1. True or false?: According to the control ladder diagram in Figure 1–39, relay 1 is latched on by one of its own NO contacts.
2. In Figure 1–39, what switching conditions must be met in order for the motor to be energized?
3. True or false?: In Figure 1–39, the motor and solenoid 1 may both be energized at the same time.
4. Describe the purpose of S_4 in Figure 1–39. Where should this switch be located to function most effectively?

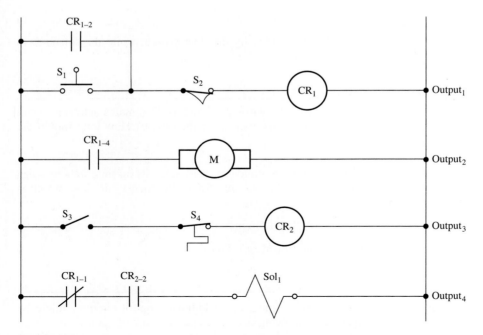

Figure 1–39
Control ladder diagram

5. Provided S_4 is in the closed condition, what must be the condition of the two relay coils for the solenoid to be actuated?
6. A dc relay coil has 247 Ω of resistance and requires 21 mA of current for actuation of the armature. What is the approximate value of its pick-up voltage?
7. Prior to actuation, an ac solenoid with a coil voltage of 125 V_{rms} offers only 107 Ω of impedance. After actuation of the plunger, solenoid current equals 255 mA. What is the approximate value of inrush current? What is the impedance offered by the coil after displacement of the plunger is accomplished?
8. Assuming that an ac solenoid is to be actuated for nearly 50 sec, what should be its maximum duty cycle? What should be its minimum off time?

1-3 THE TRANSISTOR AS A SWITCH

In recent years, solid-state devices have rapidly replaced mechanical and electromechanical switches devices in many industrial applications. If operated only in the saturation or cutoff conditions, the action of the **bipolar junction transistor** (BJT) closely resembles that of an ideal mechanical switch. Although a mechanical switch is likely to produce arcing as it is opened and closed, a transistor switch is capable of energizing or de-energizing a load through a virtually instantaneous rise or fall in current.

A basic transistor switch is illustrated in Figure 1–40. This transistor configuration is often referred to as a *buffer amplifier* because it boosts the level of current from a low-power control source to a higher power output load. When designing such a switch with a resistive load, it is customary to drive the transistor into saturation when a high pulse occurs at the base input. When switched on in saturation, the BJT is operating at maximum power efficiency, allowing virtually all of the power drawn from the source to be dissipated by the load.

The reason for the excellent power efficiency of the BJT switch may be better understood through examination of Figure 1–41(a). Ideally, for a transistor driven into saturation, 0 V would develop between the collector and emitter, resulting in the equivalent switching condition represented in Figure 1–41(b). Although this ideal condition will never

Figure 1–40
A basic transistor switch with a collector load

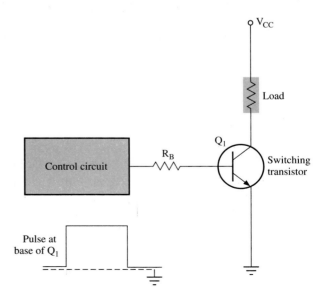

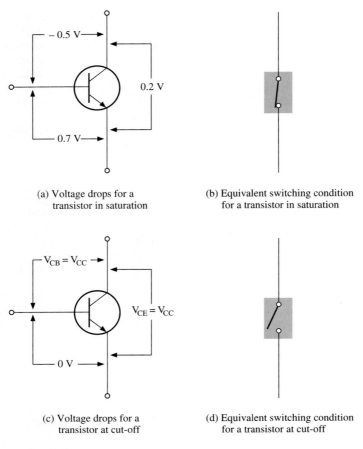

(a) Voltage drops for a
transistor in saturation

(b) Equivalent switching condition
for a transistor in saturation

(c) Voltage drops for a
transistor at cut-off

(d) Equivalent switching condition
for a transistor at cut-off

Figure 1–41

occur, the saturation level of V_{CE} will be quite low, due to a process called **conductivity modulation.** Recall that, when a BJT is operating in the active (linear) region of its load line, the collector current is a product of the base current and the β_{dc} of the device, being determined as follows:

$$I_C = I_B \times \beta_{dc}$$

However, for a BJT in deep saturation, such as the common-emitter device in Figure 1–40, collector current is limited to the following value:

$$I_C \text{ sat} = \frac{[V_{CC} - V_{CE} \text{ sat}]}{R_L} \tag{1–7}$$

Thus, during saturation, the ratio of I_C/I_B is likely to be much less than the actual β_{dc}.

With the BJT in this conductivity modulation state, an excess of minority carriers (holes for an NPN device) accumulates at the collector-base junction, resulting in a reverse electrical field at that junction, as shown in Figure 1–41a. Since the reverse voltage developed at this point will be at least -0.5 V, the maximum value of $V_{CE}(\text{sat})$ may be determined using a derivation of Kirchhoff's law:

$$V_{CE}(\text{sat}) = V_{BE} + V_{CB} = 0.7\ \text{V} + -0.5\ \text{V} = 0.2\ \text{V} \qquad (1\text{--}8)$$

The ideal amount of power consumption for the transistor switch in saturation, as with the mechanical switch, would be 0 W. Although this ideal level of power consumption is never attained, power dissipation for a BJT in deep saturation is quite low, since it is the product of $V_{CE}(\text{sat})$ and $I_C(\text{sat})$:

$$P_Q(\text{sat}) = V_{CE}(\text{sat}) \times I_C(\text{sat}) \qquad (1\text{--}9)$$

Example 1–6

Assume that, during deep saturation, a BJT switch has a V_{CE} of 124 mV and an $I_C(\text{sat})$ of 98 mA. How much power is the device dissipating when in this condition?

$$P_Q(\text{sat}) = V_{CE}(\text{sat}) \times I_C(\text{sat}) = 124\ \text{mV} \times 98\ \text{mA} = 12.15\ \text{mW}$$

As represented in Figure 1–41(c), a BJT switch in the cutoff condition will develop virtually the entire supply voltage between its collector and emitter terminals. This condition may be compared to that of an open switch, as illustrated in Figure 1–41(d). Even though this cutoff state would represent a maximum value for V_{CE}, only a reverse leakage current would flow between the main terminals of the transistor in this condition. Thus, power dissipation for the device would be determined as:

$$P_{CO} = V_{CO} \times I_R \qquad (1\text{--}10)$$

Example 1–7

Assume a transistor in the cutoff condition has a reverse leakage current of 37 μA while the supply voltage is at 12 V_{dc}. How much power is the transistor consuming in this condition?

$$P_{CO} = V_{CO} \times I_R = 12\ \text{V} \times 37\ \mu\text{A} = 444\ \mu\text{W}$$

Ideally, if the base input signal for the transistor switch in Figure 1–40 were a rectangle wave, the current pulse resulting at the collector would also be rectangular. In reality, the response to the input waveform is not instantaneous. The transistor will, as shown in Figure 1–42, have a **turn-on time** (t_{on}) that is the sum of the **delay time** (t_d) and the **rise time** (t_r). The delay time may be described as the amount of time required for the transistor to begin to respond to a transition at its base. The delay time is measured from the leading edge of the input signal to the point where the collector has attained 10% of its maximum forward amplitude. Rise time is that time required for the collector current to rise from 10% to 90% of its maximum value.

The most obvious response problem associated with the BJT switch is **storage time** (t_s). This problem becomes particularly crucial when the transistor is expected to switch in and out of saturation at a high speed. As illustrated in Figure 1–42, storage time extends from the falling edge of the input signal to the point where the collector current is reduced to 90% of its maximum amplitude. Storage time is associated with the collector-base junction capacitance, representing the amount of time required for the reverse electrical field developed at that junction to be discharged. **Fall time** (t_f) is that time required for the collector current to decrease from 90% to 10% of its maximum value. **Decay time** is that time required for the collector current to decrease from 10% of its peak amplitude

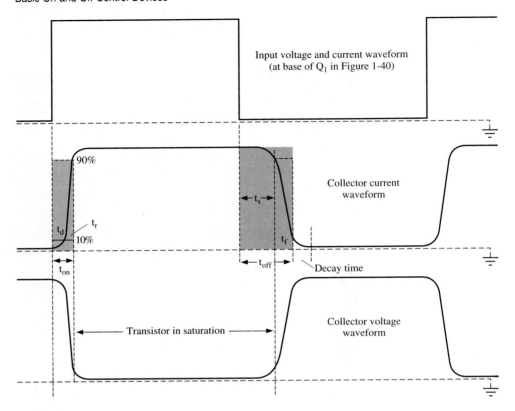

Input voltage and current waveform
(at base of Q_1 in Figure 1-40)

Collector current
waveform

Collector voltage
waveform

Transistor in saturation

Decay time

Figure 1–42
Transistor switching times

to its minimum level. The problem of storage time may be minimized by preventing the transistor from being driven into deep saturation. This may be easily accomplished by following the design procedures outlined next.

The first consideration in designing the transistor switch is the amount of current that the input control circuit may safely provide. Assuming the control source to be a digital circuit, the pulse occurring at the base will be a rectangular waveform, such as that illustrated in Figure 1–42. Referring to the transistor switch in Figure 1–43, once the peak amplitude of this input signal is known, a value of base resistance can then be selected, allowing the flow of base current to be held at the desired value.

For a BJT not yet in saturation, the collector load current is a product of the β_{dc} of the transistor and the base current. When determining this value of collector load current for a BJT switch, the minimum specified value of β_{dc} should be used:

$$I_C = \beta_{dc} \times I_B \qquad (1\text{–}11)$$

Once the collector load current has been determined, a minimum value of load resistance may be approximated by dividing V_{CC} by the minimum value of load current. (With the transistor in saturation, the voltage developed over the load resistor will nearly equal V_{CC}.)

$$R_L = \frac{V_L}{I_L} \cong \frac{V_{CC}}{I_L} \qquad (1\text{–}12)$$

Figure 1–43
Transistor switch serving as an output buffer
for a logic circuit

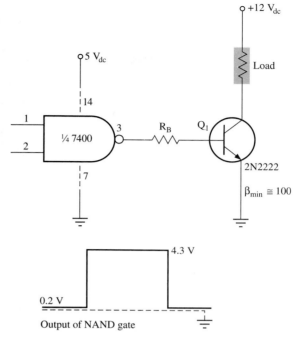

Example 1–8

For the transistor switch in Figure 1–43, assume that the output current from the transistor transistor logic (TTL) NAND gate must be limited to 200 μA, whereas the logic 1 produced at the output of this gate is nearly 4.3 V. Assume that the minimum value of β_{dc} for the BJT is 100. Given these conditions, solve for the minimum value of load resistance and the value of base resistance that would ensure saturation.

Solution:
Accounting for V_{BE}, V_{RB} is derived as follows:

$$V_{RB} = V_{in}(\text{peak}) - V_{BE} = 4.3 \text{ V} - 0.7 \text{ V} = 3.6 \text{ V}$$

$$R_B = \frac{V_B}{I_B} = \frac{3.6 \text{ V}}{200 \text{ }\mu\text{A}} = 18 \text{ k}\Omega \quad \text{(which is a standard value)}$$

The value of collector current is then determined:

$$I_L = I_C = \beta_{dc} \times I_B = 100 \times 200 \text{ }\mu\text{A} = 29 \text{ mA}$$

A minimum value of load resistance may now be derived:

$$R_L = \frac{V_{CC}}{I_L} = \frac{12 \text{ V}}{20 \text{ mA}} = 600 \text{ }\Omega$$

In Figure 1–44, the collector load for the transistor switch is now the coil of a DIP relay, which would utilize the same 5-V power source as the TTL circuitry. With a dc resistance of 250 Ω, the relay coil would require 20 mA of collector current in order to

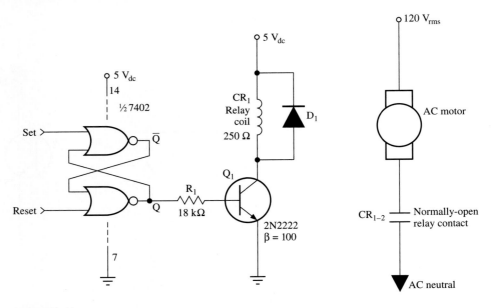

Figure 1–44
Transistor switch with DIP relay coil as the collector load

actuate its armature:

$$I_{CR_1} = \frac{V_{CR_1}}{R_{CR_1}} = \frac{5\ V}{250\ \Omega} = 20\ mA$$

Assuming a beta of 100, a base current of 200 μA would then be required:

$$I_B = \frac{I_{CR_1}}{\beta_{dc}} = \frac{20\ mA}{100} = 200\ \mu A$$

Thus, assuming a logic 1 from the NOR gate of 4.3 V, an 18-kΩ base resistor would be selected (as in the previous example).

In Figure 1–44, with the reset input at a logic 1 and the set input at a logic 0, the Q output of the NOR latch would be kept low. With this condition, the transistor switch would be biased off and the relay coil would be de-energized. This would cause the ac motor, which is in series with the relay's NO contact, to be held in the off condition. Assuming that the reset input to the NOR latch has already fallen low, at the instant the set input receives a high pulse, the Q output will be latched at a logic 1. As a result, the transistor switch will be biased on and the relay coil will be energized, causing the NO contacts to be closed and the motor to be turned on.

The purpose of D_1 in Figure 1–44 is to protect the transistor switch from the transient response of the relay coil. When the NOR latch is again reset and the transistor switch is suddenly biased off, the relay coil will attempt to induce a counter emf within itself, opposing the sudden change in collector current. With the diode connected as shown, as the magnetic field of the relay coil collapses, the induced current will be shunted to the 5-V source rather than through the transistor toward the ground. Since its function is to

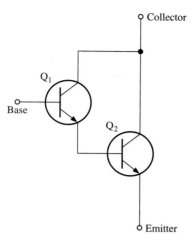

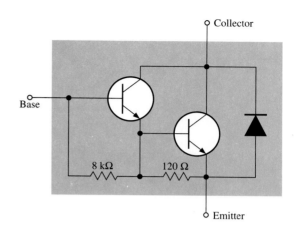

(a) Basic Darlington pair made with two NPN transistors

(b) Internal circuitry of the TIP120 Darlington power transistor circuit

Figure 1–45
The Darlington pair

protect the transistor from the inductive kick, D_1 is sometimes referred to as a **kick-back diode.**

For many industrial switching applications, a single transistor buffer may not provide an adequate boost in current. Where an extremely high current gain is necessary, two NPN or two PNP transistors may be coupled to form a **Darlington pair.** An NPN example of a Darlington pair is shown in Figure 1–45(a). As shown in Figure 1–45(b), the Darlington pair is available in a monolithic package. (Note that a kick-back diode is provided as part of the integrated circuit package.)

For a Darlington pair, current gain becomes the product of the individual dc current gains of the two transistors. Therefore,

$$\beta' = \beta_1 \times \beta_2$$

where β' represents the overall current gain.

If the two transistors are evenly matched and have nearly the same current gains, then

$$\beta' = \beta^2$$

Example 1–9

Assume for the Darlington pair in Figure 1–46 that the minimum value of beta for each of the transistors is 85. Determine the minimum values of base and load resistance that may be used if the output current from the control circuit must be no greater than 200 μA. Also determine the maximum power dissipations of the load and the Darlington pair.

Solution:
With a peak input amplitude of 9 V, the voltage developed over the base resistance would be determined as follows:

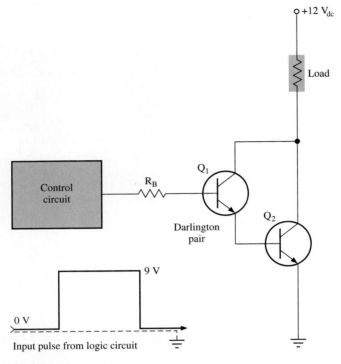

Figure 1–46
Darlington pair used to boost the flow of current for a dc load

$$V_{RB} = V_{in}(\text{peak}) - (V_{BE_1} + V_{BE_2}) = 9\text{ V} - 1.4\text{ V} = 7.6\text{ V}$$

$$R_B = \frac{V_{RB}}{I_{RB}} = \frac{7.6\text{ V}}{200\ \mu\text{A}} = 38\text{ k}\Omega \quad (\text{Use a 39-k}\Omega \text{ standard value.})$$

Current gain for the Darlington pair may be easily approximated as follows:

$$\beta' = 85^2 = 7225$$

The value of load current may now be determined:

$$I_L = \beta' \times I_{RB} = 7225 \times 200\ \mu\text{A} = 1.445\text{ A}$$

Solving for a minimum value of load resistance:

$$R_L = \frac{V_{CC} - V_{CC}(\text{sat})}{I_L} = \frac{12\text{ V} - 0.2\text{ V}}{1.445\text{ A}} = 8.17\ \Omega$$

The excellent efficiency of the Darlington pair becomes evident when its power consumption is calculated and compared to that of the load. For the resistive load:

$$P_{RL} = V_{RL} \times I_{RL} = 11.8\text{ V} \times 1.445\text{ A} = 17.05\text{ W}$$

For the Darlington pair:

$$P_D = V_{CE}(\text{sat}) \times I_{RL} = 0.2\text{ V} \times 1.445\text{ A} = 289\text{ mW}$$

The last result in Example 1–9 indicates that two 250 mW transistors forming a Darlington pair could safely control a load of nearly 20 W. (This, of course, is with the condition that the transistors will be operated only in the saturation or cutoff state). Because of this excellent power-handling capability, Darlington switching transistors are often used as the final output stages in dc-motor, stepper-motor and servomotor control circuitry, which are covered in detail later in this text.

Figure 1–47 demonstrates the ability of a Darlington pair to provide an extremely high impedance buffer between a low-power control circuit and a dc load, which in this example is a 157-Ω relay coil. Assuming the β_{dc} for each transistor to be 100, the amount of base current required for actuation of the relay would be extremely small:

$$\beta' = \beta_1 \times \beta_2 = 100 \times 100 = 10{,}000$$

$$I_C = \frac{V_{CC} - V_{CC}(\text{sat})}{r_{\text{coil}}} = \frac{11.8\text{ V}}{157\ \Omega} = 75.16\text{ mA}$$

The minimum value of base current necessary for saturation can be easily determined as follows:

$$I_B = \frac{I_C}{\beta'} = \frac{75.16\text{ mA}}{10{,}000} = \text{nearly } 7.5\ \mu\text{A}$$

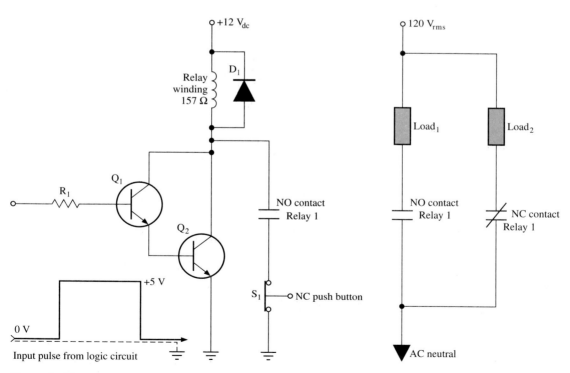

Figure 1–47
Application of a Darlington pair as the current buffer for a relay circuit

The maximum value of base resistance can then be determined:

$$R_\text{B} = \frac{V_\text{in}(\text{peak}) - (V_{\text{BE}_1} + V_{\text{BE}_2})}{I_\text{B}} = \frac{5\,\text{V} - 1.4\,\text{V}}{7.5\,\mu\text{A}} = 480\,\text{k}\Omega$$

(Use a 470-kΩ standard value for the base resistor.)

Note that in Figure 1–47 the relay is latched on by one of its own NO contacts. For this reason, the pulse width of the signal occurring at the base of the Darlington pair would need to be of sufficient length to allow the armature of the relay to be fully actuated. Once the relay is latched on, the base signal could return to 0 V. When the transistors are biased off, the relay coil may be de-energized through pressing the NC push button S_1. (Also note that the Darlington pair is protected by a kick-back diode.)

In addition to the BJT, the field effect transistor (FET) is often used in switching applications. Whereas the Darlington pair is characterized by an input impedance of several thousand ohms, the input impedance of a typical junction FET (JFET) is likely to be thousands of mega-ohms. For the metallic oxide semiconductor FET (MOSFET), input impedance approaches infinity. While JFETs are used prevalently in relatively low-power instrumentation devices such as voltmeters, **power MOSFETs** are used in many of the same circuit applications as the Darlington pair.

Figure 1–48(a) illustrates the internal structure of an **n-channel enhancement-mode** MOSFET (E MOSFET). The schematic symbol for this device is shown in Figure 1–48(b). For the E MOSFET, a channel does not exist until the gate terminal is made more positive than the source. When this occurs, as shown in Figure 1–49, a path for minority carriers is induced between the source and the drain terminals. For the n-channel E MOSFET, the channel is created within the p-type substrate as electrons are attracted toward the silicon dioxide insulating material. As the positive voltage at the gate is increased, this concentration of electrons increases, as does the drain current.

With the gate voltage at zero, the resistivity between the source and drain would be extremely high. As shown in Figure 1–50(a), this condition could be compared to that

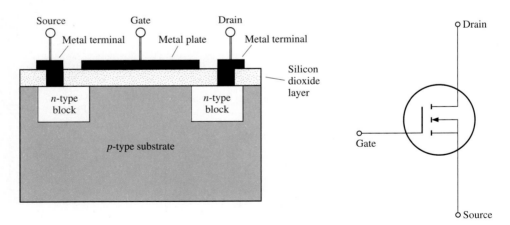

(a) Structure of n-channel enhancement mode MOSFET

(b) Schematic symbol for the n-channel enhancement mode MOSFET

Figure 1–48
The n-channel E MOSFET

Figure 1–49
Operation of the n-channel E MOSFET

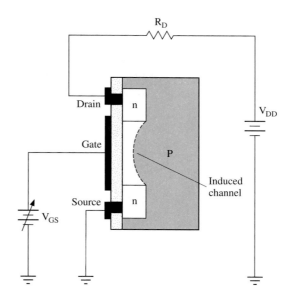

of an open switch or to that of a BJT in the cutoff condition. However, if the gate voltage of the *n*-channel E MOSFET is made sufficiently positive, the resistivity between the drain and source will be greatly reduced, due to the presence of the induced channel. With this condition, as represented in Figure 1–50(b), the operation of the E MOSFET could be compared to that of either a closed switch or to that of a BJT in saturation.

Figure 1–51 illustrates how a single power E MOSFET can be used to energize directly a dc motor. If the input signals to the gate of the E MOSFET are rectangular, simple on/off control of the motor can be maintained. Also, linear changes in gate voltage

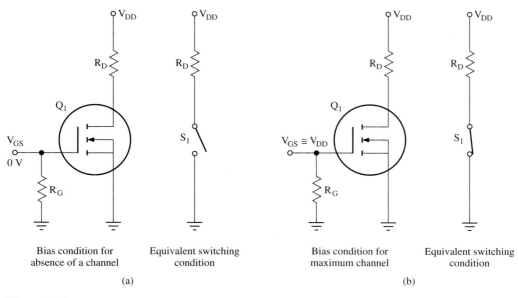

| Bias condition for absence of a channel | Equivalent switching condition | Bias condition for maximum channel | Equivalent switching condition |

(a) (b)

Figure 1–50

Figure 1–51
Power E MOSFET used to energize a dc motor

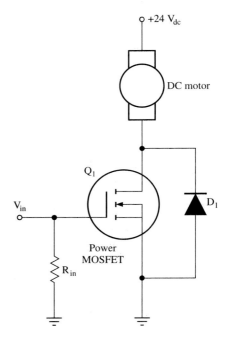

would allow proportional changes in motor speed. In applications where motor speed must be precisely controlled, a linear amplifier could precede the gate input. Such motor control circuitry is covered in detail later in this text.

SECTION REVIEW 1–3

1. In Figure 1–52, for the transistor to be in a saturated condition, collector current would equal _____ mA.
2. To ensure that the transistor in Figure 1–52 is held in a saturated condition, base current should equal at least _____ μA.
3. Assuming a logic high at point A in Figure 1–52 that equals nearly 4.1 V, which of the following values of R_1 would ensure saturation of Q_1?:
   a. 10 kΩ
   b. 22 kΩ
   c. 100 kΩ
   d. 4.7 kΩ
4. Briefly explain the purpose of D_1 in Figure 1–52.
5. Assume for the NAND latch in Figure 1–52 that point C is low, point D is high, and both inputs are at a logic high. With these logic conditions, load 1 will be _____ and load 2 will be _____.
6. Given the logic conditions specified in Review Problem 5, assume input A pulses low while input B remains at a logic high. As a result of this action, load 1 will be _____ and load 2 will be _____.
7. For the Darlington pair of Figure 1–53, the effective beta would equal nearly _____.

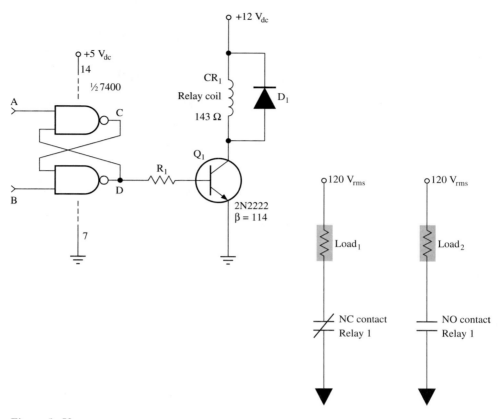

Figure 1–52

8. Assume that the motor in Figure 1–53 will draw 800 mA. To ensure that the Darlington pair switches on in saturation, the base current should equal at least _____ μA.

9. Assume a logic high from the NAND gate in Figure 1–53 will attain a level of 4.2 V. To ensure that the Darlington pair turns on in saturation, R_1 should equal
   a. 68 kΩ
   b. 56 kΩ
   c. 82 kΩ
   d. 100 kΩ

10. When both inputs to the NAND gate in Figure 1–53 are at a logic high, the Darlington pair will be biased _____ and the motor will be in the _____ condition.

1–4   HALL EFFECT AND OPTOELECTRONIC SWITCHING DEVICES

The microswitch, mercury switch, and float switch have already been introduced as position- and level-sensing devices. A disadvantage of mechanical proximity sensors, such as limit switches, is the fact that they must actually make physical contact with

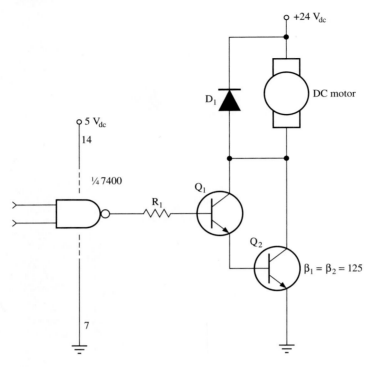

Figure 1–53

moving objects in order to sense their position. For this reason, such switches are subject to frequent adjustment and eventual replacement. Although mercury switches are position sensors that do not require physical contact with an object, the fact that they are contained in glass capsules makes them highly susceptible to breakage. The ideal position-sensing device would not require physical contact with a moving object in order to determine its proximity. Also, it would be virtually unbreakable and be able to operate at very high speeds. The **Hall effect sensor** is a solid-state electromagnetic device that satisfies these requirements and is, therefore, being used increasingly as a proximity sensor in both industrial and automotive electronics.

The **Hall effect,** which was observed as early as 1879 by the physicist E. H. Hall, occurs as a result of a current-carrying conductor being placed perpendicular to a magnetic field. The Hall effect may be defined as the development of an actual voltage across the conductor as a result of the deflection of charge carriers as they pass through the magnetic field. This slight difference in potential induced across the conductor is called the **Hall voltage** (V_H).

In recent years, scientists experimenting with various crystalline semiconductor materials discovered that the Hall effect could be greatly enhanced. This led to the development of the Hall effect sensor. As illustrated in Figure 1–54, if magnetic flux lines cut across the plate of indium arsenide at the same instant current flows through it, a Hall voltage develops. One of the most important features of the Hall sensor is the fact that the Hall voltage is directly proportional to the amount of magnetic flux (β) cutting across the semiconductor. This linear relationship of V_H and β has led to the use of the Hall effect sensor within the **gaussmeter,** which is an electronic instrument used to detect the strength of a magnetic field.

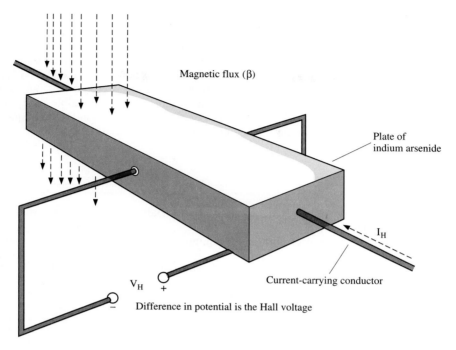

Magnetic flux (β)

Plate of
indium arsenide

I_H

Current-carrying conductor

V_H +

−

Difference in potential is the Hall voltage

Figure 1–54
The Hall effect in a piece of indium arsenide crystal

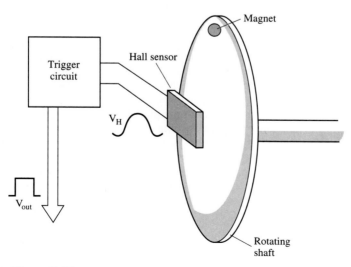

Magnet

Trigger
circuit

Hall sensor

V_H

V_{out}

Rotating
shaft

Figure 1–55
Hall effect sensor used to measure rpms

Hall effect sensors are especially effective as angular position sensors, serving as input transducers for monitoring the speed of rotating shafts within electrical motors and automobile engines. A simple method of measuring shaft rpms with a Hall effect sensor is illustrated in Figure 1–55. As the magnet mounted on the rotating shaft approaches the sensor, the magnetic flux approaches maximum. Therefore, the Hall voltage will also approach a peak level. If additional triggering circuitry is present, the pulses provided by the Hall sensor can be made rectangular in order to be compatible with digital control circuitry. A microcomputer could be easily programmed to monitor the frequency of these pulses and thus determine the angular velocity of the rotating shaft.

SECTION REVIEW 1–4

1. Describe the Hall effect. Explain how a Hall voltage may be produced within a semiconductor.
2. Identify a type of electronic test equipment in which a Hall effect sensor is likely to be found.
3. Describe a common industrial application of a Hall effect sensor.

SUMMARY

Manually operated mechanical switches exist in a variety of forms, including SPST, SPDT, DPST, and DPDT. Basic mechanical switching actions include toggle, rocker, rotary, and slide actions. Push-button switches are also prevalently used in manual control applications. Methods of actuation for these devices include push-button and rotary actuation.

Microswitches are mechanical switching devices commonly used in industrial control. They are often referred to as limit or proximity switches, because they often sense the horizontal position of moving machine parts. Mercury bulb switches are also used as angular position sensors.

Electromechanical relays are commonly used switching devices that are capable of providing electrical isolation between low-power control sources and high-power loads. Since the typical relay has multiple contacts, it is capable of controlling several loads. DIP relays are especially designed for mounting on printed circuit boards. Reed relays are high-speed switching devices controlled by electromagnets.

Solenoids, similar to relays, are electromechanical devices containing armatures that are actuated by the flow of current through a coil. Solenoids are manufactured in a variety of sizes and are designed to operate from either dc or ac power sources. These devices are used to perform such tasks as opening and closing valves or operating clutches, or opening and closing diverter gates in conveyance systems.

Solid-state devices are rapidly replacing mechanical and electromechanical switches in many industrial switching applications. Switching rapidly between saturation and cutoff, BJTs efficiently control loads placed in series with their main terminals. Transistors are often used to drive relay coils, which in turn control high-power loads. Darlington pairs and power MOSFETs may be used to control directly medium-power dc loads.

Hall effect sensors, which are transducers that convert changes in magnetic flux to changes in voltage, are used as high-speed angular position sensors. They are now used widely in both automotive and industrial electronics to measure the rpms of rotating shafts.

SELF-TEST

1. Draw the schematic symbols representing the SPST, SPDT, DPST, and DPDT manually operated switches.
2. Draw the schematic symbols representing the NO and NC forms of limit switch.
3. For the control circuit in Figure 1–56, determine the condition of the 12-V lamp during the time in which the enable input is at a logic low.
4. When a logic high occurs at point A in Figure 1–56, what would be the value of the base current entering the Darlington pair?
5. What is the effective beta for the Darlington pair in Figure 1–56? Is it sufficient to allow saturation of the Darlington pair?
6. In Figure 1–56, how much current will flow through the relay coil when it is energized?
7. What is the purpose of D_1 in Figure 1–56?
8. In Figure 1–56, what is the purpose of the relay contact that is placed in parallel with Q_2?
9. When a logic high occurs at point A in Figure 1–56, what will be the condition of the motor and the 12-V lamp?
10. In Figure 1–56, what will happen to the motor and lamp at the instant the pulse at point A falls to a logic low?

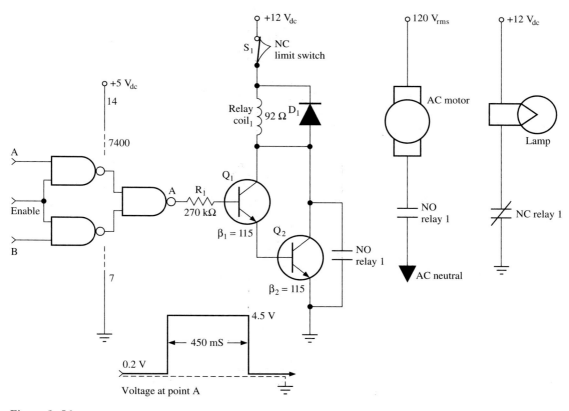

Figure 1–56

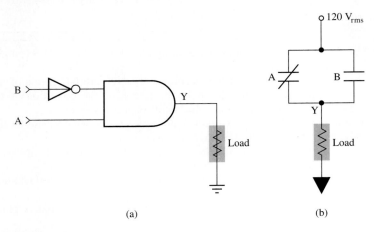

(a)   (b)

Figure 1–57

11. Once the motor in Figure 1–56 is energized, how will it be switched off?

12. Sketch the arrangement for the contacts of two relays and a resistive load that would satisfy the logic diagram shown in Figure 1–57(a). Write the Boolean expression for the logic diagram.

13. Sketch the logic diagram and write the Boolean expression for the relay contact arrangement in Figure 1–57(b).

14. For the load in Figure 1–57(b) to be energized, what should be the condition of the coils of the two relays?

2 OPERATIONAL AMPLIFIERS AND COMPARATORS

CHAPTER OUTLINE

LEARNING OBJECTIVES

On completion of this chapter, the student should be able to:

- Describe the overall structure and operation of a voltage-differencing op amp.
- Describe the overall structure and operation of a current-differencing op amp.
- Identify the inverting, noninverting, summing, and differential amplifier configurations for the op amp. Describe circuit operation and perform gain calculations for these four basic op amp configurations.
- Describe the operation of an instrumentation amplifier. Explain the purpose of the op amp voltage-follower within that device.
- Explain the operation of op amp low-pass, high-pass, and bandpass active filters. Explain their classification with regard to roll-off and frequency response.
- Explain the operation of op amp integrators and differentiators.
- Explain the basic operation of op amp comparators and Schmitt triggers.

INTRODUCTION

The operational amplifier may well be the most versatile form of integrated circuit in use in modern electronic systems. It has found a place in almost all forms of analog devices, including dc and ac amplifiers, filters, oscillators, pulse circuits, and many forms of signal conditioning and instrumentation circuits. This chapter serves as a concise introduction to the **op amp,** with emphasis placed on those applications that relate most directly to industrial electronics. (The subject of operational amplifiers and their nearly limitless applications could easily become the subject of an entire text or even a set of texts.) Also in this chapter, the comparator (sometimes referred to as a level detector) is introduced and a few simple circuits demonstrating its operation are presented.

2-1   INTERNAL STRUCTURE OF THE OP AMP AND OP AMP PARAMETERS

The idea of creating an all-purpose amplifier that could be adapted to many linear and nonlinear applications came about during World War II, as an effort of the National Defense Research Council. Credit for much of the work in this area is given to George A. Philbrick, who carried the concept of the ''operational amplifier'' into the commercial arena. This original unit, of course, was quite large, consisting primarily of vacuum tubes. As the electronic industry progressed, the op amp too continued to evolve. By the early 1960s, as the op amp was made available in a convenient modular package, it began to enjoy increased attention from design engineers. The op amp of today exists in a monolithic form and is available in integrated circuit (IC) packages that may contain one, two, or even four devices.

The op amp is basically a **differential amplifier.** It has two inputs, which are identified as inverting and noninverting. Differential amplifiers (often simply called diff amps) are easily fabricated monolithically, provide isolation of input and output dc reference levels, and have excellent noise immunity. Figure 2–1(a) contains the schematic symbol for the op amp. The noninverting input is symbolized by the + sign and the inverting input by the − sign. (The triangle symbol is generally used to represent a buffer or amplifier). Most op amps require two power supply connections, which are normally equal voltages of opposite polarity. The positive supply is represented as $+V_{CC}$, while the negative supply is identified as $-V_{EE}$.

Most op amps are **voltage-differencing amplifiers** (VDAs). That is, they will amplify a difference in voltage that exists between the two input terminals. As indicated in Figure 2–1(b), this **differential input voltage** may be symbolized as V_{id}. For a VDA, a change in the output voltage (V_o) would be directly proportional to the changing difference between the input voltages. **Current-differencing amplifiers** (CDAs) also exist, an example of

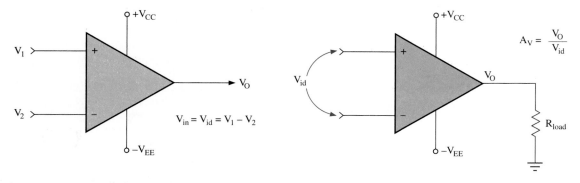

(a) Schematic symbol for the voltage-differencing op amp   (b) Function of voltage-differencing op amp

Figure 2–1

which would be the **Norton op amp.** For this device, a change in the output voltage is directly proportional to the changing difference between the two input currents.

For the student of electronics whose experience with amplifiers has been limited to discrete transistor circuits with only one input, the differential amplifier will require some preliminary explanation. For single-input linear amplifiers, regardless of the transistor configuration being used, the amplitude of the output signal is the product of the input voltage and the factor of voltage gain:

$$V_o = V_{in} \times A_v \tag{2-1}$$

With a VDA, however, the output voltage, whether ac or dc, is a product of the algebraic difference between the two input voltages and the factor of voltage gain:

$$V_o = (V_1 - V_2) \times A_v \tag{2-2}$$

As an example, consider a condition where V_1 is 35 mV, V_2 is 50 mV, and A_v is 100. As shown in Figure 2–2(a), V_1 is occurring at the noninverting input, whereas V_2 appears at the inverting input. The amplitude and polarity of the output signal may now be determined as follows:

$$V_o = (35\,\text{mV} - 50\,\text{mV}) \times 100$$
$$= -15\,\text{mV} \times 100 = -1.5\,\text{V}$$

Next, as shown in Figure 2–2(b), assume that both inputs equal exactly 100 mV. According to the formula for voltage gain, the output voltage would, ideally, become zero:

$$V_o = (100\,\text{mV} - 100\,\text{mV}) \times 100$$
$$= 0 \times 100 = 0\,\text{V}$$

Figure 2–3 contains a functional block diagram of a typical VDA. Although op amps vary greatly in the design of their internal circuitry, they generally follow the basic format illustrated here. The **MC1435** VDA, the internal circuitry of which is shown in Figure 2–4, closely adheres to this design concept. Transistors Q_1, Q_2, and Q_3 comprise the input stage, which consists of a balanced input/balanced output differential amplifier. Since Q_1 and Q_2 represent a matched pair of transistors, they would have nearly the same β. Thus, if the device is operating in the common mode (with the same voltage present at both

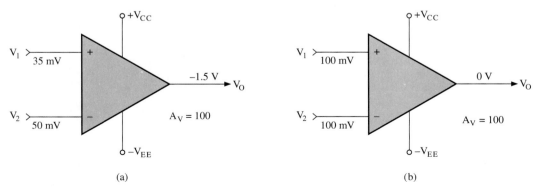

Figure 2–2
Operation of the voltage-differencing op amp

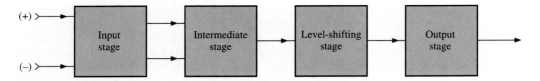

Figure 2–3
Functional block diagram of a typical op amp

inputs), the voltage at the collector of Q_1 should equal that present at the collector of Q_2, causing the differential input to the next stage of the device to equal virtually 0 V. The collector of Q_3 serves as a constant current source, enhancing the operating stability of the input stage.

The intermediate stage of the op amp, containing Q_4 and Q_5, consists of a balanced input/unbalanced output differential amplifier. Most of the actual voltage gain for the device has been achieved by the time the input signal has reached the collector of Q_5. Here, Q_6 is in an emitter-follower configuration and serves as the level-shifting stage; Q_7 and Q_8 form a complementary pair and serve as the output stage.

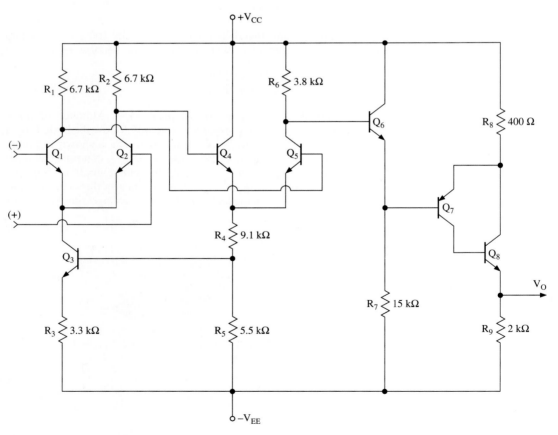

Figure 2–4
Internal circuitry of MC1435 op amp

Prior to examining the operating parameters of a typical op amp, we will consider the characteristics of the ideal op amp. The ideal op amp, of course, does not exist. However, the concept does serve as a target point for op amp design. Also, in many op amp applications, within certain limitations of frequency and voltage gain, ideal operation of the device may be assumed. As an example, consider the parameter of voltage gain. For the ideal op amp, open-loop voltage gain would approach infinity. That is, unless an external path (or loop) for negative feedback exists, regardless of how slight the difference in voltage between the two input terminals might be, the output voltage would saturate toward the level of either $+V_{CC}$ or $-V_{EE}$. For example, if both the inverting and noninverting inputs to the ideal op amp are at exactly the same voltage level, the output would be exactly 0 V. However, if the voltage at the noninverting input were to become slightly more positive than that present at the inverting input, the output would immediately saturate in the positive direction. For the ideal op amp, this rise in output voltage would be instantaneous. Conversely, if the voltage present at the inverting input were to exceed the level present at the noninverting input, the output would instantly saturate in the negative direction. The open loop action of the ideal op amp is illustrated in Figure 2–5.

The ideal op amp would have no loading effect with respect to the circuitry preceding it or following it. Therefore, the input impedance of the ideal op amp would approach infinite ohms, whereas its output impedance would be 0 Ω. (The concept of infinite input impedance applies to both the inverting and noninverting inputs of the device.) Although the real-world op amp does have finite values of input and output resistance, these parameters are close enough to the ideal that they are not even considered when designing many basic op amp circuits.

Finally, since the output of the ideal op amp has virtually instantaneous rise and fall times, it would also have virtually limitless frequency response. Therefore, we could assume that the ideal op amp would have infinitely high open-loop voltage gain, regardless of the frequency of an input signal. However, it is in this area of response time that the real-world op amp deviates furthest from the ideal. Therefore, particular care should be taken when selecting an op amp for circuit applications that require fast switching or high-frequency operation. Extensive technical information about various types of linear ICs is made available by their manufacturers. A basic understanding of this information is necessary for the technician who works with industrial control circuits. For the real-world op amp, these parameters may be considered as representing deviations from those ideal op amp characteristics that we just presented.

Input Resistance

Input impedance for any op amp may be expected to be very high. However, the actual value of this parameter varies greatly among op amps, due mainly to differences in structure. As an example, for the **μA741** op amp, input resistance may vary between 300 kΩ and 1 MΩ. For the **μA714** precision op amp, input resistance is higher yet—at least 15 MΩ. For op amps manufactured with bi–FET technology, input impedance most closely approaches the ideal level of infinity. As an example, with the **μAF771,** input resistance is likely to be 1,000,000,000,000 Ω! The input resistance of an op amp is sometimes specified as its **differential input resistance,** which is defined as the resistance that could be measured directly between either one of the input terminals and circuit ground. Input resistance may also be defined as the ratio of input voltage change to input current change, again measured at a single input terminal. This relationship of input current and voltage may be shown as:

$$R_{id} = \frac{\Delta V_{\text{in}}}{\Delta I_{\text{in}}} \qquad (2\text{--}3)$$

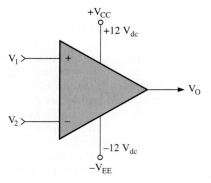

Ideal op amp

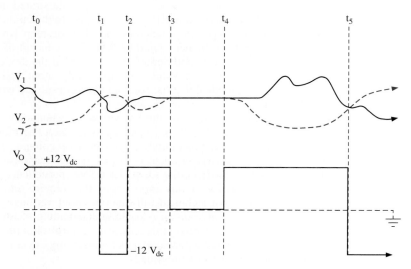

Figure 2–5
Operation of the ideal op amp where A_v equals ∞

Output Resistance

The output resistance of an op amp is likely to be very low, usually less than 100 Ω. (As an example, for the μA741 it is typically 75 Ω.) Output resistance is measured between the output terminal and circuit ground and may also be considered as a ratio of output voltage change to output current change:

$$R_o = \frac{\Delta V_o}{\Delta I_o} \tag{2–4}$$

Large Signal Voltage Gain

For a VDA, voltage gain is defined as the ratio of the output voltage to the difference in potential present between the input terminals:

$$A_v = \frac{V_o}{V_{id}} \qquad (2\text{--}5)$$

The large signal voltage gain is measured with an output load resistance of at least 2 kΩ but without a path for feedback from the output terminal to either of the input terminals. For this reason, the parameter is often referred to as the **open-loop voltage gain.** Since, for the ideal op amp, voltage gain would approach infinity, the output voltage would saturate toward the level of V_{CC} or V_{EE}, no matter how slight the value of V_{id}. (Only when V_{id} is zero will the output voltage be zero.) The voltage gains of actual op amps vary greatly but are always extremely high. For the **μA741,** minimal A_v is 25 k. For the **μA714,** A_v is typically 300 k. The action of an op amp in the open-loop configuration is similar to that of a **comparator** (level detector). In fact, an op amp in this open-loop configuration could be used as a level detector but should not be expected to perform as reliably as a precision comparator such as the **LM311.** (Comparators are studied in detail at the end of this chapter.)

Slew Rate

The slew rate (SR) of an op amp is simply its response time. More precisely, it may be expressed in the form of a differential, which is defined as the rate of change of the output voltage within a specified period of time:

$$\text{SR} = \frac{\Delta V_o}{\Delta t} \qquad (2\text{--}6)$$

In an actual op amp circuit, where a path for negative feedback is likely to exist, the slew rate of an op amp will vary with the circuit voltage gain. For this reason, a slew rate is usually specified where the voltage gain has been adjusted to be unity.

Closely associated with the problem of slew rate are the op amp's **frequency response** and **bandwidth.** Op amps with a low slew rate should never be used in amplifiers subjected to high-frequency signals. For example, the **μA741C** op amp has a slew rate of only 0.5 V/μsec and, for that reason, is likely to perform very badly as the voltage amplifier in a function generator. For such an application, a high-speed op amp such as the **LM318,** which has a slew rate of around 70 V/μsec, would perform far better.

Figure 2–6(a) contains an experimental circuit for testing the slew rate of an op amp. First, a **μA741C** could be used, with a squarewave (±5-V peak) being fed to the noninverting input. As frequency is gradually increased from 1 to 100 k pulses per second (pps), the output signal deteriorates, going through the stages represented in Figure 2–6(b). Next, we could replace the **μA741C** with an **LM318** and repeat the test procedure. Since this second op amp has a far higher slew rate, the output waveform undergoes far less distortion as the input frequency approaches 100 k pps.

Input Offset Voltage and Current

For the ideal op amp, if there is no difference in potential occurring between the two input terminals, the output voltage must be zero. For a real-world op amp, however, even if the same voltage were applied to the two input terminals simultaneously, the output voltage would not be exactly zero. Rather, it is likely to settle at a dc level slightly above or below 0 V. The **input offset voltage** (V_{io}) is the difference in potential that must occur between the two input terminals in order to nullify the output voltage (i.e., hold it at 0 V). Since the polarity of this required difference in potential between the inputs depends on the polarity of the dc level to which the output tends to deviate, input offset voltage is expressed as an absolute value. For the **μA741,** the input offset voltage could be as

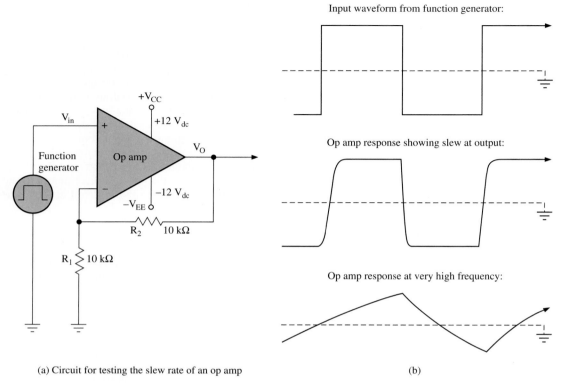

(a) Circuit for testing the slew rate of an op amp   (b)

Figure 2–6
Testing the slew rate of an op amp

much as 5 mV. However, a typical value for this device would equal 1 mV. For a precision op amp such as the μA714C, the input offset voltage would have a maximum level of only 150 μV.

The input offset voltage is actually an indication of the symmetry of the differential amplifier stages of an op amp. For example, referring back to Figure 2–4, assume that 3 V_{dc} is applied to both input terminals. If transistors Q_1 and Q_2 were perfectly matched with regard to *all* operating parameters and both of their collector resistors were exactly the same ohmic value, then the difference in potential between the collectors of Q_1 and Q_2 would be exactly 0 V. The closer this internal difference in potential is to 0 V when a common input voltage is applied, the smaller the required level of input offset voltage.

Figure 2–7(a) illustrates the concept of input offset voltage. With this op amp test circuit, once R_1 and R_2 are made exactly equal, either V_1 or V_2 is adjusted until the output voltage becomes exactly zero. The slight difference in potential which then occurs between the two input terminals represents the input offset voltage.

Closely associated with the input offset voltage is the **input offset current (I_{io}),** which is simply the difference between the two inputs currents that would exist in a condition where the input offset voltage is equal to zero. The input offset current could be determined by connecting both inputs to a common, low-amplitude voltage source, as illustrated in Figure 2–7(b). After the ohmic values of R_1 and R_2 are made exactly equal, both input currents would then be carefully measured. The absolute value of the *difference* between the two currents would represent the input offset current.

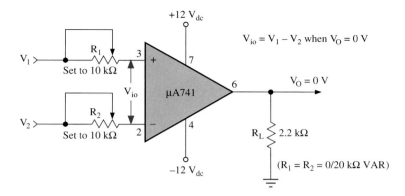

$V_{io} = V_1 - V_2$ when $V_O = 0$ V

(a) Test circuit for input offset voltage

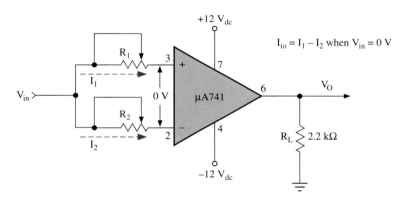

$I_{io} = I_1 - I_2$ when $V_{in} = 0$ V

(b) Test circuit for input offset current

Figure 2–7
Test circuits

While input offset current is a parameter that is seldom directly measured, input offset voltage must often be monitored and controlled because it represents a potential source of error in many forms of op amp instrumentation circuitry.

Example 2–1

Assume for the circuit in Figure 2–8 that both R_1 and R_2 are adjusted to exactly 10 kΩ and that V_1 and V_2 are both adjusted to exactly +3 V_{dc}. With these input conditions, the measured voltage between the output and circuit ground is slightly above 0 V. After the level of V_2 is slowly adjusted, the output voltage is brought to exactly zero. However, in making this correction, the value of V_2 was changed to 3.073 V. What is the input offset voltage for the op amp?

Solution:
With V_o at zero:

$$V_{io} = V_1 - V_2 = 3 \text{ V} - 3.073 \text{ V} = -0.073 \text{ V}$$

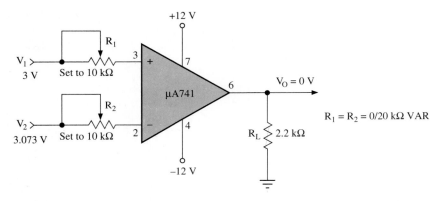

Figure 2–8
Op amp test circuit with input offset voltage

As an absolute value:

$$V_{io} = 73 \text{ mV}$$

Since, in Example 2–1, the impedance of either the inverting or noninverting inputs of the op amp itself is likely to be much greater than the ohmic value of R_1 or R_2, it may be assumed that V_1 will virtually equal the voltage at pin 3 and V_2 will virtually equal the voltage present at pin 2.

A few op amps have the built-in capability of an offset null, the μA741 being one such device. As shown in Figure 2–9, the offset null condition may be easily achieved by connecting the main terminals of a 10-kΩ potentiometer between pins 1 and 5 and the wiper to the negative power supply. With R_1 and R_2 preadjusted to exactly the same ohmic value, a common dc input voltage is now applied to the circuit. Proper adjustment of the wiper of R_3 should cause the output voltage to settle at zero volts while V_{id} is still held in a null condition. Offset nulling may be achieved with op amps that do not have

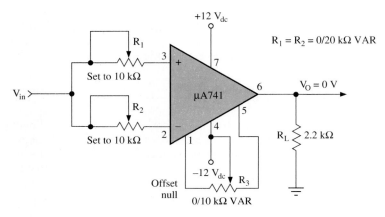

Figure 2–9
Op amp test circuit with offset null capability

the built-in capability, but an extra resistive network must be created just for that purpose. Thus, for precision instrumentation circuits, where offset problems must be eliminated, it would be far better to choose an op amp with the offset null capability.

Common-Mode Rejection Ratio

The **common-mode rejection ratio** (CMRR) is the ability of the op amp to attenuate a signal that is present at both input terminals simultaneously. As stated earlier in this chapter, one of the major advantages of the op amp is its excellent noise immunity. One reason an op amp is able to reject unwanted noise signals is, of course, its very high input impedance. Another equally important reason, however, lies in the fact that any noise signal present at one of the inputs of the device will almost certainly be present at the complementary input. We can also assume that these common signals would be perfectly in phase. For this reason, the input stage of an op amp, because it is a differential amplifier, attempts to cancel this undesired common-mode portion of an input signal. At the same time, any desired input signal, since it is usually fed to one of the two input terminals, is greatly amplified. For the ideal op amp, the CMRR would approach infinity, since that imaginary device would be able to attenuate completely any signal that is occurring in the common mode.

In mathematical terms, CMRR is expressed as the ratio of the open-loop voltage gain to the common-mode voltage gain. Note that open-loop voltage gain is indicated as A_d in order to specify a differential concept, while the common-mode voltage gain is indicated as A_{CM}.

$$\text{CMRR} = \frac{A_d}{A_{CM}} \tag{2-7}$$

Even for a real-world op amp, this ratio would represent an extremely large number and, for this reason, it is often expressed in decibels. As an example, assume for the test circuit in Figure 2–10 that the open-loop voltage gain A_d has already been determined as 32 k. If, as shown, a large sinusoidal signal is fed to both inputs, the output signal should be of a much lower amplitude. This is true due to the attenuating action of the op amp operating in the common mode. For this example, with a common input voltage of ± 10 V, we assume that the output signal is only ± 800 mV. Thus, A_{CM} would be only 0.08, determined as:

$$A_{CM} = \frac{V_o}{V_{in}} = \frac{800 \text{ mV}}{10 \text{ V}} = 0.08$$

As a ratio, CMRR may then be expressed as follows:

$$\text{CMRR} = \frac{A_d}{A_{CM}} = \frac{32 \text{ k}}{0.08} = 400 \text{ k} \tag{2-8}$$

CMRR is expressed more conveniently in decibels, being easily converted from the ratio form:

$$\text{CMRR} = 20 \log(A_d/A_{CM}) \tag{2-9}$$

Using the above example:

$$\text{CMRR} = 20 \log(32 \text{ k}/0.08) = 20 \log(400 \text{ k}) = 112 \text{ dB}$$

For the μA741C op amp, the CMRR is normally around 90 dB, while for the μA714 precision op amp, the CMRR is typically as high as 126 dB. The CMRR, input offset

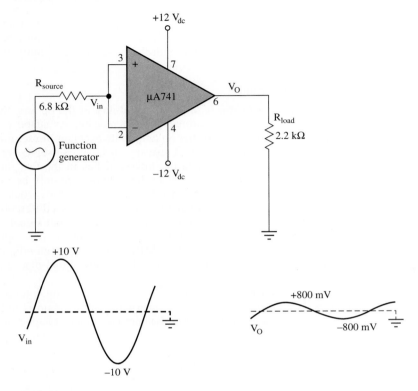

Figure 2–10
Test circuit for CMRR

voltage and input offset current are similar in that they all represent a means of testing the symmetry of the differential amplifier portion of the op amp. In many industrial control applications, op amps such as the μA741 may perform quite satisfactorily. However, in some instrumentation applications, where exact measurements must be made, precision op amps are required. With their typically higher ratings of V_{io}, I_{io}, and CMRR, they may be expected to perform more accurately and reliably.

Example 2–2

For the op amp test circuit in Figure 2–11, assume that the open-loop voltage gain is 53 k. Solve for the CMRR of the op amp. Express your answer both as a ratio and as a decibel rating.

Solution:

$$A_{CM} = \frac{V_o}{V_{in}} = \frac{435 \text{ mV}}{7.8 \text{ V}} = 0.05577$$

As a ratio:

$$\text{CMRR} = \frac{A_d}{A_{CM}} = \frac{53 \text{ k}}{0.05577} = 950.332 \text{ k}$$

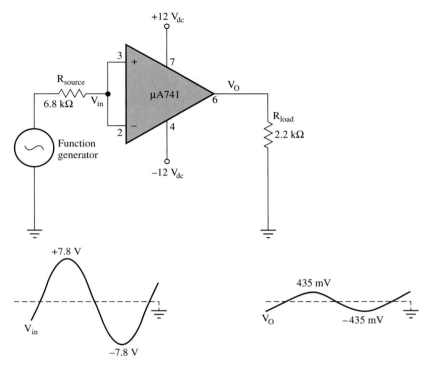

Figure 2–11
Test circuit for CMRR

In decibel form:

$$\text{CMRR} = 20\log(A_d/A_{CM}) = 20\log(950.332\,\text{k}) = 119.56\,\text{dB}$$

SECTION REVIEW 2–1

1. Explain the function of a current-differencing op amp. What type of op amp falls into this category?
2. List three advantages of the differential amplifier (as compared to a single-input device).
3. Explain the effect of the slew rate of an op amp on its frequency response.
4. With the input conditions shown in Figure 2–12, what would be the ideal value of V_{id}? What would be the ideal value of V_o?
5. Describe the purpose of R_3 in Figure 2–12.
6. For the op amp test circuit shown in Figure 2–13(a), given the input and output waveforms of Figure 2–13(b), what is the common-mode voltage gain (A_{CM})?
7. For the op amp test circuit shown in Figure 2–13(a), assume the differential-mode (open-loop) voltage gain (A_d) is 37 k. Using this value and the result of the previous question, determine the CMRR. Express your result both as a simple ratio and in decibels.

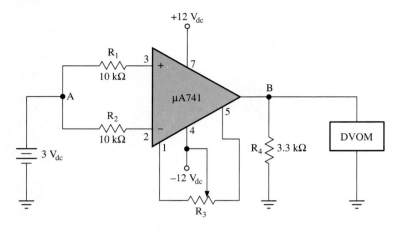

Figure 2–12

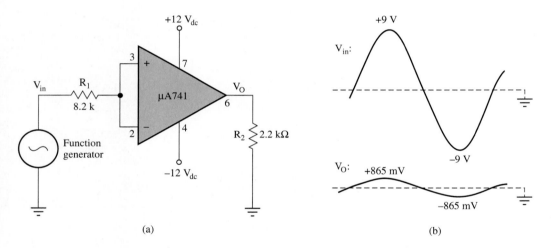

(a)   (b)

Figure 2–13

2-2   OP AMP INVERTING AND NONINVERTING AMPLIFIERS

In this section, a few of the basic configurations commonly used in the design of op amp circuitry are introduced. As these fundamental circuit designs are examined, the convenience and versatility of op amps should become evident. In most dc and low-frequency applications of op amps, ideal conditions may often be assumed. Thus, circuit design becomes very easy, since the internal structure and operation of the op amp may be virtually disregarded. The design process may then be limited to the determination of a few external component values. In higher frequency applications, however, the characteristics of the real-world op amp must be considered, since parameters such as slew rate, frequency response, and bandwidth become crucial. Mastery of the basic op amp circuit configurations presented here is a prerequisite to understanding the more complex forms of op amp signal conditioning circuitry that are now an integral part of industrial process control systems.

The Op Amp Inverting Amplifier

The simplest op amp circuit to fabricate and test is the **inverting amplifier,** which is shown in Figure 2–14(a). With this basic design, only an input resistor (R_{in}) and a feedback resistor (R_f) are required for control of voltage gain. You will recall that an op amp in the open-loop configuration has an extremely high voltage gain. With the circuit design illustrated in Figure 2–14(a), through the application of **negative** (or **degenerative**) **feedback,** this gain may be reduced to a practical value.

Notice in Figure 2–14(a) that the noninverting input to the op amp is tied directly to circuit ground, allowing only the inverting input to be active. Also notice that the inverting input (designated as point A) serves as a junction point for R_{in} and R_f. With this circuit configuration, point A functions as a **virtual ground.** That is, regardless of the large changes in voltage that would occur at both the input and output, point A will tend to hold at nearly 0 V. Assuming for now that U_1 is an ideal op amp, virtually all of the current flowing through R_{in} will bypass the inverting input and flow through R_f. Thus, these two resistors are effectively in series, with point A being equivalent to the circuit ground. This equivalent circuit condition is illustrated in Figure 2–14(b). Assuming point A to be at nearly 0 V, input current may now be estimated through simple application of Ohm's law:

$$I_{in} = \frac{V_{in}}{R_{in}} \qquad\qquad (2-10)$$

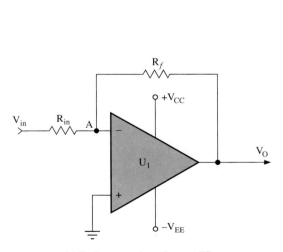

(a) Basic op amp inverting amplifier

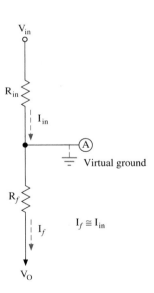

(b) Equivalent circuit for the basic op amp inverting amplifier

Figure 2–14
The op amp inverting amplifier

Remembering that feedback current virtually equals input current and that point A is at nearly 0 V, the approximate value of feedback current may be easily determined as:

$$I_{in} \cong I_f \cong \frac{V_o}{R_f} \tag{2–11}$$

As shown in Figure 2–15(a), assume that R_{in} equals 1 kΩ, R_f equals 10 kΩ, and V_{in} equals +500 mV. With these input conditions, the path for conventional current flow would be from the input to the virtual ground point. Applying Ohm's law, input and feedback current may be approximated as:

$$I_{in} = \frac{V_{in}}{R_{in}} = \frac{500 \text{ mV}}{1 \text{ k}\Omega} = 500 \text{ }\mu\text{A} \qquad \text{Thus, } I_f \cong 500 \text{ }\mu\text{A}$$

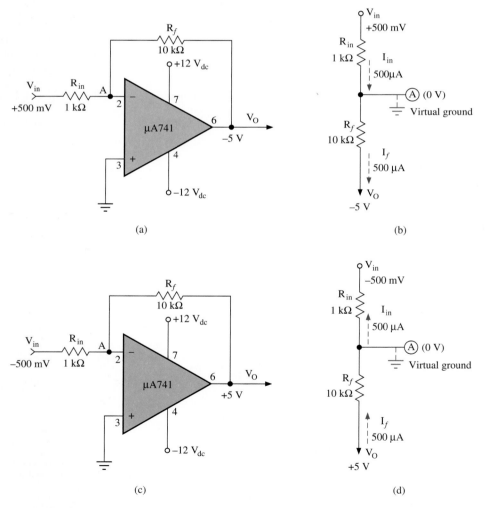

(a)

(b)

(c)

(d)

Figure 2–15
Function of the op amp inverting amplifier

Since virtually all the input current flows through R_f, the amplitude and polarity of V_o can now be determined. As shown in Figure 2–15(b), the voltage developed over R_{in} is positive with respect to the virtual ground point. Threfore,

$$0\,\text{V} - V_{R_f} = V_o$$

$$V_{R_f} \cong I_{in} \times R_f \cong 500\ \mu\text{A} \times 10\ \text{k}\Omega \cong 5\ \text{V}$$

$$V_o = 0\,\text{V} - 5\,\text{V} = -5\,\text{V}$$

Now assume, as shown in Figure 2–15(c), that V_{in} equals -500 mV. The path for conventional current flow would then be from the output toward point A and from point A to the input. Therefore,

$$I_{in} = \frac{V_{in}}{R_{in}} = \frac{-500\,\text{mV}}{1\,\text{k}\Omega} = -500\ \mu\text{A}$$

Since, as illustrated in Figure 2–15(d), point A is now negative with respect to the output, the amplitude and polarity of the output voltage are determined as follows:

$$V_{R_f} = I_{in} \times R_f = -500\ \mu\text{A} \times 10\ \text{k}\Omega \cong -5\,\text{V}$$

$$V_o = 0\,\text{V} - -5\,\text{V} = +5\,\text{V}$$

With these two examples of equivalent circuit operation, the inverting action of this first op amp configuration should become evident. Also at this juncture, a practical formula for voltage gain may be derived. First, remember that for any linear amplifier:

$$A_v = \frac{V_o}{V_{in}}$$

Thus, with reference to the previous two examples:

$$A_v \cong \frac{-5\,\text{V}}{+500\,\text{mV}} \cong -10 \quad \text{and} \quad A_v \cong \frac{+5\,\text{V}}{-500\,\text{mV}} \cong -10$$

Now substituting the literal expressions for V_{in} and V_o:

$$A_v \cong \frac{-I_{in} \times R_f}{I_{in} \times R_{in}} \quad \text{and} \quad A_v = \frac{I_{in} \times R_f}{-I_{in} \times R_{in}}$$

Simplifying:

$$A_v = \frac{-R_f}{R_{in}} \tag{2–12}$$

Applying this formula to the previous examples:

$$A_v = \frac{-R_f}{R_{in}} = \frac{-10\,\text{k}\Omega}{1\,\text{k}\Omega} \cong -10$$

Ideally, if a sine wave is occurring at the input to an op amp inverting amplifier, as shown in Figure 2–16(a), the resultant output signal would be shifted exactly 180° with respect to the source. This relationship of the input and output signals for the op amp inverting amplifier is illustrated in Figure 2–16(b). Also, the peak amplitude of the output signal would be determined as:

$$V_o(\text{peak}) = V_{in}(\text{peak}) \times A_v = V_{in} \times \frac{-R_f}{R_{in}}$$

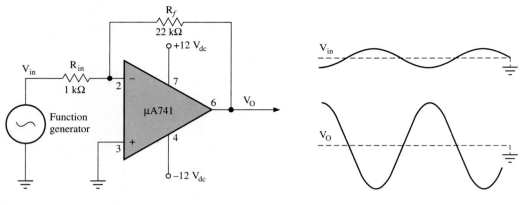

(a) Op amp inverting amplifier with AC input   (b) Input and output voltage waveforms 180° out of phase

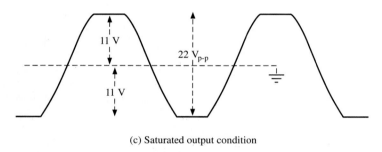

(c) Saturated output condition

Figure 2–16
AC operation of the op amp inverting amplifier

Thus, for the amplifier in Figure 2–16(a):

$$V_o(\text{peak}) = V_{in}(\text{peak}) \times \frac{-22 \text{ k}\Omega}{1 \text{ k}\Omega} = V_{in} \times -22$$

An op amp may be forced into saturation if the input signal is made too large for the given factor of A_v. For the ideal op amp, an output signal should be able to swing between the exact levels of $+V_{CC}$ and $-V_{EE}$. However, since a typical op amp is likely to drop around 2 V of the supply voltage, the output signal may be expected to saturate within 1 V of the level of $+V_{CC}$ or $-V_{EE}$. As an example, assume an op amp inverting amplifier is operating with supply voltages of ± 12 V. The output signal from this amplifier would then be limited to a peak-to-peak amplitude of only 22 V (assuming a dc reference of 0 V). A maximum peak-to-peak amplitude for the input signal may now be determined simply by dividing the 22 V by the factor of voltage gain. Assuming an R_{in} of 1 kΩ and an R_f of 22 kΩ, the gain factor would be -22. Therefore, V_{in} would be limited to a peak-to-peak amplitude of 1 V. If the input signal were to exceed this limit, saturation would occur, as represented in Figure 2–16(c).

With an op amp inverting amplifier, if R_{in} is made larger than R_f, an input signal may actually be attenuated. As an example, for the circuit in Figure 2–17(a), assume the input

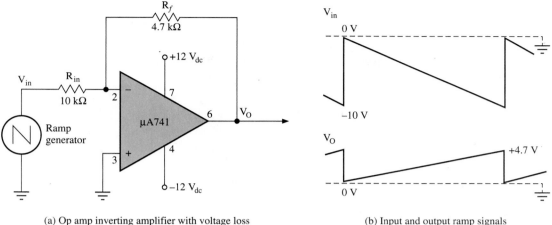

(a) Op amp inverting amplifier with voltage loss

(b) Input and output ramp signals

Figure 2–17
Op amp inverting amplifier used for conditioning of a ramp signal

to be the negative-going ramp signal in Figure 2–17(b). With this circuit condition:

$$V_o = V_{in} \times A_v = V_{in} \times \frac{-R_f}{R_{in}}$$

$$= -10 \text{ V peak} \times \frac{-4.7 \text{ k}\Omega}{10 \text{ k}\Omega} = 4.7 \text{ V peak}$$

Note that the resultant output signal, as shown in Figure 2–17(b), is a positive-going ramp signal of slightly less than half the peak amplitude of the input. Also, since the input signal does not exceed the level of 0 V, the output signal does not fall below that level. A circuit such as that shown in Figure 2–17(a) could be used where a ramp signal must be inverted then brought within a 0- to 5-V range. This process might be necessary in order to make the signal compatible with a digital control system operating from a single 5-V supply. This type of treatment of a waveform, especially when it occurs as part of an industrial control process, is generally referred to as **signal conditioning.**

Example 2–3

Assuming that R_3 is set at its maximum ohmic value, what would be the voltage gain for the op amp circuit shown in Figure 2–18(a). Given this circuit condition, determine the output voltage if V_{in} equals -270 mV.

Solution:

$$A_v = \frac{-R_f}{R_{in}} = \frac{-(R_2 + R_3)}{R_1} = \frac{-(2.2 \text{ k}\Omega + 50 \text{ k}\Omega)}{4.7 \text{ k}\Omega} = -11.1$$

$$V_o = V_{in} \times A_v = -270 \text{ mV} \times -11.1 = 3 \text{ V}$$

For the op amp circuit in Figure 2–18(b), when V_{in} attains -800 mV, V_o should equal 5 V. What ohmic setting of R_2 would allow this output condition to occur?

Figure 2–18

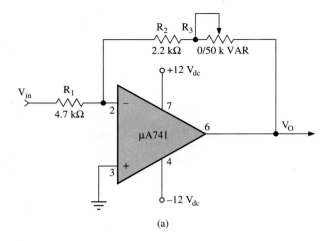

(a)

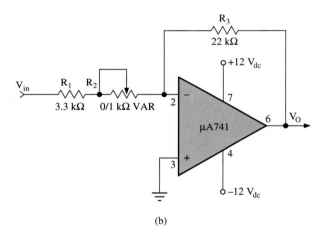

(b)

Solution:

$$A_v = \frac{V_o}{V_{in}} = \frac{5 \text{ V}}{-800 \text{ mV}} = -6.25$$

$$\text{Thus,} \frac{R_3}{R_1 + R_2} = 6.25$$

Solving for the setting of R_2:

$$\frac{R_2 + R_1}{R_3} = \frac{1}{6.25} = 0.16$$

$$\frac{R_2 + 3.3 \text{ k}\Omega}{22 \text{ k}\Omega} = 0.16$$

$$R_2 + 3.3 \text{ k}\Omega = 0.16 \times 22 \text{ k}\Omega = 3.52 \text{ k}\Omega$$

$$R_2 = 3.52 \text{ k}\Omega - 3.3 \text{ k}\Omega = 220 \text{ }\Omega$$

The Op Amp Noninverting Amplifier

Figure 2–19(a) shows a simple op amp noninverting amplifier. In this basic circuit example, the input signal is fed directly to the noninverting input of the op amp, while the inverting input functions as the virtual ground point.

With the noninverting amplifier, voltage gain is determined as:

$$A_v = \frac{R_{in} + R_f}{R_{in}}$$

which may be simplified as follows:

$$A_v = \frac{R_{in}}{R_{in}} + \frac{R_f}{R_{in}}$$
$$= 1 + \frac{R_f}{R_{in}} \tag{2–13}$$

This resultant expression for A_v indicates that, with the noninverting amplifier, voltage gain will never be less than unity. Therefore, this configuration would be used where voltage gain of at least unity is desirable and phase inversion is undesirable.

The basic amplifier shown in Figure 2–19(a) would be very difficult to control under experimental conditions, due to the fact that a signal source would be connected directly to the noninverting input of the op amp. If this input signal was a low-voltage waveform originating from a function generator, it would tend to be noisy and difficult to adjust with regard to amplitude. Figure 2–19(b) illustrates a method of alleviating these problems. Here, an input voltage divider, consisting of R_1 and R_2, provides attenuation of an input signal produced by a function generator. (Note that the actual input signal to the amplifier itself is considered to occur at pin 3 of the μA741 rather than at point A.) An input configuration such as that shown in Figure 2–19(b) might be necessary for laboratory demonstration of the noninverting amplifier. It would prevent destruction of the op amp due to excessive input voltage, attenuate input noise, and allow stable operation of an amplifier having a high voltage gain.

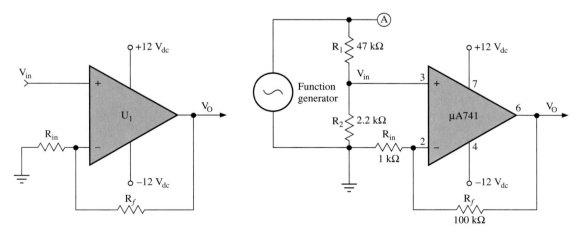

(a) Basic op amp noninverting amplifier   (b) Op amp noninverting amplifier with input attenuation

Figure 2–19
The op amp noninverting amplifier

Example 2–4

For the op amp circuit in Figure 2–19(b), assume the output signal produced by the function generator is a sine wave with a peak amplitude of ± 346 mV. Given this condition, solve for the peak voltage of the signal at pin 3 of the op amp and peak voltage of the output of the op amp. What is A_v for this op amp circuit? What would be the phase relationship of the signals occurring at point A and the output of the op amp?

Solution:

Applying the voltage divider theorem to the input section of the circuit:

$$V_{in} = \frac{R_2}{R_1 + R_2} \times V_A = 346 \text{ mV} \times \frac{2.2 \text{ k}\Omega}{47 \text{ k}\Omega + 2.2 \text{ k}\Omega} = 15.47 \text{ mV}$$

Determining voltage gain:

$$A_v = 1 + \frac{R_f}{R_{in}} = 1 + \frac{100 \text{ k}\Omega}{1 \text{ k}\Omega} = 1 + 100 = 101$$

Determining output voltage:

$$V_o = A_v \times V_{in} = 101 \times 15.47 \text{ mV}_{peak} = 1.562 \text{ V}_{peak}$$

Since Figure 2–19(b) does contain a noninverting amplifier, the voltage at point A and the voltage at the output of the op amp will, ideally, be exactly in phase.

SECTION REVIEW 2–2

1. For the op amp circuit in Figure 2–20(a), assume that R_3 is adjusted to 29.5 kΩ. If V_{in} equals $+153$ mV, what is the amplitude and the polarity of V_o? What would be the voltage gain for the circuit?

2. For the op amp circuit in Figure 2–20(a), assume that R_3 is adjusted to 14.8 kΩ. If waveform 1 in Figure 2–20(b) occurs at the input to the amplifier in Figure 2–20(a), what would be the positive and negative peak amplitudes of the waveform occurring at the output? What would be the voltage gain for the circuit? Sketch the input and output waveforms, carefully showing the phase relationship of the two signals.

3. Assume that waveform 2 in Figure 2–20(b) is now occurring at the output of the amplifier in Figure 2–20(a). If the input waveform has a peak amplitude of ± 32.5 mV, what must be the ohmic setting of R_3?

4. Describe the purpose of R_1, R_2, and R_3 in Figure 2–21(a).

5. What is the actual input resistor for the amplifier in Figure 2–21(a)? What is the actual feedback resistor? What is the voltage gain for the amplifier?

6. Assuming that $+37$ mV is occurring at point B in Figure 2–21(a), what would be the level of voltage at point C?

7. Assume that V_o in Figure 2–21(b) is occurring at point C in Figure 2–21(a). If, at the same time, V_{in} is seen at point B, what would be the peak amplitude of that signal? Is the phase relationship of these two signals represented correctly in Figure 2–21(b)? Explain your answer.

8. If, in Figure 2–21(a), the resistance between point B and ground is 9.58 kΩ and the instantaneous amplitude at point A is 57 mV, what would be the values of V_B and V_C?

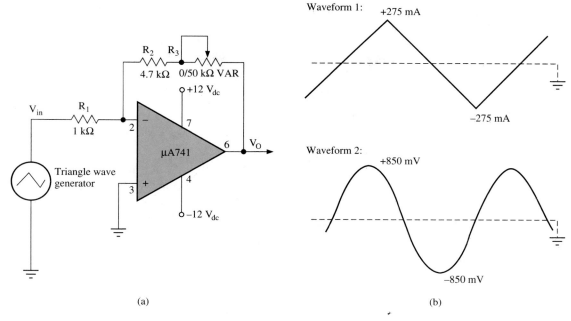

Figure 2–20

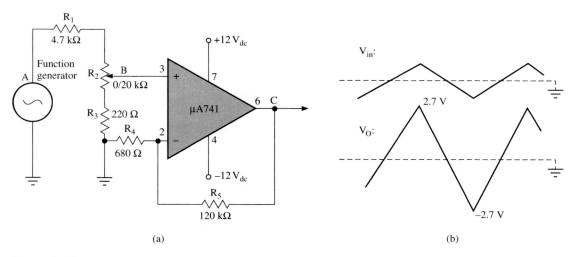

Figure 2–21

2-3   Op Amp Summer Amplifiers

The op amp summer amplifier receives its name from the fact that it is capable of performing analog addition. It is found within various forms of analog computing (or function) circuits, digital-to-analog converters, and analog-to-digital converters. An example of a very basic op amp summing circuit is shown in Figure 2–22. For this summer amplifier, assume that R_a, R_b, and R_f all equal 10 kΩ. Given these resistive values, the output voltage would simply be the negative equivalent to the sum of the two input voltages:

$$V_o = -(V_a + V_b) \qquad (2-14)$$

where $R_a = R_b = R_f$. Assume for the summer amplifier in Figure 2–22 that V_a equals 2 V_{dc} and V_b equals 4.5 V_{dc}. As with the basic inverting amplifier, the inverting input to the op amp is functioning as a virtual ground point. Therefore, the individual input currents may be approximated as follows:

$$I_a = \frac{V_a}{R_a} = \frac{2\text{ V}}{10\text{ k}\Omega} = 200\ \mu\text{A}$$

$$I_b = \frac{V_b}{R_b} = \frac{4.5\text{ V}}{10\text{ k}\Omega} = 450\ \mu\text{A}$$

Figure 2–23 illustrates the current and voltage conditions for the summer amplifier in Figure 2–22. According to Kirchhoff's current law, the summing current leaving node A must be the algebraic sum of the input currents. Thus,

$$I_a + I_b - I_s = 0\text{ A}$$

which becomes

$$I_s = (I_a + I_b) \qquad (2-15)$$

With the input current values previously calculated, summing current would then be calculated as follows:

$$I_s = 200\ \mu\text{A} + 450\ \mu\text{A} = 650\ \mu\text{A}$$

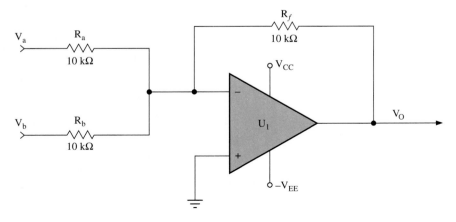

Figure 2–22
Basic op amp summer amplifier

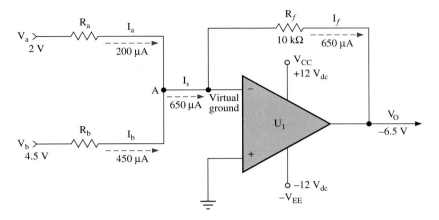

Figure 2–23
Currents and voltages for the basic op amp summer amplifier

Since virtually all of the summing current becomes feedback current, the output voltage for the summer amplifier may be derived as:

$$V_o = -(I_s \times R_f) = -(I_f \times R_f) \qquad (2\text{–}16)$$

Thus, for the example in Figure 2–23:

$$V_o = -(650 \ \mu A \times 10 \ k\Omega) = -6.5 \ V$$

Using these calculated values, Eq. (2–14) may now be proven:

$$V_o = -(V_a + V_b)$$

substituting:

$$-6.5 \ V = -(2 \ V + 4.5 \ V)$$

To bring this resultant voltage back to a positive level, it would be necessary to attach an inverting amplifier with unity gain as a second stage. A complete circuit design that would accomplish this task is illustrated in Figure 2–24. The mathematical expression for the operation of the circuit in Figure 2–24 may be stated as follows:

$$V_o = -(V_a + V_b) \times -1$$

which simplifies to:

$$V_o = (V_a + V_b)$$

Using the format established in Figure 2–24, other mathematical functions may easily be created through changing resistive values. Assume, for example, that R_3 is now made 5 kΩ, while all other resistive values remain at 10 kΩ. With this combination of resistors, the output level will become the average of the two input voltages. This mathematical function may be proven as follows:

$$\begin{aligned}
V_o &= -(I_s \times R_3) \times -(R_5/R_4) \\
&= -(650 \ \mu A \times 5 \ k\Omega) \times -(10 \ k\Omega/10 \ k\Omega) \\
&= -3.25 \ V \times -1 = 3.25 \ V
\end{aligned}$$

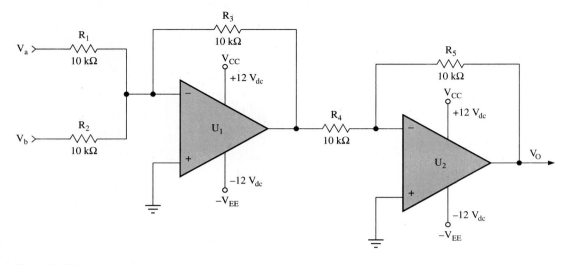

Figure 2–24
Op amp summer amplifier with unity-gain inverting output stage

The result obtained from the mathematical operation actually performed by the dual op amp circuit is equivalent to that derived by the conventional method of averaging two numbers:

$$\frac{(2 + 4.5)}{2} = 3.25$$

Example 2–5

For the dual op amp circuit in Figure 2–25, assume that R_4 is adjusted to exactly 3.333 kΩ. Given this condition and the three input voltages, determine the values of the three input currents. Also solve for the value of I_{R_4}, I_{R_5}, and I_{R_6}. What voltage would be present at the output of U_1? What would be the value of V_o? Determine the gain of the second stage of the circuit. Finally, determine the mathematical function that this circuit performs.

Solution:
Solving for the input currents:

$$I_{R_1} = \frac{V_1}{R_1} = \frac{6.3\ \text{V}}{10\ \text{k}\Omega} = 630\ \mu\text{A}$$

$$I_{R_2} = \frac{V_2}{R_2} = \frac{8.4\ \text{V}}{10\ \text{k}\Omega} = 840\ \mu\text{A}$$

$$I_{R_3} = \frac{V_3}{R_3} = \frac{7.5\ \text{V}}{10\ \text{k}\Omega} = 750\ \mu\text{A}$$

Since I_{R_4} is the summing current for U_1:

$$I_{R_4} = I_{R_1} + I_{R_2} + I_{R_3} = 630\ \mu\text{A} + 840\ \mu\text{A} + 750\ \mu\text{A} = 2.22\ \text{mA}$$

The voltage at the output of U_1 (point A) may now be determined:

$$V_o = -(I_s \times R_4) = -2.22\ \text{mA} \times 3.333\ \text{k}\Omega = -7.4\ \text{V}$$

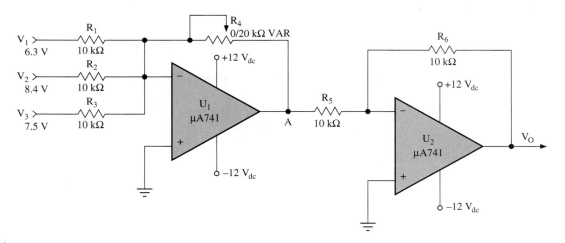

Figure 2–25

Once the voltage at point A is determined, the current flowing through R_5, which is virtually equal to that flowing through R_4, may be determined:

$$I_{R_4} = I_{R_5} = \frac{V_A}{R_5} = \frac{-7.4 \text{ V}}{10 \text{ k}\Omega} = -740 \text{ } \mu a$$

With R_5 equal to R_6, the gain for the output stage of the circuit is of course -1. Thus, V_o would be determined as:

$$V_o = V_A \times -1 = -7.4 \text{ V} \times -1 = 7.4 \text{ V}$$

The mathematical function being performed by the circuit becomes evident through determining the relationship of the output voltage to those present at the input. The sum of the input voltages is first determined:

$$(V_1 + V_2 + V_3) = 6.3 \text{ V} + 8.4 \text{ V} + 7.5 \text{ V} = 22.2 \text{ V}$$

The average of these three voltages would be determined as:

$$V_{AV} = \frac{(V_1 + V_2 + V_3)}{3} = \frac{22.2 \text{ V}}{3} = 7.4 \text{ V}$$

Thus, the circuit in Figure 2–25 is designed to derive the average of three input voltages. The actual operation being performed by the circuit is as follows:

$$V_o = (V_1 + V_2 + V_3) \times -0.3333 \times -1$$
$$= (6.3 \text{ V} + 8.4 \text{ V} + 7.5 \text{ V}) \times 0.3333 = 7.4 \text{ V}$$

Within many forms of industrial control circuitry, op amps are used in the process of **signal conditioning.** This rather general term is used to describe the process of developing an analog input signal in order to make it compatible with further stages of circuitry. Active signal conditioning may involve such processes as amplification, level shifting, and filtering. It is likely that a signal from an input device might be required to undergo all three of these processes before being sent to an analog-to-digital (A/D) converter.

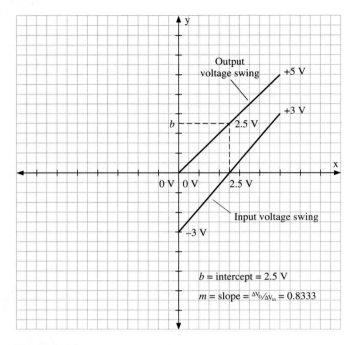

Figure 2–26
Graphic representation of relationship of input and output voltages for op amp signal condition-ing circuit

This device would derive a binary number (or several binary numbers) representing its analog value.

The subject of signal conditioning is now introduced with a design problem involving two op amps. Assume that the input signal to be conditioned has a range of 6 V, extending from −3 V to +3 V. Also assume that signal conditioning circuitry is required to make this signal compatible with an A/D conversion circuit having an input range of 0 to 5 V. Figure 2–26 graphically represents the relationship of the input and output signals for the signal conditioning circuit we want to design. The slope may be considered as the output voltage swing divided by that of the input:

$$m = \frac{5\,V - 0\,V}{3\,V - -3V} = \frac{5\,V}{6\,V} = 0.83333$$

For the graph in Figure 2–26, a literal equation for V_o would be as follows:

$$V_o = (m \times V_{in}) + b \qquad\qquad (2–17)$$

This expression may now be used to develop a simpler expression for output voltage. From the graph in Figure 2–26, it becomes evident that V_{in} is at −3 V when V_o equals 0 V. Also, when V_{in} is at +3 V, V_o must equal +5 V. Substituting these values in the preceding literal equation:

$$0\,V = (m \times -3\,V) + b$$
$$5\,V = (m \times +3\,V) + b$$

Thus, since the slope has already been determined:

$$5\text{ V} = (0.8333 \times 3\text{ V}) + b$$
$$= 2.5\text{ V} + b$$
$$b = 5\text{ V} - 2.5\text{ V} = 2.5\text{ V}$$

Now that the values of the slope and intercept have both been determined, further algebraic manipulation may be performed in order to derive an equation that may be applied directly to the design of the signal conditioning circuit:

$$V_o = (0.8333 \times V_{in}) + 2.5\text{ V}$$

Thus,

$$\frac{V_o}{0.8333} = \frac{(0.8333 \times V_{in})}{0.8333} + \frac{2.5\text{ V}}{0.8333}$$

Simplifying,

$$\frac{V_o}{0.8333} = (V_{in} + 3\text{ V})$$

Finally,

$$V_o = (V_{in} + 3\text{ V}) \times 0.8333$$

An op amp circuit may now be designed that performs this exact mathematical function. The $(V_{in} + 3)$ portion of the expression may be implemented through use of a summer amplifier that has a gain of unity. A possible solution is shown in Figure 2–27. Here, V_A would be equivalent to $-(V_{in} + 3\text{ V})$.

Since this single-stage summer amplifier inverts the sum of the input voltages, a subsequent inverting amplifier stage must be used if the final output is to be positive. The gain of this second stage will be the negative equivalent of the slope m. Thus, with the combination of R_{in} and R_f used in Figure 2–28, the second portion of the equation for V_o may be satisfied, where V_B is equivalent to $(V_A \times -0.8333)$.

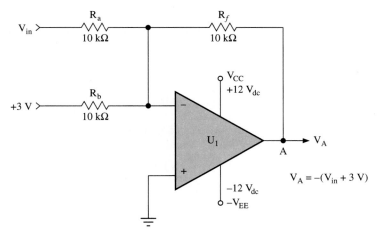

Figure 2–27
First stage of op amp signal conditioning circuit

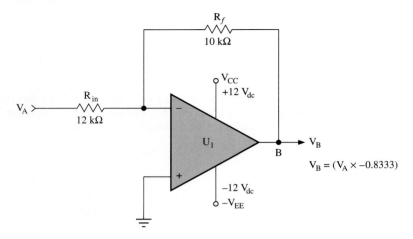

Figure 2–28
Second stage of op amp signal conditioning circuit

A design for the complete signal conditioning circuit is shown in Figure 2–29. The actual mathematical function performed by this circuit is as follows:

$$V_o = (V_{in} + 3\ \text{V}) \times -1 \times -0.8333$$

We have already established that, when the input is at its lower limit of -3 V, V_o must equal 0 V. It is easy to visualize how this condition is achieved in Figure 2–29, considering that R_2 functions as a simple voltage divider. With -3 V at the input and $+3$ V present at the wiper of R_2, exactly 0 V will be present at the inverting input of **1458a**. Thus,

$$V_o = (-3\ \text{V} + 3\ \text{V}) \times -1 \times -0.8333 = 0\ \text{V}$$

To achieve this actual null condition at point B with a summing input of 0 V, we might need to adjust the wiper of R_7. Since neither op amp of the 1458 IC has the internal offset null capability of the μA741, it is necessary to create an external resistive network to perform that function. Assuming in Figure 2–29 that V_o is slightly above ground level when the input voltage to the summer amplifier is at -3 V, the wiper of R_7 would then be adjusted in order to make the noninverting input of 1458b more negative. This action will cause V_o to move toward the desired level of 0 V. If, with V_{in} at -3 V, the output voltage settles slightly below the ground level, the wiper of R_7 may be turned in the opposite direction in order to bring the output to the desired 0-V level.

Thus far, the signal conditioning circuit in Figure 2–29 has been analyzed with respect to its maximum input and output voltages. Now, we prove that a change in the level of the input voltage results in a proportional change in voltage at the output. As an example, assume V_{in} to be at $+1.4$ V. Then V_o would be determined as follows:

$$V_o = (+1.4\ \text{V} + 3\ \text{V}) \times 0.8333 = 3.667\ \text{V}$$

Next, assume that V_{in} decreases to -2.8 V. Then V_o would be derived as:

$$V_o = (-2.8\ \text{V} + 3\ \text{V}) \times 0.8333 = 166.7\ \text{mV}$$

The percentage of change in the input voltage can now be determined as follows:

$$\frac{(V_{in1} - V_{in2})}{\Delta V_{in}(\text{max})} \times 100\% = \frac{(1.4\ \text{V} - -2.8\ \text{V})}{6\ \text{V}} \times 100\% = 70\%$$

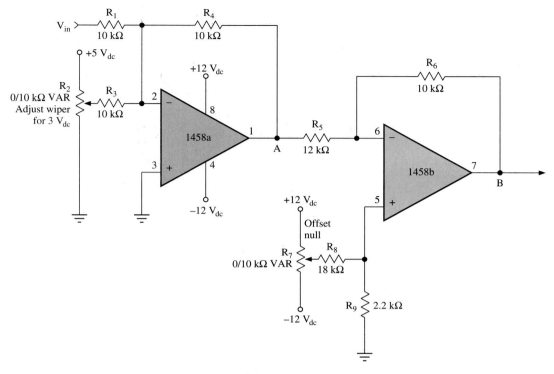

Figure 2–29
Dual op amp used as summer and inverting amplifier

An equivalent percentage of change also results at the output of the circuit, proving the linearity of its operation:

$$\frac{(V_{o1} - V_{o2})}{\Delta V_o(\text{max})} \times 100\% = \frac{(3.667 \text{ V} - 0.167 \text{ V})}{5 \text{ V}} \times 100\% = 70\%$$

SECTION REVIEW 2-3

For the op amp circuit in Figure 2–30, assume that the maximum swing of the input signal is ± 2 V. Also assume that the maximum swing of the output voltage ranges from 0 to 5 V.

1. Given the conditions just specified, graph the operation of the circuit in Figure 2–30. (Figure 2–26 should be used as a model.)
2. Given the conditions just specified, determine the slope and intercept.
3. To achieve the output voltage swing specified, what dc voltage must be constantly present at point A?
4. What must be the voltage gain of the circuit in Figure 2–30 in order to achieve the desired output voltage swing?
5. To achieve this voltage gain, what must be the setting of R_6 in Figure 2–30?
6. If V_{in} in Figure 2–30 equals -687 mV, what must be the level of output voltage?
7. If V_o in Figure 2–30 equals $+3.7$ V, what must be the level of input voltage?

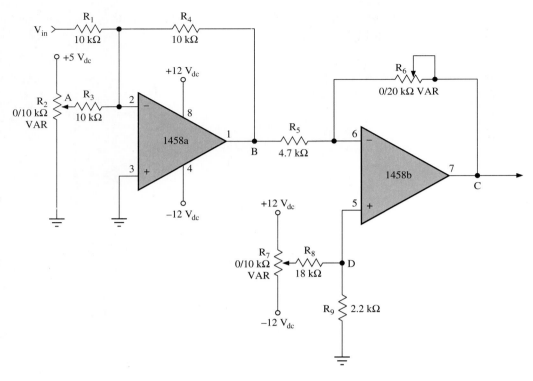

Figure 2–30

8. In Figure 2–30, what is the purpose of the resistive network consisting of R_7, R_8, and R_9?

9. In Figure 2–30, if V_{in} equals -2 V, what is the ideal value of V_o?

10. Assume that R_7 in Figure 2–30 is adjusted such that point D becomes more negative. What effect would this action have on the output voltage?

2–4   OP AMP INSTRUMENTATION AMPLIFIERS

A common characteristic of the op amp circuits studied thus far is the fact that they have used only one input terminal to receive an active signal. (Even with the summer amplifier, which may have several inputs, the summing currents actually converge on the inverting input, creating only one composite signal.) **Instrumentation amplifiers** are distinctively different in that they exploit the voltage-differencing capability of the op amp.

Recall that the input stage to a typical op amp is, in itself, a **differential amplifier** and that the major advantages of such an amplifier are high voltage gain and the ability to reject signals that are common to both inputs. The voltage-differencing amplifier is ideally suited for industrial instrumentation applications (hence, the term *instrumentation amplifier*). It is often used in conjunction with a bridge circuit containing input transducers such as thermocouples or load cells. When placed in bridge circuits, these devices generate small error signals, which usually ride on a relatively large dc component.

The purpose of the instrumentation amplifier is to amplify the error signal while rejecting the common-mode (dc) component of the signal. Because of their frequent use in conjunction with such circuits, instrumentation amplifiers are also referred to as **bridge** or **error amplifiers.**

The Basic Differential Amplifier

Figure 2–31 contains a simple differential amplifier. The device is capable of amplifying the difference in potential that exists between the two input signals, V_1 and V_2. If R_4/R_1 is equivalent to R_3/R_2, then a balanced condition exists and the output voltage can be easily determined as follows:

$$V_o = (V_1 - V_2) \times \frac{-R_f}{R_{in}} \qquad (2\text{–}18)$$

where $R_{in} = R_1$ and $R_f = R_4$. As an example of circuit operation, assume V_1 equals 235 mV and V_2 equals -115 mV. With this input condition, V_o may be derived as follows:

$$V_o = (V_1 - V_2) \times \frac{-R_f}{R_{in}}$$

$$= (235\text{ mV} - -115\text{ mV}) \times \frac{-4.7\text{ k}\Omega}{1\text{ k}\Omega} = -1.645\text{ V}$$

Next, assume V_1 is at -125 mV and V_2 is at $+87$ mV. The output voltage would then be determined as:

$$V_o = (-125\text{ mV} - 87\text{ mV}) \times \frac{-4.7\text{ k}\Omega}{1\text{ k}\Omega} = 996.4\text{ mV}$$

The High-Impedance Instrumentation Amplifier

Today, instrumentation amplifiers are readily available in monolithic form. Unlike op amps, these devices contain their own resistive feedback networks. Also, these devices

Figure 2–31
The op amp differential amplifier

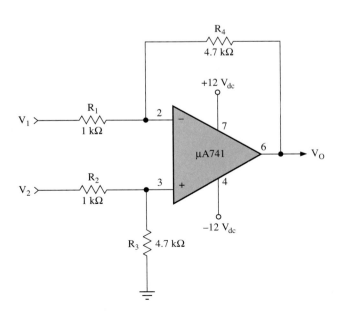

are specially designed to offer extremely high input impedance and exhibit a very high CMRR. Although a single op amp differential amplifier, such as that in Figure 2–31, may not be satisfactory for extremely precise industrial applications, combinations of op amps may be used to create specialized forms of instrumentation amplifiers. An example of such a circuit is contained in Figure 2–32. In Figure 2–32, resistors 1, 2, 3, and 4 form a **Wheatstone bridge.** This form of resistive network is often used to develop a signal from some form of transducer (such as a resistive thermal detector or a load cell). In this circuit, variable resistor R_4 is used to simulate the action of a transducer. (Various forms of input transducer will be studied in detail during the next two chapters.) The instrumentation amplifier would detect the signal generated by the transducer in the form of an **error voltage** between points A and B. During a balanced or **null** condition for the Wheatstone bridge, $(R_1 \times R_4)$ would equal $(R_2 \times R_3)$ and the error voltage, measured between points A and B, would be zero. For many forms of industrial control circuits, this null condition represents a **set point.** That is, if R_4 were a thermal transducer, the circuit could be calibrated such that V_o would equal zero at some target value of environmental temperature. Deviations from this desired temperature would create an error signal across the Wheatstone bridge. The instrumentation amplifier would then develop this difference in potential. The output from this device would then be sent to further signal conditioning or control circuitry.

In Figure 2–32, if the Wheatstone bridge is to be in a null condition, R_4 must be adjusted to 10 kΩ. With this circuit condition, each resistor in the bridge would drop 12 V and the difference in potential across the bridge (between points A and B) would

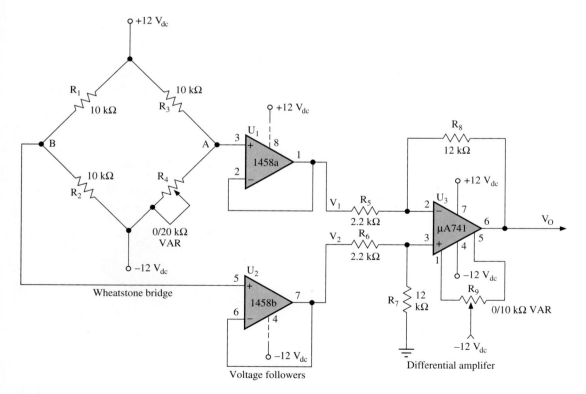

Figure 2–32
Op amp instrumentation amplifier with Wheatstone bridge

be 0 V. If, however, R_4 were adjusted to a value greater than 10 kΩ, the voltage at point A would become positive with respect to point B. If R_4 were then adjusted to less than 10 kΩ, the voltage at point A would be negative with respect to point B. Also, note that, with R_1 and R_2 being equal, point B will always be at virtually ground potential.

Assume now that R_4 in Figure 2–32 has been adjusted to 10.4 kΩ. The amount of voltage developed over R_3 and R_4 could then be determined as follows:

$$V_{R_3} = \frac{R_3}{R_3 + R_4} \times (12\text{ V} - -12\text{ V})$$

$$= \frac{10\text{ k}\Omega}{10\text{ k}\Omega + 10.4\text{ k}\Omega} \times 24\text{ V} = 11.765\text{ V}$$

$$V_{R_4} = \frac{R_4}{R_3 + R_4} \times (12\text{ V} - -12\text{ V})$$

$$= \frac{10.4\text{ k}\Omega}{10\text{ k}\Omega + 10.4\text{ k}\Omega} \times 24\text{ V} = 12.235\text{ V}$$

With respect to the circuit ground, the voltage at point B is derived as follows:

$$V_A = V_{R_4} - 12\text{ V} = 12.235\text{ V} - 12\text{ V} = 235\text{ mV}$$

The difference in potential between points A and B, which represents the **error signal,** is easily determined:

$$V_{A-B} = 235\text{ mV} - 0\text{ V} = 235\text{ mV}$$

Next assume that R_4 has been reduced to 9.8 kΩ. Following the same procedure used earlier, the new error voltage may be determined as shown:

$$V_{R_3} = \frac{10\text{ k}\Omega}{10\text{ k}\Omega + 9.8\text{ k}\Omega} \times 24\text{ V} = 12.121\text{ V}$$

$$V_{R_4} = \frac{9.8\text{ k}\Omega}{10\text{ k}\Omega + 9.8\text{ k}\Omega} \times 24\text{ V} = 11.879\text{ V}$$

$$V_A = 11.879\text{ V} - 12\text{ V} = -121\text{ mV}$$

$$V_{A-B} = -121\text{ mV} - 0\text{ V} = -121\text{ mV}$$

Op amps 1 and 2 in Figure 2–32 are both connected in a manner referred to as a **voltage follower** configuration. With this configuration, the voltage gain of each op amp is slightly less than unity, but input resistance is at maximum. The need for the voltage followers should become apparent, considering that the instrumentation amplifier itself behaves as a load across the Wheatstone bridge. Therefore, if the input impedance of the instrumentation amplifier is made very high with respect to the resistive values used within the Wheatstone bridge, its loading effect will be minimal.

Returning to the condition in Figure 2–32 in which R_4 was adjusted to 10.4 kΩ, the output voltage for the instrumentation amplifier will now be determined. Assuming the voltage gain of U_1 and U_2 to be unity, $(V_1 - V_2)$ would then virtually equal $(V_A - V_B)$. Therefore,

$$V_o = (V_A - V_B) \times \frac{-R_f}{R_{in}} = 235\text{ mV} \times \frac{-12\text{ k}\Omega}{2.2\text{ k}\Omega} = -1.282\text{ V}$$

Assuming that a null condition exists across the Wheatstone bridge, V_o should ideally equal zero. Once the bridge circuit is placed in this balanced condition, calibration of the circuit could be achieved through the adjustment of R_9, which performs the offset null function. If precision op amps are used and resistive values are closely matched, we should be able to bring V_o to 0 V whenever V_{A-B} equals 0 V.

Example 2–6

For the instrumentation amplifier in Figure 2–32, assume R_4 is now adjusted to 9.37 kΩ. What is the value of V_{A-B}? What is the value of V_o? What would be the value of V_o if R_4 were adjusted to 10 kΩ? What condition should then exist across the Wheatstone bridge?

Solution:

$$V_{R_4} = \frac{R_4}{R_3 + R_4} \times 24\ V = \frac{9.37\ k\Omega}{10\ k\Omega + 9.37\ k\Omega} \times 24\ V = 11.601\ V$$

$$V_A = 11.61\ V - 12\ V = -390\ mV$$

$$V_{A-B} = -390\ mV - 0\ V = -390\ mV$$

$$V_o = -390\ mV \times \frac{-12\ k\Omega}{2.2\ k\Omega} = 2.127\ V$$

If R_4 is adjusted to 10 kΩ, a balanced, or null, condition exists across the Wheatstone bridge ($V_{A-B} = 0$ V). Thus V_o should also equal 0 V.

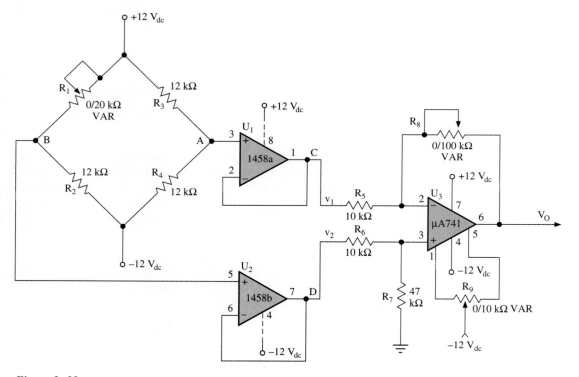

Figure 2–33

SECTION REVIEW 2-4

1. For the op amp circuit in Figure 2–33, if R_1 is adjusted to 10.39 kΩ, what would be the voltages developed over R_1 and R_2? What would be the difference in potential between points A and B?
2. Ideally, what would be the input impedance for **1458a** and **1458b** in Figure 2–33? What would be the ideal voltage gain for either of these op amps?
3. With R_1 in Figure 2–33 still set at 10.39 kΩ, what would be ideal difference in voltage between points C and D?
4. Assume that R_8 in Figure 2–33 is adjusted to exactly 47 kΩ. If R_1 is still set at 10.39 kΩ, what would be the ideal value of output voltage?
5. If R_1 in Figure 2–33 is now adjusted to 14.57 kΩ, what would be the voltages developed over R_1 and R_2? What would be the difference in potential between points A and B?
6. With R_1 in Figure 2–33 still set at 14.57 kΩ, what would be the ideal difference in voltage between points C and D?
7. With R_1 still set at 14.57 kΩ and R_8 still set at 47 kΩ, what would be the value of V_o for the circuit in Figure 2–33?
8. For the circuit in Figure 2–33, if R_1 is adjusted to exactly 12 kΩ, what is the ideal value of output voltage?

2-5   OP AMP FILTER CIRCUITS

Electronic filters are devices that have the capability to select signals within a specific range of frequencies, while rejecting signals that occur outside of that range. Filters fall into four major classifications: **high-pass, low-pass, bandpass,** and **band-reject** filters. Filters are further classified as being either *passive* or *active*. Passive filters are comprised of RC, RL, or RLC networks, whereas active filters usually contain RC networks and at least one op amp.

Active filter circuits are used extensively in industrial electronics, not only for the purpose of noise elimination but also for active signal conditioning. A major advantage of incorporating op amps into active filter design lies in the fact that reliable bandpass and band-reject filters may be created without the use of inductors. This represents a significant design advantage, considering the fact that many filters used in industry are required to operate at extremely low frequencies. An excellent example of such a circuit would be the anti-aliasing filter used in the process of A/D conversion. Essentially an active low-pass filter, this circuit might be required to attenuate unwanted signals above frequencies as low as 60 Hz. If the filtering of undesired signals at such low frequencies required the use of inductors, the values of inductance would need to be very large in order to generate the required levels of reactance. The elimination of the need for large inductors represents a significant reduction in circuit size, weight, and cost.

Figure 2–34(a) contains an RC low-pass filter. For this passive circuit, X_{C_1} increases as frequency decreases, allowing more of the ac input signal to develop over the capacitor. That frequency at which X_{C_1} equals R_1 (and capacitive voltage equals resistive voltage) is referred to as the cutoff frequency (f_{co}). For a low-pass filter, signals that occur below the cutoff frequency are considered to be within the pass band of the circuit. Signals occurring above the cutoff frequency are part of the stop band and are considered as being rejected by the filter.

For a capacitor, at any given frequency:

$$X_C = \frac{1}{2\pi f C}$$

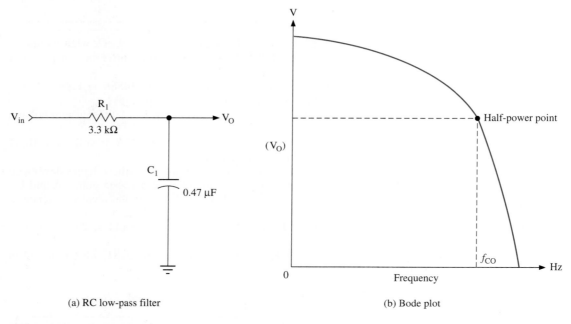

(a) RC low-pass filter

(b) Bode plot

Figure 2–34
Operation of a passive low-pass RC filter

Thus, at the half-power point, where X_C equals R:

$$f_{CO} = \frac{1}{2\pi RC} \qquad (2\text{–}19)$$

Figure 2–34(b) contains a Bode plot of the low-pass filter's output voltage versus input frequency. (It is assumed that the input voltage is being held constant.) The portion of the curve that extends from the half-power point to a frequency equivalent to ($f_{CO} \times$ 10) is referred to as the **roll-off** of the filter. The response curve of a filter is usually specified in terms of decibels of attenuation. For the passive filter in Figure 2–34(a), between the half-power point (at which an input sine wave would be attenuated by 3 dB) and the point equivalent to ($f_{CO} \times$ 10), the amount of attenuation would be 20 dB. This may be determined by means of the following example.

Example 2–7

Assume the input sine wave for the filter in Figure 2–34(a) has an amplitude of 5 V_{rms} and is occurring at the cutoff frequency. Solve for the value of f_{CO}.

With X_{C_1} equalling R at the cutoff frequency:

$$f_{CO} = \frac{1}{2\pi RC} = \frac{1}{2\pi \times 3.3 \text{ k}\Omega \times 0.47 \text{ }\mu\text{F}} = 102.6 \text{ Hz}$$

The total impedance for the filter at the half-power point is determined as follows:

$$Z_T = 1.414 \times R_1 = 1.414 \times 3.3 \text{ k}\Omega = 4.667 \text{ k}\Omega$$

The amplitude of the output signal being developed over the capacitor at the half-power point may now be determined as:

$$V_{C_1} = 0.7071 \times V_{in} = 0.7071 \times 5\,V = 3.536\,V$$

Next, assuming the filter's input signal to be occurring at a frequency equivalent to $(f_{CO} \times 10)$, the new values of capacitive reactance and total impedance are determined as follows:

$$f_{CO} \times 10 = 102.6\,Hz \times 10 = 1.026\,kHz$$

$$X_{C_1} = \frac{1}{2\pi \times 1.026\,kHz \times 0.47\,\mu F} = 330\,\Omega$$

The output voltage occurring at this higher frequency is then determined:

$$Z_T = \sqrt{R_1^2 + X_{C_1}^2} = \sqrt{3.3\,k\Omega^2 + 330\,\Omega^2} = 3.3165\,k\Omega$$

$$V_{C_1} = \frac{X_{C_1}}{Z_T} \times V_{in} = \frac{330\,\Omega}{3.3165\,k\Omega} \times 5\,V = 498\,mV \cong 500\,mV$$

This decrease in output voltage may now be expressed as decibels of attenuation:

$$dB = 20\log(V_{out}/V_{in}) = 20\log(498\,mV/5\,V) = -20\,dB$$

Thus, the roll-off for the passive RC filter occurs at a rate of -20 dB per decade. Here the term *decade* is used to describe the difference in frequency between the f_{CO} and $(f_{CO} \times 10)$.

If the positions of the capacitor and resistor are reversed, as shown in Figure 2–35(a), the circuit will function as a high-pass filter. For this device, the pass band will consist of the set of frequencies above the f_{CO}. If the input frequency is decreased below the value of f_{CO}, the roll-off will occur as before, at a rate of 20 dB per decade. The bode plot for an RC high-pass filter is illustrated in Figure 2–35(b).

A simple type of active low-pass filter is shown in Figure 2–36. Here, with the op amp in a voltage-follower configuration, the gain factor is limited to slightly less than unity. The purpose of the op amp within this circuit is to serve as a high-impedance buffer between the low-pass RC filter and an output circuit. Since the gain capability of the op amp is not being used in this simple design, the calculations for cut-off frequency and roll-off would be the same as for the passive RC filter in Figure 2–34. An active filter for which the roll-off characteristic is equivalent to that of a passive device is referred to as a **first-order filter.**

An op amp's high voltage gain and feedback capabilities may be exploited in order to sharpen the roll-off characteristic of an active filter. With a second-order filter, the roll-off would be increased to nearly -40 dB per decade. A third-order filter would have a roll-off of around -60 dB, while, for a fourth-order filter, roll-off would be increased to nearly -80 dB. Still sharper roll-off characteristics of -100 and -120 dB per decade are obtainable with fifth- and sixth-order active filters. With such higher order op amp active filters, circuit performance will approach the ideal level, with the voltage gain for input signals that occur within the pass band being held virtually constant. At the cutoff frequencies, the roll-off will be nearly instantaneous as the voltage gain for signals occurring outside of the pass band approaches zero.

Op amp filters are further classified by their design format. The three major classifications include **Butterworth, Chebyshev,** and **Bessel.** These design formats may be imple-

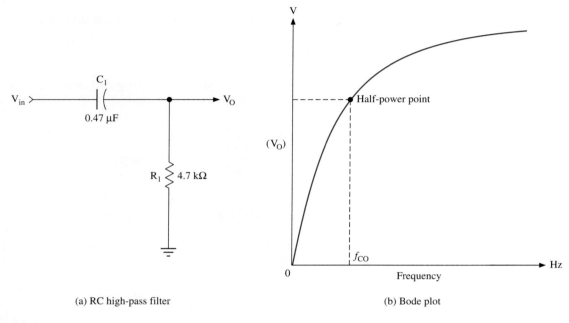

(a) RC high-pass filter   (b) Bode plot

Figure 2–35
Operation of a passive high-pass RC filter

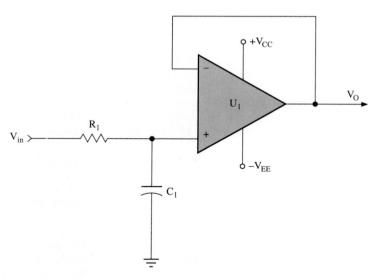

Figure 2–36
Active (first-order) low-pass filter

mented for all types and orders of active filters. To determine which design format is best suited for a particular circuit application, we would first need to know the operating characteristics of each type of filter.

The Butterworth design format allows for a very flat frequency response within the pass band. For this reason, the Butterworth characteristic is often referred to as **maximally flat.** A problem associated with the Butterworth filter is the fact that its phase response is nonlinear. This would mean that the propagation delay of signals passing through the filter would vary with frequency. This problem would become evident if a pulse waveform (nonsinusoidal) were applied to the input of the Butterworth filter. Since the various harmonic components of the signal would pass through the filter at different rates of speed, overshoots would occur at the output. These overshoots would become more prominent with higher order Butterworth filters. The major advantage of the Chebyshev filter design is the fact that steeper roll-off characteristics may be achieved with less circuitry. However, overshoots and ripples are likely to occur within the pass band of this type of filter. Also, the phase response of a Chebyshev filter is less linear than that of a Butterworth filter. The major advantage of the Bessel design format is its highly linear phase response. This characteristic makes it ideally suited for the processing of pulse signals.

An example of a second-order Butterworth low-pass filter is given in Figure 2–37. This circuit introduces an active filter design format known as either **Sallen and Key**

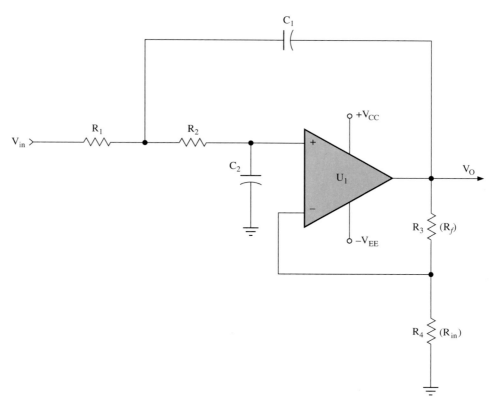

Figure 2–37
Sallen and Key (Butterworth) second-order low-pass filter

or **voltage-controlled voltage source** (VCVS). The Sallen and Key configuration is characterized by the use of one of the components of the RC network in a direct feedback path. For the device in Figure 2–37, the roll-off beyond the cutoff frequency will occur at a rate of −40 dB per decade. There are now two RC networks operating within the circuit. R_1 is combined with C_1, while R_2 is paired with C_2. Since this circuit is determined by the resistor pair R_3 and R_4. Since the circuit is essentially a noninverting amplifier configuration, the gain factor for signals occurring below the cutoff frequency approaches a value determined as follows:

$$A_v = 1 + \frac{R_f}{R_{in}}$$

where $R_3 = R_f$ and $R_4 = R_{in}$. For the filter design represented in Figure 2–37, the value of R_3/R_4 should be as near as possible to **0.586,** which is the ideal **damping factor** for ensuring a Butterworth response.

For the device in Figure 2–37, the cutoff frequency would be approximated as follows:

$$f_{CO} \frac{1}{2\pi \times \sqrt{R_1 \times R_2 \times C_1 \times C_2}} \tag{2-20}$$

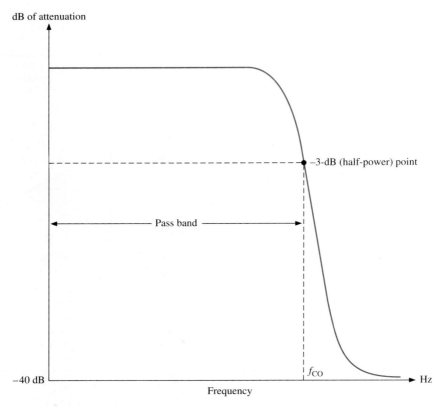

Figure 2–38
Bode plot for a Butterworth second-order low-pass filter

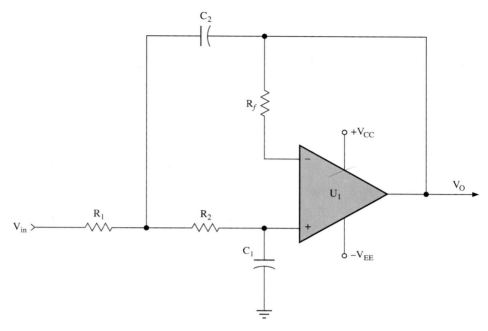

Figure 2–39
Unity-gain second-order low-pass filter

When R_1 equals R_2 and C_1 equals C_2, Eq. (2–20) may be simplified to

$$f_{CO} = \frac{1}{2\pi RC}$$

where $R = R_1 = R_2$ and $C = C_1 = C_2$. The Bode plot of a second-order low-pass filter is illustrated in Figure 2–38.

Figure 2–39 contains a second-order Butterworth low-pass filter for which voltage gain will approach unity below the cutoff frequency. Providing R_1 equals R_2, R_f nearly equals $(R_1 + R_2)$, and C_2 is nearly twice the value of C_1, the cutoff frequency for this unity-gain low-pass filter may be approximated:

$$f_{CO} = \frac{0.707}{2\pi R_1 C_1} \qquad (2–21)$$

where $R_1 = R_2 = R_f/2$ and $C_1 = C_2/2$.

Example 2–8

For the low-pass filter in Figure 2–40, what would be the cutoff frequency? What would be the frequency at $(f_{CO} \times 10)$? What is the roll-off for the circuit? What is the voltage gain of the filter for frequencies occurring within its pass band? If the peak voltage of an input signal occurring within the pass band is 470 mV, what would be the value of output voltage? What would be the values of V_o if the frequency of this input signal is increased to f_{CO} and $(f_{CO} \times 10)$? What design format does this filter represent?

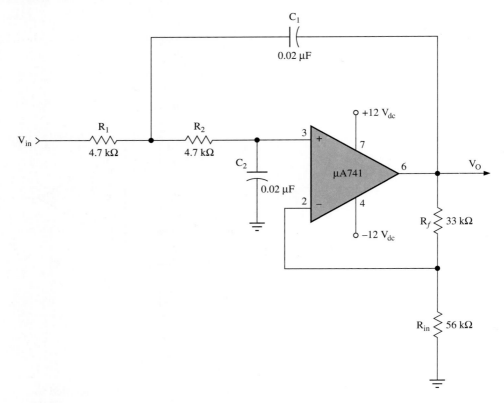

Figure 2–40

Solution:

$$f_{CO} = \frac{1}{2\pi \times RC} = \frac{1}{2\pi \times 4.7\,k\Omega \times 0.02\,\mu F} = 1.693\,\text{kHz}$$
$$f_{CO} \times 10 = 1.693\,\text{kHz} \times 10 = 16.93\,\text{kHz}$$

Within the pass band:

$$A_v = 1 + \frac{R_f}{R_{in}} = 1 + \frac{33\,k\Omega}{56\,k\Omega} = 1.589$$

If $V_{in} = 470\,\text{mV}$, then within the pass band:

$$V_o = V_{in} \times A_v = 470\,\text{mV} \times 1.589 = 747\,\text{mV}$$

At the cutoff frequency:

$$V_o = 0.707 \times V_{in} = 0.707 \times 470\,\text{mV} = 332.3\,\text{mV}$$

At $f_{CO} \times 10$, since the circuit is a second-order filter:

$$-40\,\text{dB} = 20\log(V_o/V_{in})$$
$$-2\,\text{dB} = \log(V_o/470\,\text{mV})$$

$$\text{antilog} -2 = (V_o/470\,\text{mV})$$
$$0.01 = V_o/470\,\text{mV}$$
$$V_o = 0.01 \times 470\,\text{mV} = 4.7\,\text{mV}$$

The filter in Figure 2–40 represents a Sallen and Key or VCVS form of Butterworth filter.

Active high-pass filters may be utilized to provide a sharp attenuation of signals occurring below a given cutoff frequency. At the same time, within the practical limits of the op amp, a constant gain factor may be maintained for signals above the cutoff frequencies. With the active high-pass filter, however, the deteriorating frequency response of the op amp poses a natural upper limit to the pass band. An op amp high-pass filter is often used to attenuate either a dc or a very low-frequency ac component of an input signal. At the same time, a desired higher frequency component of that same signal is easily passed with virtually no distortion.

A second-order Sallen and Key high-pass filter may be created by arranging the resistors and capacitors as shown in Figure 2–41. With the values of input and feedback resistance shown, voltage gain will approach 1.589 above the cutoff frequency and the circuit will exhibit a Butterworth response. Providing C_1 equals C_2, and R_1 equals R_2, the

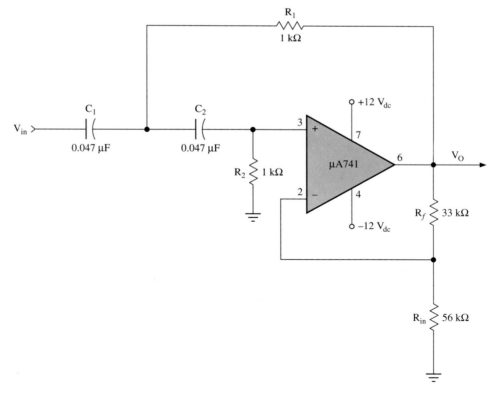

Figure 2–41
Sallen and Key (Butterworth) second-order high-pass filter

cutoff frequency for this circuit may be determined as with its low-pass counterpart:

$$f_{CO} = \frac{1}{2\pi\sqrt{R_1 \times R_2 \times C_1 \times C_2}}$$

which simplifies to:

$$f_{CO} = \frac{1}{2\pi \times RC}$$

Second-order op amp filters can be easily cascaded. One purpose for this action would be the creation of higher order low-pass or high-pass filters. As an example, two second-order low-pass filters could be cascaded, resulting in a fourth-order low-pass filter. Also, a second-order high-pass filter and a second-order low-pass filter may be cascaded, resulting in a bandpass filter. An example of such a filter is shown in Figure 2–42.

In the first stage of the bandpass filter, which is a Sallen and Key second-order Butterworth high-pass filter, C_{A_1} and R_{A_1} comprise the first RC pair, with R_{A_1} being placed in a feedback position. The R_{B_1} and C_{B_1} form the second RC pair, while R_{f_1} and R_{in_1} control the gain factor for the high-pass stage. (Note that the selected values of these two components ensure a nearly ideal Butterworth response.)

For the second stage of this circuit, which is comprised of a Sallen and Key low-pass filter, C_{A_2} now serves as the feedback element of the RC network. (Again, note that

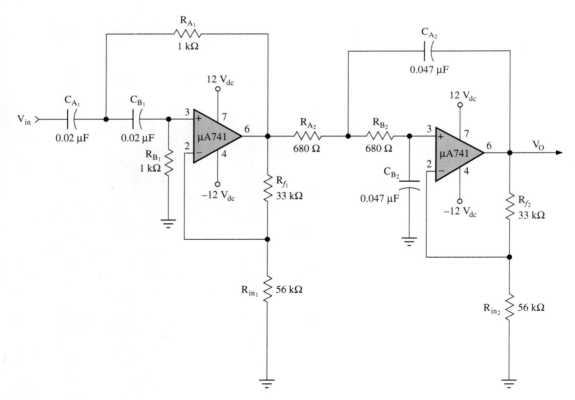

Figure 2–42
Bandpass filter formed through cascading high- and low-pass Butterworth filters

the values of R_{f_2} and R_{in_2} correspond with R_{f_1} and R_{in_1}, thus allowing a Butterworth response for this second stage of the circuit.)

A bandpass filter has a center, or resonant, frequency (f_o), as well as an upper and a lower cutoff frequency. For the bandpass design represented in Figure 2–42, the upper cutoff frequency is equivalent to that of the high-pass portion of the circuit. Since R_{A_1} equals R_{B_1} and C_{A_1} equals C_{B_1}, the upper cutoff frequency is easily calculated as:

$$f_{CO_2} = \frac{1}{2\pi \times RC} = \frac{1}{2\pi \times 1\,k\Omega \times 0.02\,\mu F} = 7.958\,kHz$$

The lower cutoff frequency can be determined by performing the same calculation for the second stage of the circuit. Since R_{A_2} equals R_{B_2} and C_{A_2} equals C_{B_2}:

$$f_{CO_1} = \frac{1}{2\pi \times RC} = \frac{1}{2\pi \times 680\Omega \times 0.047\,\mu F} = 4.98\,kHz$$

Once these two edge frequencies have been determined, the center frequency for the bandpass filter is easily determined as follows:

$$\begin{aligned} f_o &= \sqrt{f_{CO_1} \times f_{CO_2}} \\ &= \sqrt{4.98\,kHz \times 7.958\,kHz} = 6.295\,kHz \end{aligned} \qquad (2\text{–}22)$$

For the circuit design represented in Figure 2–42, the Q factor and the bandwidth are actually controlled by the proximity of the lower and upper cutoff frequencies. This should become evident with a review of these basic concepts for the bandpass filter:

$$Q = \frac{f_o}{BW} \qquad (2\text{–}23)$$

where $BW = f_{CO_2} - f_{CO_1}$. Thus, for the circuit in Figure 2–42:

$$Q = \frac{6.295\,kHz}{7.958\,kHz - 4.98\,kHz} = 2.114$$

The Bode plot in Figure 2–43, which represents the operation of the circuit in Figure 2–42, should aid in clarification of these concepts. It should now be evident that bringing the lower and upper cutoff frequencies closer together during the design process will allow a narrower pass band and, consequently, a higher value of Q. Bandpass filters for which the Q factor is at least 10 are classified as narrow-band filters. For this type of filter, the bandwidth would be no more than one-tenth the value of the center frequency. Bandpass filters with a Q factor of less than 10 are classified as wide-band filters. (For such devices, bandwidth would exceed one-tenth the value of the resonant frequency.)

Band-stop filters are widely used within industrial electronic systems. One of their primary purposes is to attenuate the noise induced by high-voltage power sources such as transformers and motor generators. The noise produced by such devices is persistent and usually occurs at very predictable frequencies. An example is the 60-cycle interference induced by an ac power line. Another industrial noise source would be a motor generator, which can produce high-voltage sine waves at a frequency of 400 Hz.

Special forms of op amp active filters may be utilized to block such unwanted noise signals. These filters, which are designed to have a very narrow stop band, are referred to as **notch filters.** The resonant frequency for such a filter is often referred to as the **notch frequency** f_n. This frequency would be equivalent to that of an undesired noise signal that is prominent in the area where a piece of electronic equipment must operate. To ensure reliable operation of the equipment, its input signal would be channeled through

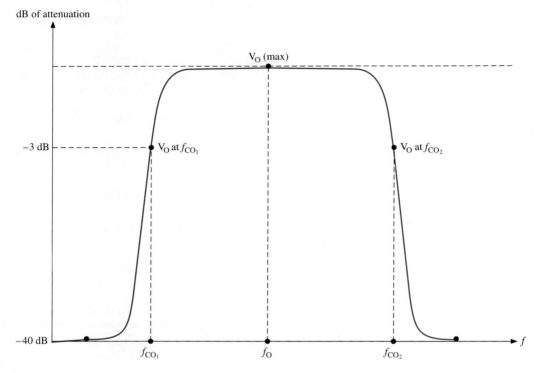

dB of attenuation

V_O (max)

-3 dB V_O at f_{CO_1} V_O at f_{CO_2}

-40 dB

f_{CO_1} f_O f_{CO_2} f

Figure 2–43
Bode plot (response curve) for Sallen and Key bandpass filter

a notch filter. Providing the desired input signal occurs at a much higher or lower frequency than that of the induced noise, the notch filter removes the unwanted noise component from that signal while allowing the desired information to pass to its destination with minimal distortion.

Figure 2–44 illustrates how a design format known as **multiple-feedback** might be used to create a notch filter. The design process may be greatly simplified if a common value of capacitance is selected for C_1 and C_2. Assuming that the notch frequency and the required value of Q are already known, the first step in circuit design is to solve for the value of the feedback resistor R_f, which is equivalent to R_4 in Figure 2–44.

$$R_f = \frac{Q}{\pi \times f_n \times C} \qquad (2\text{–}24)$$

Once this ohmic value is determined, the value of R_1 (which is the same as R_2) may be calculated:

$$R_1 = \frac{R_f}{4Q^2} \qquad (2\text{–}25)$$

Finally, the ohmic value of R_3 may be obtained as follows:

$$R_3 = 2 \times Q^2 \times R_1 \qquad (2\text{–}26)$$

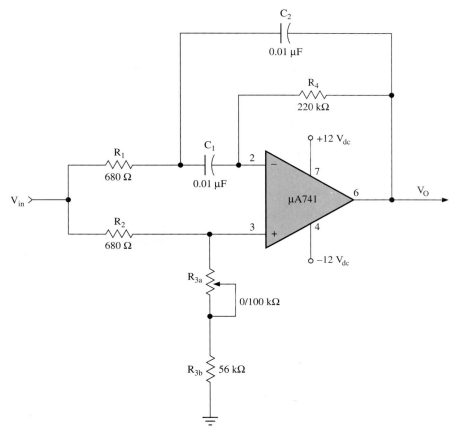

Figure 2–44
Multiple-feedback second-order notch filter

Example 2–9

Using Eqs. (2–24) through (2–26), determine the notch frequency and the setting of R_{3a} for the filter of Figure 2–44, assuming the actual value of Q is 8.5.

Solution:
Solving first for the notch frequency:

$$\text{If } R_f = \frac{Q}{\pi \times f_n \times C} \quad \text{then } f_n = \frac{Q}{\pi \times R_f \times C}$$

$$f_n = \frac{8.5}{\pi \times 220 \text{ k}\Omega \times 0.01 \text{ }\mu\text{F}} = 1.23 \text{ Hz}$$

Solving for the setting of R_{3a}:

$$R_3 = 2 \times Q^2 \times R_1$$
$$R_3 = 2 \times 8.5^2 \times 680 \text{ }\Omega = 98.26 \text{ k}\Omega$$
$$R_{3a} = R_3 - R_{3b} = 98.26 \text{ k}\Omega - 56 \text{ k}\Omega = 42.26 \text{ k}\Omega$$

SECTION REVIEW 2–5

1. What type of filter is contained in Figure 2–45? Identify the design format being utilized. With regard to frequencies within its pass band, what is the ideal voltage gain for this device? What is the ideal roll-off for this type of filter? What is its cutoff frequency?
2. For the waveform in Figure 2–46(a), assume its low-frequency component occurs at 665 Hz and its high-frequency component occurs at 10 kHz. If this waveform is being fed to the input of the circuit in Figure 2–45, the output would appear as the waveform shown in which of the following figures?:
   a. Figure 2–46(a)
   b. Figure 2–46(b)
   c. Figure 2–46(c)
3. What type of filter is contained in Figure 2–47? What design format is being utilized? What is the cutoff frequency of the filter? With regard to frequencies within the pass band, what is the ideal voltage gain for the device?
4. Assume the waveform in Figure 2–46(a) is now occurring at the input to the filter in Figure 2–47. The output waveform would now appear as that illustrated in which of the following figures?:
   a. Figure 2–46(a)
   b. Figure 2–46(b)
   c. Figure 2–46(c)

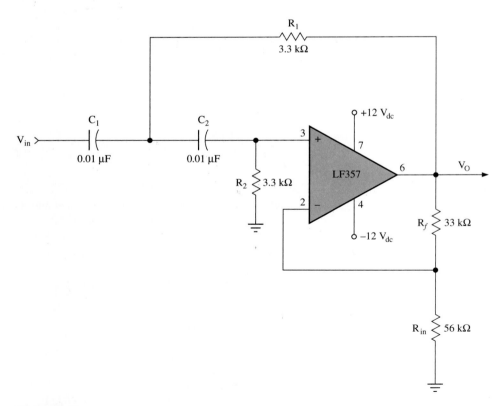

Figure 2–45

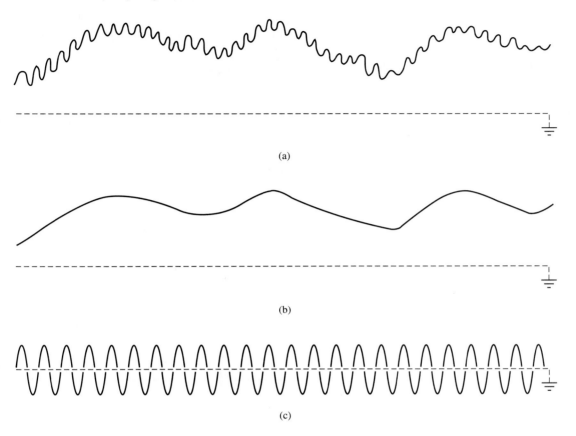

(a)

(b)

(c)

Figure 2–46

5. Identify both the design format and the function for the op amp filter contained in Figure 2–48. Assuming the actual Q of the filter to be 8.7, determine its resonant frequency. What is another name for this frequency? Explain why such a filter might be necessary within a piece of electronic equipment used at an industrial site.

2–6　　OP AMP INTEGRATORS AND DIFFERENTIATORS

Op amp integrators are used frequently in many forms of signal conditioning circuitry. Before beginning the study of the op amp integrator, it is advantageous to review the operation of the simple RC circuit shown in Figure 2–49(a). (This form of circuit has already been introduced with regard to its function as a low-pass filter.) The response of this device to high- and low-frequency sine waves is illustrated in Figure 2–49(b). This circuit is sometimes referred to as an **integrator** because the amplitude of the output signal, when developed over the capacitor, is proportional to the time integral of the input signal. Note that, as shown in Figure 2–49(b), the amplitude of the output is inversely proportional to frequency. (This is true due to the fact that frequency is inversely proportional to time.) Although the response of this simple integrator to a sine wave is characterized by varying amplitude and phase, note that the output waveform remains sinusoidal. Figure 2–49(c) represents the response of an integrator to a square wave. Where pulse

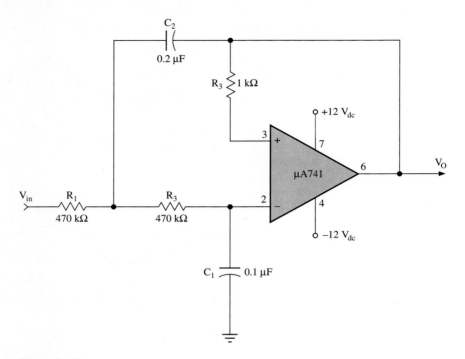

Figure 2–47

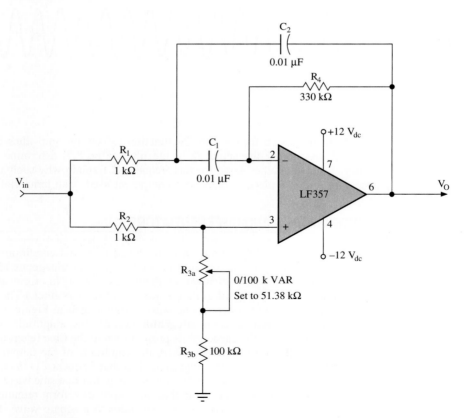

Figure 2–48

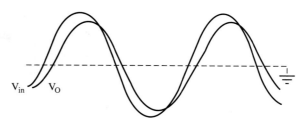

Close phase relationship of V_{in} and V_O at low f

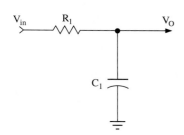

(a) Simple passive integrator

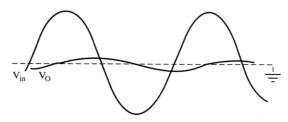

V_{in} V_O

Phase relationship of V_{in} and V_O at a much higher f

(b)

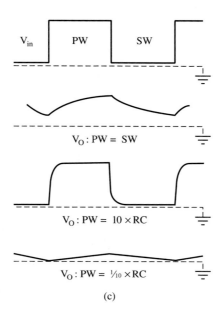

(c)

Figure 2–49
Operation of passive RC integrator

width is 10 times greater than the RC time constant, the output is still "square-like" due to the fact that the capacitor has 10 time constants in which to charge toward the peak level of the source waveform. When pulse width equals the RC time constant, an exponential wave appears at the output, the peak-to-peak amplitude of which is likely to be only 0.63 times that of the source signal. Where the pulse width of the input is only one-tenth the RC time constant, the output is a low-amplitude triangle wave. The slopes of this signal are nearly linear due to the fact that, at the instant the capacitor begins to charge (when it is behaving as a virtual short with respect to the source) the flow of charging current is nearly constant.

According to the law of time constants, if for the integrator in Figure 2–49(a), a steady dc voltage is applied at the input, the capacitor will begin to charge to that input level but never actually attain it. With the op amp integrator shown in Figure 2–50(a), however, if a constant positive dc voltage is applied to the input, the capacitor will begin to charge toward the level of $-V_{EE}$. This would be true even if the dc input voltage were much lower in amplitude than either supply voltage. If a negative dc voltage were applied to the input, the capacitor would begin to charge toward the level of $+V_{CC}$. Also, due to the action of the virtual ground at the inverting input, providing the input voltage remains at a constant dc level, the flow of charging current will also be constant. Thus, the resultant voltage change at the output would be a linear ramp. Figure 2–50(b) illustrates this relationship of input and output signals.

Remembering that the capacitor tends to behave as an open switch with respect to a dc input and that the open-loop gain of the op amp is very large should help you to understand how, through time, the capacitor could charge to a level of nearly V_{CC} or $-V_{EE}$. This ability of the open-loop gain of the op amp to enhance the charging of the capacitor is referred to as the **Miller effect** and, for that reason, the op amp circuit in Figure 2–50(a) is sometimes referred to as the **Miller integrator.** The Miller effect can be explained mathematically as shown here:

$$Q = V_{id} \times (1 + A_v) \times C_1$$

where Q is the charge on the capacitor, V_{id} is the slight difference in potential between the input terminals of the op amp, and A_V is the open-loop gain of the device.

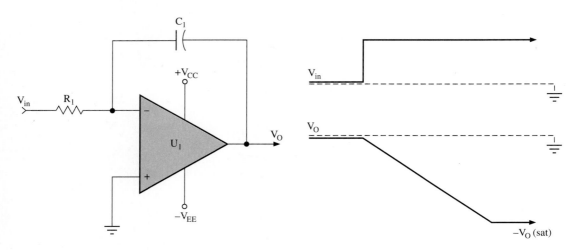

(a) Basic op amp integrator   (b) Response of the op amp integrator to a constant DC input level

Figure 2–50

Since Q is equivalent to $(C_1 \times V_{C_1})$:

$$C_1 \times V_{C_1} = V_{id} \times (1 + A_V) \times C_1$$

which simplifies to:

$$V_{C_1} = V_{id} \times (1 + A_V) \qquad (2\text{--}27)$$

For the Miller integrator, if the input is a square wave, the output will be a triangle wave. A thorough analysis of this particular circuit application provides an excellent foundation for understanding the operation of the integrator in more advanced signal conditioning circuitry.

We review the **coulomb,** the basic unit of charge, first. The charge held by a capacitor may be considered as the product of its capacitance and voltage, stated as follows:

$$Q = C \times V$$

Also, when the charging current is constant, the growth of charge over the capacitor will be the product of the flow of this current through time. This relationship may be stated simply as:

$$Q = I \times T$$

Therefore, the following relationship may be stated:

$$(C \times V) = (I \times T)$$

With the values of charging current and capacitance being held constant, the variable elements in the equation would then be voltage and time. Stated in a simple differential form, the equation would then be:

$$(C \times \Delta V) = (I \times \Delta t)$$

Therefore,

$$\frac{\Delta V}{\Delta t} = \frac{I}{C}$$

A ratio has now been established, which is referred to as the **slope** for the output signal. The slope can be defined simply as the rate of voltage change through time. Also, consider the expression for the change in voltage:

$$\Delta V = \frac{I \times \Delta t}{C} \qquad (2\text{--}28)$$

where ΔV represents the difference in output voltage and I represents the charging current.

It should now be evident that the amount of change in the output voltage of the integrator is directly proportional to both the value of the input current and the amount of time this current is allowed to flow. Also, the growth of this voltage is indirectly proportional to the value of capacitance. Or simply, if the value of the capacitor is increased, the slope will decrease.

Example 2–10

For the Miller integrator in Figure 2–51(a), assume the input to be the square wave shown in Figure 2–51(b). Determine the amplitude and polarity of input current that occurs during the positive alternation of the input square wave. Determine the amplitude and

Figure 2–51

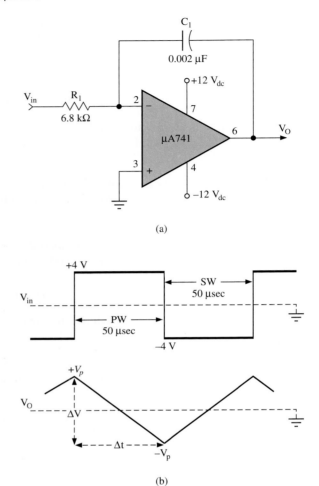

(a)

(b)

polarity of the input current that occurs during the negative alternation of the input square wave. What would be the value of ΔV_o? Determine the positive and negative peak amplitudes of this output signal. At what frequency would the input and output signals occur?

Solution:

During the pulse width of the input signal:

$$I = I_{in} = \frac{+4 \text{ V}}{6.8 \text{ k}\Omega} = 588 \text{ } \mu\text{A}$$

During the space width of the input signal:

$$I = I_{in} = \frac{-4 \text{ V}}{6.8 \text{ k}\Omega} = -588 \text{ } \mu\text{A}$$

$$\Delta V_o = V_{o\,p\text{-}p} = \frac{I \times \Delta t}{C} = \frac{588 \text{ } \mu\text{A} \times 50 \text{ } \mu\text{sec}}{0.002 \text{ } \mu\text{F}} = 14.7 \text{ V}$$

$$+V_p = \frac{14.7 \text{ V}}{2} = 7.35 \text{ V} \qquad (-V_p = -7.35 \text{ V})$$

$$f = \frac{1}{T} = \frac{1}{PW + SW} = \frac{1}{100 \text{ }\mu\text{s}} = 10 \text{ kpps}$$

Unless the pulse width and space width of the square wave in Figure 2–51(b) are *exactly* equal, a residual dc charge is likely to develop over C_1, causing the output to drift slowly toward the level of V_{CC} or V_{EE}. If pulse width is greater than space width, the output would clamp in the negative direction. If the space width were longer than the pulse width, the output would drift toward V_{CC}. For the integrator in Figure 2–52, the presence of R_2 dampens the Miller effect, providing a path for feedback current. Thus C_1 is able to release any excess charge during each cycle of the input square wave. (A value of R_2 should be selected such that the time constant for R_2 and C_1 is at least 10 times the pulse width of the input square wave.)

For many function generator applications, it is necessary to vary the dc reference level of an output signal. This action is made possible through the addition of the offset adjustment circuitry as shown in Figure 2–52. Adjustment of potentiometer R_3 should allow the dc reference of the output signal to be altered without changing the signal's waveshape or peak-to-peak amplitude. If the wiper of R_3 is moved in the direction of the 12-V supply, the output signal will also shift in the positive direction. The opposite effect will occur if the voltage at the wiper of R_3 is made more negative.

Figure 2–53 serves to summarize the operation of a Miller integrator. When the input is a cosine wave, the output will be an inverted sine wave. That is, the output signal will

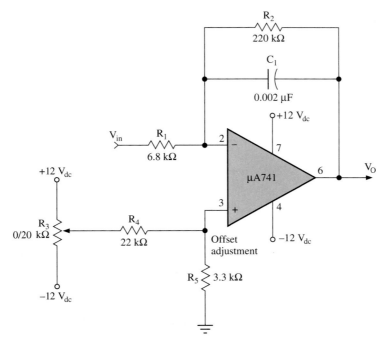

Figure 2–52
Op amp integrator with feedback resistor and offset adjustment

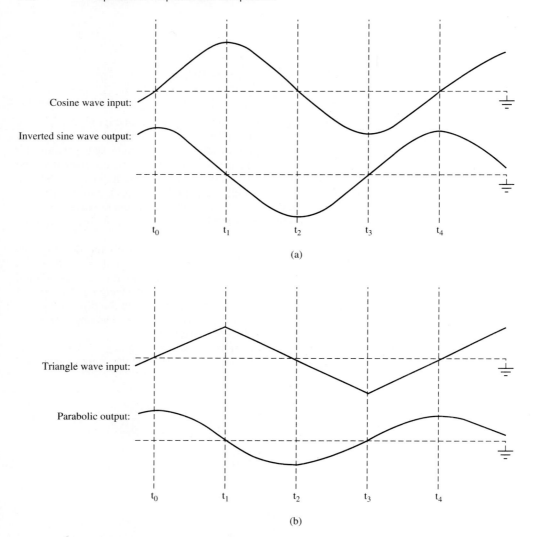

Cosine wave input:

Inverted sine wave output:

t_0 t_1 t_2 t_3 t_4

(a)

Triangle wave input:

Parabolic output:

t_0 t_1 t_2 t_3 t_4

(b)

Figure 2–53

actually lead the source by 90°. If the input is a triangle wave, the output will be a parabolic function.

Figure 2–54(a) shows a simple RC differentiator. With the output taken between the resistor and ground, the amplitude of this signal is now proportional to the time derivative of the input. As shown in Figure 2–54(b), when the input is a sine wave, the amplitude of the output will be directly proportional to the input frequency. (Also note that, with the sine wave input, as frequency increases, the phase angle decreases.)

Since the output of the differentiator is developed over the resistor, according to Kirchhoff's law, this output signal must consist of the algebraic difference between the source voltage and that developed over the capacitor. Figure 2–54(c) illustrates the response of the simple RC differentiator to a square wave at three significant ratios of pulse width to RC time. Careful comparison of these signals to the integrator outputs in Figure 2–49(c)

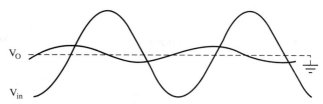

Phase relationship of V_{in} and V_O at low f

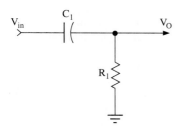

(a) Simple passive differentiator

Close phase relationship of V_{in} and V_O at high f

(b)

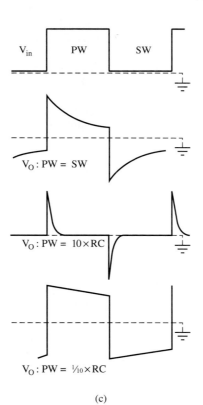

(c)

Figure 2–54
Operation of passive RC differentiator

should help you visualize how the resistor actually develops the "missing piece" of the source waveform.

As shown in Figure 2–55(a), a basic op amp differentiator may be created simply by reversing the positions of the resistor and capacitor within the op amp integrator. Like the Miller integrator, the op amp differentiator is capable of producing a larger output voltage than is present at the input. Also, due to the low output impedance of the op amp, the loading effect of the differentiated output will be minimized.

As a signal conditioning tool, the op amp differentiator is not as accurate as the op amp integrator and, therefore, is not as prevalent. Also, the basic circuit in Figure 2–55(a) has an inherent tendency to develop high-frequency noise. This is understandable considering that the simple RC differentiator of Figure 2–54(a) is essentially a high-pass filter. Due to the relatively low value of capacitive reactance offered by C_1 to high-frequency noise, such unwanted signals could easily pass this input capacitor and be developed over the feedback resistor. Also, as the frequency of an input sine wave is increased, additional phase delay will gradually be added to the 90° phase shift already present due to the circuit's inverting action. Finally, a frequency would be attained where enough regenerative feedback is occurring to cause oscillation. Such problems as these render the basic op amp differentiator of Figure 2–55(a) impractical for most circuit applications.

Figure 2–55(b) contains a Bode plot representing the operation of the basic differentiator. Here the frequency response curve of the differentiator is plotted against the open-loop frequency response curve of a typical op amp. Note that, for an ideal op amp differentiator, the circuit gain factor would continue to increase as frequency is increased and X_C is decreased. However, with an actual differentiator, a frequency is eventually reached at which gain begins to decrease due to the degenerating high-frequency response of the op amp itself. On the Bode plot, the difference between this point (f_b) and an origin point representing unity gain (f_a) reveals a closure rate of 20 dB per decade. With the voltage gain rising this sharply, overemphasis of high-frequency harmonics and noise could occur to the point of obscuring the desired output signal.

An op amp differentiator design that provides the operating stability and noise immunity necessary for most practical circuit applications is shown in Figure 2–56(a). With

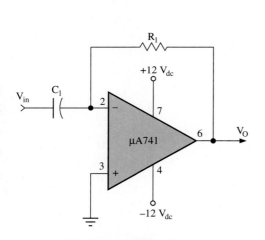

(a) Basic op amp differentiator

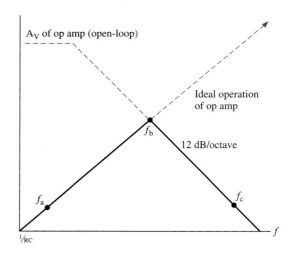

(b) Bode plot of basic op amp differentiator

Figure 2–55
Operation of basic op amp differentiator

this circuit design, **stop resistor** R_s is placed in series with the input capacitor. A value of R_s should be chosen that would be much greater than the value of capacitive reactance occurring at the anticipated frequency of oscillation. If this frequency has not been determined, then a value of stop resistor should be chosen that is no less than $1/100 \times R_1$. This should ensure stable circuit operation at higher frequencies because, when X_C becomes minimal, voltage gain will still be limited by a factor of $(-R_1/R_s)$. The circuit would then begin to function as a simple inverting amplifier, as represented by its Bode plot in Figure 2–56(b).

As shown in Figure 2–57(a), a feedback capacitor may be used to further enhance the operating stability of the differentiator. If a value of C_f is chosen such that $(R_1 \times C_f)$ nearly equals $(R_s \times C_1)$, then the Bode plot would appear as shown in Figure 2–57(b). The relative values of f_a and f_b for this circuit would be derived as follows:

$$f_a = \frac{1}{2\pi \times R_1 \times C_1} \quad f_b = \frac{1}{2\pi \times R_s \times C_1} = \frac{1}{2\pi \times R_1 \times C_f}$$

As shown in Figure 2–57(b), once the frequency of f_b has been attained, the closed-loop response of the differentiator should reveal a decrease in voltage gain. Remembering that component values have been chosen such that $(R_1 \times C_f)$ nearly equals $(R_s \times C_1)$ and that the ohmic value of R_s will be much less than the ohmic value of R_1, C_f must then be much smaller in value than C_1. Thus, for any given value of input frequency, the value of capacitive reactance for C_1 will be much less than that of C_f. Ideally, as input frequency increases, the X_C of the feedback capacitor will continue to decrease such that the response curve for voltage gain beyond f_b will be a mirror image of that leading up to this frequency. Therefore, the overall response curve of this final circuit will become a dampened version of the one in Figure 2–55(b), which represented unstable circuit operation.

Designing a practical op amp differentiator, such as that in either Figure 2–56(a) or 2–57(a) involves the manipulation of a few basic formulae. As with the Miller integrator, the inverting input of the op amp will still function as a virtual ground. Thus, at any given

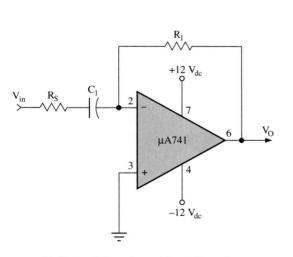

(a) Op amp differentiator with stopping resistor

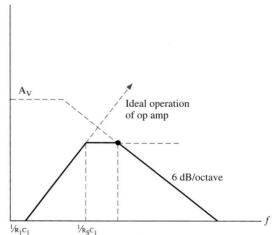

(b) Bode plot for differentiator with stopping resistor

Figure 2–56
Operation of op amp differentiator with stopping resistor

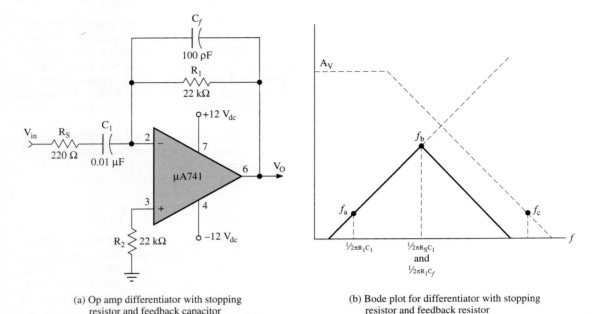

(a) Op amp differentiator with stopping
resistor and feedback capacitor

(b) Bode plot for differentiator with stopping
resistor and feedback resistor

Figure 2–57
Operation of op amp differentiator with stopping resistor and feedback capacitor

instant, the output voltage could be estimated as follows:

$$V_o = -RC \times \frac{\Delta V_{in}}{\Delta t} \qquad (2\text{–}29)$$

According to this formula, the more rapid a change in amplitude of the input signal, the greater the amplitude of the output. As an example, assume the triangle wave shown in Figure 2–58(a) is serving as the input to the differentiator in Figure 2–57(a). Notice that this signal begins at a constant frequency of 1 k pps, but with a constantly varying amplitude. As shown in Figure 2–58(b), the resultant output signal will be a square wave. Changes in the pulse amplitude of this output signal will be proportional to changes in the slope of the input signal.

At first you may have difficulty visualizing how a triangle wave might be differentiated to form a square wave. If you recall that the slope of a linear ramp signal is actually a constant you should understand how this occurs. From t_0 to t_1 in Figure 2–58, during the positive slope of the triangle wave, the input voltage is actually changing at a rate of 8000 V/sec, calculated as

$$\text{Slope} = \frac{\Delta V_{in}}{\Delta t} = \frac{4 \text{ V} - 0 \text{ V}}{500 \text{ } \mu\text{sec}} = 8000 \text{ V/sec}$$

Since this **rate of change** is constant, the output voltage level for the differentiator in Figure 2–57a will remain constant, being calculated as:

$$V_o = -R_1 \times C_1 \times \frac{\Delta V}{\Delta t} = -22 \text{ k}\Omega \times 0.01 \text{ } \mu\text{F} \times 8000 = -1.76 \text{ V}$$

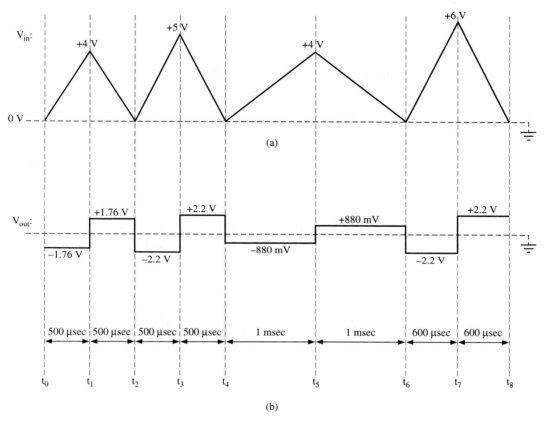

Figure 2–58
Response of the practical op amp differentiator to a varying ramp input

From t_1 to t_2, while the input is ramping at the same rate but in the negative direction, the output voltage is determined to be

$$V_o = -22 \text{ k}\Omega \times 0.01 \ \mu\text{F} \times \frac{0 \text{ V} - 4 \text{ V}}{500 \ \mu\text{sec}} = +1.76 \text{ V}$$

If the input to the op amp differentiator in Figure 2–57(a) is a sine wave then, as shown in Figure 2–59, the output will be an inverted cosine wave. This is true due to the fact that, as the source waveform crosses the zero-degree point in the positive direction, the rates of input current and voltage change are at maximum. Therefore, since the amplitude of the output signal will be proportional to the time derivative of the input, V_o will then be at its maximum negative value. Due to the inverting action of the circuit, as the input sine wave develops from the 0-V level to its positive peak, the output waveform is moving from its negative peak potential toward 0 V. (Notice that the 0-V points in the output signal correspond with the points at which input current and voltage change are at minimum.)

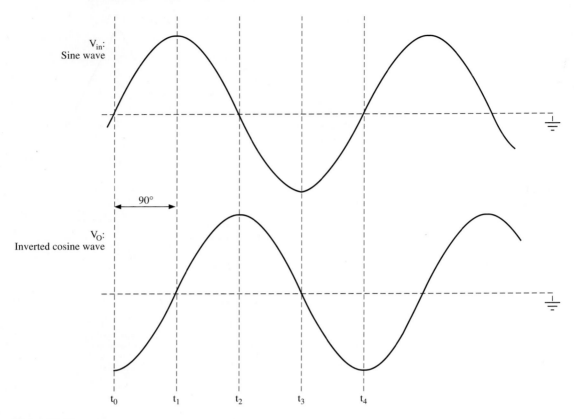

Figure 2–59
Response of differentiator to sinusoidal input

Example 2–11

Solve for the values of f_a and f_b for the op amp differentiator in Figure 2–57(a). Assuming the input voltage is a linear ramp that changes from 0 to 3.5 V in a period of 450 μsec, what would be the value of output voltage during that time? Next, assuming the input ramps from 3.5 to 0 V in a period of 225 μsec, what would be the value of V_O?

Solution:
Solving for f_a:

$$f_a = \frac{1}{2\pi \times R_1 \times C_1} = \frac{1}{2\pi \times 22\ \text{k}\Omega \times 0.01\ \mu\text{F}} = 723.4\ \text{Hz}$$

Solving for f_b:

$$f_b = \frac{1}{2\pi \times R_s \times C_1} = \frac{1}{2\pi \times 220\ \Omega \times 0.01\ \mu\text{F}} = 72.34\ \text{kHz}$$

or

$$f_b = \frac{1}{2\pi \times R_1 \times C_f} = \frac{1}{2\pi \times 22\ \text{k}\Omega \times 100\ \text{pF}} = 72.34\ \text{kHz}$$

Solving for V_o with a positive-going ramp at the input:

$$V_o = -\text{RC} \times \frac{\Delta V}{\Delta t} = -22 \text{ k}\Omega \times 0.01 \text{ } \mu\text{F} \times \frac{3.5 \text{ V}}{450 \text{ } \mu\text{sec}} = -1.711 \text{ V}$$

Solving for V_o with a negative-going ramp at the input:

$$V_o = -22 \text{ k}\Omega \times 0.01 \text{ } \mu\text{F} \times \frac{-3.5 \text{ V}}{225 \text{ } \mu\text{sec}} = +3.422 \text{ V}$$

Differentiators have a wide variety of applications, which extend from the creation of simple trigger or timing pulses to complex analog computations. The processes of integration and differentiation are both incorporated into industrial electronic system design. Specific applications of these processes are analyzed in those sections of the text dedicated to signal conditioning and closed-loop control systems.

SECTION REVIEW 2–6

1. Describe the Miller effect. How does this effect enhance the performance of an op amp integrator?
2. True or False?: For the op amp integrator in Figure 2–52, changing the value of C_1 to 0.01 μF would increase the slope of its output signal.
3. Explain the purpose of R_2 in Figure 2–52.
4. Assume the square wave in Figure 2–51(b) is now serving as the input to the circuit in Figure 2–52. If R_1 in Figure 2–52 is now replaced with a 12 kΩ resistor, what is the peak-to-peak amplitude of the output signal? What is the slope of the output signal?
5. True or False?: The output amplitude of a differentiator is directly proportional to the time derivative of its input signal.
6. Explain the problem inherent to the basic op amp differentiator in Figure 2–55(a).
7. Describe the purpose of the stop resistor in Figure 2–57(a). What is the purpose of the feedback capacitor contained in this same circuit?
8. For the circuit in Figure 2–57(a), assume the input is a linear ramp signal that changes from 352 mV to 28 mV in 36 μsec. What is the amplitude of the output during this time?

2–7   COMPARATORS AND SCHMITT TRIGGERS

A **comparator** is a voltage-differencing device that is commonly used in the process of **A/D conversion**. Like the op amp, the comparator has both an inverting and a noninverting input. In fact, as shown in Figure 2–60, the schematic symbol for the comparator is identical to that for the op amp. The operation of the ideal comparator may be compared to that of the ideal op amp in the open-loop condition. This is due to the fact that the ideal comparator, like the ideal op amp, is considered to have infinite voltage gain. Therefore, regardless of how slight the difference in voltage between the two input terminals, the output of the ideal comparator will approach the level of the positive or negative supply voltage.

Figure 2–61 represents the operation of the ideal comparator. Note that in Figure 2–61(a) a constant **reference voltage** V_{ref} is applied to the inverting input of the comparator.

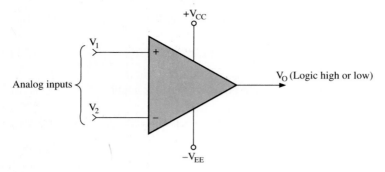

Figure 2–60
Schematic symbol for the comparator

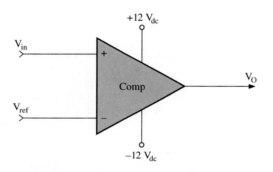

(a) Comparator with active signal at the noninverting input

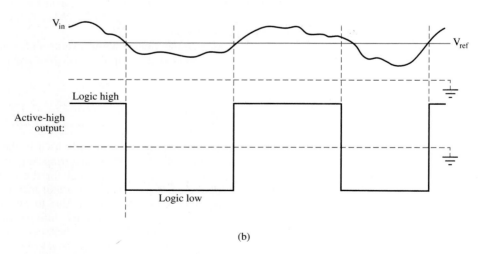

(b)

Figure 2–61
Operation of a comparator with the active signal occurring at the noninverting input

Table 2–1
Operation of comparator with active-high output

Input	Output
$V_{in} > V_{ref}$	High (1)
$V_{in} < V_{ref}$	Low (0)

This input configuration allows the voltage present at the noninverting input to be compared continually to a reference level. As illustrated in Figure 2–61(b), when the active input signal V_{in} becomes more positive than the reference voltage, the output will instantly rise toward the level of V_{CC}. When the active signal falls below the reference level, the output instantly falls toward the level of $-V_{EE}$. Note that, while changes in the amplitude of the active input signal may be gradual, transitions occurring at the output of the comparator are virtually instantaneous. Thus, its output signal, which appears as a rectangular pulse,

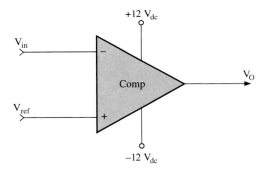

(a) Comparator with active signal at the inverting input

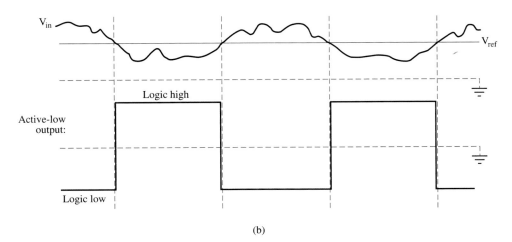

(b)

Figure 2–62
Operation of a comparator with the active signal occurring at the inverting input

Table 2–2
Operation of comparator with active-low output

Input	Output
$V_{in} < V_{ref}$	High (1)
$V_{in} > V_{ref}$	Low (0)

can be considered to be a digital, or discrete, response to changes in the analog input voltage. With the input configuration of Figure 2–61(a), note that the output toggles in the positive direction when V_{in} is greater than V_{ref}. This high output condition could be considered as a logic 1. When V_{in} falls below the level of V_{ref}, the output condition may be considered as a logic 0. The operation of the comparator in Figure 2–61(a) may be summarized as shown in Table 2–1.

In Figure 2–62(a), the active input signal occurs at the inverting input of the comparator, whereas the reference voltage is placed at the noninverting input. Thus, as illustrated in Figure 2–62(b), the output of the comparator will toggle to a logic 1 when the input signal falls below the reference level. When V_{in} exceeds the level of V_{ref}, the output of the comparator will toggle to a logic 0.

Table 2–2 represents the operation of the comparator in Figure 2–62(a). Note that the function of this device is complementary to that for the comparator in Figure 2–61(a). The output conditions represented in Table 2–1 may be considered as active-high, since the output is at a logic 1 when the input condition ($V_{in} > V_{ref}$) is met. Table 2–2 represents the operation of a comparator with an active-low output, since V_o will be at a logic 0 when the input condition ($V_{ref} > V_{in}$) is met.

Figure 2–63 illustrates how a reference voltage might be established for a simple comparator circuit. If the extremely high input impedance of the comparator is ignored during the calculation process, the reference voltage in Figure 2–63 may be determined

Figure 2–63
Comparator with active-high output and adjustable reference voltage

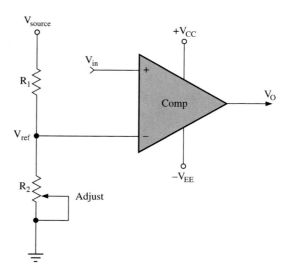

as follows, based on the assumption that R_1 and R_2 form a simple series voltage divider:

$$V_{\text{ref}} = V_{\text{source}} \times \frac{R_2}{R_1 + R_2} \qquad (2\text{-}30)$$

Thus, increasing the ohmic value of R_2 will result in a larger value of V_{ref}.

Example 2–12

For the comparator circuit in Figure 2–64, assume that R_2 is adjusted to 7.46 kΩ. What is the value of V_{R_2}? If V_{in} equals 6.3 V? What is the approximate output voltage? What digital logic level could this output voltage represent?

Assuming the ohmic setting of R_2 remains the same while V_{in} becomes 2.78 V, what is the approximate value and polarity of V_o? What digital logic level could this output voltage represent?

Solution:

$$V_{\text{ref}} = V_{\text{source}} \times \frac{R_2}{R_1 + R_2} = 12\text{ V} \times \frac{7.46\text{ k}\Omega}{10\text{ k}\Omega + 7.46\text{ k}\Omega} = 5.127\text{ V}$$

With V_{in} equal to 6.3 V, $V_{\text{in}} > V_{\text{ref}}$. Since V_{in} is occurring at the noninverting input of the comparator:

$$V_o \cong +V_o(\text{sat})$$

or approximately $+11$ V for a logic 1 (allowing a 2-V internal voltage drop for the comparator). With V_{in} equal to 2.78 V, $V_{\text{in}} < V_{\text{ref}}$. Thus,

$$V_o \cong -V_o(\text{sat})$$

or approximately -11 V for a logic 0.

Figure 2–65(a) illustrates how an **LM311,** which is a high-speed precision comparator, may be used as a **zero-crossing detector.** This form of circuit is commonly used to

Figure 2–64
Comparator with adjustable reference voltage

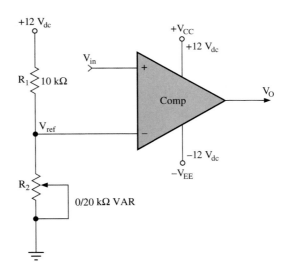

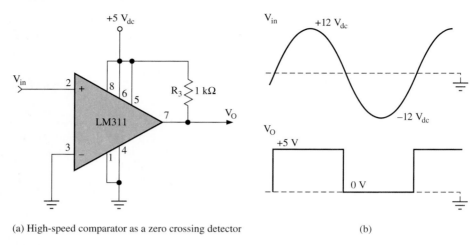

(a) High-speed comparator as a zero crossing detector   (b)

Figure 2–65
Operation of comparator as a zero-crossing detector

convert a sine wave into a square wave. Such a process may be necessary where a sinusoidal signal is to be used as a clock (timing source) for digital circuitry. With the circuit configuration shown in Figure 2–65(a), the output square wave is made compatible with a digital system operating from a single 5-V power supply. (Note that since the output of the **LM311** is an open-collector, a pull-up resistor must be attached between pin 7 and the 5-V source.) The phase relationship of the input sine wave and output square wave is illustrated in Figure 2–65(b).

The operation of the comparator circuit in Figure 2–65(a) can be briefly summarized as follows. Since the inverting input to the comparator is connected directly to circuit ground, 0 V is established as the reference level. Therefore, due to the comparator's extremely high voltage gain, the output signal will rise toward V_{CC} at the instant the active input attempts to exceed 0 V. This action is seen at the 0° point in Figure 2–65(b). The output of the comparator will remain high until the 180° point in the cycle, where the input signal again approaches the 0-V reference level. At the instant the input signal falls below 0 V, the output will toggle to 0 V, remaining at this level until the input sine wave again approaches 0 V.

A problem inherent to the high-speed comparator is that of switching noise. Assume, as illustrated in Figure 2–66(a), that the active input signal to the comparator circuit in Figure 2–65 is a sine wave containing high-frequency noise distortion. As the instantaneous amplitude of the sine wave approaches the value of the reference voltage, the high-frequency noise riding on this waveform will rapidly switch above and below this threshold level. As a result, the output of the comparator will rapidly toggle between its minimum and maximum switching levels. This spurious switching action, which is illustrated in Figure 2–66(b), cannot be tolerated, especially if the comparator output is to serve as a trigger or timing pulse for digital ciruitry.

One method to prevent the erratic switching action shown in Figure 2–66(b) would be to feed the sine wave through an active low-pass filter, attempting to eliminate any high-frequency noise prior to its arrival at the active input of the comparator. Although this method might be effective, it requires the addition of at least one op amp to the existing circuit design.

The noise problem could be eliminated without the use of any additional ICs by abandoning the high-speed comparator in favor of a device called the **Schmitt trigger,**

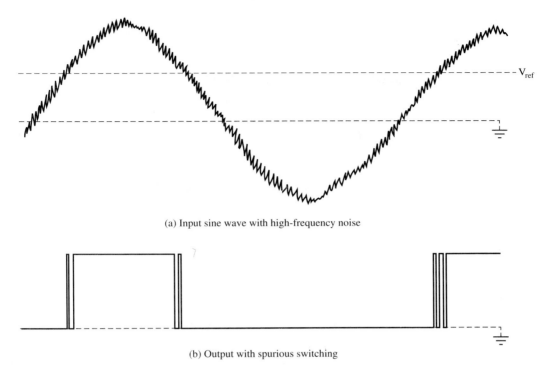

(a) Input sine wave with high-frequency noise

(b) Output with spurious switching

Figure 2–66
Response of a high-speed comparator to a noisy input signal

an example of which is shown in Figure 2–67. An op amp with a high slew rate may be used effectively as a Schmitt trigger, if the frequency of the input signal is well within the bandwidth of the device. The Schmitt trigger in Figure 2–67 provides switching stability through its *hysteresis,* which is established by the voltage divider action of R_1 and R_2. Since the junction of these two resistors occurs at the positive input of the op amp, R_2 actually functions as a positive feedback resistor. Note that R_1 and R_2 *do not* limit the voltage gain of the device. Thus, the op amp, when placed in this configuration, may still be considered as having an open-loop value of A_v.

In electronics, the term **hysteresis** is generally used to describe a lag or gap between a changing input condition and a resultant change at an output. The ideal comparator would have no hysteresis since the output of the device would toggle instantaneously as the active input voltage deviated above or below the established reference level. However, it is this nearly instantaneous switching action that leads to the undesired output noise occasionally exhibited by the high-speed comparator. With the Schmitt trigger, an intentional hysteresis gap, or **deadband,** is created through establishing **upper** and **lower trigger points** (UTP and LTP, respectively).

For the Schmitt trigger in Figure 2–67, the UTP and LTP are determined through simple voltage divider calculations:

$$\text{UTP} \cong +V_o(\text{sat}) \times \frac{R_1}{R_1 + R_2} \tag{2–31}$$

$$\text{LTP} \cong -V_o(\text{sat}) \times \frac{R_1}{R_1 + R_2} \tag{2–32}$$

Figure 2–67
Op amp Schmitt trigger

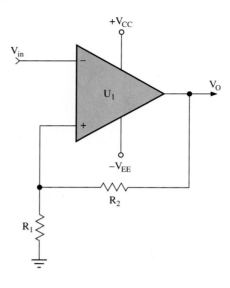

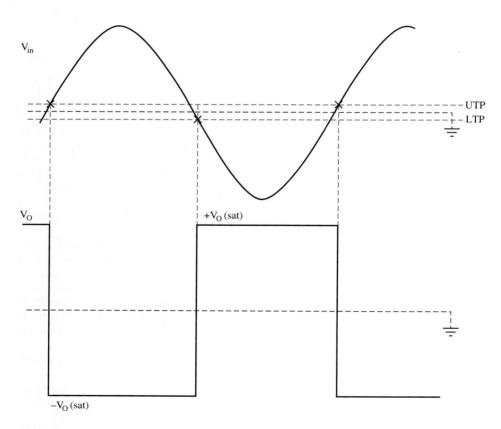

Figure 2–68
Schmitt trigger input and output waveforms

The **hysteresis gap** for the Schmitt trigger would then be the difference in voltage between the two trigger points:

$$\text{Hysteresis gap} = (\text{UTP} - \text{LTP})$$

Figure 2–68 illustrates the relationship of the input and output waveforms for the Schmitt trigger in Figure 2–67. Note that since the active signal occurs at the inverting input, the comparator action of the device may be considered as active-low. Therefore, when the signal present at the inverting input is below the UTP, the output voltage will remain at its positive saturation level, allowing the UTP to be maintained at the inverting input. At the instant the input signal exceeds the level of the UTP, the output rapidly toggles to its negative saturation level and remains there until the input signal falls below the level of the LTP, which is now present at the noninverting input. It should become evident that the Schmitt trigger in Figure 2–67 is functioning as a zero-crossing detector. The smaller the value of R_1 becomes with respect to R_2, the closer the upper and lower trigger points will be to 0 V.

Example 2–13

For the op amp Schmitt trigger in Figure 2–69, determine the upper and lower trigger points as well as the hysteresis gap. Assume the positive and negative saturation levels of output voltage to be +11 and −11 V, respectively.

Solution:

$$\text{UTP} = +V_o(\text{sat}) \times \frac{R_1}{R_1 + R_2} = 11 \text{ V} \times \frac{12 \text{ k}\Omega}{12 \text{ k}\Omega + 680 \text{ k}\Omega} = 191 \text{ mV}$$

Figure 2–69
Schmitt trigger

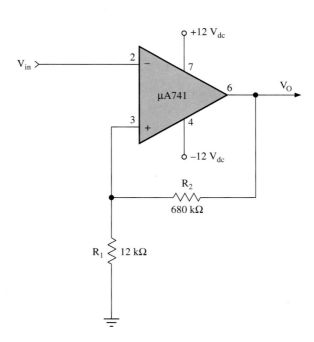

$$\text{LTP} = -V_o(\text{sat}) \times \frac{R_1}{R_1 + R_2} = -11 \text{ V} \times \frac{12 \text{ k}\Omega}{12 \text{ k}\Omega + 680 \text{ k}\Omega} = -191 \text{ mV}$$

$$\text{Hysteresis gap} = \text{UTP} - \text{LTP} = 191 \text{ mV} - -191 \text{ mV} = 382 \text{ mV}$$

SECTION REVIEW 2–7

1. For the comparator circuit in Figure 2–70, what is the reference voltage?
2. For the comparator in Figure 2–70, is the output of the comparator active-high or active-low?
3. For the comparator circuit in Figure 2–70, assume the input is at 2.38 V. What would be the value of V_o?
4. For the comparator in Figure 2–70, assume that V_{in} now equals 8.49 V. What would be the value of V_o?
5. If the sine wave in Figure 2–71(a) is occurring at the input to the circuit in Figure 2–71(b), what type of waveform would be occurring at the output?
6. What would be the upper and lower trigger points for the circuit in Figure 2–71(b)? What would be the hysteresis gap?
7. For the circuit in Figure 2–71(b), if the instantaneous amplitude of the input signal is −1.86 V, what would be the approximate value of V_o?
8. For the circuit in Figure 2–71(b), if the instantaneous amplitude of the input signal is now +837 mV, what would be the approximate amplitude of the output signal?
9. What type of op amp circuit is represented in Figure 2–71(b)? What function is this device performing?
10. What advantage does the circuit in Figure 2–71(b) have over the type of comparator circuit shown in Figure 2–70?

Figure 2–70

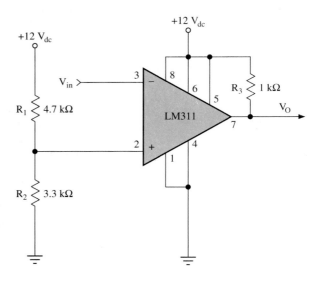

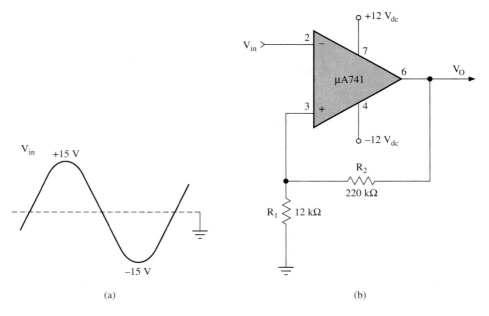

Figure 2–71

SUMMARY

Op amps are linear integrated circuits that serve a wide variety of purposes in industrial electronics. Being manufactured with differential inputs (either voltage-differencing or current-differencing), op amps have major advantages over single-port amplifiers. The strengths of the op amp include high voltage gain and excellent noise immunity. To better understand the function of the op amp within industrial circuitry, a technician must become thoroughly familiar with the operating parameters of the device. These parameters include open-loop voltage gain, slew rate, common-mode rejection ratio, and bandwidth.

Two commonly used op amp configurations are the inverting and noninverting amplifier. Both are effectively single-port amplifiers that are used in various low-power linear applications. The inverting configuration is the most versatile of the two designs, since it is capable of very high voltage gains as well as gains of less than unity.

The op amp summer amplifier, which is actually a modified form of inverting amplifier, is capable of performing simple analog computations. It is found in both D/A and A/D conversion systems. Summer circuits are often used in conjunction with inverting amplifiers for the purpose of level shifting and gain adjustment.

The differential, or instrumentation, amplifier exploits the op amp's voltage-differencing capability. Having two active input signals, the device is often used to amplify the error or differential signal developed over a resistive bridge circuit. Op amps in a voltage-follower configuration are often placed at each input to the instrumentation amplifier in order to increase the input impedance of the device and thus decrease its loading effect.

Op amp active filters are used prevalently in both signal conditioning and noise elimination. The Butterworth design format is used quite frequently due to its virtually flat frequency response within a given pass band. A low-pass and a high-pass active filter may be combined to form a bandpass filter. Of particular importance for industrial applications is the notch filter, which is used to attenuate a particular noise frequency. A notch filter may be created with a single op amp using the multiple-feedback design format.

Op amp integrators and differentiators are frequently used in the process of active signal conditioning. The integrator will produce the time integral of an input signal, whereas a differentiator will produce its time derivative. Of the two types of device, the op amp integrator is used more frequently, due to the fact that it is more accurate and less noise prone than the op amp differentiator.

The comparator is a device that is used frequently in the process of A/D conversion. It makes either a high- or a low-output decision based on the comparison of an active input signal to an established reference level. High-speed precision comparators tend to generate output noise, especially if there is noise riding on the input signal. In a noise-prone environment, a Schmitt trigger may be used instead of the comparator. The op amp Schmitt trigger is designed to have a hysteresis that reduces the problem of spurious switching at its output. The comparator and the Schmitt trigger are both used as zero-crossing detectors, converting input signals into rectangular waveforms that are compatible with digital circuitry.

SELF-TEST

1. For the op amp circuit in Figure 2–72, what would be the maximum possible voltage gain?
2. For the op amp circuit in Figure 2–72, what would be the required setting of R_2 in order to achieve a voltage gain of -14 V?
3. For the op amp circuit in Figure 2–72, assume V_{in} equals 173 mV and V_o equals -2.738 V. What is the voltage gain? What is the ohmic setting of R_2?
4. Describe the purpose of R_8, R_9, and R_{10} in Figure 2–73. If all inputs to the circuit are temporarily at 0 V, what should be the value of V_o? If, with this condition, V_o is slightly positive, what adjustment should be made to bring the output to the 0 V level?

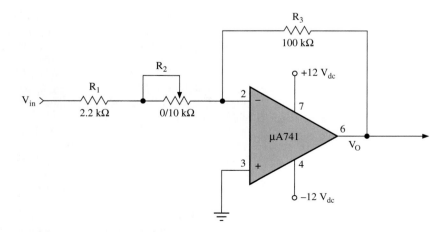

Figure 2–72

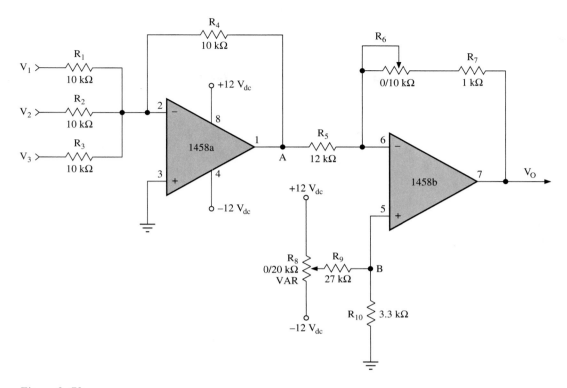

Figure 2–73

5. In Figure 2–73, assume $V_1 = -1.743$ V, $V_2 = -863$ mV, and $V_3 = -562$ mV. What should be the value of I_s? What should be the voltage level at point A? Assuming the offset null adjustment has already been performed, what should be the value of V_o if R_6 is set at 2.85 kΩ?

6. True or false?: In Figure 2–73, if point B is made more positive, V_o will become more negative.

7. In Figure 2–73, assume that the offset null adjustment has already been made. If $V_1 = -625$ mV, $V_2 = +2$ V, and $V_3 = -425$ mV, what must be the ohmic setting of R_6 in order to allow an output of $+475$ mV?

8. What type of amplifier is represented in Figure 2–74? What type of op amp configuration is represented by U_1 and U_2?

9. For the op amp circuit in Figure 2–74, assume $V_A = 1.36$ V and $V_B = 0$ V. What is the ideal level of V_C? What is the ideal level of V_D? If R_8 is set to 34.5 kΩ, what would be the value of V_o? With these conditions, what must be the ohmic setting of R_1? What is the ohmic setting of R_4?

10. In Figure 2–74, assume R_1 is adjusted to 94.7 kΩ, R_4 is adjusted to 73.4 kΩ, and R_8 is set at 42.7 kΩ. What must be the values of V_A and V_B? What is the value of V_o?

11. What type of circuit is contained in Figure 2–75? Identify both the order and design format represented by that device.

12. What is the cutoff frequency of the circuit in Figure 2–75? What is the roll-off for the device?

13. What type of filter is contained in Figure 2–76?

14. What two types of individual filters are contained within the circuit in Figure 2–76? What design format is represented by U_1? What design format is represented by U_2?

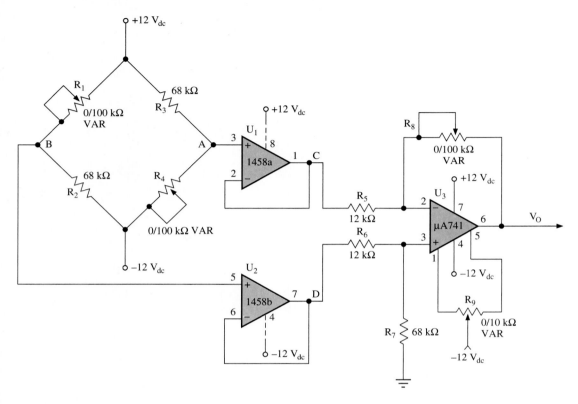

Figure 2–74

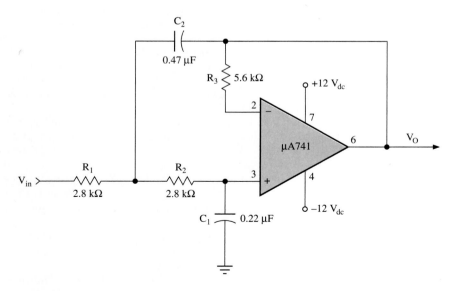

Figure 2–75

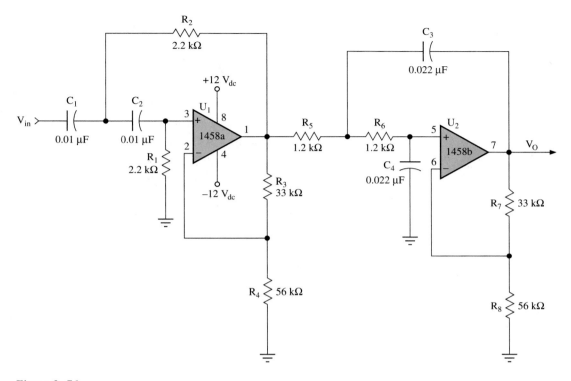

Figure 2–76

15. What is the center (resonant) frequency for the filter in Figure 2–76? What are the upper and lower cutoff frequencies of the device? What is the bandwidth of the circuit? What is the value of Q?

16. What type of filter is represented in Figure 2–77? Describe a probable application of this device within an industrial environment.

17. Assuming a Q value of 8.7, what would be the resonant frequency for the filter in Figure 2–77? What is another term for this frequency?

18. What type of op amp circuit is represented in Figure 2–78?

19. Providing the op amp in Figure 2–78 is not being driven into saturation, if a cosine wave is occurring at its input, what waveform would be occurring at its output?

20. For the circuit in Figure 2–78, if the input signal is that contained in Figure 2–79(a), what type of waveform would be developed at the output? What would be the positive and negative peak amplitudes of that signal (assuming offset nulling has already been performed)?

21. For the circuit in Figure 2–78, describe the purpose of R_2.

22. True or False?: The output of the circuit in Figure 2–78 is directly proportional to the time derivative of its input.

23. If the waveform in Figure 2–79(b) occurs at the input of the circuit in Figure 2–80, what type of waveform would occur at its output? What would be the positive and negative peak amplitudes of that waveform?

24. What type of device is contained in Figure 2–80? What is its reference voltage? Is the operation of the device active-high or active-low?

25. What type of device is contained in Figure 2–81? What are its upper and lower trigger points? Is the operation of the device active-high or active-low?

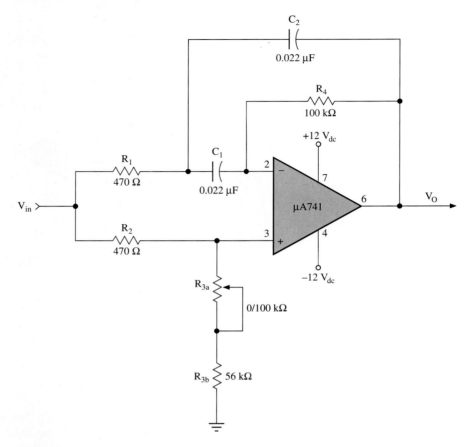

Figure 2–77

Figure 2–78

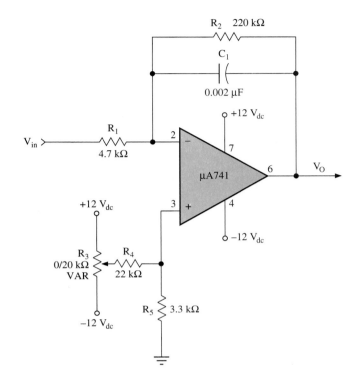

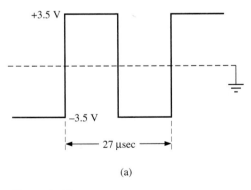

(a)

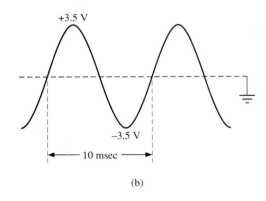

(b)

Figure 2–79

Figure 2–80

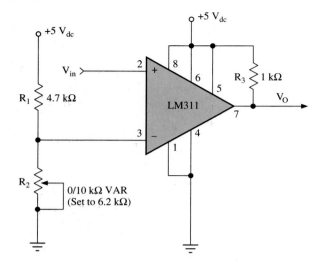

Figure 2–81

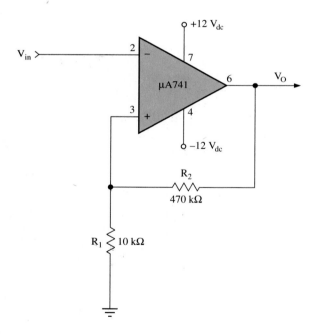

3 THERMAL TRANSDUCERS/ INTRODUCTION TO SIGNAL CONDITIONING

LEARNING OBJECTIVES

On completion of this chapter, the student should be able to:

- Identify the various types of thermal transducers commonly used in industrial electronics.
- Describe basic passive and active signal conditioning methods involving thermal transducers.
- Describe the temperature and electrical characteristics of thermistors. Explain how these characteristics may be exploited in various industrial measurement and control processes.
- Describe the operation of the resistive temperature detector. Compare the operation of this device to that of the thermistor.
- Describe the operation of the thermocouple. Identify advantages and disadvantages of this device as compared to the thermistor and the resistive temperature detector.
- Describe the operation of various forms of solid-state temperature detectors. Explain how they may be used in place of or in conjunction with thermal transducers.

INTRODUCTION

In this chapter, those transducers which are capable of converting changes in temperature to changes in an electrical property are introduced. Also, basic signal conditioning circuitry used in conjunction with these devices is analyzed. Passive signal conditioning circuits, consisting of simple resistive voltage dividers and bridge circuits, are introduced first. Next, active signal conditioning circuitry, containing op amps and comparators, is analyzed. Examples from actual industrial control circuits are used to introduce the specialized functions of each of the thermal transducers introduced here.

3-1   INTRODUCTION TO THERMAL TRANSDUCERS

The term **transducer** can be generally applied to any device that is capable of converting energy from one form to another. With regard to industrial electronics, the term transducer is used to describe a device that converts changes in some physical parameter to changes in voltage and current. The term **thermistor,** which is a contraction of the words ''thermal resistor,'' is used to describe a common form of temperature transducer. This semiconductor device exhibits a change in resistance as a result of a change in its own body temperature. Other forms of temperature transducer include the **resistance temperature detector** (RTD) and the **thermocouple.** The thermocouple is markedly different from the thermistor and RTD with regard to both structure and operation. This device, which consists of bimetallic junctions, is actually capable of producing a small **thermoelectric voltage.** This potential will vary as a function of temperature.

Thermistors are created by mixing either ceramic-based or metallic-oxide semiconductor materials. Materials commonly used in the manufacture of thermistors include oxides of iron, nickel, and titanium. Through a process known as **sintering,** these materials are permanently bonded together. (Sintered materials are subjected to intense heat, but are not actually melted together during the manufacturing process.) The resultant semiconductors exhibit various temperature and electrical effects, based on the proportions of sintered materials. The mixture of materials established for each type of thermistor is referred to as its **material system.**

Because the thermistor has a high sensitivity to changes in temperature, it can be connected directly into a resistive network in order to produce significant changes in voltage and current. The schematic symbol for a thermistor is shown in Figure 3–1(a),

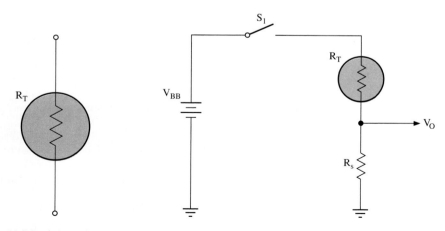

(a) Schematic symbol   (b) Simple passive signal conditioning application of the thermistor
    for the thermistor

Figure 3–1

and a simple method of passive signal conditioning using the device is shown in Figure 3–1(b).

A thermistor is usually characterized by a **negative temperature coefficient** (NTC), which means that its resistance is inversely proportional to its body temperature. Although it exhibits less sensitivity than the thermistor, an RTD has a **positive temperature coefficient** (PTC). This implies that changes in resistance for the RTD will be proportional to changes in temperature, while representing less of a range in ohmic value than that of a typical thermistor. This relationship between the two types of devices is represented graphically in Figure 3–2. Note that the thermistor is capable of a much greater change in resistance for a given range of temperatures. However, the RTD is capable of a more linear response. If the thermistor represented in Figure 3–2 were placed within the voltage divider in Figure 3–1(b), its resistance would decrease with an increase in its body temperature. Thus, due to the voltage divider action, V_{RT} would decrease, resulting in an increase in the value of V_o. If the RTD were placed in the circuit, the opposite effect would occur, due to the PTC of the device. However, the change in the value of V_o would be less significant, due to the lower sensitivity of the RTD (assuming the value of R_s remains constant).

While the thermistor has greater sensitivity to changes in temperature, its nonlinearity places a practical limit on its operating range. For example, as illustrated by triangulation in Figure 3–3(a), linear portions of the operating range of a thermistor may be approximated. For these linear sections of the curves, the slope of $\Delta R/\Delta T$ may be assumed to

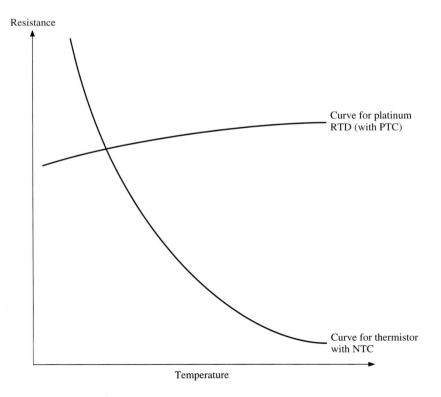

Figure 3–2
General comparison of the operating curves (resistance versus temperature) for a typical thermistor and an RTD

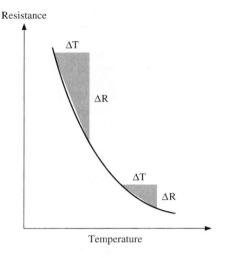

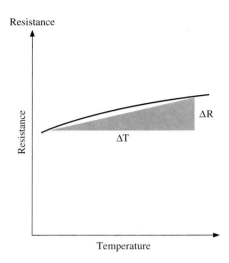

(a) Estimation of linear portions of thermistor curve

(b) Estimation of linear portion of curve for RTD

Figure 3–3
Estimation of linear portions

be fairly constant. Note that, while the variation in resistance for the RTD is much less than that of the thermistor, the RTD is actually capable of monitoring a much greater range of temperatures with a nearly constant value of $\Delta R/\Delta T$.

SECTION REVIEW 3–1

1. Define the term *material system* as it applies to the manufacture of thermistors. Identify some materials commonly used in the production of thermistors.
2. Define the term *positive temperature coefficient*. What type of transducer is likely to exhibit this characteristic?
3. Define the term *negative temperature coefficient*. What type of transducer is likely to exhibit this characteristic?
4. Define the term *sintering*. What form of temperature transducer is manufactured by this process?
5. True or false?: A thermistor is likely to have a wider practical temperature range than an RTD.
6. True or false?: The temperature response curve for an RTD is likely to be more linear than that of the thermistor.
7. True or false?: The thermistor is less sensitive than the RTD with respect to changes in temperature.
8. Which form of temperature transducer is capable of generating its own thermoelectric voltage?

3–2   THERMAL AND ELECTRICAL CHARACTERISTICS OF THERMISTORS

The operation of a thermistor may be analyzed with regard to both its **thermal** and **electrical** properties. While a detailed understanding of the thermal characteristics is not a requirement for electronics technicians, a basic knowledge of the heating effects of the device is essential to an understanding of its electrical characteristics. Therefore, we briefly explain the temperature characteristics of the thermistor prior to the study of its electrical operation.

When a thermistor is conducting current, it dissipates energy in the form of heat. Thus, at any given instant, the temperature of the device may be slightly higher than that of the medium in which it is placed. Expressed mathematically, thermistor temperature T is the sum of the ambient temperature of the environment T_A and the **self-heating effect** ΔT of the device itself:

$$T = T_A + \Delta T \tag{3-1}$$

The rate at which energy is supplied to the thermistor is equivalent to the rate at which energy is dissipated plus the rate at which energy is being absorbed. Thus,

$$\frac{dH}{dt} = \frac{dH_L}{dt} + \frac{dH_A}{dt} \tag{3-2}$$

where
dH/dt = the rate at which energy is being supplied
dH_L/dt = the rate at which energy is being dissipated
dH_A/dt = the rate at which energy is being absorbed.

For a thermistor conducting current, the rate at which energy is being supplied may also be expressed as follows:

$$\frac{dH}{dt} = I^2 \times R = V \times I = P \tag{3-3}$$

The rate at which heat is dissipated by the thermistor is proportional to the self-heating effect (or temperature rise) of the device. Energy loss is actually the product of the self-heating effect and the **dissipation constant** of the thermistor. (The dissipation constant for a thermistor is symbolized by δ and is typically expressed in milliwatts per degree Celsius.) The dissipation constant depends on several factors, including the thermal conductivity of the medium in which the device is placed and the rate at which heat is being transferred via its metal leads. Accounting for the dissipation constant, the rate of energy loss for the thermistor may be expressed as follows:

$$\frac{dH_L}{dt} = \delta \Delta T = \delta(T - T_A) \tag{3-4}$$

The rate of energy absorption for the device may be stated in terms of its **specific heat** s and mass m. The product of these two factors is referred to as the **heat capacity** C, which is actually a function of the thermistor's physical structure. The rate of absorption for a given thermistor may be expressed as follows:

$$\frac{dH_A}{dt} = sm \times \frac{dT}{dt} = C \times \frac{dT}{dt} \tag{3-5}$$

where dT/dt represents the rate of change in thermistor temperature.

Through substitution, we can create an expression for thermistor energy consumption that accounts for both the dissipation constant and heating capacity of the device. This **heat transfer equation** may be written as follows:

$$P = \frac{dH}{dt} = \delta(T - T_A) + C \times \frac{dT}{dt} \tag{3-6}$$

When dT/dt approaches zero, a state of **equilibrium** is assumed to exist, where the rate of heat dissipation for the thermistor is virtually equal to the power being drawn from the electrical source. Thus, with this condition, the expression for energy transfer may be simplified as follows:

$$P = V_T \times I_T \cong \delta(T - T_A) \tag{3–7}$$

In Eq. (3–7) V_T and I_T represent static values of thermistor voltage and current.

Dividing the heating capacity by the dissipation constant yields the **thermal time constant** τ of the thermistor. The thermal time constant is that time period required for the temperature of the thermistor to attain 63.2% of the difference between its original value and a step increase in temperature to which the device is being subjected. This step increase in temperature is assumed to be occurring while thermistor power dissipation is approaching zero. The concept of the thermal time constant is illustrated in Figure 3–4, which shows the response of the thermistor to a step increase in temperature through a period of five time constants. In this example, at t_0, the thermistor is subjected to an instantaneous increase in the temperature of the medium in which it is placed. By t_1, the thermistor temperature has increased by 63.2% of the difference between T_1 and T_A. Thus, by the end of the first thermal time constant, the difference in temperature between the thermistor and the medium is only 36.8% of the original step difference, which occurred

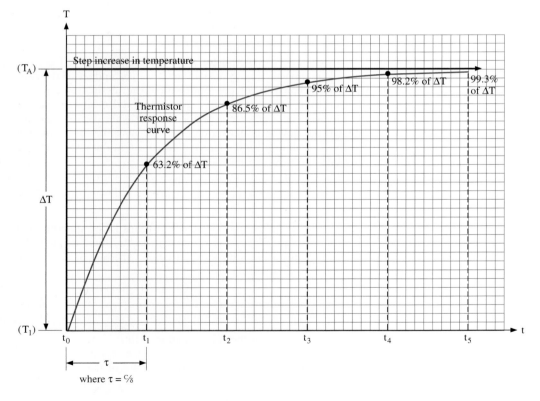

Figure 3–4
Response of a thermistor to a step increase in temperature

at t_0. During the second time constant (between t_1 and t_2), thermistor temperature increases by 63.2% of the remaining difference. Thus by t_2, thermistor temperature has increased by a factor of nearly 86.5% of the original difference between T_1 and T_A. After five time constants, thermistor temperature is considered to be virtually equal to the value of T_A and an equilibrium condition is again considered to exist.

When thermistor power consumption is reduced to a point where the self-heating effect is minimal, the heat transfer equation may be expressed as follows:

$$\frac{dT}{dt} = \frac{-\delta}{C} \times (T - T_A) = \frac{-(T - T_A)}{\tau} \tag{3-8}$$

This expression indicates that, for a given difference in temperature between a thermistor and the medium in which it is placed, the rate of temperature change is inversely proportional to the thermal time constant of the device. (This concept should become clear after considering that C represents the ability to store energy, whereas δ represents the ability to dissipate energy in the form of heat.)

Example 3-1

The time constant specified for a certain thermistor is 4.5 sec. If the dissipation constant for the device is 280 mW/°C, what is the heating capacity for the device?

Solution:
If $\tau = C/\delta$, then $C = \tau \times \delta$. Thus,

$$C = 4.5 \text{ sec} \times 280 \text{ mW} = 1.26$$

Note that the rate of thermistor temperature change through time is not constant (as would be the slope for a linear function). Referring back to Figure 3-4, at the instant the thermistor is subjected to a step change in medium temperature, the rate of temperature change is at its maximum. However, as the exponential curve for temperature growth extends into the fifth time constant, (dT/dt) decreases toward zero. A practical expression for thermistor temperature is given in Eq. (3-9). With this equation, providing that the thermal time constant of the device has already been derived, the instantaneous thermistor temperature may be calculated for any point on the exponential curve:

$$T = T_A + (T_1 - T_A)\varepsilon^{-t/\tau} \tag{3-9}$$

where
$T =$ instantaneous value of thermistor temperature
$T_1 =$ starting temperature
$T_A =$ ambient temperature of the medium.

Example 3-2

The time constant of a given thermistor is 7.4 sec. Assume the starting temperature for the device is 29°C. If the device is initially in an equilibrium condition, what would be the temperature of the thermistor 12 sec after being subjected to a step increase in medium temperature from 29°C to 36°C?

Solution:

$$T = T_A + (T_1 - T_A)\varepsilon^{-t/\tau}$$
$$= 36°C + (29°C - 36°C)\varepsilon^{-12/7.4}$$

$$= 36°C + (-7°C \times 0.1976)$$
$$= 36°C - 1.383°C = 34.617°C$$

A thermistor exhibits three basic electrical characteristics: the **current-time charac-teristic,** the **voltage-current characteristic,** and the **resistance-temperature characteris-tic.** All three of these electrical properties are exploited in the various applications of the device. The current-time characteristic of the thermistor is utilized in such industrial applications as sequential switching and overload protection. Applications of the voltage-current characteristic include flowmeters and temperature controllers. The resistance-temperature characteristic is exploited in such varied applications as thermostats, industrial process controllers, and medical electronics instrumentation.

The current-time characteristic of a thermistor is closely associated with the self-heating effect of the device. When a significant amount of power is dissipated by the thermistor, the increase in its body temperature due to self-heating can be determined as follows:

$$\Delta T = \frac{P}{\delta} \times (1 - \varepsilon^{-t/\tau}) \qquad (3-10)$$

where P is the static value of $(V_T \times I_T)$ as in Eq. (3–7). This equation clearly indicates that, as thermistor body temperature increases above environmental temperature due to I^2R loss within the device, the difference in temperature between the device and the medium in which it is placed (ΔT) becomes a function of the applied power and the amount of time in which the power is applied.

Example 3–3

Assume that a thermistor is connected as shown in Figure 3–5(a). The dissipation constant for the device is 126 mW/°C, while its heating capacity is 1.35. Also, the static-state value of dc current through the thermistor is assumed to be 65 mA. At t_0 in Figure 3–5(b),

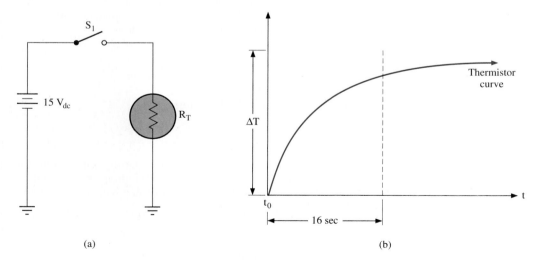

(a)   (b)

Figure 3–5

the thermistor temperature equals the ambient temperature of the medium. However, as current begins to flow, the thermistor temperature increases due to self-heating. Using the information given, solve for the thermal time constant and P. What would be the value of ΔT 16 sec past t_0?

Solution:

Solving for the thermal time constant:

$$\tau = \frac{C}{\delta} = \frac{1.35}{126 \text{ mW/}^\circ\text{C}} = 10.714 \text{ sec}$$

Solving for power:

$$P = V_T \times I_T = 15 \text{ V} \times 65 \text{ mA} = 975 \text{ mW}$$

Determine ΔT at 16 sec past t_0:

$$\Delta T = \frac{975 \text{ mW}}{126 \text{ mW/}^\circ\text{C}} \times (1 - \varepsilon^{-16/10.714})$$

$$= 7.738 \times (1 - 0.2246) = 7.738 \times 0.7754$$

$$= 6^\circ\text{C}$$

For Example 3–3, after a time lapse of approximately 54 sec (representing five time constants), a state of equilibrium will be achieved where ΔT virtually equals (P/δ). Recall that, according to Eq. 3–7, for an equilibrium condition:

$$V_T \times I_T = \delta(T - T_A)$$

where $(T - T_A)$ is equivalent to ΔT and $(V_T \times I_T)$ equals P.

Figure 3–6 contains the voltage-current characteristic curve for a thermistor, indicating the negative temperature coefficient of the device. When thermistor current is held at a very low value, the power consumption of the device will be negligible. Consequently, the device will behave as an ordinary linear resistor. The actual ohmic value of the resistor in this condition, referred to as its **zero-power resistance,** may vary, depending on the ambient temperature of the environment. That is, the zero-power resistance of a thermistor at 26°C will be less than its zero-point resistance at 14°C, assuming the device has a negative temperature coefficient.

When thermistor resistance is close to the zero-power point and the device is behaving as a linear resistance, it will conform to Ohm's law. This effect is exhibited at the beginning of the curve in Figure 3–6, where thermistor voltage increases sharply in proportion to the increase in current. This linear relationship of ΔV and ΔI continues until the curve approaches the peak point. Beyond this maximum voltage point, within the **negative resistance** portion of the curve, thermistor current increases rapidly. At the same time, the voltage developed over the device actually begins to decrease. This action occurs as a result of the self-heating effect, which causes the thermistor's resistance to decrease. As current increases, resistance decreases still further, causing further self-heating action. Finally, an equilibrium condition is attained where the power supplied by the source virtually equals that being dissipated by the thermistor.

An application of the current-time characteristic of a thermistor is illustrated in Figure 3–7. Here a thermistor with a negative temperature coefficient is placed in series with a relay coil and a current-limiting resistor. Assuming that R_T is initially in a cool condition, the combined resistance of R_T, R_S, and the relay coil will be high enough to hold the coil current below the minimum value required for actuation of the relay contacts. Thus, when

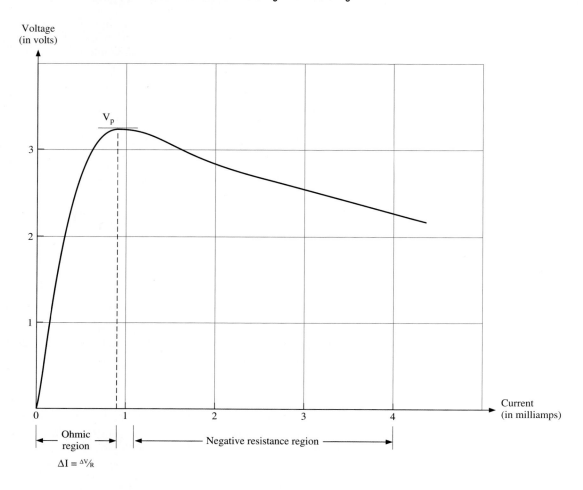

Figure 3–6
Voltage-current characteristic curve for a thermistor with an NTC

S_1 is closed, the relay contacts will remain in their normally open positions and the load will not be energized.

As thermistor resistance decreases due to the self-heating effect, current flow through the relay winding increases. Once this current attains a level that allows actuation of the relay contacts, the load will be energized. At the same time, the normally open contacts in parallel with the thermistor will close. The purpose of this action is twofold. First, with current being shunted past the thermistor, the voltage and current for that device will equal zero. Consequently, the power consumption of the thermistor will be zero and it will be able to return to its zero-power resistance. Second, with the thermistor being shorted, the relay coil will receive a steady dc current. This current will remain nearly constant, ensuring that the load will remain energized until S_1 is opened.

When S_1 is again closed, assuming the thermistor to be in a cool state, a delay cycle of nearly the same period could be repeated, providing that the ambient temperature of the environment has not changed significantly. While the delay circuit in Figure 3–7 may not be acceptable for precision timing applications, it is suitable for use within a sequential

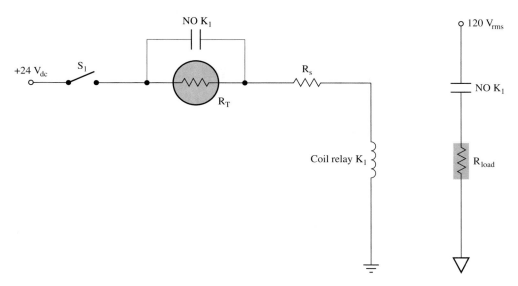

Figure 3–7
Self-heating thermistor used in a relay switching delay circuit

switching circuit for which delay time may deviate beyond some minimum value. An advantage of such a circuit is the fact that an *RC* timing network may be replaced by a single thermistor.

Example 3–4

For the circuit in Figure 3–8, assume that, with S_1 open, the thermistor has a resistance of 417 Ω. At the instant S_1 is closed, what will be the value of current and voltage for the relay coil? What will be the condition of the standby indicator lamp and the load at that time? To what ohmic value must R_T be reduced in order to allow actuation of the relay? Once actuation of the relay does occur, what will be the values of I_s, V_{R_s}, and V_L? What happens to thermistor voltage and current as a result of the actuation of the relay? What is the purpose of this action?

Solution:
    Solving for the conditions at the instant S_1 is closed:

$$R_T + R_s + r_L = 417\ \Omega + 220\ \Omega + 316\ \Omega = 953\ \Omega$$

$$I_s = \frac{24\ \text{V}}{953\ \Omega} = 25.18\ \text{mA}$$

$$V_L = I_s \times r_L = 25.18\ \text{mA} \times 316\ \Omega = 7.958\ \text{V}$$

At this time the relay will not be actuated. Thus the load will be de-energized and the standby lamp will be on.
    To solve for the value of R_T at the point of actuation of the relay: Since the relay coil requires 12 V_{dc} for actuation:

$$I_s = \frac{V_L}{r_L} = \frac{12\ \text{V}_{dc}}{316\ \Omega} = 37.975\ \text{mA}$$

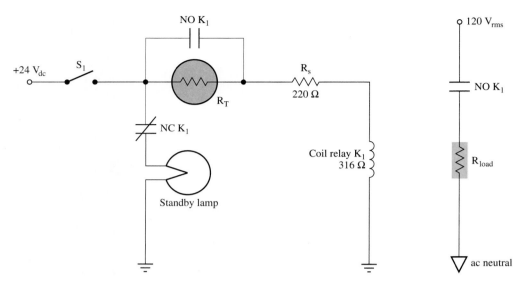

Figure 3–8

$$\text{Total series resistance} = \frac{24\ V_{dc}}{37.975\ mA} = 632\ \Omega$$

$$R_T = 632\ \Omega - (220\ \Omega + 316\ \Omega) = 96\ \Omega$$

To solve for the conditions following actuation: When the relay armature is actuated, current is shunted around the thermistor. Thus,

$$I_s = \frac{24\ V_{dc}}{(220\ \Omega + 316\ \Omega)} = 44.776\ mA$$

$$V_L = 44.776\ mA \times 316\ \Omega = 14.149\ V$$

While relay current is shunted around the thermistor, the device will begin to cool and R_T will increase toward its starting value. However, relay current will remain constant and the load will be energized until S_1 is opened.

Several types of thermistor are manufactured especially for the purpose of protecting individual components or entire circuits from adverse temperature effects. With such applications (just as in the previous example) the current-time characteristic of the thermistor is being utilized. An example of such a device would be the **tempsistor.** At the initial power-on of a circuit, this component would offer its highest resistance to current flow. Depending on the type of tempsistor, this initial ohmic value might range from as low as 56 Ω to as much as 39 kΩ. However, as the device begins to self-heat, due to its steep negative resistance characteristic, its ohmic value will begin to decrease sharply. When in the self-heated condition, its voltage will drop to its minimum value, allowing virtually all of the source voltage to be developed over a series load. This common application of a tempsistor with a negative temperature coefficient is illustrated in Figure 3–9(a). The load represented in Figure 3–9(a) might actually be a power supply containing at least one large electrolytic filter capacitor. At the instant the circuit is energized, these capacitors

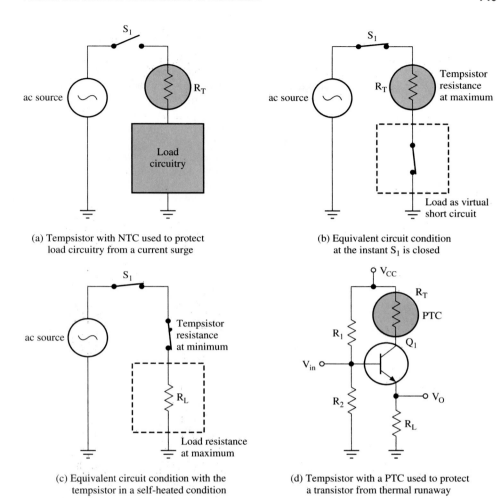

(a) Tempsistor with NTC used to protect
load circuitry from a current surge

(b) Equivalent circuit condition
at the instant S_1 is closed

(c) Equivalent circuit condition with the
tempsistor in a self-heated condition

(d) Tempsistor with a PTC used to protect
a transistor from thermal runaway

Figure 3–9
Tempsistors used in overcurrent protection circuitry

will behave as virtual short circuits. This would normally result in a large in-rush of current. However, with the tempsistor present between the source and the load, this initial value of current will be limited to a safe level, as illustrated by the equivalent circuit in Figure 3–9(b).

As the power supply capacitors begin to charge toward their maximum voltage levels, the tempsistor will approach a self-heated condition. Thus, as the load impedance approaches a maximum level, the tempsistor resistance will fall rapidly, allowing virtually all of the source potential to be transferred to the load. This stable condition for the load and tempsistor is illustrated in Figure 3–9(c).

Various forms of thermistors exhibiting a positive temperature coefficient may also be used for current-limiting purposes. These devices are ideally suited for the protection of individual components such as power transistors. By limiting the flow of current from the transistor, an overheated condition, which could lead to thermal runaway and eventual destruction of the component, might be prevented.

The action of the tempsistor in Figure 3–9(d) might best be understood by first considering how thermal runaway could occur if this PTC device were not present in the circuit. As the body temperature of the transistor begins to increase with current flow, thermal agitation occurs, resulting in still more current. An avalanche effect could ensue, which, without a reduction in temperature, will lead to the demise of the transistor. However, with the tempsistor placed in the circuit as shown, its resistance will increase sharply as the operating temperature of the transistor exceeds a safe level. Prior to the overheated condition, the tempsistor functions as a collector resistance of only a few ohms. However, as transistor temperature begins to exceed a safe level, the ohmic value of the tempsistor will rise, driving the transistor toward saturation. With a V_{CE} now approaching 0 V, its power dissipation will be minimized, thus protecting the device from thermal runaway.

Along with self-heating thermistors, **temperature-sensing reed switches** may also be used to protect various loads from excess current. A cutaway view of the temperature-sensing reed switch is shown in Figure 3–10(a), and the schematic symbol for this device is given in Figure 3–10(b). Though not actually an input transducer, this device may still

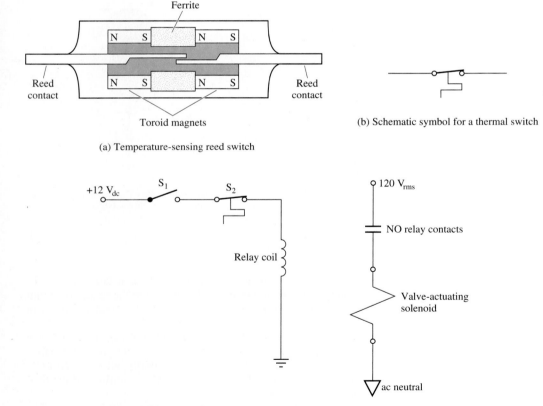

(a) Temperature-sensing reed switch

(b) Schematic symbol for a thermal switch

(c) Temperature-sensing reed switch used to protect a solenoid
(the reed switch is in close thermal contact with the solenoid)

Figure 3–10
Operation of the temperature-sensing reed switch

be considered a heat sensor due to the fact that it is able to toggle as temperature fluctuates above or below a threshold level.

As illustrated in Figure 3–10(a), the temperature-sensing reed switch contains two pieces of ferrite core material that are surrounded by small toroid magnets. With these magnetic assemblies placed on either side of the switching contacts, magnetic flux lines will pass through them, thus holding them in the closed position. However, as the switch begins to heat toward a point called the **Curie temperature,** the strength of the magnetic field that is holding its contacts in the closed condition will rapidly dissipate, due to an increasing magnetic reluctance within the ferrite material. Consequently, the reed contacts will separate and remain in the open position until the body temperature of the switch falls below the threshold (or Curie) point.

A possible application of the temperature-sensing reed switch is illustrated in Figure 3–10(c). Here, the switch would be placed in close thermal contact with a solenoid-controlled valve, in order to protect its windings from an overheated condition. Remember that, for an ac solenoid, the inductive reactance of the coil will be relatively low prior to actuation of the plunger. Therefore, a relatively large value of current will flow through the windings until actuation is achieved. Consequently, if the valve were stuck in the closed position, this large in-rush of current would continue to flow indefinitely, causing the windings to overheat and even burn. If, however, the Curie temperature of the reed switch is low enough, its contacts will open well before the solenoid is damaged. Only when the solenoid windings have been cooled sufficiently will the valve again be operable.

It has been demonstrated that thermistor applications involving the current-time and voltage-current characteristics exploit the self-heating effect of the device. However, in applications involving the resistance-temperature characteristic of the thermistor, the self-heating effect must be avoided. This is due to the fact that, in instrumentation circuitry designed for the purpose of monitoring and controlling environmental temperature, the self-heating effect represents a significant source of error. In such applications, the thermistor must operate as closely as possible to its zero-power condition.

SECTION REVIEW 3–2

1. Describe the **self-heating effect** as it applies to the operation of a thermistor.
2. The rate at which energy is being dissipated by a thermistor is actually the product of what two parameters?
3. What is the symbol for the dissipation constant of a thermistor? In what units is this parameter expressed?
4. Identify two factors that influence the dissipation constant of a thermistor.
5. What condition is said to exist for a thermistor when dT/dt approaches zero? With this condition, the energy being dissipated by a thermistor is virtually equal to what other parameter?
6. What thermistor property is the product of its specific heat and mass? What is the symbol for this parameter?
7. What thermistor parameter is derived through dividing the heating capacity of the device by its dissipation constant? What is the symbol for this parameter?
8. The heating capacity of a thermistor equals 1.24, whereas its time constant is 7.43 sec. What is the dissipation constant for the device?
9. Assume the static dc voltage for a thermistor is 12 V_{dc} and its static current is 37.5 mA. Also assume that the time constant for the device is 8.8 sec and its heat capacity is 1.79. Given this information, what would be the self-heated temperature of the device 23 sec after power is applied?

10. List the three basic electrical characteristics of a thermistor. Which two characteristics are closely associated with the self-heating effect of the device?
11. Describe a possible application of a tempsistor with an NTC. Also describe an application of such a device having a PTC.
12. Explain the operation of the temperature-sensing reed switch. Define the term *Curie temperature* as it applies to the function of this device.

3-3   INTRODUCTION TO SIGNAL CONDITIONING USING THERMISTORS

The goal in designing instrumentation circuitry involving thermistors is to establish a linear relationship between the output voltage or current and the temperature of the environment being monitored. Since the zero-point resistance of a thermistor will vary as a function of its body temperature, the change in environmental temperature may be detected as either a change in the voltage developed across the device or a deviation in the value of its current.

In Figure 3–11(a), a thermistor has been placed in series with a linear resistance, over which the output signal is being developed. For this simple voltage divider network, since the thermistor exhibits a negative temperature coefficient, the output voltage will increase as environmental temperature rises.

In Figure 3–11(b), the output voltage (V_o) is plotted against increasing temperature. At any given point on this curve:

$$V_o = V_s \times \frac{R_1}{R_1 + R_T} \tag{3-11}$$

where R_T represents the zero-power resistance of the thermistor at any point within the practical range of the device.)

For the circuit in Figure 3–11(a), the practical range for temperature measurement is limited to the portion of the thermistor curve that is approximately linear. Ideally, a reference temperature (symbolized as T_o) is established at the middle of the curve. (The

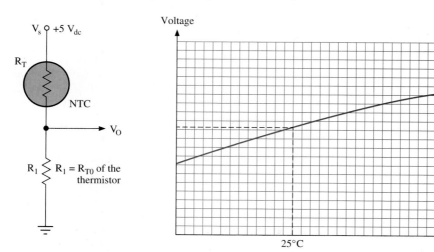

(a) Voltage divider containing a thermistor with an NTC

(b) Nearly linear curve for output voltage

Figure 3–11
Passive signal conditioning using a thermistor with an NTC in a voltage divider

reference level specified for most thermistors is 25°C.) The zero-power resistance that occurs at the specified reference temperature (symbolized as R_{T0}) should thus represent the midpoint of the zero-power resistance range of the thermistor. Manufacturers do provide listings of temperature versus resistance for various types of thermistors. An example of such a listing is found in Table 3–1, which represents the operation of an A919a fluid temperature sensor, which is manufactured by Thermometrics Inc.

Figure 3–12 represents the physical structure of two contrasting forms of thermistor. A simple, relatively inexpensive, bead-type thermistor is shown in Figure 3–12(a), while the A919 fluid temperature sensor (a relatively complex, state-of-the-art device) is illustrated in Figure 3–12(b). While the type of thermistor contained in Figure 3–12(a) is suitable for such applications as monitoring gas flow or measuring temperature within a limited area, the device in Figure 3–12(b) is especially designed to monitor the temperature of liquid flowing through a tube. The actual thermistor is in the form of a tiny probe, contained within a plastic Y assembly, which is easily placed in-line with plastic tubing used to convey chemicals. The size and shape of the thermal sensor, as well as its position within the Y assembly, allow for minimal disturbance of the flow of liquid through the tubing. At the same time, due to its short thermal time constant, this thermistor is able to respond rapidly to fluctuations in temperature. Note that in Table 3–1, a precise ohmic value is provided for every half degree Celsius. Such detailed information is often necessary for chemical processing and medical laboratory applications. The A919a is suitable for laboratory use, while the A919b, due to its rugged design, is better suited for industrial applications.

Table 3–1
Temperature versus thermistor resistance for standard A919 sensors

Degrees Celsius	Resistance (Ω)	Degrees Celsius	Resistance (Ω)	Degrees Celsius	Resistance (Ω)	Degrees Celsius	Resistance (Ω)
0.00	56341.02	12.50	32931.96	25.00	20000.00	37.50	12579.79
0.50	55100.98	13.00	32258.53	25.50	19619.70	38.00	12357.11
1.00	53891.82	13.50	31600.79	26.00	19247.70	38.50	12139.00
1.50	52712.66	14.00	30958.34	26.50	18883.81	39.00	11925.35
2.00	51562.71	14.50	30330.79	27.00	18527.82	39.50	11716.05
2.50	50441.16	15.00	29717.76	27.50	18179.56	40.00	11511.00
3.00	49347.23	15.50	29118.86	28.00	17838.82	40.50	11310.12
3.50	48280.17	16.00	28533.75	28.50	17505.43	41.00	11113.29
4.00	47239.27	16.50	27962.07	29.00	17179.21	41.50	10920.44
4.50	46223.79	17.00	27403.47	29.50	16859.98	42.00	10731.47
5.00	45233.07	17.50	26857.63	30.00	16547.59	42.50	10546.29
5.50	44266.43	18.00	26324.21	30.50	16241.87	43.00	10364.82
6.00	43323.24	18.50	25802.91	31.00	15942.65	43.50	10186.97
6.50	42402.85	19.00	25293.41	31.50	15649.79	44.00	10012.67
7.00	41504.66	19.50	24795.43	32.00	15363.13	44.50	9841.83
7.50	40628.07	20.00	24308.67	32.50	15082.52	45.00	9674.38
8.00	39772.52	20.50	23832.85	33.00	14807.83	45.50	9510.24
8.50	38937.45	21.00	23367.70	33.50	14538.91	46.00	9349.34
9.00	38122.30	21.50	22912.94	34.00	14275.62	46.50	9191.60
9.50	37326.56	22.00	22468.33	34.50	14017.84	47.00	9036.95
10.00	36549.71	22.50	22033.61	35.00	13765.44	47.50	8885.34
10.50	35791.26	23.00	21608.53	35.50	13518.29	48.00	8736.68
11.00	35050.72	23.50	21192.86	36.00	13276.27	48.50	8590.92
11.50	34327.63	24.00	20786.37	36.50	13039.25	49.00	8447.99
12.00	33621.52	24.50	20388.82	37.00	12807.13	49.50	8307.83
						50.00	8170.38

Interchangeability $\pm$ 0.25°C from 0°C to 30°C.

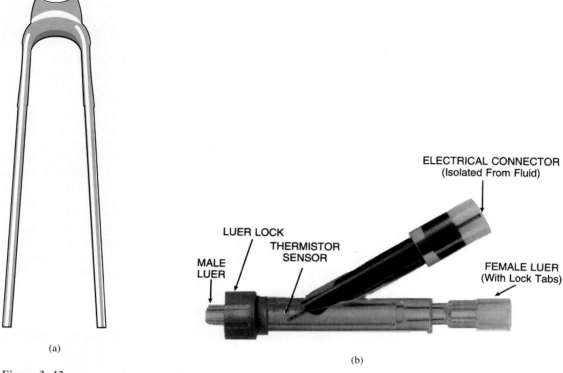

(a)

(b)

Figure 3–12
(a) Typical bead-type thermistor, (b) A919 fluid temperature sensor (Courtesy of Thermomet-
rics, Inc.)

Note also that in Table 3–1, the midpoint temperature is the standard reference level of 25°C, which corresponds with 20 kΩ of zero-power resistance. According to this table, thermistor resistance may be as high as 56.34 kΩ when the body temperature of the device is at the freezing point (0°C). When body temperature has increased to 50°C, thermistor resistance will have decreased to 8.17 kΩ. A curve of temperature versus resistance for the A919a is contained in Figure 3–13.

In Figure 3–14(a), the A919a is placed in a voltage divider network, such that changes in output voltage will represent fluctuations in the temperature of the fluid medium in which the device is placed. Using the curve in Figure 3–13 as a guide, the values of R_T may be substituted into Eq. (3–11) in order to plot a curve for V_o as thermistor temperature varies between 0°C and 50°C.

For example, using thermistor resistance at 0°C:

$$V_o = V_s \times \frac{R_1}{R_1 + R_T} = 5\,\text{V} \times \frac{20\,\text{k}\Omega}{20\,\text{k}\Omega + 56.34\,\text{k}\Omega} = 1.31\,\text{V}$$

At 5°C:

$$V_o = 5\,\text{V} \times \frac{20\,\text{k}\Omega}{20\,\text{k}\Omega + 45.23\,\text{k}\Omega} = 1.533\,\text{V}$$

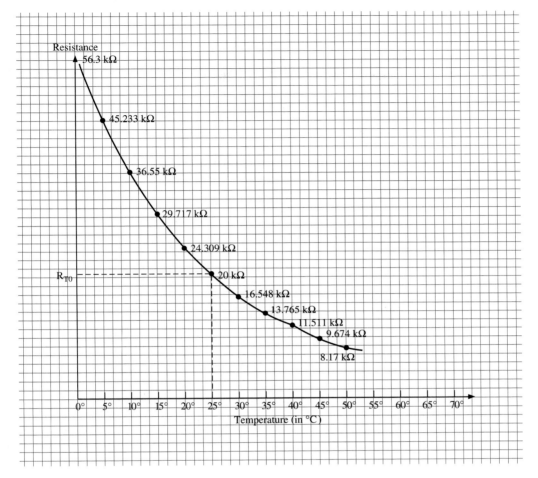

Figure 3–13
Plot of temperature versus resistance for the A919a fluid temperature sensor

At 10°C:

$$V_o = 5 \text{ V} \times \frac{20 \text{ k}\Omega}{20 \text{ k}\Omega + 36.55 \text{ k}\Omega} = 1.768 \text{ V}$$

As the values of output voltage are calculated, they may be plotted as shown in Figure 3–14(b), resulting in a nearly linear slope of temperature versus voltage. During the first half of the curve, $\Delta V_o/\Delta T$ is approximately 48 mV/°C, which is determined as follows:

$$\Delta V_o/\Delta T = \frac{2.5 \text{ V} - 1.31 \text{ V}}{25°C - 0°C} = \frac{1.19 \text{ V}}{25°C} = 47.6 \text{ mV/°C}$$

During the second half of the curve, the slope decreases slightly and is determined as follows:

$$\Delta V_o/\Delta T = \frac{3.55 \text{ V} - 2.5 \text{ V}}{50°C - 25°C} = \frac{1.05 \text{ V}}{25°C} = 42 \text{ mV/°C}$$

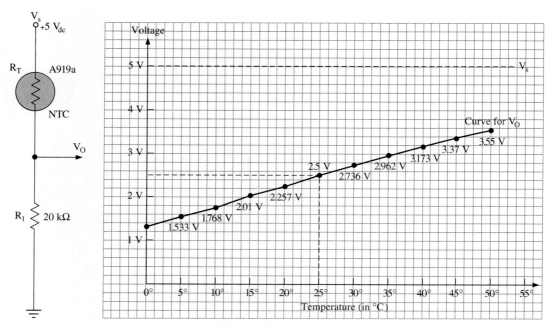

(a) Fluid temperature-sensor
in a voltage-divider

(b) Nearly linear curve for output voltage

Figure 3–14
Operation of the A919a fluid temperature sensor in a voltage-divider

Finally, the overall slope or average voltage change per degree Celsius may be determined:

$$\Delta V_o / \Delta T = \frac{3.55 \text{ V} - 1.31 \text{ V}}{50°\text{C} - 0°\text{C}} = \frac{2.24 \text{ V}}{50°\text{C}} = \text{nearly } 45 \text{ mV/°C}$$

Table 3–1, along with the graphs in Figures 3–13 and 3–14(b), can now be used in the design of active signal conditioning circuitry involving the A919a thermistor. Assume, for example, that a circuit must be designed that senses an overheated condition for a liquid material being used in an industrial process. When the temperature of the fluid exceeds 104°F, an active-low signal must be sent to digital control circuitry for which 5 V represents a logic high and 0 V represents a logic low.

First, the specified maximum temperature must be converted to degrees Celsius in order to utilize Table 3–1. Using the standard conversion method,

$$°\text{F} = (1.8 \times °\text{C}) + 32° \tag{3–12}$$

Substituting and solving for degrees Celsius:

$$104°\text{F} = (1.8 \times °\text{C}) + 32°$$
$$72°\text{F} = 1.8 \times °\text{C}$$
$$\frac{72°\text{F}}{1.8} = 40°\text{C}$$

According to Table 3–1, at 40°C, the value of R_T for the A919a should be 11.511 kΩ. Since the near linearity of the voltage divider network containing this device has already been proven, it will be incorporated into the circuit design. Thus a value for V_o must be determined that corresponds with 40°C.

$$V_o = V_s \times \frac{R_1}{R_1 + R_T} = 5\,\text{V} \times \frac{20\,\text{k}\Omega}{20\,\text{k}\Omega + 11.511\,\text{k}\Omega} = 3.173\,\text{V}$$

As shown in Figure 3–15, an LM311 comparator with an adjustable reference voltage will be used to develop the input signals for the digital circuitry. Since the overtemperature condition is represented as an active-low, V_{in} from the voltage divider must occur at the inverting input of the comparator. The reference voltage, which occurs at the noninverting input of the comparator, should be adjusted to equal V_{in} of the voltage divider at 40°C. The required setting of trim pot R_1 would be determined as follows. Referring back to Eq. 2–29:

$$V_{\text{ref}} = V_s \times \frac{R_2}{R_1 + R_2} \qquad \text{Thus, } 3.173\,\text{V} = 5\,\text{V} \times \frac{10\,\text{k}\Omega}{R_1 + 10\,\text{k}\Omega}$$

Solving for R_1:

$$\frac{3.173\,\text{V}}{5\,\text{V}} = \frac{10\,\text{k}\Omega}{R_1 + 10\,\text{k}\Omega} \qquad \text{Thus, } 1.5758 = \frac{R_1 + 10\,\text{k}\Omega}{10\,\text{k}\Omega}$$

$$R_1 + 10\,\text{k}\Omega = 15.758\,\text{k}\Omega \qquad \text{Thus, } R_1 = 5.758\,\text{k}\Omega$$

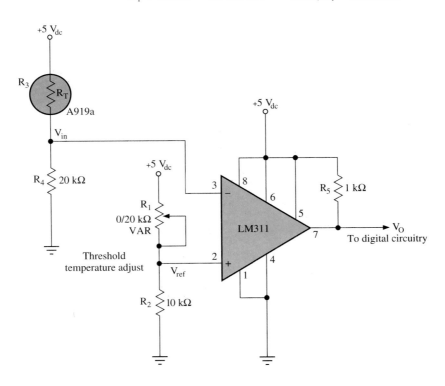

Figure 3–15
Over-temperature detection circuit using a fluid temperature sensing thermistor and a comparator

With potentiometer R_1 adjusted to the value calculated here, the output of the comparator will toggle to a logic low at the instant the body temperature of the thermistor exceeds 40°C. This action would indicate an overheated condition for the liquid material being monitored. As temperature decreases, the value of V_{in} will fall below the reference level, allowing the output of the comparator to toggle back to a logic high. If a new set-point temperature is to be established, R_1 would simply be adjusted to allow the new reference voltage to occur at the noninverting input of the comparator. This voltage would, of course, be equivalent to the value of V_{in} occurring at the new set point. The operation of the circuit in Figure 3–15 is represented in Figure 3–16.

Within a limited temperature range, the output of the voltage divider in Figure 3–14(a) could serve as the input to **thermometric** (temperature-measuring) circuitry. For example, as shown in Figure 3–17, the slope of temperature versus voltage between 15°C and 35°C is sufficiently linear for this purpose.

Op amp circuitry may be designed for the purpose of expanding the range and shifting the level of the V_{in} signal. As an example, such active signal conditioning circuitry would be required to make the signal from the voltage divider compatible with an A/D converter having an input range of 0 to 5 V. A type of circuit that could accomplish such a task was introduced in Section 2–3 of Chapter 2. (Review the design process pertaining to Figures 2–26 through 2–29.) When adapting the design shown in Figure 2–29 for the purpose of conditioning the V_{in} signal from the thermistor voltage divider, m (or slope) becomes $\Delta V_o/\Delta V_{in}$:

$$m = \frac{\Delta V_o}{\Delta V_{in}} = \frac{5\,V - 0\,V}{2.962\,V - 2.01\,V} = \frac{5\,V}{952\,mV} = 5.2521$$

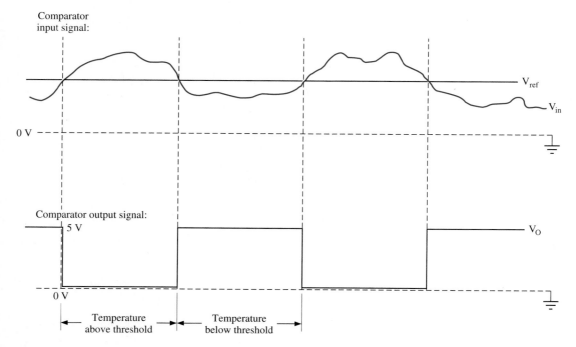

Figure 3–16
Operation of over-temperature detection circuit in Figure 3–15

Figure 3–17
Slope of temperature versus voltage between
15°C and 35°C

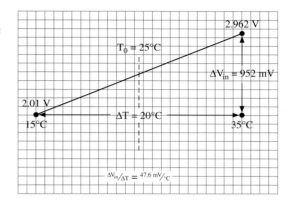

(Note that this result represents the overall voltage gain of the op amp circuit.) For the signal conditioning circuit being designed:

$$V_o = (m \times V_{in}) + b$$

Using the minimum and maximum values of output voltage, the value of b may be derived:

$$0 \text{ V} = 5.2521 \times 2.01 \text{ V} + b = 10.556721 \text{ V} + b$$
$$b = -10.556721 \text{ V}$$

or

$$5 \text{ V} = 5.2521 \times 2.962 \text{ V} + b = 15.556721 \text{ V} + b$$
$$b = 5 \text{ V} - 15.556721 \text{ V} = -10.556721 \text{ V}$$

Simplifying:

$$\frac{V_o}{m} = \frac{(m \times V_{in})}{m} + \frac{b}{m} = V_{in} + \frac{b}{m}$$

Thus,

$$V_o = (V_{in} + b/m) \times m$$

Substituting:

$$V_o = \left(V_{in} + \frac{-10.556721 \text{ V}}{5.2521} \right) \times 5.2521$$
$$= (V_{in} + -2.01 \text{ V}) \times 5.2521$$

The intercept may also be determined by graphic means, as shown in Figure 3–18. (This process was first illustrated in Figure 2–26.)

Using the design format introduced in Figure 2–29, a dual op amp circuit may now be created. In Figure 3–19, V_{in} of the thermistor voltage divider will now serve as the input to the dual op amp circuitry. Due to the high input impedance of the 1458 op amp, its loading effect on the voltage divider is negligible.

For the first stage of the signal-conditioning circuitry, which is a summer amplifier, point A must be adjusted to −2.01 V in order to allow the summing current to approach 0 A when V_{in} is at +2.01 V. With this condition, which represents the lowest temperature within the specified range, the output would be at 0 V. With V_{in} at its maximum value

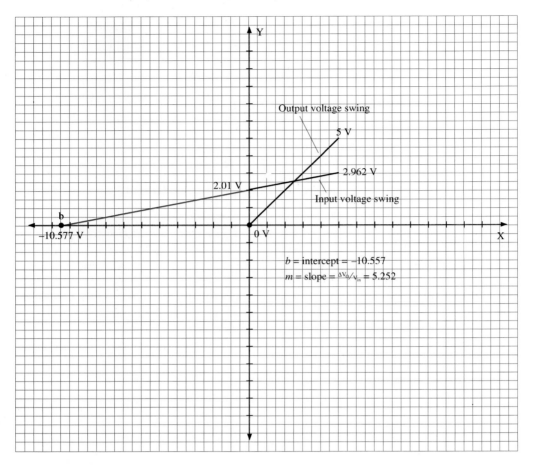

Figure 3–18
Graphic representation of solution for op amp circuit

for the given range (2.962 V), V_o should equal 5 V. This is easily proven as follows: According to Eq. (2–17),

$$V_o = (m \times V_{in}) + b$$
$$= (5.2521 \times 2.962 \text{ V}) + -10.556721 \text{ V} = 5 \text{ V}$$

As illustrated graphically in Figure 3–18, $+2.5$ V represents the midpoint for both the input (V_{in}) and output (V_o) voltage swings. This intersect point on the graph represents the midpoint temperature of 25°C. Thus, when 25°C occurs, V_o should equal 2.5 V. Again using Eq. (2–17), the expected output voltage representing T_o may be determined:

$$V_o = (5.2521 \times 2.5 \text{ V}) + -10.556721 \text{ V} = 2.574 \text{ V}$$

Note that a slight error exists in this result. If this deviation is to be expressed as a percentage of error, its value would be determined as follows:

$$\frac{2.574 \text{ V} - 2.5 \text{ V}}{5 \text{ V} - 0 \text{ V}} \times 100\% = \frac{74 \text{ mV}}{5 \text{ V}} \times 100\% = 1.48\% \text{ deviation}$$

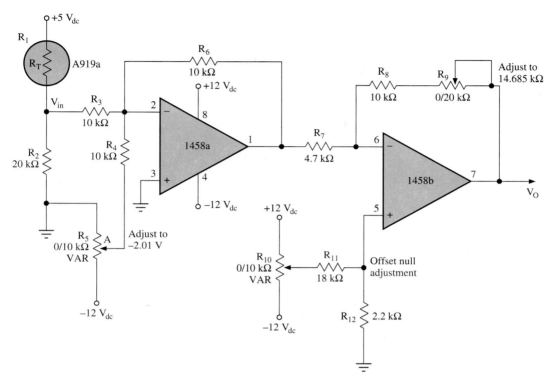

Figure 3–19
Op amp signal conditioning circuit that will fulfill the graphic equation of Figure 3–18

This degree of error may be acceptable in some industrial control applications. However, in precise thermometric circuitry, it would not be tolerable. In such highly linear instrumentation circuitry, temperature transducers other than thermistors must be used in order to achieve a greater degree of accuracy.

Example 3–5

For the circuit in Figure 3–19, what would be the values of V_{in} and V_o if the environmental temperature is 17°C? What would be the values of V_{in} and V_o if the temperature increases to 29°C? What percentage of change does this rise in temperature represent? Does V_o increase by nearly the same percentage?

Solution:
At 17° C: according to Table 3–1, R_T for the A919a = 27.403 kΩ. Thus,

$$V_{in} = 5 \text{ V} \times \frac{20 \text{ k}\Omega}{20 \text{ k}\Omega + 27.403 \text{ k}\Omega} = 2.1096 \text{ V}$$

$$V_o = (5.2521 \times 2.1096 \text{ V}) + -10.5567 \text{ V} = 523.1 \text{ mV}$$

At 29°C: According to Table 3–1, R_T for the A919a = 17.179 kΩ. Thus,

$$V_{in} = 5 \text{ V} \times \frac{20 \text{ k}\Omega}{20 \text{ k}\Omega + 17.179 \text{ k}\Omega} = 2.6897 \text{ V}$$

$$V_o = (5.2521 \times 2.6897 \text{ V}) + -10.5567 \text{ V} = 3.5699 \text{ V}$$

For the given range of temperature:

$$\frac{29°C - 17°C}{20°C} \times 100\% = 60\%$$

$$\frac{3.5699 \text{ V} - 523.1 \text{ mV}}{5 \text{ V}} \times 100\% = 60.9\%$$

Thus, the circuit is functioning with an error in linearity of less than 1%.

One alternative form of passive signal conditioning circuit that may be used in temperature sensing is the Wheatstone bridge. This resistive network is often used to develop a differential error signal that could be sent to an instrumentation amplifier, as shown in Figure 3–20. The Wheatstone bridge is especially suitable for developing signals produced by highly resistive, low-power transducers.

Figure 3–21 illustrates various thermistor assemblies belonging to the A990 family of devices, manufactured by Thermometrics Inc. An example of such a thermistor assembly

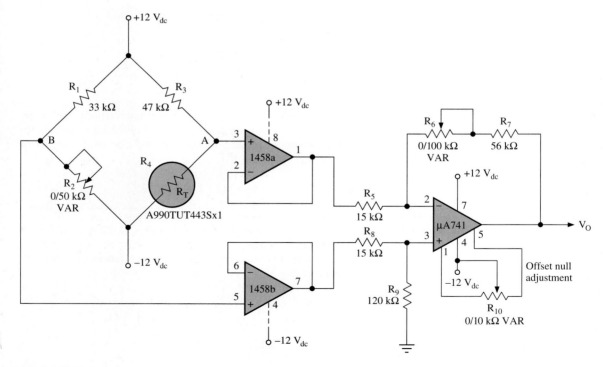

Figure 3–20
Highly resistive thermistor with Wheatstone bridge and instrumentation amplifier

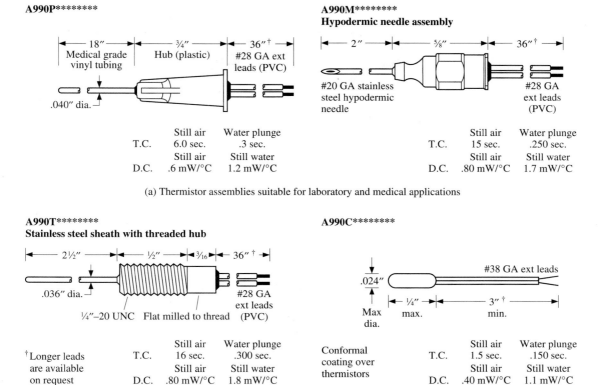

A990P********

	Still air	Water plunge
T.C.	6.0 sec.	.3 sec.
	Still air	Still water
D.C.	.6 mW/°C	1.2 mW/°C

A990M********
Hypodermic needle assembly

	Still air	Water plunge
T.C.	15 sec.	.250 sec.
	Still air	Still water
D.C.	.80 mW/°C	1.7 mW/°C

(a) Thermistor assemblies suitable for laboratory and medical applications

A990T********
Stainless steel sheath with threaded hub

†Longer leads
are available
on request

	Still air	Water plunge
T.C.	16 sec.	.300 sec.
	Still air	Still water
D.C.	.80 mW/°C	1.8 mW/°C

(b) Thermistor assembly suitable for
severe industrial environments

A990C********

Conformal
coating over
thermistors

	Still air	Water plunge
T.C.	1.5 sec.	.150 sec.
	Still air	Still water
D.C.	.40 mW/°C	1.1 mW/°C

(c) Thermistor assembly suitable for
immersion in conductive fluids

Figure 3–21
A990 thermistor assemblies (Courtesy of Thermometrics, Inc.)

is represented in the Wheatstone bridge circuit of Figure 3–20. The temperature sensors contained within the assemblies shown in Figure 3–21 consist of matched pairs of small, glass-encapsulated bead-type thermistors. Within the thermistor packages, these sensors may be connected in either series or parallel, depending on the resistance levels and performance characteristics established by the manufacturer. The two thermistor assemblies contained in Figure 3–21(a) have various medical and laboratory applications, while the more rugged device of Figure 3–21(b) is suitable for monitoring temperature in severe industrial environments. The thermistor assembly shown in Figure 3–21(c) may be safely immersed in conductive fluids. Common applications of this device would include monitoring the temperature of chemicals stored in vats or being mixed during industrial processes.

Each of the thermistor packages represented in Figure 3–21 is made available in the resistances listed in Table 3–2(a). Also, these devices may be ordered in accordance with the curve tolerances specified in Table 3–2(b). The exact part number for a desired thermistor assembly is compiled from the information contained in these tables, as well as Figure 3–21. For example, the number designation of the thermistor contained in Figure 3–20 is a compilation of the assembly prefix A990T, the resistance code UT443, and the tolerance code S×1. Thus, the complete part number A990TUT443S×1 indicates a stainless steel sheath assembly with a thermistor resistance ranging from 257.5775 kΩ at 0°C

Table 3–2(a)
Resistance versus temperature characteristics

Temperature (°C)	Resistance Code (Data in Ohms)					
	UN 103	**UT 103**	**UN 223**	**UT 223**	**UN 443**	**UT 443**
0	14129.9	56519.5	31452.1	125808.0	64394.4	257577.5
5	11335.1	45340.5	25168.0	100672.0	51167.4	204669.5
10	9152.8	36611.2	20273.6	81094.6	40931.5	163726.2
15	7437.4	29749.5	16435.7	65742.7	32956.3	131825.2
20	6080.3	24321.1	13406.3	53625.1	26701.3	106805.3
25	5000.0	20000.0	11000.0	44000.0	21764.2	87056.8
30	4134.9	16539.7	9077.0	36307.9	17843.2	71372.8
35	3438.1	13752.5	7531.1	30124.5	14710.6	58842.5
40	2873.8	11495.1	6281.4	25125.7	12193.5	48774.1
45	2414.2	9656.7	5265.6	21062.5	10159.7	40638.8
50	2038.0	8151.9	4435.6	17742.3	8507.6	34030.3
55	1728.4	6913.8	3753.9	15015.5	7154.4	28617.7
60	1472.6	5890.2	3191.3	12765.2	6043.0	24171.9
65	1260.0	5040.0	2724.8	10899.2	5126.1	20504.4
70	1082.7	4330.6	2336.2	9344.9	4366.4	17465.8
75	934.02	3736.1	2011.1	8044.5	3734.4	14937.8
80	808.93	3235.7	1738.0	6952.0	3206.5	12825.9
85	703.23	2812.9	1507.6	6030.3	2763.7	11054.7
90	613.55	2454.2	1312.4	5249.6	2390.8	9563.4
95	537.18	2148.7	1146.5	4586.0	2075.7	8303.0
100	471.90	1887.6	1004.9	4019.6	1808.5	7233.9
105	415.90	1663.6	883.67	3534.7	1580.9	6323.7

Courtesy of Thermometrics, Inc.

Table 3–2(b)
Curve tolerances

Tolerance Code	Temperature Tolerance in °C (±)			
	0°C–25°C	**25°C–50°C**	**50°C–70°C**	**70°C–105°C**
S×1	0.1	0.05	0.1	0.2
S×2	0.1	0.1	0.1	0.2
S×3	0.2	0.1	0.2	0.3
S×4	0.2	0.2	0.2	0.3

Courtesy of Thermometrics, Inc.

to 6.3237 kΩ at 105°C. Also, between 0°C and 25°C, the actual temperature of the device will be within 0.1°C of accuracy. That is, if the thermistor resistance is 106.8053 kΩ, its temperature must then be no higher than 20.1°C and no lower than 19.9°C. [The remaining tolerances within the temperature range of the device are easily derived from Table 3–2(b).]

It is now evident that the value of R_T for the thermistor in Figure 3–20 approaches the value of input impedance for a typical op amp. This could result in a considerable loading effect if the bridge output at point A in Figure 3–20 were fed to a single-input circuit such as that in Figure 3–19. However, in Figure 3–20, since an op amp in the voltage-follower configuration is used to buffer each of the inputs from the Wheatstone bridge, the impedance across the terminals of the instrumentation amplifier is likely to be several megohms. Since the amplifier functions as an extremely high-resistance load across the Wheatstone bridge, the bridge may be considered to be in an unloaded or open condition. This greatly simplifies both circuit design and analysis.

An important feature of the instrumentation amplifier is its ability to attenuate common-mode signals. Since thermistors are likely to require leads of considerable length, they tend to pick up high-frequency noise present within an industrial environment. However, since these noise signals are likely to be present at both inputs of the instrumentation amplifier, they will be greatly attenuated. At the same time, the differential error signal from the Wheatstone bridge will be amplified. For the circuit in Figure 3–20, the gain factor will be nearly equal to $-(R_6 + R_7)/R_5$.

Example 3–6

For the circuit in Figure 3–20, assume that the set-point temperature is to be 45°C. Given this condition, what would be the value of R_T for the thermistor? What should be the ohmic setting of R_2 if a null condition is to exist across the Wheatstone bridge at the set-point temperature?

Solution:
According to Table 3–2(a), at 45°C, R_T equals 40.639 kΩ. For a null condition to occur at this temperature:

$$R_1 \times R_4 = R_2 \times R_3$$

Thus,

$$R_2 = \frac{R_1 \times R_4}{R_3} = \frac{33 \text{ k}\Omega \times 40.639 \text{ k}\Omega}{47 \text{ k}\Omega} = 28.534 \text{ k}\Omega$$

If the environmental temperature increases to 50°C, what will be the new error voltage across the bridge? What will be the value of output voltage if R_6 is adjusted to 64 kΩ?

Solution:
According to Table 3–2(a), at 50°C, R_T equals 34.03 kΩ. Thus,

$$V_B = \left(\frac{28.534 \text{ k}\Omega}{28.534 \text{ k}\Omega + 33 \text{ k}\Omega} \times 24 \text{ V} \right) + -12 \text{ V} = -871 \text{ mV}$$

$$V_A = \left(\frac{34.03 \text{ k}\Omega}{34.03 \text{ k}\Omega + 47 \text{ k}\Omega} \times 24 \text{ V} \right) + -12 \text{ V} = -1.921 \text{ V}$$

$$V_{A-B} = -1.921 \text{ V} - -871 \text{ mV} = -1.05 \text{ V}$$

$$V_0 = V_{A-B} \times -(R_6 + R_7)/R_5$$

$$= -1.05 \text{ V} \times -(64 \text{ k}\Omega + 56 \text{ k}\Omega)/15 \text{ k}\Omega = 8.4 \text{ V}$$

Thus far, applications of the thermistor based on its current-time and resistance-temperature characteristics have been examined. The voltage-current characteristic of the device may also be used in the design of thermistor circuitry. An example of such an application is found in Figure 3–22. Here, two thermistors of the same type are connected in a Wheatstone bridge configuration in order to monitor the flow of gas through a pipe.

When conducting current, thermistor R_2 will be in a self-heated condition, thus having a body temperature significantly higher than the ambient temperature of the environment. As illustrated in Figure 3–23, R_2 is a bead-type thermistor, which is placed directly in the center of the gas pipe. **Reference thermistor R_1** is housed within a small metal block,

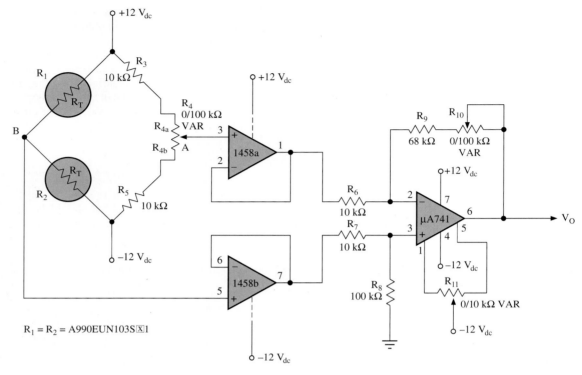

Figure 3–22
Two thermistors used in a gas-flow measurement system

which is mounted on the inside surface of the pipe. Since R_1 is able to dissipate heat rapidly via the metal block, its body temperature will virtually equal the ambient temperature of the gas within the pipe; R_4 is a trim pot used to establish the null condition for the Wheatstone bridge, which should exist whenever there is no gas flow within the pipe. Since R_2 is in a self-heated state, its ohmic value will be lower than that of R_1. Thus, for

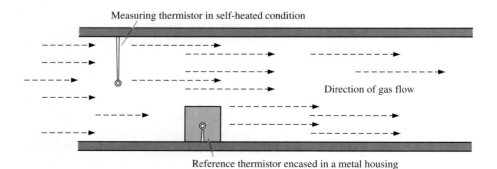

Figure 3–23
Placement of the measuring and reference thermistors within the gas pipe

a null condition (with no gas flowing), R_4 must be adjusted such that:

$$R_1 \times (R_{4b} + R_5) = R_2 \times (R_{4a} + R_3)$$

To understand the operation of the circuit in Figure 3–22, we must first refer back to the discussion of the thermistor dissipation constant. It has already been stated that the dissipation constant of a thermistor depends on the thermal conductivity of the medium in which the device is placed. Since R_2 is placed in the direct path of the gas flow, it will experience convective heat loss. Thus, the self-heated temperature of R_2 will vary as a function of changes in the rate of the flow of gas. Assuming that the temperature of the gas within the pipe remains the same, the reference value of R_T (zero-power resistance) will remain the same. Thus, as the flow of gas increases, the dissipation constant increases, causing the self-heated temperature (ΔT) to decrease. This decrease in temperature causes an increase in the resistance of R_2, which causes the voltage at point B on the Wheatstone bridge to become more positive.

As the temperature within the pipe fluctuates, the zero-power resistance of the two thermistors must also change. However, since R_1 and R_2 are identical, any change in R_T for R_1 will also occur in R_2. Thus, the effect of changes in ambient temperature on the measurement of gas flow will be negated. Assuming the effect of R_T to be minimized, the change in self-heated temperature for R_2 becomes proportional to the square root of the rate of gas flow. The output signal from the Wheatstone bridge (V_{A-B}) is developed by the instrumentation amplifier. The output of this circuit could be rapidly processed by an A/D converter and sent to a microcomputer. This device could then perform an algorithm or access a look-up table in its memory to derive the flow rate (expressed in cubic feet per minute).

SECTION REVIEW 3–3

1. For the voltage divider in Figure 3–14(a), assume the body temperature of the A919a is 7°C. Using Table 3–1 as a guide, determine the value of V_o.
2. If V_o in Figure 3–14(a) now equals 1.97 V, what is the body temperature of the thermistor in °F?
3. For the comparator circuit in Figure 3–15, assume the body temperature of the thermistor is 116°F. If the potentiometer is now adjusted to 12.75 kΩ, what is the logic condition of the comparator output?
4. For the dual op amp circuit in Figure 3–19, assume the body temperature of the thermistor to be at 69.8°F. What would be the equivalent temperature in °C? What would be the values of V_{in} and V_o with this temperature condition?
5. What form of resistive network is especially suitable for developing signals produced by low-power, highly resistive temperature transducers?
6. For the circuit in Figure 3–20, what must be the setting of R_2 if a null condition is to occur at 40°C?
7. Assume that R_1 in Figure 3–20 is replaced by an A990ES×3UN223 thermistor assembly. Briefly describe this device with regard to general function and resistance range. Would the circuit function be less accurate as a result of this substitution (assuming the bridge has been calibrated for the new set-point value of R_4)?
8. For the thermistor assembly identified in the previous question, according to its tolerance code, what would be its lowest body temperature with a body resistance of 7.5311 kΩ?

9. True or false?: For the circuit in Figure 3–22 to have a positive-going output voltage swing, thermistor R_2 should be kept in the self-heated condition, while thermistor R_1 is operated as closely as possible to the zero-power point.
10. For the circuit in Figure 3–22, what parameter of the measuring thermistor is actually being changed as a result of fluctuations in the rate of gas flow?

3–4   INTRODUCTION TO RESISTIVE TEMPERATURE DETECTORS

In addition to the thermistor, the **resistive temperature detector** is frequently used in industrial instrumentation circuitry. It has already been stated that the RTD, while having a lower resistivity than the thermistor, is capable of a more linear response to changes in temperature. The operation of the RTD is based on the highly predictable characteristic of temperature versus resistance exhibited by pure metals. In spite of its high cost and relatively low sensitivity, platinum is commonly used to manufacture RTDs. Platinum exhibits greater linearity than nickel or tungsten and has an operating range extending from nearly $-200°C$ to $+600°C$.

The sensitivity of an RTD is often expressed in terms of its **temperature coefficient** (symbolized as α), which is the fractional change in the resistance of the device that occurs with an incremental change in temperature. For a platinum RTD with a resistance of $100\ \Omega$ at the freezing point ($0°C$), resistance is likely to change by nearly $0.4\ \Omega/°C$. The temperature coefficient for the device is actually specified as the ratio of the fractional change in resistance and the zero-degree resistance per $°C$. Thus,

$$\alpha = \frac{\Delta R_o/R_o}{°C} \tag{3–13}$$

where ΔR_o = the fractional change in resistance per $°C$
R_o = the resistance of the RTD at the $0°C$ reference point.

Therefore, with a resistance of $100\ \Omega$ at $0°C$, the temperature coefficient for the RTD can be determined:

$$\alpha = \frac{\Delta R_o/R_o}{°C} = \frac{0.4\ \Omega/100\ \Omega}{°C} = 0.004/°C$$

Due to the near linearity of the temperature versus resistance characteristic of pure platinum, the temperature coefficient for the platinum RTD may now be used to estimate quickly its resistance at any given value of body temperature. This is easily accomplished using the following formula:

$$R_T = R_o \times (1 + \alpha \times T) \tag{3–14}$$

where R_T is the body resistance of the platinum RTD being approximated for the given value of T, which is its body temperature.)

Example 3–7

A platinum RTD has a resistance of $100\ \Omega$ at $0°C$. Assuming the temperature coefficient of the device to be of $0.004/°C$, what would be the resistance of the device at $47°C$?

Solution:

$$R_T = R_o \times (1 + \alpha \times T)$$
$$= 100\ \Omega \times (1 + 0.004 \times 47°C) = 118.8\ \Omega$$

Table 3–3 contains a detailed listing of temperature versus resistance for a platinum RTD. The device's positive temperature coefficient becomes obvious as its body resistance increases with temperature. Also, careful analysis of the table would reveal that the increases in resistance, for each incremental change in °C, are nearly constant (varying from 0.39 to 0.35 $\Omega/°C$ as temperature increases from 0°C to 299°C).

As shown in Figure 3–24, a typical RTD is likely to consist of a thin platinum wire wrapped around a ceramic core material. This assembly is encased in a stainless steel sheath in order to isolate it from the medium in which it is immersed. The stainless steel sheath also enhances the ability of the RTD to dissipate heat. Since the RTD is likely to be larger than the typical bead-type thermistor, it will have a significantly longer thermal time constant. Due to the relatively large cross-sectional area of the RTD, it is well suited for monitoring temperature within a large area.

RTDs used in temperature measurement are likely to have much lower resistive ranges than thermistors used for the same purpose. Although wire resistance may not be an important consideration in designing thermistor circuits, it does become significant in the design of RTD circuitry. This is especially true when the RTD is contained within a Wheatstone bridge, as part of a precise temperature measurement system. In such circuitry, the RTD is likely to be in a remote location, several feet away from the electronic control source. The resistance of the wire could then represent a significant source of error in calibrating the bridge circuitry.

As shown in Figure 3–25, a **compensation line** may be used to reduce the effect of the lead resistance. The compensation line consists of wire of the same length and gauge as that used to access the RTD. Thus, any error resulting from wire resistance introduced on the right side of the Wheatstone bridge is compensated for on the left side. For the circuit in Figure 3–25, since the effect of wire resistance has been minimized, a null condition may still be considered to exist where $(R_1 \times R_4)$ equals $(R_2 \times R_3)$. (This is based on the assumption that the ohmic value of R_2 and the compensation line equals that of the RTD and its leads.) The compensation lead must, of course, be subjected to the same environmental conditions as those leads connected to the RTD. This ensures that any fluctuations in wire resistance due to changes in the temperature of the medium will occur on both sides of the bridge and thus be negated.

Example 3–8

Assume that the circuit in Figure 3–25 is being used to monitor the temperature within an oven and that the oven temperature is to be maintained at 275°F. If a null condition occurs across the Wheatstone bridge with the oven temperature at this set-point value, what must be the ohmic setting of R_2? (Assume that R_4 is a platinum RTD with a temperature versus resistance characteristic conforming to Table 3–3.)

Assume that R_2 has already been adjusted for the set-point temperature specified. If, when the oven is first turned on, its temperature is only 59°F, what would be the value of V_A, V_B, V_{A-B}, and V_o? (Assume that the resistance of the thermistor leads, as well as the compensation line, is 1.4 Ω.)

Solution:
Convert the set-point temperature to °C:

$$\text{Set point} = \frac{5}{9} \times (275°F - 32°F) = 135°C$$

Select the resistance value from Table 3–2 that corresponds with the set-point temperature:

$$\text{At } 135°C, R_T = 151.7 \ \Omega$$

Table 3-3
Resistance versus temperature for a platinum RTD

°C	Ohm	Diff.	°C	Ohm	Diff.	°C	Ohm	Diff.	°C	Ohm	Diff.	°C	Ohm	Diff.	°C	Ohm	Diff.
±0	100.00	0.39	+60	123.24	0.38	+120	146.06	0.38	+180	168.47	0.37	+240	190.46	0.36	+300	212.03	0.36
+1	100.39	0.39	61	123.62	0.39	121	146.44	0.37	181	168.84	0.37	241	190.82	0.36	301	212.39	0.35
2	100.78	0.39	62	124.01	0.38	122	146.81	0.38	182	169.21	0.37	242	191.18	0.37	302	212.74	0.36
3	101.17	0.39	63	124.39	0.38	123	147.19	0.37	183	169.58	0.37	243	191.55	0.36	303	213.10	0.35
4	101.56	0.39	64	124.77	0.39	124	147.56	0.38	184	169.95	0.37	244	191.91	0.36	304	213.45	0.36
5	101.95	0.39	65	125.16	0.38	125	147.94	0.38	185	170.32	0.36	245	192.27	0.36	305	213.81	0.35
6	102.34	0.39	66	125.54	0.38	126	148.32	0.37	186	170.68	0.37	246	192.63	0.36	306	214.16	0.36
7	102.73	0.39	67	125.92	0.38	127	148.69	0.38	187	171.05	0.37	247	192.99	0.37	307	214.52	0.35
8	103.12	0.39	68	126.30	0.39	128	149.07	0.37	188	171.42	0.37	248	193.36	0.36	308	214.87	0.36
9	103.51	0.39	69	126.69	0.38	129	149.44	0.38	189	171.79	0.37	249	193.72	0.36	309	215.23	0.35
10	103.90	0.39	70	127.07	0.38	130	149.82	0.38	190	172.16	0.37	250	194.08	0.36	310	215.58	0.36
11	104.29	0.39	71	127.45	0.38	131	150.20	0.37	191	172.53	0.37	251	194.44	0.36	311	215.94	0.35
12	104.68	0.39	72	127.83	0.39	132	150.57	0.38	192	172.90	0.36	252	194.80	0.37	312	216.29	0.36
13	105.07	0.39	73	128.22	0.38	133	150.95	0.37	193	173.26	0.37	253	195.17	0.36	313	216.65	0.35
14	105.46	0.39	74	128.60	0.38	134	151.32	0.38	194	173.63	0.37	254	195.53	0.36	314	217.00	0.36
15	105.85	0.39	75	128.98	0.38	135	151.70	0.37	195	174.00	0.37	255	195.89	0.36	315	217.36	0.35
16	106.23	0.38	76	129.36	0.38	136	152.07	0.38	196	174.37	0.37	256	196.25	0.36	316	217.71	0.36
17	106.62	0.39	77	129.74	0.39	137	152.45	0.37	197	174.74	0.36	257	196.61	0.37	317	218.07	0.35
18	107.01	0.39	78	130.13	0.38	138	152.82	0.38	198	175.10	0.37	258	196.98	0.36	318	218.42	0.36
19	107.40	0.39	79	130.51	0.38	139	153.20	0.37	199	175.47	0.37	259	197.34	0.36	319	218.78	0.35
20	107.79	0.39	80	130.89	0.38	140	153.57	0.38	200	175.84	0.37	260	197.70	0.36	320	219.13	0.35
21	108.18	0.39	81	131.27	0.38	141	153.95	0.37	201	176.21	0.36	261	198.06	0.36	321	219.48	0.36
22	108.57	0.38	82	131.65	0.38	142	154.32	0.38	202	176.57	0.37	262	198.42	0.36	322	219.84	0.35
23	108.95	0.39	83	132.03	0.38	143	154.70	0.37	203	176.94	0.37	263	198.78	0.36	323	220.19	0.35
24	109.34	0.39	84	132.41	0.39	144	155.07	0.38	204	177.31	0.37	264	199.14	0.36	324	220.54	0.36
25	109.73	0.39	85	132.80	0.38	145	155.45	0.37	205	177.68	0.36	265	199.50	0.36	325	220.90	0.35
26	110.12	0.39	86	133.18	0.38	146	155.82	0.38	206	178.04	0.37	266	199.86	0.36	326	221.25	0.35
27	110.51	0.39	87	133.56	0.38	147	156.20	0.38	207	178.41	0.37	267	200.22	0.36	327	221.60	0.35

#	Value		#	Value		#	Value		#	Value		#	Value		#	Value	
28	110.89	0.38	88	133.94	0.38	148	156.57	0.37	208	178.78	0.37	268	200.58	0.36	328	221.95	0.35
29	111.28	0.39	89	134.32	0.38	149	156.95	0.38	209	179.14	0.36	269	200.94	0.36	329	222.31	0.36
30	111.67	0.39	90	134.70	0.38	150	157.32	0.37	210	179.51	0.37	270	201.30	0.36	330	222.66	0.35
31	112.06	0.39	91	135.08	0.38	151	157.69	0.37	211	179.88	0.37	271	201.66	0.36	331	223.01	0.35
32	112.44	0.38	92	135.46	0.38	152	158.07	0.38	212	180.24	0.36	272	202.02	0.35	332	223.36	0.35
33	112.83	0.39	93	135.84	0.38	153	158.44	0.37	213	180.61	0.37	273	202.37	0.36	333	223.72	0.36
34	113.22	0.39	94	136.22	0.38	154	158.81	0.37	214	180.97	0.36	274	202.73	0.36	334	224.07	0.35
35	113.61	0.39	95	136.60	0.38	155	159.19	0.38	215	181.34	0.37	275	203.09	0.36	335	224.42	0.35
36	113.99	0.38	96	136.98	0.38	156	159.56	0.37	216	181.71	0.37	276	203.45	0.36	336	224.77	0.35
37	114.38	0.39	97	137.36	0.38	157	159.93	0.37	217	182.07	0.36	277	203.81	0.35	337	225.12	0.35
38	114.77	0.39	98	137.74	0.38	158	160.30	0.37	218	182.44	0.37	278	204.16	0.36	338	225.48	0.36
39	115.15	0.38	99	138.12	0.38	159	160.68	0.38	219	182.80	0.36	279	204.52	0.36	339	225.83	0.35
40	115.54	0.39	100	138.50	0.38	160	161.05	0.37	220	183.17	0.37	280	204.88	0.36	340	226.18	0.35
41	115.93	0.39	101	138.88	0.38	161	161.42	0.37	221	183.54	0.37	281	205.24	0.36	341	226.53	0.35
42	116.31	0.38	102	139.26	0.37	162	161.79	0.37	222	183.90	0.36	282	205.60	0.35	342	226.88	0.35
43	116.70	0.39	103	139.63	0.38	163	162.16	0.37	223	184.27	0.37	283	205.95	0.36	343	227.23	0.35
44	117.08	0.38	104	140.01	0.38	164	162.53	0.37	224	184.63	0.36	284	206.31	0.36	344	227.58	0.36
45	117.47	0.39	105	140.39	0.38	165	162.91	0.38	225	185.00	0.37	285	206.67	0.36	345	227.94	0.35
46	117.86	0.39	106	140.77	0.38	166	163.28	0.37	226	185.36	0.36	286	207.03	0.36	346	228.29	0.35
47	118.24	0.38	107	141.15	0.37	167	163.65	0.37	227	185.73	0.37	287	207.39	0.35	347	228.64	0.35
48	118.63	0.39	108	141.52	0.38	168	164.02	0.37	228	186.09	0.36	288	207.74	0.36	348	228.99	0.35
49	119.01	0.38	109	141.90	0.38	169	164.39	0.37	229	186.46	0.37	289	208.10	0.36	349	229.34	0.35
50	119.41	0.39	110	142.28	0.38	170	164.76	0.37	230	186.82	0.36	290	208.46	0.36	350	229.69	0.35
51	119.78	0.39	111	142.66	0.38	171	165.13	0.37	231	187.18	0.37	291	208.82	0.35	351	230.04	0.35
52	120.17	0.38	112	143.04	0.37	172	165.50	0.37	232	187.55	0.37	292	209.17	0.36	352	230.39	0.35
53	120.55	0.39	113	143.41	0.38	173	165.87	0.37	233	187.91	0.36	293	209.53	0.36	353	230.74	0.35
54	120.94	0.39	114	143.79	0.38	174	166.24	0.37	234	188.28	0.37	294	209.89	0.36	354	231.09	0.35
55	121.32	0.38	115	144.17	0.38	175	166.62	0.38	235	188.64	0.36	295	210.25	0.35	355	231.44	0.35
56	121.70	0.38	116	144.55	0.38	176	166.99	0.37	236	189.00	0.37	296	210.60	0.36	356	231.79	0.35
57	122.09	0.39	117	144.93	0.37	177	167.36	0.37	237	189.37	0.36	297	210.96	0.36	357	232.14	0.35
58	122.47	0.38	118	145.39	0.37	178	167.73	0.37	238	189.73	0.36	298	211.32	0.35	358	232.49	0.35
59	122.86	0.39	119	145.68	0.38	179	168.10	0.37	239	190.10	0.37	299	211.67	0.37	359	232.84	0.35

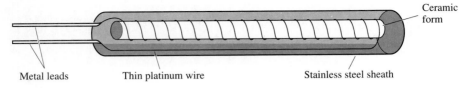

Figure 3–24
Physical structure of a resistive temperature detector

With R_1 and R_3 both equaling 120 Ω, R_2 must be adjusted to 151.7 Ω in order to create the null condition at the set-point temperature of 135°C.

Converting 59°F to °C:

$$T = \frac{5}{9} \times (59°F - 32°F) = 15°C$$

At 15°C, the resistance of the RTD equals 105.85 Ω. With this condition:

$$I_B = \frac{5\ \text{V}}{R_1 + R_2 + R_L} = \frac{5\ \text{V}}{120\ Ω + 151.7\ Ω + 1.4\ Ω} = 18.308\ \text{mA}$$

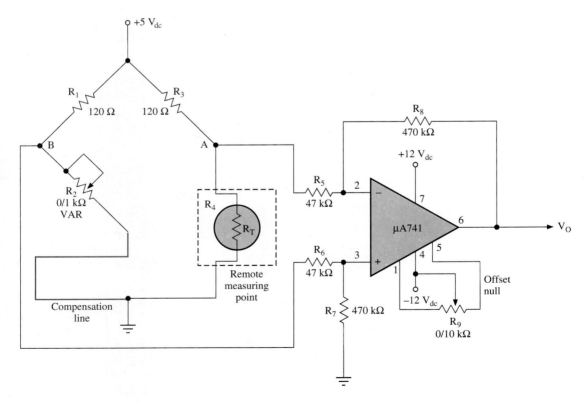

Figure 3–25
RTD in a Wheatstone bridge configuration with compensation line

$$I_A = \frac{5\ V}{R_3 + R_4 + R_L} = \frac{5\ V}{120\ \Omega + 105.85\ \Omega + 1.4\Omega} = 22.002\ mA$$

$$V_B = I_B \times (R_2 + R_L) = 18.308\ mA \times (151.7\ \Omega + 1.4\ \Omega)$$

$$V_B = 2.803\ V$$

$$V_A = I_A \times (R_4 + R_L) = 22.002\ mA \times (105.85\ \Omega + 1.4\ \Omega)$$

$$= 2.36\ V$$

$$V_{A-B} = V_A - V_B = 2.36\ V - 2.803\ V = -443\ mV$$

$$V_o = V_{A-B} \times \frac{-R_8}{R_5} = -443\ mV \times \frac{-470\ k\Omega}{47\ k\Omega} = +4.433\ V$$

SECTION REVIEW 3–4

1. Explain why, in spite of its cost, platinum is a highly preferred material for manufacturing RTDs.
2. True or false?: Platinum is characterized by a negative temperature coefficient.
3. A platinum RTD has a resistance of 100 Ω at 0°C. If the temperature coefficient of the device is 0.004/°C, what is its resistance at 132.8°F?
4. True or false?: RTDs used in thermometric circuitry generally have much lower values of resistance than thermistors used for the same purpose.
5. True or false?: Because of their larger cross-sectional area, RTDs are better suited than bead-type thermistors for monitoring temperature within large areas.
6. Since RTDs are larger in size than typical bead-type thermistors, they tend to have much shorter time constants.
7. Explain the purpose of the compensation line contained in Figure 3–25.
8. For the circuit in Figure 3–25, the oven set-point temperature is to be established at 325°F. What must be the setting of R_2? Assuming R_2 is set at this ohmic value, what would be the value of V_o if the body temperature of the RTD is at 392°F?

3–5   INTRODUCTION TO THERMOCOUPLES

Although they have different physical compositions, thermistors and RTDs are similar in that changes in their body temperatures result in changes in their electrical resistance. Only when placed in a resistive network with power applied is either of these two devices capable of translating temperature fluctuations into voltage changes. The **thermocouple,** which is a third major form of temperature transducer, has the ability to translate changes in temperature directly to changes in voltage.

Figure 3–26 illustrates the principle of operation for the thermocouple. As shown here, when two different types of metals are joined, a small amount of electromotive force (referred to as **thermoelectric** or **Seebeck voltage**) is developed across their point of contact. Although changes in the value of this EMF are limited to the microvolt range, they may be fairly linear, depending on the types of metals used to form the junction and the span of temperatures being monitored. All thermocouples exhibit a positive temperature coefficient, which may vary from 10 to 60 μV/°C, again depending on the type of bimetallic junction and the point in the given temperature range. Generally speaking, the incremental change in thermoelectric voltage per °C tends to become greater as temperature

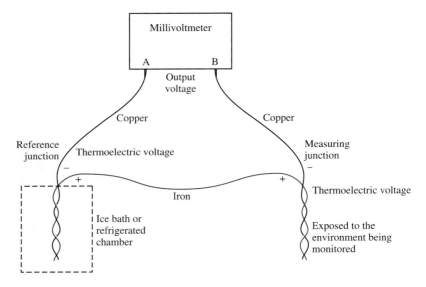

Figure 3–26
Principle of operation for the thermocouple

increases. The usable temperature range for a typical thermocouple is greater than that of either a thermistor or an RTD, extending from around −150°C to as high as 1500°C.

As symbolized in Figure 3–27, a complete thermocouple measurement system consists of at least one reference junction T_r and a measuring junction T_m. Note that at the points where the copper leads from the meter are joined with the iron strips within the

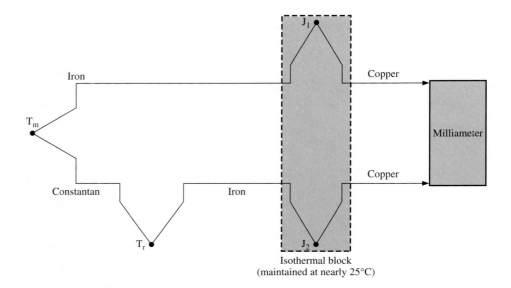

Figure 3–27
Symbolic representation of a thermocouple connected to an analog meter movement

thermocouple, two additional thermal junctions are effectively formed (designated in Figure 3–27 as J_1 and J_2). If these two junction points are contained within a common isothermal block, and thus held at the same temperature, their effect on the operation of the thermocouple is negated. This is true due to the fact that the thermoelectric voltages developed over J_1 and J_2, like those developed at the reference and measuring junctions, are series-opposing. Thus, if J_1 and J_2 are held at the same temperature, their thermoelectric voltages will cancel each other within the series loop.

Because, for the thermocouple in Figure 3–27, both the measuring and the reference junctions are identical, their thermoelectric voltages will be equal and opposite when their temperatures are the same. Thus, if the reference junction is kept at the ice-point temperature of 0°C, a meter reading of 0 V will represent the freezing point.

A bimetallic junction may be made from iron and constantan (which is an alloy of copper and nickel). Such a junction is referred to as **J-type.** A higher grade of bimetallic junction, **S-type,** may be created using pure platinum and an alloy of platinum-rhodium. Although more costly than the J-type thermocouple, this noble-metal junction is highly resistive to oxidation. While J-type thermocouples have practical temperature ranges extending through nearly 900°C, S-type devices have usable ranges extending through nearly 1600°C.

Table 3–4 contains a listing of voltage versus temperature for a J-type thermocouple (made from iron and constantan). For the sake of simplicity, this listing is made in increments of 5°C. However, more detailed listings, in which thermoelectric voltages are specified at each °C or °F, can be easily obtained from manufacturers.

To derive a listing of voltage versus temperature such as that in Table 3–4, a milli-voltmeter would be connected as illustrated in Figure 3–28. The voltage levels recorded in the table are obtained with the positive lead of the meter connected to the J_1 terminal of the isothermal block and the negative lead connected to the J_2 terminal. With the reference junction held at 0°C, a balanced condition should exist when the measuring junction is also at the freezing point. At this time, the thermoelectric voltages of the measuring and reference junctions will be equal in value and opposite in polarity. Thus, as indicated in both Table 3–4 and Figure 3–28(a), the output voltage from the thermocouple would be at zero. Below this 0 V/0°C reference point, the iron would become more negative with respect to the constantan, resulting in the voltage condition shown in Figure 3–28(b). Below the freezing point, the meter readings would become more negative as temperature continues to decrease. With temperature increasing above the 0 V/0°C point, the iron would become more positive than the constantan. Thus, the polarity of the meter voltage would be reversed. As represented in Figure 3–28(c), the output voltage of the thermocouple would continue to become more positive with increases in temperature.

Example 3–9

Using Table 3–4, determine the difference in thermocouple output voltage that would occur between −195°C and −190°C. Determine the approximate difference in voltage per °C represented by this span of temperatures. Next, determine the difference in voltage occurring between 740°C and 745°C. Again determine the approximate difference in voltage per °C.

Solution:
According to Table 3–4, at 195°C, thermocouple voltage would be −7.78 mV. At −190°C, thermocouple voltage would equal −7.66 mV. Thus, the rate of voltage increase would be 120 μV per 5°C change in temperature:

$$\Delta V = (-7.66 \text{ mV} - -7.78 \text{ mV})/5°C = 120 \, \mu V/5°C$$

Table 3–4
Voltage versus temperature for a J-type thermocouple

°C	0°	5°	10°	15°	20°	25°	30°	35°	40°	45°
					(in millivolts)					
−150°	−6.5	−6.66	−6.82	−6.97	−7.12	−7.27	−7.4	−7.54	−7.66	−7.78
−100	−4.63	−4.83	−5.03	−5.23	−5.42	−5.61	−5.8	−5.98	−6.16	−6.33
−50°	−2.43	−2.66	−2.89	−3.12	−3.34	−3.56	−3.78	−4.00	−4.21	−4.42
−0°	0.0	−0.25	−0.5	−0.75	−1.0	−1.24	−1.48	−1.72	−1.96	−2.2
+0°	0.0	0.25	0.5	0.76	1.02	1.28	1.54	1.8	2.06	2.32
50°	2.58	2.85	3.11	3.38	3.65	3.92	4.19	4.46	4.73	5
100°	5.27	5.54	5.81	6.08	6.36	6.63	6.9	7.18	7.45	7.73
150°	8.0	8.28	8.56	8.84	9.11	9.39	9.67	9.95	10.22	10.5
200°	10.78	11.06	11.34	11.62	11.89	12.17	12.45	12.73	13.01	13.28
250°	13.56	13.84	14.12	14.39	14.67	14.94	15.22	15.5	15.77	16.05
300°	16.33	16.6	16.88	17.15	17.43	17.71	17.98	18.26	18.54	18.81
350°	19.09	19.37	19.64	19.92	20.2	20.47	20.75	21.02	21.3	21.57
400°	21.85	22.13	22.4	22.68	22.95	23.23	23.5	23.78	24.06	24.33
450°	24.61	24.88	25.16	25.44	25.72	25.99	26.27	26.55	26.83	27.11
500°	27.39	27.67	27.95	28.23	28.52	28.8	29.08	29.37	29.65	29.94
550°	30.22	30.51	30.8	31.08	31.37	31.66	31.95	32.24	32.53	32.82
600°	33.11	33.41	33.7	33.99	34.29	34.58	34.88	35.18	35.48	35.78
650°	36.08	36.38	36.69	36.99	37.3	37.6	37.91	38.22	38.53	38.84
700°	39.15	39.47	39.78	40.10	40.41	40.73	41.05	41.36	41.68	42.0

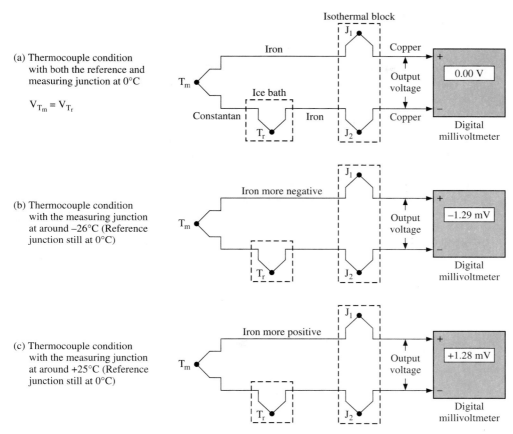

(a) Thermocouple condition with both the reference and measuring junction at 0°C

$V_{T_m} = V_{T_r}$

(b) Thermocouple condition with the measuring junction at around −26°C (Reference junction still at 0°C)

(c) Thermocouple condition with the measuring junction at around +25°C (Reference junction still at 0°C)

Figure 3–28
Operation of a T-type thermocouple

The rate of change in voltage per °C could also be estimated as follows:

$$\Delta V = \frac{-7.66 \text{ mV} - -7.78 \text{ mV}}{-190°C - -195°C} = \frac{120 \ \mu V}{5°C} = 24 \ \mu V/°C$$

According to Table 3–4, at 740°C, the thermocouple voltage would equal 41.68 mV. At 745°C, this voltage would have increased to 42 mV. Thus, the new rate of voltage change would have increased to 320 μV per 5°C change in temperature, which is determined as follows:

$$\Delta V = (42 \text{ mV} - 41.68 \text{ mV})/5°C = 320 \ \mu V/5°C$$

The rate of change in voltage per °C could also be determined as follows:

$$\Delta V = \frac{42 \text{ mV} - 41.68 \text{ mV}}{745°C - 740°C} = \frac{320 \ \mu V}{5°C} = 64 \ \mu V/°C$$

With a thermocouple, changes in temperature are directly translated into small, yet predictable, changes in voltage. Analog thermometers can be created that exploit this property. These devices are capable of accurately monitoring a wide variety of temperature ranges. An example of such an instrument is illustrated in Figure 3–29. Here, a **D'Arsonval movement** (identical to those found in millivoltmeters) is used to indicate temperature. If a low-resistance meter movement is used (one requiring only 250 μA for full-scale deflection), the thermoelectric voltage present at the output terminals of the thermocouple will actually function as the source of power for the meter.

Careful examination of the voltage and current scales on the face of an analog meter will indicate that these scales are not perfectly linear. That is, for any given difference in voltage, the degree of arc (or deflection) of the D'Arsonval movement will vary, depending on the point within the voltage range at which this change takes place. The scale of a high-quality analog meter must, therefore, compensate for these deviations from linearity in order to maintain accuracy. (This is also true with regard to the scale of the analog thermometer.) The fact that the changes in thermoelectric voltage for the thermocouple tend to increase with temperature further complicates the problem of providing compensation in the incrementation of the meter scale.

As an example, consider the operation of the J-type thermocouple represented in Table 3–4. Between 0°C and 5°C, which represents the low end of the meter scale, the thermoelectric voltage would be changing at a rate of nearly 50 μV/°C:

$$\Delta V/°C = \frac{0.25 \text{ mV} - 0.00 \text{ mV}}{5°C} = 50 \; \mu V/°C$$

However, between 495°C and 500°C, which represents the high end of the meter scale, the rate of thermoelectric voltage change per °C will have increased to 56 μV and is determined as follows:

$$\Delta V/°C = \frac{27.39 \text{ mV} - 27.11 \text{ mV}}{5°C} = 56 \; \mu V/°C$$

As evidenced by the readout meters shown in Figure 3–30, the meter scales must compensate for the fact that a change in temperature at its upper end will cause a greater amount

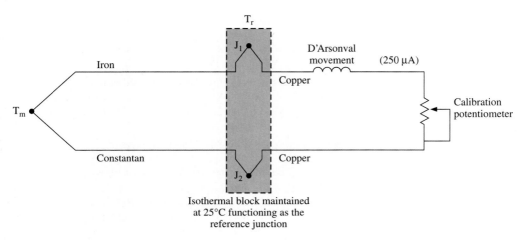

Figure 3–29
Analog thermometer with D'Arsonval meter movement

Figure 3-30
Omega Series 7000 Readout Meter (© Copyright 1992 Omega Engineering, Inc. All rights reserved. Reproduced with permission of Omega Engineering, Inc., Stamford, CT 06907)

of needle deflection than a temperature change at its lower end. However, once the nonlinearities of the thermocouple and the D'Arsonval movement are compensated for, a highly accurate temperature meter may be created. In addition to the fact that the temperature meter in Figure 3-29 requires no external power source, an advantage of this analog device is the fact that it is easily calibrated. This is accomplished through adjustment of a small trim pot. Also, as illustrated in Figure 3-30, both degrees Celsius and degrees Fahrenheit may be indicated simultaneously.

Example 3-10

Through examining the meter face in Figure 3-30, it becomes evident that, when the needle is at 150°C, it will be close to the 300°F mark. Determine if this would be an accurate indication of temperature in °F.

Solution:

$$°F = 9/5 \times °C + 32 = (1.8 \times 150°) + 32° = 302°F.$$

This result proves the temperature scale to be sufficiently accurate.

SECTION REVIEW 3-5

1. Define *thermoelectric voltage* as it applies to the operation of the thermocouple. What is another term used to designate this voltage?
2. True or false?: If a thermocouple is to operate properly, the voltages developed at its reference and measuring junctions must be series-aiding.
3. Describe the purpose of the reference junction of a thermocouple. What particular circuit design problem is associated with this junction?
4. What type of thermocouple junction is particularly resistive to oxidation? What materials may be used to form this type of junction? What is the practical range of a thermocouple made from these metals?
5. What materials are used to form the J-type thermocouple? What is the practical temperature range of this form of thermocouple?
6. True or false?: For a thermocouple, the difference in output voltage per °C becomes greater toward the upper end of its temperature range.
7. Explain how a temperature meter consisting of a thermocouple and a D'Arsonval movement is able to operate without the aid of a battery.

3-6   SOLID-STATE TEMPERATURE DETECTING AND SIGNAL CONDITIONING DEVICES

The thermocouple has many advantages over other forms of temperature transducers, including wide operating ranges, durability, and relatively low cost. The primary disadvantage of the thermocouple is the requirement of maintaining the reference junction at an absolutely constant temperature. Maintaining the necessary controlled environment for the reference junction may be quite difficult, possibly posing an even greater design problem than the physical placement of the measuring junction. Fortunately, through the utilization of specialized ICs, the task of holding the reference junction of the thermocouple at a constant temperature may be rendered unnecessary. A circuit containing such special-purpose ICs is contained in Figure 3–31.

The AD590 IC is a **temperature-sensitive current source,** for which changes in output current are directly proportional to changes in temperature. Note that in Figure 3–31 this device is placed in close thermal contact with the reference junction of the thermocouple. Thus, any fluctuations from the actual reference temperature that occur at the reference junction of the thermocouple will also be sensed by the AD590.

At the desired reference temperature of 0°C, an offset current of 273 μA will be flowing from the AD590. This is due to the fact that the AD590 is designed to operate in accordance with the **Kelvin** system. As defined by the **International System of Units**

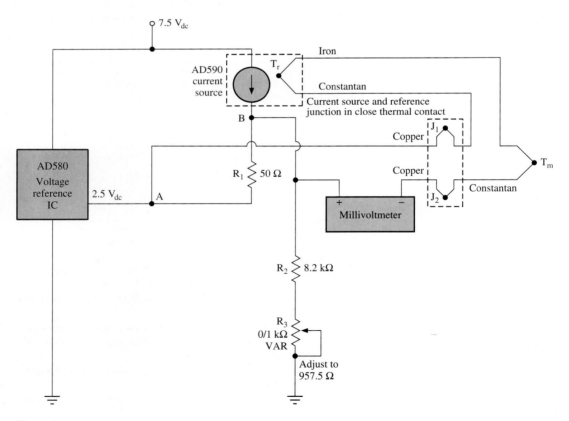

Figure 3–31
Thermocouple reference junction temperature compensation circuit that uses two ICs

(SI), an object has attained the absolute zero point, or 0 K, when it is completely devoid of thermal energy. Within the Kelvin scale, each additional unit of temperature actually represents a specific amount of molecular energy being acquired by the object. Even at the freezing point (either 0°C or 32°F) an object has considerable thermal energy, equivalent to 273.15 K. Therefore,

$$T\,(°C) = T\,(K) - 273.15° \qquad (3–15)$$

where $T\,(°C)$ represents temperature in degrees Celsius and $T\,(K)$ represents temperature in Kelvin. (Note that degrees are not used when indicating the Kelvin scale.)

Not until its body temperature approaches 0 K does the output current from the AD590 equal 0 A. As temperature begins to rise above the 0 K point, current from the AD590 will begin to increase at a rate of 1 μA/°C. According to Eq. (3–15), 0 K must be equivalent to nearly -273°C. Thus, when the body temperature of the AD590 approaches 0°C, its output current will already equal an offset value of nearly 273 μA, representing the difference between the 0 K point and the freezing point.

To compensate for this offset current, the AD580 **voltage reference IC** is connected as shown, providing a constant 2.5 V_{dc} at point A. Assuming that R_3 is adjusted to the value indicated, 0 V will be present over R_1 when the output current (I_{CS}) from the AD590 IC is at 273 μA. Thus, when the temperature at the reference junction of the thermocouple equals 0°C, V_{R_1} will equal 0 V, reflecting the **difference in thermoelectric voltage,** which should exist between the reference and measuring junctions at that temperature. Remember that, with the thermocouple, output voltage is assumed to equal 0 V at 0°C. Since the voltages developed over the reference and measuring junctions are series-opposing, they will be allowed to cancel each other only at the freezing point. Thus, for that condition to occur, V_{R_1} must equal 0 V at 0°C.

As the temperature increases, current from the AD590 also increases, at a rate of 1 μA/°C. With R_1 equaling nearly 50 Ω, its voltage will be virtually equal and opposite in polarity to the thermoelectric potential of the reference junction. R_1 will, thus, continue to compensate for these voltage increases, effectively holding the reference voltage at the value corresponding to the freezing point. This action allows the measuring junction to operate as if the reference junction were being held at a constant 0°C!

The operation of the circuit in Figure 3–31 should become clear with the following analysis, considering first the condition where temperature is 0°C and the current source is providing 273 μA. At this time, assuming virtually all of the current from the AD590 will be flowing through R_2 and R_3, the voltage that develops over this combined resistance would be determined as follows:

$$V_B = I_B \times (R_2 + R_3)$$
$$V_B = 273\ \mu A \times (8.2\ k\Omega + 957.5\ \Omega) = 2.5\ V$$

Thus, with both points A and B at 2.5 V, no compensation voltage will develop over R_1.

Now consider the condition where the temperature of the environment being monitored has increased to 10°C. Since the reference junction of the thermocouple is not being held at the freezing point, it must experience the same increase in voltage as the measuring junction. It is at this point that the function of R_1 becomes apparent. According to Table 3–4, between 0°C and 10°C, the output of the thermocouple should increase from 0 V to 0.5 mV. To hold the reference junction effectively at its freezing point potential, the voltage over R_1 should also increase to 0.5 mV. With R_1 equaling 50 Ω, this action will, in fact, occur, as illustrated in Figure 3–32(a).

At 10°C, the current-sourcing capability of the AD590 will increase to 283 μA. Thus, the voltage at point B must increase slightly. Since R_1 represents the path of least resistance for current leaving the AD590, it may be assumed that virtually all of the 10 μA represent-

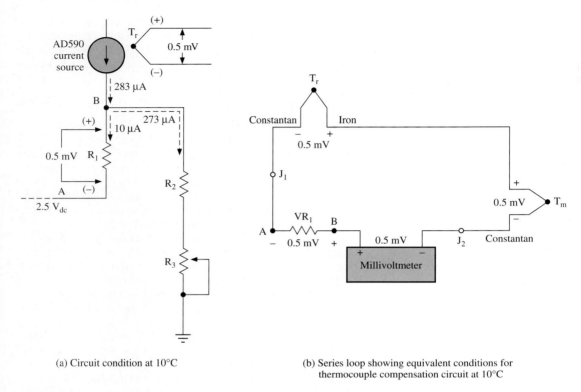

(a) Circuit condition at 10°C

(b) Series loop showing equivalent conditions for thermocouple compensation circuit at 10°C

Figure 3–32
Operation of thermocouple temperature compensation circuits

ing the 10°C increase in temperature will flow through this resistor. Meanwhile, the current flow through the combined resistance of R_2 and R_3 may be assumed to remain at its offset value of 273 μA. Based on these assumptions, the voltage developed over R_1 at 10°C may be determined as follows:

$$V_{R_1} = 10 \ \mu A \times 50 \ \Omega = 500 \ \mu V \text{ or } 0.5 \text{ mV}$$

Note that, according to Table 3–4, this value of V_{R_1} corresponds to the thermocouple voltage occurring at 10°C. As illustrated in Figure 3–32(b), since the voltages developed over the reference junction of the thermocouple and R_1 are series-opposing, the voltage between point A and the positive terminal of the T_r junction will be virtually 0 V, simulating a freezing-point condition. Thus the output of the thermocouple will be able to develop +0.5 mV, accurately representing 10°C.

Example 3–11

For the circuit in Figure 3–31, assume that the temperature at both the measuring and reference junctions of the thermocouple is 35°C. According to Table 3–4, what voltage should be present at the output of the thermocouple? What would be the value of current provided by the AD590 at that temperature? What would be the voltage developed over R_1 at this time? Does the absolute value of V_{R_1} nearly equal that specified for the output of the thermocouple?

Solution:
According to Table 3–4, at 35°C, V_{R_1} for the thermocouple should equal 1.8 mV. With current from the AD590 increasing at a rate of 1 μA/°C, I_{CS} would equal the sum of the offset current (representing 0°C) and the increase in current (representing 35°C). Thus, at 35°C,

$$I_{CS} = 273 \ \mu A + 35 \ \mu A = 308 \ \mu A$$

Still assuming virtually all of the current above the offset value to flow through R_1:

$$V_{R_1} = 35 \ \mu A \times 50 \ \Omega = 1.75 \ mV$$

This approximated value of V_{R_1} nearly equals the 1.8 mV obtained from Table 3–4.

Since the AD590 exhibits excellent linearity, it can be used in a variety of temperature sensing applications where it is desirable to convert a change in temperature directly to a proportional change in current. Other linear temperature sensing ICs are also available. An example of such a device would be the **LM335** temperature sensor, manufactured by National Semiconductor. This device, which is able to produce linear changes in voltage in response to changes in temperature, is shown in Figure 3–33.

As implied by the schematic representation of Figure 3–33(a), the LM335 operates in a manner similar to a Zener diode, having a dynamic resistance of less than 1 Ω and a current range of 400 μA to 5 mA. Thus, the output voltage of the device changes as a function of temperature rather than current, increasing at a rate of 10 mV with each temperature increment of 1 K. Since the LM335, like the AD590, is designed to operate in accordance with the Kelvin system, at 0 K, the output of the device will be 0 V. However, its practical range is typically −10°C to 100°C. An alternative device, the **LM135,** has a wider practical range, extending from −55°C to 150°C.

The connection of the LM335 shown in Figure 3–34 allows optimal operation of the device for thermometric applications. When, with a body temperature of 25°C, the output is adjusted to 2.982 V, the device will exhibit its greatest linearity. In fact, when calibrated

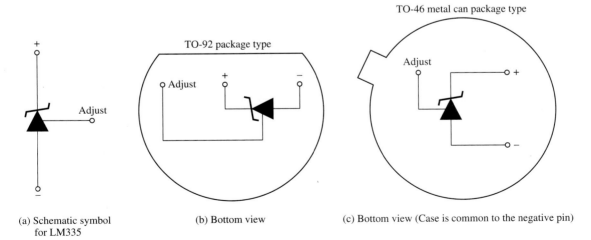

(a) Schematic symbol for LM335

(b) Bottom view

(c) Bottom view (Case is common to the negative pin)

Figure 3–33
LM335 solid-state temperature sensor

Figure 3–34
Calibration of LM335 for maximum linearity

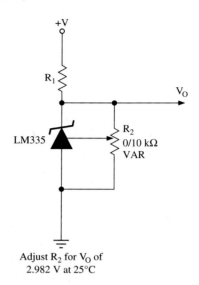

Adjust R_2 for V_O of
2.982 V at 25°C

in this manner, the LM335 will be able to sense temperature changes within a 100°C range with less than a 1% margin of error! Like a thermistor, the LM335 is susceptible to the self-heating effect, which represents a major source of error in thermometric applications. For this reason, the value of series current, as controlled by R_1 in Figure 3–34, should be kept at a minimal value. However, as with the thermistor, the self-heating property of the LM335 might be exploited in the detection and measurement of gas flow.

Example 3–12

For the circuit in Figure 3–35, determine the value of V_{in} when the body temperature of the LM335 attains 49°C. What ohmic setting of R_2 would establish a reference voltage equaling this value of V_{in}? Once this reference voltage is established, what would be the output condition of the comparator if the body temperature of the LM335 exceeds 49°C? What would this logic condition be at temperatures below 49°C?

Solution:
To solve for the value of V_{in} at 49°C: Assuming V_{in} equals 2.982 V at 25°C, then

$$\Delta T = 49°C - 25°C = 24°C$$
$$\Delta V = 10\,mV \times 24°C = 240\,mV$$
$$V_{in} = 2.982\,V + 240\,mV = 3.222\,V$$

Solving for the adjusted value of R_2:

$$\frac{3.222\,V}{5\,V} = \frac{R_2}{R_1 + R_2} = \frac{R_2}{10\,k\Omega + R_2}$$

Thus,

$$1.551831 = \frac{10\,k\Omega}{R_2} + 1 \quad \text{and} \quad \frac{10\,k\Omega}{R_2} = 0.551831$$

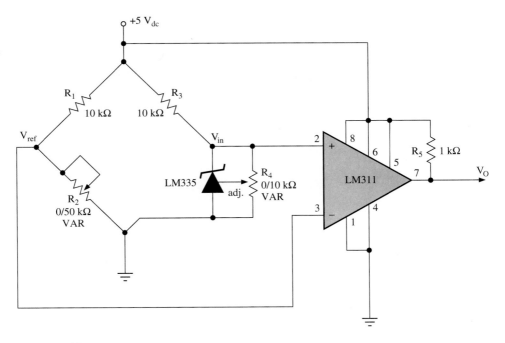

Figure 3–35

Finally,

$$\frac{R_2}{10 \text{ k}\Omega} = 1.81215 \quad \text{and} \quad R_2 = 18.1215 \text{ k}\Omega$$

With V_{in} being fed to the noninverting input of the LM311 comparator, the output of that device will toggle to a logic high when the body temperature of the LM335 exceeds 49°C. When the body temperature of the device falls below 49°C, the output of the comparator will switch to a logic low.

___

In addition to ICs, which actually contain temperature sensing elements, solid-state signal conditioning devices are also made available. Such special-purpose ICs are manufactured to develop the output signals from various forms of transducers. Figure 3–36 illustrates an application of the 2B31 IC, which is designed to function as a bridge amplifier for various forms of resistive transducers. In this example, an RTD with an R_o (freezing-point resistance) of 500 Ω is placed in a Wheatstone bridge, which provides the differential voltage input to the 2B31.

Note in Figure 3–36 that the positive side of the Wheatstone bridge is connected to pins 20 and 18 of the IC rather than directly to the external DC voltage source. The **+ Sense** terminal is capable of monitoring the excitation voltage being provided by the **+ V$_{exc}$** output. This action allows the bridge voltage source to be held at a constant +5 V, in spite of wide variations in the ohmic value of the RTD brought about by fluctuations in temperature. As a result, greater accuracy is achieved with respect to the differential input signal being developed at pins 12 and 15. The desired value of bridge

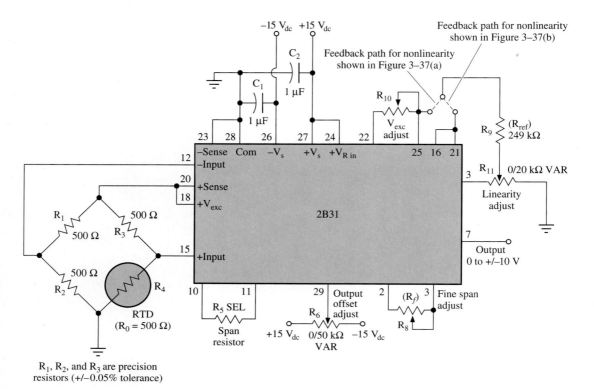

Figure 3–36
Operation of the 2B31 signal conditioning IC

excitation voltage may be made adjustable by connecting a potentiometer, as shown, between pins 22 and 25 of the IC. In addition to this voltage-regulating capability, the 2B31 provides a low-pass filtering action, greatly attenuating any noise present at the differential input. This filtering action requires the connection of capacitors C_1 and C_2 as shown. The internal instrumentation amplifier of the 2B31 also exhibits an excellent noise immunity, having a CMRR of 140 dB. External resistors R_5 and R_8 are used to control the voltage gain of the internal amplifiers of the IC, while potentiometer R_6 allows large changes in the dc offset of the output signal (between +10 V and −10 V).

Another important feature of the 2B31 signal conditioning IC is its ability to linearize, as well as amplify, the output signal. Figure 3–37 contains two typical input voltage curves that could be produced by a transducer placed within the external bridge circuit. Through the introduction of a small amount of negative feedback from the output of the internal instrumentation amplifier, the nonlinearities exhibited by these two curves might be easily corrected. The amount of this feedback is controlled through the adjustment of R_{11}. As shown in Figure 3–36, if the input response curve of the transducer resembles that contained in Figure 3–37(a), the feedback path should be connected to pin 25. If, however, the response curve of the input transducer is like that shown in Figure 3–37(b), the feedback path should be terminated at pins 16 and 21.

In addition to such multipurpose signal conditioning ICs as the 2B31, highly special-ized solid-state devices are also made available. They are designed to amplify and linearize

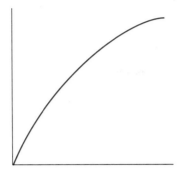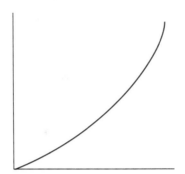

Figure 3–37
Typical transducer nonlinearities

the output signal of one particular form of transducer. Examples of this type of IC are the AD594 and the AD595, which function as thermocouple amplifiers. The AD594 is specially designed to be used in conjunction with a J-type thermocouple, whereas the AD595 is calibrated to work with the K-type device. The internal architecture of these two ICs, as well as their external connection to a thermocouple, is shown in Figure 3–38. These ICs contain both an instrumentation amplifier and the compensation circuitry necessary to counteract temperature changes at the reference junctions of the thermocouple. As indicated by the block diagram of the circuitry within the IC, the signals from the input differential amplifier and the ice point compensation circuitry are algebraically summed. The resultant signal is then fed to an additional amplifier stage. A valuable troubleshooting aid is provided by the output circuitry of the IC in that an overload condition resulting from a defective thermocouple junction may be sensed by level-detection circuitry. If either the reference or measuring junction is opened, this condition is immediately sensed by the internal circuitry of the IC. A high signal then occurs at the base of the output transistor, the open collector of which is accessed at pin 12. In Figure 3–38, this internal transistor functions as a current sink for an LED in series with a limiting resistor. Thus, during a fault condition for the thermocouple, the LED will be illuminated.

When using either of these thermocouple amplifiers, the thermocouple leads must be carefully soldered to pins 1 and 14 as shown in Figure 3–38. A special type of solder must be used for this purpose, consisting of one of the following three combinations of materials:

1. 95% tin and 5% antimony
2. 95% tin and 5% silver
3. 90% tin and 10% lead.

Also, for reliable operation, care must be taken to select a noncorrosive core material. (Note that the solder typically used for electronic circuits is 60% tin and 40% lead.) An advantage of this method of connection lies in the fact that the solder connections made at pins 1 and 14 effectively become the reference junctions for the thermocouple. Also, another design advantage of this IC is the fact that the input signal from the thermocouple is linearized, resulting in output voltage changes that are directly proportional to fluctuations in temperature. Whereas the thermocouple voltage changes are likely to be only a few microvolts per °C, the output of either the AD594 or AD595 will vary at a nearly linear rate of approximately 10 mV/°C.

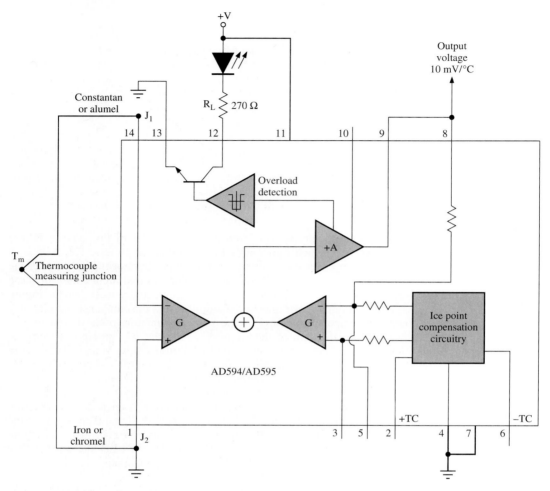

Figure 3–38
AD594/AD595 thermocouple amplifier

SECTION REVIEW 3–6

1. Briefly explain the function of the AD590 IC in Figure 3–31.
2. For the circuit in Figure 3–31, if the temperature being detected by the thermocouple is 29°C, what should be the value of I_{R_1}? What should be the value of V_{R_1}?
3. Describe the difference between 0°C and the 0 K point.
4. A temperature of 327 K is equivalent to what temperature in °C?
5. Ideally, what current would be flowing from the AD590 IC at a temperature of 0 K?
6. For the comparator circuit in Figure 3–35, assume the wiper of R_4 is still adjusted for a V_o of 2.982 V at 25°C. What must be the ohmic setting of R_2 in order to cause the output of the LM311 comparator to toggle as temperature fluctuates above or below a threshold level of 96.8°F?
7. Describe a valuable troubleshooting aid provided by the AD594 and the AD595 ICs.

SUMMARY

Transducers are devices capable of translating one form of energy to another. Temperature transducers include thermistors, RTDs, and thermocouples. Thermistors and RTDs both function as variable resistances in response to changes in temperature. Thermistors are typically characterized by a negative temperature coefficient, whereas RTDs, which are metallic devices, exhibit a positive temperature coefficient.

Although RTDs are less sensitive than thermistors with regard to changes in temperature, they do tend to be more linear and, consequently, may be used over a wider range of temperatures for signal conditioning applications. Thermistors, however, typically have higher values of impedance than RTDs. For this reason, lead resistance becomes a less significant factor in the design of thermistor instrumentation circuitry, especially where the temperature transducer must be placed in a remote location. While small, bead-type thermistors are well suited for monitoring temperature changes at a specific point, RTDs are better suited for monitoring temperature changes within a large area. When serving as input transducers for signal conditioning purposes, both the thermistor and the RTD are often placed in voltage divider or Wheatstone bridge configurations. In addition to their wide use within temperature control systems, industrial applications of thermistors also include switching delay, overcurrent protection, and gas flow monitoring.

Thermocouples are temperature sensitive devices capable of converting changes in temperature directly to small, but highly predictable, changes in voltage. These devices operate on the principle that a small electromotive force is generated across a junction formed by two different forms of metal. This thermoelectric (or Seebeck) voltage tends to become greater with increasing temperature. To function properly, a thermocouple must consist of both a measuring junction and a reference junction. Ideally, the reference junction of the thermocouple should be kept at the freezing point, allowing the output of the thermocouple to be at 0 V when both the reference and measuring junctions of the thermocouple are at 0°C. In an industrial environment, however, attempting to hold the reference junction at the freezing point may not be practical. For this reason, compensation circuitry is often used to simulate a constant freezing-point condition at the reference junction.

Specialized analog ICs have also been developed for the purpose of sensing changes in temperature. For such devices, fluctuations in temperature are translated into directly proportional changes in current or voltage. Also, specialized solid-state devices have been designed for the purpose of amplifying and linearizing the output signals from various temperature transducers. One type of IC serves as a general-purpose bridge amplifier, while another functions strictly as a thermocouple amplifier. With their linearizing capability, these ICs greatly expedite the design of signal conditioning circuitry.

SELF-TEST

1. Match the device listed in column A with the pertinent information listed in column B.

Column A	Column B
1. thermocouple	a. amplifier used with K-type device
2. 2B31	b. −10°C to 100°C range
3. AD580	c. Seebeck voltage
4. thermistor	d. −55°C to 150°C range
5. reed switch	e. amplifier for J-type device

6. AD590   f. provides a voltage reference
7. LM335   g. transducer signal conditioning device
8. AD595   h. normally has an NTC
9. LM135   i. Curie temperature
10. AD594   j. temperature-sensitive current source

2. Which three devices listed in column A of Problem 1 are calibrated to operate according to the Kelvin system?
3. For which device in column A of Problem 1 should output current equal 273 μA at 0°C?
4. What type of thermal transducer is characterized by a positive temperature coefficient? Name a material that is highly preferred for the manufacture of this device.
5. Assume that a thermistor with a time constant of 8.6 sec is subjected to an instantaneous environmental temperature increase of 14°C. If the original body temperature of the device was 27°C, what would be the temperature 14 sec past the time of the step increase?
6. True or false?: Increasing the dissipation constant of a thermistor will result in a decrease in its thermal time constant.
7. True or false?: RTDs typically operate at higher ohmic ratings than thermistors.
8. At 32°F, the resistance of an RTD is 500 Ω. At 117°F, its resistance has increased to 602 Ω. What is the temperature coefficient for this device?
9. Define *sintering*. What thermal transducer is manufactured by means of this process?
10. Explain a major design challenge involving the reference junction of the thermocouple. Briefly describe a practical solution to this problem.
11. A gas flowmeter utilizing two thermistors is likely to exploit the
    a. resistance-temperature characteristic of the measuring thermistor

Figure 3–39

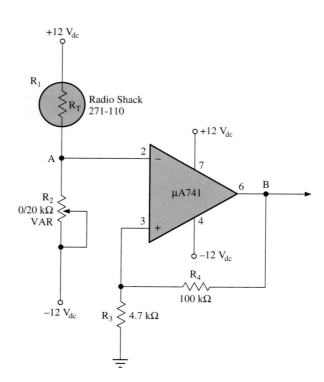

    b. resistance-temperature characteristic of the reference thermistor
    c. voltage-current characteristic of the measuring thermistor
    d. characteristics identified in all of the above choices
12. The self-heating property of a thermistor may be exploited in
    a. gas-flow measurement systems
    b. accurate temperature measurement systems
    c. delay circuitry for electromechanical relay switching
    d. those applications identified in choices a and c
13. True or false?: When a thermistor is used to measure gas flow, the parameter of the device that is actually being varied is its dissipation constant.
14. An RTD is to be placed in a Wheatstone bridge and used for the purpose of accurately measuring temperature. If the RTD must be located several feet away from the rest of the instrumentation circuitry, what design consideration becomes important? What action should be taken to ensure accurate operation of the circuit? Why might this consideration be less critical if a thermistor were used instead of the RTD?

Table 3–5
Temperature versus resistance for a Radio Shack 271-110 NTC thermistor

Temperature (°C)	Resistance (Ω)
−50	329.2 kΩ
−45	247.5 kΩ
−40	188.4 kΩ
−35	144 kΩ
−30	111.3 kΩ
−25	86.39 kΩ
−20	67.74 kΩ
−15	53.39 kΩ
−10	42.25 kΩ
−5	33.89 kΩ
0	27.28 kΩ
5	22.05 kΩ
10	17.96 kΩ
15	14.68 kΩ
20	12.09 kΩ
25	10.00 kΩ
30	8.313 kΩ
35	6.941 kΩ
40	5.828 kΩ
45	4.912 kΩ
50	4.161 kΩ
55	3.537 kΩ
60	3.021 kΩ
65	2.589 kΩ
70	2.229 kΩ
75	1.924 kΩ
80	1.669 kΩ
85	1.451 kΩ
90	1.366 kΩ
95	1.108 kΩ
100	973.5 kΩ
105	856.5 kΩ
110	757.9 kΩ

15. Identify the op amp configuration being used in Figure 3–39. What type of feedback is occurring in this example? Assuming the internal circuitry of the op amp drops nearly 2 V, what would be the minimum and maximum levels of output voltage for the circuit?

16. Assuming that point A in Figure 3–39 is to be at 0 V when the body temperature of the thermistor is 176°F, what would be the necessary ohmic setting of R_2? (Refer to Table 3–5 for information on the thermistor being used.)

17. For the circuit in Figure 3–39, what is the hysteresis gap with respect to input voltage? What are the upper and lower trigger points? What temperature does each of these voltages represent? What is the hysteresis gap in terms of °C?

18. What valuable troubleshooting aid is provided with the AD594 and AD595 IC's? What external components could be connected to either of these ICs to exploit this capability?

19. For the circuit in Figure 3–40, what must be the adjusted voltage at point B in order to obtain an output of 0 V while the thermistor body temperature is 0°C? (Refer to Table 3–5 for information about the thermistor.)

20. For the circuit in Figure 3–40, assume that point B has already been adjusted for a 0-V output at 0°C. If the ohmic value of R_9 is adjusted to 4.75 kΩ, what would be the output voltage at 25°C?

21. Assume that the 0 V/0°C adjustment is being maintained in Figure 3–40. What must be the ohmic setting of R_9 in order to have an output of 4.3 V at 100°C?

22. Explain the purpose of op amp 1458a in Figure 3–40. What op amp configuration does it represent? What is its ideal voltage gain?

23. What op amp configuration is represented by op amp 1458b in Figure 3–40? What is its voltage gain?

24. Describe the difference between 0°C and the 0 K point. Which point represents the total absence of thermal energy?

25. A temperature of 42°F is equivalent to what temperature on the Kelvin scale?

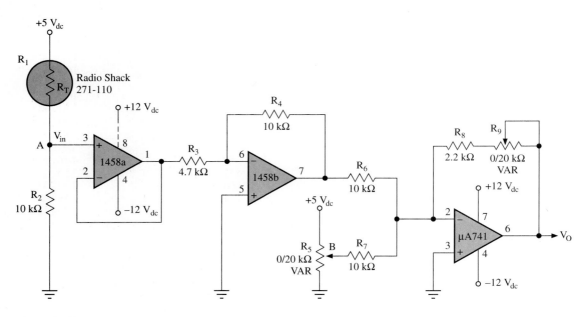

Figure 3–40

4 OTHER TRANSDUCERS AND SIGNAL CONDITIONING CIRCUITRY

LEARNING OBJECTIVES

On completion of this chapter, the student should be able to:

- Describe the operation of a displacement potentiometer. Explain how such a device might be used in conjunction with industrial instrumentation circuitry.
- Explain the function of a linear variable differential transformer. Describe how this device could be used in conjunction with signal conditioning circuitry.
- Identify two types of magnetic proximity sensors commonly used within modern industrial systems.
- Describe the operation of capacitive level-detection transducers. Explain methods by which these devices might be connected with signal conditioning circuitry.
- Describe the operation of ultrasonic level-sensing transducers. Explain how these devices might be utilized to monitor the level of both stationary and moving materials.
- Explain the operation of the Bourdon tube and the bellows as pressure-sensing transducers. Describe how such devices might be used in conjunction with displacement transducers in order to convert changes in pressure to electrical signals.
- Explain how a bellows might be used in conjunction with a linear variable differential transformer to measure gas flow.
- Explain the operation of turbine, nutating disk, and magnetic flow measurement devices.
- Describe the operation of humidity-sensing transducers, including the psychrometer, hygrometer, and resistive humidity sensor.
- Describe the structure and operation of strain gauges. Explain how these devices might be used to measure tensile and compressive stress. Explain how the strain gauge might be used to measure pressure indirectly.

- Describe the structure and operation of piezoresistive pressure-sensing transducers. Explain advantages these devices might have over strain gauges.
- Explain the structure and operation of piezoelectric force-sensing transducers.
- Describe the structure and operation of various forms of load cells. Explain how these devices might be used in the measurement of force, weight, and flow.

INTRODUCTION

In the previous chapter, thermal transducers were examined in detail, along with a few basic signal conditioning circuits that utilize op amps. In this chapter, several more forms of transducers are introduced as we continue our study of signal conditioning circuitry. The various transducer functions studied in this chapter include position sensing, proximity detection, level measurement, pressure sensing, flow measurement, humidity detection, and the measurement of force and weight.

The two forms of position-sensing transducers we introduce in this chapter are the potentiometer and the linear variable differential transformer. These devices are frequently used in conjunction with some form of mechanical linkage in order to detect the position of a moving machine part. Also, these two devices may be used along with other transducers in such applications as temperature and pressure measurement.

The Hall effect device, which is commonly used as a position sensor, was introduced in Chapter 1. Two types of proximity sensors, which also operate on magnetic principles, are studied in this chapter. These devices are commonly used in conveyance systems in order to detect possible jamming conditions.

During industrial process control, we often need to monitor the level of fluids being stored in sealed vats or open tanks. While the simple float switch, which was introduced in Chapter 1, might be actuated when fluid attains a maximum level, capacitive level detectors may be used to monitor actively the fluid level within a container. These devices, along with their associated signal conditioning circuitry, are examined in detail here. Also, the operation of the ultrasonic level detector is discussed.

Pressure, flow, and humidity sensing transducers serve important roles in industrial process control. The operation of a few of these devices will be briefly introduced. Also, methods of using force-summing devices as the Bourdon tube and the bellows in conjunction with displacement transducers is explained.

The operation of the strain gauge is then examined. The special problems involved with these relatively low-amplitude transducers are examined in detail. Also, strain gauge applications, including the indirect measurement of pressure, are explained. In addition, the operation of piezoresistive pressure-sensing and piezoelectric force-sensing transducers is introduced.

Finally, the construction and applications of various forms of load cells are investigated. These versatile devices may be used to measure the weight of stationary material and to monitor the flow rate of material along conveyance systems. Also, the possible use of load cells to measure indirectly the level of material stored in large containers is introduced.

4–1   DISPLACEMENT POTENTIOMETERS

Most electronics technicians are already familiar with the operation of potentiometers, since a vast amount of the routine maintenance performed on electronic control circuitry requires the adjustment of trim pots. However, not all potentiometers used in industrial control systems are manually adjusted. The fact that a potentiometer is actually a variable voltage divider may be exploited for the purpose of position sensing. For example, mechanical linkage could be provided between a moving machine part and the wiper of a linear potentiometer. As illustrated in Figure 4–1, movement of the shaft would result in a corresponding motion of the wiper along the body of the resistor. Thus, with the output being taken at the wiper, changes in the position of the shaft may be translated to changes in voltage. This output signal could be developed from point B to point A or from point B to point C.

The mechanical linkage between the shaft and the potentiometer can be designed such that a given amount of travel by the shaft will result in a proportional amount of wiper movement. As an example, assume the maximum amount of travel for the shaft in Figure 4–1 is 5 in. Given this condition, 1 in. of shaft motion would represent 20% of its total travel. Thus, for linear operation, this motion should result in a 20% change in resistance, which would be measurable either from point B to point A or from point B to point C.

Now assume the maximum resistance of the potentiometer in Figure 4–1 is 150 Ω (measured from point A to point C). Also assume that both the shaft and the wiper are in their midpoint positions. With these initial conditions, the shaft could move as much as 2.5 in. in either direction. Ideally, the resistance from point A to point B would be the same as that measured from point B to point C (equaling 75 Ω). If the shaft moves a distance of 1 in. to the right, representing 20% displacement, the wiper will also move to the right, causing the resistance between points A and B to increase by a factor of 20%. This increased ohmic value would be determined as follows:

$$\Delta R = 20\% \text{ of } R_T = 0.2 \times 150 \ \Omega = 30 \ \Omega$$

With the wiper originally in its midpoint position:

$$R_{A-B} = R_{T/2} + (0.2 \times R_T) = 75 \ \Omega + 30 \ \Omega = 105 \ \Omega$$

Ideally, these changes in resistance from point A to point B would be perfectly proportional to variations in shaft position. This ideal linear relationship is represented in Figure 4–2(a). Real-world displacement potentiometers are not perfectly linear, as evidenced by the graph in Figure 4–2(b). Manufacturers of displacement potentiometers usually specify a worst-case **percentage of deviation** for a given device, with 1% being a typical value. As an example, for the 150-Ω potentiometer, 20% shaft displacement could

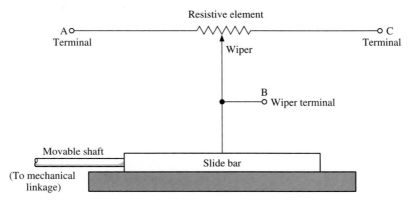

Figure 4–1
Basic linear displacement potentiometer

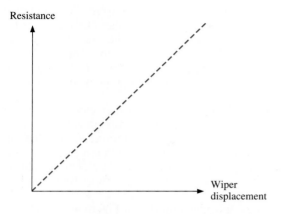

(a) Ideal potentiometer response to wiper displacement

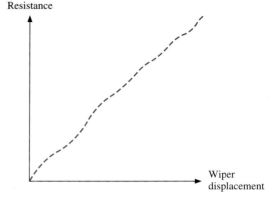

(b) Actual potentiometer response to wiper displacement

Figure 4–2

conceivably result in as much as 21% or as little as 19% change in resistance. Thus, allowing for worst-case operation of this potentiometer, minimum and maximum values of R_{A-B} would be determined as follows:

$$R_{A-B}(\text{min}) = (50\% + 19\%) \times R_T = 0.69 \times 150 \ \Omega = 103.5 \ \Omega$$
$$R_{A-B}(\text{max}) = (50\% + 21\%) \times R_T = 0.71 \times 150 \ \Omega = 106.5 \ \Omega$$

(Note that with an R_T of 150 Ω, deviation of 1% would be equivalent to 1.5 Ω.)

Another important parameter for displacement potentiometers is that of resolution. Ideally, regardless of how slight the displacement of a potentiometer wiper might be, a corresponding change would occur in the value of resistance measured from the wiper to either terminal of the device. However, if the potentiometer is wire-wound, the resolution of the device is limited by its number of turns. For example, assume the potentiometer contained in Figure 4–1 consists of 200 turns and has an R_{A-C} value of 150 Ω. The resolution of this potentiometer would then be determined as follows:

$$\Delta R = R_{A-C}/N \tag{4–1}$$

where N represents the number of turns for a wire-wound potentiometer. Thus, with the values given earlier:

$$\Delta R = 150 \ \Omega/200 \text{ turns} = 0.75 \ \Omega/\text{turn}$$

This result would indicate that, as the wiper moves along the body of the potentiometer, the smallest changes in the value of R_{A-B} and R_{B-C} are likely to occur in increments of 0.75 Ω.

When current is flowing through the potentiometer, as illustrated in Figure 4–3, a voltage-divider relationship will exist for points A, B, and C. Because R_{A-C} is not affected by the movement of the wiper, V_{A-C} will remain constant. However, if the wiper is moved toward point C, V_{A-B} will increase as V_{B-C} decreases. (Remember that, according to Kirchhoff's voltage law, at any instant, the algebraic sum of V_{A-B}, V_{B-C}, and V_{A-C} must be zero. Thus:

$$V_{A-B} + V_{B-C} = V_{A-C} \tag{4–2}$$

Also, applying the voltage divider theorem derived from Kirchhoff's law:

$$V_{A-B} = V_{A-C} \times \frac{R_{A-B}}{R_{A-C}} \qquad (4-3)$$

and

$$V_{B-C} = V_{A-C} \times \frac{R_{B-C}}{R_{A-C}} \qquad (4-4)$$

Therefore,

$$\frac{\Delta \text{Wiper travel}}{\text{Maximum wiper travel}} = \frac{\Delta R_{A-B}}{R_{A-C}} = \frac{\Delta V_{A-B}}{V_{A-C}} \qquad (4-5)$$

and

$$\frac{\Delta \text{Wiper travel}}{\text{Maximum wiper travel}} = \frac{\Delta R_{B-C}}{R_{A-C}} = \frac{\Delta V_{B-C}}{V_{A-C}} \qquad (4-6)$$

Thus, for a linear potentiometer, the following ideal relationships exist:

$$\Delta R_{A-B} \propto V_{A-B} \quad \text{and} \quad \Delta R_{B-C} \propto V_{B-C} \qquad (4-7)$$

This relationship of voltage and resistance may be proven through the following example.

Example 4–1

Assume for the potentiometer in Figure 4–3 that R_{A-C} equals 150 Ω and that the wiper is in the center position. Also assume that V_{A-C} equals 12 V_{dc}. Assuming perfectly linear

Figure 4–3
Potentiometer in a basic voltage-divider configuration

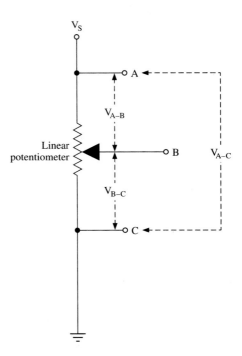

operation of the potentiometer, solve for the values of V_{A-B} and V_{B-C} with the wiper in the center position. Next, assume the wiper moves downward such that R_{B-C} equals 42 Ω. Solve for the resultant values of V_{A-B} and V_{B-C}. Does the percentage of change in R_{B-C} equal the percentage of change in V_{B-C}?

Solution:
Solving for V_{A-B} and V_{B-C} with the wiper centered:

$$R_{A-B} \text{ and } R_{B-C} \text{ both equal } 1/2R_T = 0.5 \times 150\ \Omega = 75\ \Omega$$

Thus,

$$V_{A-B} = V_{B-C} = 1/2V_{A-C} = 0.5 \times 12\ V = 6\ V$$

Applying the voltage divider theorem to obtain the new value of V_{B-C}:

$$V_{B-C} = V_{A-C} \times \frac{R_{B-C}}{R_{A-C}} = 12\ V \times \frac{42\ \Omega}{150\ \Omega} = 3.36\ V$$

Proving linear operation:

$$\Delta R_{B-C} = 75\ \Omega - 42\ \Omega = 33\ \Omega$$

Expressed as a percentage of change:

$$\frac{33\ \Omega}{150\ \Omega} \times 100\% = 22\%$$

$$\Delta V_{B-C} = 6\ V - 3.36\ V = 2.64\ V$$

Expressed as a percentage of change:

$$\frac{2.64\ V}{12\ V} \times 100\% = 22\%$$

Note that the percentages of change in resistance and voltage are exactly the same.

In determining the operation of a wire-wound displacement potentiometer, the resolution of the device is an important consideration, due to the fact that the resolution of the output signal (expressed as ΔV_{A-C}) depends on the potentiometer's number of turns. For example, assume a potentiometer with 200 turns has a V_{A-C} of 12 V_{dc}. The resolution of the output signal would then be determined as follows:

$$\Delta V_{A-C} = V_{A-C}/N \tag{4–8}$$

Thus, for the given value of N and V_{A-C}:

$$\Delta V_{A-C} = 12\ V/200\ \text{turns} = 0.06\ V/\text{turn} = 60\ mV/\text{turn}$$

This result would indicate that, regardless of how slight a change in the position of the wiper might be, the smallest amount of change in the output signal would be 60 mV. The resolution of the potentiometer thus represents a significant source of error, especially in applications where accurate detection of displacement is critical. Referring back to Example 4–1, we determined that a wiper displacement of 22% should, ideally, produce an output voltage change of 22%. However, due to the resolution limitation of the potentiometer, the actual percentage of voltage change could vary significantly from the ideal value. As an example, assume again that a potentiometer consists of 200 turns and has an R_{A-C} of

150 Ω. As was determined earlier, ΔR_{A-C} would equal 0.75 Ω and ΔV_{A-C} would be 60 mV. In either case, the possible deviation from an ideal output signal could be expressed as a percentage, which is determined as follows:

$$\% \text{ deviation} = \frac{\Delta R_{A-C}}{R_{A-C}} \times 100\%$$

$$= \frac{0.75\ \Omega}{150\ \Omega} \times 100\% = 0.5\%$$

(4–9)

Also

$$\% \text{ deviation} = \frac{\Delta V_{A-C}}{V_{A-C}} \times 100\%$$

$$= \frac{60\ \text{mV}}{12\ \text{V}} \times 100\% = 0.5\%$$

(4–10)

If the number of turns for a potentiometer is known, the percentage of deviation associated with the resolution may be determined simply as follows:

$$\% \text{ deviation} = 1/N \times 100\%$$

(4–11)

Thus,

$$\% \text{ deviation} = 1/200 \text{ turns} \times 100\% = 0.5\%$$

These calculations indicate that, if the wiper of the 200-turn potentiometer is moved a distance equivalent to 22% of its travel, the resultant change in output voltage could be as great as 22.5% or as little as 21.5%. (Note that this percentage of deviation accounts only for the resolution problem. Additional deviation could also occur due to nonlinearity.) With these given error factors, the actual output voltage in Example 4–1 could vary between 2.58 and 2.7 V.

Figure 4–4
Connection of a potentiometer in a
Wheatstone bridge

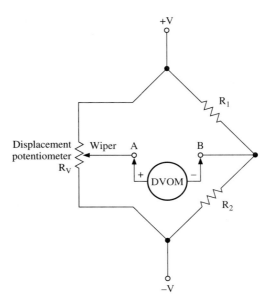

As shown in Figure 4–4, a displacement potentiometer may be connected in a Wheatstone bridge configuration. Here, if R_1 and R_2 are of equal ohmic value, a null condition exists across the bridge when the wiper of the potentiometer is in the center position. This form of resistive network is particularly useful in applications where the central position of the wiper represents the home position for a moving machine part. Movement of the wiper from the center position toward the +V side would result in a positive voltage reading from point A to point B. Wiper movement in a downward direction from the center position would result in a negative voltage reading from point A to point B. The amplitude of the error signal taken at point A would be proportional to the distance by which the wiper is displaced from its central position.

Figure 4–5 illustrates how a displacement potentiometer might be utilized within a temperature monitoring system. Such a **resistance thermometer** bridge circuit may be used where precise temperature measurement is required. Due to its high linearity, a platinum RTD would become the temperature-sensing element rather than a thermistor. Through the action of the displacement potentiometer, the circuit in Figure 4–5 has a self-nulling capability. Note that the sliding wiper of the potentiometer actually becomes the indicator for the thermometer scale.

Consider first the condition where the environmental temperature is raised above the set-point level. Since the RTD has a positive temperature coefficient, point B will attempt to become more negative than point A. However, a self-balancing action will automatically

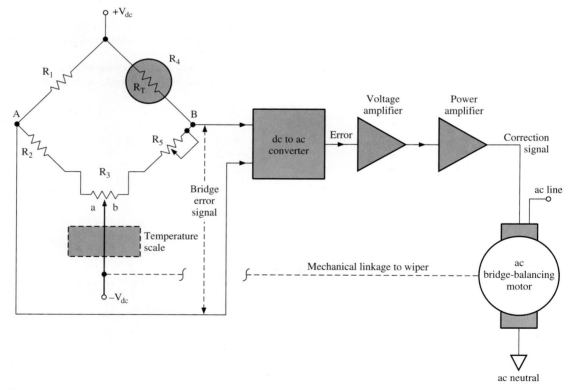

Figure 4–5
Functional diagram of a resistance thermometer

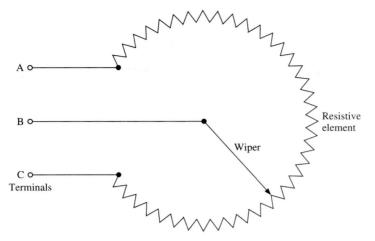

Figure 4–6
Schematic representation of rotary potentiometer

be initiated, bringing the bridge back into a null condition. This compensating action begins as the error signal starts to develop across the bridge. This error signal is fed through a dc-to-ac converter, voltage amplifier, and power amplifier in order to drive an ac motor. Through mechanical linkage, the wiper will be moved in the right-hand direction, causing R_{3a} to decrease as R_{3b} increases. When the ohmic value of $R_1 \times (R_{3b} + R_5)$ equals $R_4 \times (R_2 + R_{3a})$, the null condition will be restored. As a result, the drive signal for the motor will approach 0 V, causing the motor to stop running and the wiper of R_3 to stop moving. The meter scale is placed such that this left-hand motion of the wiper results in an increased temperature indication on the scale.

Next, assume that the environmental temperature has decreased, such that point B attempts to become more positive with respect to point A. To maintain a null condition across the Wheatstone bridge, the wiper must now be moved in the right-hand direction. For this action to take place, the motor rotation must be reversed. Since the error signal now has reversed polarity, this reversed motor rotation will, in fact, occur. As the wiper moves to the right, the ohmic value of R_{3a} will be increased while that of R_{3b} is decreased. This action will continue until a null condition is again restored across the bridge.

For some industrial processes, we want to maintain a record of changing temperature through time. Although a computer system could easily be used to collect and provide a listing of such data, the device in Figure 4–5 could also be modified to perform such a task. A strip-chart pen could be mechanically linked to the wiper of the potentiometer, allowing fluctuations in temperature through time to be graphically represented.

We should also mention that displacement potentiometers are manufactured in rotary form, as illustrated in Figure 4–6. Such devices are often used to monitor the angular displacement of a machine part.

SECTION REVIEW 4–1

1. Assume that R_{A-C} for a displacement potentiometer is 500 Ω, and its maximum wiper travel is 2 in. Given this information, a wiper displacement of 7/16 in. should result in a resistance change of how many ohms between the wiper and terminal A or C?
2. For the potentiometer in Review Problem 1, assume there are 235 turns. What would

be the minimum amount of resistance change that could result from movement of the wiper? If V_{A-C} is 15 V_{dc}, what would be the minimum value of ΔV_{A-B} or ΔV_{B-C}? What would be the percentage of deviation for this potentiometer?

3. True or false?: The percentage of nonlinearity that is considered acceptable for most displacement potentiometers used in industry is $\pm 10\%$.

4. Assume that a wire-wound potentiometer consisting of 215 turns has 24 V_{dc} applied across its main terminals. Also assume the device is 3⅝ in. long and has an R_{A-C} of 100 Ω. If the wiper moves 1/4 in. from the center position, what would be the ideal change in voltage measured between the wiper and either main terminal? Based on the number of turns specified for the device, determine the minimum values of ΔR_{A-C} and ΔV_{A-C} and the percentage of deviation.

5. For the Wheatstone bridge in Figure 4–7, assume the RTD resistance is 327 Ω. What must be the ohmic values of R_2 and R_{3a} for a null condition to occur across the bridge? What voltage would then be developed over both R_{3b} and the RTD?

6. Assume the RTD in Figure 4–7 has a positive temperature coefficient. If the body temperature of the RTD is decreased, in what direction must the wiper of R_3 be moved in order to restore a null condition across the bridge?

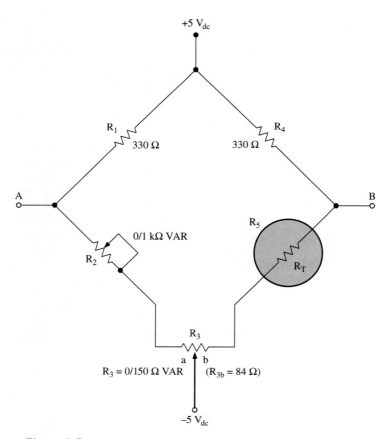

Figure 4–7

4–2   LINEAR VARIABLE DIFFERENTIAL TRANSFORMERS

While the **linear variable differential transformer** (LVDT) is a more costly form of displacement transducer than the potentiometer, it does have greater linearity and therefore is a more accurate sensing device. The LVDT is especially suited for use in ac control circuitry. As illustrated in Figure 4–8, the LVDT is basically a transformer consisting of a single primary and two secondary windings. These three windings are placed on a common cylinder, which is made of a nonferrous material such as plastic. The two secondary windings, which have the same number of turns, are wound in such a manner that the voltage developed from point A to point B will be 180° out of phase with the voltage developed from point E to point F. Thus, if the peak-to-peak amplitude of the output sine wave developed between points A and B equals that of the signal developed from point E to point F, the differential output signal taken between points A and F would remain at 0 V.

The LVDT contains a movable core, usually made of a ferrous material such as iron. With some forms of LVDT, this ferrous plunger may be moved manually by an adjustment screw. However, in applications where the LVDT is being used to detect shaft displacement, the plunger is allowed to be free moving. Through mechanical linkage, the movement of the plunger would be made proportional to the displacement of the shaft.

To achieve a condition where V_{A-B} equals V_{E-F} and V_{A-F} equals 0 V, the movable core must first be placed in its center position. This condition for the LVDT and the resultant secondary and output waveforms are represented in Figure 4–9(a). As illustrated in Figure 4–9(b), movement of the plunger toward the right will cause V_{E-F} to become greater as V_{A-B} decreases. Thus, the resultant waveform for V_{A-F} will appear as shown. Figure 4–9(c) represents the opposite effect, which occurs as the plunger is moved toward the left.

Because the two secondary voltages for the LVDT are 180° out of phase, the peak amplitude of the differential output signal may be determined simply as follows:

$$V_{A-F}(\text{peak}) = V_{E-F}(\text{peak}) - V_{A-B}(\text{peak}) \qquad (4\text{–}12)$$

Figure 4–10 illustrates a possible application of an LVDT within a gas flow measurement system. Note that the function of the LVDT and bellows mechanism is identical to

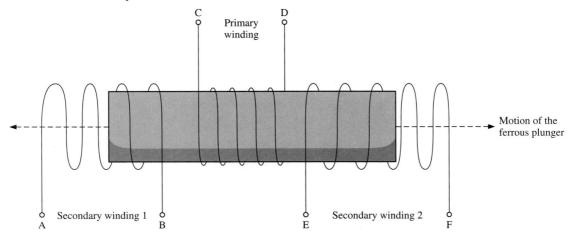

Figure 4–8
Structure of a basic LVDT

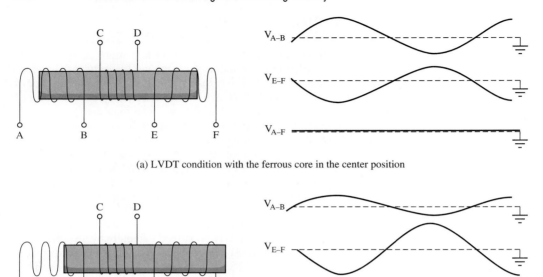

(a) LVDT condition with the ferrous core in the center position

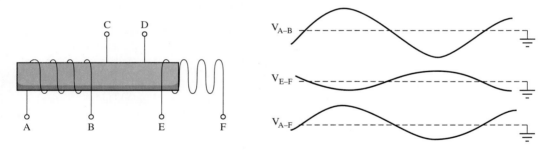

(b) LVDT condition with the plunger moved to the right ($V_{E-F} > V_{A-B}$)

(c) LVDT condition with the plunger moved to the left ($V_{A-B} > V_{E-F}$)

Figure 4–9

that of the thermistor circuitry shown in Figure 3–22. Whereas the thermistor-based system relies on the varying dissipation constant of a self-heated thermistor for sensing changes in the rate of convection, the system in Figure 4–10 accomplishes this task by detecting differences in air pressure at two different points within a pipe.

The flow measurement system in Figure 4–10 exploits a physical principle known as the **Bernoulli effect.** (This is the same principle that accounts for the ability of a rapidly moving airplane to lift into the atmosphere.) Just as the air pressure below the wing of a moving airplane becomes greater than that above, the air pressure at the **venturi** point tends to become greater than the pressure downstream. As a result, the bellows will push the plunger of the LVDT to the right. As the flow rate of the gas within the pipe increases, the difference in pressure between the venturi point and the downstream point also becomes

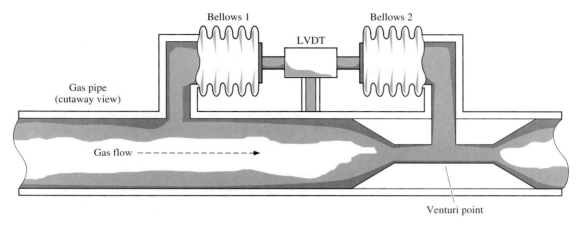

Figure 4–10
Gas flow measuring system with an LVDT and venturi

greater, resulting in further right-hand displacement of the plunger. Consequently, a greater differential voltage will be produced at the output of the LVDT.

The differential output of the LVDT in Figure 4–10 may be conditioned by an op amp instrumentation amplifier. However, since the output of the LVDT is an ac signal, rectification and filtering would be required prior to A/D conversion. The output of the A/D converter could then be fed to a microcomputer programmed to calculate and display the flow rate. It is important to remember that, although the system in Figure 4–10 senses a difference in pressure between two points in a gas pipe, the variable being monitored is flow rate. More information concerning pressure- and flow-sensing transducers is presented later in this chapter.

SECTION REVIEW 4–2

1. True or false?: Linear variable differential transformers are typically less accurate and less linear than displacement potentiometers.
2. For the LVDT in Figure 4–8, assume that the plunger is moved such that V_{E-F} equals 3.75 V(peak) and V_{A-B} equals 2.63 V(peak). What is the peak-to-peak amplitude of the differential output signal taken between points A and F?
3. What must be the position of the ferrous plunger of an LVDT in order for the differential output voltage to equal 0 V?

4–3   MAGNETIC PROXIMITY SENSORS

During Chapter 1, which served as an introduction to basic on/off devices, the concept of proximity detection was first introduced. A basic example of proximity sensing was illustrated in Figure 1–14, where a microswitch was actuated by a cam protruding from the bottom of a moving machine part. The problem associated with this method of position sensing is the fact that physical contact must be made between a moving machine part and a relatively fragile limit switch. With repeated actuation, the microswitch will eventually be destroyed, especially if the cam, possibly due to poor mechanical adjustment, strikes it with excessive force. As is explained in Chapter 5, an infrared light barrier may also be used to sense the approach of an object. However, an inherent problem with the light barrier is that of physical alignment. Failure to focus the emitted light beam directly at the lens of the detector could result in false jam indications. Such problems are all too prevalent within modern conveyance systems, which are likely to contain many light barriers and photosensors, a few of which will inevitably be knocked out of alignment.

If a sturdier form of proximity sensor is required, a device that operates on the principle of electromagnetism can be utilized. A simple example of such a device is the **reluctance proximity sensor,** which is illustrated in Figure 4–11. This sensing device consists primarily of a permanent magnet and a coil. With no ferrous material close to the sensor, the flux lines produced by the permanent magnet will emanate from the north pole and extend to the south pole in a symmetrical, undistorted pattern, as shown in Figure 4–12(a). Although the flux lines would be cutting across the windings of the permanent magnet at this time, there is no relative motion between the flux lines and the windings. Thus, during this passive condition, since there is no induced voltage or current within the coil, the output of the device is at 0 V.

When an object containing ferrous material approaches the reluctance sensor, distortion of the magnetic field results as the flux lines are bent and concentrated in order to enter the ferrous material. Since this action produces relative motion between the flux lines and the windings, a current pulse is induced within the coil and a voltage pulse is produced across the coil as shown in Figure 4–12(b).

We must stress that an output signal is present only when relative motion occurs between the coil and the flux lines. Therefore, when the object being detected becomes stationary, the output will return to 0 V. However, as illustrated in Figure 4–12(c), when the ferrous object moves away from the sensor (in the opposite direction of its approach) another output pulse is developed, due again to relative motion between the flux lines and the windings. Since, as the ferrous object moves away, the flux lines will be returning

Figure 4–11
Basic structure of a reluctance proximity sensor

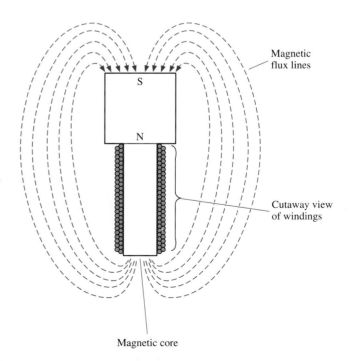

Magnetic flux lines

Cutaway view of windings

Magnetic core

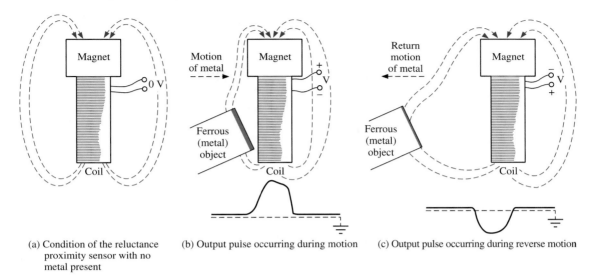

(a) Condition of the reluctance proximity sensor with no metal present

(b) Output pulse occurring during motion

(c) Output pulse occurring during reverse motion

Figure 4–12

to their original, undistorted pattern, they will be cutting across the windings in the opposite direction. Therefore, the polarity of the output signal will be reversed.

The reluctance proximity sensor is essentially a passive device that is brought temporarily into an active state by either the approach or removal of a ferrous object. However, the **eddy-current-killed oscillator** sensor (ECKO) can be considered to be an active device, because it continually produces a high-frequency output signal. This signal will be at its maximum amplitude as long as no metallic objects come close to the sensor. As a metal object approaches the ECKO, a portion of the energy contained in the rapidly expanding and contracting magnetic field of the sensor will be absorbed, producing eddy currents within the metal. This eddy current loss will be sensed as a decrease in amplitude for the output of the oscillator. If this loss in signal strength is great enough, due to extremely tight coupling between the ECKO sensor and the target object, the oscillator output may be totally dampened. (Hence, the description "eddy-current-killed" is particularly appropriate.) The operation of the ECKO is illustrated in Figure 4–13.

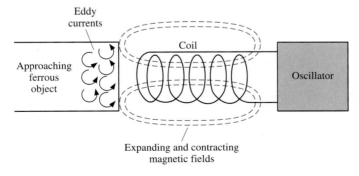

Figure 4–13
Structure and operation of the ECKO

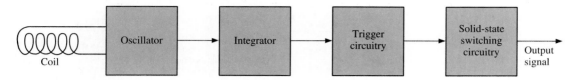

Figure 4–14
Signal conditioning circuitry for an ECKO

Both the reluctance proximity sensor and the ECKO require a considerable amount of signal conditioning. In most applications, only an on/off output is required, since the signal is likely to be processed by digital circuitry. As shown in Figure 4–14, the output of the ECKO could be fed to an integrator, which in turn would feed triggering circuitry and a solid-state switch.

A commonly used form of reluctance proximity sensor, referred to as a **proxistor,** is shown in Figure 4–15. This device contains its own switching circuitry and an LED that illuminates during the time in which the device detects the presence of metal. This LED is particularly useful during the alignment process because it allows the technician to ascertain the exact point at which a metallic object activates the sensor. Note that the body of the proxistor is threaded and has two hex nuts attached. This allows the device to be mounted easily on a plate or bracket within a piece of machinery.

The proxistor is extremely versatile, being capable of functioning at a wide variety of dc supply voltages. The operation of the device is inherently on/off. That is, its active-high output will rapidly switch to a level approaching V_{CC} when a metal object comes in close proximity. At this same instant, the LED indicator on the back of the device will illuminate. When the metal object is moved away, the output of the device will toggle to nearly 0 V and the LED will turn off.

Figure 4–16 illustrates how a proxistor might be utilized within a conveyance system. Notice here that the device is mounted on a bracket, which protrudes from the base plate. A metallic shoe, which is either spring loaded or has a leaf spring action, is mounted adjacent to the proxistor. Such an assembly would be located at a point within a conveyance system where either a jam or a backup condition is likely to occur.

When no backup or jam condition exists, the shoe will be able to extend fully outward. As illustrated in Figure 4–17(a), its metal surface will nearly touch the sensing surface of the proxistor. As a result, the LED indicator remains illuminated, indicating that a steady high signal is being maintained at the proxistor output. If the material on the conveyor becomes backed up, pressure will be exerted against the shoe, counteracting its spring force. Eventually, if the backup condition gets worse, the shoe will be displaced

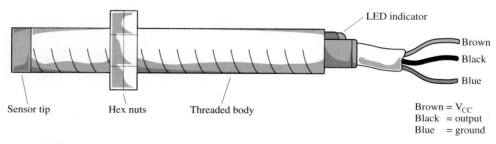

Figure 4–15
Proxistor

Figure 4–16
Conveyance system application of a proxistor

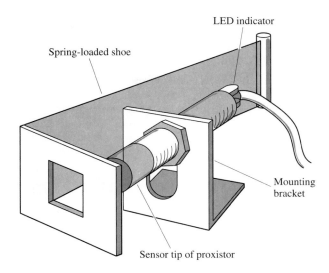

LED indicator

Spring-loaded shoe

Mounting bracket

Sensor tip of proxistor

to the position represented in Figure 4–17(b). With the opening in the shoe now positioned in front of the sensor, the output of the proxistor will fall toward the 0-V level, alerting the system control circuitry to a possible jamming condition. With most conveyance systems, a jamming condition will not actually be declared until this low signal is present for a specified minimum period of time. If the jam signal persists, part or all of the conveyance system would shut down until the jam is manually cleared. Only when the shoe is moved back to its normal position, allowing the output of the proxistor to toggle back to a logic high, will the control circuitry sense that the jam has been cleared. Even then, as a safety precaution, the machine would probably not restart automatically. The

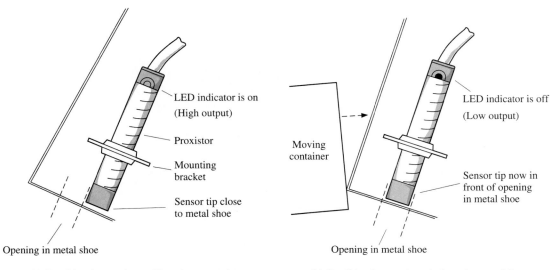

LED indicator is on
(High output)

Proxistor

Mounting bracket

Sensor tip close to metal shoe

Opening in metal shoe

(a) Condition for proxistor with no jam occurring

Moving container

LED indicator is off
(Low output)

Sensor tip now in front of opening in metal shoe

Opening in metal shoe

(b) Condition for proxistor during a jam condition

Figure 4–17

operator would, instead, be required to press the start switch in order to re-activate the system.

SECTION REVIEW 4–3

1. Explain the operating principle for the eddy-current-killed proximity sensor.
2. Describe the operation of a reluctance proximity sensor.
3. Explain at least one possible application of a reluctance proximity sensor within a solid material conveyance system.

4–4   CAPACITIVE AND ULTRASONIC LEVEL-SENSING TRANSDUCERS

Industrial process control often requires monitoring of the level of materials being stored within large containers such as hoppers, open tanks, or sealed vats. The concept of level detection was introduced in Chapter 1 with the valve control system shown in Figure 1–37. This relatively simple control circuit relies on a float switch to detect when fluid within a tank has attained a maximum allowable level. While this maximum fluid level may be associated with a known volume of liquid within the tank, interim fluid levels are not being monitored. Thus, in applications where it is necessary to monitor material levels continually, the simple float switch must be replaced by actual level-sensing transducers, which are likely to be capacitive or ultrasonic sensing devices. Pressure-sensing devices may also be utilized to determine indirectly the level of fluids stored in heavy containers. Their use is explained later in this chapter.

Figure 4–18
Sealed vat containing a bare capacitance probe

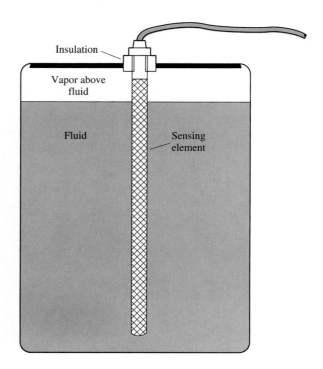

In vats that are completely enclosed, a capacitive probe may be used to measure the material level directly. If the fluid within the tank is nonconductive, a bare (noninsulated) form of probe may be used. In vats containing conductive materials, an insulated probe must be utilized. Though commonly found in sealed vats containing liquids, such probes may be used effectively in the measurement of the level of solid materials in granulated or powder form. Also, capacitive probes are used to monitor the level of **slurries,** which are insoluble mixtures of solid particles within liquid bases.

Figure 4–18 illustrates how a bare capacitance probe may be utilized within a sealed vat containing a nonconductive fluid. With this measurement scheme, three capacitances become important. In Figure 4–19, which is the equivalent circuit condition for Figure 4–18, C_V represents the capacitance occurring between the probe and the inner wall of the tank above the fluid level. C_L, which also exists between the probe and the inner wall of the vat, occurs below the fluid level. For both of these capacitances, the contents of the tank functions as the dielectric material. Thus, the effective capacitor formed by the probe and the vat is **coaxial.** The probe becomes the inner conductor, while the interior wall of the vat functions as the outer conductor. In addition to the capacitances involving the vat and the probe, the capacitance of the output cable (designated C_1) must also be considered, especially if the cable is several feet in length.

The wall of the vat is connected directly to the circuit ground, so these three capacitances are electrically in parallel. Since capacitances in parallel are additive, the sum of C_1, C_V, and C_L (designated C_O) could be measured between the output terminal and ground. Note that the resistivity of the material within the vat (designated R_L) may be considered as a single resistor in parallel with the three capacitances.

Assume that the storage tank in Figure 4–18 is approximately half full of fluid. With this condition, the value of C_V will be much less than the value of C_L. This is true due to the fact that the dielectric constant of the vapor within the upper part of the tank is

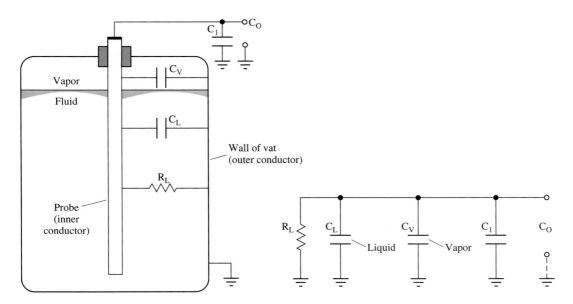

(a) Equivalent components for vat   (b) Equivalent circuit condition for vat containing bare capacitance probe
   containing bare capacitance probe

Figure 4–19

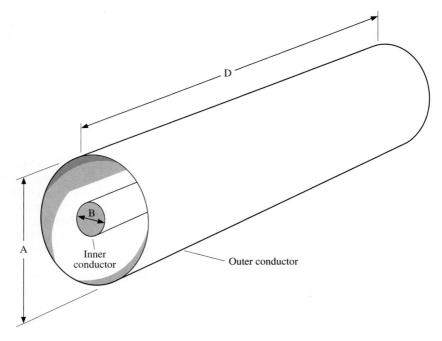

Figure 4–20
Equivalent coaxial structure of a vat containing a bare capacitance probe

much lower than that of the material (solid or liquid) being stored in the lower portion of the vat. For example, the dielectric constant of air, which serves as a reference for other materials, is considered to be unity. (Most other gases also have a dielectric constant approaching 1.) By contrast, water may have a dielectric constant as low as 50 or as high as 80, depending on the various minerals it contains. For a solid material such as grain, which is likely to be stored in a silo containing a level-detection system, the dielectric constant may fall between 3.0 and 4.0. This value will vary significantly as the humidity within the tank fluctuates.

The equivalent structural condition for the bare capacitance probe is represented in Figure 4–20. For such a device, the value of capacitance may be determined as follows:

$$C = 7.32 \, \kappa \, \frac{D}{\log A/B} \, \text{pF} \qquad (4\text{–}13)$$

where D equals probe length in feet, A equals the diameter of the interior of the vat (outer conductor), B equals the diameter of the probe (inner conductor), and κ equals the dielectric constant of the material being stored within the vat. Providing A and B are both in the same units, A/B becomes a simple ratio.

Example 4–2

Assume that the length of a bare capacitance probe is 57 in. The probe is immersed in a vat filled with a nonconductive liquid having a dielectric constant of 32. The diameter of the interior of the tank is 28¼ in., and the diameter of the probe is 5/8 in. (The vapor above the fluid level may be assumed to have a dielectric constant of unity.) What are the values of C_L and C_V when the vat is completely full? What are the values of C_L and C_V when the vat is completely drained?

Solution:

To determine capacitance values with the vat full, first convert dimensions to decimal values:

$$A = 28.25 \text{ in.} \quad \text{and} \quad B = 0.625 \text{ in.}$$

Converting D to feet:

$$57 \text{ in.}/12 = 4.75 \text{ ft}$$

then

$$C_L = C_L(\text{max}) = 7.32 \, \kappa \, \frac{D}{\log A/B} \text{ pF} = 7.32 \times 32 \times \frac{4.75}{\log 28.25/0.625} \text{ pF}$$

$$C_L(\text{max}) = 672.2 \text{ pF}$$

(With the vat completely full, C_V will approach zero.)

To determine capacitances with the vat empty:

$$C_V \cong C_V(\text{max}) = 7.32 \times 1 \times \frac{4.75}{\log 28.25/0.625} \text{ pF} = 21 \text{ pF}$$

(With the vat completely empty, C_L approaches zero.)

The key to understanding the operation of the bare capacitance probe lies in the ability to perceive C_V and C_L as two variable capacitors that share common coaxial conductors. As the fluid level within the vat in Figure 4–18 is increased, the value of C_L will increase significantly because of the relatively high dielectric constant of the liquid. Meanwhile, the value of C_V will decrease, because less vapor will now be present within the tank. Thus, the slight decrease in C_V will be overcompensated by the relatively larger increase in C_L. Consequently, the overall capacitance (C_O) will be increased. In fact, the increase in C_O will be nearly proportional to the increase in fluid level. If the fluid level decreases, C_V will increase. However, the decrease in C_L will be far more significant, resulting in an overall reduction in the value of C_O.

For processing by signal conditioning circuitry, the varying output capacitance (C_O) must be converted to variations in voltage. Two possible circuits for accomplishing this initial step in signal conditioning are shown in Figure 4–21. A fully capacitive bridge is represented in Figure 4–21(a). With the vat in an empty condition, C_4 would be adjusted such that the differential signal developed between points A and B would equal 0 V. As the vat begins to fill, the value of C_O will increase. Since C_1 and C_O are effectively in series, they will both be charged by the same current. As a result, at any given instant, the charge on C_O (in coulombs) will virtually equal the charge on C_1. Remember that:

$$Q = I \times T \quad \text{and} \quad V = Q/C$$

Thus, as the value of C_O increases, due to the rising fluid level in the vat, the value of V_B decreases. As a result, the value of V_{A-B} will become greater with the rise of fluid level within the vat. In Figure 4–21(b), a resistive voltage divider is used to establish the value of V_A. However, this bridge circuit will produce the same differential output signal as that in Figure 4–21(a).

The frequency of the bridge excitation signal must be high enough to ensure that the value of R_L (the resistivity of the material within the vat) is much higher than the capacitive

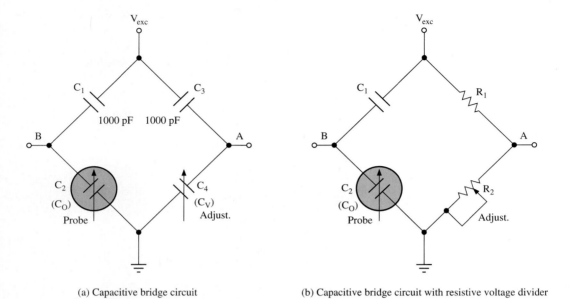

(a) Capacitive bridge circuit   (b) Capacitive bridge circuit with resistive voltage divider

Figure 4–21

reactance of C_O. Thus, the frequency of the sine wave being developed over the bridge circuit is often set as high as 100 kHz.

The output of the bridge circuit, taken between points A and B, is amplitude modulated, with the peak-to-peak voltage varying in proportion to the changing fluid level of the vat. For most applications of the capacitive level detector, it is necessary to convert the bridge output signal to a varying dc voltage. At first, it may seem that the easiest method of converting the output of the bridge circuit to dc would be through simple diode rectification and filtering. However, this would lead to a nonlinear response at lower voltage levels, which occur as the vat nears an empty condition. For this reason, the rectifier concept must be abandoned in favor of a more complex, but highly linear, **demodulation system.** A block diagram showing this design alternative is shown in Figure 4–22.

As represented in Figure 4–22, the differential output from the capacitive bridge is fed to an instrumentation amplifier. The ac output of this circuit is then fed to the demodulator, which converts the peak amplitude of this signal to a varying dc voltage. This dc output signal may require additional amplification and/or level shifting, which could be accomplished by a final output stage. The output of such amplifier circuitry could be fed to the analog indicator as shown. This D'Arsonval indicator might, for example, require an input signal ranging from 0 to 5 V_{dc}. The scale for the meter movement could simply show the amount of material within the vat in terms of a percentage (rather than an exact level or volume).

A detailed schematic of the instrumentation amplifier is shown in Figure 4–23. Assuming the vat is empty, a null condition would exist across the bridge because V_A would equal V_B. As the vat begins to fill, C_O increases, causing V_B to decrease. Consequently, a sine wave develops between points A and B. This signal will be amplified by a factor approximately equal to $-R_4/R_3$. The relationship of the bridge excitation signal, V_A, V_B, V_{A-B}, and V_C, is represented in Figure 4–24.

The actual demodulator circuit, which processes the ac output signal from the instrumentation amplifier, is represented in Figure 4–25. For stable operation, the oscillator

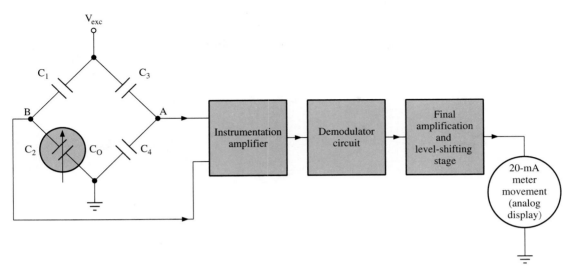

Figure 4–22
Block diagram of capacitive bridge and signal conditioning circuitry

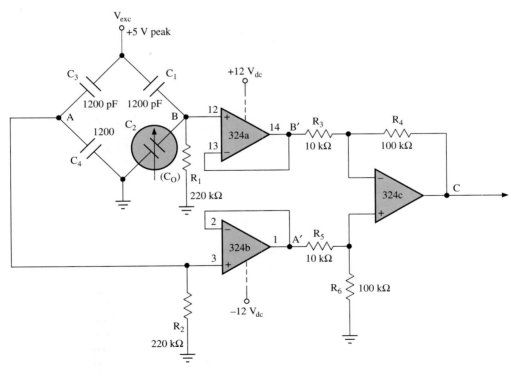

Figure 4–23
Capacitive bridge with instrumentation amplifier

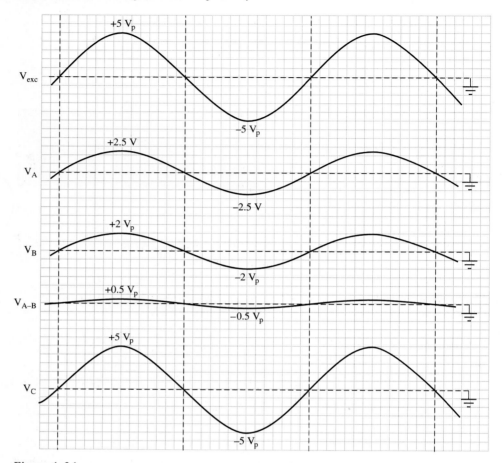

Figure 4–24

Sample waveforms for the capacitive bridge and instrumentation amplifier in Figure 4–23 (The liquid level in the vat is assumed to be at a point where V_B attains $2V_p$.)

controlling this stage of the circuit should be the same one used for bridge excitation. Note that, whereas the bridge excitation signal is sinusoidal, the oscillator signal feeding the demodulator is a square wave. (These two waveforms are easily produced simultaneously by a function generator IC.)

In Figure 4–25, after the signal from the instrumentation amplifier passes the voltage-follower stage, it is fed simultaneously to complementary JFET sampling gates. Also note that the square wave output of the oscillator is fed to the gates of both JFETs. Thus, at any given instant, one JFET switch will be conducting while the other is in the pinch-off condition. During the pulse width of the oscillator square wave, the sine wave at the input of the demodulator will be in its positive alternation. With this condition, Q_1 will be conducting, allowing C_3 to charge in the positive direction. Thus, C_3 will be able to attain a voltage level nearly equal to the peak amplitude of the input sine wave.

During the space width of the oscillator square wave, Q_2 will be enabled to conduct. At this time, the input sine wave will be in its negative alternation, allowing C_4 to charge toward the negative peak amplitude of this signal. Thus, the output differential amplifier

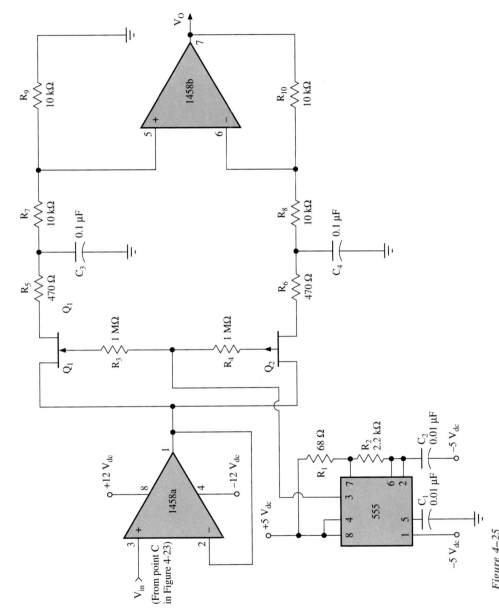

Figure 4–25
Demodulator circuitry for the signal from the capacitive bridge and instrumentation amplifier

217

will develop a dc signal representing the difference in amplitude between the two capacitor voltages. As the level of material within the vat increases, the peak-to-peak amplitude of the input sine wave increases. This action produces an increase in the amplitude of the negative dc voltage (V_0) present at the output of the differential amplifier in Figure 4–25.

In many signal conditioning applications, we might want to remove the slight ripple riding on the output signal. As seen in Figure 4–26, this may be accomplished through the use of an active low-pass filter. Final level shifting and gain control is accomplished by the inverting output amplifier. Here, for example, the output voltage range might be adjusted such that a 0% full condition for the vat would produce a 0-V output level. This condition could be achieved through fine adjustment of the offset-nulling resistor. The gain adjustment resistor could then be set to provide an output of +5 V when the vat is 100% full.

Figure 4–27 illustrates an alternative form of display for a level-detecting system. Here, a resistive voltage divider is used to establish reference voltages representing 25%, 50%, 75%, and 100% of the 5-V maximum output level. These reference voltages are sensed at the inverting inputs of four comparators. The active input of these comparators is taken from the final output of the circuit in Figure 4–26. When the level of fluid rises within the vat, the input voltage to the comparators increases proportionally. As the vat level exceeds the 25% point, the output of comparator 1 goes high, producing a logic high at point A. This condition produces a logic high at the output of XOR_1. As a result,

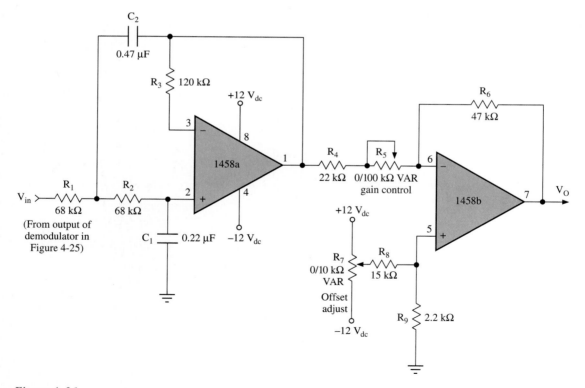

Figure 4–26
Final stage of signal conditioning circuitry consisting of a unity-gain low-pass filter and an inverting amplifier

the Darlington pair consisting of Q_1 and Q_2 will be biased on, illuminating the 25% indicator lamp. As the input voltage continues to increase, exceeding the 50% threshold, the output of both comparators 1 and 2 will be at a logic high. When this occurs, the 25% light will turn off and the 50% lamp will be illuminated. The comparator and logic gate conditions for each of the five possible display lamp combinations are shown in Figure 4–28.

Example 4–3

For the circuit in Figure 4–23, assume the fluid in the vat has risen to a level where C_O equals 1386 pF. Given this condition, determine the peak amplitude of V_A, V_B, and V_{A-B}. What would be the peak amplitude of the output of the instrumentation amplifier in Figure 4–23? What would be the ideal dc output for the demodulator circuit in Figure 4–25?

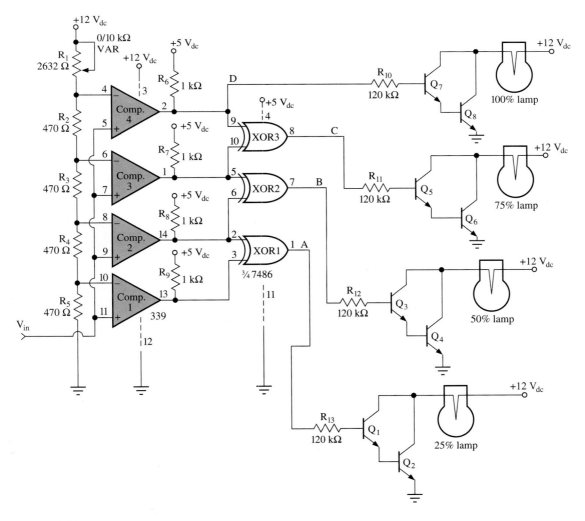

Figure 4–27
Display circuitry for capacitive level-detection system

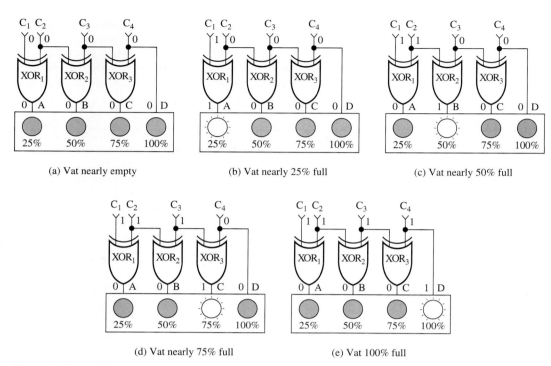

(a) Vat nearly empty   (b) Vat nearly 25% full   (c) Vat nearly 50% full

(d) Vat nearly 75% full   (e) Vat 100% full

Figure 4–28
Five possible logic and display lamp conditions for the circuit in Figure 4–27

Also assume that the output of the demodulator stage is fed to the low-pass filter and level-shifting circuitry in Figure 4–26. (Since the filter design is unity gain, A_V for that part of the circuit should be considered to be $+1$.) Also assume that R_7 in Figure 4–26 is adjusted for a 0-V output when the capacitive bridge circuit is in a null state. If R_5 is set at 34.5 kΩ, what would be the value of V_O?

Finally, if the output of the level-shifting amplifier is fed to the display circuit in Figure 4–27, what would be the condition of the indicator lamps?

Solution:
Step 1: Solve for the peak values of V_A, V_B, and V_{A-B}. Since C_1 and C_2 (C_O) are effectively in series, Q_1 will equal Q_2. Since C_3 equals C_4, $V_{C3} = V_{CA} = 1/2 V_{exc} = 2.5$ V. Thus,

$$Q/C_1 + Q/C_2 = V_{exc} \text{ (peak)} \quad \text{(where } Q = Q_1 = Q_2)$$
$$Q/1200 \text{ pF} + Q/1386 \text{ pF} = 5 \text{ V (peak)}$$
$$1/1200 \text{ pF} + 1/1386 \text{ pF} = 5 \text{ V}/Q$$
$$833.333 \times 10^6 + 721.5 \times 10^6 = 5 \text{ V}/Q$$
$$Q = 5 \text{ V}/1.5548 \times 10^9 = 3.2158 \times 10^{-9} \text{ C}$$
$$V_{C_1} = Q/C_1 = \frac{3.2158 \times 10^{-9}}{1200 \times 10^{-12}} = 2.68 \text{ V}$$
$$V_B = V_{C_2} = Q/C_2 = \frac{3.2158 \times 10^{-9}}{1386 \times 10^{-12}} = 2.32 \text{ V}$$
$$V_{A-B} = V_A - V_B = 2.5 \text{ V} - 2.32 \text{ V} = 180 \text{ mV}$$

Step 2: Solve for V_0 of the instrumentation amplifier.

$$V_0 = V_{B-A} \times -R_4/R_3$$
$$V_0 \text{ (peak)} \cong -180 \text{ mV} \times -100 \text{ k}\Omega/10 \text{ k}\Omega = +1.8 \text{ V (peak)}$$

Step 3: Determine the dc output voltage for the demodulator, low-pass filter, and level-shifting amplifier. Ideally, for the differential amplifier in the demodulator circuit:

$$V_0 = [V \text{ (peak)} - -V \text{ (peak)}] \times -R_{10}/R_8$$

Thus,

$$V_0 = (1.8 \text{ V} - -1.8 \text{ V}) \times -10 \text{ k}\Omega/10 \text{ k}\Omega = -3.6 \text{ V}$$

Since the low-pass filter is in a unity-gain configuration:

$$V_0 = -3.6 \text{ V} \times +1 = -3.6 \text{ V}$$

For the level-shifting amplifier:

$$V_0 = V_{in} \times \frac{-R_6}{R_4 + R_5} = -3.6 \text{ V} \times \frac{-47 \text{ k}\Omega}{22 \text{ k}\Omega + 34.5 \text{ k}\Omega} = 2.995 \text{ V}$$

Since the preceding result is greater than 2.5 V but less than 3.75 V, the 50% lamp will be illuminated.

Many of the fluids stored in tanks and vats at industrial sites are likely to be conductive. For the level measurement of such materials, a bare capacitance probe would prove unsatisfactory, due to the fact that a path for current flow would exist between the bare inner conductor and the metallic wall of the container. To allow accurate and safe measurement of conductive fluids, an **insulated capacitance probe** would be required.

Figure 4–29 illustrates an application of an insulated capacitance probe, being used to measure the level of a conductive fluid within a sealed vat. The equivalent circuit for this measurement system is more complex than that of the bare capacitance example in Figure 4–19, due to the fact that, in addition to C_V, C_L, and C_1, two other variables become important. The capacitance between the metal conductor within the probe and the outside of the probe above the fluid level (represented as C_A) can be considered to be in series with C_V. The variable C_B represents the capacitance occurring between the metal conductor within the probe and the outside of the probe below the fluid level. The very low resistance of the material within the vat (represented as R_L) is effectively in parallel with C_L. For this reason, we must choose an operating frequency that will ensure that the reactance of C_B will be much greater than the ohmic value of R_L. Assuming this condition is met, the equivalent circuit conditions for an empty and a partially filled vat may be represented as shown in Figure 4–30.

When the vat is empty, the probe capacitance consists essentially of the series capacitances C_A and C_V. Thus, the resultant capacitance may be calculated as follows:

$$C_0 = \frac{C_A \times C_V}{C_A + C_V} \tag{4–14}$$

Since, during an empty condition, the value of C_V will be much greater than C_A, the equivalent probe capacitance will approximately equal C_V. As the vat begins to fill, the value of C_B will increase, becoming far greater than C_V. Since, at the desired operating frequency, X_{C_B} will be a much greater ohmic value than the combined impedance of R_L and X_{C_L}, the effective impedance of branch 2 will nearly equal X_{C_B}. Also, since C_B is becoming much greater than C_V, the parallel capacitance of C_V and C_B will approach the value of C_B as the fluid level approaches maximum.

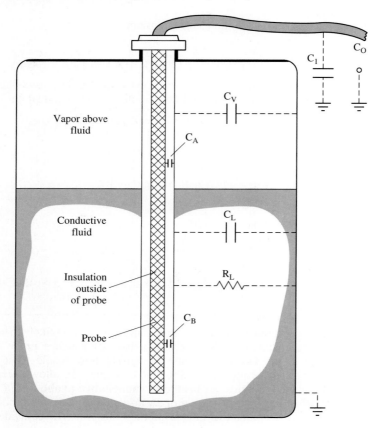

Figure 4–29
Sealed vat containing an insulated capacitance probe

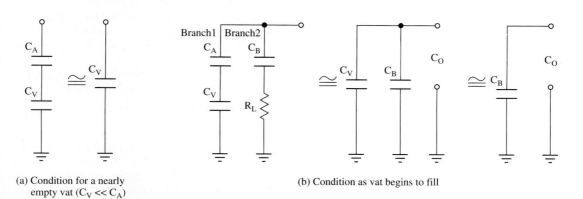

(a) Condition for a nearly
   empty vat ($C_V \ll C_A$)

(b) Condition as vat begins to fill

Figure 4–30
Changing equivalent circuit conditions for the vat and insulated probe in Figure 4–29

The signal conditioning circuitry for the insulated capacitance probe could be identical to that shown for the bare capacitance probe. However, care must be taken to select the proper oscillator frequency for bridge excitation and demodulation. Remember that, for the bare capacitance probe, the operating frequency may approach 100 kHz in order to ensure that X_{C_0} is much less than R_L. With the insulated capacitance probe, X_{C_B} must be much greater than R_L. Thus, the operating frequency for a circuit containing the insulated form of probe is likely to be much lower. For either type of probe, the amplitude of the bridge excitation signal must be very stable. Any deviation in this voltage would result in an error in the measurement process.

In addition to capacitive level detectors, a wide variety of ultrasonic level detectors is also used in industrial control systems. These transducers may be considered as active devices, since they both transmit and detect energy pulses at frequencies just above the audio range. Operating on the same principle as depth-sounding sonar transducers, they send out a burst of energy then await a return echo.

As illustrated in Figure 4–31, the difference in time between the initial pulse and the return echo is measured in order to determine the distance between the transducer and

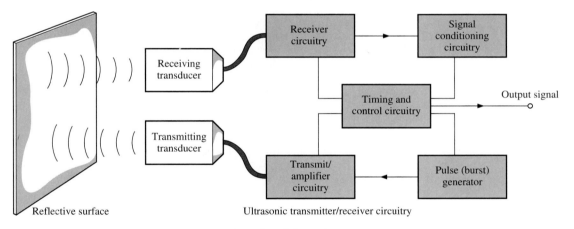

(a) Operation of ultrasonic sensor

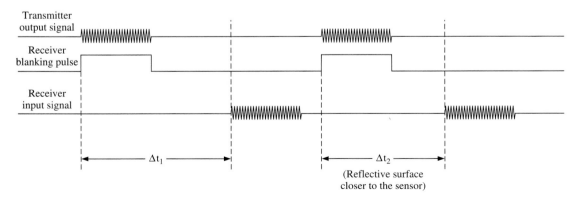

(b) Ultrasonic sensor waveforms

Figure 4–31

the surface from which the ultrasonic pulse is reflected. Ultrasonic waves travel at nearly 340 m/sec. Thus, if the time between the initial pulse and return echo is precisely measured, distance can be accurately determined as follows:

$$D = \frac{\Delta t \times 340}{2} \tag{4-15}$$

where D equals the distance in meters and Δt represents the time lapse between the initial and return pulses.

According to Eq. (4–15), the distance between the transducer and the reflecting surface is directly proportional to the time required for the return echo. Digital counting circuitry could be enabled to operate at the instant the initial pulse is transmitted. At the instant the return echo is detected, the counting action would halt. The highest binary number attained by the counter could undergo further processing in order to derive and display an exact distance. However, in many industrial applications, it is not necessary to calculate exact distances. Instead, we might want to determine only relative changes in distance. In such applications, the only requirement might be to indicate variations in distance as a percentage of the maximum. Thus, a simple D'Arsonval movement could again serve as the display.

The primary advantage of ultrasonic level-sensing transducers is the fact that these devices need not come in direct contact with the materials being monitored. This represents a distinct advantage over capacitive probes, especially when these materials include corrosive chemicals or abrasive slurries. Since ultrasonic transducers do not require immersion, they may be placed directly above conveyor belts used to transport granulated materials. Here, they are able to monitor continually the level of the material traveling along the conveyance path. The information thus obtained may be used to control the flow rate of these materials. An application of an ultrasonic level detector within a conveyance system is shown in Figure 4–32.

In some industrial applications, including filtration and sewage treatment systems, it is necessary to monitor the difference in fluid level between two points. The **ultrasonic differential level detector,** which is represented in Figure 4–33, is ideally suited for such a task. As shown here, both transducers 1 and 2 transmit ultrasonic pulses simultaneously. However, with transducer 1 being closer to the surface of the liquid than transducer 2, its return echo will occur sooner. Through signal conditioning circuitry, Δt_1 and Δt_2 are translated to dc voltage levels, with the difference between these two potentials being amplified to become the output signal. If the distance between the two fluid levels increases, the difference between Δt_1 and Δt_2 will increase. Consequently, the difference between the dc voltages representing Δt_1 and Δt_2 becomes greater, causing the dc output voltage to increase. As with the single-channel ultrasonic detector in Figure 4–31, the output of the differential detector may be fed to a D'Arsonval movement, with the difference in fluid levels being represented as a percentage rather than in specific units.

As illustrated in Figure 4–33, an ultrasonic level detector may be used to monitor the difference in fluid level on either side of a filtration system. Such a detection system is likely to be equipped with an audible alarm, which would be activated when the difference in fluid level deviates from a minimum or maximum allowable range. Thus, the differential level detector may be used to sense possible clogged filters or drains, which could lead to costly damage if not quickly remedied.

Example 4–4

Assume for the differential level detector in Figure 4–33 that Δt for transducer 1 is 8.2 msec, while Δt for transducer 2 is 14.5 msec. What is the distance (in meters) between

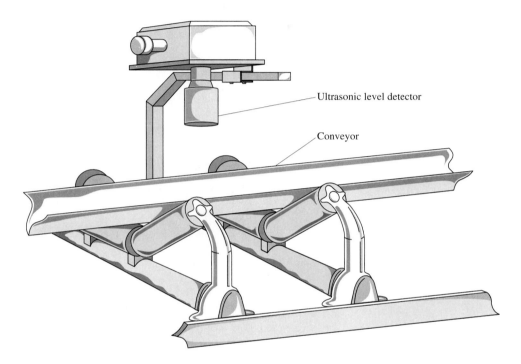

Figure 4–32
Ultrasonic level detector used within a conveyance system

transducer 1 and the fluid ahead of the filter? What is the distance between transducer 2 and the fluid downstream from the filter? What is the difference between the two fluid levels?

Solution:
Solving for D_2:

$$D_2 = \frac{\Delta t_2 \times 340}{2} = \frac{14.5 \text{ msec} \times 340}{2} = 2.465 \text{ m}$$

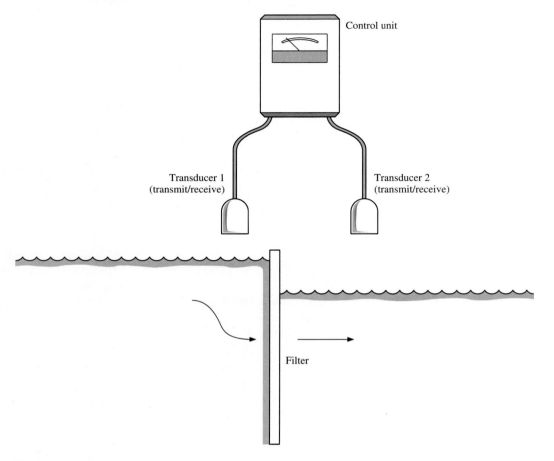

Figure 4–33
Application of a differential ultrasonic level detector within a filtration system

Solving for D_1:

$$D_1 = \frac{\Delta t_1 \times 340}{2} = \frac{8.2 \text{ msec} \times 340}{2} = 1.394 \text{ m}$$

Also,

$$\Delta t_2 - \Delta t_1 = 14.5 \text{ msec} - 8.2 \text{ msec} = 6.3 \text{ msec}$$

$$\Delta D = \frac{6.3 \text{ msec} \times 340}{2} = 1.071 \text{ m}$$

SECTION REVIEW 4–4

1. True or false?: For the vat in Figure 4–18, as the fluid level rises, the decrease in C_V will be greater than the increase in C_L, causing the output capacitance (C_O) to decrease.
2. True or false?: Dry grain has a higher dielectric constant than water.

3. The interior of a sealed vat containing a bare capacitance probe is 37½ in. in diameter. The bare capacitance probe is 3/4 in. in diameter and 63 in. long. Assuming the tank is filled with a material having a dielectric constant of 4.7, what is the approximate value of C_O.

4. For the capacitive bridge circuit in Figure 4–21(a), assume the value of C_2 (C_O) is now 1483 pF and that C_4 is adjusted to 1200 pF. Assuming the peak amplitude of the bridge excitation signal is 12 V, solve for the values of V_A, V_B, and V_{A-B}.

5. Explain why the excitation frequency for a bridge circuit containing a bare capacitance probe is likely to be much higher than the excitation frequency for a bridge containing an insulated capacitance probe.

6. Explain why the demodulator system contained in Figure 4–25 is preferred over a simple diode rectifier in converting the output of the level-detecting circuitry to a dc signal.

7. For the ultrasonic differential level detector in Figure 4–33, assume that Δt for transducer 1 is 7.5 msec. If the fluid level ahead of the filter is 1.83 m higher than that downstream, what would be the value of Δt for transducer 2? What would be the distance between transducer 2 and the fluid downstream from the filter?

4–5   PRESSURE AND FLOW-SENSING TRANSDUCERS

Pressure is generally defined as the application of force by one body directly on another. In more specific terms, pressure is defined as the amount of force being applied over a surface within a defined area. The scientific international (SI) unit for pressure is the **pascal** (after the seventeenth-century French mathematician). This unit, abbreviated ''Pa,'' is defined as a force of 1 newton applied within an area of 1 cm². The **newton** (N) is the unit of force in the meter-kilogram-second (mks) system, being equal to nearly 0.225 pounds. A force of 1 N would be capable of accelerating a body with a mass of 1 kg at a rate of 1 m/sec during a period of 1 sec.

While the pascal may serve as the worldwide standard for the measurement of pressure, the units most commonly used within the industrial community are **pounds per square inch absolute** (psia) and **pounds per square inch gauge** (psig). Psia units are referenced to a perfect vacuum and, therefore, are unaffected by changes in altitude. Psig units, however, are referenced to atmospheric pressure. Thus, when determining air pressure within a pneumatic system, the ambient atmospheric pressure functions as an offset variable. Air pressure in psig would be determined by subtracting the ambient atmospheric pressure from the absolute (psia) value. Occasionally, it may be necessary to convert psi units to pascal units. The conversion factor for this process is 6.8948×10^3. Thus,

$$Pa = 6.8948 \times 10^3 \, psi \qquad (4–16)$$

Example 4–5

Assume the gauged air pressure for a relieving regulator indicates 47.5 psig. The device is located at sea level, where the ambient air pressure is nearly 14.7 psi. What is the absolute air pressure for the regulator? Express this absolute pressure in terms of pascal units.

Solution:

$$psig = psia - ambient \ air \ pressure$$

Thus,

$$psia = psig + \text{ambient air pressure}$$
$$\text{Absolute pressure} = 47.5\ psig + 14.7\ psi = 62.2\ psia$$

Converting to pascal units:

$$Pa = 62.2\ psia \times 6.8948 \times 10^3 = 4.28857 \times 10^5\ Pa$$

All transducers introduced thus far in this chapter have been capable of converting a change in a physical parameter directly to a change in an electrical parameter such as current, voltage, or resistance. For the measurement of low values of pressure, however, two transducers are used that are **force-summing.** These two devices, the **Bourdon tube** and the **bellows,** convert changes in pressure to changes in angular or linear displacement. Four basic types of Bourdon tubes are illustrated in Figure 4–34.

To create an electrical signal representing variations in pressure, a displacement transducer, either a potentiometer or an LVDT, is linked mechanically to the Bourdon tube. Such an arrangement is shown in Figure 4–35, where a C-shaped Bourdon tube is connected to an LVDT.

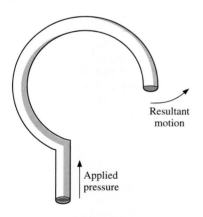

(a) C-shaped Bourdon tube

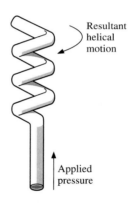

(b) Helical-shaped Bourdon tube

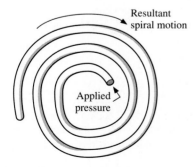

(c) Spiral-shaped Bourdon tube

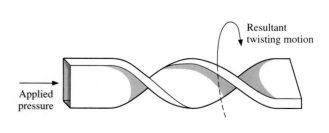

(d) Twisting-type Bourdon tube

Figure 4–34
Four basic types of Bourdon tubes

Figure 4–35
C-shaped Bourdon tube mechanically linked
to the ferrous plunger of an LVDT

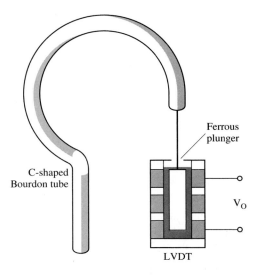

In Figure 4–10, a gas flow measurement system utilizing a bellows and an LVDT was introduced. Figure 4–36, which contains the same two devices, shows more details about the structure and operation of the bellows. As pressure increases within the housing, the pressure spring inside the bellows is contracted, causing the output boss to move upward. This action causes the swing arm to move, which, in turn, pushes the ferrous plunger of the LVDT. The differential output signal should then increase proportionally with the rise in pressure.

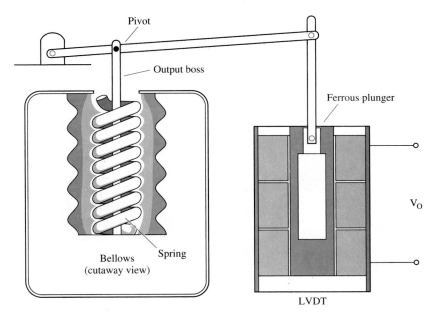

Figure 4–36
Bellows mechanically linked to an LVDT

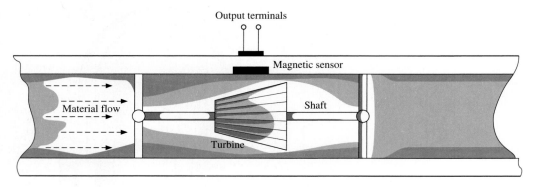

Figure 4–37
Cutaway view of a pipe containing a turbine flow sensor

Thus far in this text, two methods of gas flow measurement have been presented. One utilized a self-heated thermistor, while the other employed a bellows and venturi. Complex mechanical transducers are also used for detection of the flow rate of liquids and slurries. Though not designed for use with slurries, the **turbine** flow measurement system in Figure 4–37 is very effective in measuring the flow of liquid materials. With the turbine placed as shown, its speed of rotation will vary as a function of the rate of

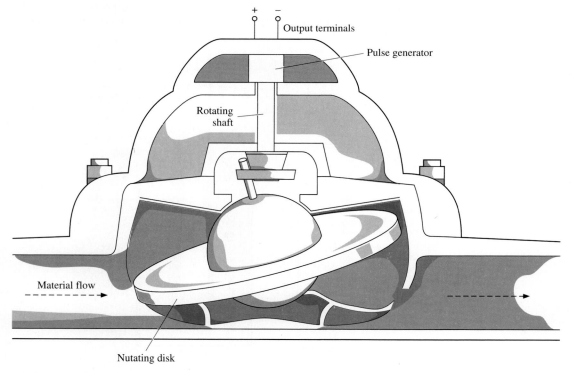

Figure 4–38
Nutating disk and pulse generator used to measure flow

flow. Through use of a magnetic pick-up, the transducer is able to generate an output signal, the frequency of which is proportional to the angular velocity of the turbine.

Figure 4–38 illustrates an alternative form of mechanical flow transducer. Here, a **nutating,** or wobbling, disk is used to divide the material being transported into measured segments. As with the turbine, rotation is produced and a magnetic pick-up is used to create an output pulse. The frequency of this output signal is directly proportional to the flow of liquid material through the device. The nutating disk has the same limitation as the turbine, in that it is unable to measure the flow of slurries. A venturi system, which has already been introduced, would be effective in that regard.

A major problem with either the turbine or the nutating disk method of measuring liquid flow lies in the fact that they both must be placed directly in the path of the material being transported. In a sense, therefore, they become impediments, disrupting the flow that they are actually attempting to measure. Also, they may be damaged or destroyed if they are placed in direct contact with conductive or corrosive fluids. The **magnetic flow-sensing transducer,** therefore, has a distinct advantage over the two mechanical devices just described in that it doesn't directly contact the material being monitored. The operation of this transducer is illustrated in Figure 4–39.

The design of the device shown in Figure 4–39 capitalizes on the conductive property of the material within the pipe. Note that an electromagnetic field is produced by a saddle coil placed on the outside of the pipe through which the conductive material must pass. Excitation for this electromagnet must, of course, be provided by an ac source. The field strength will vary as a function of the amount of conductive material moving within the pipe. Since an ac voltage may be induced within a conductor that passes through an electromagnetic field, the amplitude of the voltage picked up between the output terminals (V_0) will become a function of the rate of flow.

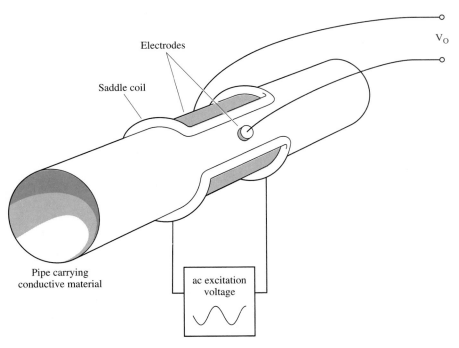

Figure 4–39
Structure of magnetic flow-sensing transducer

SECTION REVIEW 4–5

1. 6.895×10^5 pascals is equivalent to how many psia units?
2. Describe the difference between psia and psig units.
3. The air pressure indicated by a gauge on a relieving regulator in a pneumatic system is 34.5 psi. The ambient air pressure at this time is 12.63 psi. What is the air pressure in psia units?
4. Explain how output signals are produced for the nutating disk and turbine flow transducers.
5. Identify a limitation shared by the nutating disk and turbine flow transducers.
6. Describe the operation of the magnetic flow-sensing transducer. What advantage does this device have over the turbine and nutating disk devices? What limitation does this transducer have with regard to the types of material it is able to monitor?

4–6   STRAIN GAUGES AND OTHER FORCE-SENSING TRANSDUCERS

Strain gauges are transducers capable of sensing minute changes in the physical displacement of an object. The basic resistance strain gauge illustrated in Figure 4–40 operates according to the principle that the resistance of a conductive material will become greater if either its length is increased or its cross-sectional area is decreased. When this conductive material is forced to stretch, both of these conditions will in fact occur, resulting in an

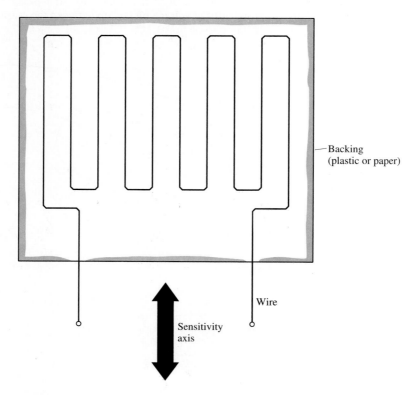

Figure 4–40
Bonded-wire–type strain gauge

increase in the resistivity of the material. When the stretching force is alleviated, the length and cross-sectional area of the conductor will tend to return to their original dimensions, thus allowing the resistivity of the material to be reduced to its original value. The function of the strain gauge just described may be exploited in the measurement of acceleration, pressure, or weight.

As seen in Figure 4–40, a basic strain gauge may consist of several loops of a fine wire, which are permanently bonded to a paper or plastic backing. Resistance strain gauges are more likely to consist of a metal foil (rather than wire) conductor, which is bonded to a flexible plastic backing material. Since such a device may be as small as a postage stamp and as thin as 0.003 mm, it can be easily cemented to the surface of the object undergoing strain. The zigzag or serpentine arrangement of the foil conductor allows maximum gauge length to be achieved within a limited surface area. As shown in Figure 4–41, arrow marks may be provided on the surface of the strain gauge. These indications may be used as alignment aids when mounting the device. Also, these markings are used by some manufacturers to represent the **sensitive axis** of the strain gauge. If a strain gauge

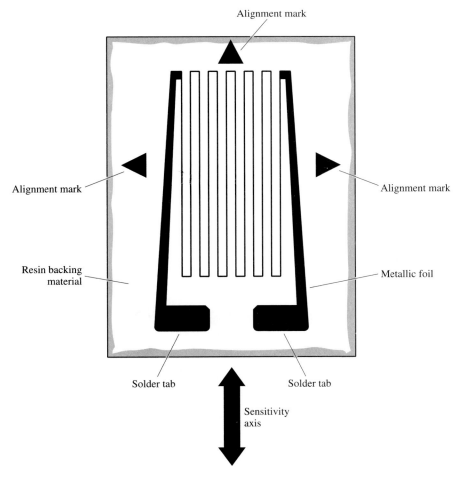

Figure 4–41
Foil-type strain gauge

is to operate correctly, its sensitive axis must be directly aligned with the stretching or compressing force being monitored.

The need for such a thin, small-size strain-sensing transducer should become apparent as the phenomenon of strain itself is analyzed. When, for example, force is applied to a steel bar, the length of the bar will be varied by some small, yet predictable amount. This variation in length resulting from strain is referred to as **deformation** and is symbolized as ΔL (representing difference in length).

Strain on an object may be either **compressive** or **tensile.** Tensile strain, as illustrated in Figure 4–42(a), results from stretching or pulling on an object. With this condition, the body under stress will become slightly longer than normal. Providing the material is not stretched beyond its limit of elasticity, it will tend to return to its normal length as soon as the pulling force is removed. As seen in Figure 4–42(b), compressive strain produces an opposite effect, with the length of the body being slightly reduced as force is applied. Again assuming that the amount of strain has not exceeded the elasticity limit, the body should return to its normal length when the compressive force is removed.

Strain, which is symbolized as ε, is determined as the ratio of the amount of deformation to the normal length of the body under stress. Strain may be expressed fractionally or as a percentage and is determined as follows:

$$\varepsilon = \Delta L/L \times 100\% \qquad (4\text{–}17)$$

where ε represents strain, ΔL equals the degree of deformation, and L is the normal length of the body under stress.

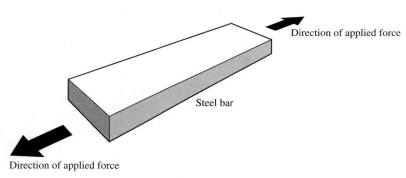

(a) Tensile (stretching) strain

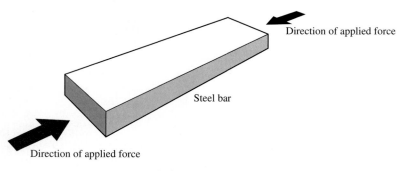

(b) Compressive strain

Figure 4–42

Whether expressed as a ratio or percentage, strain represents a very small value, typically much less than 0.01 or 1%. For this reason, strain is more conveniently represented as **microstrains** (microinches per inch), which are symbolized as $\mu\varepsilon$. Expressed mathematically:

$$\mu\varepsilon = (\Delta L/L)/\mu \text{ in./in.} \tag{4–18}$$

Example 4–6

Assume that a metal bar normally 53³⁄₁₆ in. long is compressed such that its length is reduced to 53¹⁄₁₆ in. What is the amount of deformation in inches? What is the amount of strain as a ratio? What is the amount of strain in percentage form? Finally, express the degree of strain in microstrains.

Solution:
Step 1: Solve for the deformation.

$$\text{Deformation} = \Delta L = L_1 - L_2$$
$$= 53\tfrac{3}{16} \text{ in.} - 53\tfrac{1}{16} \text{ in.} = 1/8 \text{ in.} = 0.125 \text{ in.}$$

Step 2: Solve for strain as a ratio. (Converting 53³⁄₁₆ in. to decimal form $= 53.1875$ in.)

$$\varepsilon = \Delta L/L = 0.125 \text{ in.}/53.1875 \text{ in.} = 2.35018 \times 10^{-3}$$

Step 3: Solve for strain as a percentage.

$$\varepsilon = \Delta L/L \times 100\% = 2.35018 \times 10^{-3} \times 100\% \cong 0.235\%$$

Step 4: Solve for strain in $\mu\varepsilon$.

$$\mu\varepsilon = (\Delta L/L)/(\mu\text{in./in.}) = (\Delta L/L)/1 \times 10^{-6} = (\Delta L/L) \times 1 \times 10^{6}$$
$$= 2.30518 \times 10^{-3} \times 1 \times 10^{6} = 2.35018 \times 10^{3} \ \mu\varepsilon$$

To utilize a strain gauge effectively for the purposes of monitoring force and pressure, engineers must have a clear understanding of the relationships of stress, strain, and resistance. According to **Hooke's law,** stress is directly proportional to the force being applied to a body under strain and inversely proportional to the cross-sectional area of that body. Thus,

$$\text{Stress} = F/A \tag{4–19}$$

where F equals force in pounds and A equals the cross-sectional area, in square inches, of the body under strain.

The convenient units for stress would be, as with pressure, pounds per square inch (psi).

Example 4–7

If a force of 1245 pounds is exerted over an area of 47 square inches, what would be the amount of stress?

$$\text{Stress} = F/A = 1245 \text{ lbs}/47 \text{ in.}^2 = 26.49 \text{ psi}$$

Another factor that must be considered in the analysis of strain is the **modulus of elasticity,** or **Young's modulus,** which is simply the ratio of the stress applied to a given body and the resultant strain. Expressed mathematically, Young's modulus (symbolized as Y) would be as follows:

$$Y = \text{stress/strain} = (F/A)/\varepsilon \qquad (4\text{--}20)$$

Although Young's modulus varies significantly for different materials, it tends to remain constant for any one material within its limits of elasticity.

Example 4–8

Assume that a metallic block is 14 in. tall, 23 in. wide, and 7 in. thick. As a compressive force of 8.75 tons is applied to the block, its thickness is reduced to $6^{31}\!/_{32}$ in. What is the amount of strain (expressed in $\mu\varepsilon$)? What is the stress being incurred by the block? What is the modulus of elasticity for the material?

Solution:
Step 1: Solve for the deformation. Converting L_2 to decimal format: $6^{31}\!/_{32}$ in. = 6.96875 in.

$$\text{Deformation} = \Delta L = L_1 - L_2$$
$$= 7 \text{ in.} - 6.96875 \text{ in.} = 0.03125 \text{ in.}$$

Step 2: Solve for the cross-sectional area.

$$A = 14 \text{ in.} \times 23 \text{ in.} = 322 \text{ in.}^2$$

Step 3: Solve for the stress. Converting tons to pounds: 8.75 tons $\times$ 2000 lb = 17,500 lb

$$\text{Stress} = F/A = 17{,}500 \text{ lb}/322 \text{ in.}^2 = 54.348 \text{ psi}$$

Step 4: $\varepsilon = \Delta L/L = 0.03125 \text{ in.}/7 \text{ in.} = 4.4643 \times 10^{-3}$

$$\text{Strain} = 4.4643 \times 10^{-3} \times 1 \times 10^6 = 4.4643 \times 10^3 \ \mu\varepsilon$$

Step 5: Solve for the modulus of elasticity.

$$Y = (F/A)/\varepsilon = 54.348 \text{ psi}/4.4643 \times 10^{-3} = 12.1734 \times 10^3$$

With a strain gauge mounted as shown in Figure 4–43, it will undergo the same amount of tensile or compressive stress as the body to which it is attached. As a result, the resistance of the strain gauge will increase with tensile stress and decrease with compressive stress. The proportional relationship of changing body length and changing gauge resistance may be developed as follows. For a resistive material with uniform cross-sectional area:

$$R = \rho \times (L/A) \qquad (4\text{--}21)$$

where L and A represent length and area, respectively, and ρ stands for the resistivity of the given material.

Assuming that the body in Figure 4–43 undergoes tensile stress, the length of the gauge material will increase by a factor of ΔL. At the same time, also due to stretching, the cross-sectional area A of the strain gauge decreases. As a result, the resistance of the strain gauge increases by an amount specified as ΔR. Letting R_0 represent the original

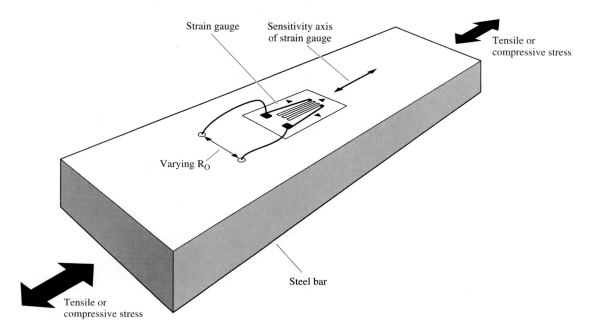

Figure 4–43
Function of strain gauge attached to a steel bar under stress

(normal) value of strain gauge resistance and L_O represent the original strain gauge length, the relationship of strain gauge resistance, length, and cross-sectional area may be expressed as:

$$R = R_0 + \Delta R = \rho \times \frac{(L_O + \Delta L)}{(A_O - \Delta A)} \qquad (4\text{–}22)$$

Note that, while ΔL represents deformation, ΔA represents the difference in cross-sectional area resulting from the applied force.

For most metallic materials, $\Delta L/L$ and $\Delta A/A$ are nearly equal to each other. Also, $\Delta R/R_0$ is nearly twice the value of $\Delta L/L$ or $\Delta A/A$. Thus, a practical statement for the relationship of these variables may be developed:

$$\Delta R/R_0 \cong \Delta L/L_O + \Delta A/A_O \text{ where } \Delta L/L_O \cong \Delta A/A_O$$

Therefore,

$$\Delta R/R_0 \cong 2 \times \Delta L/L_O = 2\varepsilon \qquad (4\text{–}23)$$

Thus, when a strain gauge undergoes tensile or compressive stress, its ratio of resistive change may be expected to be twice the value of the strain incurred by the body to which it is attached. The ratio of $\Delta R/R_0$ to $\Delta L/L$ is referred to as the **gauge factor** (GF). Expressed mathematically,

$$GF = (\Delta R/R_0)/(\Delta L/L) \qquad (4\text{–}24)$$

Thus, for a wire or metallic foil-type of strain gauge, the gauge factor would be approximately 2. Providing the strain gauge is securely bonded to the surface of the metal body,

$\Delta L/L$ for that device will be equivalent to $\Delta L/L$ (or ε) of the body to which it is attached. Thus,

$$\text{GF} = (\Delta R/R_O)/\varepsilon \tag{4-25}$$

The amount of strain incurred by the metal body may then be determined indirectly, simply by detection of the degree of change in strain gauge resistance. Once ΔR is known, strain for the metal body may easily be determined through manipulation of Eq. (4-24). Assuming the gauge factor to equal nearly 2, an expression for ε may be developed as follows:

$$2 \cong (\Delta R/R_O)/\varepsilon$$

Thus,

$$\varepsilon \cong (\Delta R/R_O)/2 \cong (\Delta R/R_O) \times 0.5$$

Values of R_O for a strain gauge range from as low as 30 Ω to as high as 3 kΩ, with typical values lying between 100 and 400 Ω. As a strain gauge undergoes stress, ΔR is likely to be varied by only a few ohms. However, the response of the strain gauge will be highly linear.

Example 4–9

Assume that a strain gauge with an R_O value of 347.6 Ω is attached to a steel bar 14 in. long, as illustrated in Figure 4–44. Assume that, after tensile stress is exerted on the bar, the resistance of the strain gauge increases to 348.3 Ω. What is the value of ΔR? Assuming a gauge factor of 2, what would be the amount of strain expressed in micro-strains? What would be the amount of deformation for the steel bar? Finally, what would be the approximate increased length of the steel bar as it undergoes tensile stress?

Solution:
Step 1: Solve for ΔR.

$$\Delta R = R_2 - R_1 = 348.3 \ \Omega - 347.6 \ \Omega = 0.7 \ \Omega$$

Step 2: Solve for strain.

$$2\varepsilon = \Delta R/R_O = 0.7 \ \Omega/348.3 \ \Omega = 2.0098 \times 10^{-3}$$
$$\varepsilon = 2.0098 \times 10^{-3}/2 = 1.0049 \times 10^{-3}$$

Expressed in $\mu\varepsilon$:

$$\text{Strain} = 1.0049 \times 10^{-3} \times 1 \times 10^6 \cong 1{,}005 \ \mu\varepsilon$$

Step 3: Solve for deformation.

$$\Delta L = \varepsilon \times L = 1{,}005 \times 10^{-3} \times 14 \text{ in.} = 0.0141 \text{ in.}$$

Step 4: Solve for the increased length of the bar.

$$L_2 = L + \Delta L = 14 \text{ in.} + 0.0141 \text{ in.} = 14.0141 \text{ in.}$$

Signal conditioning of the output from a strain gauge, as with a thermistor or an RTD, is accomplished through sensing a change in the output of a voltage divider. A problem associated with strain gauges lies in the fact that such changes in output voltage are likely to be very slight. For example, consider the simple voltage divider in Figure

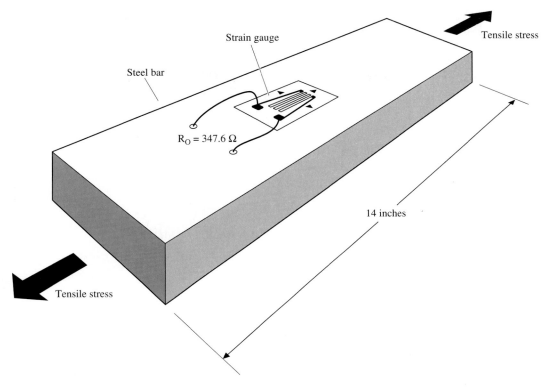

Figure 4–44

4–45. Assume for the strain gauge in this circuit that ΔR is equivalent to that specified in the previous example. With $R\varepsilon$ in the unstressed condition, its exact value of R_O would be as indicated in Figure 4–44. Then the V_O for this condition would be determined as follows:

With no stress: $\quad V_{O_1} = V_S \times \dfrac{R_G}{R_G + R_1} = 12 \text{ V} \times \dfrac{347.6 \ \Omega}{347.6 \ \Omega + 330 \ \Omega} = 6.1558 \text{ V}$

Under stress: $\quad V_{O_2} = 12 \text{ V} \times \dfrac{348.3 \ \Omega}{348.3 \ \Omega + 330 \ \Omega} = 6.1619 \text{ V}$

Note that a relatively significant amount of strain produces a relatively small change in output voltage, which is determined as follows:

$$\Delta V_O = V_{O_2} - V_{O_1} = 6.1619 \text{ V} - 6.1558 \text{ V} = 6.1 \text{ mV}$$

With so small a change in the value of V_O being produced by the voltage divider in Figure 4–45, precise signal conditioning would be quite difficult. This is true for several possible reasons. One source of error could be fluctuations in the level of the dc supply voltage. Unless the source voltage is precisely regulated, deviations in that voltage could easily exceed the 6-mV change brought about by the variation in strain gauge resistance. Also, with such a small change in signal voltage, even the slightest amount of noise within a system could further disrupt the signal conditioning process. For these reasons, the simple

Figure 4–45
Strain gauge in a simple voltage-divider con-figuration

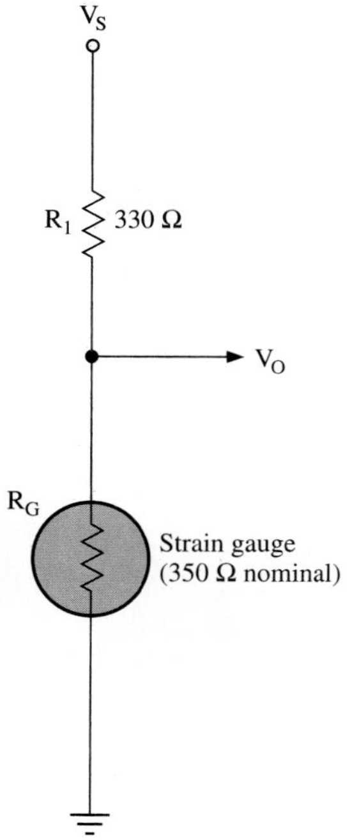

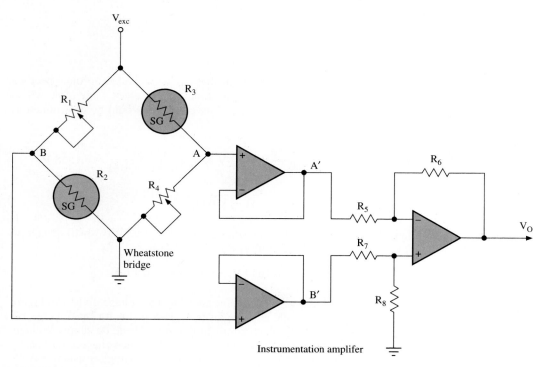

Figure 4–46
Basic strain gauge signal conditioning circuitry

voltage-divider concept must be abandoned in favor of the Wheatstone bridge and differential amplifier contained in Figure 4–46.

The advantages of the design scheme represented here are numerous. First, through utilization of two strain gauges, the range of the strain gauge output signal, taken between points A and B, is effectively doubled. Second, since $(R_1 + R_2)$ will, at any instant, nearly equal $(R_3 + R_4)$, fluctuations in supply voltage and current will be felt equally on either side of the Wheatstone bridge. Therefore, these fluctuations will be nullified as the desired output signal is developed between points A and B. Also, due to the common-mode rejection ratio of the instrumentation amplifier, any noise present at its inputs will be attenuated even further.

As an example of bridge circuit operation, consider the condition represented in Figure 4–47. The two strain gauges could be mounted on either side of the body under stress, as shown in Figure 4–47(a). Ideally, both strain gauges would experience the same stress at the same time. Thus, if both gauges are subjected to compressive stress, the voltage measured from point A to ground would increase, whereas the voltage between point B and ground would decrease. With no stress occurring, both strain gauge resistances would be at their quiescent value of nearly 350 Ω. With R_1 and R_4 both adjusted to nearly 350 Ω, a null condition would exist across the bridge during a nonstress condition.

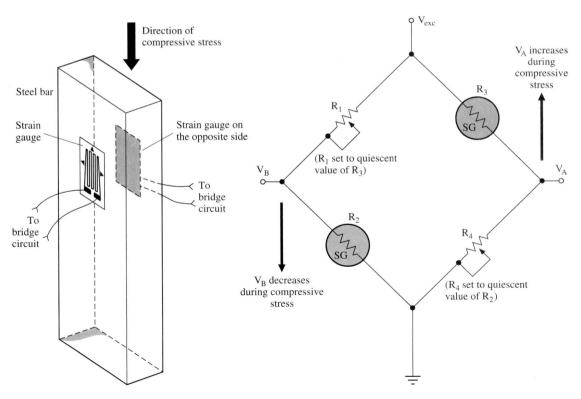

(a) Placement of two active strain gauges on steel bar (Sensitivity axes are identical)

(b) Wheatstone bridge configuration for two active strain gauges

Figure 4–47

Example 4–10

For the circuit in Figure 4–48, assume that the value of R_O for both strain gauges is exactly 350 Ω. Next, assume that both strain gauges undergo 5,245 με of strain as a result of compressive stress. What would be the new ohmic value for R_2 and R_3? What would be the new values of V_A, V_B, and V_{A-B}? Also solve for the values of V_C and V_O.

Solution:
Step 1: Solve for the ohmic values of the strain gauges. Assuming a gauge factor of 2:

$$(\Delta R/R_O)/\varepsilon = 2$$

Substituting

$$(\Delta R/350\ \Omega)/5{,}245 \times 10^{-6} = 2$$

Solving for ΔR:

$$\Delta R = 2 \times 5{,}245 \times 10^{-6} \times 350\ \Omega = 3.6715\ \Omega$$

Since the stress is compressive:

$$R_2 = R_3 = R_G - \Delta R = 350\ \Omega - 3.6715\ \Omega = 346.3285\ \Omega$$

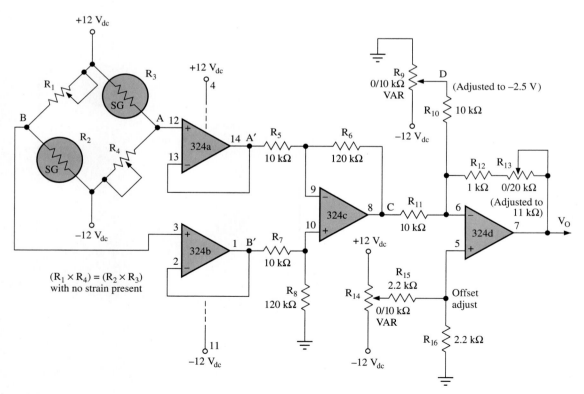

Figure 4–48
Strain gauge bridge and signal conditioning circuitry

Step 2: Solve for the values of V_A, V_B, and V_{A-B}.

$$I_{R_3} \cong I_{R_4} = \frac{12 \text{ V} - -12 \text{ V}}{R_3 + R_4} = \frac{24 \text{ V}}{696.3285 \text{ }\Omega} = 34.467 \text{ mA}$$

$$V_A = -12 \text{ V} + V_{R_4} = -12 \text{ V} + (34.467 \text{ mA} \times 350 \text{ }\Omega) \cong 63 \text{ mV}$$

$$I_{R_1} \cong I_{R_2} = 34.467 \text{ mA}$$

$$V_B = -12 \text{ V} + V_{R_2} = -12 \text{ V} + (34.467 \text{ mA} \times 346.3285 \text{ }\Omega) \cong -63 \text{ mV}$$

$$V_{A-B} = V_A - V_B \cong 63 \text{ mV} - -63 \text{ mV} = 126 \text{ mV}$$

Step 3: Solve for V_C and V_O. Since $R_8/R_7 = R_6/R_5$:

$$V_C = V_{A-B} \times -R_6/R_5 = 126 \text{ mV} \times -120 \text{ k}\Omega/10 \text{ k}\Omega = -1.512 \text{ V}$$

$$I_{R_{10}} = -2.5 \text{ V}/10 \text{ k}\Omega = -250 \text{ }\mu\text{A}$$

$$I_{R_{11}} = -1.512 \text{ V}/10 \text{ k}\Omega = -151.2 \text{ }\mu\text{A}$$

$$I_{FB} = -250 \text{ }\mu\text{A} + -151.2 \text{ }\mu\text{A} = -401.2 \text{ }\mu\text{A}$$

$$V_O = -I_{FB} \times (R_{12} + R_{13}) = 401.2 \text{ }\mu\text{A} \times (1 \text{ k}\Omega + 11 \text{ k}\Omega)$$

$$= 401.2 \text{ }\mu\text{A} \times 12 \text{ k}\Omega = 4.814 \text{ V}$$

As with other resistive devices, strain gauges are affected by changes in temperature. In fact, extreme fluctuations in temperature are likely to produce larger resistance changes than those resulting from variations in strain.

A remedy to the problem of ambient temperature is illustrated in Figure 4–49. Note that this orientation of the two gauges causes their sensitivity axes to be perpendicular. Thus, only strain gauge 1 will actually respond appreciably to changes in stress. (Because strain gauge 2 is mounted perpendicular to the line of force, it would show virtually no response to stress.) However, if the strain gauges are identical, they would exhibit the same thermal response, regardless of their physical orientation. Thus, within the Wheatstone bridge configuration shown in Figure 4–50, strain gauge 1 would be responding to changes in both stress and temperature, while strain gauge 2 would be responding only to temperature fluctuations. Ideally, since both gauges would respond identically to variations in temperature, changes in voltage resulting from ambient temperature effects would be nullified between points A and B.

Even though the bridge circuit in Figure 4–50 contains two strain gauges, only R_4 is being used to detect applied force. (R_2, which is present only for the purpose of temperature compensation, is appropriately referred to as the **dummy** gauge.) Thus, the circuit in Figure 4–50 would be referred to as a 1/4 bridge, since just one of its four resistive components is active. Figure 4–46, however, contains a 1/2 bridge, since both R_2 and R_3 are actually detecting force. A full bridge is represented in Figure 4–51. In such a circuit, strain gauges R_1 and R_4 could be subjected to tensile stress while R_2 and R_3 undergo compressive stress. Thus, the ohmic value of R_1 will increase as that of R_2 decreases. The result of this combined action is a drop in the voltage level at point B. The opposite action occurs at point A, resulting from the increasing ohmic value of R_4 and the decreasing ohmic value of R_3. The full bridge circuit configuration allows for the greatest possible differential output signal while, at the same time, providing for excellent temperature compensation. An application of this form of strain gauge circuit is covered in detail in the next section of this chapter.

In addition to foil-type strain gauges, force-sensing transducers also exist that are made from special types of semiconductor material. These semiconductors exhibit a

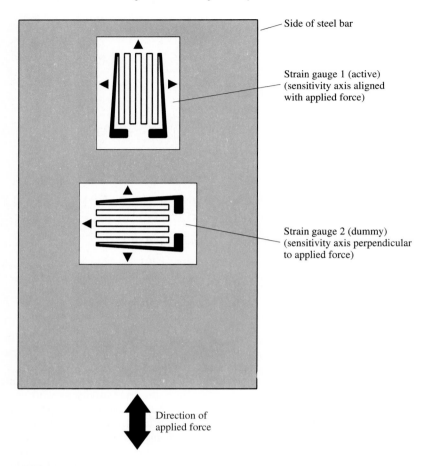

Figure 4–49
Alignment of active and dummy strain gauges on a steel bar undergoing stress

property called **piezoresistance.** Within such materials, resistivity varies as a function of strain. The primary advantage of piezoresistive transducers over conventional strain gauges is their significantly higher gauge factors. While the bonded-wire and foil-type strain gauges typically have a gauge factor of nearly 2, piezoresistive devices have GF values ranging from around 50 to as high as 175. Semiconductor pressure sensors are usually made of silicon. However, germanium may also be used. These devices exhibit excellent temperature stability as long as their operating temperatures are held below 200°C. Due to their high resistivity, semiconductor pressure transducers may be fabricated in the form of very small, thin wafers, which are easily mounted on bodies being subjected to tensile or compressive stress.

The **piezoelectric property** of crystalline materials is widely exploited, especially in such familiar applications as the phonographic needle and the microphone. The piezoelectric effect is also utilized in a few, highly specialized, industrial applications, including the measurement of minute changes in pressure. **Piezoelectric pressure transducers** are extremely sensitive and, therefore, are not found in typical strain measurement applications. Rather, they are often used to detect slight changes in the density of gases and liquids.

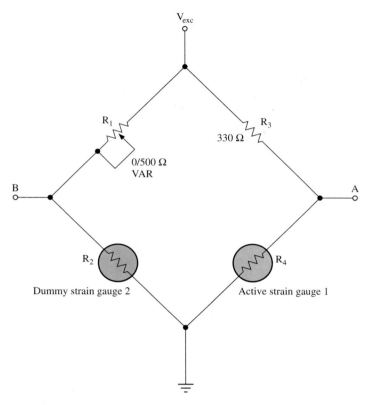

Figure 4–50
Connection of dummy and active strain gauges in a Wheatstone bridge

Also, piezoelectric sensors, commonly made of quartz and barium titanate, are found within ultrasonic sensors.

Piezoelectric transducers are often used to measure force. The **load washer** type of transducer, shown in Figure 4–52(a), is designed to detect axial force. The more complex **dynamometer** shown in Figure 4–52(b) is capable of sensing force from three directions. Note that these lines of force share an **orthogonal** (90°) relationship. Three-element piezoelectric transducers such as that in Figure 4–52(b) are used in milling machine operations to monitor the amount of force being exerted on a workpiece. The transducer shown here is an inherently ac device, having a resonant frequency of around 4 kHz. The piezoelectric pressure sensor is capable of sensing a force of up to 5000 N. However, its response is essentially dynamic. Thus, the device is better suited for detecting rapidly changing forces than monitoring steady-state conditions.

SECTION REVIEW 4–6

1. Assume that a metal beam 15¾ in. long is subjected to tensile stress resulting in a strain factor of 0.00632. What is the amount of deformation? What is the length of the beam while under stress? Express the amount of strain both as a percentage and in microstrains.

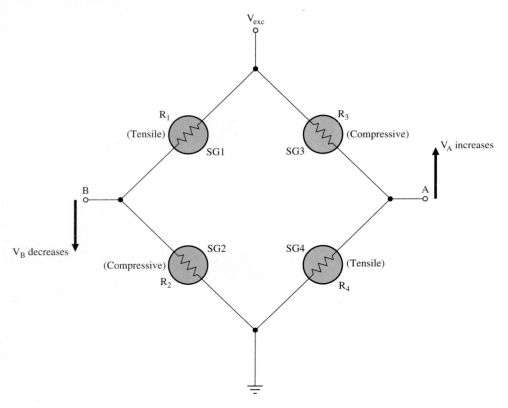

V_{exc}

R_1 (Tensile) SG1

R_3 (Compressive) SG3

V_A increases

B

A

V_B decreases

SG2 (Compressive) R_2

SG4 R_4 (Tensile)

Figure 4–51
Operation of four strain gauges in a full bridge

2. Assume that the metal beam described in Review Problem 1 now has a single strain gauge bonded to its surface with its sensitivity axis aligned in the direction of tensile stress. If the quiescent resistance (R_O) of the device is 1.2 kΩ, what would be the value of ΔR, assuming that the strain factor is still 0.00632 and the gauge factor is 2? What would be the resistance value of the strain gauge while under stress?

3. Assume the strain gauge is still bonded to the metal beam as described in Review Problem 2. If its resistance value is now 1.193 kΩ, what form of stress is being exerted on the beam? What is the amount of stress (expressed in microstrains)? What is the degree of deformation for the metal beam? What is the length of the beam while under stress?

4. True or false?: Figure 4–53 represents a 1/2 bridge circuit.

5. True or false?: The two strain gauges represented in Figure 4–53 would be bonded very close together on the body under stress, with the sensitivity axis of the dummy gauge being perpendicular to the sensitivity axis of the active gauge.

6. Describe the exact function of the dummy gauge in Figure 4–53.

7. For the circuit in Figure 4–54, assume that R_2 and R_3 are both bonded to a metal block. The sensitivity axes of the gauges are both aligned to detect compressive force, as shown in Figure 4–47(a). The block is 11¾ in. thick, 18½ in. wide, and 8³⁄16 in. long. Assume that block is being subjected to 2.43 tons of pressure in the direction

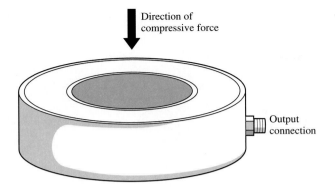

(a) Load washer form of piezoelectric force transducer

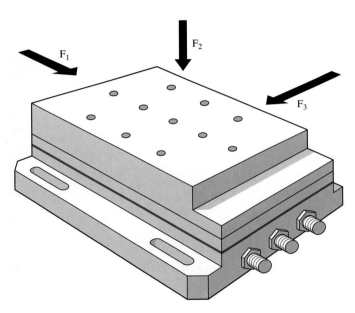

(b) Dynamometer form of piezoelectric force transducer

Figure 4–52

of its 11¾ in. dimension. Also assume the gauge factor of R_2 and R_3 equals 2. Given a Young's modulus of 1.24×10^4, solve for the following:

Stress = _____ psi $\Delta R_2 = \Delta R_3$ = _____ Ω

ε = _____ V_{A-B} = _____ V

με = _____ V_C = _____ V

ΔL = _____

8. True or false?: Foil-type strain gauges tend to have much higher gauge factors than semiconductor force transducers.

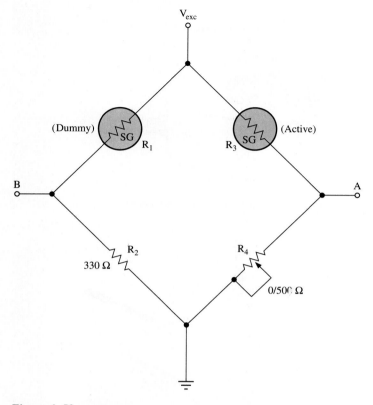

Figure 4–53

9. True or false?: Semiconductor force transducers exhibit a property known as piezoelectricity.

10. Briefly explain a few applications of piezoelectric pressure-sensing transducers.

4–7   LOAD CELLS AND LOAD CELL APPLICATIONS

The term **load cell** is generally used to describe an industrial weighing device. Load cells are commonly used to monitor the weight of materials stored in vats, hoppers, and other large stationary containers. Also, load cells may be mounted under conveyance systems in order to monitor the flow of material.

A basic load cell may be formed by combining four strain gauges into a full-bridge configuration, as represented in Figure 4–55(a). The physical structure of such a **tension load** cell is illustrated in Figure 4–55(b). As the body being weighed pulls downward, the **proving ring** is elongated. As a result of this distortion, R_1 and R_4 undergo compressive stress, which causes a decrease in their ohmic values. At the same time, R_2 and R_3 undergo tensile stress, causing an increase in their ohmic values. Therefore, the differential voltage developed from point A to point B will vary as a function of the weight of the body being suspended from the load cell. (The method for determining the value of V_{A-B} was demonstrated in the previous section of this chapter.)

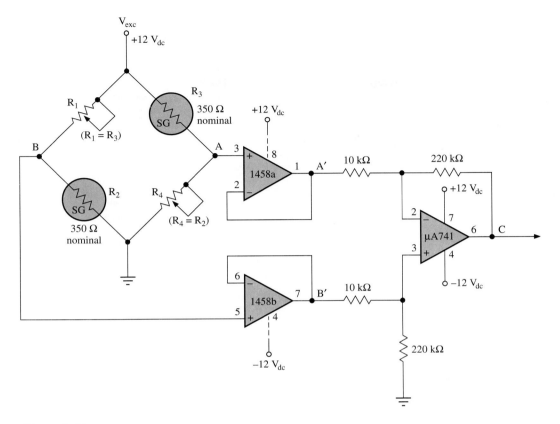

Figure 4–54

A sample application of the tension load cell is shown in Figure 4–56. Note that the load cell is positioned such that it will sense the changing tensile force being exerted by the suspended hopper. The **gross weight** of the suspended hopper, which includes both the weight of the hopper and its contents, may be detected simply by connecting the differential output of the load cell directly to an instrumentation amplifier. However, in most load cell applications, the **tare weight,** which is the weight of the container itself, must be negated in order to determine the actual weight of the material it holds. The weight of the material within the container, which is referred to as the **net weight,** may be determined as follows:

$$\text{Net weight} = \text{gross weight} - \text{tare weight} \qquad (4-26)$$

Negation of the effect of the tare weight may be easily accomplished by connecting compensation circuitry as shown in Figure 4–57. Through a one-time manual adjustment of **tare pot** R_5, Eq. (4–26) may be implemented. Note that V_{A-B}, developed directly across the bridge, represents the combined weight of the suspended hopper and its contents. Since V_{B-C} is series opposing with respect to V_{A-B}, the following relationship may be established using Kirchhoff's law:

$$-V_{A-B} + V_{B-C} + V_{C-A} = 0$$

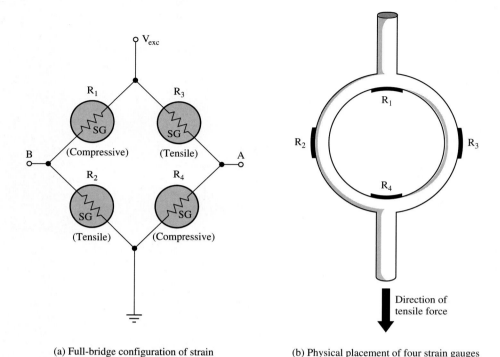

(a) Full-bridge configuration of strain gauges for a tension load cell

(b) Physical placement of four strain gauges on the proving ring of a tension load cell

Figure 4–55

Thus,

$$V_{C-A} = V_{A-B} - V_{B-C} \qquad (4\text{--}27)$$

where V_{C-A} represents net weight, V_{A-B} represents gross weight, and V_{B-C} represents the tare weight.

While V_{A-B} will vary with the changing weight of the contents of the hopper, V_{B-C}, once adjusted manually, will remain constant. With the hopper empty, the tare pot would be adjusted until a null condition exists between points A and C. At this time, V_{A-B} and V_{B-C} would be of equal value and opposite polarity. Thus:

$$V_{A-B} - V_{B-C} = 0$$

(Note that this condition indicates a net weight of 0.)

The signal developed between points A and C in Figure 4–57, though highly linear, is likely to be only a few millivolts in amplitude. Therefore, it must undergo extensive signal conditioning. After being processed by an instrumentation amplifier, this signal would then be sent to system control circuitry, which would be responsible for operation of both the input conveyance system to the hopper and the solenoid, which actuates the hopper's output control valve. Through continual monitoring of the weight of the contents of the hopper, the control system would attempt to maintain a constant level of material within the hopper. In a modern industrial facility, such a system is likely to be controlled by a microcomputer. Therefore, the final stage of signal conditioning would require A/D conversion.

Figure 4–56
Application of a tension load cell

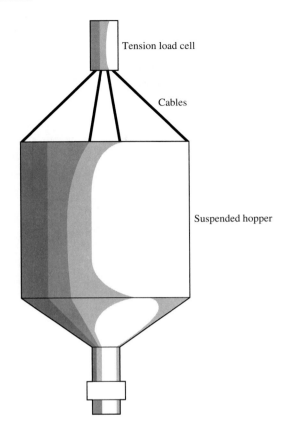

Tension load cell

Cables

Suspended hopper

The major problem confronting the designer of signal conditioning circuitry for the strain gauge form of load cell lies in the fact that the differential input signal is in the millivolt range. Commercially manufactured load cells are usually rated in terms of millivolts per volt of bridge excitation. For example, if a load cell is specified as having an absolute full-scale output of 2 mV/V and the bridge excitation voltage is 10 V, the maximum differential voltage produced at the output of the load cell would be only 20 mV. With so small an initial signal, an instrumentation amplifier exhibiting both a high input impedance and a high common-mode rejection ratio (CMRR) must be created. Such a circuit is shown in Figure 4–58. The μA725 op amp used in Figure 4–58 is especially designed for instrumentation applications. Though characterized by a high input impedance and an especially high CMRR, this first generation op amp lacks the internal frequency compensation provided within many second-generation devices such as the μA741. Therefore, external frequency compensation circuitry may be necessary in ac signal conditioning applications to counteract the effects of the internal capacitance of the device. At higher operating frequencies, the gain of a typical op amp will begin to deteriorate. Also, with sinusoidal waveforms, considerable phase-shifting occurs. With rectangular waveforms, distortion is particularly noticeable, due to the lack of higher frequency components within the output waveform. (See Figure 2–6 in Chapter 2.) In system applications where the bridge excitation voltage consists of a high-frequency ac signal, frequency compensation within the instrumentation amplifier would be essential.

For the amplifier in Figure 4–58, to obtain maximum CMRR, the ratio of (R_9/R_8) should equal (R_{11}/R_{10}) and (R_5/R_8) should equal (R_7/R_{10}).

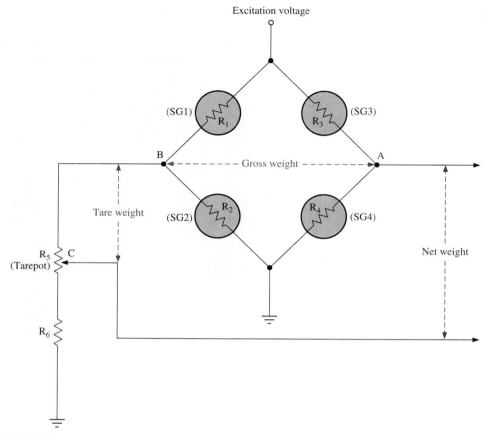

Figure 4–57
Full-bridge and tare circuitry for a tension load cell

Example 4–11

For the load cell in Figure 4–59, assume the bridge excitation voltage is a precisely regulated 12 V_{dc}. If the absolute full-scale output is 2.67 mV/V, what would be the maximum differential voltage developed between points A and B? If the maximum weight being detected by the load cell is 4000 lb and the hopper itself weighs 830 lb, what is the maximum value of net weight? What should be the value of V_{A-C} during this maximum net weight condition? With this maximum value of V_{A-C}, assume V_O must equal +2 V. What should be the ohmic setting of R_{14}? What must be the overall gain of the amplifier circuitry in order to achieve this output voltage?

Solution:
Step 1: Solve for maximum V_{A-B}.

$$V_{A-B}(\text{max}) = 12 \times -2.67 \text{ mV} = -32.04 \text{ mV}$$

Step 2: Solve for the maximum net weight.

$$\text{Net weight} = \text{gross weight} - \text{tare weight}$$
$$= 4000 \text{ lb} - 830 \text{ lb} = 3170 \text{ lb}$$

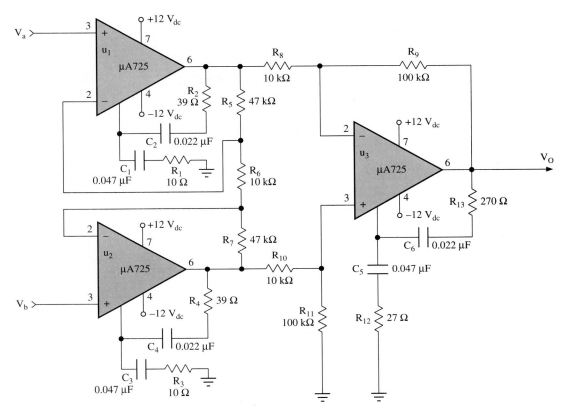

Figure 4–58
Instrumentation amplifier with high CMRR

Step 3: Solve for the maximum value of V_{A-C}.

$$V_{A-C}(\text{max}) = (3170 \text{ lb}/4000 \text{ lb}) \times -32.04 \text{ mV} = -25.392 \text{ mV}$$

Step 4: Solve for the ohmic setting of R_{14}.

$$A_{V_1} = -(R_9/R_8) = -10$$

Thus,

$$V_{A'-C'} = -10 \times -25.392 \text{ mV} = 253.92 \text{ mV}$$

$$A_{V_2} = 1 + \frac{R_{16}}{(R_{14} + R_{15})} = V_O/V_D = 2 \text{ V}/253.92 \text{ mV} = 7.8765$$

$$1 + \frac{68 \text{ k}\Omega}{R_{14} + 2.2 \text{ k}\Omega} = 7.8765 \qquad \text{Thus,} \frac{68 \text{ k}\Omega}{R_{14} + 2.2 \text{ k}\Omega} = 6.8765$$

$$R_{14} + 2.2 \text{ k}\Omega = (68 \text{ k}\Omega/6.8765) = 9.8888 \text{ k}\Omega$$

Thus,

$$R_{14} = 9.8888 \text{ k}\Omega - 2.2 \text{ k}\Omega = 7.6888 \text{ k}\Omega$$

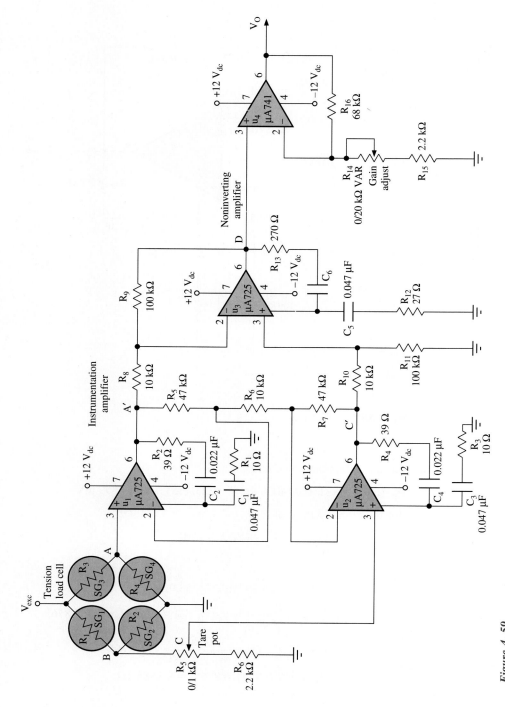

Figure 4–59
Tension load cell with instrumentation and non-inverting amplifier

Step 5: Solve for overall gain:

$$A_{V'} = A_{V_1} \times A_{V_2} = -10 \times 7.8765 = -78.765 \text{ or}$$

$$A_{V'} = V_O/V_{A'-C'} = 2\,V/-25.392\,mV = -78.765$$

Thus far, the operation of the load cell has been explained from the standpoint of sensing tensile stress. Load cells also exist that operate on the principle of compression. For lightweight to mediumweight applications, load cells containing strain gauges may still be found. However, load cells containing LVDTs are also prevalently used for this purpose. The detailed structure of a load cell containing an LVDT is illustrated in Figure 4–60. This versatile weighing device has a platform at the top, on which could be mounted the actual platform of a scale or a container such as a hopper. The resultant line of force would then be downward toward the yoke. A strong range spring provides counterpressure, constantly exerting force upward against the yoke. A bracket that is welded to the inside of the housing holds the windings of the LVDT in a fixed position. Since the ferrous plunger of the LVDT is attached to the yoke, its movement will be in proportion to the amount of applied force. Thus the differential output signal will vary as a function of weight.

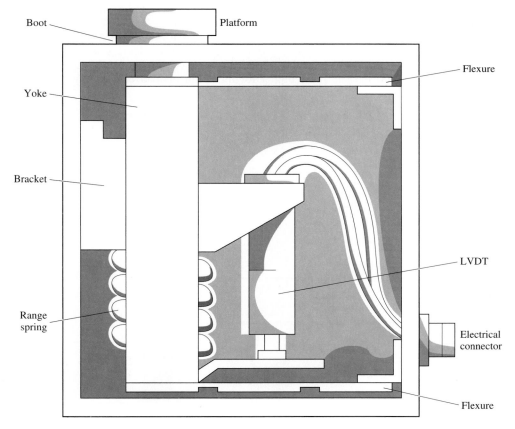

Figure 4–60
Cutaway view of the internal structure of an LVDT load cell

A dynamic application of the LVDT form of load cell is shown in Figure 4–61. Here, the load cell continually monitors the weight of material as it is moved along a conveyance system. Providing the velocity of the conveyor belt is also being sensed, the flow rate of the moving material can be easily determined. The flow rate is directly proportional to both the weight and the velocity of the material being transported. Thus, the flow rate may be obtained by deriving the product of the load cell output and the output of the velocity sensing device. (Methods of detecting conveyor belt speed are covered in detail later in this text.) This multiplication process could be performed directly by analog function circuitry. However, a more likely approach would involve the conversion of both outputs to binary data via A/D conversion circuitry. A microcomputer could then perform a multiplication algorithm, thus deriving the flow rate. This same microprocessor-based system, through use of the derived information, could also control the speed of the conveyor, attempting to maintain a constant flow rate.

Load cells may even be used to determine indirectly the level of materials contained in large storage vats. As shown in Figure 4–62, three load cells may be used to measure simultaneously the weight of the vat and its contents. For such applications, strain gauge load cells capable of withstanding extremely large weights must be utilized. As with ultrasonic level detectors, the advantage of the system represented in Figure 4–62 lies in the fact that the transducers do not contact the material being monitored. Assuming the tare weight (weight of the container) is already known and compensated for, this measuring method may be highly accurate, since deviations in force detected by the load cell will correspond to the changing level of material within the vat.

Commercially manufactured load cells are available in a variety of designs and weight ranges. Such load cells are shown in Figure 4–63. The **S-type** devices shown in Figure 4–63(a) are similar in structure, having standard capacities ranging from as low as 50 lb to ͭs high as 10,000 lb. The rated output for this form of load cell is only 3.0 to 3.6

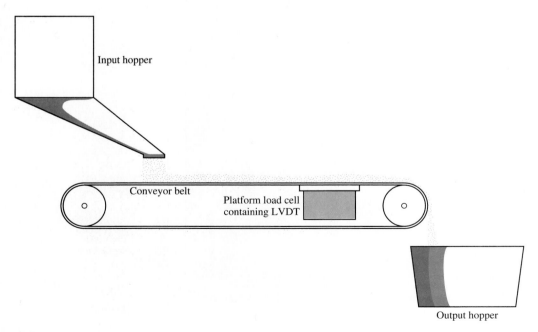

Figure 4–61
Dynamic application of a load cell within a conveyance system

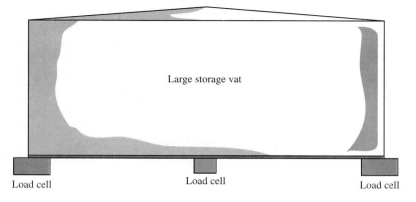

(a) Three load cells used to monitor weight of material in a storage vat

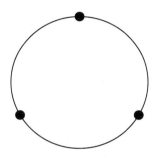

(b) Bottom view: placement of load cells

Figure 4–62

Figure 4–63
(a) S-type load cells and (b) Load cell designed for measurement of extremely heavy weights load cell (Courtesy of Revere Transducers)

(expressed in mV/V). However, they are highly accurate, having a maximum **combined error** of only 0.03%. The combined error is expressed as a percentage of the full-scale differential output voltage, measured between a no-load condition and the maximum weight condition.

Example 4–12

Assume an S-type load cell has an actual full-scale output of 3.35 mV/V and is given an excitation voltage of exactly 12 V_{dc}. What would be the ideal maximum value of output voltage? What would be the maximum allowable amount of error, based on a combined error rating of only ±0.03%?

Solution:
At full-scale weight:

$$V_{A-B} = 12 \times 3.35 \text{ mV} = 40.2 \text{ mV}$$
$$0.03\% = 0.03/100 = 3 \times 10^{-4}$$
$$\text{Maximum error voltage} = 3 \times 10^{-4} \times 40.2 \text{ mV} = 12.06 \text{ μV}$$

The S-type load cell shown in Figure 4–63(a) is ideally suited for use either in a platform scale system or within a conveyance system, as shown in Figure 4–61. The device in Figure 4–63(b), however, is designed for much heavier weighing applications. While the rated output of this device may be only 1.75 mV/V, transducers of this type are available that are capable of safely withstanding and accurately measuring up to 500,000 lb! The relatively low output voltage rating of this load cell is compensated for by the fact that it may accept an excitation voltage (dc or ac) of up to 25 V. [The device in Figure 4–63(a) might have a maximum excitation voltage of only 15 V.] The form of load cell shown in Figure 4–63b, due to its extreme accuracy and wide weight range, is ideal for use in truck weighing systems and in heavy storage vat systems, as shown in Figure 4–62.

---

1. Define the terms *gross weight, tare weight,* and *net weight* as they apply to industrial weighing systems.
2. The combined weight of a hopper and its contents is 1432 lb. Assuming that the hopper weighs 312 lb when totally empty, determine the net weight and the tare weight.
3. A load cell used to measure the weight of a very heavy, suspended hopper is likely to consist of
   a. an LVDT
   b. a proving ring
   c. four strain gauges in a full-bridge configuration
   d. both choices b and c.
4. The maximum weight rating for an S-type load cell is 500 lb. When a load with a gross weight of only 300 lb is placed on the device, its differential output voltage equals 19.04 mV. Assuming that the excitation voltage is exactly 12 V_{dc}, what is the full-scale output voltage of the device in mV/V?
5. For the weighing system shown in Figure 4–59, assume that the voltage measured from point C to point A is now the 5-kHz sine wave shown in Figure 4–64(a). If the output V_O of the signal conditioning circuitry in Figure 4–59 is the sine wave in Figure 4–64(b), what must be the ohmic setting of potentiometer R_{14}?

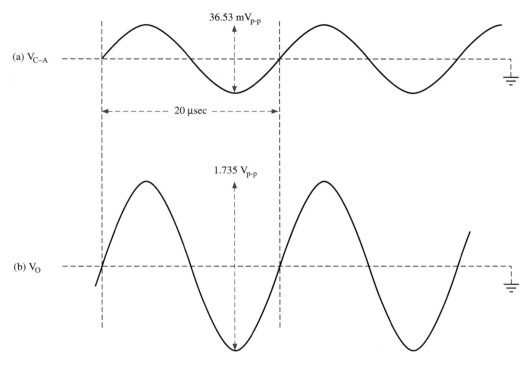

Figure 4–64
Waveforms for load cell and signal conditioning circuitry in Figure 4–59

6. Describe at least two applications of the load cells shown in Figure 4–63(a). Explain a possible application of the load cell in Figure 4–63(b).

SUMMARY

The displacement potentiometer and the linear variable differential transformer are capable of converting differences in physical position to proportional changes in voltage. These transducers may be linked with moving machine parts in order to monitor changes in their positions. Either the displacement potentiometer or the LVDT may be linked with such pressure-sensing transducers as the Bourdon tube and the bellows. These force-summing transducers require mechanical linkage with displacement transducers in order to convert changes in pressure to changes in voltage. While potentiometers are less costly than LVDTs and operate from either dc or ac sources, LVDTs are more accurate and, therefore, more likely to be used in precision measurement and control applications.

Magnetic proximity sensors are capable of detecting the approach of ferrous objects. They are commonly used in electromechanical and conveyance system control for the detection of the arrival of a part in the proper location or detection of a possible jamming condition. Two basic forms of magnetic proximity sensor are the reluctance and the ECKO devices. Magnetic proximity sensors are sturdy and quite durable. For these reasons, they are often used in applications where microswitches and optical sensors could possibly be knocked out of alignment or destroyed.

Capacitive level detectors are often used to monitor the level of liquids, slurries, and granulated materials being stored within sealed vats. Bare capacitance probes may be used safely to monitor nonconductive materials. However, for corrosive and conductive materials, insulated probes must be used. In some measurement applications, the concept of immersion of the transducer may be abandoned entirely in favor of an ultrasonic form of sensor. Also, heavy vats may be placed on top of load cells, which, through force detection, are able to accurately sense the level of the material stored within.

Flow sensors are important types of input transducers that are able to translate the changing rate of material flow into a varying electrical signal. Turbines and nutating disks are placed directly in the path of moving fluids. Changes in their rotational velocity are translated into changes in pulse output frequency. Magnetic flow-sensing transducers have the advantage of being mounted entirely on the outside of a conduit. Therefore, they do not impede the flow of material. Venturi points, used in conjunction with bellows and displacement transducer assemblies, detect changes in flow rate by actually sensing differences in pressure on either side of the measurement point.

Strain gauges are resistive devices capable of detecting the effects of tensile and compressive stress on metals. They may be combined into bridge networks to form load cells. Though load cells produce differential output signals limited to only a few millivolts, they are highly linear. If these signals are properly conditioned, highly accurate weight and force measurement is possible. Load cells may also be created with LVDTs. Such devices are effective in lightweight and mediumweight measurement applications. Pressure-sensing transducers may also be fabricated from piezoresistive and piezoelectric substances. Piezoresistive transducers have much higher gauge factors than foiltype strain gauges. Piezoelectric pressure-sensing transducers, such as the dynamometer, are used in such industrial applications as machine tooling.

SELF-TEST

1. For the bridge circuit in Figure 4–65, assume that V_{A-B} equals -423 mV. What would be the amount of resistance between the wiper and the positive voltage source? What would be the resistance between the wiper and the circuit ground? In order for this differential voltage to have occurred across the bridge, how far (as a percentage of total resistance length) would the wiper have moved from the central (null-state) position?

2. Assume that displacement potentiometer R_1 in Figure 4–65 consists of 225 turns. What is its percentage of deviation? Based on the bridge excitation voltage shown, what would be the resolution of the potentiometer output?

3. Identify two forms of transducer that an LVDT could be mechanically linked with in order to detect changes in pressure.

4. True or false?: LVDTs are normally less expensive than displacement potentiometers but tend to be less accurate.

5. True or false?: The secondary voltages of an LVDT are series-opposing.

6. ECKO is the abbreviated name for what form of transducer? Explain the operation of this device.

7. True or false?: For a bare capacitance probe, the dielectric is actually formed by the material being stored within the container in which the probe is immersed.

8. A bare capacitance probe is immersed in a sealed vat that is completely full of a nonconductive fluid having a dielectric constant of 47. The length of the probe is 4

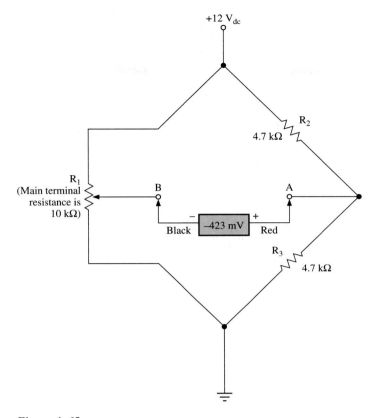

+12 V$_{dc}$

R$_2$
4.7 kΩ

R$_1$
(Main terminal
resistance is
10 kΩ)

B

A

Black   –423 mV   Red
−   +

R$_3$
4.7 kΩ

Figure 4–65

ft, 9 in., and its diameter is 3/4 in. If the capacitance of the probe is now 960 pF, what would be the diameter of the vat?

9. For the probe mentioned in Problem 8, explain why capacitance would decrease as the vat is emptied.

10. For the capacitive bridge network of Figure 4–21a, assume that C_0 for the bare capacitance probe is 678 pF. Assuming C_4 is adjusted to 875 pF and V_{exc} equals 5 V, what would be the value of V_{A-B}?

11. Explain the operation of the demodulator in Figure 4–25. Why is this circuit better suited for signal conditioning than a simple diode bridge rectifier and filter?

12. True or false?: Excitation signals for bridges containing bare capacitance probes typically occur at around 1000 Hz.

13. For an ultrasonic level detector, the difference in time between the output burst and return echo is 34 msec. What is the distance between the transducer and the reflecting surface?

14. Explain the operation of an ultrasonic differential level detector. Identify a possible industrial application of this device.

15. How many psia units is 5.875×10^4 pascal units equivalent to?

16. Explain the operation of a nutating disk. What are two limitations of this device?

17. Describe the operation of the tension load cell represented in Figure 4–57. Assuming the four strain gauges are mounted on the proving ring in Figure 4–55(b), identify

which undergo tensile stress and which experience compressive stress. What is the exact purpose of potentiometer R_5?

18. For the weighing system in Figure 4–59, assume that the bridge excitation voltage is now 10 V_{dc}. Also assume that the tension load cell is being subjected to a maximum-weight condition. With this maximum load, the output voltage equals 2.75 V_{dc}. If R_{14} is set to 5.985 kΩ, what must be the value of V_{A-C}?

19. Explain the structure and operation of a compressive load cell containing an LVDT. Describe a dynamic application of this device.

20. Assume that, while subjected to maximum weight, the output of a load cell is 27.45 mV. If the excitation voltage being applied to the device is 12 V_{dc}, what would be the absolute output rating in mV/V?

5 OPTOELECTRONIC DEVICES AND SIGNAL CONDITIONING CIRCUITRY

CHAPTER OUTLINE

LEARNING OBJECTIVES

On completion of this chapter, the student should be able to:

- Explain the characteristics of infrared, incandescent, ultraviolet, and phosphorescent light.
- Explain the difference between the terms *photometric* and *radiometric* as they pertain to the operation of optoelectronic devices.
- Explain how infrared, incandescent, and ultraviolet energy can be exploited in the operation of industrial control systems.
- Describe the operation of industrial, incandescent, and ultraviolet light sources.
- Explain the special power requirements as well as the safety precautions involving halogen and ultraviolet lamps.
- Describe in detail the structure and operation of the LED. Explain the structure and operation of both the common-anode and the common-cathode seven-segment display.
- Explain the operation of the photoconductive cell as a linear transducer.
- Describe the structure of the photodiode. Explain both its photoconductive and photo-voltaic modes of operation.
- Describe the structure of the phototransistor. Explain its operation as a linear transducer.
- Explain how photoconductive cells and phototransistors may be connected to op amp signal conditioning circuitry.

- Describe the advantages of using optoisolators when interfacing between printed circuit boards.
- Explain how an optocoupled interrupter module is used within a shaft encoder to monitor either rotational velocity or conveyor belt travel.
- Explain how infrared light barriers and photosensors may be used in conjunction with shaft encoders to monitor gap, length, and distance within conveyance systems.
- Explain how the elements of bar code may be used to form binary numbers.
- Explain the basic structure and operation of bar code readers. Describe how these devices may be used within sorting systems.
- Explain the basic structure and operation of the diode arrays used within video acquisition systems. Explain the signal conditioning process required for these devices.
- Explain the basic structure and operation of a photomultiplier tube.
- Explain the basic structure and operation of a laser.

INTRODUCTION

The purpose of this chapter is to explain the structure and operation of various light sources and optical transducers commonly found in modern industrial control systems. After the on/off function of simple infrared light barriers and photosensors is introduced, this chapter progresses into the study of more complex optoelectronic devices and their applications. Included in this study are systems that utilize optoelectronic devices to measure light intensity, monitor and control motor speed, and detect the presence of special stamps or labels on envelopes and packages. After bar code readers and basic video acquisition systems are presented, industrial applications of photomultiplier tubes and lasers are briefly surveyed.

5–1   INTRODUCTION TO LIGHT

Light may be generally described as **electromagnetic radiation,** a form of energy that is characterized by simultaneous periodic variations in the intensity of both an electrical and a magnetic field. The various types of electromagnetic energy are organized by wavelength into the **electromagnetic spectrum,** which includes radio waves, infrared energy, visible light, ultraviolet energy, x-rays, and gamma rays. The common characteristic of all forms of electromagnetic radiation is the fact that, within a perfect vacuum, it is propagated at a rate of 186,000 miles per second.

Technically speaking, the term *light* pertains only to that narrow portion of the electromagnetic spectrum that is visible to human beings. The term **photometry** refers to the measurement of visible light. Thus, a photometric sensing device would be one that has a sensitivity range similar to that of the human eye. **Radiometry,** however, may be defined as the measurement of electromagnetic energy within a much broader range, including the infrared, visible, and ultraviolet bands. Thus, a radiometric sensor would, ideally, be able to detect the presence of all three types of energy with equal sensitivity. In reality, optoelectronic devices have been developed that are sensitive to specific types of radiant energy. An example of such a device would be the infrared phototransistor.

Thus, the term *optoelectronic* is not confined to only those devices that emit or detect visible light. Rather, it applies equally to those semiconductors that produce or detect infrared, visible, or ultraviolet radiation. Table 5–1 summarizes the audio-frequency band and the electromagnetic spectrum, showing the relationship of audio, radio, and light frequencies. As the frequency of the energy form increases, its wavelength decreases. Beyond the extra-high radio-frequency (RF) range, the familiar unit of the hertz is no

Table 5–1
Audio and electromagnetic frequency spectrum

Frequency/Wavelength	Form of Energy
0 Hz	Purely dc current or voltage
16 to 16 kHz	Audio-frequency band
16 kHz to 30 kHz	Ultrasonic frequency band
30 kHz to 30,000 MHz	Radio-frequency band
30,000 MHz to 300,000 MHz	Extra-high frequency band
300,000 MHz to 7600 Å	Infrared radiation band
7600 Å to 3900 Å	Visible light radiation band
320 Å to 0.1 Å	Ultraviolet radiation band
0.1 Å to 0.006 Å	Gamma-ray band
0.006 Å →	Cosmic rays

longer practical. It is at this point in the electromagnetic spectrum that **wavelength** becomes the more practical unit. The units commonly used to designate wavelengths of light energy are the **nanometer** (nm), the **micrometer** (μm), and the **angstrom** (Å). One nanometer is equivalent to 1×10^{-9} m. A micrometer is equal to 1×10^{-6} m, and an angstrom represents a wavelength of 1×10^{-10} m. Note that, as shown by equation 5–1, frequency and wavelength have a reciprocal relationship. Thus, wavelength will decrease as frequency increases.

$$\lambda = \frac{c}{f} \tag{5–1}$$

(Where λ is wavelength, c is the speed of light, and f is frequency.)

According to quantum theory, light exists in the form of photons, which may be described as tiny particles of light energy. The energy contained by a photon is directly proportional to its frequency. Thus:

$$W = \hbar f \tag{5–2}$$

(Where W is energy and $\hbar$ is Planck's constant, equivalent to 6.624×10^{-34} joules per second.)

The color red represents the lowest frequency and, hence, longest wavelength within the spectrum of visible light. The prefix *infra* is the Latin term for "under" or "below." Thus, the term **infrared** would describe the energy band occurring just below the band of visible light. The color violet represents the highest frequency and, hence, shortest wavelength within the spectrum of visible light. The prefix *ultra* is the Latin term for "beyond." Thus, the term **ultraviolet** would describe the energy band occurring just above the band of visible light. Infrared light is used extensively in the operation of industrial light barriers and photosensors, especially those found in automated conveyance systems. Ultraviolet light is used in a variety of specialized industrial processes.

The visible light spectrum is divided into several distinct colors. However, the form of visible light most commonly associated with industrial process control is diffused or white light, which is a mixture of virtually all of the colors within the visible spectrum. White light produced by the conventional light bulb may also be referred to as **incandescent** light, because it is produced by sources that must become hot prior to radiating energy. (The term *incandescent* is derived from the Latin *incandescere,* meaning to become hot and glow.) Thus, incandescent light sources, in addition to white light, also produce considerable heat. Sources of infrared and ultraviolet radiation, by contrast, produce virtually no heat.

While incandescence may be defined as the ability of a material to radiate light as a result of being heated, **phosphorescence** is defined as the ability of a material to emit light both during and after exposure to radiation. Phosphorescent material is actually ionized by the radiation source, resulting in a glowing effect referred to as **scintillation.** Truly phosphorescent materials exhibit an afterglow, being able to sustain their luminescence for an appreciable amount of time after the radiant source has been removed. **Fluorescence** is defined as the ability of a material to emit light during exposure to radiation. However, with fluorescent materials, luminescence occurs only while the source of radiation is actually present. The properties of phosphorescence and fluorescence are both used in industrial process control, with ultraviolet light usually serving as the radiant source. Phosphorescent and fluorescent light produced through exposure to ultraviolet radiation is characterized by the virtual absence of heat. It is this attractive feature that has led to the predominant use of fluorescent lighting.

SECTION REVIEW 5–1

1. Define the term *incandescent.* What is a by-product of virtually all incandescent light sources?
2. Define the term *ultraviolet.* Where does this form of radiation occur within the electromagnetic spectrum?
3. Define the term *infrared.* Where does this form of radiation occur within the electromagnetic spectrum?
4. True or false?: Electromagnetic energy is radiated through the variation of either a single electrical or a single magnetic field.
5. True or false?: An infrared phototransistor would be an example of a photometric sensor.
6. Define the process of phosphorescence.
7. A wavelength of 17,582 Å is equal to how many micrometers?
8. Explain the difference between fluorescence and phosphorescence.
9. Define scintillation.
10. What form of radiant energy is often used to induce fluorescence and phosphorescence?

5–2   INDUSTRIAL LIGHT SOURCES

Various kinds of lamps serve important functions within industrial process control. Simple applications of incandescent lamps include heating and drying of granulated or powdered materials. Also, incandescent and ultraviolet (UV) lamps have gained increasing importance in modern conveyance systems, which rely heavily on optoelectronics. Specially designed incandescent lamps and lens assemblies are used to provide extremely concentrated beams of white light for the operation of bar code readers and video acquisition systems. UV lamps are also used extensively in automated document sorting systems. This is especially true with regard to postal sorting machines, which rely heavily on the properties of fluorescence and phosphorescence in the recognition of valid indicia (a stamp or meter mark).

Incandescent lamps are the oldest and most familiar form of electric lighting. The structure of a typical incandescent lamp is illustrated in Figure 5–1. When power is applied to the lamp and current begins to flow through its filament, the filament emits light as a result of heating. Incandescent light bulbs manufactured for home use may contain either a vacuum or gas. In either case, however, their operating principle is basically the same.

Figure 5–1
Typical incandescent light source

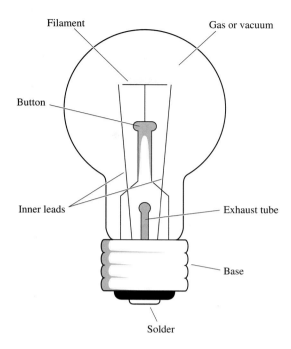

When light bulbs are selected for general use, the primary consideration is the wattage rating, which simply represents the amount of electrical power the bulb is likely to dissipate as it converts electricity to light and heat. However, for industrial applications such as video acquisition systems and bar code readers, the precise amount of light produced by the incandescent source becomes an extremely important factor. The unit of luminous flux (visible light) is the **lumen.** To define the lumen, two important terms must first be introduced. A **blackbody** is actually an ideal object, one that absorbs all forms of radiation and, therefore, exhibits absolutely no reflectivity. The concept of the blackbody serves as a basis for determining the intensity of incandescent light, since none of the light emanating from its surface would have originated from another source. The **candle** is a specific unit of light intensity, being equivalent to 1/16 of the luminous intensity of a square centimeter of the surface of a blackbody at the solidification temperature of pure platinum (1755°F). With the unit of the candle now established, the unit lumen may be specified. A lumen is equal to the light emitted in a straight line by a point source of one candle intensity. Incandescent light sources are likely to be rated by lumens as well as wattage. For instance, a typical 60-W light bulb manufactured for home use could produce between 850 and 900 lumens.

Incandescent lamps used for home and commercial lighting are usually powered directly from an ac source. (Methods of controlling the intensity of such lighting, through various forms of dimmer circuitry, are addressed in Chapters 7 and 8.) For bar code readers and video acquisition systems, however, powering and controlling the required incandescent lamps directly from an ac source would prove unsatisfactory. The ac line voltage, especially if unfiltered, could easily introduce noise into the video signal conditioning circuitry. In fact, the 60-cycle alternations would be reflected in the video signal, interfering greatly with the data acquisition process. Thus, for reliable operation, incandescent lighting within such systems is powered from a dc source.

Halogen lamps are used extensively in bar code readers and video acquisition systems, due to their relatively small size and high luminescence. In such applications, where the level of light must remain absolutely constant, these lamps must be operated from well-regulated dc power supplies. Due to their relatively small size, halogen lamps are easily placed in the various housings and assemblies associated with video acquisition systems. However, halogen lamps become *extremely* hot, requiring heat-sinking and possibly even air cooling (from a fan or pneumatic system). Unless adequate heat escape is provided for halogen lamps contained within enclosed housings, costly optical devices such as lenses, beamsplitters, and filters may eventually crack. However, assuming that adequate cooling is provided, halogen lamps are likely to last far longer than other forms of incandescent light. When removing a burned out halogen lamp, several minutes should be allowed for the bulb to cool. Also, when replacing a halogen lamp, care should be taken not to touch the bulb directly. Active chemicals emitted from the human skin gradually weaken the glass housing of the lamp causing it to explode while it is energized, resulting in damage to surrounding hardware.

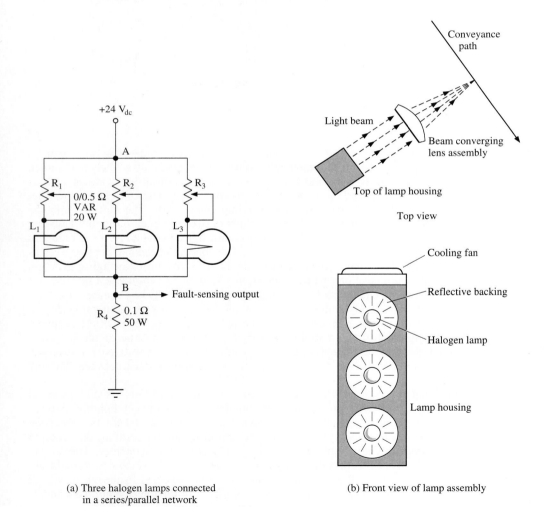

(a) Three halogen lamps connected in a series/parallel network

(b) Front view of lamp assembly

Figure 5–2
Halogen lighting system

Figure 5–2(a) contains a diagram of a halogen lighting system; Figure 5–2(b) illustrates how the three lamps involved are likely to be mounted. This lamp assembly is typical of those found in video acquisition systems, possibly being placed along the conveyance path of an automated sorting machine. The video circuitry would quickly scan the surface of a passing container, obtaining an image of the destination label. For optimal performance, the beam-concentrating lens assembly must be aligned so that the light beam is as concentrated as possible, converging precisely at the point being viewed by the scanning circuitry. This image could then be digitized and sent to processing circuitry, which would read the digitized information and determine the routing for the given container. Further control circuitry would then ensure that the package is routed to the proper storage location.

Failure of any one of the lamps could result in the misreading of a significant number of destination labels and, consequently, several missort conditions. For this reason, a voltage divider and comparator circuitry are included within the system, as illustrated in Figure 5–3. If one of the lamps fails, the effective resistance between point A and point B will increase because the resistance of a burned out lamp approaches infinity.

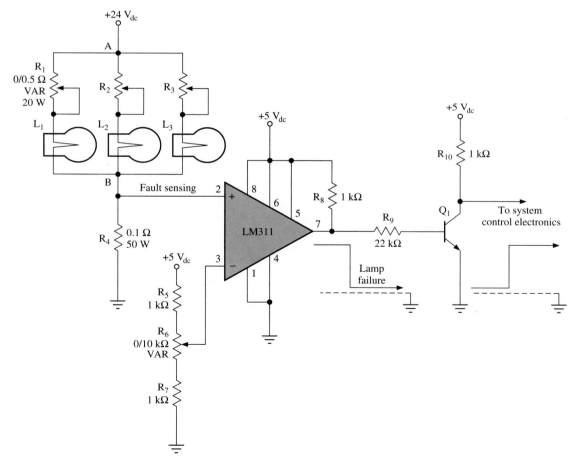

Figure 5–3
Halogen lamp-failure sensing circuit

Consequently, the voltage between point B and the ground will decrease, causing the active comparator input voltage to fall below the threshold level. This results in an active-low lamp-failure signal at the comparator output. The error message could be sent to system control, which would react by shutting down the conveyance motors and displaying a lamp failure message at the operator station. Only when the lamp is replaced will the error message terminate, allowing the motors to be re-energized. To avoid a shut down of the conveyance system due to lamp failure, halogen bulbs are replaced periodically, as part of the machine's maintenance cycle.

Ultraviolet and fluorescent lamps operate on an entirely different principle than the incandescent bulb. These lamps contain no filament. Rather, they rely on ionization of gas within a glass tube to produce energy. As illustrated in Figure 5–4, the cathodes within the lamp are heated by ac current, resulting in **thermionic emission** of electrons. After power is applied to the lamp, the mercury vapor inside gradually ionizes due to the presence of the free electrons. These mercury atoms, now in an excited condition, produce ultraviolet radiation. The inside of the fluorescent light bulb is coated with a phosphorescent material such as magnesium tungstate. This powdered material is excited by the presence of the UV radiation and, in turn, emits a bluish-white light. As the term fluorescent implies, the light immediately dissipates when the cathodes are deenergized and the UV radiation stops.

Fluorescent lamps are perfectly safe because virtually all of the energy actually leaving the lamp is in the form of visible light. However, in actual UV lamps, there is no protective coating of phosphorescent material. Thus, the energy leaving the lamp remains in the form of ultraviolet radiation. UV lamps are dangerously deceptive since the mercury vapor within a UV lamp emits a soft, pleasant bluish light. Also, the UV lamp radiates virtually no heat. These two characteristics of the UV lamp could cause unwary technicians to look directly at the lamp and even handle it while it is energized. In actuality, prolonged exposure to ultraviolet radiation can result in burning of the skin and damage to the eyes. For this reason, UV lamps should never be handled when energized. Special filtered sunglasses should also be worn if it is necessary to work around energized UV lamps when they are illuminated.

Figure 5–5 illustrates a common method of powering UV lamps. Note that the regulated 24 V_{dc} is actually chopped (possibly via an astable multivibrator) at a relatively high frequency of 20 kHz. This signal is then fed to a step-up transformer, which converts the astable signal to a high-amplitude sine wave of around 900 V_{p-p}. As the UV lamp warms up and begins to glow, it will have a slight loading effect on the power source, causing the cathode voltage to drop to about 600 V_{p-p}. For reliable operation, UV lamps must be allowed around 10 to 15 minutes to warm up. Automated document handling machines may contain sensing circuitry that won't allow the conveyance motors to be

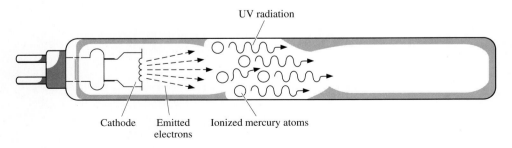

Cathode   Emitted   Ionized mercury atoms
           electrons

Figure 5–4
Structure and operation of a UV lamp

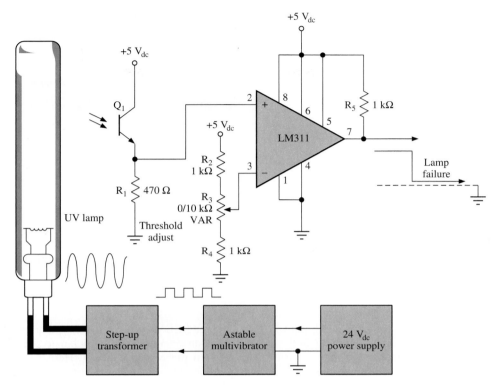

Figure 5–5
UV lamp with power supply and failure detection circuit

started until the UV lamps are fully operative. The same circuit could also shut down the system in the event of a lamp failure. The sensing circuit required for this function, as shown in Figure 5–5, consists essentially of a phototransistor and a comparator.

The **light-emitting diode** (LED) is also an important light source within industrial control systems. LEDs operating in the infrared band serve as the emitters in photosensors and light barriers. Red, yellow, and green LEDs are used prevalently as machine status indicators, often serving as invaluable aids in troubleshooting. Complex industrial control systems may contain cabinets filled with several complex circuit cards. LEDs mounted on the edges of these cards are often used to indicate the status of signals being routed through a system.

Figure 5–6(a) shows the schematic symbol for the LED, and Figure 5–6(b) illustrates its structure and operation. The LED, like a typical rectifier or signal diode, consists basically of a *p-n* junction. However, the LED is manufactured from special materials such as gallium arsenide (GaAs), gallium arsenide phosphide (GaAsP), or gallium phosphide (GaP). The LED radiates light through a process known as **electroluminescence.** As conduction electrons leave the *n*-type material and cross the depletion region into the *p*-type material, they become valence electrons. During this recombination process, they release energy in the form of heat and light. As illustrated in Figure 5–6(b), a gold film at the cathode terminal of the LED allows for optimal intensification of the light beam. At the anode terminal of the device, placement of the metallic film as close as possible to the perimeter allows for optimal radiation of this intensified light.

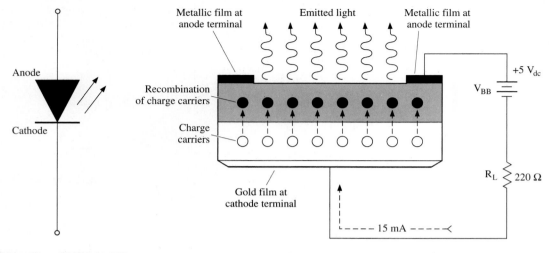

(a) Schematic symbol for the LED   (b) Structure and operation of an LED

Figure 5–6

As shown in Figure 5–7, a single red LED may serve an extremely important function as a system fault indicator, allowing an operator or technician to ascertain quickly a problem that has occurred with a given piece of machinery. Here, an active-low lamp failure signal (possibly originating at the output of the comparator in Figure 5–3 or 5–5) forward biases the indicator LED, which could be mounted on the edge of the circuit card containing the comparator circuit.

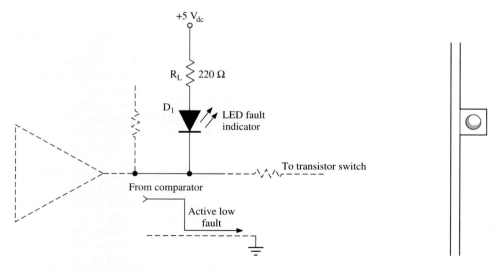

(a) Indicator LED, which illuminates when a lamp failure occurs in Figure 5-3   (b) Placement of LED on edge of a circuit card

Figure 5–7

A more elaborate LED display system is shown in Figure 5–8. Here, a 42-V_{dc} power line is sent to circuitry used to drive four solenoids, each of which has its own circuit branch containing a 4-A slow-blow fuse. The condition of each of the fuses is represented by an LED. As illustrated in Figure 5–9, to expedite troubleshooting, the LEDs could be mounted on a maintenance panel, directly below their corresponding fuses. Providing that all four of the fuses remain intact, the four LEDs will be biased on. In the event that any one of the solenoids malfunctions and draws excessive current, its series fuse will blow. As a result, the associated LED will be biased off due to the loss of its current source. For example, assume F_1 blows. Since D_1 now functions as an open switch, no current will flow through R_2, resulting in a logic low condition at pin 5 of the 7420. If one or more of the four inputs to the NAND function fall to 0 V, the output of that gate will go to a logic high. This action forward biases the Darlington pair, turning on a 24-V incandescent lamp. Such an indicator lamp, often referred to as a **beacon,** could be mounted on a pole far enough above the given machine module that it might be seen from a distance.

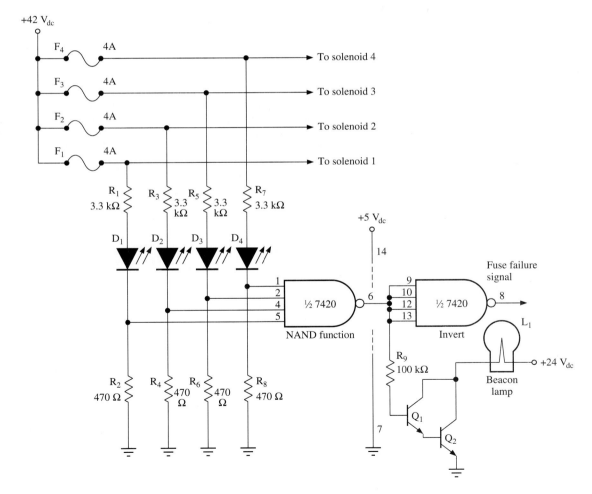

Figure 5–8
LEDs used to indicate the condition of four fuses

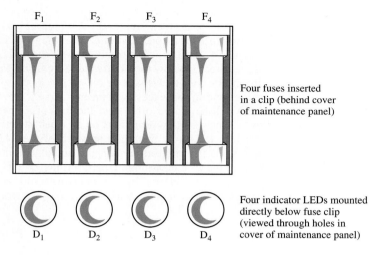

F_1 F_2 F_3 F_4

Four fuses inserted
in a clip (behind cover
of maintenance panel)

D_1 D_2 D_3 D_4

Four indicator LEDs mounted
directly below fuse clip
(viewed through holes in
cover of maintenance panel)

Figure 5–9
Physical placement of LEDs and fuses in failure detection system

Alerted by the beacon, the technician or operator could then look at the LED indicators on the maintenance panel to ascertain which specific fuse has opened. The logic low present at the output of the invert function could serve as the actual fault message, alerting system control that a fuse has opened at the module containing the four solenoids. System control would immediately shut down the machine until the fault is remedied.

In addition to individual LEDs, **seven-segment displays** are also important industrial indicator devices. Figure 5–10(a) illustrates the physical placement of the seven individual LEDs (designated a through g) within the device. Figure 5–10(b) shows the arrangement of the diodes within a **common-cathode** display, while Figure 5–10(c) represents a **common-anode** configuration. The common-cathode display is sometimes referred to as a *pull-up device*. The reason for this should become evident through examination of Figure 5–11(a). Note here that all of the cathodes are connected directly to circuit ground, and each of the seven anodes has an individual positive current source. The common-anode display may be considered as a pull-down device. As illustrated in Figure 5–11(b), all of the anodes are connected directly to a positive current source, and each of the cathodes is connected to the collector side of a current-sinking transistor switch.

Example 5–1

For the common-cathode configuration shown in Figure 5–11(a), what must be the logic conditions at points a through g in order for the number 5 to be displayed? For the common-anode configuration in Figure 5–11(b), what must be the logic conditions at points a through g for the letter F to be displayed?

Solution:

Step 1: For the common-cathode device forming the number 5, the display would appear as shown in Figure 5–12(a). Since the device requires active-high inputs, the logic conditions for segments a through g would be as represented below.

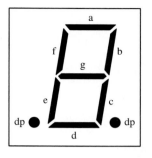

(a) Seven-segment display package

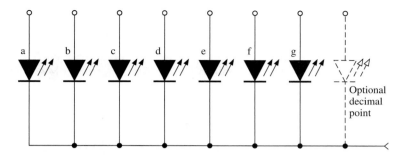

(b) Common-cathode display format

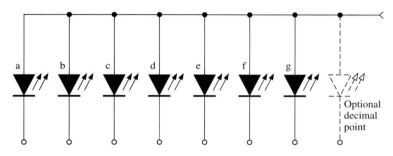

(c) Common-anode display format

Figure 5–10

Condition of inputs for common-cathode display:

a	b	c	d	e	f	g
1	0	1	1	0	1	1

Step 2: For the common-anode device in Figure 5–11(b) to display the letter F, the segments must be illuminated as shown in Figure 5–12(b). Since the display requires active-low inputs, the logic levels for segments a through g would be as represented below.

Condition of inputs for common-anode display:

a	b	c	d	e	f	g
0	1	1	1	0	0	0

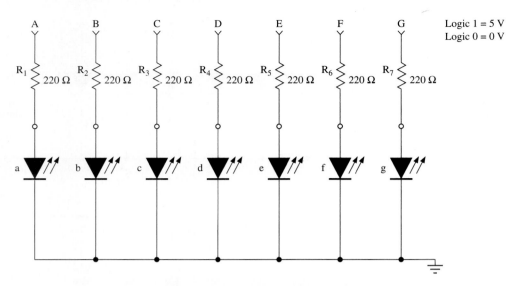

(a) Electrical connection of a common-cathode LED display

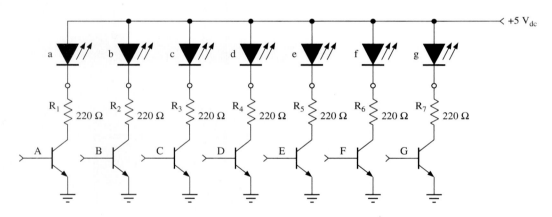

(b) Electrical connection of a common-anode LED display

Figure 5–11

Seven-segment displays are quite versatile devices, since they are able to represent clearly the numbers 0 through 9 and the letters A through F. However, they do require complex decoder/driver circuitry to control the logic levels of each of the segments. Because seven-segment displays are used so prevalently in industry, a wide variety of decoder/driver ICs are available for their control. The internal logic circuitry of a relatively simple TTL decoder device, the 7447, is shown in Figure 5–13. This device is designed for binary-coded decimal (BCD) operation. Thus, it is able to represent clearly the decimal equivalent of the nibbles 0000 through 1001. The outputs of the 7447 are open-collector, making it compatible with common-anode LED displays. The **lamp-test** feature allows all seven of the segments to be tested simultaneously, while **ripple-blanking** allows the

Figure 5–12

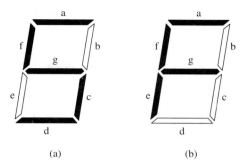

(a)   (b)

most significant displays to be totally blanked rather than display the number 0. For example, assume two of these ICs are used together to display the numbers 0 through 99. If the ripple-blanking output of the least significant decoder is connected to the ripple-blanking input of the most significant device, single-digit values would be displayed simply as 3 or 4, rather than 03 or 04.

In recent years, much more elaborate LED display devices have been developed, examples of which are the 4 × 7 and 5 × 7 dot matrices shown in Figure 5–14. These

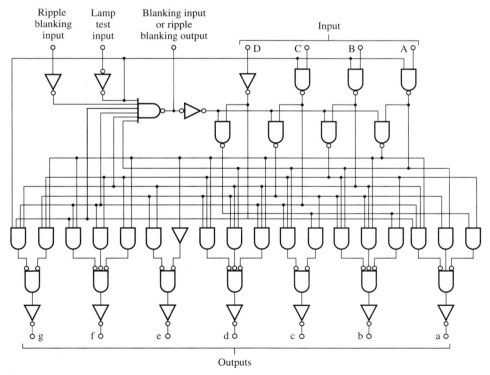

Figure 5–13
Internal circuitry of the 7447 BCD to seven-segment decoder

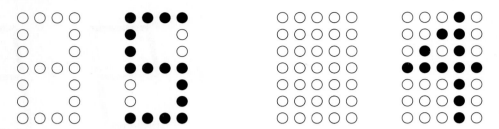

(a) Structure and operation of
a 4 × 7 dot matrix display

(b) Structure and operation of
a 5 × 7 dot matrix display

Figure 5–14

devices are ideal for use with microprocessor systems used to control machinery. Their compact size allows them to be mounted easily on the edge of printed circuit boards. A single display matrix might be all that is required when operating and maintaining a small control system or a specific module within a larger system. The control microcomputer could generate single-character codes through the display matrix, informing the operator or technician of the current machine status. For example, the letter T could be used to indicate that the system is in a test mode, while R could represent a ready condition. Single-digit numbers could then be used to represent common faults detected by the system control. A technician familiar with the system would, of course, recognize these codes and know where to begin to look for the source of a machine malfunction.

SECTION REVIEW 5–2

1. What is a blackbody? Why is the concept of such an ideal surface necessary in defining light intensity?
2. Define the term candle as it applies to the measurement of light intensity.
3. What is the exact definition of the lumen?
4. Describe the operation of an incandescent light source.
5. Describe two advantages of the halogen lamp over conventional incandescent light sources.
6. Explain two precautions which should be observed when replacing halogen lamps.
7. Explain the structure and operation of the ultraviolet lamp.
8. Explain the structure and operation of the fluorescent lamp.
9. Describe a safety hazard involved with UV lamps.
10. Define electroluminescence. What optoelectronic device works on this principle?

5–3   PHOTOCONDUCTIVE CELLS, PHOTODIODES, AND PHOTOTRANSISTORS

Solid-state optoelectronic devices may be generally classified as **photoemissive, photoconductive,** or **photovoltaic.** With photoemissive semiconductors, such as the LED, light is produced as the result of current flow. Recall that with the LED, photons are released as conduction band electrons, cross a PN junction, and recombine with atoms to become valence band electrons. For photoconductive devices, including the **photoconductive cell** (or **photoresistor**), **photodiode,** and **phototransistor,** an opposite effect occurs. With these devices, current flow is induced due to the presence of light. The photodiode is unique in that it is capable of both a photoconductive and a photovoltaic response to light energy. During this section of the chapter, after the structure and operation of

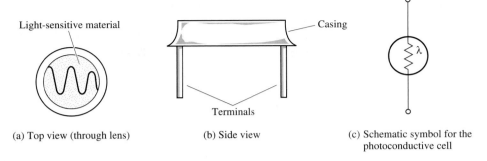

(a) Top view (through lens)   (b) Side view   (c) Schematic symbol for the
                                                                                       photoconductive cell

Figure 5–15
The photoconductive cell

photoconductive cell, photodiode, and phototransistor are investigated, both discrete (on/off) and linear industrial applications of these devices are introduced. Also, a common photovoltaic application of the photodiode is examined in detail.

   The structure of the photoconductive cell, which may be considered as a light-sensitive resistor, is represented in Figure 5–15(a) and (b). The schematic symbol for the device is shown in Figure 5–15(c). The materials commonly used in the manufacture of photoconductive cells are cadmium selenide (CdSe) and cadmium sulfide (CdS). A relatively fast response time is the primary advantage of the cadmium selenide type of photoconductive cell, which is able to change from maximum to minimum resistivity in around 10 msec. A disadvantage of the cadmium selenide device is that its resistance will vary significantly with changes in temperature. Although the cadmium sulphide device exhibits a slower response time (around 100 msec), it does have significantly greater temperature stability than the cadmium selenide photoconductive cell.

   The resistivity of the photoconductive cell is inversely proportional to the intensity of the light beam striking it. This property may be exploited in simple passive signal-conditioning applications, as shown in Figure 5–16. With the photoconductive cell placed

Figure 5–16
Operation of a photoconductive cell within a
voltage divider

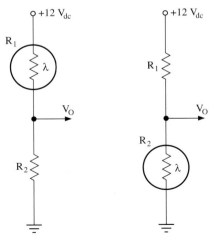

(a) V_O increases with   (b) V_O decreases with
       light intensity   light intensity

within a voltage divider as shown in Figure 5–16(a), an increase or decrease in light energy may be sensed as a corresponding increase or decrease in V_O. Through reversal of the components, as illustrated in Figure 5–15(b), the circuit response is reversed. V_O will now decrease with an increase in light intensity and increase as light energy is reduced.

As shown in Figure 5–17, a photoconductive cell may be used in discrete switching applications. Placed in series with a relay coil, the photoconductive cell controls the switching conditions of two loads. With no light penetrating the lens of the photoconductive cell, its resistance, combined with that of R_1, should be able to hold the voltage and current of the relay coil well below the levels specified for actuation of the relay armature. At the same time, the ohmic setting of R_1 should be low enough to allow actuation of the relay armature once the light penetrating the lens of the photoconductive cell has reached a desired threshold. The circuit in Figure 5–17 could represent a means of automatically controlling the operation of a lamp. The lamp, equivalent to load 1, would be placed in series with the normally closed contacts of the relay. Thus, when the ambient light of the environment reaches a desired threshold, the relay armature is actuated, causing the lamp to switch off.

Figure 5–18 represents an op amp configuration referred to as a **current-to-voltage converter,** or simply a **current amplifier.** This form of circuit is commonly used in signal conditioning applications involving photoconductive cells and photodiodes. When connected as shown here, the photoconductive cell behaves as a nearly ideal current source, exhibiting extremely high output impedance and providing an output current that is independent of the load resistance. Due to the fact that a photocell is a passive transducer (unable to generate its own voltage directly from light), an external biasing source (V_B) must be provided. The circuit in Figure 5–18 behaves essentially like an inverting amplifier, with the inverting input to the first op amp functioning as a virtual ground. Thus, the current flow through the feedback resistance ($R_2 + R_3$) is nearly equal to that flowing through the photoconductive cell ($I\lambda$). Since photoconductive cell resistance decreases as light intensity increases, feedback current will vary as a function of light intensity. For

Figure 5–17
Use of a photoconductive cell in a relay circuit

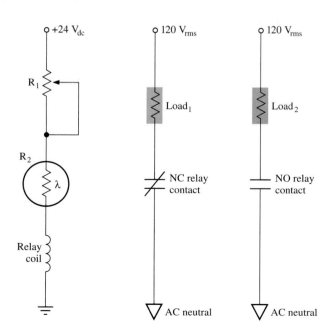

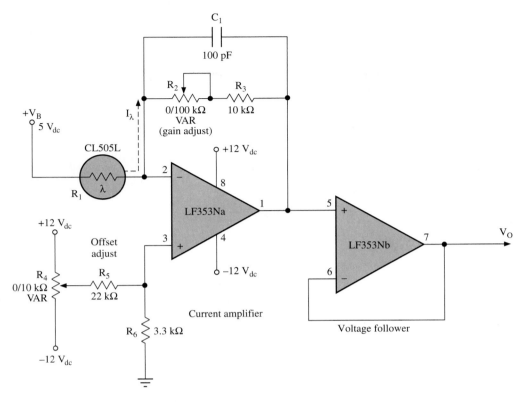

Figure 5–18
Active signal conditioning technique using a photoconductive cell and current amplifier

the CL505L, the minimum **dark resistance** is at least 100 kΩ. With an illumination of 0.61 lux, photocell resistance could be as low as 1.5 kΩ. (The **lux** is a unit used to measure surface illumination. One lux of illumination exists if a surface 1 m distant from a point source is receiving luminous flux with an intensity of 1 lumen/m².) Provided V_B remains constant, the output voltage of the current amplifier may be determined as follows:

$$V_O = I\lambda \times -R_f \qquad (5-3)$$

The resolution, or sensitivity, of the current amplifier may be greatly increased if the input impedance of the op amp is extremely high. For this reason, a dual bi-FET op amp is chosen. Since each of the two op amps within the LF353N is likely to have an input impedance in excess of 10^{12} Ω, extremely small changes in feedback current will be sensed and converted to changes in output voltage. The placement of the voltage-follower stage at the output of the current amplifier virtually eliminates the loading effect of any subsequent signal conditioning circuitry. With the positive dc biasing source, the output signal for the circuit in Figure 5–18 is a negative dc voltage, the amplitude of which varies in proportion to changing light intensity. Because this circuit is likely to be operating in an industrial environment containing fluorescent lighting, filter capacitor C_1 is provided for the purpose of high-frequency noise elimination.

Figure 5–19(a) represents the photodiode, which may be used in many of the same industrial applications as the photoconductive cell. For a photodiode operating in the

Figure 5–19
The photodiode

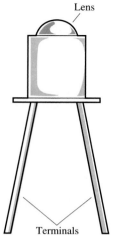
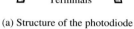
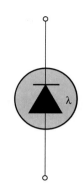

(a) Structure of the photodiode

(b) Schematic symbol
for the photodiode

photoconductive mode, it is the reverse current that becomes a function of light energy. Thus, when operating as a light-variable resistance, the photodiode must be reverse-biased, as implied by the orientation of its schematic symbol in Figure 5–19(b). During photoconductive operation, an increase in the level of light energy entering the lens of the photodiode results in a decrease in resistivity, accompanied by an increase in reverse current. When virtually no light is entering the lens of the photodiode, the reverse leakage current will be at a minimum value, referred to as **dark current** (I_{dark}). This current, although usually in the nanoampere range, must be considered in small-signal applications of the photodiode. However, it is the reverse current resulting from **photonic emission** (I_P) that actually varies in proportion to light intensity. I_{dark} may be considered as a constant or offset value of reverse current that must be added to I_P. Thus, at any time, the total value of reverse current for the photodiode operating in the photoconductive mode may be determined as follows:

$$I_R = I_{dark} + I_P \qquad (5\text{–}4)$$

Figure 5–20 shows the operation of the photodiode within a passive voltage divider. With placement of the diode as shown in Figure 5–20(a), circuit operation is linear, with V_O becoming greater as light intensity increases. As light intensity is reduced, the resistivity of the photodiode increases, resulting in a decrease in V_O. If the placement of the components in the voltage divider is reversed, as shown in Figure 5–20(b), an opposite effect is produced at the output. V_O will now decrease as light intensity is increased.

An active signal conditioning circuit utilizing a photodiode in the photoconductive mode is shown in Figure 5–21. The operation of this circuit is essentially the same as that of the current amplifier in Figure 5–18, with the inverting input of the first op amp still functioning as a virtual ground. However, with a negative biasing source present at the anode of the photodiode, feedback current flows in the opposite direction of that in Figure 5–18. Thus, the output voltage increases in the positive direction as light intensifies. With this circuit configuration, feedback current is virtually equal to the reverse current of the diode. Thus, output voltage may be determined as follows:

$$V_O = (I_p + I_{dark}) \times R_f \qquad (5\text{–}5)$$

Figure 5–20
Operation of a photodiode within a voltage divider

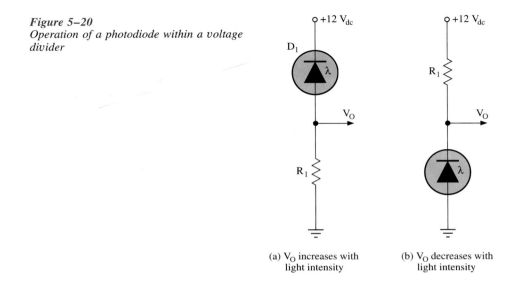

(a) V_O increases with light intensity

(b) V_O decreases with light intensity

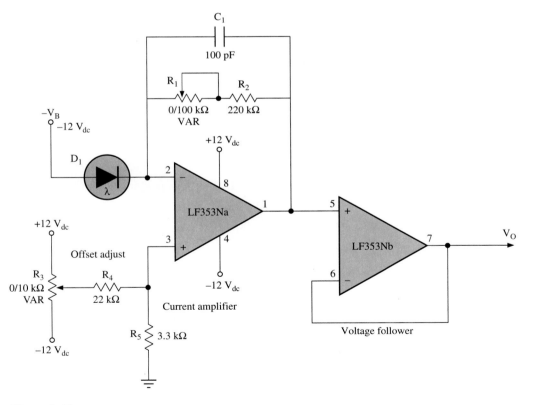

Figure 5–21
Active signal conditioning using a current amplifier and a photodiode operating in the photoconductive mode

Figure 5–22
Photodiode functioning in the photovoltaic mode

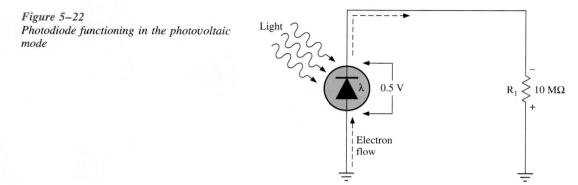

Figure 5–23
Preamplifier circuit for developing a signal produced by a p-i-n photodiode operating in the photovoltaic mode

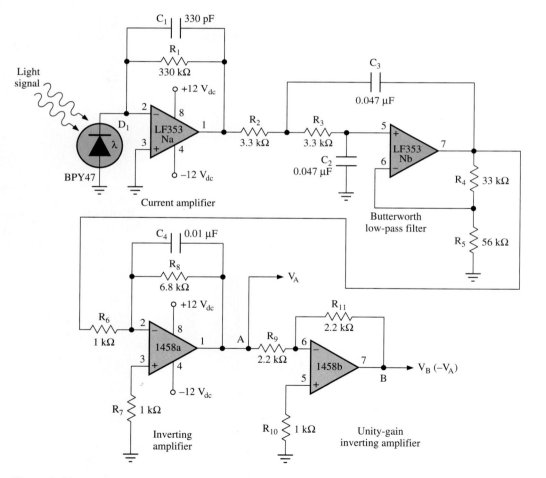

The photodiode, when connected as shown in Figure 5–22, operates like a miniature solar cell in that it is capable of generating electrical energy directly from light. In this **photovoltaic mode** of operation, no biasing source is necessary because the photodiode itself is functioning as the voltage source! As the photodiode is excited by the presence of light, electrons begin to flow across the depletion region and are emitted as free electrons via the cathode. These electrons flow through the load resistance in the direction shown, then recombine with the *p*-type material of the anode. This photovoltaic action will be sustained indefinitely as long as a light source is present. Note that, with a typical photodiode, the photovoltaic action generates only about 0.5 V from anode to cathode. Also, in the basic circuit configuration of Figure 5–22, the load resistance must be several megohms since the current sourcing capabilities of a single photodiode are quite limited. If the load resistance is kept below 10 MΩ, the photovoltaic operation of the photodiode will be linear. Above 10 MΩ, its operation becomes logarithmic.

A major problem associated with a *p-n* junction photodiode operating in the photovoltaic mode is that of slow response time. In fact, the time constant for a circuit such as that in Figure 5–22 could be as long as 100 msec. For this reason, in industrial applications requiring rapid processing of optical signals, the *p-i-n* form of photodiode is often used instead of the conventional *p-n* type. The term *p-i-n* is an acronym describing the basic structure of the device, which contains heavily doped *p* and *n* regions separated by an undoped intrinsic region (hence *p-i-n*). In addition to a response time of as little as 1 nsec, *p-i-n* photodiodes are characterized by dark current values in the picoampere range. Such low values of dark current allow the *p-i-n* photodiode to operate with greater noise immunity than the conventional *p-n* device.

The circuit in Figure 5–23 shows a common application of the *p-i-n* photodiode. This application is typical of preamplifiers found in various forms of optical recognition systems. Note that, since the photodiode is now operating in the photovoltaic mode, no bias source is required at the input. Here, the operation of the photodiode and feedback resistor R_1 is similar to that of the circuit in Figure 5–22. The cathode of the *p-i-n* diode functions as the current source for the initial amplifier stage. As light intensity increases, the ability of the photodiode to supply current to the feedback resistor is increased. This results in a positive voltage increase at the output of the current amplifier. A second-order Butterworth low-pass filter forms the subsequent amplifier stage. As illustrated in Figure 5–24, the low-pass filter removes high-frequency noise from the developing optical

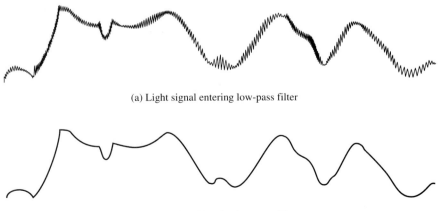

(a) Light signal entering low-pass filter

(b) Light signal leaving low-pass filter

Figure 5–24

signal. Further amplification and filtering is provided by the second inverting amplifier stage. Note that, at point B, a third inverting amplifier having a voltage gain of -1 provides a positive equivalent of the negative output taken at point A. Either signal V_A or V_B could function satisfactorily as a single output signal. However, in applications where the optical signal requires greater amplification, these two equal and opposite signals could be used simultaneously to form a differential output signal V_{A-B}. This differential signal could then be fed to an instrumentation amplifier, which would further reduce common-mode noise and bring the signal to a level required for processing by an A/D converter.

Example 5–2

For the preamplifier in Figure 5–23, determine the value of V_{A-B}, assuming I_{dark} equals 18 pA and $I\lambda$ equals 156.34 nA.

Solution:

Step 1: Solve for the output amplitude of the current amplifier:

$$V_O = (I\lambda + I_{dark}) \times R_f$$
$$= (156.34 \text{ nA} + 18 \text{ pA}) \times 330 \text{ k}\Omega = 51.6 \text{ mV}$$

Step 2: Solve for the output amplitude of the low-pass filter: For the second-order Butterworth filter, with the given values of R_4 and R_5, A_V approaches 1.586. Therefore,

$$V_O = 1.586 \times V_{in} = 1.586 \times 51.6 \text{ mV} = 81.83 \text{ mV}$$

Step 3: Solve for V_A:

$$V_A = V_{in} \times -R_8/R_6 = 81.83 \text{ mV} \times (-6.8 \text{ k}\Omega/1 \text{ k}\Omega) = -556.44 \text{ mV}$$

Step 4: Solve for V_{A-B}: Since the final stage of the circuit is an inverting amplifier with a gain of -1:

$$V_{A-B} = V_A - V_B = -556.44 \text{ mV} - 556.44 \text{ mV} = -1.1129 \text{ V}$$

Since it does not require a base terminal, the phototransistor is likely to be identical in appearance to the photodiode, as shown in Figure 5–19(a). However, the operation of the phototransistor is markedly different from that of the photodiode. As illustrated in Figure 5–25(a), the base of the phototransistor represents a relatively large surface area. It is the base-emitter junction of the phototransistor that is sensitive to light. If no light is entering the lens of the device, only a slight reverse leakage current, again referred to as dark current, will flow between the collector and emitter terminals. However, as photons strike the base-emitter junction, the phototransistor becomes forward-biased and a forward current (symbolized as $I\lambda$) begins to flow across the base-emitter junction.

A major advantage of the phototransistor is its high linearity. Ideally, changes in the amplitude of its base current are directly proportional to changes in the intensity of the light entering its lens aperture. Thus:

$$\Delta I\lambda \, \alpha \, \Delta \hbar f \tag{5–6}$$

Recall that for any BJT operating in the active (linear) region of its load line (between saturation and cut-off):

$$I_C = I_B \times \beta$$

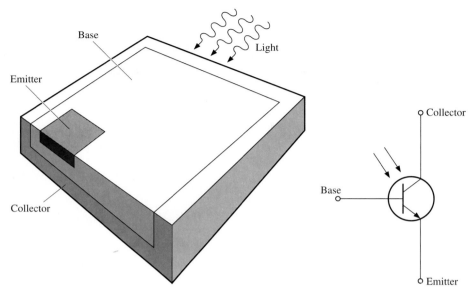

(a) Basic structure of the phototransistor   (b) Schematic symbol for the phototransistor

Figure 5–25
The phototransistor

Thus, for a phototransistor during linear operation:

$$I_C = I\lambda \times \beta \tag{5–7}$$

Therefore,

$$\Delta I_C \; \alpha \; \Delta \hbar f \tag{5–8}$$

Figure 5–26 contains a phototransistor in an emitter-follower configuration. Since emitter current is virtually equal to that of the collector, any change in light intensity sensed at the base will result in a proportional change in voltage at the emitter. In the absence of light, with only dark current flowing, the transistor will be in a near cutoff state, behaving as an open switch. With this condition, V_O would be limited to ($I_{dark} \times R_L$), which represents only a few millivolts. However, as light intensifies and emitter current increases, Q_2 will approach a saturation condition. Between these two operating extremes, the operation of the phototransistor would be linear, being expressed mathematically as follows:

$$\Delta I_E \; \alpha \; \Delta \hbar f \quad \text{Thus,} \; \Delta V_E \; \alpha \; \Delta \hbar f$$

In Figure 5–27, two phototransistors in an emitter-follower configuration are used in conjunction with a comparator for the purpose of detecting a difference in light intensity between two surfaces. Such circuits are used in automated sorting systems, since they are able to discern quickly if the surface of a container passing along a conveyance pass contains a glossy area. This gloss-detection process often precedes that of video acquisition. In fact, it actually expedites the video acquisition process, since the destination address on many containers and envelopes is located beneath a cellophane-covered window. Another interesting application of gloss-detection circuitry would be within automated currency sorting systems. One of the functions of such systems is to remove worn bills from circulation. Thus, if a glossy area is detected on the surface of a bill passing through

Figure 5–26
Phototransistor in emitter-follower configu-
ration

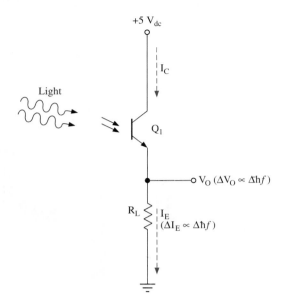

the system, it is likely to be a piece of clear tape. Thus, the bill is assumed to be worn and is, therefore, rejected. (Electronics technicians who service these systems are not allowed to keep the worn bills.)

Although the function of the circuit in Figure 5–27 is quite simple from the electronic standpoint, its operation becomes far more complex when the optical theory of its operation is considered. In order for it to respond to a glossy area on a passing container, an elaborate optical assembly such as that shown in Figure 5–28 must be utilized. In Figure 5–28, note that Q_1 represents the diffused light detector in Figure 5–27, and Q_2 represents the incident light detector. For the circuit in Figure 5–27 to operate reliably, the incandescent light source must be aligned such that its beam reflects off of mirror 1, glances off of the glossy surface of an object present behind a slot in the face plate, reflects off of mirror 2, then directly enters the lens of Q_2. Once this physical alignment of the mirrors and light source is achieved, electronic calibration of the circuit may be accomplished through the adjustment of potentiometers R_2 and R_4. The voltage at point A functions as the threshold level. This is because it develops in response to diffused or scattered light. With a glossy-surfaced piece of white cardboard placed behind the slot, R_2 should be adjusted to establish a reference voltage of around 1.5 V at point A. With the card still present, R_4 should be adjusted until the voltage at point B attains around 4 V. Since the voltage at the inverting input of the comparator is now higher than the reference level, the output of the comparator will be at a logic low, signaling the presence of a glossy surface.

The effectiveness of the circuit in Figure 5–27 becomes apparent when the glossy cardboard is replaced by a white piece of cardboard that has a nonreflective surface. The voltage at point A will remain virtually constant, since the amount of diffused light within the detection assembly is altered very little by the differences in surface texture. Also, R_1 functions as a shunt resistor, serving to dampen the sensitivity of Q_1. However, Q_2 will be greatly affected by the change in surface texture, due to the fact that the light beam reflected from mirror 1 will be scattered by the rough cardboard. Consequently, the amount of light detected by Q_2 will be greatly diminished. As a result, the voltage at point B will drop to a level well below the reference (0.5 V is typical). The output of the comparator will then toggle to a logic high, signaling the absence of a glossy surface.

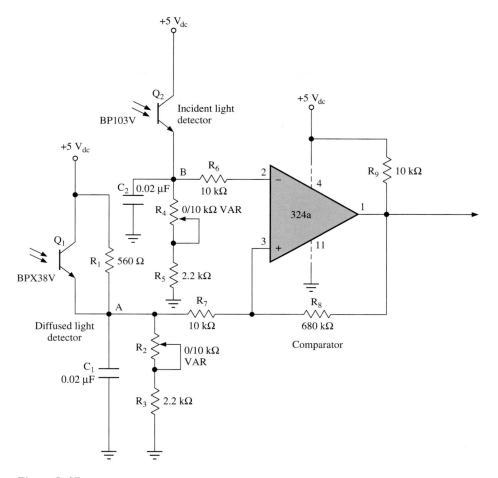

Figure 5–27
Differential light-sensing circuit used to detect the presence of glossy surface on a passing object

Another factor to consider in the operation of the circuit in Figure 5–27 is the fact that several detectors, as well as lamps, would need to be stacked within the assembly in order to scan large surfaces. For an assembly consisting of eight detector circuits, eight lamps and sixteen phototransistors would be stacked in the arrangement shown in Figure 5–28. However, since the 324 IC contains four comparators, only two of these ICs would be needed. The eight comparator outputs could form a byte of data, representing the surface of a passing container or piece of currency.

SECTION REVIEW 5–3

1. The photoconductive cell in Figure 5–18 is functioning
   a. photoemissively
   b. photovoltaically
   c. as a current source
   d. as a constant voltage source

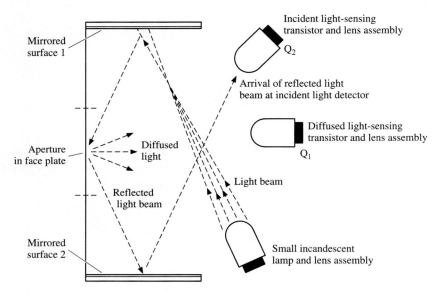

Figure 5–28
Gloss-detection assembly

2. The op amp configuration used in Figures 5–18 and 5–21 is referred to as
   a. a voltage-to-current converter
   b. a current-to-voltage converter
   c. a current amplifier
   d. both b and c
3. The photodiode in Figure 5–21 is functioning in the
   a. photovoltaic mode
   b. photoconductive mode
   c. electroluminescent mode
   d. photoemissive mode
4. The photodiode in Figure 5–23 is functioning in the
   a. photovoltaic mode
   b. photoconductive mode
   c. photometric mode
   d. phototropic mode
5. Define the term *dark current* as it applies to the operation of the photodiode and the phototransistor.
6. Describe the structure of the *p-i-n* photodiode. What main advantage does this device have over the *p-n* photodiode?
7. Explain why the photodiode in Figure 5–23 is able to operate without a bias voltage.
8. True or false?: The comparator in Figure 5–27 is actually connected in a Schmitt trigger configuration.
9. What is the function of R_1 in Figure 5–27? Why is it placed in parallel with the diffused light detector?
10. Explain how the circuit in Figure 5–27, operating in conjunction with the optical assembly in Figure 5–28, is able to detect a glossy area on the surface of a passing object. Identify two possible applications of this type of system.

5-4   OPTOISOLATORS, OPTOCOUPLERS, AND INTERRUPTER MODULES

A photoconductive and a photoemissive device may be contained in a single package to form an **optocoupler** (also referred to as an **optoisolator**). Such devices are used in a wide variety of industrial applications, especially where the input, or **emitter,** portion of a circuit must be electrically isolated from the output, or **detector,** side. Optocouplers represent an extremely low-noise medium for data transfer. Therefore, they are ideally suited for use in microcomputer-controlled industrial systems that contain such noise-generating components as relays, motors, and switching power supplies. Three commonly used forms of optocoupler are represented in Figure 5–29. Note that in all three examples, the light-emitting element is an LED, which is likely to produce infrared light. For low and medium power applications, the detector may be a photodiode or a single phototransistor. Where greater current amplification is necessary, a **photodarlington** is used as the detector, as shown in Figure 5–29(c).

At this juncture, a clarification of the terms **optocoupler** and **optoisolator** is necessary. While these words are sometimes used interchangeably, the term optoisolator specifically

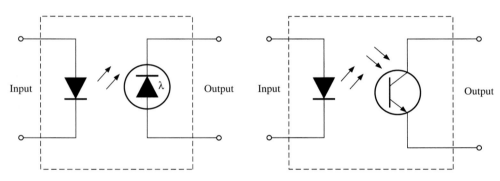

(a) Optocoupler with photodiode as detector   (b) Optocoupler with phototransistor as detector

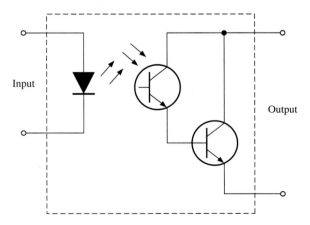

(c) Optocoupler with photodarlington as detector

Figure 5–29
Common forms of optocoupler

refers to a device often used as an interface between two circuit boards. When signals are being sent from one circuit board to another, especially if these boards are separated by a considerable amount of cabling, a device such as the PC-849 quad optoisolator is often utilized, serving as an input device for the circuit board at the receiving end. Where digital signals are involved, the output stage for the board sending data is usually in an open-collector configuration. The internal structure of the PC-849, which contains four separate optoisolators, is shown in Figure 5–30.

In large pieces of machinery, where signals must often pass along several feet of cabling, optoisolators are utilized for a variety of reasons. First, light is a virtually noise-free medium for sending signals. While long wire runs tend to pick up a considerable amount of noise from such sources as fluorescent lamps and motors, much of this noise may be attenuated when signals are exchanged by optical means. Also, long wire runs tend to have a loading effect on the signal source. This loading effect is minimized by an optoisolator, since the receiving circuit draws no current from the signal source. Finally, if the two circuit boards are a large distance from each other, it is very likely they are probably being operated from different dc power sources that do not share the same ground reference. Direct coupling of signals between two such circuit boards would result in unstable operation at best. Optoisolators are also ideal for coupling between low-power control circuits and high-power driver circuits. The wide differences in voltage that could occur between the two sides of the device are not a major design concern, due to the fact that a typical optoisolator is able to withstand extremely large differences in voltage between its emitter and detector. This **isolation voltage** is typically in excess of 5000 V!

Figure 5–30
Internal structure of PC-849 quad opto-coupler

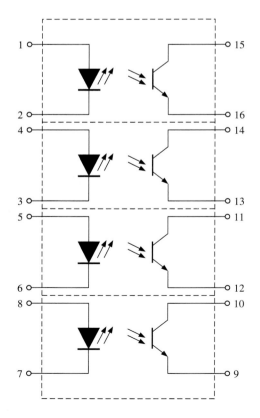

Figure 5–31 illustrates a method of transmitting signals via an optoisolator that is commonly used in industrial control systems. Since the number of terminals available on the edge of a circuit board is limited, a technique called **multiplexing** is being represented. Here, eight bits of data are being sent from one circuit card to another as two nibbles (groups of four bits). Using this method of data transmission, only four output pins are required instead of eight. The internal schematic diagram of the 8234 multiplexer is provided in Figure 5–32. Careful analysis of the operation of this device will help you understand the multiplexing process.

Data bits A_0 through A_3 form one nibble, while bits B_0 through B_3 form another. The condition of the two control inputs determines which of the two nibbles passes through the multiplexer. Because the outputs of the 8234 are open-collector outputs, it is actually the logic complement of the two nibbles that is being transmitted. If, for example, S_0 is

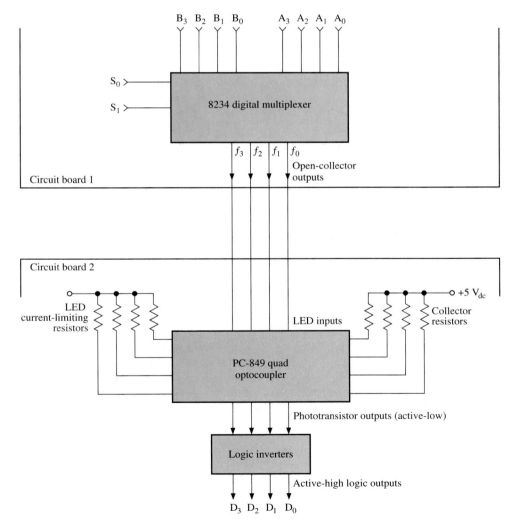

Figure 5–31
Block diagram of typical optoisolator system

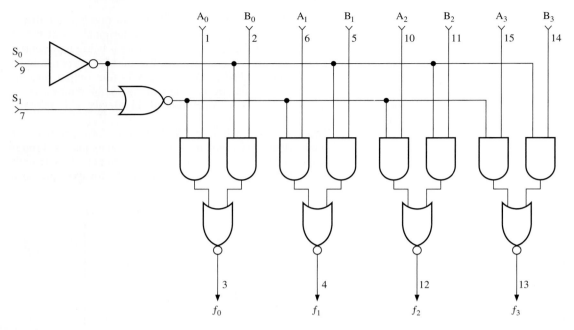

Figure 5–32
Internal logic circuit of the 8234 digital multiplexer with open-collector outputs

high while S_1 is low, the output of the NOR gate in Figure 5–32 will be high while the output of the inverter is low, resulting in an enable condition for the complements of the A bits. Thus the logic conditions at pins 3, 4, 12, and 13 will be the opposite of those present at pins 1, 6, 10, and 15. If S_0 is low while S_1 is high, the complement of the B nibble will be present at the outputs of the multiplexer. This output condition also exists if both S_0 and S_1 are low. If both S_0 and S_1 are high, all four outputs will be high regardless of the conditions of the input variables.

Figure 5–33 illustrates how, for the circuit in Figure 5–31, a single logic level is actually transmitted from one circuit board to the next. Assume, for example, that one of the A bits is at a logic high. Its corresponding output (represented as f_n) will then be low. Thus, due to the open-collector action of the output transistor, the voltage at point A will be pulled low as current flows through the LED of the optoisolator and R_1. This action causes a light beam to excite the base of the detector phototransistor within the optoisolator. This phototransistor, in turn, switches on in saturation, bringing the voltage at point B to a logic low. As a result, the output of the inverter goes to a logic high, representing the high condition of the original input variable.

Assuming that the original A bit was at a logic low, the output transistor Q_1 in Figure 5–33 would then be biased off, allowing only a minute amount of dark current to flow through the emitter side of the optoisolator. The phototransistor would then function as an open switch, with a logic high being present at point B. Capacitor C_1 would then quickly charge toward the 5-V supply level, causing the output of the inverter to toggle to a logic low. Resistor R_1 and capacitor C_1 perform a filtering action, removing any high-frequency noise from the signal that might still be present after passing through the optoisolator. As a result of this filtering action, high-frequency harmonics within the digital signal are attenuated, resulting in an undesired increase in its rise and fall time.

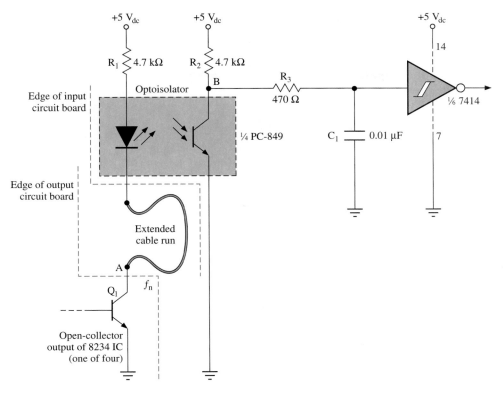

Figure 5–33
Typical circuit configuration for sending a digital signal from one circuit card to another using an optoisolator

The Schmitt-triggering action of the inverter restores the rectangular shape of the digital signal, greatly reducing its rise and fall time.

Optocouplers are readily available in a variety of specialized forms, often having signal conditioning circuitry built into the IC package. Two examples of such devices are shown in Figure 5–34. These devices are sometimes referred to as **OPIC photocouplers.** (OPIC is an abbreviation for optical integrated circuit. The PC901V, illustrated in Figure 5–34(a), is designed especially for digital switching applications. The photodiode on the detector side of the device operates in the photovoltaic mode, providing current to an amplifier that, in turn, feeds a Schmitt trigger with an open-collector transistor. In addition, the device contains a voltage regulator. This important feature allows it to be powered by a source higher than 5 V_{dc} and still provide a TTL compatible output signal.

The PC902 in Figure 5–34(b) is designed for the optocoupling of ac input signals. Through utilization of two LEDs in its emitter section, the device is able to transmit light to the detector during either alternation of an ac input signal. The photodiode on the detector side of the optocoupler, operating photovoltaically, is able to function as a current source, regardless of the direction of the input current. As a result, the output becomes a rectified, pulsating dc signal. The current provided by the photodiode is converted to a voltage, buffered, and then fed to an output transistor. Note that the pin-out of the device allows direct access to the output of the internal amplifier and also to the collector of the output transistor.

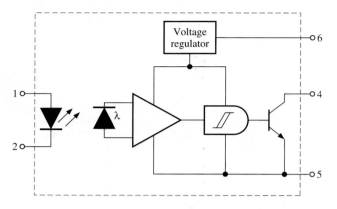

(a) The PC901V digital OPIC photocoupler

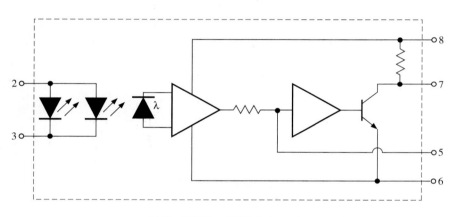

(b) The PC902 ac OPIC photocoupler

Figure 5–34

Another optoelectronic device commonly used in modern industrial systems is the **optocoupled interrupter module** (or simply **photointerrupter**). The structure of this switching device is nearly identical to that of the optoisolator. However, as illustrated in Figure 5–35(a), it is not totally encapsulated. Rather, the emitter and detector are placed on either side of a notched sensing area through which a form of motion transducer known as a **shaft encoder** may rotate. As shown in Figure 5–35(b), the CNY37 represents the most basic form of interrupter module, consisting of a single LED and phototransistor. Also, the shaft encoder represented in Figure 5–29(c) is of the most basic type, consisting only of a timing disk, possibly mounted on a rotating motor armature or pulley. The interrupter module is precisely placed such that the holes of the shaft encoder pass directly through the sensing area.

Figure 5–36 illustrates the operation of the CNY37. Because this device contains no signal conditioning circuitry, external components must be provided to shape its output signal for digital control applications. Here, since it is biased on continually, the LED is able to provide a continuous beam of light. However, the phototransistor is enabled to conduct only when a hole passes between the emitter and detector. The ohmic values of R_1 and R_2 allow the phototransistor to conduct at near saturation when the light beam

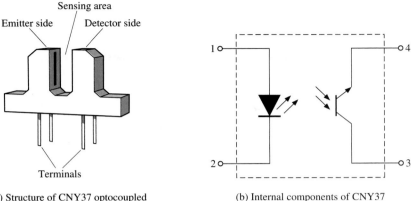

(a) Structure of CNY37 optocoupled
　　interrupter module

(b) Internal components of CNY37

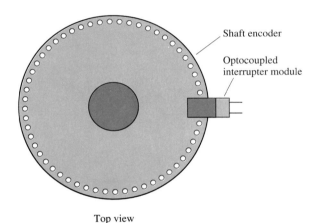

Top view

(c) Typical arrangement of shaft encoder
　　and optocoupled interrupter module

Figure 5–35

passes through a hole in the shaft encoder. At this instant, the voltage at point A will be near ground level. When the light beam is momentarily blocked, the phototransistor approaches cutoff, with only dark current flowing through its main terminals. With this condition, the voltage at point A approaches the level of V_{CC}.

The op amp in the signal conditioning circuitry, functioning as a Schmitt trigger, transforms the rough signal from the interrupter module into a rectangular waveform. Potentiometer R_4, by varying the levels of UTP and LTP, allows adjustment of the duty cycle of the Schmitt trigger output. Two Schmitt-triggered inverters provide TTL compatible clock signals, available at points D and E. These signals are in phase with each other but are complementary to the output of the op amp. The signal at point E serves as the clock input to a divide-by-four ripple counter, consisting of two JK flip-flops. With their J and K inputs tied to V_{CC} via R_8, these devices will function as T flip-flops, toggling on each negative transition that occurs at their clock inputs. If we assume that the shaft encoder is rotating at a steady speed, the squarewave at point F will occur at one-half the frequency of the signal at point E. In turn, since the signal at point F

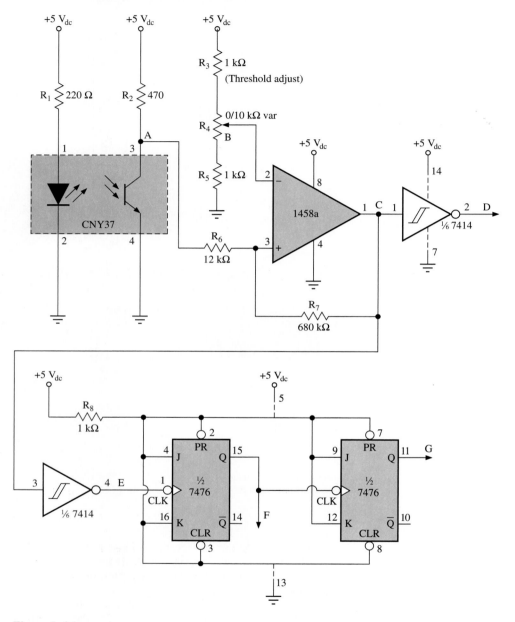

Figure 5–36
Optocoupled interrupter module and typical signal conditioning circuitry

serves as the clock input to the second JK flip-flop, the squarewave at point G will occur at one-quarter the frequency of the signal at point E. The waveforms contained in Figure 5–37 serve to clarify the operation of the shaft encoder and its signal conditioning circuitry.

One problem associated with the type of shaft encoder illustrated in Figure 5–35 is the fact that it is highly susceptible to damage. Both the interrupter module and the timing disk are likely to be exposed to moving machine parts. Also, the dirt and dust of a typical

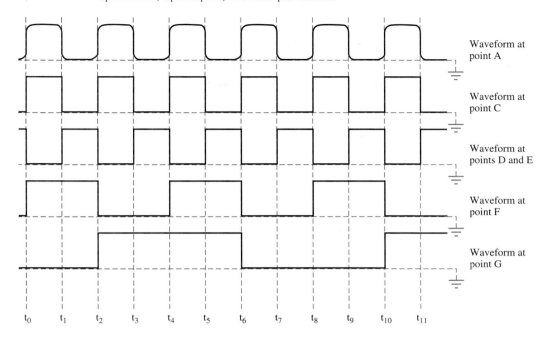

Figure 5–37
Waveforms generated by optocoupled interrupter module and signal conditioning circuitry

industrial environment tend to clog the holes in the timing disk and block the sensing area of the interrupter module. These problems are virtually eliminated by the type of shaft encoder shown in Figure 5–38. Although such devices are far more expensive than the basic shaft encoder system of Figure 5–36, they do have the advantage of being fully encapsulated. Also, many of these devices contain their own signal conditioning circuitry

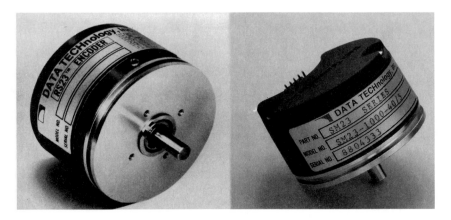

Figure 5–38
Encapsulated high-resolution shaft encoders (From Data Technology, Inc., used by permission)

and are thus able to provide high-resolution digital outputs in binary, BCD, and even gray code.

Although the type of device shown in Figure 5–38 contains just one LED, it utilizes an array of several photosensors. Through the use of a large number of holes on the timing disk, such a device is capable of extremely high output resolution. For example, the RS23 model shown in Figure 5–38, manufactured by Data Technology, is capable of a resolution of 2000 pulses per revolution (ppr). Much higher resolutions are possible with such devices, especially those operated by means of a laser. Applications of such high resolution shaft encoders are covered in detail in Chapter 14.

SECTION REVIEW 5–4

1. Define the term *isolation voltage* as it applies to the operation of an optoisolator.
2. If, for the 8234, inputs S_0 and S_1 are both high, the outputs of the device will form
   a. the A nibble
   b. the B nibble
   c. the complement of the B nibble
   d. 1111
3. Explain at least three reasons why it is advisable to use optoisolators when interfacing between circuit boards placed a considerable distance apart.
4. Describe how the optocoupled interrupter module in Figure 5–35 operates in conjunction with a shaft encoder.
5. What is the purpose of the op amp Schmitt trigger shown in Figure 5–36?
6. What is the purpose of R_4 in Figure 5–36?
7. If the shaft encoder used with the interrupter module in Figure 5–36 has a resolution of 512 pulses per revolution, how many pulses would occur at point G during one revolution of the shaft encoder disk?
8. Describe the advantages of the encapsulated form of shaft encoder shown in Figure 5–38.
9. How does an OPIC differ from a basic optocoupler?
10. What is the basic advantage of using multiplexers when interfacing digital logic signals between circuit boards?

5–5   APPLICATIONS OF LIGHT BARRIERS, INTERRUPTER MODULES, AND PHOTOSENSORS

The term **light barrier** is often used to describe a pair of optical components, consisting of an emitter and a detector, that are placed on either side of a conveyor. The emitter element is usually an LED that emits an infrared beam. This LED is carefully aligned with the detector element, which is often a photodarlington. Thus, as an object passes between the light barrier components, it breaks the infrared beam, allowing the detector to send a pulse to system control circuitry indicating the arrival of the object. In this text, the term **photosensor** shall be used to describe a proximity sensing device that contains encapsulated emitter and detector elements. The photosensor operates on the same principle as the light barrier, relying on the presence or absence of a light beam at its detector to sense the arrival of an object. However, since the photosensor is encapsulated, the emitter within the device must produce a beam that is reflected off a surface and returned to the detector. Light barriers and photosensors, in conjunction with shaft encoders, are often used to measure the length of objects moving through conveyance systems. These same devices also allow for precise tracking of the progress of these objects through the conveyance path. Light barriers may be used, either in arrays or in conjunction with shaft encoders, to monitor the gap between such objects.

In conveyance system engineering, the term *transport clock* is commonly used to describe a signal taken from the output of a shaft encoder (as opposed to that produced by a crystal-controlled clock or astable multivibrator). In conveyance applications, the shaft encoder is often used to monitor belt travel. In such systems, the shaft encoder is connected to a pulley within the belt system rather than directly to the shaft of a drive motor. Since the rotation of the pulley varies in proportion to changes in belt speed, any such deviations are reflected as changes in the frequency of the transport clock signal. This deviation in the transport clock signal is actually desirable, because the purpose of the signal is to monitor accurately the progress of objects along the conveyance path rather than to maintain an absolutely constant frequency.

As an example of the operation of a **measuring light barrier,** consider the circuit in Figure 5–39. When the light barrier is unobstructed, current flows through the photodarlington, which is connected in an emitter-follower configuration. The values of R_1 and R_2 allow Q_1 and Q_2 to approach saturation, pulling V_{R_2} to a logic high. With this condition, the output of the NOR gate is held at a logic low, regardless of the activity of the transport clock signal. However, when the light barrier is momentarily blocked by a passing object, V_{R_2} will fall to nearly 0 V. During this condition, the complement of the transport clock signal will be enabled to pass through the NOR gate. Thus, the number of transport clock pulses passing through the NOR gate will vary as a function of the length of an object on the conveyance path.

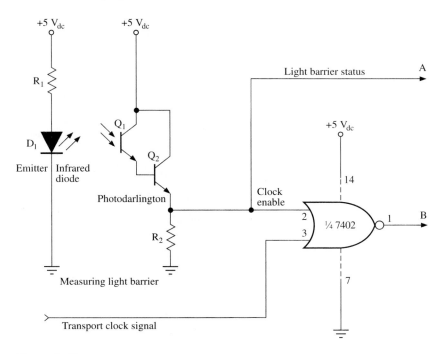

Figure 5–39
Measuring light barrier and clock enable

The transport clock signal arriving at the NOR gate in Figure 5–39 could be generated by the circuit shown in Figure 5–36. Depending on the resolution required, the clock signal could be taken at point D, E, F, or G. In most applications for which transport clock signals are used, time is not as important a variable as the distance that a conveyor belt has traveled. For instance, one cycle of the signal taken at point D or E in Figure 5–36 could represent 5 mm of belt travel. Based on this value, the signal at point F would represent 10 mm of belt travel and that taken at point C would represent a distance of 20 mm.

Assume that a container is moving along a conveyor belt as illustrated in Figure 5–40. Also assume that the output of the light barrier is taken from the NOR gate in Figure 5–39 and that the transport clock signal is derived from point D or E of the shaft encoder system in Figure 5–36. Finally, assume that the output of the NOR gate is now feeding the 8-bit binary counter in Figure 5–41. Prior to the time at which the container enters the measuring light barrier, the system control electronics would have reset the counter to 00000000. Because the output of the NOR gate (point B) remains low while the light barrier is unobstructed, the counter stays in the reset condition until the container breaks the infrared beam. At this instant, the status signal at point A falls to a logic low indicating that a container has just entered the measuring light barrier. Also at this time, clock transitions will begin to occur at point B, causing the counter to increment with each cycle of the transport clock. Once the container clears the measuring light barrier, the counting sequence is halted and the light barrier status signal returns to a high level. At this point, the microprocessor within the system control circuitry reads the contents of the counter and quickly determines the length of the container. One meter is equivalent to approximately 39.37 in. Thus, a resolution of 5 mm could be converted to inches as

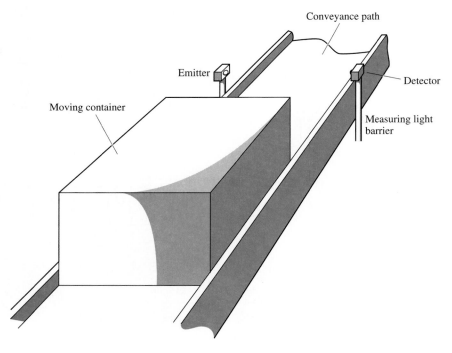

Figure 5–40
Container on a conveyance system approaching a measuring light barrier

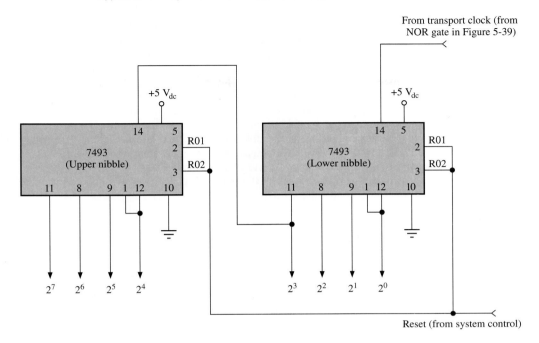

Figure 5–41
Eight-bit binary counter used as part of a measuring light barrier system

shown:
$$5 \text{ mm} = 0.005 \text{ m}$$
$$0.005 \times 39.37 \text{ in.} = 0.197 \text{ in. or nearly } 2/10 \text{ in.}$$

Example 5–3

Assume that during the time a container blocks the measuring light barrier in Figure 5–39, the binary counter in Figure 5–41 increments from 00000000 to 10111101. Also assume that each cycle of the transport clock signal in Figure 5–39 represents 5 mm of conveyor belt travel. What is the approximate length of the object passing the light barrier? Express this result in both millimeters and inches.

Solution:

Step 1: Convert the highest binary count to a decimal number:
$$10111101 = 2^0 + 2^2 + 2^3 + 2^4 + 2^5 + 2^7$$
$$= 1 + 4 + 8 + 16 + 32 + 128 = 189$$

Step 2: Convert to millimeters:
$$189 \times 5 \text{ mm} = 945 \text{ mm}$$

Step 3: Convert to inches:
$$945 \text{ mm} = 0.945 \text{ m} \quad \text{Thus, } 0.945 \times 39.37 \text{ in.} = 37.2 \text{ in.}$$

Assume the purpose of the measuring light barrier just described is to verify that a container passing through a conveyance system is small enough to be processed at a later point in an automated sorting system. Such a container might be diverted to another path, leading to a temporary storage area for oversized containers. In this case, the light barrier and shaft encoder could be accessed by a second counter, the purpose of which would be to track the progress of the container toward a diverter gate, which could be located several feet downstream from the measuring light barrier. Through the use of simple comparative logic circuitry, the control electronics could detect when the count produced in Figure 5–41 exceeded a maximum allowable value, which would be represented as a binary value (possibly encoded by DIP switches). If this maximum count is exceeded, the control logic enables operation of the diverter gate. This gate is driven by a rotary solenoid, which is actuated just as the oversized container approaches the diverting station. As shown in Figure 5–42, the precise timing of this function is made possible through use of the 20-mm transport clock, taken at point F in Figure 5–36, which serves as the clocking source for the second binary counter.

Assume that the diverting station is to be located 12 ft beyond the measuring light barrier. The system designers then need to ascertain the number of cycles of the 20-mm clock that would occur within 12 ft of belt travel. Once this number of clock cycles is determined, it could be encoded by a DIP switch and fed, along with the output of the second counter, to a logic comparator function as shown in Figure 5–42. The advantage of using DIP switches in this control application is that they allow flexibility in adjustment of both the maximum container length and the solenoid triggering time. The initial setting of the DIP switch would be determined as follows. Convert feet to meters:

$$12 \text{ ft} = 12 \times 12 \text{ in.} = 144 \text{ in.}$$
$$144 \text{ in.}/39.37 = 3.6576 \text{ m}$$

Then determine the number of clock cycles:

$$3.6576 \text{ m}/20 \text{ mm} = 3.6576/0.02 = \text{nearly } 183 \text{ 20-mm clock cycles}$$

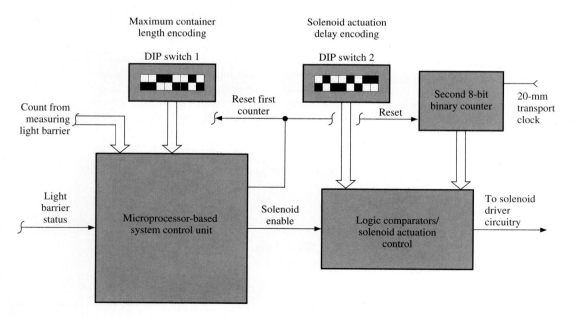

Figure 5–42
Block diagram of measuring light barrier and solenoid control

Thus, the second DIP switch is set to binary 10110111. At the instant the second counter increments to this value, the belt system will have conveyed the container the 12-ft distance from the measuring light barrier to the diverter gate. If the container had been identified as oversized, the gate would be enabled to operate at the instant the count of 10110111 was attained. If the first counter determined the length of the container to be within tolerance, the diverter would not be enabled. The second counter would still track the progress of that container, but the solenoid would not be activated as it passed the diverting station.

Light barriers are often arranged in close arrays for the purpose of measuring the gap between small containers entering a narrow conveyance path. As shown in Figure 5–43, the emitter and detector elements are often inserted at close intervals within bars of synthetic material (sometimes referred to as **phenolic blocks**) and placed on either side of a conveyance path. One block contains an array of infrared LEDs, and the second contains an array of photodarlingtons. The two blocks are aligned such that the infrared beam from each emitter is focused directly into its corresponding detector. Indicator LEDs are often provided as part of the detector block, serving as alignment and troubleshooting

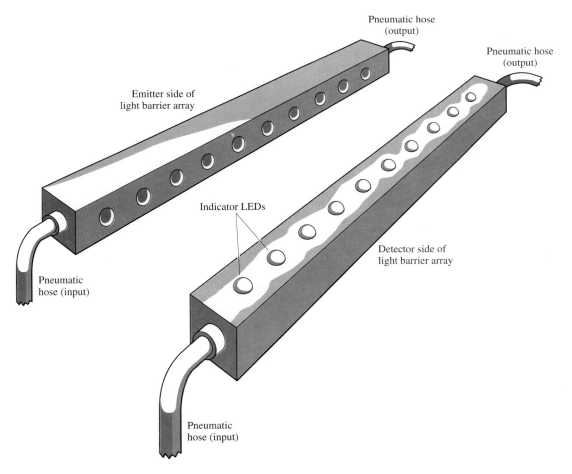

Figure 5–43
Light barrier array formed by two phenolic blocks

aids. They allow defective emitter or detector elements to be identified quickly. Dust accumulation is a constant problem with light barrier arrays. For this reason, phenolic blocks are likely to have pneumatic hose connections, allowing for continuous cleaning of the arrays during machine operation. The importance of keeping the light barrier array free of dust and debris becomes evident as the details of its operation are explained.

The schematic diagram for one of the ten light barriers within the array is shown in Figure 5–44. Although the output of the photodarlington is in an emitter-follower configuration, the first logic inverter causes the output of the detector circuitry to be effectively active-low. Thus, when the light barrier is unobstructed, the voltage at point A should be well above the level of a minimum logic high. The output of the first inverter would then be at a logic low. This low voltage pulls the cathode voltage for D_2 toward 0 V, forward biasing the LED and causing it to illuminate.

The Schmitt-triggered inverter in Figure 5–44 serves as an interface between the light barrier and subsequent digital control circuitry. Not only does it sharpen the rise and fall time of the output signal produced by the light barrier, it also prevents the logic output signal L_1 from floating within the TTL noise margin. If the two phenolic blocks are properly aligned, the logic outputs of this circuit should change states both cleanly and reliably. Thus, with D_2 placed at the output of the first inverter, its on/off condition will clearly indicate the logic level being sent to the control circuitry.

Remembering that the circuit shown in Figure 5–44 is only one of ten contained in the light barrier array, consider the condition shown in Figure 5–45. Assume that the light barriers are spaced 10 mm apart, with an approximate 125 mm gap between each emitter and detector. This system could then effectively monitor the progress of small containers from the feeder mechanism to the exit point of the array. In this example, the sensing system is used by the control circuitry to maintain a spacing of approximately 60 mm between the containers as they enter the conveyance path. To accomplish this

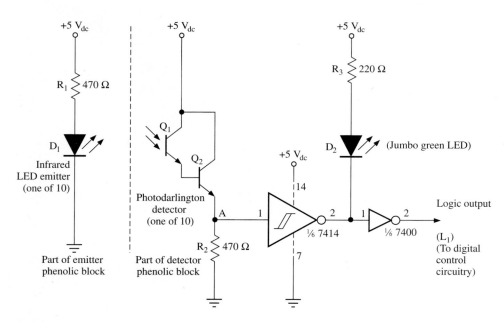

Figure 5–44
Detail of emitter and detector circuitry for the light barrier array

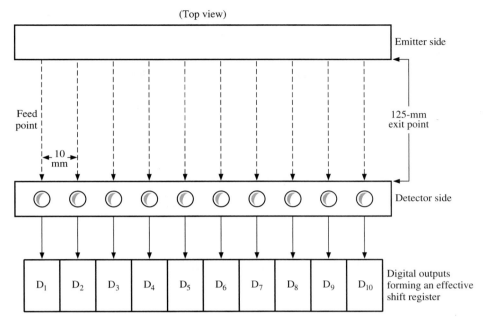

Figure 5–45
Structure and operation of a light barrier array

task, the outputs of the ten light barriers are treated as if they form a shift register in that they are fed simultaneously (in parallel) to the control circuitry.

As illustrated in Figure 5–46(a), until the first container clears the fifth light barrier, the feeder remains in the off condition, which prevents the next container from entering the conveyance path. However, at the instant the logic outputs start to form the number 1111100000, the control circuit senses that the first container has just cleared the fifth light barrier. In response, the microcomputer immediately activates the feeder mechanism, placing the next container onto the conveyor. The containers would then appear as illustrated in Figure 5–46(b), with the light barrier outputs forming the number 0111110000. Finally, Figure 5–46(c) indicates an intermediate condition, in which the first container has nearly cleared the array, resulting in the output variable 0000111110. One of the tasks of the system control circuitry is to detect quickly a possible jamming condition and immediately remove power from the conveyor and feeder motors. One method of doing this is to sense if any one of the logic outputs remains low beyond a maximum allowable time, since such a condition would indicate a possible obstruction within the conveyance path. While this method of jam-sensing is reliable, it may also cause the system to shut down intermittently if any one of the light barriers is operating in a marginal condition as a result of dust accumulation or poor alignment.

Encapsulated **photosensors,** such as that illustrated in Figure 5–47, have one major advantage over light barriers. This advantage is the fact that emitter and detector elements are housed within the same container. These versatile devices are utilized in a variety of industrial applications, operating either in the **proximity-sensing mode** or **reflex mode.** The proximity-sensing mode involves the reflection of either a visible or infrared light beam off of an approaching object, while the reflex mode involves the breaking of a reflected light beam. Photosensors are often used in conveyance applications where utilization of typical light barriers would not be practical, either due to limited space or likelihood of damage.

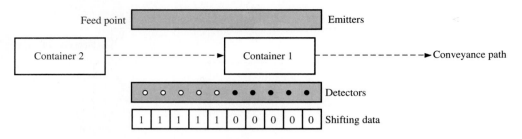

(a) Logic conditions for container 1 just clearing the fifth light barrier

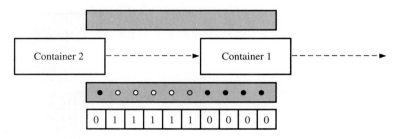

(b) Logic conditions for container 2 just entering the array

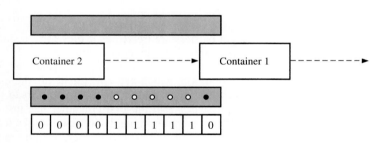

(c) Logic conditions for container 1 just leaving the array

Figure 5–46

Although simple light barriers are basically on/off devices, most commercially available photosensors have the capability of voltage gain or **sensitivity** adjustment. As shown in Figure 5–47, these photosensors might also have a light/dark adjustment. When this control is adjusted fully to the light position, the output voltage of the device will increase as it senses light. If the light/dark control is adjusted fully to the dark position, the output of the device will remain low while light is being detected. Not until the source of light is interrupted does the output voltage increase toward the supply level. Note that, as a positioning aid, the device in Figure 5–47 contains a beam status/alignment indicator. The intensity of the light from this miniature LED is proportional to the amount of reflected light being received by the photosensor. Thus, when aligning the sensor, this indicator should be closely observed, since the point at which it glows the brightest represents optimal alignment.

Figure 5–48 is a simplified schematic diagram of the type of photosensor shown in Figure 5–47. For added versatility, the device has both a **sink** and a **source** output. The

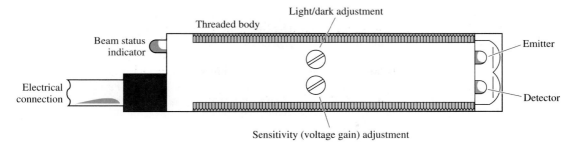

Figure 5–47
Encapsulated photosensor (side view)

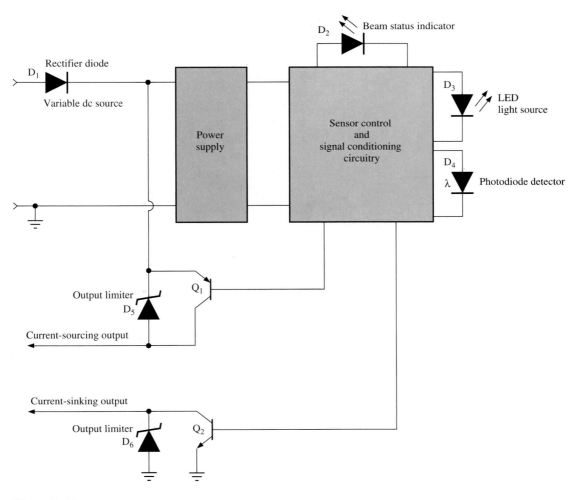

Figure 5–48

source output contains a *p-n-p* transistor, which provides current to a load connected between its collector and the ground. The sink output is in the more familiar open-collector format. The two Zener diodes provide overvoltage protection for these output transistors.

Figure 5–49 illustrates a common conveyance application of a proximity-sensing photosensor. Such a device is designed to have an extremely sharp drop-off in voltage gain beyond a specified range, which might be only a few inches. The operation of this form of sensor may be considered the opposite of that of a light barrier. A light barrier relies on the breaking of a through-beam as the result of a passing object; a proximity-sensing device, however, actually depends on the presence of the passing object in order to create a reflected beam.

The performance curves of two forms of photosensor designed to operate in the proximity-sensing mode are shown in Figure 5–50. For the photosensors represented here, the term **excess gain** refers to the ratio of the amount of light that should be reflected at a given range to a minimum reference value. An excess gain of at least 8 is recommended for reliable operation. That is, the photosensor should be adjusted such that the amount of light being returned by an object entering the established field of view is at least eight times greater than the minimal value required for producing an output signal. The graph in Figure 5–50(a) represents the performance of a proximity-sensing photosensor that has a visible red LED as its light source. Note that its excess gain actually increases slightly between 0.1 and 1 in., then rolls off markedly beyond the 2-in. point. Such a performance

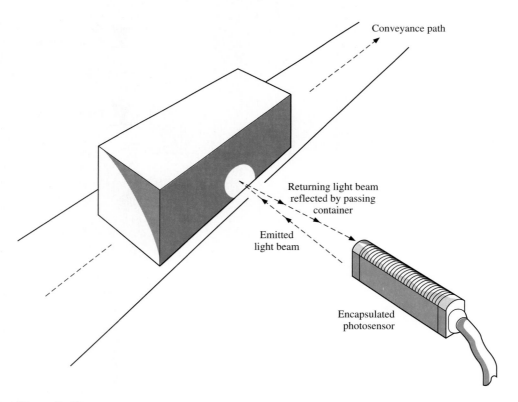

Figure 5–49
Operation of proximity-sensing photosensor

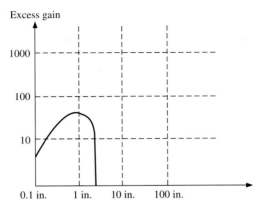

(a) Performance curve for proximity-sensing photosensor with red LED emitter

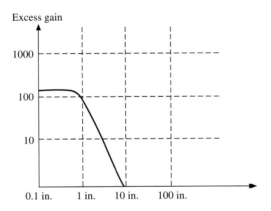

(b) Performance curve for proximity-sensing photosensor with infrared LED emitter

Figure 5–50

curve is actually quite desirable, because it greatly reduces the possibility of the device sending out false signals in response to shiny surfaces appearing in the background. Figure 5–50(b) represents the operation of the same form of photosensor in which the light source is an infrared LED. The advantage of the infrared device is, obviously, an increase in sensing range. However, its roll-off is not as sharp as that of the sensor operating in the visible red range.

Figure 5–51 illustrates a conveyance application of a photosensor designed to operate in the reflex mode. In this mode of operation, a reflective surface must be present on the opposite side of the conveyance path to provide a return light beam. Thus, the photosensor detects the approach of an object such as a container by the interruption of the return light beam. In some applications, where a shiny or burnished metal surface is already present, no special reflective device is necessary. However, for more reliable operation, **polarized reflex sensors** are often utilized. Special reflective devices, sometimes called **retroflectors,** are used to condition the light beam produced by the polarized reflex sensor. The retroflector rotates the light beam produced by the polarized reflex sensor by 90°, and the sensor, in turn, only detects the returning rotated light. Thus, false triggering of

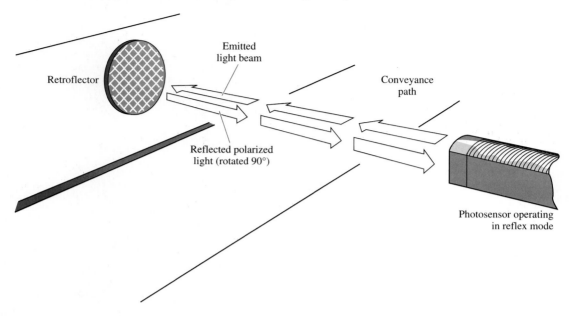

Figure 5–51
Reflex mode operation of an encapsulated photosensor

the sensor is greatly reduced. For both the basic and polarized reflex sensors, the source of light is usually a visible red LED. The primary advantage of photosensors operated in the reflex mode is the fact that they have a much larger field of detection than devices operated in the proximity-sensing mode. While the operating range of a proximity-sensing device is only a few inches, the sensing field of a reflex sensor may extend to as much as 10 ft. As represented in Figure 5–52, the performance curve of a photosensor operating in the reflex mode appears to be nearly linear as excess gain decreases gradually with distance.

Figure 5–52
Performance curve for a photosensor operating in the reflex mode

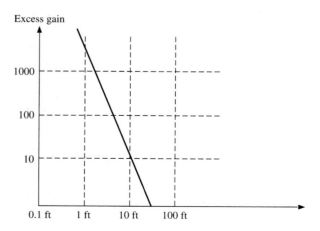

SECTION REVIEW 5–5

1. Assume that the transport clock input to the circuit in Figure 5–39 is taken from point G in Figure 5–36 and that one cycle of the waveform taken at either point D or point E in Figure 5–36 represents 10 mm of conveyor belt movement. If, in Figure 5–39, 21 square wave cycles occur during the time that the clock enable is low, what must be the approximate length of an object passing through the measuring light barrier? Express your answer in both inches and metric units.

2. Explain an application of the light barrier array in Figure 5–43. What form of light is likely to be used to operate this device?

3. What is the purpose of the pneumatic portion of the light barrier array in Figure 5–43?

4. For the circuit in Figure 5–44, what would be the condition of the indicator LED (D_2) when the light barrier is unobstructed?

5. What is the purpose of the Schmitt-triggered inverter in Figure 5–44? Why is it especially important when interfacing with TTL circuitry?

6. Describe two possible modes of operation for the encapsulated photosensor.

7. True or false?: Infrared proximity-sensing photosensors have a greater range than those that utilize red LEDs as their light source.

8. Define the term *excess gain* as it applies to the operation of a photosensor.

9. Describe the operation of a polarized reflex sensor. What advantage does this device have over conventional photosensors?

10. True or false?: A proximity-sensing form of photosensor typically has a much greater operating range than a polarized reflex sensor.

5–6   INTRODUCTION TO BAR CODE AND BAR CODE READERS

Bar code is a method of symbolizing numbers and letters that has evolved from the underlying on/off concept associated with the binary number system and digital logic. Thus, bar code is directly compatible with microcomputer-based control systems. Bar code consists of two basic elements, **bars** and **spaces.** Bars are the vertical black lines, which may be either wide or narrow; spaces are simply the vertical white gaps, either wide or narrow, that occur between the black bars. During this section of the chapter, to aid in understanding the operation of **bar code scanners,** two common types of bar code shall be introduced. These bar codes are used prevalently in warehouse inventory control, conveyance and sorting systems, and shipping applications. The first bar code, **uniform symbol description-1 (USD-1),** is purely numeric and thus rather limited, being able to represent only 10 digits and 2 codes for start and stop. Nevertheless, USD-1 symbology is widely used, a common application being the representation of zip codes on shipping labels. The second, more complex bar code, **uniform symbol description-3 (USD-3) code 39,** is alphanumeric and, thus, more versatile than USD-1, representing 44 characters in addition to its start and stop codes.

Table 5–2 contains the character set for USD-1 bar code. Before examining the actual bar code, we must first note some deviations between the notation shown in this table and the true binary-coded decimal (BCD) number system. While BCD numbers consist of nibbles, or groupings of four bits, the USD-1 characters are represented as five bits, always consisting of two high bits and three low bits. As shown in Table 5–2, the fifth bit is a parity bit. The parity bit is high as necessary to ensure that a two high-bit format is maintained for all characters. (Scanning through Table 5–2 at this time should help to reinforce this important concept.) Also note that the numeric weights of the bit positions

Table 5–2
USD-1 Bar Code Character Set

1	2	4	7	Parity	Data
0	0	1	1	0	0
1	0	0	0	1	1
0	1	0	0	1	2
1	1	0	0	0	3
0	0	1	0	1	4
1	0	1	0	0	5
0	1	1	0	0	6
0	0	0	1	1	7
1	0	0	1	0	8
0	1	0	1	0	9

are 1, 2, 4, and 7 rather than 1, 2, 4, and 8. Also, these numeric weights extend from left to right, not right to left. With the exception of zero, which appears to equal 11, logic ones may be summed from left to right (ignoring the parity bit) to determine the numeric weight of each character.

USD-1 bar code is referred to as **interleaved two of five.** The reason for the description "two of five" should already be clear from examining Table 5–2, since two of every five elements in a character code are always high bits. To understand interleaving, however, we must first learn more about the bar code. In USD-1 symbology, a wide element, whether a bar or a space, represents a logic high, while a narrow bar or space represents a logic low. (This may dispel a common misconception that a black bar always represents a high bit and a space represents a logic zero.) USD-1 symbology has distinct start and stop characters that allow for bidirectional scanning of a bar code cluster. A narrow bar, narrow space, narrow bar, and narrow space comprise the start character, while the stop character is formed by a wide bar, narrow space, and narrow bar.

Figure 5–53(a) illustrates how the numbers 1 through 5 and the number 9 are represented as a USD-1 bar code cluster. First, note the presence of the start and stop characters, which clearly indicate to the bar coder reading device where the cluster begins and ends. The process of interleaving should now become apparent. For the numbers 1, 3, and 5, wide black bars form high bits and narrow black bars form low bits. For the numbers 2, 4, and 9, however, a complementary scheme is used, with wide spaces representing high

Figure 5–53

(a) Example of USD-1 bar code cluster

(b) USD-3 code 39 representation of letter *T*

bits and narrow spaces representing low bits. It should also be clear at this point how interleaving saves space, allowing characters to be overlapped.

Table 5–3 contains the complete character set for USD-3 code 39. In addition to the 44 alphanumeric characters, this symbology has distinct start and stop characters, allowing for bidirectional decoding. As may be seen in Table 5–3, each code 39 character consists of nine elements. Three of these elements are wide and six are narrow, as implied by "code 39," which actually stands for the concept of "three out of nine" elements. As with USD-1 bar code, a wide element represents a high bit and a narrow element represents a low bit. However, because of its much larger character set, code 39 is not interleaved.

Code 39 might best be understood by comparing one of the codes contained in Table 5–3 with the corresponding bar code representation of the letter T in Figure 5–53(b). Scanning the bars from left to right, the pattern is narrow, narrow, wide, wide, and narrow. This pattern corresponds with 00110, as specified in the Bars column next to the letter *T* in Table 5–3. Looking at the same bar code, we now scan the spaces from left to right. The resultant pattern, narrow, narrow, narrow, and wide, corresponds with 0001, as specified in the Spaces column next to the letter *T* in Table 5–3.

A bar code reader can be a relatively simple device. As shown in Figure 5–54, it consists of a light source, a lens assembly, and a single photodetection element. For the simple hand-held bar code reader shown here, the light source consists only of an LED. The first lens focuses the light from this source, concentrating it on that portion of the bar code currently being read. The second lens focuses the light returning from the surface being scanned. This returning light is concentrated on the small photodetection element within the bar code reader. This device can be either a single photodiode or phototransistor. The bar code is read as relative motion takes place between the hand-held device and the object being scanned.

Bar code readers are also used in conveyance systems. A bar code is either sprayed on a container by a special ink-jet printer, or a label containing the code is placed on the package. The bar code reader scans the surface of the container as it passes a small window in the detection assembly.

For a bar code reader used in a conveyance system, the light source is likely to be a halogen lamp rather than an LED. A functional diagram of a typical bar code reader assembly is contained in Figure 5–55. Note that the halogen lamp is placed within a housing that functions as a heat sink. The light from the halogen lamp passes through a glass window prior to entering lens 1. The focused light from lens 1 then passes through a beamsplitter. Also, the illuminated image of the surface being scanned passes through a second lens, approaching the beamsplitter from the opposite direction. With the beam-

Table 5–3
Character set for OSD-3 code 39

Ch.	Bars	Sp.	Ch.	Bars	Sp.	Ch.	Bars	Sp.	Ch.	Bars	Sp.
1	10001	0100	B	01001	0010	M	11000	0001	X	00101	1000
2	01001	0100	C	11000	0010	N	00101	0001	Y	10100	1000
3	11000	0100	D	00101	0010	O	10100	0001	Z	01100	1000
4	00101	0100	E	10100	0010	P	01100	0001	–	00011	1000
5	10100	0100	F	01100	0010	Q	00011	0001	.	10010	1000
6	01100	0100	G	00011	0010	R	10010	0001	SP	01010	1000
7	00011	0100	H	10010	0010	S	01010	0001	*	00110	1000
8	10010	0100	I	01010	0010	T	00110	0001	$	00000	1110
9	01010	0100	J	00110	0010	U	10001	1000	/	00000	1101
0	00110	0100	K	10001	0001	V	01001	1000	+	00000	1011
A	10001	0010	L	01001	0001	W	11000	1000	%	00000	0111

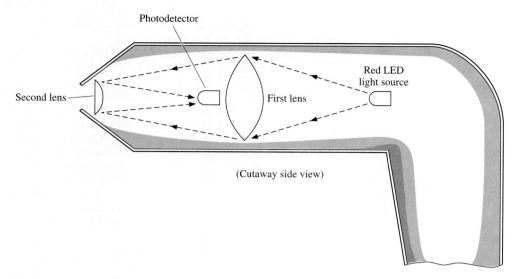

Figure 5–54
Operation of a hand-held bar code reader

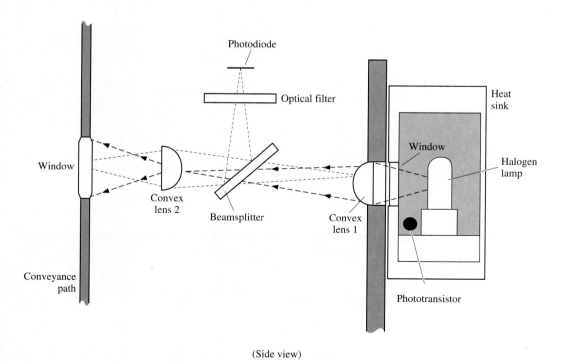

(Side view)

Figure 5–55
Operation of a conveyance form of bar code reader

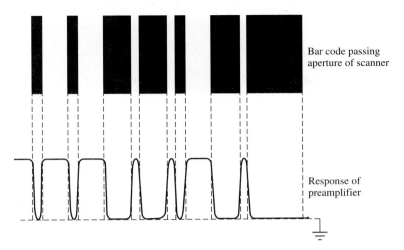

Bar code passing
aperture of scanner

Response of
preamplifier

Figure 5–56
Waveform produced by preamplifier of bar code reader

splitter placed at the angle shown, the light beam containing the image of the scanned surface is reflected upward to be picked up by the photosensing circuitry. After passing through an optical filter, this image is sensed by a *p-i-n* photodiode operating in the photovoltaic mode. The circuit configuration for this photodiode and preamplifier would be identical to that contained in Figure 5–23.

Figure 5–56 represents the response of the preamplifier circuitry of a bar code reader as it scans the surface of a passing container. The analog electronic operation of the circuit is relatively simple, with fluctuations in reflected light intensity being transformed to variations in output voltage. Since the bar code being scanned consists of either light or dark areas on the surface of a passing object, the output consists of either high or low voltages, which are easily transformed to binary code. The output of the analog amplification circuitry is fed to an A/D converter. The output of this device is then fed to a microprocessor system. The microprocessor system analyzes the data received from the A/D converter in order to determine a code representing the object passing on the conveyance system. For a container, this code might represent a destination within a warehouse facility. Thus, the system control circuitry would track the progress of the container and route it to the proper storage location.

SECTION REVIEW 5–6

1. Explain why the USD-1 bar code shown in Figure 5–53(a) is described as interleaved two of five.
2. Explain why the example of USD-3 bar code in Figure 5–53(b) is referred to as code 39.
3. True or false?: For all industrial bar codes, a black bar represents a high bit and a space represents a low bit.
4. Explain the purpose for having unique start and stop characters in USD-1 and USD-3 bar codes.
5. Explain the function of the beamsplitter in Figure 5–55. Why are the photodetector and preamplifier placed directly above the beamsplitter?

6. For the bar code reader represented in Figure 5–55, what would be the function of the phototransistor contained within the housing of the halogen lamp? What action should take place if this device fails to detect the presence of light?

5-7   INTRODUCTION TO VIDEO ACQUISITION SYSTEMS

The hand-held and conveyance forms of bar code reader covered thus far are similar in that they require the use of just one photodetector element. This is possible since the presence of the bar code is detected as a sudden transition in voltage at a single analog input. In video acquisition systems, which electronically obtain an image of a large surface area of a passing container, it may be necessary to process analog inputs rapidly from more than 1000 photodetectors! Such systems are likely to contain at least one array of photodiodes, similar to that shown in Figure 5–57(a).

The RL512D is an IC manufactured especially for the processing of video signals. The IC contains 512 active photodiodes and 10 inactive black reference diodes. The purpose of these 10 inactive (sometimes referred to as "blindfolded") diodes is to serve as a black reference for adjustment of the subsequent signal-conditioning circuitry. The 522 photodiodes within the array, usually referred to as **pixels,** are arranged in a straight line and placed behind a small window on the bottom of the device. As illustrated in

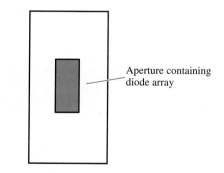

(a) Basic structure of RL512D diode array

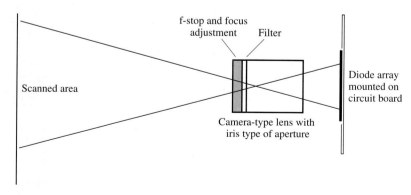

(b) Typical application of the RL512D diode array

Figure 5–57

Figure 5–57(b), the RL512D diode array is designed with the intent that it be mounted on a circuit board that is placed behind a lens assembly. This lens assembly would be identical to that of a camera, having both an adjustable shutter and a focus capability. The lens and circuit board assembly would be placed at an optimal distance for the given video-lift application. If the assembly is moved closer to the conveyance path, the field of view decreases. However, this action would also increase the resolution of the image being scanned, since each of the photodiodes would be seeing a smaller area of the image. Moving the lens assembly further away from the conveyance path would allow a greater area to be scanned. However, the resolution would be decreased.

At any instant, the voltage developed over each of the 512 active pixels is proportional to the intensity of the light it is sensing. However, these voltages are not readily accessible. Two complex processes are required in order to read the voltage levels developed over the pixels and to condition these analog signals for use by subsequent circuitry. First, intricate timing control circuitry must be developed, allowing the voltage level of each pixel to be read out of the array. As shown by the block diagram in Figure 5–58, video signals from odd-numbered pixels are processed by one amplifier channel, while signals produced by even-numbered pixels are processed by an identical amplifier circuit. The timing circuit also controls two **sample-and-hold** gates. These analog switches are triggered in such a way that the recombined video signal enters a final amplifier stage. In applications where the video output must be sent a considerable distance, a coaxial cable is frequently used. For such a coupling configuration, the output impedance of the final amplifier stage is matched to the impedance of the cable, which is typically 50 Ω.

In addition to the 522 pixels, the RL512D contains the necessary digital control circuitry required to read sequentially the contents of the analog data held by the array. This internal circuitry must in turn be controlled by an external timing circuit such as that contained in Figure 5–59. Here, a 10-MHz clock signal arrives at the clock input of a JK flip-flop, which is configured to function as a free-running T-type flip-flop, dividing the input clock frequency by two. The input clock signal and the Q output of the JK flip-flop are represented as waveforms B and D in Figure 5–60. Note that the JK flip-flop is toggled on the negative transition of the input clock signal if the inhibit clock signal at point B is in its inactive logic high condition. The JK and D flip-flops remove noise from

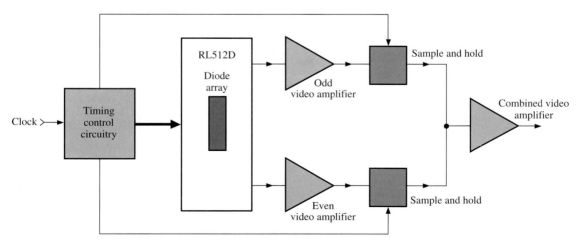

Figure 5–58
Block diagram for a video acquisition system containing an RL512D diode array

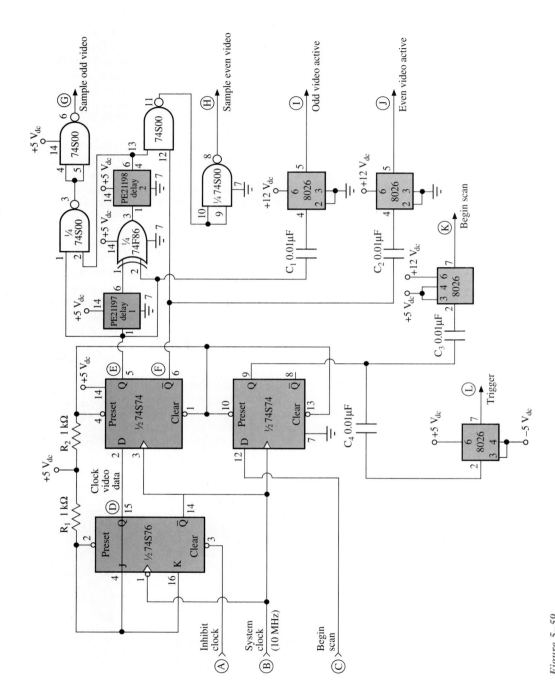

Figure 5–59
Timing control circuitry for the RL512D diode array and sampling gates

320

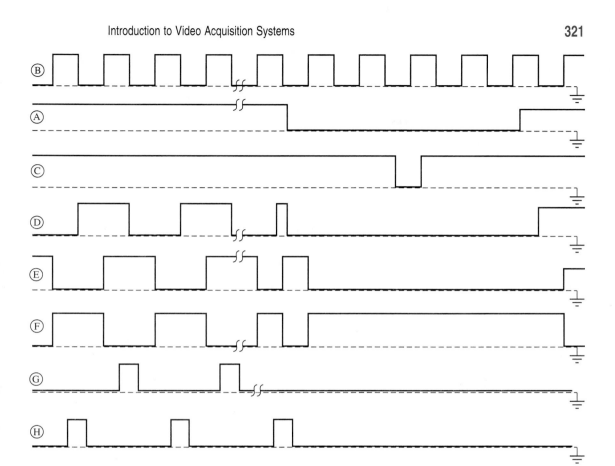

Figure 5–60
Waveforms for the timing control circuit in Figure 5–58

the input clock signal, which is likely to be sent via a cable from a remote circuit board. Also note that the complementary outputs of the first D flip-flop are nearly perfect square waves. As will become evident later, this is essential for reliable operation of the RL512D. (The functions of the delayed sampling signals developed at points G and H are also covered later in this analysis.)

On the positive transition of the 10-MHz system clock signal, through the action of the first D flip-flop, the logic condition at point D is latched at point E. Thus, at the same instant, the complement of the logic condition at point D will be latched at point F. These complementary signals are fed to metal-oxide semiconductor (MOS) drivers, which boost their logic high levels to 12 V_{dc}. This increase in logic level is necessary for reliable clocking of the array. The waveform at point I, which is virtually in phase with the signal at point E, serves to clock the analog information from the odd-numbered pixels within the array. The waveform at point J, which is virtually in phase with the signal at point F, serves as the clock for acquiring information from the even-numbered pixels. Note that the signal at point I is fed to pins 3 and 10 of the array IC, as indicated in Figure 5–61. With each positive transition of the clocking signal at input 01 of the array, analog information from an odd numbered pixel is made available at pin 14 as Video 1. The clocking signal from point J is being fed to pins 1 and 20 of the array. Thus, with each

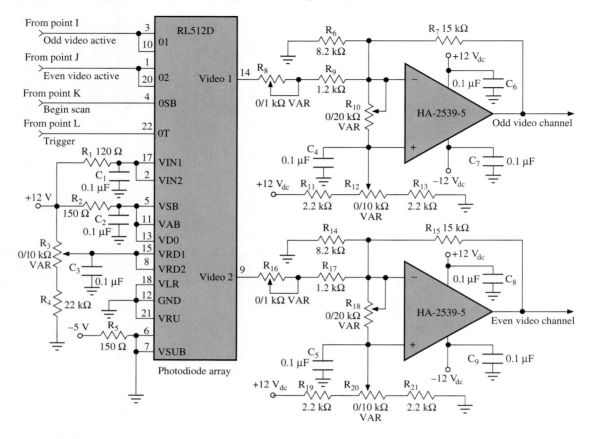

Figure 5–61
Photodiode array with odd and even video amplifiers

positive transition of input 02, information from an even-numbered pixel is made available at pin 9 of the array as Video 2.

Referring back to Figure 5–59, we now explain the purpose of the second D flip-flop. The timing control circuitry shown here is driven by a 10-bit binary counter (not shown), which defines the extent of the scanning cycle. (This counter could easily be constructed by cascading three high-speed 4-bit binary counters such as the 74S163.) At count 522, with completion of the scanning cycle, clock pulses are temporarily removed from point D in Figure 5–59. This action is accomplished by bringing the signal at point A to a logic low at that instant. This signal would remain low for several clock states, to allow subsequent circuitry to process the data received during the previous scanning cycle. Within this dormant period, possibly between counts 522 and 540, the scan pulse would occur. This active-low signal arrives at the D input of the second D flip-flop and is latched at the Q output on the positive transition of the system clock. As shown in Figure 5–61, this scan signal is fed to pins 4 and 22 of the diode array via two MOS buffers. As a result of this signal, the instantaneous voltages developed by the 512 active pixels (as well as the black reference diodes) will be latched into analog switches. Also, the internal scan control circuitry of the array will point to the first pixel (diode 000).

At the end of this dormant state, the counter is reset to 000 and the clock inhibit signal at point A in Figure 5–59 returns to its inactive high condition. With the first positive transition of the signal at input 02 of the RL512D, the analog voltage level acquired from pixel 000 during the scan pulse appears at the Video 2 output. With the first positive transition of the pulse at input 01, the voltage obtained from pixel 001 occurs at the Video 1 output. The next positive transition at input 02 yields the voltage from pixel 002 at output Video 2, and the following positive transition at input 01 yields the voltage from pixel 003. This odd/even acquisition pattern continues until all the voltages from all 522 pixels have been obtained.

As shown in Figure 5–61, two symmetrical amplifier stages are used for conditioning the odd and even video signals obtained from the photodiode array. The HA-2539-5 op amp used in these amplifier stages is especially designed for high-frequency signal conditioning applications. Although a relatively costly device, this IC has a bandwidth of 600 MHz and a slew rate of 600 V/μsec! These characteristics allow the device to process pulse waveforms reliably within a video acquisition system, which is likely to have a basic clocking frequency of at least 10 MHz.

At this juncture, the advantage of using the odd/even clocking format should become apparent. If a single op amp were made responsible for processing the analog information from all 512 pixels, it would have to operate at a rate of 10 MHz. However, for the circuit in Figure 5–61, the work load is literally divided in half. At any given time, only one of the op amps is actually processing information from the array. Thus, the operating frequency for a single op amp is only 5 MHz, allowing more time for developing the signals obtained from the separate pixels. Figure 5–62 illustrates the timing relationships for the signals occurring at the 01 and 02 inputs of the array and the outputs of the two op amps. Note that, while the logic level at the 01 input is a logic low, the output of the odd video channel is low, indicating that no odd pixel signal is being processed at that time. Also,

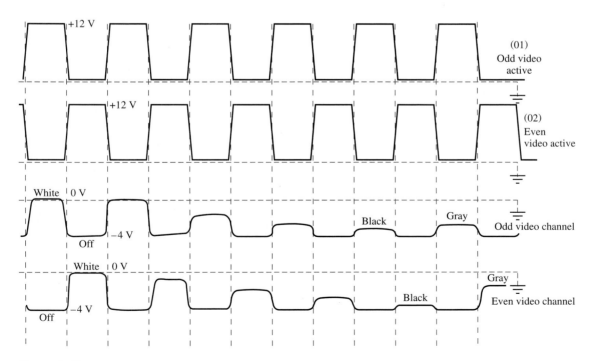

Figure 5–62
Waveforms present at the inputs and outputs of a photodiode array

when the 02 input is at a logic low, the even video channel is dormant. As explained in detail later in this analysis, the use of odd and even channels also allows for more reliable operation of the sample-and-hold circuitry.

As illustrated in Figure 5–62, the operation of the diode array and inverting op amp channels is such that the presence of a white surface at the video acquisition point should cause the voltage level of either channel to approach nearly 0 V. By contrast, the presence of a black surface should cause the video signal to fall to approximately -4 V. Gradations between black and white would, of course, result in an output voltage between 0 and -4 V. If the circuitry in Figure 5–61 is to function reliably, the voltage gain, frequency compensation, and dc offset must be exactly the same for both channels. Resistors R_8 and R_{16} allow gain adjustment for both channels, while R_{12} and R_{20} allow adjustment of dc offset. By effectively changing the input impedance to the op amps, R_{10} and R_{18} provide frequency compensation.

Decreasing the ohmic value of either compensation resistor causes the output wave-form of the op amp to become increasingly rectangular. However, this action also intro-duces noise into subsequent signal conditioning circuitry. Increasing the ohmic value of R_{10} and R_{18} reduces noise by attenuating high-frequency harmonics. However, an adverse result of this action is the deterioration of the rectangular quality of the output waveform of the op amp. The exponential waveform in Figure 5–63(a) is the result of a value for the compensation resistance that is too high. By contrast, Figure 5–63(b) illustrates overshoots and ringing, which result from an ohmic value for R_{10} and R_{18} that is too low. The preferred output signal is shown in Figure 5–63(c). While the rise and fall times of this properly conditioned signal are slightly longer than those of the waveform in Figure 5–63(b), the flattened peak allows for reliable sampling and formation of the aggregate (combined) video signal.

The operation of the timing circuitry involved with the sample-and-hold gates in Figure 5–64 is now explained. Referring back to Figure 5–59, note that clocking signals for the odd and even pixels, which originate at points E and F, branch to delay line and gating circuitry, which is used for precise operation of the sample-and-hold circuitry found in Figure 5–64. The first delay line, the PE21197 IC found between point E and pin 1 of the XOR gate in Figure 5–59, provides a time delay of approximately 30 nsec. The second time delay IC (PE21198) provides an additional 20 nsec of delay. These ICs allow for the development of the intricate wave patterns illustrated in Figure 5–65.

Referring to the first D flip-flop in Figure 5–59, assume that the outputs are toggling as shown at t_0 in Figure 5–65. Although a high condition will immediately be present at pin 2 of the 7486, the Q condition will not be sensed at pin 1 of that gate until 30 nsec have passed. Prior to that time, an odd-parity condition exists at the inputs of the XOR gate, causing its output to be at a logic high. However, after the 30-nsec delay, the output of the PE21197 will go to a logic high, as seen at t_1 in Figure 5–65. As a result of this even-parity condition, the output of the XOR gate falls to a low level. This action produces a rapid 20-nsec pulse at the output of the PE21198. It is this signal that actually causes pixel information to be transferred to the sample-and-hold circuit in Figure 5–64. During the time in which the clocking signal at point E is high, note that a logic 1 is present at pin 1 of the 74S00 IC. Thus, during the 20-nsec output pulse of the PE21198, pin 3 of the 74S00 will be at a logic low. This results in a 20-nsec high pulse at point G, which is connected to the sampling control circuitry in Figure 5–64.

During the time in which odd pixel information is being accessed, a logic low is present at point F in Figure 5–59. Consequently, point H is held at a logic low, preventing sampling of even pixel information. However, at t_3 in Figure 5–65, the logic conditions at points E and F are reversed. Since the negative transition at point E is not sensed at pin 1 of the 7486 until 30 nsec have passed, the output of the XOR gate remains high

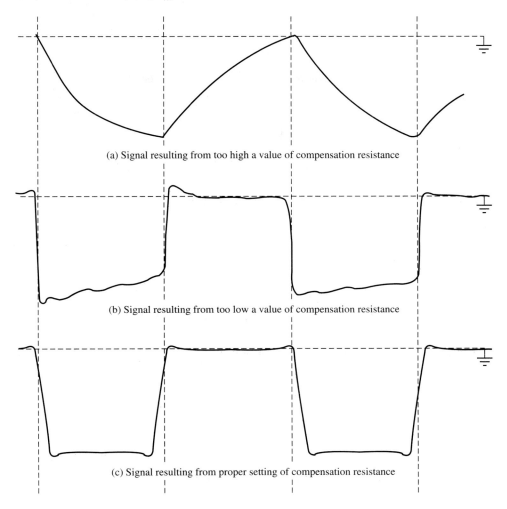

(a) Signal resulting from too high a value of compensation resistance

(b) Signal resulting from too low a value of compensation resistance

(c) Signal resulting from proper setting of compensation resistance

Figure 5–63

due to odd parity. However, at the end of the delay time, the output of the XOR falls low, again triggering the 20-nsec pulse from the PE21198. With point E now low and point F at a logic high, pin 11 of the 7400 will go low for 20 nsec. This action produces a 20-nsec high pulse at point H, allowing the sampling of even pixel information.

Through the action of the 7486 in Figure 5–64, a 20-nsec high condition from point G in Figure 5–59 produces a momentary low pulse at pin 2 of the first 8026 and a momentary high pulse at pin 4. A similar action occurs with the second 8026 as a result of a high pulse at point H in Figure 5–59. The switching action of the two level-shifting buffers allows for reliable sampling of the odd and even video signals arriving from the signal conditioning circuitry in Figure 5–61. As should be made evident by the waveforms in Figure 5–65, the odd and even video information has considerable rise and fall times, even when the frequency-compensation potentiometers in Figure 5–61 are properly adjusted. The purpose of the timing circuitry just analyzed is to delay the sampling of this odd and even pixel information until it has settled at the desired level. It is imperative

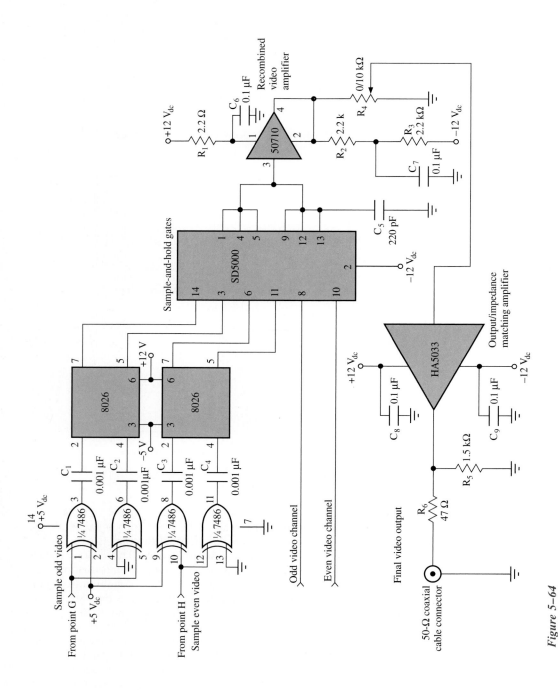

Figure 5–64
Sample-and-hold circuitry recombination amplifier, and output stage for photodiode array signal conditioning system

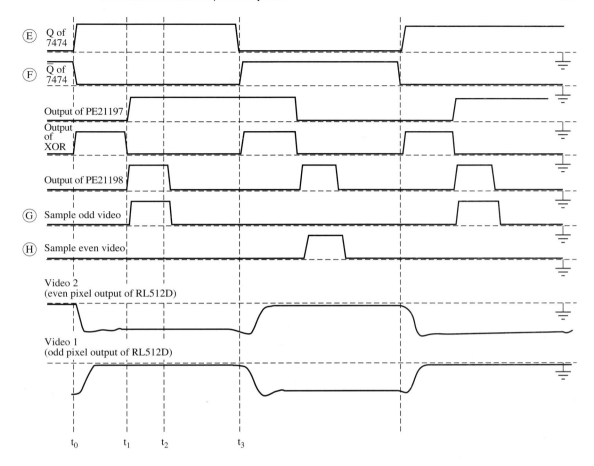

Figure 5–65
Detailed timing diagram for sampling control circuitry

that the actual sampling does not occur until the rise time and the initial ringing of these pulses have already passed.

As shown in Figure 5–64, the sampled odd and even pixel information is sent to a recombined video amplifier, where R_4 allows for simultaneous adjustment of both the amplitude and the negative dc reference of the output signal. The final amplifier stage allows for impedance matching with a 50-Ω coaxial cable, which would carry the combined video signal to subsequent processing circuitry. This circuitry would contain an A/D converter and possibly a microprocessor-based system, which could create a digitized image of the surface being scanned. The operation of such A/D circuitry is analyzed later in this text.

In addition to the complex signal conditioning circuitry just analyzed, another important consideration of the video acquisition system is the illumination of the surface being scanned. A lighting system that could be used in conjunction with video acquisition circuitry was introduced early in this chapter. As illustrated in Figure 5–2 a rectangular lens assembly is used to concentrate light on the video pick-up area. Since this lens is flat on the side where the light enters and convex on the opposite side, the light from the three powerful halogen lamps converges on the surface being scanned. This narrow light

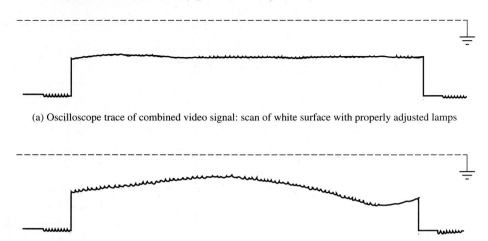

(a) Oscilloscope trace of combined video signal: scan of white surface with properly adjusted lamps

(b) Oscilloscope trace of combined video signal: scan of white surface with overly bright middle lamp and weak bottom lamp

Figure 5–66

beam is then reflected at an angle, directly entering the aperture of the lens assembly used in conjunction with the RL0512D diode array. For optimal performance of the acquisition system, both the diode array circuit board and the rectangular lens assembly should be adjustable. Also, as shown in Figure 5–2(a), high-wattage potentiometers should be used to allow slight variation in the light intensity for each of the halogen lamps.

To allow for even distribution of light over the scanned surface, it is likely that current for the middle lamp will need to be reduced slightly. Otherwise, the middle of the scanning area is likely to be much brighter than the upper and lower edges, resulting in the undesirable combined video signal shown in Figure 5–66(b). To avoid this bulging effect in the middle of the scan, it may be necessary to increase R_2 toward its maximum value, while reducing the ohmic values of R_1 and R_3. This adjustment should be made by placing a nonglossy white card in the video acquisition area and viewing the final output signal (from Figure 5–64) with an oscilloscope. The adjustment should be made until the flattened white signal appears, as shown in Figure 5–66(a). Note that the lens assembly in front of the diode array actually inverts the image being scanned. Therefore, the top of the image appears at the left-hand side of the oscilloscope display, even though the surface of the white test card is being scanned from the bottom to the top!

SECTION REVIEW 5-7

1. Describe the function of the 10 inactive photodiodes contained within the RL0512D.
2. Describe the advantage of using an odd/even pixel clocking format for the diode array, as represented in Figure 5–58.
3. Why is the complex timing control circuitry found in Figures 5–59 and 5–64 necessary for the reliable operation of the sample-and-hold gates contained within the SD5000? Why are the two delay-line ICs found in Figure 5–59 especially important within this process?
4. Explain the function of R_8 and R_{16} in Figure 5–61.
5. Explain the function of R_{10} and R_{18} in Figure 5–61.
6. Explain the function of R_{12} and R_{20} in Figure 5–61.

7. Describe the actions that occur within the RL0512D when logic lows are present at inputs 0T and 0SB.

8. True or false?: Adjustment of R_4 in Figure 5–64 affects only the dc reference of the final video output signal.

5–8   INTRODUCTION TO PHOTOMULTIPLIER TUBES AND LASERS

Photomultiplier tubes and **lasers** are optoelectronic devices that are becoming increasingly prevalent in industry. Both types of device rely on the process of light amplification for their operation. As will be explained, photomultiplier tubes are input devices, amplifying a low amplitude signal from a light source and converting it to proportional changes in output current. Lasers, however, function as light sources, using the same principle of light amplification to produce a coherent beam of light. Photomultiplier tubes and lasers could thus be found within the same industrial system. For example, within a bar code scanning system, lasers could function as light sources and photomultiplier tubes could serve as the detector elements, converting a light signal representing bar code clusters to changes in output current.

The photomultiplier tube (**PMT**) functions as a high gain amplifier for low power optical signals. PMTs exploit an effect called **photoelectric emission.** To initiate this process, as illustrated in Figure 5–67, a **photocathode** is connected to a highly negative voltage source. As the photocathode is bombarded by photons from an incident light source, free electrons are emitted. These free electrons are attracted toward another electrode called

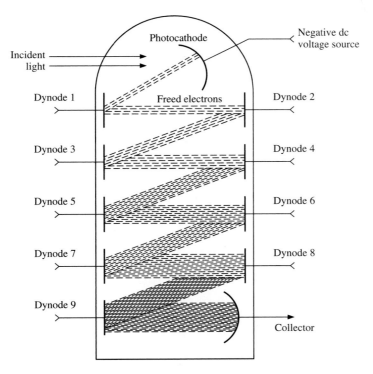

Figure 5–67
Operation of a PMT

a **dynode,** which has a more positive potential than the photocathode. As the free electrons strike the first dynode, they dislodge additional electrons. This intensified beam of free electrons is attracted toward the second dynode, which has a more positive potential than the first dynode. As these electrons bombard the second dynode, they dislodge still more free electrons. As shown in Figure 5–67, this process may involve several more dynodes, until finally, the freed electrons accumulate at the **collector,** or anode, terminal of the PMT. Here, an output current may be accessed and converted to a voltage signal.

A method of applying the increasingly positive voltages to the dynodes of the PMT is shown in Figure 5–68. Here, a simple resistive voltage divider network is connected between the -800 V photocathode voltage source and the circuit ground. With R_1 through R_{10} all equalling 100 kΩ, each resistor develops 80 V. Thus, the potential for dynode 1

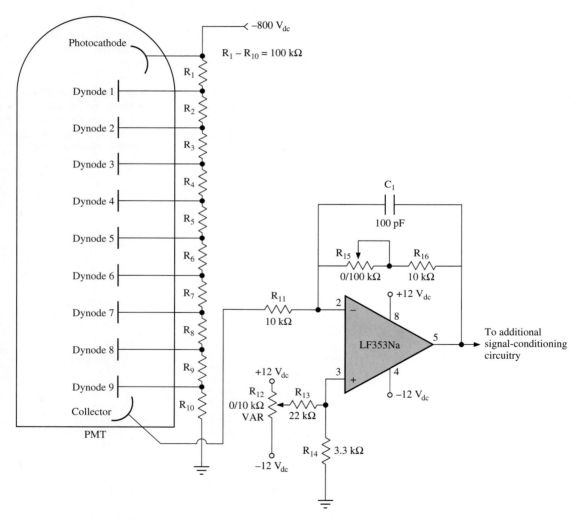

Figure 5–68
PMT with associated voltage divider and current to voltage converter

becomes -720 V, and the potential at dynode 2 is -640 V. This voltage division pattern continues through to dynode 9, which is biased at -80 V. The collector terminal of the PMT is connected to the input of an op amp current to voltage converter. This device, as in Figure 5–23, could feed additional filtering and amplification circuitry.

The number of free electrons available at the collector terminal of the PMT is a function of the current gain (G) of the device. This amplification factor is dependent on other factors, including the number of dynodes (N) and the types of surface material used in fabricating the photocathode and dynodes. The angle of incidence of the **primary electrons** (those electrons striking a dynode) is important, since it affects the number of **secondary electrons** (those electrons dislodged from the dynode by the primary electrons). The ratio of secondary to primary electrons is referred to simply as the **secondary emission ratio (δ)**. Given these important factors, the overall current gain for the PMT may be determined as follows:

$$G = \delta^N \tag{5-9}$$

Example 5–4

Assume a PMT contains nine dynodes and has a secondary emission ratio of 5.4. What is the overall current gain of the device?

Solution:

$$G = \delta^N = 5.4^9 \cong 3.9 \times 10^6$$

As with the photodiode and the phototransistor, dark current becomes an important factor in determining the minimum amplitude of an incident light beam entering the PMT. For the PMT, I_{dark} is a small current that flows from the collector terminal of the device when no light is present at the photocathode.

The term laser is an acronym that stands for **light amplification by stimulated emission of radiation.** While low to medium power lasers may operate as light sources in bar code scanners and inspection systems, more powerful lasers are used in cutting tools, drills, and welders. Types of lasers commonly used in modern industrial systems include the ruby or solid-state laser, the semiconductor laser, the gas laser, and the organic dye laser. (At this juncture, it should be stressed that lasers could easily comprise the subject of an entire textbook. The purpose of the final section of the chapter is to briefly introduce laser structure and operation.)

The basic structure and operation of a ruby laser is shown in Figure 5–69. The original light source for this device is a **xenon flash tube,** the energy source for which is a high voltage DC power supply. As shown in Figure 5–69, a large capacitance (either a single capacitor or a bank of capacitors) is placed in parallel with the DC power supply. The output voltage developed over this power supply typically ranges from 2000 to 5000 V for a small ruby laser. When a trigger pulse arrives at the transformer of the xenon flash tube, the energy stored by the large reservoir capacitor is released via the electrodes of the xenon flash tube, producing intense radiation over a wide range of wavelengths.

The radiation from the xenon flash tube is absorbed by the ruby laser crystal. This crystal consists of crystalline aluminum oxide. To form the ruby laser crystal, less than 1% of the aluminum is replaced by chromium. The intense light produced by the xenon flash tube pumps the chromium atoms within the ruby laser from the G (ground) energy band to the unstable E and F bands. From here, the atoms jump momentarily to the M

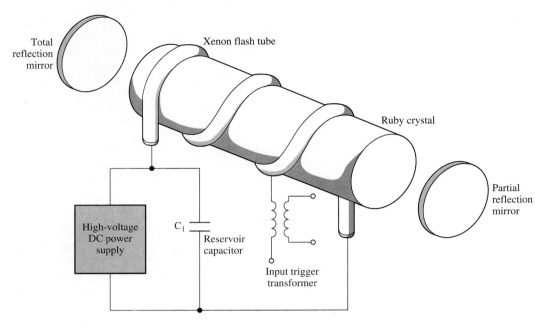

Figure 5–69
Structure of ruby laser

band, then return to the G level, producing both incident and emitted photons. These photons produce still further photon emission as the optical pumping process continues.

At first, the emission of photon energy within the ruby laser crystal is random. Eventually, however, several photons begin to bombard the total reflection mirror at one end of the laser housing and the partial reflection mirror at the output end. It is at this point that the desired laser action actually begins. The photons striking the total reflection mirror are reflected back to the partial reflection mirror. From here, these same electrons return to the total reflection mirror. This back and forth reflection may occur thousands of times, with more and more photons joining in this perpendicular beam formation. Finally, the photons gain sufficient energy to escape out of the laser housing via the partial reflection mirror. The light that leaves the laser through the partial reflection mirror is **coherent,** consisting of virtually a single wavelength. Such coherent light is **monochromatic** (consisting of one distinct color).

SECTION REVIEW 5–8

1. Define photoelectric emission. Describe how this effect is produced in a PMT.
2. What are dynodes? Describe a typical method of biasing these electrodes within a PMT.
3. Define secondary emission ratio as it pertains to the operation of a PMT.
4. Assume that the overall current gain for a PMT is 6.351×10^6. If the device contains nine dynodes, what is the secondary emission ratio?
5. For the PMT in Figure 5–68, what would be the voltage developed between the sixth dynode and circuit ground?
6. What is an alternative term for the ruby laser?

7. What materials comprise the ruby laser crystal? Which of these materials represents a slight impurity that must be added to produce the laser effect?

8. Describe in detail what action must occur within the ruby laser for laser light to escape through the partial reflection mirror. What are the characteristics of this laser light?

SUMMARY

Light is a form of electromagnetic energy that is becoming increasingly important as a means of signal propagation within industrial control systems. In addition to the visible light band, both ultraviolet and infrared light are frequently used within industrial process control. Visible light is used in both bar code readers and video acquisition systems. UV light is used in bar code detection systems, which are designed to detect identifying marks or stamps on containers. In such systems, the phenomena of phosphorescence and fluorescence are often exploited.

A common source of incandescent light used within industrial recognition systems is the halogen lamp. These compact light sources operate easily from dc sources and allow for low-noise operation. However, they do produce considerable heat, thus requiring cooling fans and/or heat sinking. Although UV lamps produce a pleasant blue light and are virtually heatless, they also produce harmful radiation. For this reason, UV lamps should be turned off if possible when performing maintenance on systems in which they are contained. Individual LEDs are often used as system-status indicators within industrial machinery to serve as troubleshooting aids. LEDs grouped into the familiar seven-segment and dot-matrix display modules are also common types of industrial indicators. Infrared LEDs are commonly used as the emitter element within light barriers and encapsulated photosensors. This is especially true for conveyance systems that must operate in brightly lit industrial facilities.

Light-detecting devices commonly found within industrial control systems include the photoconductive cell, photodiode, and phototransistor. Photoconductive cells and photodiodes are often used in linear light-detecting applications. For example, a single photodiode is often used as the sensing element within a bar code reader. In such applications, p-i-n photodiodes are often used, due to their high-speed operation. Photodiodes may be operated either photoconductively or photovoltaically, often being connected to op amps in a current-to-voltage converter configuration. Specialized ICs containing photodiodes are often used in video acquisition systems. Such a device is the RL0512D, which contains 512 photodiodes.

Phototransistors are used in a variety of industrial applications. These devices are commonly used as the detector element within light barriers and encapsulated photosensors. They are also used as detector elements within optoisolators and optocouplers for either switching or linear applications. Phototransistors are commonly used in conjunction with LEDs to form interrupter modules. Such devices are used to form shaft encoders, which function in a variety of conveyance applications. Precise measurement of motor and belt speeds is made possible by such devices. Shaft encoders are often used in conjunction with measuring light barriers to monitor and control the progress of objects through conveyance systems.

PMTs and lasers are devices that operate through the process of light amplification. Through photoelectric emission, the PMT amplifies changes in a low amplitude optical signal, converting them to proportional changes in output current. Op amp signal conditioning circuitry is required to convert this current signal from the PMT to proportional changes in voltage. The laser is a device that produces a coherent, monochromatic light beam. With the ruby, or solid-state, laser, a xenon flash tube elevates the energy level of chromium atoms that are added to the crystalline material.

This optical pumping action eventually causes a coherent light beam to leave through the partial reflection mirror of the device. Lasers are used in industrial bar code scanners and inspection equipment. More powerful lasers are used in cutting, drilling, and welding operations.

SELF-TEST

1. How many micrometers is represented by 14,532 Å?
2. An afterglow is associated with the phenomenon of
   a. phosphorescence
   b. fluorescence
   c. scintillation
   d. ultraviolet radiation
3. What device operates on the principle of electroluminescence?
4. Explain the difference between phosphorescence and fluorescence.
5. What is a disadvantage of halogen lamps?
6. Explain the advantages of using halogen lamps in video acquisition systems.
7. What optoelectronic device operates through the process of photoelectric emission?
8. Explain the difference between photoconductive and photovoltaic operation of a photodiode.
9. Describe the function of the *p-i-n* photodiode. What is the primary advantage of this device over a conventional photodiode?
10. The op amp configuration commonly used in linear applications of photodiodes and photocells is the
    a. Schmitt trigger
    b. voltage-to-current converter
    c. noninverting amplifier
    d. current-to-voltage converter
11. Define the term *dark current* as it applies to the operation of photodiodes, phototransistors, and PMTs.
12. Explain at least two advantages of using the optoisolator as an interface device.
13. What is an interrupter module?
14. True or false?: Optocouplers are limited to the propagation of digital signals.
15. Explain the operation of the shaft encoder. Describe at least two conveyance applications of this device.
16. Identify the primary structural advantage of the encapsulated photosensor. Describe two adjustment capabilities that such a device is likely to have.
17. What are the two operating modes of the encapsulated photosensor represented in Figure 5–47? Describe the operation of the device in each mode.
18. True or false?: Encapsulated photosensors may contain either infrared or red LEDs.
19. Describe the operation of a hand-held bar code reader, such as that shown in Figure 5–54. What form of light source is it likely to contain?
20. Describe the operation of the bar code reader assembly shown in Figure 5–55.
21. True or false?: USD-3 code 39 bar code derives its name from the fact that it uses 39 characters.
22. Describe the operation of the RL512D photodiode array. Explain how this device might be used in conjunction with a lens assembly in video acquisition systems.
23. Explain the advantages of using odd/even pixel clocking for the RL512D.
24. What is the purpose of the 10 blindfold pixels within the RL512D?
25. What special types of op amps must be used within the signal conditioning circuitry of the RL512D? Why must such high-performance devices be used?

26. What types of lamps are likely to be used within the illumination assembly of a video acquisition system? Why is it important for each of these lamps to be adjusted separately?
27. What optoelectronic device may be described as being optically pumped?
28. Identify the three types of electrode contained within the PMT. Explain the function of each electrode.

6 INTRODUCTION TO PULSE AND TIMING-CONTROL CIRCUITS

CHAPTER OUTLINE

LEARNING OBJECTIVES

On completion of this chapter, the student should be able to:

- Explain the operation of the discrete transistor one-shot.
- Describe industrial applications of the discrete transistor one-shot.
- Explain the operation of the discrete transistor astable multivibrator.
- Describe the internal structure and operation of various forms of TTL one-shots.
- Explain the difference between a retriggerable and nonretriggerable one-shot.
- Describe the internal structure and operation of the 555 IC timer.
- Explain how the 555 IC timer may be connected to function as either a one-shot or an astable multivibrator.
- Describe the structure and operation of the D flip-flop.
- Describe the structure and operation of the JK flip-flop.
- Explain how D and JK flip-flops are used in frequency-division circuitry.
- Explain how a dual 555 timer can be made to operate as a pulse width modulator.
- Describe the operation of a voltage-controlled oscillator. Explain how a single 555 timer may function as a VCO.
- Explain the operation of a phase-locked loop. Describe an industrial application of this circuit.

INTRODUCTION

In this chapter, we introduce the pulse circuitry commonly used within industrial control systems. This study begins with monostable and astable multivibrators, which serve as a basis for many other forms of industrial pulse circuitry. Bistable multivibrators, including the D and JK flip-flops, are studied. Frequency-division applications of the D and JK flip-flop are also examined. Other devices covered in this chapter include the pulse width modulator and the voltage-controlled oscillator. The phase-locked loop, which serves in a wide variety of closed-loop control applications, is studied in detail. Various devices and integrated circuits (ICs) commonly used within industrial pulse circuitry are surveyed, including transistor–transistor logic (TTL) multivibrators, counters, and the 555 IC timer.

6-1   INTRODUCTION TO MONOSTABLE AND ASTABLE MULTIVIBRATORS

Multivibrators are basic forms of pulse circuits used in a wide variety of timing, counting, and frequency-division applications. Multivibrators exist in three basic forms: **monostable, astable,** and **bistable.** The operation of the monostable multivibrator, often referred to as a **one-shot,** is shown in Figure 6–1. As implied by its name, the monostable multivibrator has one stable condition. The one-shot is in a stable condition when its $\overline{Q}$ output is high

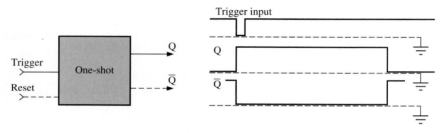

(a) Operation of basic one-shot (monostable multivibrator)

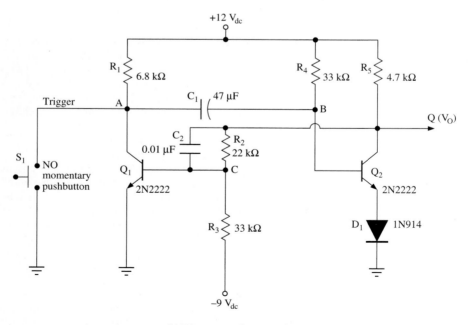

(b) Discrete transistor one-shot

Figure 6–1

and its Q output is low. The one-shot remains in the stable condition until the arrival of a trigger signal from an external source. At the instant the trigger signal occurs, the Q output of the device goes to a high condition. The duration of this unstable high condition at the Q output is definite, being determined by an RC combination contained within the one-shot. At the end of this period, the Q output falls low, returning to its stable state. A $\overline{Q}$ output terminal is not always provided for a one-shot. If this output is made accessible, its logic level will always complement that of the Q output. Thus, as seen in Figure 6–1(a), the $\overline{Q}$ signal is a mirror image of the Q output, having an active-low pulse of the same duration as the active-high Q output. The trigger input to the one-shot in Figure 6–1 is shown as an active-low pulse. However, this input could just as easily be made active-high. One-shots are **edge triggered.** Thus, for a device with an active-low trigger, the Q output goes high on the negative transition of the input signal. For a one-shot with an active-high trigger, Q goes high on the positive transition of the input.

Figure 6–1(b) represents a discrete transistor one-shot. Prior to the closure of S_1, which provides an active-low trigger input, the one-shot is in its stable condition. At this time, Q_1 is in a cutoff state and Q_2 is biased on in saturation. Closure of S_1 causes the conditions of these two transistors to reverse momentarily, with Q_1 entering saturation and Q_2 going into the cutoff state. The duration of this unstable state is determined by the values of R_4 and C_1. During the stable condition of the one-shot, R_4 simply functions as the base resistor for Q_2. However, when the one-shot is triggered, Q_2 functions as an open switch, allowing R_4 to become the path of the charging current for C_1. Resistors R_2 and R_3 form a voltage divider that is salient to the proper operation of the one-shot. With the one-shot in the stable condition, the voltage divider maintains a negative voltage at point C, holding Q_1 in the cutoff state. At the instant S_1 closes, the potential at point C rapidly changes to 0.7 V. This action holds Q_1 in a saturated condition during the unstable period. Capacitor C_2 is simply a commutating (or speed-up) capacitor, which improves the switching time for the one-shot.

During the stable state for the one-shot in Figure 6–1(b), the voltage conditions would be as shown in Figure 6–2(a). Note that, at this time, C_1 is impressed with a charge equivalent to the difference in potential between the collector voltage of Q_1 and the base voltage of Q_2. Due to the presence of D_1, which protects the base-emitter junction of Q_2 from the negative charge developed by C_1, V_B is 1.4 V during the stable condition of the one-shot. Since Q_1 is in cutoff, V_A is virtually at the level of V_{CC}. Thus, the charge over C_1 is 10.6 V, negative with respect to point B:

$$V_{A-B} = V_A - V_B = 12\ \text{V} - 1.4\ \text{V} = 10.6\ \text{V}$$

With Q_2 in saturation, the voltage at the Q output of the one-shot is less than 1 V, being equal to the sum of V_{D1} and $V_{CE}(\text{sat})$ of Q_2:

$$V_{out} = V_{D1} + V_{CE}(\text{sat}) = 0.7\ \text{V} + 0.2\ \text{V} = 0.9\ \text{V}$$

With this low voltage at the Q output, a negative potential is present at point C, being determined as follows:

$$V_{out} - -V_{BB} = 0.9\ \text{V} - -9\ \text{V} = 9.9\ \text{V}$$

$$I_{R2} = I_{R3} = \frac{9.9\ \text{V}}{(R_2 + R_3)} = \frac{9.9\ \text{V}}{(22\ \text{k}\Omega + 33\ \text{k}\Omega)} = 180\ \mu\text{A}$$

$$V_C = V_{out} - (I_{R3} \times R_2) = 0.9\ \text{V} - (180\ \mu\text{A} \times 22\ \text{k}\Omega) = -3.06\ \text{V}$$

This negative potential at point C holds Q_1 in the cutoff condition until S_1 is pressed. At the instant S_1 closes, point A is brought to ground potential. As represented in Figure 6–2(b), this sudden drop in voltage at point A clamps point B momentarily at -10.6 V.

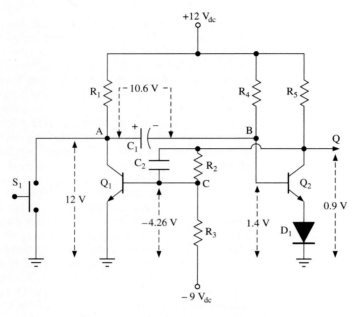

(a) Voltage conditions for one-shot during stable state

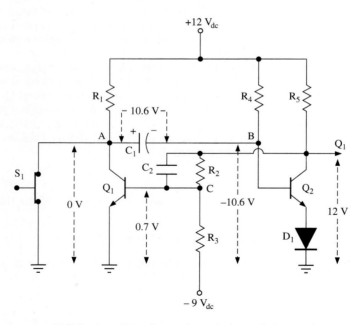

(b) Voltage conditions for one-shot at the instant S₁ is pressed

Figure 6–2

Since C_1 was already charged to 10.6 V, positive with respect to point A, this negative potential is present for an instant directly at the base of Q_2. (Normally, this negative potential could destroy the base-emitter junction of an *n-p-n* transistor. However, the presence of D_1 prevents this from occurring.) This negative base potential immediately places Q_2 in a cutoff condition, which brings the output of the one-shot temporarily to the level of V_{CC}. Since the collector of Q_2 is now at nearly 12 V, Q_1 is forward biased, with 0.7 V being present at point C. Q_1 remains in saturation and Q_2 remains in cutoff until the voltage at point B attains the level of 0.7 V. This increase in voltage at point B occurs as the negative charge on C_1 dissipates via R_4. The rate of discharge for C_1 is based on the following formula:

$$t = (RC) \ln \left(\frac{V - V_0}{V - V_C} \right) \tag{6-1}$$

where V represents the value of V_{CC}, V_0 represents the charge already present on the capacitor, and V_C represents the threshold or target value of capacitor voltage.

Thus, for the one-shot in Figure 6–1, the duration of the pulse width at the Q output is determined as follows:

$$PW = 33 \text{ k}\Omega \times 47 \text{ }\mu\text{F} \times \ln \frac{(12 \text{ V} - -10.6 \text{ V})}{(12 \text{ V} - 1.4 \text{ V})}$$

$$= 1.174 \text{ sec}$$

Figure 6–3 illustrates the pulses generated by the one-shot during its active state. The rate of growth for the charge over C_1 is decreased if a larger value of capacitance is utilized. Consequently, the duration of the pulse width at the Q output would be extended.

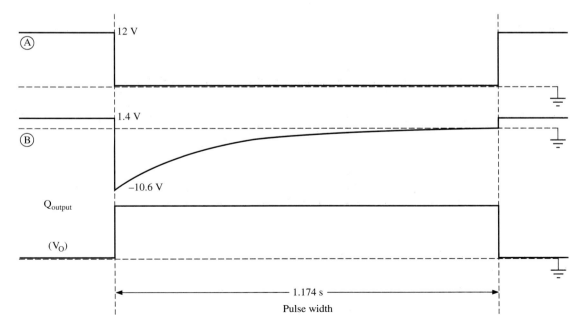

Figure 6–3
Operation of the one-shot during its unstable condition

For example, a capacitance of 100 μF would allow a pulse width of nearly 2.5 sec, whereas a capacitance of 1000 μF would allow a pulse width in excess of 20 sec.

A possible application of the discrete one-shot in Figure 6–1(b) is illustrated in Figure 6–4. Here, a motor within a piece of machinery is energized from a control panel in a remote location. Assuming that the Q output of the one-shot is connected to point A as shown, the photodiode and, consequently, the phototransistor within the optocoupler are biased on for approximately 1 sec at the instant S_1 is pressed. This provides ample time for the relay coil to energize and be locked on by its own holding contacts. As a result, the motor energizes and remains on after the output of the one-shot falls low. It should now be evident that the one-shot allows for reliable operation of the motor-start circuit, because it causes the motor to be energized regardless of how quickly S_1 in Figure 6–1(b) is pressed and released. If the one-shot were not present in the circuit, it would be possible to release S_1 before the holding contacts were able to engage, resulting in a possibly dangerous false-start condition for the motor. The combined operation of the

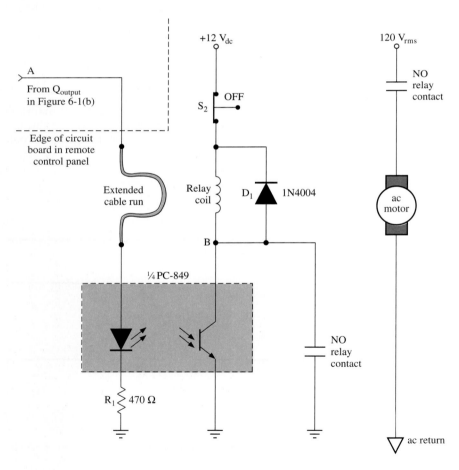

Figure 6–4
Relay actuating circuit controlled by the manually triggered one-shot

circuitry in Figures 6–1(b) and 6–4 represents a vast improvement over the simple relay-latching circuit contained in Figure 1–24.

Figure 6–5 illustrates how the operating principle of the one-shot may be incorporated into control circuitry for a solenoid. Such circuitry is often used in industrial applications where rapid actuation of the solenoid is necessary. Assume, for example, that a rotary solenoid is being used to open a diverter gate within a conveyance system. Also assume that this gate must actuate very rapidly and remain energized for approximately 1 sec. To ensure that the gate is able to overcome inertia and instantly deflect to its fully open position, the solenoid winding is excited by a short burst of very high current. While this **excitation current** is several amperes, it lasts only a few milliseconds. When the excitation current is past, only a holding current flows, sufficient for keeping the gate in its fully deflected position. Where the rotary solenoid must overcome the force of a retaining spring, the high-current/low-current method of actuation just described is very efficient. Enough energy is provided for initial motion, but the source energy is greatly reduced once this movement is achieved. This method of actuation prevents excess heating of the solenoid during prolonged operation. As a result, the lifetime of the device is greatly increased.

Prior to actuation of the solenoid, a logic low is present at point A in Figure 6–5. With this condition, power transistors Q_1 and Q_3 are biased off, and switching transistor

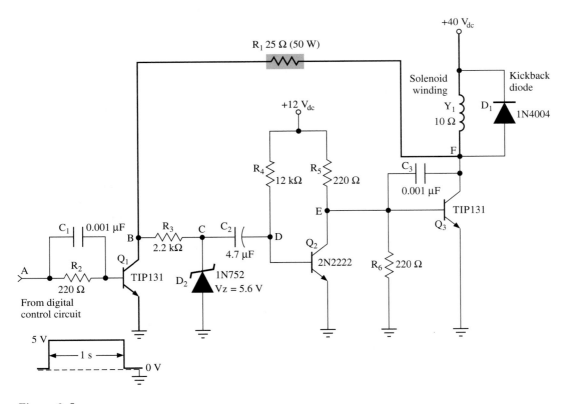

Figure 6–5
Solenoid-actuation circuit that uses the operating principle of a one-shot to provide approxi-mately 20 msec excitation current

Q_2 is biased on in saturation. A slight current does flow through the solenoid winding at this time, the path for current flow being from the 40-V source, continuing through R_1, R_3, and D_2. Although this current is sufficient for operating D_2 in its Zener region, it is far too small for actuation of the solenoid. This quiescent current is estimated as follows:

$$V_{F-C} = 40 \text{ V} - V_Z = 40 \text{ V} - 5.6 \text{ V} = 34.4 \text{ V}$$

$$I_Z = \frac{34.4 \text{ V}}{(r_{dc} + R_1 + R_3)} = \frac{34.4 \text{ V}}{(10 \ \Omega + 25 \ \Omega + 2.2 \text{ k}\Omega)} = 15.4 \text{ mA}$$

With 5.6 V present at point C and Q_2 conducting, C_2 charges to 4.9 V, negative with respect to point D. Since Q_2 is in saturation at this time, point E is at nearly ground potential, keeping Q_3 in a cutoff state. This quiescent circuit condition continues until a triggering pulse arrives from system control.

As shown in Figure 6–6, a rectangular pulse occurs at point A at the instant the solenoid is to be actuated. The duration of this control pulse is approximately 1 sec, which is equivalent to the amount of time the diverter gate is to be closed. On the rising edge of this input signal, due to the low ohmic value of R_2, Q_1 rapidly switches into saturation. This action brings the voltage levels at points B and C to near ground potential. Since C_2 is already charged negative with respect to point D, the sudden drop in potential at points B and C causes the base voltage for Q_2 to be clamped momentarily at −4.9 V. Consequently, Q_2 immediately goes into cutoff, resulting in a forward-biased condition for Q_3. Since current flows via the path of least resistance, solenoid current begins to flow

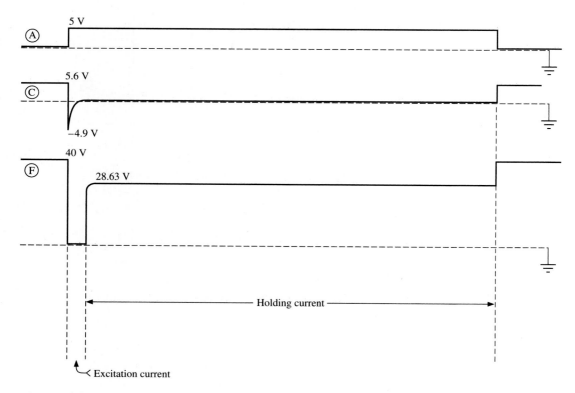

Figure 6–6
Timing diagram for solenoid-actuation circuit

from the 40-V source directly through Q_3 toward circuit ground. At this time, solenoid current is limited only by the dc resistance of the winding. Thus,

$$I_{Y1} = \frac{[40 \text{ V} - V_{CE}(\text{sat})]}{r_{dc}} = \frac{39.8 \text{ V}}{10 \text{ } \Omega} = 3.98 \text{ A}$$

This high level of excitation current is present only while Q_3 is conducting. The voltage at point D increases in the positive direction as C_2 charges via R_4. At the instant V_D attains 0.7 V, Q_2 switches on into saturation. This action returns Q_3 to a cutoff condition, terminating the flow of excitation current. However, as shown in Figure 6–6, the logic high still remains at point A, keeping Q_1 in a saturation condition. Thus, an alternative path for solenoid current still exists, via current-limiting resistor R_1. Through the action of R_1, solenoid voltage and current are greatly reduced. However, current flow is sufficient for holding the gate in the open condition. The values of solenoid current and voltage during this holding condition are determined as follows:

$$I_H = \frac{[40 \text{ V} - V_{CE}(\text{sat})]}{(r_{dc} + R_1)} = \frac{39.8 \text{ V}}{35 \text{ } \Omega} = 1.137 \text{ A}$$
$$V_{Y1} = I_H \times r_{dc} = 1.137 \text{ A} \times 10 \text{ } \Omega = 11.37 \text{ V}$$

An important consideration in the design of this solenoid control circuit is the power dissipation of the limiting resistor, which dissipates considerable heat. During the time in which holding current is flowing, power dissipation for the limiting resistor is determined as follows:

$$P_{R1} = I_H^2 \times R_1 = 1.137 \text{ A}^2 \times 10 \text{ } \Omega = \text{nearly } 12.93 \text{ W}$$

Another important design concern is, of course, the exact time period for the excitation current. Once the optimal duration is known, a suitable value of capacitance is then selected. Formula (6–1) applies directly to this calculation process. Since the value of C_2 has already been determined, the pulse width for the high signal at point E is determined as follows:

$$\text{PW} = R_4 \times C_2 \times \ln \left(\frac{12 \text{ V} - -4.9 \text{ V}}{12 \text{ V} - 0.7 \text{ V}} \right)$$
$$= (12 \text{ k}\Omega \times 4.7 \text{ } \mu\text{F}) \times 0.4025 = 22.7 \text{ msec}$$

If the desired pulse width is known, and the values of R_4 and V_Z have already been determined, the preceding expression can be readily manipulated to solve for a value of capacitance.

Example 6–1

For the solenoid control circuit in Figure 6–5, assume that excitation current is to flow for nearly 50 msec. What standard value of capacitance would allow this to occur?

Solution:
For the circuit in Figure 6–5:

$$\text{PW} = RC \times 0.4025$$

Substituting:

$$50 \text{ msec} = 12 \text{ k}\Omega \times C_1 \times 0.4025$$

Solving for C_2:

$$C_2 = \frac{50 \text{ msec}}{(12 \text{ k}\Omega \times 0.4025)} = 10.352 \text{ } \mu F$$

Choose 10 μF as a standard value.

The **astable multivibrator,** like the monostable device, exploits the principles of RC timing and transistor switching. However, the astable multivibrator has no stable condition. Rather, it functions as a free-running oscillator, producing complementary waveforms as shown in Figure 6–7(a). Although, in some applications, a synchronizing input is applied to the astable device, it requires no triggering input, as does the one-shot. The discrete astable multivibrator shown in Figure 6–7(b) consists of symmetrical transistor switches. If the two capacitors were not present in the circuit, these switches would go immediately into saturation at the instant power is applied. However, due to slight differences in resistance values, as well as differences in β_{dc}, one switch will almost certainly go into

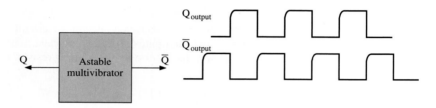

(a) Operation of the basic astable multivibrator

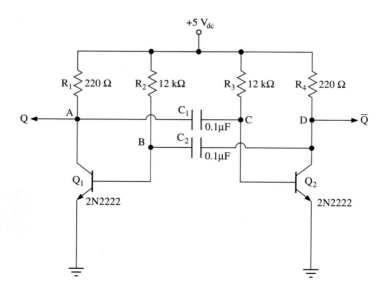

(b) Discrete transistor astable multivibrator

Figure 6–7

saturation a few nanoseconds before the other. It is this inherent imperfection within the circuit that actually allows it to operate.

Assume, for example, that Q_1 switches on in saturation prior to Q_2. With this condition, as illustrated in Figure 6–8(a), point A is at nearly ground potential due to the saturation of Q_1. At the same time, point B is at 0.7 V. Thus, a charge of 4.3 V is impressed over C_2, negative with respect to point B. At the instant of turn on, with C_1 momentarily behaving as a dead short, point C is at near ground potential. Thus, point D is at 5 V due to the momentary cutoff condition of Q_2. With R_3 serving as a path for current, C_1 begins to charge in the positive direction with respect to point C. As the charge at point C attains 0.7 V, Q_2 begins to switch on. As a result, the voltage at point D drops rapidly to near ground potential. Thus, as seen in Figure 6–8(b), point B is momentarily clamped at -4.3 V, resulting in an immediate cutoff condition for Q_1. With point A now at 5 V, a 4.3-V charge is impressed over C_1, negative with respect to point C. It is at this point (t_1) that the flip-flop action of the astable multivibrator actually begins. Capacitor C_2 must now charge via R_2 until the voltage at point B attains the threshold level of 0.7 V. Once this occurs, the circuit condition changes rapidly to that illustrated in Figure 6–8(c). With Q_1 able to switch on again into saturation, the voltage at point A returns to near ground potential. As a result, point C is clamped momentarily at -4.3 V. Consequently, Q_2 switches into cutoff, with point D rising to 5 V. Now C_1 must charge via R_3 until the voltage at point C attains the 0.7-V threshold. It should now be evident that this free-running switching action will continue as long as a dc source is present.

The waveforms generated by the astable multivibrator in Figure 6–7(b) are illustrated in Figure 6–9. When the base voltage of either transistor is below 0.7 V, its collector output approaches the level of V_{CC}. When the base voltage of either transistor attains 0.7 V, the collector output falls to near ground potential. Thus, the pulse width and space width of the signal developed at either the Q or the $\overline{Q}$ output depends on the two RC combinations contained within the circuit. As seen in Figure 6–9, where R_2 equals R_3 and C_1 equals C_2, the pulse widths and space widths of both signals should be nearly equal.

An expression for the frequency of the astable multivibrator in Figure 6–7(b) can be developed using formula (6–1). As with the discrete transistor one-shot, a capacitor is impressed with a starting voltage level. Then, through transistor switching, the positive terminal of the capacitor is clamped at near ground potential, causing the voltage at its negative terminal to clamp below 0 V. The negative side of the capacitor must then charge in the positive direction until a switching threshold is attained. For the circuit in Figure 6–7(b), the initial capacitor voltage (V_0) is -4.3 V. This is the potential momentarily clamped at the base of either transistor as it is switched into cutoff. The threshold voltage is simply the switching level of V_{BE}. Thus, the pulse width of the signal developed at either the Q or $\overline{Q}$ output is determined as follows:

$$ PW = t = (RC) \ln \left(\frac{V - V_0}{V - V_C} \right) $$

$$ PW = 12 \text{ k}\Omega \times 0.1 \text{ }\mu F \times \ln \left(\frac{5 \text{ V} - -4.3 \text{ V}}{5 \text{ V} - 0.7 \text{ V}} \right) = 925.7 \text{ }\mu\text{sec} $$

As illustrated by the waveforms contained in Figure 6–9, the Q and $\overline{Q}$ outputs of the astable multivibrator are always in complementary conditions. Thus, the pulse width of the Q output is equal to the space width of the $\overline{Q}$ output. Conversely, the space width of the Q output is equal to the pulse width of the $\overline{Q}$ output. Where R_2 equals R_3 and C_1

Figure 6–8

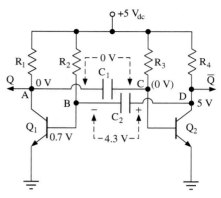

(a) Multivibrator conditions at instant of turn on (assuming Q_1 switches on before Q_2)

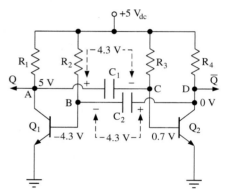

(b) Multivibrator conditions at instant when Q_2 turns on, forcing Q_1 into cutoff

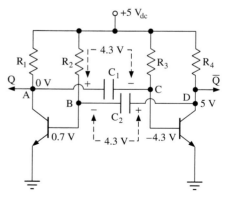

(c) Multivibrator conditions at instant when Q_1 turns on, forcing Q_2 into cutoff

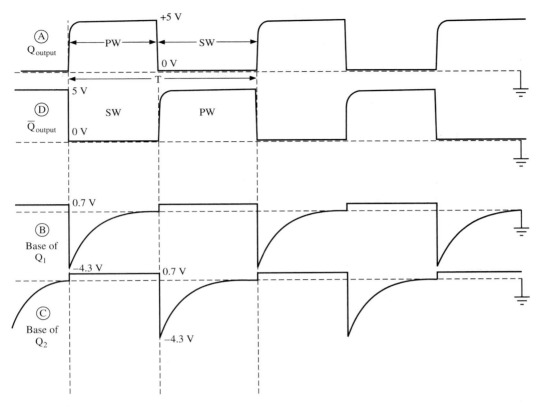

Figure 6–9
Waveforms produced by the discrete transistor astable multivibrator

equals C_2, frequency may be determined simply as follows:

$$f = \frac{1}{(2 \times PW)} \tag{6–2}$$

Thus, for the circuit in Figure 6–7(b), the frequency of the signal at either the Q or $\overline{Q}$ output is determined as follows:

$$f = \frac{1}{(2 \times 925.7 \ \mu sec)} = \text{nearly 540 pps}$$

where pps is pulses per second.

As seen in Figure 6–9, the output signals produced by the discrete transistor astable multivibrator are not true square waves. Since the transistors do not instantly switch into the cutoff state, the Q and $\overline{Q}$ waveforms contain a slight exponential curve at their leading edges. These signals can be made more rectangular by feeding them through Schmitt-triggered inverting buffers, as illustrated in Figure 6–10(a). Since the astable multivibrator is not crystal controlled, its frequency is likely to deviate slightly as temperature changes. However, the buffered output signal shown in Figure 6–10(b) is still suitable to serve as a clock source for TTL counting circuitry, as long as extreme precision is not required.

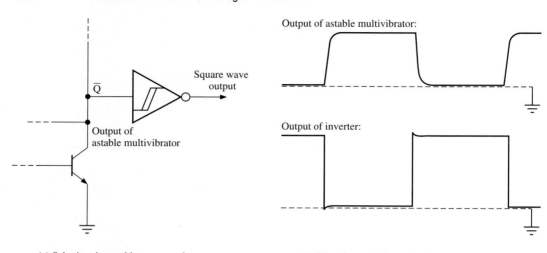

(a) Schmitt-triggered inverter used to
buffer output of astable multivibrator

(b) Waveform relationship of astable multivibrator
and inverting Schmitt trigger

Figure 6–10

In the previous chapter, a method of detecting the failure of fuses within machine control circuitry was introduced. As illustrated in Figure 5–8, a beacon lamp could be illuminated at the instant one of four fuses was blown. For machinery that is quite large, a more effective method of fault indication would be that shown in Figure 6–11. Large machinery is likely to contain several beacons, indicating system status for various operating zones. If the system is functioning properly at a given point, the green lamp within the beacon would be continuously illuminated. If a fault occurs within that zone, the green lamp would switch off and the red emergency lamp would begin to flash. A horn could also sound, switching on and off at the same rate as the emergency lamp.

The large values of capacitance used in Figure 6–11 allow for an operating frequency of less than 1 pps. This frequency is determined as follows, using formulas (6–1) and (6–2):

$$PW = (22 \text{ k}\Omega \times 47 \text{ }\mu\text{F}) \times \ln \frac{(5 \text{ V} - -4.3 \text{ V})}{(5 \text{ V} - 0.7 \text{ V})} = 798 \text{ msec}$$

$$f = \frac{1}{2 \times 798 \text{ msec}} = 0.627 \text{ pps}$$

This slow operating frequency for the astable multivibrator creates the desired visual effect. In the event of a system failure, the attention of an operator or technician would be drawn to the location of the flashing beacon. There, further visual indications, such as LEDs on control panels or circuit card edges, would assist in leading the technician to the exact fault.

The astable multivibrator in Figure 6–11 begins to operate at the instant power is applied. Thus, the two LEDs, which could be mounted on a control panel, flash continually. As long as the system being monitored is operating correctly, the input at point A is at a logic low. This input condition brings point B to a logic high, forward biasing the Darlington pair consisting of Q_3 and Q_4. As a result, the green lamp is illuminated. The low at point A also causes a logic high to be present at point D. This holds point E at a

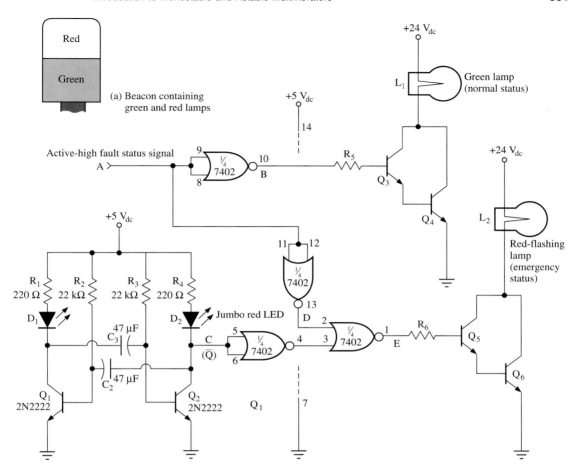

(a) Beacon containing green and red lamps

(b) Astable multivibrator-controlled flasher

Figure 6–11

logic low, regardless of the alternating logic states at point C. Thus, the red lamp remains in the off condition. When a fault condition has occurred, point A switches to a logic high, causing point D to fall to a logic low. This condition allows the alternating logic levels at point C to pass to point E. Thus, when the $\overline{Q}$ output is high, point E is also high, causing the Darlington pair consisting of Q_5 and Q_6 to be forward biased. As a result, the red lamp turns on. During the time in which the $\overline{Q}$ output is low, point E is also low, switching off the emergency lamp. Thus, when a fault has occurred, the red lamp flashes on and off, while the green lamp is held in the off condition.

Example 6–2

System designers have decided that a red flashing beacon would be easier to see if it flashes at a rate of 1½ times per second. What value of capacitance should then be used for C_2 and C_3 in Figure 6–11?

Solution:

Step 1: Derive an expression for capacitance:

$$f = \frac{1}{(2 \times PW)} = \frac{1}{2 \times RC \times 0.7714} = \frac{0.6482}{RC}$$

Step 2: Solve for capacitance:

$$C = \frac{0.6482}{22 \text{ k}\Omega \times 1.5 \text{ pps}} = 19.64 \text{ }\mu F$$

Use 22 μF as a standard value.

SECTION REVIEW 6–1

1. True or false?: An astable multivibrator has two stable output conditions.
2. For the discrete one-shot in Figure 6–1(b), explain the functions of C_2 and D_1.
3. For the one-shot in Figure 6–1(b), assume that the resistive values are kept the same. What standard value of capacitance used for C_1 would allow a pulse width of approximately 25 sec?
4. For the circuit in Figure 6–5, what is the voltage level at point F when point A is at 0 V? What is the voltage level at point F at the instant point A switches to a logic high? What is the voltage level at point F when point E is at a logic low, but point A is still at a logic high?
5. For the circuit in Figure 6–5, what value of C_2 would allow high current to flow through the solenoid for nearly 1/10 of a second?
6. True or false?: For both the astable and the monostable multivibrator, the Q and $\overline{Q}$ outputs are always in complementary logic states.
7. For the astable multivibrator in Figure 6–7(b), what standard value of capacitance would be used for C_1 and C_2 in order to obtain an operating frequency of 245 pps?

6–2   INTEGRATED CIRCUIT TIMERS

Monostable and astable multivibrators are so prevalent within industrial control systems that the electronic circuitry required to create these devices is made available in monolithic form. Such ICs contain the basic components of the multivibrator, to which external resistors and capacitors are connected. Those ICs most commonly used as part of industrial control systems include the **74121,** the **74122,** and the **74123,** which are three forms of TTL one-shots. The **555 IC timer,** which may be configured to function either in a monostable or astable mode, is also widely used for industrial control applications.

Figure 6–12 shows the structure of the 74121, which functions as a single monostable multivibrator. This IC contains its own timing resistor R_{int}. This resistance may be unused, as shown in Figure 6–12(a), or placed in series with an external resistance (R_{ext}) as shown in (b). The designer also has the option of using only R_{int} as the resistive element for RC timing. In this case, only the external capacitor is connected to the 74121. For standard TTL, R_{int} equals 2 kΩ, while for low-power TTL, this internal resistance is 4 kΩ. For the

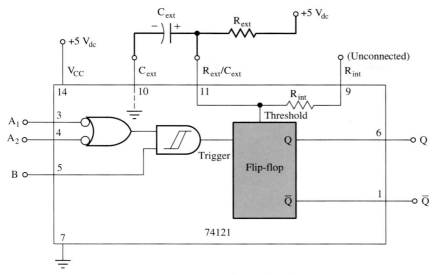

(a) TTL one-shot with fixed pulse width

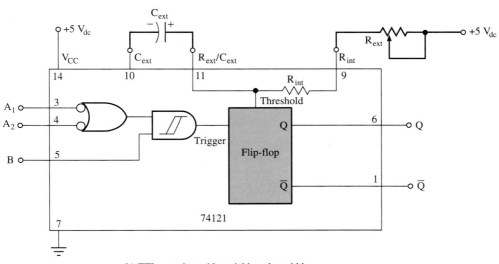

(b) TTL one-shot with variable pulse width

Figure 6–12

74121, the duration of the active-high Q output is determined as follows:

$$PW = R \times C_{ext} \times \ln 2 \cong 0.7 \times RC \qquad (6\text{–}3)$$

where R may equal R_{ext}, R_{int}, or $(R_{ext} + R_{int})$.

The 74121 is capable of producing pulse widths ranging from 40 nsec to 28 sec. The most accurate operation is achieved if the value of C_{ext} is kept between 10 pF and 10 μF. Also, the resistive component within the RC timing combination should not fall below

1.4 kΩ. Where the exact duration of the pulse width is not critical, C_{ext} may be as high as 1000 μF.

The design of the internal circuitry of the 74121 allows for negative-edge triggering at either input A_1 or A_2, while positive edge triggering is possible at point B. To bring the Q and $\overline{Q}$ outputs into their active states, a positive transition must occur at the trigger input of the internal flip-flop of the 74121. Providing input B is held at a logic high, the output of the Schmitt-triggered AND function will go to a logic high as either A_1 or A_2 is brought to a logic low. If either input A_1 or A_2 is held low, the output of the internal NAND function will be at a logic high. Thus, as input B goes from a low to a high condition, a positive transition also occurs at the output of the Schmitt-triggered AND function. The versatility of the 74121 is further enhanced by the availability of the $\overline{Q}$ output. Figure 6–13 illustrates the operation of the 74121. In (a), input B is assumed to be high prior to the occurrence of the active low trigger pulse. In (b), either A_1 or A_2 must be made low prior to the arrival of the active-high trigger pulse.

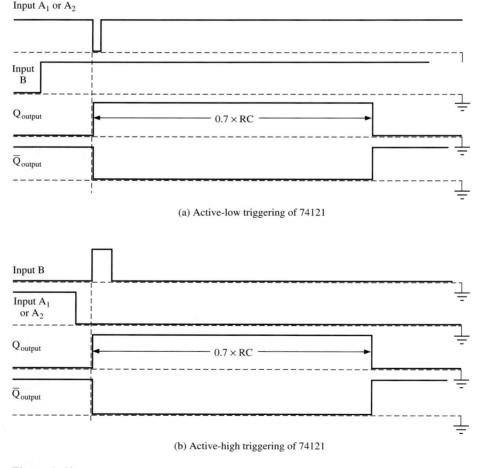

(a) Active-low triggering of 74121

(b) Active-high triggering of 74121

Figure 6–13

Example 6–3

Assume that the 74121 monostable multivibrator in Figure 6–12(b) is a low-power TTL device. Also assume that R_{ext} is a 0/20-kΩ potentiometer and C_{ext} is a 1-μF capacitor. What must be the ohmic setting of R_{ext} in order to obtain a pulse width of 10 msec?

Solution:

Step 1: Derive an expression for R_{ext}: Since the 74121 is specified as a low-power device, R_{int} equals 4 kΩ. Thus,

$$PW = 0.7 \times (R_{ext} + 4\,k\Omega) \times C_{ext}$$

Step 2: Substitute known values:

$$10\,\text{msec} \cong 0.7 \times (R_{ext} + 4\,k\Omega) \times 1\,\mu F$$

Step 3: Solve for R_{ext}:

$$(R_{ext} + 4\,k\Omega) = \frac{10\,\text{msec}}{0.7 \times 1\,\mu F} = 14.286\,k\Omega$$

$$R_{ext} = 14.286\,k\Omega - 4\,k\Omega = 10.286\,k\Omega$$

Set R_{ext} to 10.29 kΩ prior to installing it into the circuit.

Figure 6–14 illustrates a possible application of the 74121 one-shot. Note that, because the device is to be triggered by a single active-low control signal, the B input is permanently tied to 5 V_{dc}. Also note that A_1 and A_2 are tied together to form a common input terminal. The device being controlled in Figure 6–14 is a small linear solenoid. Unlike the device in Figure 6–5, this solenoid requires only a single current pulse for reliable actuation. At the instant a trigger pulse arrives from system control, the Q output of the 74121 goes to a logic high, remaining in that state for nearly 200 msec. During this time, the Darlington pair is forward biased, energizing the linear solenoid. As a result, the plunger pushes forward, overcoming the pressure of the detent spring. The device in Figure 6–14(b) could be used to operate a needle valve or a small punching or stamping mechanism.

A problem with the form of solenoid control illustrated in Figure 6–14 is the fact that pulse width is easily affected by deviations in operating temperature. Thus, in applications where exact timing becomes critical, the one-shot is likely to be abandoned in favor of a microprocessor-based timing system. At the instant the solenoid is to be actuated, a microcomputer sends a triggering signal to the Darlington pair. As illustrated in Figure 6–15, this trigger signal is likely to be stored in a buffer while the microcomputer executes a delay algorithm. At the end of this delay time, the microcomputer simply removes this trigger signal from the buffer. Since the microcomputer is driven by a high-frequency, crystal-controlled clock, it is capable of extremely precise timing control. Crystal-controlled clocking also allows the microcomputer to be immune to temperature effects.

The microcomputer initiating the actuating signal for the linear solenoid in Figure 6–15 is likely to be in a remote location. For this reason, the output of the tri-state buffer enters a Schmitt-triggered inverter prior to leaving the microcomputer circuit board. The inverted control signal, now active-low, arrives at the cathode terminal of the optocoupler on the solenoid drive circuit board. When point A is at a logic high, point B is low, forward biasing the LED within the optocoupler. The phototransistor then goes into saturation, bringing point C to a logic low. Point D then goes high, forward biasing the

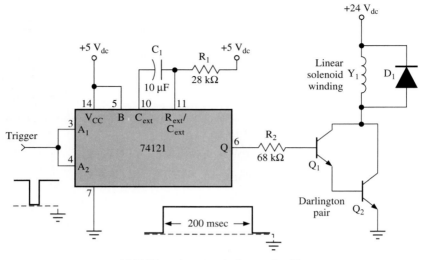

(a) 74121 used to control a linear solenoid

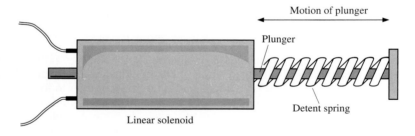

(b) Operation of the linear solenoid

Figure 6–14

Darlington pair. When the output of the tri-state buffer falls low, the Darlington pair is reverse biased, de-energizing the solenoid.

Although the one-shot in Figure 6–15 does not directly control the pulse width of the solenoid control signal, it does perform a necessary function in limiting the amount of time in which current may flow to the solenoid winding. The tri-state buffer actually has three possible output conditions. When enabled, its output is either a valid logic high or a logic low. When not enabled, the device functions as an open switch, with its output effectively removed from any subsequent circuitry. Thus, during the time in which a logic low is present at its enable input, the output of the tri-state buffer will be equivalent to the signal present at the data input. When a logic high is present at the enable input, the output of the tri-state buffer actually functions as an open switch, exhibiting an extremely high impedance to current flow in either direction at point A. When the solenoid is to be actuated, the address-decoding circuitry of the microcomputer sends an active-high pulse to the chip-select input of the 74LS373, which contains eight tri-state buffers. This action is likely to occur a few nanoseconds after the microprocessor has sent a high bit to the data input of the tri-state buffer via the microcomputer data bus. This chip-select pulse inverts to provide an active-low trigger to the 74121 one-shot. At the instant this low

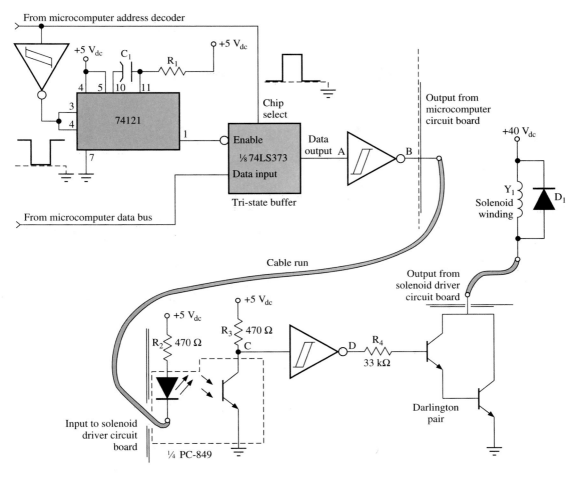

Figure 6–15
74121 IC controlling maximum duration of solenoid-actuation pulse

pulse arrives at pins 3 and 4 of the 74121, the $\overline{Q}$ output of that device goes to a logic low, enabling the tri-state buffer to pass the high present at the data input to the data output terminal (point A).

If the microcomputer is functioning correctly, it performs a delay algorithm, which precisely times the period for which point A is held at a logic high. At the end of this delay routine, the microcomputer places a logic low at the data input of the tri-state buffer. Since, at that time, the $\overline{Q}$ output of the 74121 is still low and the chip-select input is still high, the logic low immediately passes to point A. This causes the Darlington pair to be reverse biased, de-energizing the solenoid. The need for the one-shot should now be apparent. If the microcomputer malfunctions, the logic condition that could exist at point A becomes unpredictable. If the one-shot were not in the circuit, it is possible that the logic high would be present at point A indefinitely, allowing a high level of current to keep flowing through the solenoid winding. This condition would be extremely dangerous, since the prolonged current flow could easily cause the solenoid to overheat and be destroyed. Also, the prolonged actuation of the solenoid could result in costly damage to

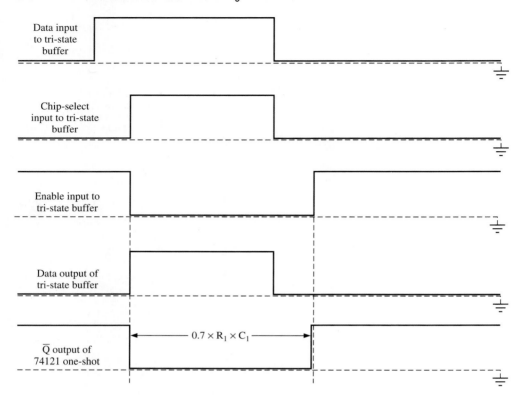

Figure 6–16
Pulses showing operation of solenoid-control circuit in Figure 6–15

machinery and to the materials being processed. However, as shown by the pulse train in Figure 6–16, the low pulse at the $\overline{Q}$ output times out soon after the delay algorithm, causing the tri-state buffer to go to its high-impedance output condition. This ensures that the solenoid winding will remain energized no longer than the pulse width of the one-shot, regardless of the condition of the data input to the tri-state buffer.

Example 6–4

For the solenoid-control circuitry in Figure 6–15, assume that the duration of the output pulse produced at point A is exactly 125 msec, when the microcomputer is functioning correctly. The active-low enable pulse produced at the $\overline{Q}$ output of the 74121 is to be approximately 25% longer than the signal produced at point A. If the largest capacitance value recommended for accurate operation is to be used for C_1, what ohmic value for R_1 would produce the desired signal at the $\overline{Q}$ output of the one-shot?

Solution:

Step 1: Determine the duration of the $\overline{Q}$ signal:

$$\overline{Q} = 1.25 \times PW \text{ at } A = 1.25 \times 125 \text{ msec} = \text{nearly } 156 \text{ msec}$$

Step 2: Solve for the combined value RC: For the 74121 one-shot in Figure 6–15,

$$PW \cong 0.7 \times R_1 \times C_1$$

Thus,

$$RC \cong \frac{156 \text{ msec}}{0.7} \cong 223 \text{ msec}$$

Step 3: Solve for the value of R_1: Since the largest value of capacitance recommended for accurate operation is 10 μF, then

$$R_1 \times (10 \text{ μF}) \cong 223 \text{ msec}$$

Thus,

$$R_1 \cong \frac{223 \text{ msec}}{(10 \text{ μF})} \cong 22.3 \text{ k}\Omega$$

Choose 22 kΩ as a standard value.

The structure of the 74122 monostable multivibrator is shown in Figure 6–17. This device has four trigger inputs: A_1 and A_2 allow negative-edge triggering, whereas positive-edge triggering is possible with inputs B_1 and B_2. Also, like the 74121, the 74122 has an internal timing resistor. The 74122 has two distinct design advantages over the 74121.

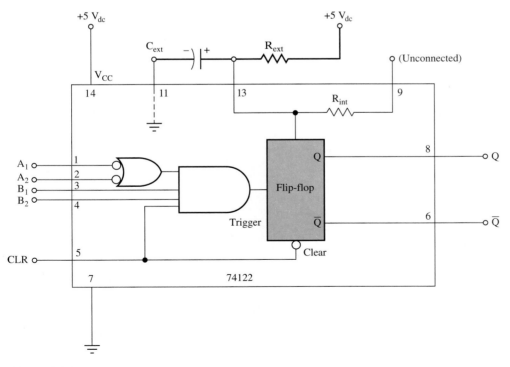

Figure 6–17
Internal structure of TTL retriggerable one-shot with clear input

The first advantage is the fact that the 74122 is **retriggerable.** This means that the pulse width of the Q output may be greatly extended by sending multiple trigger signals to the device. The timing diagram in Figure 6–18 serves to demonstrate the operation of the 74122. Note that, for the sake of simplicity, the B_2 input is tied to a logic high. This allows A_1, A_2, and B_1 to become the effective trigger inputs. At t_0, the active-low input at A_1 initiates the extended high pulse at the Q output. At t_1, prior to the termination of the first pulse width, a second trigger pulse arrives at the A_2 input. This retriggering action (which could also have occurred a second time at A_1) prevents the return of the one-shot to its quiescent state. Before the termination of the second pulse width, with the A_2 input remaining low, a second retriggering pulse arrives at t_2, as the B input returns to a logic high. This action extends the output pulse width even further. With no additional retriggering, the final portion of the output pulse, extending from t_2 to t_3, is equivalent to $(0.7 \times R_{ext} \times C_{ext})$. Thus, through the retriggering process, a single output pulse is created, extending from t_0 through t_3. Such a process allows relatively long output pulses to be obtained while utilizing relatively small values of R_{ext} and C_{ext}.

The second advantage of the 74122 over the 74121 is the presence of the **clear input** (CLR). This active-low input resets the internal flip-flop of the device. Thus, as shown at t_5 in Figure 6–18, the CLR signal is able to prematurely terminate the high condition at the Q output of the one-shot. Since CLR also serves as an input to the internal AND gate of the 74122, it has priority over all four trigger inputs. As illustrated in Figure 6–18, between t_5 and t_9, the logic low condition at the CLR input inhibits any trigger signals present at A_1, A_2, or B_1.

Figure 6–19 shows the internal structure of the 74123 IC, which contains two retriggerable one-shots. These devices share the same V_{CC} and ground, but are operated indepen-

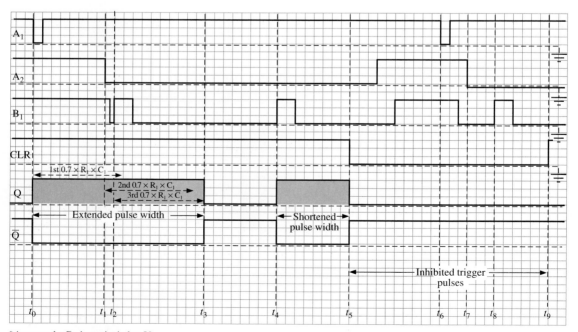

*Assume the B_2 input is tied to V_{CC}

Figure 6–18
Timing diagram demonstrating the operation of the 74122 retriggerable one-shot

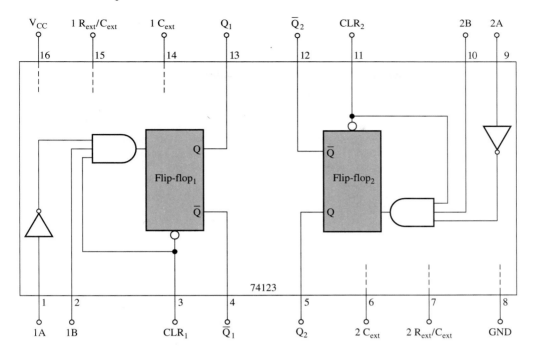

Figure 6–19
Internal structure of TTL dual retriggerable one-shots with clear inputs

dently. Note that each multivibrator is provided with terminals for its own R_{ext} and C_{ext}. However, no R_{int} is made available for either multivibrator. The obvious advantage of the 74123 over the 74121 or 74122 is the availability of two one-shots, but it does have a slight disadvantage, however, in that it has only two trigger inputs for each device. For the first multivibrator, 1A serves as the active-low input, while 1B is active-high. If input 1B is held at a logic high, internal flip-flop 1 is triggered as a negative transition occurs at the 1A input. If the 1A input is held low, flip-flop 1 is triggered as a positive transition occurs at the 1B input. Inputs 2A and 2B function identically to inputs 1A and 1B, providing negative-edge and positive-edge triggering for the second monostable multivibrator. The two CLR inputs, of course, have priority over the trigger inputs. A low condition at pin 3 holds the Q output low and the $\overline{Q}$ output high for the first one-shot, regardless of the activity at inputs 1A or 1B. A low at pin 11 holds $\overline{Q}$ high and Q low for the second one-shot, regardless of any activity at inputs 2A or 2B.

In determining the duration of the output pulse for the 74122 or the 74123, a different formula must be utilized than for the 74121. For values of C_{ext} exceeding 1000 pF, the output pulse width is determined by the following formula:

$$PW = \kappa \times (R_T \times C_{ext}) \times \left(1 + \frac{0.7}{R_T}\right) \qquad (6\text{–}4)$$

where R_T represents $R_{int} + R_{ext}$ for the 74122 and just R_{ext} for the 74123.

For Eq. (6–4), κ is a constant that varies slightly with each type of device. For a standard TTL 74122, κ is approximately 0.32, whereas for the low-power equivalent of the same device, κ increases to 0.37. For a standard TTL 74123, κ is 0.28, whereas for the low-power form of the same device, κ becomes 0.33. An examination of Eq. (6–4) shows that, for practical values of R_{ext}, the expression for pulse width may be simplified to

$$PW \cong \kappa \times (R_T \times C_{ext}) \qquad (6\text{–}5)$$

Figure 6–20 illustrates an application of a retriggerable one-shot. Here the device is used to control the operation of the solenoid-actuation circuit first presented in Figure 6–5. The use of the 74122 retriggerable one-shot in this control circuit results in more efficient and reliable operation of the diverter gate than would be possible with the 74121. Remember that the purpose of such a diverter gate is to route rapidly moving containers to an alternative conveyance path or a storage bin. The decision to divert would be based on previously obtained information about each container. Such information might include length and height, as well as destination data obtained via bar code reading, indicia-detection, or image-lift circuitry.

At the input to the conveyance path, a light barrier array such as that illustrated in Figures 5–43 through 5–46 would be used to establish a uniform gap between the arriving containers. For optimal performance of most high-speed sorting machines, the gap between containers is kept as narrow as possible. Consequently, diverter gates within such systems must operate rapidly, with extremely precise timing. The high-current/low-current actuation technique, which has already been explained in detail, results in both rapid response and energy efficiency for the rotary solenoid that drives the diverter gate. In even the best designed conveyance systems, however, diverter gates are points where jamming is likely

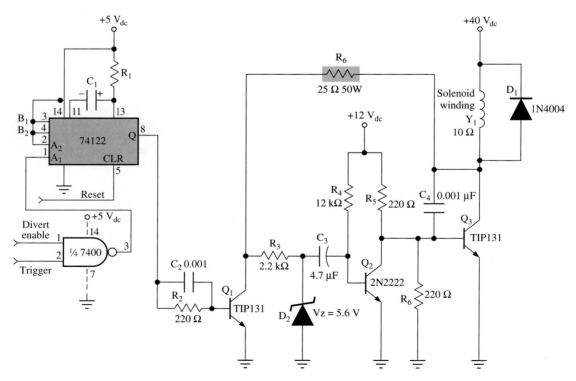

Figure 6–20
Retriggerable one-shot controlling the operation of a solenoid

to occur. To minimize the occurrence of this problem, any unnecessary motion of the gate must be avoided.

Consider the possible conditions that could exist for two containers rapidly approaching the diverting station, as represented in Figure 6–21. It is possible that both containers are destined to bypass the diverting path. In that case, the gate simply does not actuate when the red light beam of the reflex photosensor is broken by either container. It is also possible that the first container is destined to bypass the diverter, but the second container must enter the diverting path. Since the system control microcomputer is tracking the progress of each container, it would withhold the trigger signal for the one-shot controller until the second container trips the reflex photosensor. An interesting design problem arises for a condition where both containers are to be diverted. In this situation, to avoid unnecessary motion, the gate should actuate to divert the first container and remain actuated until the second container has also been routed to the diverting path. This is easily accomplished through the action of the retriggerable one-shot.

As illustrated by the timing diagram in Figure 6–22, if a container is to be diverted, the system control circuitry brings the divert enable signal to a logic high. This occurs just before that container approaches the reflex photosensor. At the instant the light beam is broken, a trigger pulse is initiated by the system control circuitry. The resultant active-low output of the NAND gate triggers the 74122. The Q output of that device then goes to a logic high, actuating the diverter gate. Since system control microcomputer already knows that the second container is to be diverted, it holds the divert enable signal high. As represented in Figure 6–22(a), the speed of the conveyor belt and the gap between the containers is such that the second container is likely to break the light beam and initiate a second trigger pulse prior to the termination of the pulse width initiated by the arrival of the first container. Thus, the Q output of the 74122 remains high, prolonging

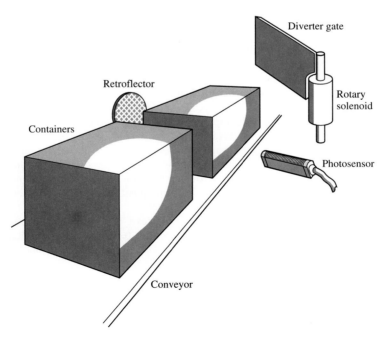

Figure 6–21
Two containers in close proximity approaching diverter gate

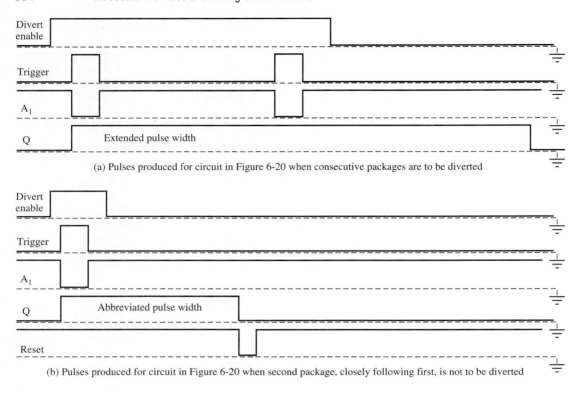

(a) Pulses produced for circuit in Figure 6-20 when consecutive packages are to be diverted

(b) Pulses produced for circuit in Figure 6-20 when second package, closely following first, is not to be diverted

Figure 6–22

the holding current for the diverter gate. As a result, the gate remains actuated and the second container is diverted.

A final condition must be considered for the two containers in Figure 6–21. Assume that the first container is to be diverted and the second must bypass the diverting path. Providing that the second container is far enough behind the first, system control could safely let the Q output of the one-shot continue through its complete pulse width. However, if the second container is too close on the heels of the first, the control circuitry must prematurely terminate the Q output of the one-shot through the initiation of an active-low reset signal. As illustrated in Figure 6–22(b), since the reset signal arrives at the CLR input of the 74122, the Q output of the one-shot immediately terminates. This action ensures that the diverter gate returns to its home position prior to the arrival of the second container.

The action just described, of course, requires extremely precise timing. If the reset signal is sent too soon, the first container may not be completely diverted, resulting in a jam condition. If the reset signal is sent too late, the gate could block the path of the second container, accidentally sending it to the wrong location. To ensure proper timing, the control circuitry often uses the output of a transport clock, such as that shown in Figure 5–36, to monitor the gap between the arriving containers. (Remember that, in conveyance systems, such clocking circuitry is often used to convert angular rotation to belt travel.) By counting the number of transport clock transitions that occur between obstructions of the acceptance light barrier, the system control microcomputer is able to determine if it is necessary to initiate a reset pulse. The microcomputer would be programmed to compare the measured gap to a binary value representing a minimal acceptable distance. If, during measurement of the gap between two containers, the count of transducer clock cycles exceeds this reference value, no reset pulse is initiated. If the count falls below this number, a reset pulse is immediately sent to the one-shot.

Although the light barrier array at the input to the conveyor ensures that proper spacing is maintained between the containers, it is possible through slippage that the second container is too close for proper operation of the diverter gate. The microcomputer could also be programmed to recognize this condition. To prevent damage to both the gate and the container, it could possibly initiate the necessary signals to shut down the conveyance motors and declare a fault. However, a more efficient course of action would be to allow the conveyance system to continue operating and simply send both containers to a reject bin. Packages arriving at the reject bin could either be hand sorted or possibly processed by the conveyance system a second time.

Although the function of the 74121, 74122, and 74123 is limited to that of a monostable multivibrator, the 555 timer is utilized in a wide variety of pulse applications. In Figure 6–23, which shows the operation of the 555 timer as a basic one-shot, the internal structure of the device is represented. Before analyzing the monostable operation of the IC, we must first become acquainted with its internal circuitry. The 555 contains a voltage divider consisting of three 5-kΩ resistors (hence, 555). Note that this voltage divider provides reference levels for two internal comparators. The comparators, in conjunction with the three resistors, form a voltage level-sensing device known as a **window detector.** For comparator 1, the reference voltage (which is equal to 2/3 V_{CC}) is present at the inverting input. For comparator 2, the reference voltage present at the non-inverting input is 1/3 V_{CC}. Since the non-inverting input is the active input terminal for comparator 1, the ouptut of this device goes high at the instant the voltage at pin 6 exceeds 2/3 V_{CC}. With the inverting input of comparator 2 serving as the active terminal, the output of that comparator goes high at the instant the voltage at pin 2 falls below 1/3 V_{CC}.

The central switching device of the 555 IC timer is a **set/reset** (SR) **flip-flop,** which is triggered on the positive edge of pulses arriving from the two comparators. When the voltage at pin 2 falls below 1/3 V_{CC}, providing the output of comparator 1 is already low, the SR flip-flop goes to a set condition. The logic low now present at the $\overline{Q}$ output creates a reverse-biased condition for Q_1, which is a current-sinking transistor switch. As a result, the discharge of the external timing capacitor via pin 7 is temporarily inhibited. The low condition for $\overline{Q}$ inverts to a logic high at pin 3. This high output voltage approaches the level of V_{CC}. At the instant the voltage at pin 6 exceeds the level of 2/3 V_{CC}, a positive transition occurs at the reset input of the flip-flop. If the set input is already at a logic low, this action brings the $\overline{Q}$ output to a logic high. As a result, pin 3 falls to nearly 0 V. Also, the current-sinking transistor becomes forward biased, allowing the external timing capacitor to discharge via pin 7. As with the 74122 and 74123, the active output pulse of the 555 may be prematurely terminated by a priority clearing signal. The reset input at pin 4 of the 555 timer functions identically to the CLR input of either the 74122 or 74123. As long as pin 4 is held at a logic low, pin 3 also remains low. The availability of the threshold input at pin 5 further enhances the operation of the 555 IC timer. If this terminal is left unconnected, the switching threshold for comparator 1 remains at 2/3 V_{CC}. However, an external voltage source, connected directly to pin 5, may be used to override the 2/3 V_{CC} threshold. An example of this capability of the 555 IC timer is introduced later in this chapter.

The timing diagram in Figure 6–24 illustrates the operation of the 555 configuration in Figure 6–23. At t_0, prior to triggering of the 555, a reset pulse ensures that the $\overline{Q}$ output of the internal flip-flop is latched at a logic high. With this condition, current-sinking transistor Q_1 is switched on in saturation. At this time, with R_1 serving as the

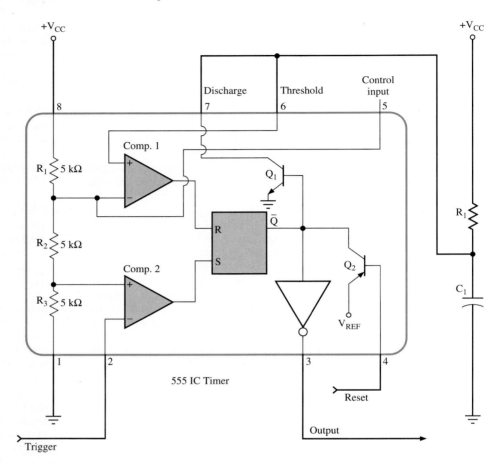

Figure 6–23
The 555 IC timer in a monostable multivibrator configuration

collector resistor for Q_1, current is shunted past C_1. Therefore, this capacitor remains in a discharged state. At t_1, the negative transition of the trigger signal causes the $\overline{Q}$ output of the flip-flop to toggle to a logic low and initiates the active output pulse at pin 3. With Q_1 now reverse biased, C_1 is able to charge via R_1. As shown in Figure 6–24, the high condition at pin 3 continues until the voltage developing over C_1 approaches 2/3 V_{CC}. When, at t_2, this threshold level is attained, the internal flip-flop resets. This action causes the $\overline{Q}$ output to go high and the voltage at pin 3 to fall to near ground potential. Also, with Q_1 now forward biased, C_1 rapidly discharges via pin 7. The capacitor remains in the discharged condition until another trigger pulse arrives at pin 2. The monostable circuit in Figure 6–23 is *not* retriggerable. Thus, any subsequent trigger pulses occurring between t_1 and t_2 would not lengthen the output pulse width. However, the high condition at pin 3 could be abbreviated at any point by a reset pulse at pin 4.

A simple formula for determining the pulse width of the 555 one-shot may be derived from formula (6–1). Since the initial charge on C_1 is equal to $V_{CE}(sat)$ of Q_1, V_0 may be assumed to be 0 V. Also, with pin 6 common to the positive terminal of C_1, as seen in

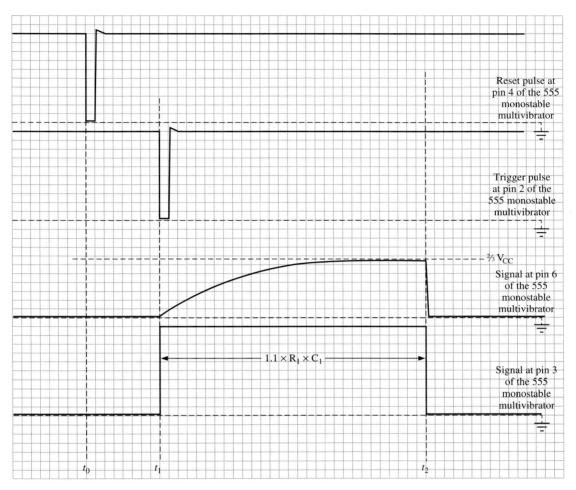

Reset pulse at pin 4 of the 555 monostable multivibrator

Trigger pulse at pin 2 of the 555 monostable multivibrator

$\frac{2}{3}V_{CC}$

Signal at pin 6 of the 555 monostable multivibrator

$1.1 \times R_1 \times C_1$

Signal at pin 3 of the 555 monostable multivibrator

t_0 t_1 t_2

Figure 6–24
Pulses produced by 555 monostable multivibrator

Figure 6–23, V_C becomes 2/3 V_{CC}. Thus, for any supply voltage:

$$PW = R_1 \times C_1 \times \ln\left(\frac{V_{CC} - 0\,V}{V_{CC} - 2/3V_{CC}}\right)$$

$$PW = R_1 \times C_1 \times \ln 3$$

$$PW \cong 1.1 \times R_1 \times C_1 \qquad (6\text{–}6)$$

Example 6–5

For the 555 one-shot in Figure 6–25, what ohmic setting of R_1 would allow an output pulse width of 175 msec?

Solution:

Step 1: Derive an expression for R_1:

$$PW \cong 1.1 \times (R_1 + R_2) \times C_1$$

$$(R_1 + R_2) \cong \frac{PW}{(1.1 \times C_1)}$$

$$R_1 = \frac{PW}{(1.1 \times C_1)} - R_2$$

Step 2: Substitute known values and solve for R_1:

$$R_1 = \frac{175 \text{ msec}}{(1.1 \times 4.7 \text{ } \mu F)} - 22 \text{ k}\Omega \cong 11.85 \text{ k}\Omega$$

Adjust R_1 to 11.85 kΩ prior to placing it in the circuit.

As shown in Figure 6–26(a), two or more one-shots may be cascaded together to form a **sequential timer.** The integrated circuit represented here is a **556 dual timer,** which contains two independently functioning 555 timers. Note that capacitive coupling is used to produce a rapid, active-low trigger pulse for the second timer at the instant the output of the first timer falls low. As shown by the timing diagram in Figure 6–26(b), the space width of the trigger pulse for the second one-shot should be made much narrower than the pulse width of the second output signal. This relationship may be achieved by ensuring that the time constant for R_2 and C_2 is much shorter than the output pulse width of the second timer. For example, if τ for R_2 and C_2 is approximately 1/10 of the output pulse width of the second one-shot, the voltage at pin 8 of the second one-shot will be at nearly V_{CC} by the time the second output signal terminates. This ensures that retriggering of the second one-shot does not occur. The time constant for the coupling capacitor and resistor should not be too short. If the trigger pulse for the second one-shot is too narrow, that device might not be activated.

Figure 6–27 illustrates the 555 circuit configuration for a basic astable multivibrator. (Here, the internal circuitry is shown in order to clarify the operation of this device.)

Figure 6–25

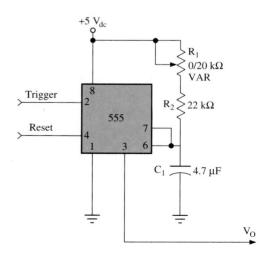

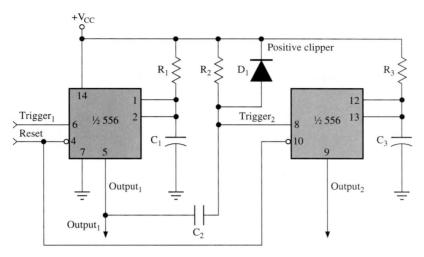

(a) 556 dual timer used to create a sequential timer

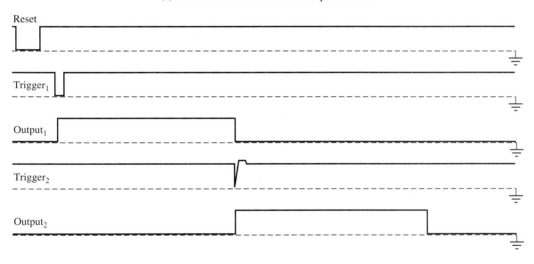

(b) Timing diagram for the sequential timer

Figure 6–26

Because the device is to be free running, no external trigger source is used. Instead, both pins 2 and 6 are made common to the positive terminal of the timing capacitor. If the device is to run continuously, pin 4 can be connected permanently to V_{CC} in order to inhibit the reset function. The threshold input at pin 5 is connected to a 0.01-μF capacitor, which is grounded for the purpose of noise elimination. At the instant power is applied to the circuit, C_1 is in the discharged state. Thus, as shown at t_0 in Figure 6–28, the internal flip-flop is set and the output at pin 3 is at a logic high. Since the internal current sink Q_1 is reverse biased at this time, C_1 is able to charge via the combined resistance of R_1 and R_2. At t_1, as the value of V_{C1} attains $2/3$ V_{CC}, the flip-flop is reset. At this time, the output at pin 3 falls to a logic low. Since Q_1 is now forward biased, C_1 begins to discharge into pin 7 via R_2. At t_2, however, as the voltage over C_1 is reduced to $1/3$ V_{CC},

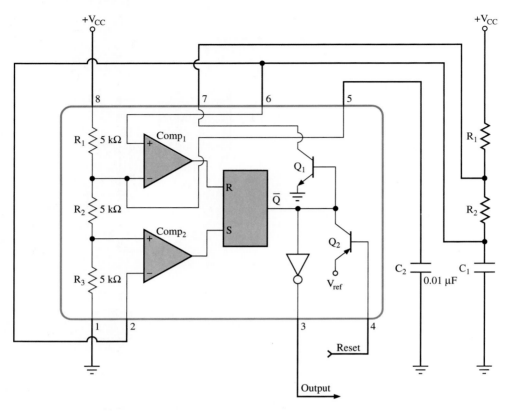

Figure 6–27
A 555 IC timer in an astable multivibrator configuration

the flip-flop sets and the output again switches to a logic high. With Q_1 now reverse biased, C_1 again charges forward 2/3 V_{CC}, again causing the flip-flop to reset. This cycle continues as long as power is applied.

A formula for the pulse width of the rectangular output of the 555 astable multivibrator may be derived from Eq. (6–1). Once the circuit has begun operating, the residual charge maintained over the timing capacitor, represented as V_0 in Eq. (6–1), is equal to 1/3 V_{CC}. The highest voltage attained by the capacitor, of course, is still 2/3 V_{CC}. Since, during the time the output is high, the timing capacitor charges via the combined resistance of R_1 and R_2, the expression for pulse width is developed as follows:

$$PW = (R_1 + R_2) \times C_1 \times \ln\left(\frac{V_{CC} - 1/3V_{CC}}{V_{CC} - 2/3V_{CC}}\right)$$
$$= (R_1 + R_2) \times C_1 \times \ln 2$$

which simplifies to:

$$PW = 0.693 \times (R_1 + R_2) \times C_1 \qquad (6-7)$$

Since the discharge path for C_1 is only through R_2, the space width of the output signal is calculated as follows:

$$SW = 0.693 \times (R_2 \times C_1) \qquad (6-8)$$

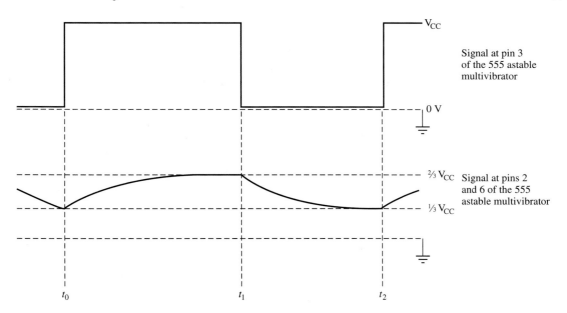

Figure 6–28
Waveforms for 555 astable multivibrator

Note that for the 555 astable configuration in Figure 6–27, since the capacitor charges via two resistors and discharges through only one, pulse width for the output signal is always greater than space width. If the ohmic value of R_1 is made much greater than that of R_2, the charge time for C_1 is much longer than the discharge time. Thus, for the rectangular waveform at pin 3, pulse width becomes much greater than space width. If the ohmic value of R_1 is made much less than that of R_2, then the charge time for C_1 nearly equals the discharge time. As a result, the output at pin 3 more closely resembles a square wave.

For any rectangular waveform, the period (represented as T) is the sum of the pulse width and the space width. Thus, for the astable 555 timer circuit in Figure 6–27:

$$T = \text{PW} + \text{SW} \tag{6–9}$$

As with any periodic signal, frequency is determined as follows:

$$f = \frac{1}{T} \tag{6–10}$$

An important parameter for rectangular wave generators used within industrial control circuitry is **duty cycle (δ).** The duty cycle of a rectangle wave is the ratio of the pulse width to the period, usually expressed as a percentage. Thus,

$$\delta = \frac{\text{PW}}{T} \times 100\% \tag{6–11}$$

The concept of duty cycle is examined in detail later in this chapter, during the discussion of pulse width modulators. Duty cycle also becomes an important factor in the study of switching regulators and power supplies, which are introduced in Chapter 9.

Example 6–6

For the 555 astable multivibrator in Figure 6–27, assuming R_1 equals 22 kΩ, R_2 equals 1.5 kΩ, and C_1 equals 0.01 μF, solve for pulse width, space width, period, frequency, and duty cycle of the output signal.

Solution:

Step 1: Solve for pulse width:

$$PW = 0.693 \times (22 \text{ k}\Omega + 1.5 \text{ k}\Omega) \times 0.01 \ \mu\text{F} = 162.9 \ \mu\text{sec}$$

Step 2: Solve for space width:

$$SW = 0.693 \times 1.5 \text{ k}\Omega \times 0.01 \ \mu\text{F} = 10.4 \ \mu\text{sec}$$

Step 3: Solve for the period:

$$T = 162.9 \ \mu\text{sec} + 10.4 \ \mu\text{sec} = 173.3 \ \mu\text{sec}$$

Step 4: Solve for frequency:

$$f = \frac{1}{173.3 \ \mu\text{sec}} = 5.77 \text{ kpps}$$

Step 5: Solve for duty cycle:

$$\delta = \frac{162.9 \ \mu\text{sec}}{173.3 \ \mu\text{sec}} \times 100\% = 94\%$$

An application of the basic 555 astable circuit is shown in Figure 6–29. Large industrial machines could contain several embedded microcomputer systems, each of which is dedicated to controlling a certain subfunction of the machine. The embedded microcomputers, in turn, are controlled by a system control microcomputer. Although this main microcomputer may have a keyboard and monitor, similar to those of the personal computer, the embedded microcomputers are usually contained on printed circuit boards and placed in card cage assemblies. Manual control of these embedded microcomputers, in many cases, is limited simply to a reset switch, and visual display of its status may consist of only a few card-edge LEDs.

The programming format for a typical embedded microcomputer normally consists of several subroutines, most of which are executed within a few milliseconds. After the execution of each subroutine, the embedded microcomputer automatically executes a routine in which it resets its own **watchdog timer** circuit. This reset pulse clears the counter within the watchdog timer, preventing the watchdog from sending a fault message to the main microcomputer system. Most watchdog timers allow only a few milliseconds to pass before sending an error message to system control. This rapid response to a failure is necessary, considering that the results of an embedded microcomputer malfunction would be unpredictable and even dangerous. For example, assume that an embedded microcomputer experienced a read/write failure immediately after sending a high-current pulse to a solenoid. A one-shot circuit such as that in Figure 6–15 would prevent immediate damage to the solenoid by automatically switching off the output buffer and thus removing the high-current signal. Since the malfunctioning embedded microcomputer is unable to send a reset pulse to the watchdog timer, the watchdog "barks" by sending a fault message to the system control microcomputer. The watchdog timer could directly reset the embedded microcomputer, bringing it back to a cold-start condition. Also, the system

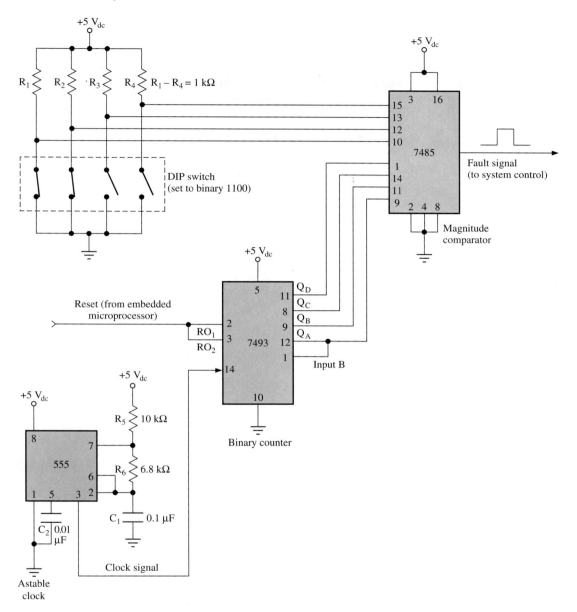

Figure 6–29
Watchdog timer circuitry containing 555 IC timer in astable multivibrator configuration

control microcomputer could initiate its own reset pulse as a response to the fault signal from the watchdog circuit.

The operation of the basic watchdog circuit in Figure 6–29 is quite simple. The astable multivibrator serves as the clocking source to a 4-bit binary counter. The four *Q* outputs of that device are fed to a logic comparator that continually compares the output number of the counter to a maximum number encoded by the dip switch. In this example, this maximum number is 1100 (equivalent to decimal 12). If the embedded processor is functioning properly, it is always able to reset the counter back to 0000 well before the maximum count is attained. If a reset pulse fails to arrive in time, the counter is able to increment beyond 1100, thus producing a fault signal. At first the use of a multivibrator as the clocking source for the watchdog might seem unnecessary, since the embedded microcomputer has its own crystal-controlled clocking source. However, a possible source of a system failure might be the embedded microcomputer clocking circuitry. If the watchdog were driven by this circuit, it would fail at the same instant as the microcomputer! It is for this reason that an independent astable clock source is used.

Example 6–7

For the watchdog timer in Figure 6–29, assume that the maximum allowable delay between reset pulses is to be approximately 18 msec. What setting of the DIP switch would most closely produce this desired delay?

Solution:
Step 1: Solve for the time period and frequency of the astable clock signal:

$$PW = 0.693 \times (10 \text{ k}\Omega + 6.8 \text{ k}\Omega) \times 0.1 \text{ }\mu F = 1.164 \text{ msec}$$
$$SW = 0.693 \times 6.8 \text{ k}\Omega \times 0.1 \text{ }\mu F = 471.24 \text{ }\mu sec$$
$$T = 1.164 \text{ msec} + 471.24 \text{ }\mu sec = 1.63524 \text{ msec}$$
$$f = \frac{1}{1.63524 \text{ msec}} = 611.5 \text{ pps}$$

Step 2: Solve for the proper setting of the DIP switch:

$$N = \frac{\text{Maximum delay}}{T} = \frac{18 \text{ msec}}{1.63524 \text{ msec}} = 11$$

Set the DIP switch to 1011 binary.

The following circuit analysis serves to clarify the important concepts introduced during this section of the chapter. As illustrated in Figure 6–30, a **clutch/brake mechanism** is mechanically linked to a continuously running belt drive system. This mechanical linkage consists of a toothed timing belt, which is wrapped firmly around a gear on the rotating drive shaft and a similar gear on the shaft of the clutch brake. When the shaft is to remain stationary, the brake winding is energized while the clutch is held in the off condition. On the arrival of a signal from the system control microcomputer, the brake is released and the clutch is energized. As the clutch engages, the shaft of the clutch/brake picks up drive from the belt system and begins to rotate. As shown in Figure 6–30, the clutch/brake contains a metal timing disk. On completion of one rotation, a magnetic sensor, in the form of a Hall-effect switch, sends a pulse to the control circuitry. This pulse indicates that the shaft has returned to its home position. A watchdog timer is operative during the time in which the shaft of the clutch/brake is turning. Failure of the clutch brake to return to its home position within a maximum period of time is assumed to indicate a fault condition. The fault message, which is likely to result from a mechanical problem such as a worn or loose-fitting timing belt, causes the system control microcomputer to inhibit further operation of the machinery.

Figure 6–31 contains a block diagram of the control circuitry for the clutch/brake mechanism. Note that, for activation of both the brake and the clutch, a high-current/

low-current (equivalent to excitation and holding current) switching technique is used. Thus, four separate output lines are required for activating the high-power transistor switches located on the driver circuit board.

The operation of the clutch/brake control circuitry is better understood by first analyzing the timing control diagram in Figure 6–32. At t_0, when power is first applied to the system, the control processor sends an active-high initialization signal to the clutch/brake controller. This pulse inhibits the operation of the clutch/brake mechanism until the circuitry on the control card is placed in the proper starting condition. This action is achieved by a reset pulse arriving at t_1. When the initialization pulse terminates at t_2, the brake mechanism is activated, with brake excitation and holding currents beginning simultaneously. The brake remains engaged until the arrival of a trigger pulse at t_3. This pulse terminates the brake low current and also activates the watchdog timer. To ensure smooth mechanical operation, a brief dead time occurs prior to the activation of the clutch, allowing complete release of the shaft by the brake. The clutch excitation current and holding current begin simultaneously at t_4. The clutch holding current terminates with the arrival of a pulse from the Hall-effect switch at t_5. If the shaft rotates with the proper angular velocity, this pulse resets the watchdog timer well ahead of its maximum count. A second dead time occurs between t_5 and t_6, allowing the clutch to be fully released before the brake excitation and holding currents begin. After t_6, the brake remains engaged indefinitely, until the arrival of another trigger signal from the system control microcomputer. A detailed schematic diagram of the clutch/brake control circuit is contained in Figure 6–33(a) and the driver circuit is shown in Figure 6–33(b).

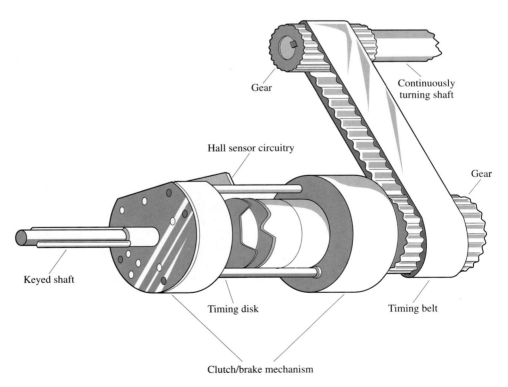

Figure 6–30
Clutch brake mechanism mechanically linked to drive source

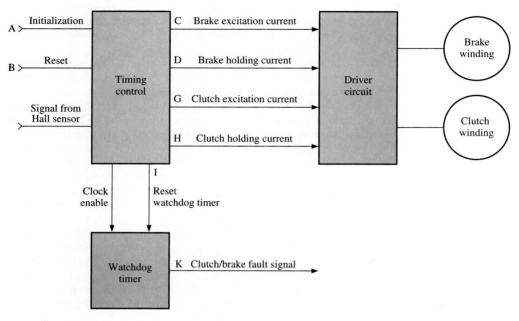

Figure 6–31
Block diagram of clutch/brake control system

At point A in Figure 6–33(a), the active-high initialization signal arrives from the system control processor. This signal branches to two different points, serving as the A input to the fifth one-shot and inverting to become the enable signal for the active-low outputs of the control circuit card. With a low condition present at point J, the outputs of the Schmitt-triggered NAND gates are all high. This condition inhibits the flow of current to the emitters of the optoisolators of the driver circuitry in Figure 6–33(b). As a result, neither the clutch or brake is energized while the initialization signal is high. The arrival of the reset pulse at point B directly resets the fourth one-shot. This signal also results in a logic low condition at point I, which resets one-shots 1, 2, and 3.

According to Figure 6–32, when the signal at point A falls low at t_2, point I has already returned to a logic high. At this juncture, it is important to note that the B input of the fifth one-shot (OS5) is common to point I, while the A inputs are common to point A. Thus, at t_2, the negative transition at point A triggers OS5. Since point J is now at a logic high, the high condition at the Q output of OS5 results in a logic low at the output of the fourth Schmitt-triggered NAND gate. This low condition forward biases the emitter LED of the first optoisolator on the driver circuit board in Figure 6–33(b), causing the collector voltage of the detector phototransistor to fall toward ground potential. The collector output of switching transistor Q_1 then rapidly switches to a logic high, forward biasing the Darlington pair and delivering excitation current directly to the brake winding. This action rapidly engages the brake, prohibiting rotation of the shaft until the clutch is activated.

Since the output of the fourth Schmitt-triggered NAND gate is common to point C, the negative transition that initiates the flow of brake excitation current also sets the NAND latch in Figure 6–33(a), bringing its Q output to a logic high. Since the enable signal at point J is also at a logic high, point D falls low, enabling the flow of holding

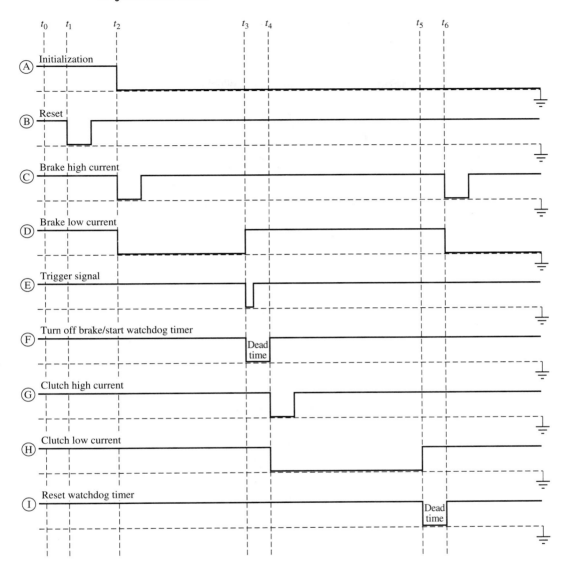

Figure 6–32
Timing diagram for clutch/brake control circuitry

current to the brake winding. The momentary low pulse at the $\overline{Q}$ output of OS5 results in a momentary high pulse at the reset input to the counter within the watchdog timer, initializing the count at 0000. Also, the low condition now present at the $\overline{Q}$ output of the NAND latch inhibits the flow of clock pulses to the counter. The control circuit remains latched in this wait state until the arrival of a trigger signal from system control.

When the clutch is to be actuated, an active-low trigger pulse arrives at point E. This signal triggers the first one-shot (OS1). With point C already at a logic high, the negative transition at the $\overline{Q}$ output of the one-shot resets the NAND latch, terminating the flow of

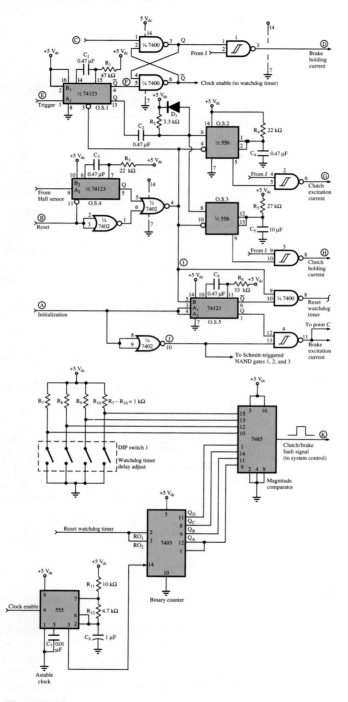

Figure 6–33

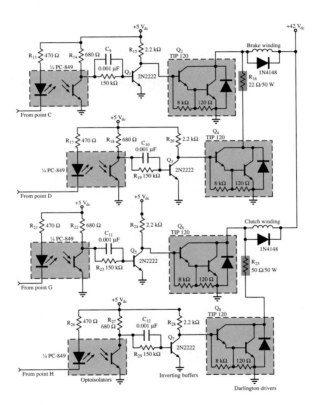

Figure 6–33
(continued)

holding current to the brake winding. With the $\overline{Q}$ output of the NAND latch now high, the counter of the watchdog timer begins to increment. The Q output of OS1 is capacitively coupled to the trigger inputs of OS2 and OS3. Since the trigger inputs to the 556 respond only to a negative transition, these devices are not activated until the Q output of OS1 falls low. This timing relationship allows a dead time to occur between t_3 and t_4, as represented in Figure 6–32. The negative transition at the Q output of OS1 initiates the excitation and holding-current signals for actuating the clutch. These active-low signals occur simultaneously at points G and H. While the excitation current signal quickly terminates, the low current signal continues until it is interrupted by the reset signal at point I. If the clutch is functioning correctly, an active-low pulse from the Hall sensor within the clutch/brake assembly triggers OS4 well before the termination of the pulse width of OS3. It is this action that results in the low pulse at point I and the high pulse at the reset input to the watchdog timer. If the clutch mechanism fails or the timing belt slips, the duration of the clutch holding current is still limited by the pulse width of the third one-shot. Also, failure of the watchdog timer to reset results in a fault signal at point K. System control responds to this signal by returning the initialization signal at point A to a logic high. This action inhibits all current flow to the clutch/brake.

During normal operation of the clutch/brake assembly, with the reset pulse at point I arriving in time to reset the watchdog timer, a second dead time begins, occurring between t_5 and t_6 in Figure 6–32. At the end of the dead time, the positive transition of the signal at point I triggers OS5. The resultant logic the high condition at the Q output produces a logic low at point C, initiating the flow of excitation current to the brake winding. The low condition at point C also sets the NAND latch, initiating the flow of holding current to the brake and inhibiting the clocking of the watchdog timer. The control circuit is again latched in a wait state, ready for another trigger pulse to arrive from system control.

The operating cycle for the clutch/brake assembly just described could easily happen three or four times per second, depending on the type of machinery in which it is contained. Although the excitation currents for both the clutch and the brake would last only a few milliseconds, the clutch brake, as well as the driver circuit card, could become quite warm with prolonged use. For this reason, the Darlington driver transistors must be protected by a large heat sink. Also, the large current-limiting resistors, which dissipate a great deal of power, are mounted on a separate board, several inches away from the driver circuit card. In spite of these precautions to avoid overheating of the components, it is still necessary to use a cooling fan in many such high-power electromechanical control applications.

In addition to its complexity, a disadvantage of the control circuit contained in Figure 6–33(a) is the fact that its performance is likely to be influenced by deviations in temperature. A major advantage of the circuit is the fact that it is able to detect quickly a mechanical failure of the clutch/brake. Also, since it initiates its own timing sequence, it requires limited interfacing with the system control microcomputer. As explained later in this text, the entire timing sequence for the clutch/brake could be controlled directly by an embedded microprocessor. An advantage of this control technique would, of course, be extreme precision of timing.

SECTION REVIEW 6–2

1. Explain the difference in operation between a retriggerable and a non-retriggerable one-shot.
2. For the monostable multivibrator in Figure 6–17, assume that inputs A_1 and A_2 are both tied directly to circuit ground, while input B_2 is tied to V_{CC}. If C_{ext} equals 0.47

μF and R_{ext} is now a 0/20-kΩ variable resistor, what must be the setting of R_{ext} in order to produce an output pulse width of 2 msec when the B_1 input goes to a logic high?

3. What is the duration of the active-low brake excitation current pulse seen at point C in Figure 6–33?

4. What is the duration of the active-low clutch excitation current pulse seen at point G in Figure 6–33?

5. If the watchdog timer in Figure 6–33 fails to reset, what would be the maximum duration of the clutch holding current?

6. For the circuit in Figure 6–33, what is the duration of the dead time occurring between t_3 and t_4?

7. What is the pulse width of the clocking signal produced by the astable multivibrator within the watchdog circuit in Figure 6–33? What is the space width of that clocking signal? What is the duty cycle of that signal? What is the frequency?

8. Assume the time required for one rotation of the shaft within the clutch/brake mechanism in Figure 6–30 is 150 msec. Also assume the DIP switch within the watchdog circuit in Figure 33 is set to 1101. If the clutch mechanism is not slipping, will the shaft be able to complete a rotation in time to reset the watchdog timer?

6–3   INDUSTRIAL APPLICATIONS OF BISTABLE MULTIVIBRATORS

The **bistable** form of multivibrator has two stable output conditions. This type of device may be latched into a set condition, where the Q output stays high and $\overline{Q}$ remains low. Also, the bistable multivibrator may be latched into a reset condition, where the $\overline{Q}$ output stays high and Q remains low. A bistable multivibrator is often referred to simply as a **flip-flop.** Flip-flops are used in a wide variety of digital circuitry. However, in this chapter, only those flip-flop applications that apply directly to industrial pulse and timing circuitry are addressed.

The simplest form of bistable multivibrator is the **T flip-flop,** which has a single trigger input and two complementary outputs. A block representation of this type of device is contained in Figure 6–34(a). As implied by the letter T, the T flip-flop **toggles** on each trigger pulse. If, at a given instant, Q is high and $\overline{Q}$ is low, Q goes low and $\overline{Q}$ goes high on the next active transition of the trigger signal. On the subsequent active transition of the input signal, toggling will again occur, with Q returning to a high state and $\overline{Q}$ going low.

Table 6–1 serves to clarify the symbols used in representing the operation of the various forms of flip-flop. From this listing it may be determined that the T flip-flop shown in Figure 6–34(a) toggles on the negative transition of the trigger signal. The operation of this device is further clarified by the waveforms in Figure 6–34(b). Here, the trigger waveform is represented as a clocking signal for which space width is much narrower than pulse width. Note that at t_0 the flip-flop is in the reset state. Thus, with the first negative transition of the input signal, occurring at t_1, the flip-flop toggles, with Q going high and $\overline{Q}$ going low. At t_2, on the subsequent negative transition at the input, the outputs again toggle, with Q going low and $\overline{Q}$ going high. At this point, it becomes obvious that, regardless of the duty cycle of the input signal, the T flip-flop produces complementary square waves. It is important to note that these output waveforms occur at one-half the frequency of the T input.

As illustrated in Figure 6–35, T flip-flops can be **cascaded** to form a binary counter. Note that the Q output of the least significant flip-flop serves as the clocking input to the next most significant device. (This pattern continues for the second and third flip-flops as well.) The relatively simple device shown here is often referred to as a **ripple counter,**

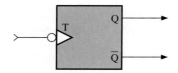

(a) Block representation of the T flip-flop

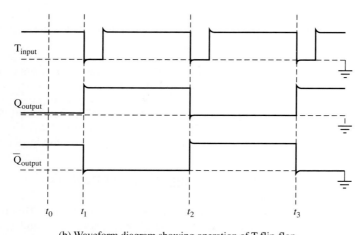

(b) Waveform diagram showing operation of T flip-flop

Figure 6–34

due to the fact that a clock transition at the input to the least significant flip-flop must "ripple" through the device before it is able to affect the output of the most significant flip-flop. This **propagation delay** problem limits the operating frequency of the ripple counter.

The waveforms in Figure 6–35(b) represent the operation of the 4-bit ripple counter. Prior to t_1, all of the flip-flops are assumed to be in the reset condition. Thus the binary number formed by the Q outputs at that time is 0000. At t_1, with the first negative clock transition, FF_0 toggles to the set condition. At this instant, the Q outputs form the number 0001. At t_2, on the next negative clock transition, FF_0 toggles a second time. The negative transition at the Q output of FF_0 causes FF_1 to toggle to the set condition. As a result, the Q outputs of the counter now form the number 0010. On the third negative clock transition, FF_0 again toggles into the set condition. However, this positive transition at the 2^0 output has no effect on FF_1, which remains in the set condition. Thus at t_3, the Q outputs of the counter form 0011. This toggling pattern continues through t_{15}, where all four Q outputs are at a logic high. At t_{16}, the negative clock transition causes all four flip-flops to reset, forming the initial number of 0000. If the ripple counter is allowed to run freely, it continuously increments from 0000 through 1111 every 16 clock states. It is interesting to note that, although a binary up-count is present at the Q outputs, a down-count would be available through accessing the four $\overline{Q}$ outputs. This becomes evident by considering the fact that when the Q outputs form 0000, the $\overline{Q}$ outputs must form the complementary number of 1111. When the Q outputs are forming 1111, the $\overline{Q}$ outputs are at 0000. Thus, at any instant, the Q and $\overline{Q}$ numbers are complementary, with the sum of the two variables always equaling 1111. For example, when the Q outputs form 0100

Table 6–1
Symbols used to indicate triggering methods for bistable multivibrators

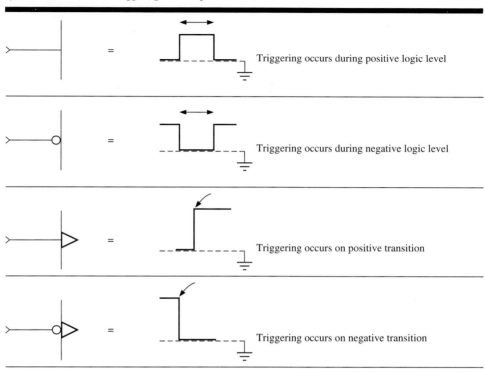

	=	Triggering occurs during positive logic level
	=	Triggering occurs during negative logic level
	=	Triggering occurs on positive transition
	=	Triggering occurs on negative transition

(equivalent to decimal 4), the $\overline{Q}$ outputs must form 1011 (equivalent to decimal 11). The sum of these two binary numbers is 1111, equivalent to decimal 15.

Of equal significance to the counting capability of the binary counter is that of **frequency division.** Through examining the waveforms in Figure 6–35, it becomes evident that FF_0 produces a square wave at one-half the frequency of the clock source. Since FF_1 is clocked by FF_0, FF_1 produces a square wave at one-half the frequency of FF_0, or one-quarter the frequency of the clock source. This **divide-by-two** pattern continues through to FF_3, where the frequency of the Q output of that device is only one-sixteenth that of the clock source. Binary ripple counters that are used for the purpose of frequency division are often referred to as **modulo-N counters,** where N is the number of T flip-flops being used within the device. The *modulo* of the counter is the actual ratio of the clock input frequency to the output frequency taken at the most significant flip-flop. Thus, for any free-running binary ripple counter:

$$f_{\text{output}} = \frac{f_{\text{input}}}{\text{Modulo}} = \frac{f_{\text{input}}}{2^N} \qquad (6\text{–}11)$$

where N represents the number of flip-flops contained within the ripple counter.

Example 6–8

Assume the clock source to a modulo-N counter is an astable multivibrator, the output signal of which has a duty cycle of 72% and a space width of 18 μsec. If the counter

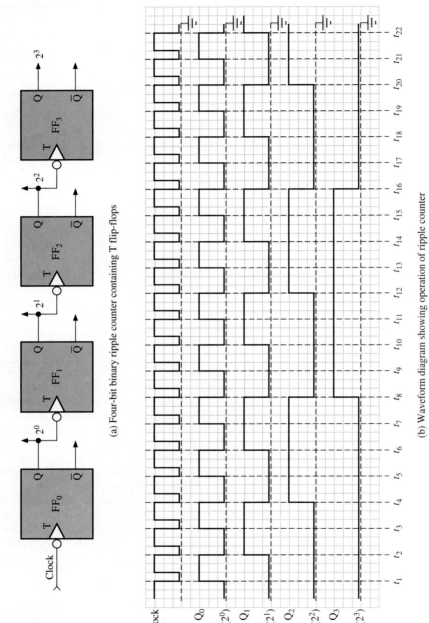

(a) Four-bit binary ripple counter containing T flip-flops

(b) Waveform diagram showing operation of ripple counter

Figure 6–35

contains five cascaded T flip-flops, what is the frequency of the signal taken at the most significant Q output?

Solution:

Step 1: Solve for the frequency of the clock input: With a duty cycle of 72%, the space width represents 28% of the period of the clock signal. Thus,

$$0.28 \times T = 18 \ \mu\text{sec}$$

$$T = \frac{18 \ \mu\text{sec}}{0.28} = 64.3 \ \mu\text{sec}$$

$$f = \frac{1}{T} = \frac{1}{64.3 \ \mu\text{sec}} = 15.552 \ \text{kpps}$$

Step 2: Solve for the output frequency: Since the counter contains five flip-flops, $N = 5$. Thus,

$$f_{\text{out}} = \frac{15.552 \ \text{kpps}}{2^5} = 486 \ \text{pps}$$

Two forms of bistable multivibrator commonly found within industrial timing circuitry are the **D flip-flop** and the **JK flip-flop.** Although the D flip-flop is often used as a data storage device, it may also be used in divide-by-two applications. The JK flip-flop, which is the most versatile form of bistable multivibrator, may be configured to function as an SR, a T, or even a D flip-flop.

Figure 6–36 shows the internal structure of the 7474 IC, which contains two D flip-flops. When one of the flip-flops contained within this IC is operating as a data storage device, the logic level present at the D input is latched at the Q output on the positive transition of the clock input. At the same instant, the $\overline{Q}$ output assumes the complementary logic condition. This action is enabled to occur whenever the **preset** (PR) and **clear** (CLR) inputs are in their inactive high states. If a logic low is present at the preset input while clear is high, the Q output is high and the $\overline{Q}$ output is low, regardless of the conditions of the D and clock inputs. If a logic low is present at the clear input while preset is high, Q is low and $\overline{Q}$ is high, regardless of the activity at the D and clock inputs. Having both the preset and clear inputs low simultaneously is a forbidden condition for the D flip-flops within the 7474. While it does not harm the device, it places both outputs in the high condition, which is logically incorrect for any form of flip-flop. Figure 6–37 illustrates how a 7474 IC may be utilized as a divide-by-four device, effectively functioning as two cascaded T flip-flops.

In the configuration shown in Figure 6–37(a), the two D flip-flops could perform the same function as the two JK flip-flops in Figure 5–36 (of the previous chapter). The input clock waveform in Figure 6–37(b) is shown as having considerable noise content, possibly the result of traveling via a lengthy cable run. The two D flip-flops could actually be serving as input buffers on a remote circuit board. Note that the waveform available at Q_A is a symmetrical square wave, having a frequency of one-half that of the input. The Q_B output has a frequency of one-half that of the Q_A signal and one-quarter that of the input. At t_0, with the arrival of the common clear pulse, both Q outputs are brought to a logic low and both $\overline{Q}$ outputs switch to a logic high. Thus, the D inputs of both flip-flops are at a high level prior to the first positive transition of the input clock. At t_1, with the arrival of a positive clock transition, the logic high present at the $\overline{Q}_A$ input becomes the Q_A condition, causing $\overline{Q}_A$ to toggle to a logic low. At t_2, the positive transition of the input clock again causes Q_A and $\overline{Q}_A$ to toggle. At this instant, since $\overline{Q}_A$ is changing in

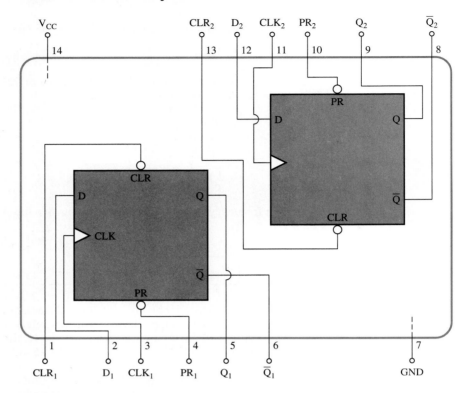

Figure 6–36
Internal structure of SN7474 dual D flip-flop with present and clear inputs

the positive direction, the logic high present at the $\overline{Q}_B$ output latches at the Q_B output, causing $\overline{Q}_B$ to fall to a logic low.

The 7476 IC, which contains two **JK flip-flops,** is represented in Figure 6–38. The two flip-flops within the 7476 package, like those in the 7474, have active low preset (PR) and clear (CLR) inputs. These inputs have priority over J, K, and clock. It is important to note that the JK flip-flops within the 7476 IC are triggered on the negative transition of the clock input. This allows the devices to be cascaded easily to create a modulo-N ripple-counter. Again referring to the interrupter module circuitry in Figure 5–36, note that the J, K, preset, and clear inputs are all tied to a logic high via a pull-up resistor. Also, the Q output of the first flip-flop becomes the clock input to the second. This configuration allows the 7476 to operate as a modulo-2 ripple counter, with the two JK flip-flops functioning identically to the T-type devices. In addition to this **toggle mode** of operation, which occurs whenever J and K are at a logic high, the JK flip-flop also has a **memory** (or **hold**) mode and a **set/reset mode.** If J and K are both low, while preset and clear are high, the device is in a memory state, holding Q and $\overline{Q}$ in their most recent conditions, regardless of activity at the clock input. If J goes to a logic high and K is at a logic low (assuming preset and clear to be high), on the next negative clock transition Q goes to a logic high and $\overline{Q}$ goes to a logic low. Thus, J effectively functions as the set input. If J is low while K goes to a logic high, on the next negative clock transition, Q goes low and $\overline{Q}$ goes high. In this instance, K would be functioning as the reset input.

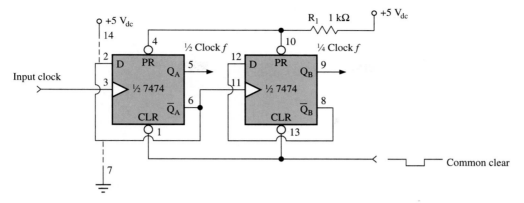

(a) 7474 dual D flip-flop IC in a modulo-2 configuration

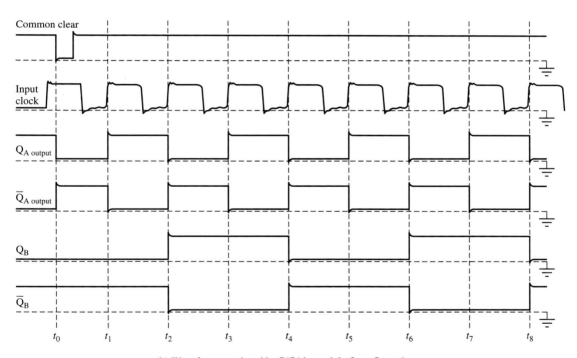

(b) Waveforms produced by 7474 in modulo-2 configuration

Figure 6–37

The 7493 is an IC ripple counter that exploits the ability of a JK flip-flop to perform in the toggle mode. The internal circuitry of this device is represented in Figure 6–39. Note that the J and K inputs to the four flip-flops are inaccessible. Instead, the J and K inputs are tied high internally, allowing the four flip-flops to function as T-type devices. The 7493 has two active-high reset terminals, RO_1 and RO_2, which serve as inputs to an internal NAND gate. The output of this gate in turn becomes the active-low clear input

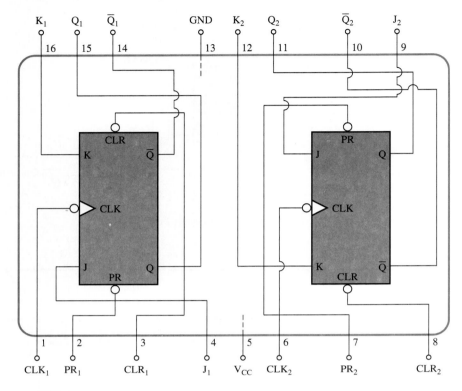

Figure 6–38
Internal structure of SN7476 dual JK flip-flop with present and clear inputs

to all four flip-flops. Thus, if either RO_1 or RO_2 is low, the counter is enabled to operate. However, if both RO_1 and RO_2 are high simultaneously, all four Q outputs are brought to a logic low.

Figure 6–40(a) illustrates how the 7493 could be made to function as an **octal** or **modulo-3** counter. Here, the first flip-flop remains unused, with the clock source being connected to input B rather than input A. Since both RO_1 and RO_2 are grounded, the counter is enabled to count continuously from 000 through 111. In Figure 6–40(b), note that the clock signal now arrives at input A, while Q_A is jumpered to input B. This configuration allows the 7493 to function as a modulo-4 counter. Since RO_1 and RO_2 are again tied to ground, the device is enabled to operate continuously, counting from 0000 through 1111. With the configuration shown in Figure 6–40(c), the 7493 is functioning as a **decade counter.** Note here that the Q_B and Q_D outputs are feeding back to the RO_1 and RO_2 inputs. Thus, at the instant the counter attempts to increment to 1010, which is the first number in the binary upcount for which Q_B and Q_D are both high at the same time, the counter resets. Since the counter attains the number 1010 for only a few nanoseconds, it effectively increments from 0000 through 1001. Such a counting action may be necessary for a ripple counter operating in a **binary-coded decimal** (BCD) format. Since the counter in Figure 6–40(c) does not have an N value that is equivalent to a power of two, it should not be referred to as a modulo-N counter. Instead, it should be referred to as a decade counter or a **divide-by-ten** device.

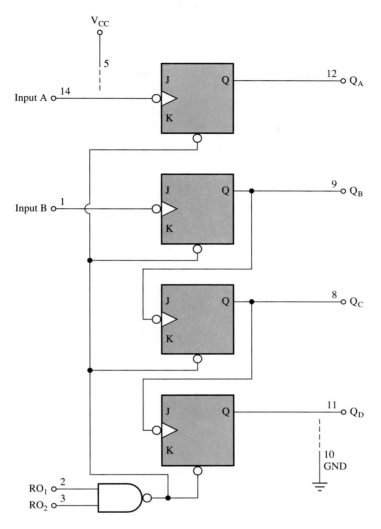

Figure 6–39
Internal circuitry of 7493A 4-bit binary ripple counter

Two problems associated with IC ripple counters are noise and propagation delay. For a TTL ripple counter, such as the 7493, noise spikes are likely to be present on the Q output signals, occurring at the active transition of the input clock signal. In relatively low-frequency applications, propagation delay is not a problem. As an example, refer once again to the interrupter module circuit in Figure 5–36. Assume the output at point F is being utilized as the clock input to a 7493 modulo-4 counter. If the period of the clock signal taken at this point represents 5 mm of conveyor belt movement, the clock cycle is likely to be at least 1 msec. Therefore, the clock frequency would not exceed 1 kpps. According to the data sheets for the 7493, a negative transition at input A should be able to produce a transition at the Q_D output in no more than 70 nsec. Thus, when compared to an input clocking cycle of 1 msec, this propagation delay is insignificant,

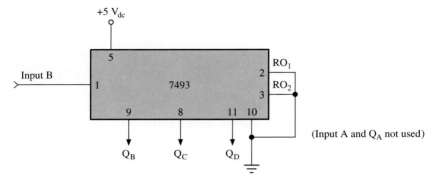

(a) 7493 configured to function as an octal counter

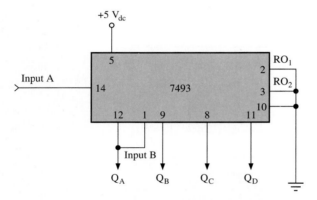

(b) 7493 configured to function as a 4-bit binary counter

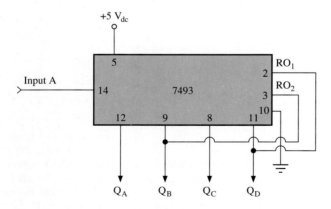

(c) 7493 configured to function as a decade counter

Figure 6–40

representing a very small fraction of the clocking period:

$$\frac{\Delta t}{T} = \frac{70 \times 1 \times 10^{-9}}{1 \times 10^{-3}} \times 100\% = 0.007\% \text{ of input clock period}$$

where Δt propagation delay and T equals $1/f$.

Thus, the 7493 ripple counter would function quite satisfactorily in most conveyance system and motor control applications. Assume, however, that an engineer attempted to use this same device within a video acquisition system, which is likely to have a clocking frequency of 10 MHz. In such a system, the propagation delay of the ripple counter would quickly prove to be unacceptable:

$$\frac{\Delta t}{T} = \frac{70 \times 1 \times 10^{-9}}{100 \times 1 \times 10^{-9}} \times 100\% = 70\% \text{ of input clock period}$$

It should also be mentioned that the propagation delay problem is compounded when two or more ripple counters are cascaded together. For example, if the Q_D output of one 7493 modulo-4 counter is fed to input A of a second 7493, a counter with a modulo of 256 would be created. However, the maximum propagation time would now be twice that of a single IC, equaling around 140 nsec. Thus, at higher operating frequencies, the ripple counter is usually abandoned in favor of a device utilizing **synchronous clocking.** An example of this more advanced form of counter is contained in Figure 6–41.

Synchronous counters derive their name from the fact that all of the internal flip-flops have a common clock source and are thus triggered simultaneously. Note that, after passing through an inverting buffer, the clock input to the 74191 connects directly to the four JK flip-flops within the device. The 74191 is actually a binary **up/down counter.** When the down/up control input at pin 5 is at a logic low, the device is programmed to increment on each positive transition of the input clock signal. If the down/up control is at a logic high, the 74191 is programmed to decrement on each positive transition of the clock pulse. Only when the enable input at pin 4 is at a logic low does the counter operate. When this control input is high, the counter simply holds its current count, regardless of the activity of the clock signal at pin 14. The method by which the enable control operates becomes evident by considering its effect on the Q_A and ripple clock outputs. When the enable signal is high, through the action of an inverting buffer, both the J and K inputs to the least significant flip-flop are low. As a result, that flip-flop is placed in a memory, or hold, condition. Also, the low produced by the inverter holds the ripple clock output high. With further analysis, it becomes evident that the other three flip-flops are also placed in a memory mode, through the action of their AND/OR arrays. When the enable input falls low, the least significant flip-flop is immediately placed in the toggle mode. The other flip-flops are enabled to toggle when necessary during the progression of a binary up- or down-count.

In addition to the four Q outputs, the 74191 has two other outputs, which allow the device to be cascaded with other counting devices. As long as the down/up and enable inputs are both at a logic low, a high pulse occurs at the **max/min** output whenever the number 1111 is attained. If the enable input remains low while the down/up control goes high, a high pulse occurs at the max/min output whenever the number 0000 is attained. The pulse width of the max/min signal is equal to the period of one clock cycle. As shown in Figure 6–42, during the time in which the max/min pulse is high and the space width of the clock signal is occurring, a low is present at the ripple clock output. The positive transition of the **ripple clock signal** occurs as the max/min pulse falls low. This ripple clock signal may be cascaded to the clock input of a second 74191, in order to form a modulo-8 counter. Thus, during an up-count, as the least significant counter

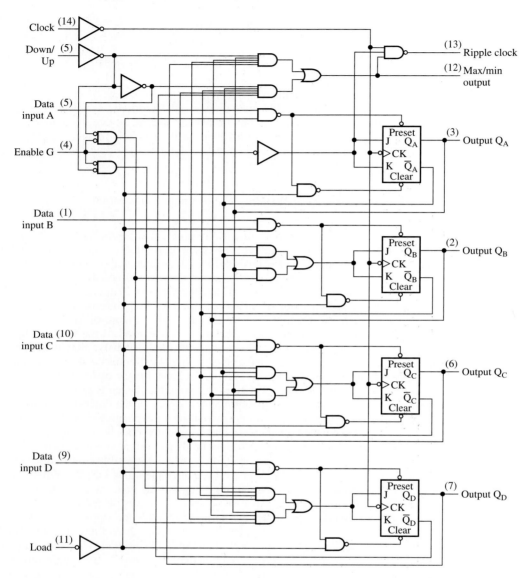

Figure 6–41
Internal circuitry of 74191 binary up/down counter

increments from 1111 to 0000, a positive transition occurs at its ripple clock output, incrementing the second 74191. During a binary down-count, a positive transition occurs at the ripple clock output of the least significant counter just as it decrements from 0000 to 1111. This action decrements the second counter.

A major advantage of the 74191 up/down counter is the fact that, whenever the load input is low, the four Q inputs may be preset to the 4-bit variable present at the data inputs. With the 74191, because this loading action involves the preset and clear inputs of the internal JK flip-flops, it occurs independently of the clock input. Whenever the load input is at a logic high, the output of an internal inverter buffer is at a logic low. With this condition, through the operation of eight NAND gates that control the loading function, the preset and clear inputs of all four flip-flops are held in their high states. Consider, however, what would happen at the instant the load input falls low if a logic high is already present at data input A. The two NAND gates controlling the load function of the least significant flip-flop cause a low to be present at preset and a high to be present at clear. Thus Q_A immediately goes high, reflecting the condition at the data input A. Assume also that data input B is low prior to the arrival of the low pulse at the load input. At the instant load goes low, the two NAND gates controlling the B flip-flop cause preset to be high and clear to be low. Thus, Q_B immediately goes low, reflecting the logic condition at data input B.

In Figure 6–43, the quick loading capability of two 74191 ICs is exploited to create a frequency-divider circuit with a variable modulo. Connection of the data inputs to a single 8-bit DIP switch allows the modulo of the two counters to be adjusted to any binary number between 00000001 and 11111111. With both of the down/up control inputs tied high via pull-up resistors R_9 and R_{10}, and the enable inputs tied to ground, the two 74191s operate continuously as binary down-counters. The load function of both 74191s is controlled by a single two-input NAND gate. Only when the two 74191s are forming 00000000 are both max/min outputs at a logic high. When this occurs, the output of the NAND gate goes low, activating the loading function of both of the counters. Consequently, the binary number formed by the DIP switch immediately appears at the Q outputs of the counters. As a result of this action (providing the DIP switch is forming a number other than 00000000), the output of the NAND gate immediately goes high, disabling the load function. Thus, the duration of the feedback pulse at the output of the NAND gate is in the nanosecond range, being dependent on the propagation delay of the counter and the gate itself. Note that this narrow negative pulse also serves as the clocking signal for the JK flip-flop. With each loading function of the two 74191s, this flip-flop toggles. An advantage of having the Q output of a JK flip-flop as the final output of the counter in Figure 6–43 is the fact that the output signal remains a square wave, despite the changes in frequency that result from varying the setting of the DIP switches.

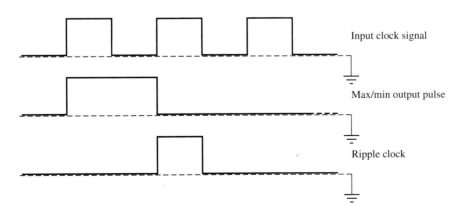

Figure 6–42
Relationship of clock and output pulses for 74191 IC counter

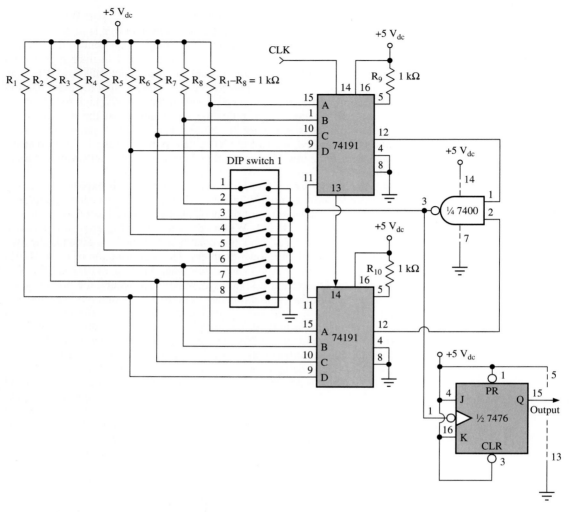

Figure 6–43
Adjustable frequency divider containing cascaded down-counters and T flip-flop

---

Example 6–9

For the frequency-divider circuit in Figure 6–43, assume the input clock signal is occurring at a rate of 116 kpps. What must be the setting of the DIP switch in order to produce an output signal with a period of nearly 3.2 msec?

Solution:
Step 1: Solve for the required overall modulo of the circuit:

$$f_{out} = \frac{1}{3.2 \text{ msec}} = 312.5 \text{ pps}$$

$$\text{Modulo} = \frac{f_{\text{input}}}{f_{\text{output}}} = \frac{116 \text{ kpps}}{312.5 \text{ pps}} \cong 371.2$$

Step 2: Solve for the required modulo of the programmable portion of the circuit: Since the JK flip-flop performs a divide-by-two operation, the modulo for the two 74191s is determined as follows:

$$\text{Modulo} = \frac{371.2}{2} \cong 186$$

Step 3: Solve for the setting of the DIP switches: The DIP switches must represent the binary equivalent of 186, which is 10111010. In Figure 6–43, closure of a DIP switch results in a logic low for a given data input, whereas leaving a DIP switch open causes the input to be pulled up to a logic high. Thus, for switches 1 through 8, respectively, the configuration should be

1	2	3	4	5	6	7	8
CLOSED	OPEN	CLOSED	OPEN	OPEN	OPEN	CLOSED	OPEN

SECTION REVIEW 6–3

1. True or false?: A modulo-4 counter divides the input clock frequency by a factor of 4.
2. Carefully sketch the circuit configuration in which a 7474 IC could function as a modulo-2 counter.
3. Sketch the circuit configuration in which a 7476 IC could function as a modulo-2 counter.
4. Describe the three basic modes of operation for the JK flip-flop. Describe the conditions that should exist for the J, K, preset, and clear inputs during each of these operating modes.
5. Sketch a possible configuration for a 7493 IC functioning as a modulo-3 counter.
6. Sketch the configuration for a 7493 IC functioning as a modulo-4 counter. If the signal at input A is occurring at 458 pps, what is the frequency of the signal available at Q_D?
7. Sketch the configuration for a 7493 IC functioning as a decade counter. Describe what happens within the device as it attempts to increment to the number 1010.
8. Describe the function of the NAND gate in Figure 6–43. Explain what happens at the instant both of its inputs are at a logic high.
9. True or false?: The output signal in Figure 6–43 remains a virtual square wave, despite changes in the setting of the DIP switch.
10. For the circuit in Figure 6–43, assume the input clock signal has a frequency of 97.8 kpps. What must be the setting of the DIP switch in order to produce a square wave at the output with a frequency of nearly 800 pps?

6–4   ADVANCED INDUSTRIAL PULSE CIRCUITS

The **pulse width modulator** (PWM) and **voltage-controlled oscillator** (VCO) are advanced forms of industrial pulse circuits. When used within industrial control circuitry, PWMs and VCOs are likely to contain astable multivibrators as their basic oscillating devices. During the operation of the PWM, the pulse width of the output signal varies in response to a change in either resistance or voltage. However, the operating frequency of the device remains constant. For the VCO, the operating frequency varies as a function of an input signal, usually referred to simply as the **control voltage.** The principle of pulse width modulation is incorporated into the operation of A/D converters, motor control circuits, switching power supplies, and switching voltage regulators. The VCO is used in a variety of signal-conditioning applications. As explained later, the VCO is an integral part of a control device called the **phase-locked loop** (PLL).

Figure 6–44 shows a simple PWM that could easily be constructed using a 556 dual timer. The first 555 functions as an astable multivibrator, and the second timer operates in the monostable mode. The values of R_1 and R_2 allow the first timer to produce a rectangular waveform for which pulse width is much greater than space width. The pulse width and space width of the signal at point A are determined as follows:

$$PW = 0.693 \times (R_1 + R_2) \times C_1 = 0.693 \times (100 \text{ k}\Omega + 2.2 \text{ k}\Omega) \times 0.01 \text{ }\mu F = 708.2 \text{ }\mu sec$$

$$SW = 0.693 \times 2.2 \text{ k}\Omega \times 0.01 \text{ }\mu F = 15.25 \text{ }\mu sec$$

$$T = PW + SW = 708.2 \text{ }\mu sec + 15.25 \text{ }\mu sec = 723.5 \text{ }\mu sec$$

$$f = \frac{1}{T} = \frac{1}{723.5 \text{ }\mu sec} = 1.382 \text{ kpps}$$

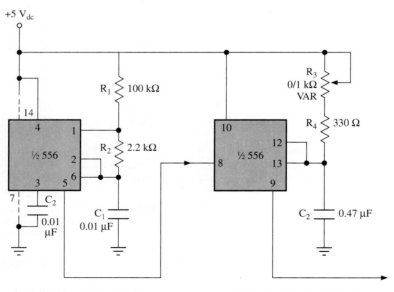

(a) Astable multivibrator stage   (b) Monostable multivibrator stage

Figure 6–44
A 556 dual timer being utilized to form a PWM

Since the ohmic value of R_1 is much greater than that of R_2, the duty cycle of the astable signal is quite long, being determined as follows:

$$\delta = \frac{PW}{T} \times 100\% = \frac{708.2 \ \mu sec}{723.5 \ \mu sec} \times 100\% = 97.9\%$$

Thus, the active-low signal that triggers the second timer is present during only about 2% of the astable clock period. This narrow pulse is ideal for triggering the monostable multivibrator, which is designed to have a variable pulse width. For reliable operation of the PWM, the maximum pulse width of the monostable section should be less than the period of the astable signal. Figure 6–45 illustrates the various waveforms produced by the PWM. In Figure 6–45(a), R_3 is assumed to be set to 0 Ω. Thus, the pulse width is at its minimum duration. In Figure 6–45(b), the value of R_3 is increased, allowing the duty cycle of the output signal to approach 50%. Finally, in Figure 6–45(c), R_3 is assumed to be set to its maximum value, resulting in maximum duty cycle.

The pulse width and duty cycle of the final output of the PWM are easily calculated as follows:

$$PW = 1.1 \times (R_3 + R_4) \times C_2$$

Assuming R_3 is set to its maximum ohmic value:

$$PW = 1.1 \times (1 \ k\Omega + 330 \ \Omega) \times 0.47 \ \mu F = 687.6 \ \mu sec$$

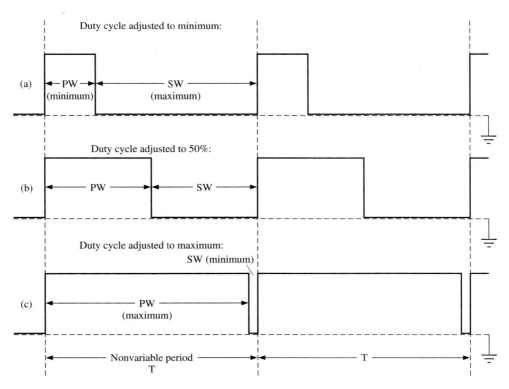

Figure 6–45
Variable duty cycle of 556 PWM

The maximum duty cycle for the output signal is then determined:

$$\text{Duty cycle} = \frac{688\ \mu\text{sec}}{723.5\ \mu\text{sec}} \times 100\% = 95\%$$

Minimum pulse width and duty cycle are then calculated:

$$PW = 1.1 \times 330\ \Omega \times 0.47\ \mu\text{F} = 170.6\ \mu\text{sec}$$

$$\text{Duty cycle} = \frac{170.6\ \mu\text{sec}}{723.5\ \mu\text{sec}} \times 100\% = 23.6\%$$

With regard to switching power supplies and regulators, as well as motor control circuitry, the average dc voltage produced by the PWM becomes an important consideration. For example, during the time in which the pulse width is present, current may be enabled to flow to a high power load such as a dc servo motor. Thus Figure 6–45(a) would represent a condition of minimum power transfer to the load, whereas Figure 6–45(c) would represent a maximum power condition. Assume, as shown in Figure 6–45, that the control signal from the PWM is a rectangular waveform alternating between V_{CC} and ground. The average voltage is then easily determined by multiplying the peak amplitude by the duty cycle. Since the peak voltage virtually equals V_{CC}:

$$V_{AV} = V_{peak} \times \delta \cong V_{CC} \times \text{Duty cycle} \qquad (6\text{–}12)$$

Note here that duty cycle is a ratio of PW/T rather than a percentage.

Example 6–10

For the PWM of Figure 6–44, assume R_3 is adjusted to 463 Ω. What is the pulse width of the output signal? Assuming V_{peak} of the output signal approaches V_{CC}, what is the average output voltage?

Solution:
Step 1: Solve for pulse width:

$$PW = 1.1 \times (463\ \Omega + 330\ \Omega) \times 0.47\ \mu\text{F} = 410\ \mu\text{sec}$$

Step 2: Solve for duty cycle:

$$\text{Duty cycle} = \frac{410\ \mu\text{sec}}{723.5\ \mu\text{sec}} \times 100\% = 56.7\%$$

Step 3: Solve for average output voltage:

$$V_{AV} \cong 0.567 \times 5\ \text{V} \times 5\ \text{V} \cong 2.833\ \text{V}$$

In many industrial control applications, the adjustment of potentiometer R_3 in Figure 6–44 would represent a one-time setting. Such an adjustment might be necessary to obtain a specific rpm for a dc motor or an exact level of reference voltage for a switching regulator. However, in other forms of industrial control circuitry, it is necessary for the duty cycle of the PWM to vary on a continual basis. For such circuitry, the manually adjusted potentiometer is abandoned in favor of a variable dc control source. An example of this more complex form of pulse width modulation is analyzed in detail in Chapters 8 and 14.

Figure 6–46 illustrates how a 555 IC timer and a D flip-flop could be utilized to form a voltage-controlled oscillator. The 555 configuration shown here is essentially that of an astable multivibrator. However, the use of the control input at pin 5 allows the upper threshold voltage (which is normally 2/3 V_{CC}) to be overridden by an input signal from an external source. Variation of this switching threshold allows both the pulse width and the space width of the output of the 555 to be altered. For example, if the voltage at pin 5 decreases, both pulse width and space width decrease due to the fact that the capacitor charging and discharging time is abbreviated. As a result, output frequency increases. An inherent problem with the 555 timer configuration in Figure 6–46 is the fact that, as the frequency of the signal at pin 3 changes, duty cycle deviates as well. In most industrial applications of a VCO, this condition is unacceptable. By connecting the D flip-flop at pin 3 of the 555 timer, the VCO output signal remains a square wave as frequency varies. However, the frequency of the signal at the Q output of the flip-flop is only one-half that of the signal at pin 3 of the 555 IC timer.

Determining the frequency of the circuit in Figure 6–46 for different values of control voltage becomes a laborious process, due to the fact that calculations for pulse width and space width are based on formula (6–1). In the following two equations, note that the instantaneous value of control voltage (V_{CONT}) replaces V_C. Since the lower threshold voltage of the timer must equal one-half that present at pin 5, V_0 is replaced by $1/2 V_{CONT}$. Remember, with the flip-flop present at the output of the astable circuit, the frequency of the signal at pin 3 of the 555 timer must be divided by 2 in order to determine the actual operating frequency.

$$PW = (R_1 + R_2) \times C_1 \times \ln\left(\frac{V_{CC} - 1/2 V_{CONT}}{V_{CC} - V_{CONT}}\right) \tag{6–13}$$

$$SW = R_2 \times C_1 \times \ln\left(\frac{V_{CC} - 1/2 V_{CONT}}{V_{CC} - V_{CONT}}\right) \tag{6–14}$$

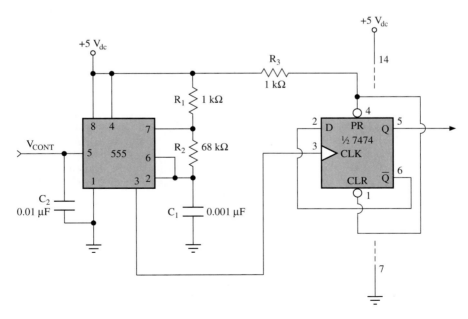

Figure 6–46
A 555 astable multivibrator and D flip-flop combined to form a VCO

Example 6–11

For the circuit in Figure 6–46, assume that the instantaneous value of control voltage is 2.7 V. What is the frequency of the signal present at the output of the D flip-flop?

Solution:
Step 1: Solve for pulse width and space width:

$$PW = 69 \text{ k}\Omega \times 0.001 \text{ }\mu F \times \ln\left(\frac{5 \text{ V} - 1.35 \text{ V}}{5 \text{ V} - 2.7 \text{ V}}\right) = 31.87 \text{ }\mu sec$$

$$SW = 68 \text{ k}\Omega \times 0.001 \text{ }\mu F \times 0.46182 = 31.4 \text{ }\mu sec$$

Step 2: Solve for output frequency: For the astable signal,

$$T = \frac{1}{31.87 \text{ }\mu sec + 31.4 \text{ }\mu sec} = 63.27 \text{ }\mu sec$$

$$f_A = \frac{1}{63.27 \text{ }\mu sec} = 15.805 \text{ kpps}$$

$$f_{out} = 0.5 \times 15.805 \text{ kpps} = 7.903 \text{ kpps}$$

The **phase-locked loop** (PLL) is a form of control circuit widely used in both industrial and communication electronics. A block diagram for the basic PLL is contained in Figure 6–47. Design details for individual PLL circuits may vary significantly, depending on specific applications. However, the basic elements contained within *all* forms of phase-locked loops are a **phase detector, low-pass filter,** and **voltage-controlled oscillator.** As indicated in Figure 6–47, the PLL is also likely to contain some form of amplification or signal conditioning circuitry. For a PLL used within communication circuitry, there could be a linear amplifier following the low-pass filter. For a PLL used within an industrial controller, there could be an integrator between the phase detector and low-pass filter.

The basic operation of the PLL is as follows. When no signal is present at the input, the device is in a free-running condition, with the VCO operating at its **center frequency.** When a signal is present at the input, the phase detector (which may be a specialized IC) continually compares the frequency and phase of that signal (symbolized as v_i) to the frequency and phase of a **servo** (or feedback) signal produced by the VCO. The output of the phase detector is an **error signal** (symbolized as v_e), the components of which include the sum of the input and servo frequencies ($f_{input} + f_{VCO}$) and the difference between these two frequencies ($f_{input} - f_{VCO}$). This error signal is fed to the low-pass filter, which is usually a passive device, consisting simply of a resistor and capacitor. This filter easily passes the dc and difference frequency components of the error signal, while attenuating the higher frequency content. Thus, the control input to the VCO (symbolized as V_{CONT}) is a dc voltage, which pulsates at the rate of ($f_{input} - f_{VCO}$).

With the servo signal from the VCO being fed back to the phase detector, the action of the PLL is such that f_{VCO} begins to approach f_{input}. Consequently, the frequency of the error signal decreases. Finally, when f_{VCO} virtually equals f_{input}, the frequency of the error signal approaches 0 Hz. With this condition, which is referred to simply as **lock,** the frequency of the VCO remains locked on the frequency of the input signal. Ideally, the self-correcting action of the PLL would continue until f_{VCO} exactly equalled f_{input}. In reality, however, a slight disparity in both phase and frequency persists between these two signals during the lock condition. Within a given frequency band, referred to as the **lock, holding,** or **tracking range,** the frequency of the VCO is able to track closely any fluctuations of

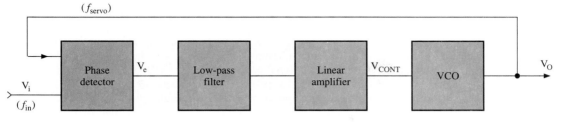

(a) Block diagram of PLL found in communication circuitry

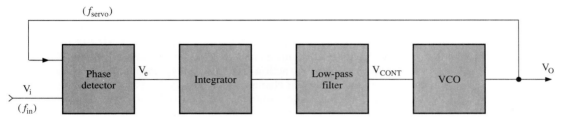

(b) Block diagram of PLL found in industrial control circuitry

Figure 6–47

the input frequency. This lock range of the PLL should not be confused with the **capture range,** which is also called the **acquisition range** or, simply, the **bandwidth.** The capture range of the PLL is that set of input frequencies within which the VCO is able to acquire a lock condition, having previously been in a free running (or center frequency) state.

Within industrial control systems, PLLs are often utilized to synchronize the operating frequency of a control circuit with the frequency of an external timing source. As an example, consider the task of the video acquisition circuitry analyzed in Section 5–7 of the previous chapter. To obtain an undistorted image of an object passing along the conveyance path, the frequency of the Begin Scan signal (seen in Figures 5–59 and 5–61) must quickly increase or decrease in response to any fluctuations in the speed of the conveyance system. Such a compensation process is necessary because the speed of any conveyor belt is likely to deviate due to variations in loading. Figure 6–48 contains a block diagram of a PLL system designed especially to allow the scan clock for a video acquisition system to respond quickly to changes in belt speed. This control circuit contains two major elaborations on the basic PLL design. First, frequency-divider circuitry is utilized to condition both the servo and the input signal prior to their arrival at the phase detector. Second, both a VCO and a crystal-controlled oscillator serve as possible sources for the system clocking signal. When the conveyance belts are stationary, the crystal oscillator is enabled to operate, providing a constant clock signal to the video processing circuitry. (The presence of this signal allows maintenance to be performed on the video system while the belts are not running.) When the conveyance system is operating, the VCO becomes the source for the system clock signal, which must now closely track deviations in the frequency of a transducer clock pulse. This signal is provided by a high-resolution shaft encoder that is connected to a pulley within the conveyance system.

A special design problem associated with the PLL represented in Figure 6–48 is the fact that the basic clock pulse for the video acquisition system and the shaft encoder clock pulse occur at widely differing frequencies. As indicated in Figure 5–59, the frequency of the System Clock for the video acquisition system is likely to approach 10 MHz. However, the shaft encoder module is likely to produce a signal with a frequency of less than 30 kpps. Thus, to achieve a condition where f_{input} and f_{VCO} (at the input to the phase detector) are nearly equal, divider circuitry is used to lower the frequency of both the shaft encoder signal and the servo signal produced by the VCO. The actual circuitry used to perform this function within the PLL is shown in Figures 6–49 and 6–50. This circuitry is designed to operate on the principle demonstrated in Figure 6–43.

Since the frequency range of the servo signal is much higher than that of the shaft encoder, much more divider circuitry is required to condition the f_{VCO} input to the phase detector. As shown in Figure 6–49, the servo signal from the VCO is fed back to the clock input of the least significant 74191. The four 74191 ICs are cascaded together and connected to a pair of 8-bit DIP switches to form a programmable 16-bit down-counter. When all four ICs are forming 0000 at the same instant, all four max/min outputs are high, causing the output of the NAND gate to go low. This action loads the 74191s with the numbers formed by the DIP switches and toggles the JK flip-flop. (The JK flip-flop provides an additional divide-by-two action and allows the f_{VCO} input to the phase comparator to remain a square wave, despite changes in frequency.)

In Figure 6–50, the input signal from the shaft encoder first enters a JK flip-flop. The Q output of this device then serves as the clock input to an 8-bit programmable down-counter. The output of this device, like the counter in Figure 6–49, enters a JK flip-flop, which allows the f_{input} signal to remain a square wave as its frequency varies. One task of a technician performing maintenance on the video acquisition system would be to ensure that the DIP switches within the PLL are set correctly.

Assume, for example, that in a quiescent condition, with the belts not running and no signal arriving from the shaft encoder, the frequency of the signal arriving at the f_{VCO}

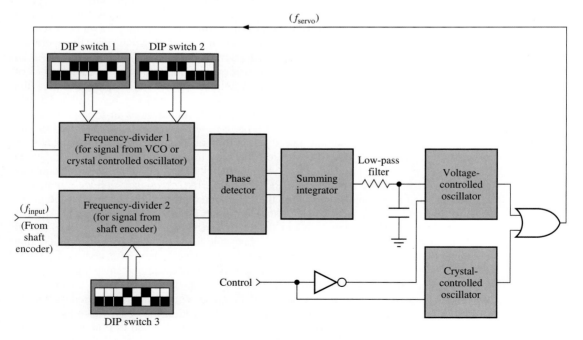

Figure 6–48
PLL circuit used to control clocking of video scanning system

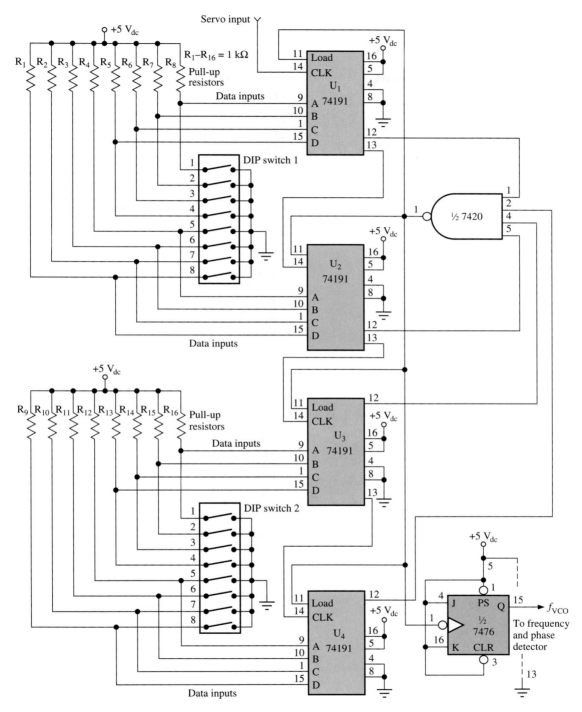

Figure 6–49
Frequency-division circuitry for servo signal from VCO

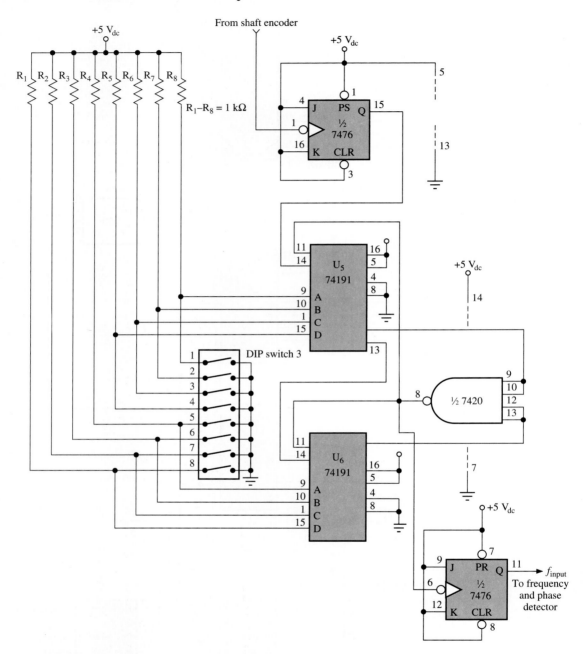

Figure 6–50
Frequency-division circuitry for signal from shaft encoder

input of the phase detector should equal 330 pps. With the crystal control oscillator now active, the servo input to the frequency divider in Figure 6–49 would be a constant 10 MHz. Thus, the modulo of the divider circuit would be determined as:

$$\text{Modulo} = \frac{10\,\text{MHz}}{330\,\text{pps}} = 30.303\,\text{k}$$

Since the JK flip-flop at the output of the divider circuit performs a divide-by-two action, the modulo of the 16-bit counter is only half that determined above. Thus,

$$\text{Modulo of counter} = \frac{30.303\,\text{k}}{2} = 15.152\,\text{k}$$

To achieve a frequency of 330 pps, DIP switches 1 and 2 would then be set to the binary equivalent of 15.152 k. To form this variable, which is 0011101100110000, DIP switch 1 would be set to closed, closed, closed, closed, open, open, closed, and closed for switches 1 through 8, respectively. DIP switch 2, following the same sequence, would be set as open, open, closed, open, open, open, closed, and closed. For proper calibration of the PLL, the f_{input} signal produced by the divider circuit in Figure 6–50 should also be adjusted as closely as possible to 330 pps. Assuming the 0.15-mm clock signal from the shaft encoder has a period of 37 μsec, the frequency of this signal would be determined as follows:

$$f = \frac{1}{37\,\mu\text{sec}} = 27.027\,\text{kpps}$$

With the two JK flip-flops present in Figure 6–50, the required modulo for the 8-bit counter would be reduced by a factor 4:

$$\frac{27.027\,\text{kpps}}{4} = 6.757\,\text{kpps}$$

$$\text{Modulo of counter} = \frac{6.757\,\text{kpps}}{330\,\text{pps}} = 20.48 \cong 20$$

Thus, DIP switch 3 must be set to produce an input variable of 00010100. This is accomplished by setting switches 1 through 8, respectively, to closed, closed, open, closed, open, closed, closed, and closed.

Figure 6–51 contains a schematic diagram of the phase detector, integrator, and low-pass filter sections of the PLL circuitry represented in Figure 6–48. The two signals from the frequency-divider circuits arrive at the inputs to the MC4044, which is a frequency and phase detector IC. Through high-speed Schmitt trigger action, this device is capable of squaring any type of input waveform. The phases of the f_{VCO} and f_{input} signals are precisely compared at the UF and DF outputs of this device. The outputs of the phase comparator are summed at point A and fed to an op amp integrator. Through the voltage divider action of R_1 and R_2, a dc reference of approximately 1.5 V is established for the averaging circuit. As phase differences occur between the UF and DF signals, the voltage at point B increases and decreases with respect to the reference level. When a lock condition is achieved, the UF and DF signals occur at virtually the same frequency and are nearly in phase. With this condition, because the difference frequency between the two signals approaches 0 Hz, the output of the passive filter, consisting of R_4 and C_2, is a virtually constant dc level.

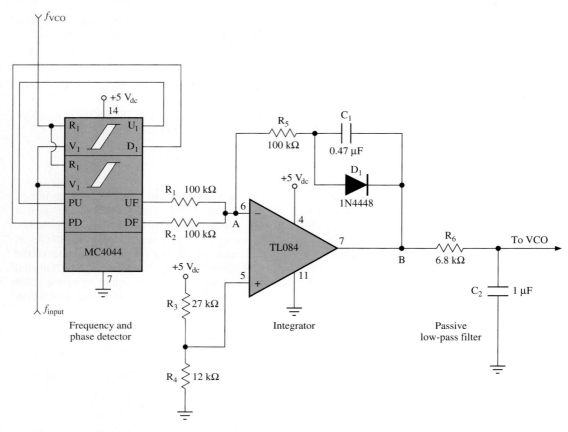

Figure 6–51
Phase-detecting, integrating, and filtering circuitry of VCO

Figure 6–52 shows the oscillator circuitry contained within the PLL. The 74LS629 is a high-frequency dual timer IC. One timer functions to develop the clocking signal from a 20-MHz crystal, and the other functions as the VCO. The outputs of both the VCO and the crystal-controlled oscillator pass through a logic OR function, consisting of three TTL NAND gates. It is important to note that, when the control input is low, the VCO is enabled and the PLL becomes active. Thus, the system control processor would bring the control input to a logic low shortly after the conveyance system is activated. When the belts are halted, the control input is brought to a logic high. As a result, the crystal-controlled clock becomes active, with both the servo and the clock data signals remaining at a constant 10 MHz.

SECTION REVIEW 6–4

1. For the pulse width modulator in Figure 6–44, assume the desired average output voltage is 3.6 V. What must be the ohmic setting of R_3? What would be the pulse width, space width, and duty cycle at this time?

2. For the voltage-controlled oscillator in Figure 6–46, assume the control voltage is 3.86 V. What is the frequency of the signal at pin 3 of the 555 timer? What is the frequency of the signal taken at the Q output of the D flip-flop?
3. Describe the purpose of the D flip-flop in Figure 6–46.
4. Identify the three major functional entities found within all phase-locked loops. Describe the task performed by each of these functional entities.
5. For the phase-locked loop represented in Figure 6–48, assume that the crystal-controlled oscillator is operating at a frequency of 18.7 MHz. What must be the settings of DIP switches 1 and 2 in Figure 6–49 in order to create a f_{VCO} with a period of 3.2 msec during the time the crystal-controlled oscillator is active?
6. Assume the period of the shaft encoder clock signal is expected to be 40 μsec when

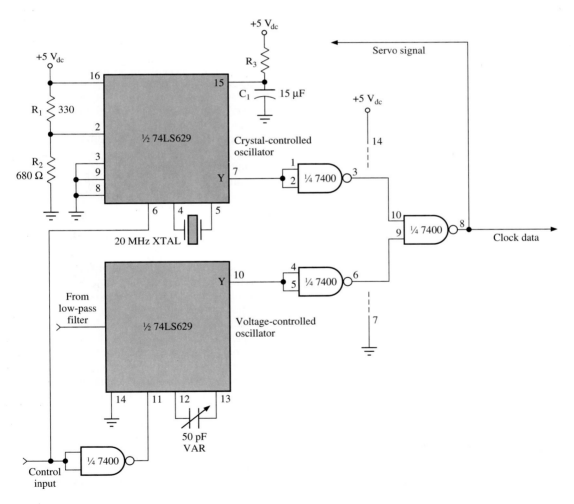

Figure 6–52
Crystal-controlled oscillator and VCO circuitry

the conveyance system is running at full speed. What must be the setting of DIP switch 3 in Figure 6–50 in order to obtain an f_{input} signal with a period of 3.2 msec?

7. Describe the function of the op amp integrator contained in Figure 6–51.

SUMMARY

The three basic solid-state timing devices used in industrial control circuitry are the monostable, astable, and bistable multivibrators. Monostable multivibrators (or one-shots) are used frequently to time the operation of such electromechanical devices as relays, solenoids, and clutch/brakes. One-shots may be used individually or cascaded to form sequential timers. Within the TTL family, commonly used monostable ICs include the 74121, 74122, and 74123. The 74123 is especially useful in industrial applications where extended pulse widths may be required, due to the fact that the two one-shots within the device are retriggerable. In addition to these TTL devices, the 555 IC timer and the 556 dual timer may also be configured to function as one-shots.

Astable multivibrators are found in a variety of industrial circuits, often serving as clock sources in circuitry where precise timing is not required. An example of such an application would be the slow-running clock source for a flashing beacon or emergency indicator. Astable clock sources are also ideal for use in the watchdog timer circuitry of embedded microprocessor systems. Astable multivibrators are often used as the basic clocking elements within more complex pulse circuitry, including pulse width modulators and voltage-controlled oscillators. The 555 IC timer is often configured to function as an astable multivibrator.

Bistable multivibrators are used prevalently in industrial frequency-division circuitry. Both the D flip-flop and the JK flip-flop, especially those contained in the 7474 and 7476 ICs, often function as T flip-flops within divide-by-two and divide-by-four circuitry. Due to their simplicity of design, ripple counters are often used in low-frequency counting and frequency-division circuitry, where propogation delay is not a significant problem. A commonly used IC ripple counter is the 7493, which may be configured for modulo-3, modulo-4, or decade operation. In high-speed frequency-division applications, synchronous IC up/down counters such as the 74191 are preferred over ripple counters. The 74191 is particularly versatile due to the fact that it may be instantly loaded with any 4-bit variable. This action allows the device to be used in conjunction with DIP switches for complex modulo operation.

Advanced forms of industrial pulse circuitry include the pulse width modulator (PWM), voltage-controlled oscillator (VCO), and phase-locked loop (PLL). A PWM may be easily fabricated using two 555 IC timers or a single 556. The first timer, which is configured to operate as an astable multivibrator with a narrow space width, triggers the second timer, which operates as a one-shot with a variable output pulse. The PWM is used prevalently in switching power supplies and regulators, as well as motor control circuitry. A VCO may be fabricated using a single 555 timer. The device operates essentially as an astable multivibrator, the operating frequency of which is controlled by a variable dc input. A T flip-flop may be placed at the output of the VCO in order to maintain the integrity of the output square wave as frequency changes. The VCO is an integral part of the PLL, a control device that is becoming increasingly important in industrial electronics. Through the action of the phase detector, low-pass filter, and VCO contained within the circuit, the PLL is capable of locking on to a variable input frequency and closely tracking that signal as it fluctuates through time. An elaboration on the basic concept of the PLL includes the use of frequency dividers to compensate for two widely differing frequency ranges of f_{in} and f_{servo}. An averaging integrator circuit may also be contained within the PLL, especially in applications where the input signals are rectangular waveforms.

SELF-TEST

1. For the circuit in Figure 6–1(b), if C_1 equals 4.7 μF, what is the pulse width of the output signal?
2. Explain how the one-shot in Figure 6–1(b) improves the operation of the relay actuating circuit in Figure 6–4.
3. For the astable multivibrator in Figure 6–7(b), assume that C_1 and C_2 both equal 0.22 μF. What is the frequency of the output signal?
4. Describe the function of the 74121 IC in Figure 6–15.
5. If, in Figure 6–15, R_1 equals 10 kΩ and C_1 is the largest value of capacitor recommended for accurate operation of the 74121 IC, what would be the maximum actuation time for the solenoid?
6. Explain the operation of a retriggerable one-shot. What is the advantage of using such a device to control a diverter gate, as shown in Figure 6–20?
7. For the clutch/brake control circuit in Figure 6–33, what factors determine the duration of the brake low-current signal?
8. For the circuit in Figure 6–33, what is the reason for the dead time occurring between t_3 and t_4?
9. For the circuit in Figure 6–33, which transistors are conducting while the clutch high current is flowing?
10. For the circuit in Figure 6–33, what is the duration of the dead time between t_5 and t_6? What is the purpose of this dead time?
11. Assume for the clutch/brake in Figure 6–30 that the minimum required velocity of shaft rotation is 350 rpm. What must be the setting of the DIP switch in Figure 6–33 in order to detect failure of the clutch brake shaft to rotate at this velocity?
12. If, for the control circuitry in Figure 6–33, both the Hall sensor and the watchdog circuitry failed, what would be the maximum duration of the clutch holding-current signal?
13. Describe exactly how a logic high condition at point A in Figure 6–33 inhibits the operation of the clutch/brake driver circuitry in Figure 6–33.
14. Describe the function of the integrator in Figure 6–51. What are the frequency components of its output signal?
15. Describe the function of the low-pass filter in Figure 6–51. Ideally, what is its output frequency when the PLL is in a lock condition?
16. For the frequency divider in Figure 6–43, assume the input clock frequency is 22.6 kpps. What must be the setting of the DIP switch in order to obtain an output signal with a period of nearly 11 msec?
17. Describe the difference in operation between a retriggerable and a nonretriggerable one-shot. Which type is likely to be preferred for operating a diverter gate within a high-speed conveyance system?
18. Explain two advantages of the high-current/low-current method for actuating solenoids and clutch/brakes.
19. Describe a common application of a watchdog timer circuit. Why is it necessary for this monitoring device to be driven by a free-running astable multivibrator rather than the crystal-controlled clock of the system in which it is contained?
20. Why does the PLL-based control circuit represented in Figure 6–48 contain both a VCO and a crystal-controlled oscillator?

7 INTRODUCTION TO UNIJUNCTION TRANSISTORS AND THYRISTORS

CHAPTER OUTLINE

LEARNING OBJECTIVES

On completion of this chapter, the student should be able to:

- Describe the internal structure and operation of the unijunction transistor.
- Explain the operation of the unijunction transistor relaxation oscillator.
- Describe the internal structure and operation of the programmable unijunction transistor.
- Define the term *thyristor*. Explain the advantages of thyristors over rheostats in controlling power transfer to ac loads.
- Describe the internal structure and operation of the silicon-controlled rectifier.
- Describe the basic forms of gate-control circuitry used for triggering SCRs.
- Describe the operation of the light-activated SCR. Explain how the sensitivity of this device is controlled.
- Explain how the UJT may be used to enhance the reliability and accuracy of SCR gate-control circuitry.
- Explain how SCRs may be used as switches in the control of various dc loads.
- Describe the operation of the silicon-controlled switch. Explain the advantages of using this device in the control of dc loads.
- Explain the structure and operation of the gate turn-off switch. Describe a common automotive application of this device.

411

- Describe the internal structure and operation of the TRIAC. Explain the major advantage of using this device for the control of ac loads.
- Identify and explain the operation of various forms of unilateral breakover devices.
- Identify and explain the operation of various forms of bilateral breakover devices.

INTRODUCTION

The **unijunction transistor** (UJT) is a semiconductor device used extensively within pulse circuitry. The UJT, which operates on the principle of avalanche breakdown, is an inherently nonlinear device and, therefore, not usable for voltage or current amplification. However, the device is ideally suited for providing trigger pulses within various forms of control circuitry.

Thyristors comprise a family of semiconductor devices that have many applications in modern industrial electronics. Their main purpose is to control the transfer of power to various ac loads. The term *thyristor* is derived from the words *thyratron* and *transistor* and is especially appropriate for describing the operation of **silicon-controlled rectifiers** (SCRs). The operation of the SCR is analogous to that of the three-element, gas-filled tube known as the **thyratron.** A small trigger current at the grid of this device is used to induce the main flow of current between the anode and cathode. The SCR is sometimes referred to as a *four-layer* device (*p-n-p-n*) because it is formed through the fusion of an *n-p-n* and a *p-n-p* transistor. SCRs and other thyristors, like UJTs, are inherently nonlinear. In this chapter, SCRs and TRIACs will be examined in detail, while breakover devices such as DIACs and silicon bilateral switches are briefly surveyed.

7–1   UNIJUNCTION TRANSISTORS AND PROGRAMMABLE UNIJUNCTION TRANSISTORS

Figures 7–1(a) and (b) represent the internal structure and operation of the UJT, and Figure 7–1(c) contains the schematic symbol for the device. NOTE: The schematic symbol for the UJT resembles that of the *n*-channel JFET. In drawing or reading schematic diagrams, one must be careful to differentiate between these two devices.

Like the bipolar transistor, the UJT has three terminals. However, the UJT consists of two bases and an emitter. As seen in Figure 7–1(a), the UJT contains a single *p-n* junction, which is the basis for the term **uni***junction*. This junction is created by connecting a heavily doped piece of *p*-type silicon to one side of a lightly doped *n*-type silicon bar. The *p*-type material forms the emitter of the UJT. It is the emitter terminal that functions as the control input to the device. The base terminals are placed at opposite ends of the *n*-type silicon bar.

Since the *n*-type bar is lightly doped, it exhibits very high resistivity. This piece of *n*-type silicon forms an unbroken channel within the UJT. However, from the standpoint of its equivalent operation, it may be represented as shown in Figure 7–1(b). Here the *n* material is divided into two internal base resistances, designated r_{B1} and r_{B2}. The junction of these two resistances occurs at the cathode of an equivalent diode, which represents the *p-n* junction. The interbase resistance, which may be measured by placing an ohmmeter between the two base terminals, is designated R_{BB}. This parameter, which is specified on the data sheet for a given device, varies significantly, even for UJTs of the same type. As an example, for the 2N4948 UJT, R_{BB} could be as low as 4 kΩ or as high as 12 kΩ. Note that in Figure 7–1(b) r_{B1} is represented as a variable resistor. This is because the resistivity of that portion of the *n*-type material is inversely proportional to the amount

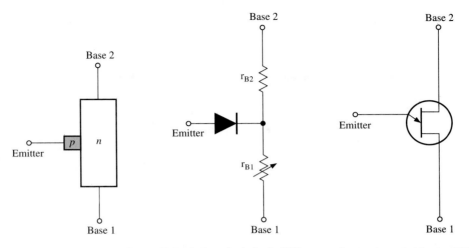

(a) Internal structure of the UJT   (b) Equivalent circuit for the UJT   (c) Schematic symbol for the UJT

Figure 7–1

of current entering the device at the emitter. The relationship between emitter current and the ohmic value of r_{B1} forms the underlying principle for the operation of the UJT.

Figure 7–2(a) indicates the proper biasing condition for the UJT. Notice that the B_2 terminal is made positive with respect to B_1. The difference in potential between the two base terminals is designated V_{B2B1}. When the emitter terminal is reverse biased, current still flows between the base terminals of the UJT, as long as a difference in potential exists between B_1 and B_2. This current is designated I_{B2}. With the emitter terminal reverse-

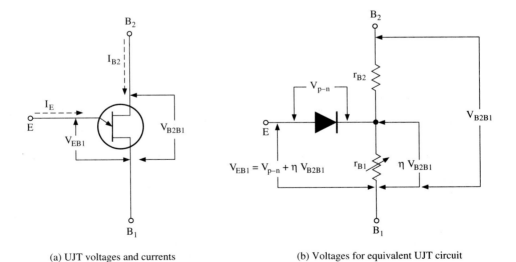

(a) UJT voltages and currents   (b) Voltages for equivalent UJT circuit

Figure 7–2

biased, r_{B1} and r_{B2} are considered to be in series. Thus, during this condition:

$$R_{BB} = r_{B1} + r_{B2} \qquad (7\text{–}1)$$

Also, with the emitter terminal reverse biased, I_{B2} could be determined using Ohm's law:

$$I_{B2} = \frac{V_{B2B1}}{R_{BB}} \qquad (7\text{–}2)$$

To clearly understand the operation of the emitter junction, it is necessary to examine the equivalent UJT circuit shown in Figure 7–2(b). A voltage-divider function occurs within the UJT, involving r_{B1} and r_{B2}. The ratio of r_{B1} to R_{BB}, which exists with the emitter in a reverse-biased condition, is an important parameter for the UJT. The symbol for this relationship, referred to as the **intrinsic-standoff ratio,** is η, which is the lowercase Greek letter eta. The intrinsic-standoff ratio may be determined as follows:

$$\eta = \frac{r_{B1}}{(r_{B1} + r_{B2})} = \frac{r_{B1}}{R_{BB}} \qquad (7\text{–}3)$$

As indicated in Figure 7–2(b), the voltage developed over r_{B1} within the UJT is easily determined using the voltage-divider equation:

$$V_{r_{B1}} = \frac{r_{B1}}{R_{BB}} \times V_{B2B1} = \eta \times V_{B2B1} \qquad (7\text{–}4)$$

Since $V_{p\text{-}n}$ represents the forward voltage of a silicon diode, its value would be nearly 0.7 V. Thus, to initiate the flow of emitter current, the minimum difference in potential that must occur between the emitter and base 1 terminals would be determined as follows:

$$V_{EB1} = V_{p\text{-}n} + (\eta \times V_{B2B1}) \qquad (7\text{–}5)$$

Until this threshold voltage is attained, the *p-n* junction at the emitter of the UJT remains reverse biased. As long as V_{EB1} is less than $(\eta \times V_{B2B1})$, the UJT would be operating in the **cutoff region** of its characteristic curve, as indicated in Figure 7–3. With this condition, only an emitter-leakage current (indicated as I_{EB20}) flows across the junction. At the point where V_{EB1} exactly equals $(\eta \times V_{B2B1})$, there is a difference of 0 V across the junction. With this condition, as shown in the curve, emitter current equals exactly 0 A.

If the voltage at the emitter is increased slightly beyond the peak point, the *p-n* junction approaches a forward-biased condition. At this point, V_{EB1} attains its peak level (designated as V_P on the characteristic curve). It should be stressed that, although η is a constant for a given UJT, the value of V_P varies as a function of the voltage applied between the two base terminals. As the emitter junction approaches a forward-biased condition, with V_{EB1} equaling V_P, a small forward current begins to flow. Since this current corresponds with the peak voltage, it is referred to as **peak current** (I_P). As I_E increases beyond the level of I_P, the UJT begins to operate in the **negative-resistance region.** Within this portion of the characteristic curve, the difference in potential between the emitter and the B_1 terminal decreases with the rapid increase in I_E.

This action might, at first, seem to defy Ohm's law. However, due to avalanche breakdown between the emitter and B_1 terminals, a large decrease in the ohmic value of r_{B1} occurs simultaneously with the increase in I_E. As I_E continues to increase, V_{EB1} approaches a saturation level and, finally, decreases to a low point referred to as the **valley voltage** (designated V_V on the characteristic curve). At the valley point, r_{B1} approaches its minimum ohmic value. This **saturation** level of resistance is designated as r_S. The valley voltage is equal to the sum of the forward voltage drop at the *p-n* junction and the voltage drop

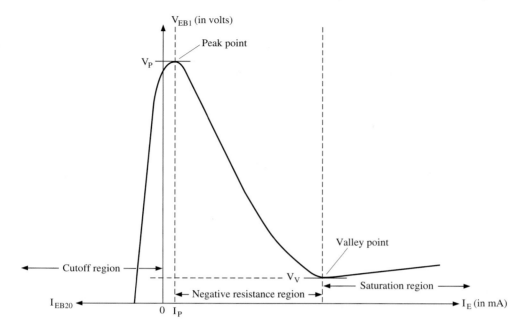

Figure 7–3
Emitter characteristic curve for UJT

over r_{B1}. Thus, at the saturation point:

$$V_V = V_{p\text{-}n} + [(I_E + I_{B2}) \times r_S] \qquad (7\text{–}6)$$

Beyond the valley point of the characteristic curve, r_{B1} remains at its saturation level. Thus, V_{EB1} increases only slightly with any further increase in I_E.

A problem associated with the UJT is that of temperature stability. The interbase resistance of the device may vary significantly with deviations in operating temperature. Thus, the **interbase resistance temperature coefficient** (αR_{BB}) becomes an important operating parameter for the UJT. On the data sheet for a given device, this parameter is likely to be listed for a minimum and maximum percentage of change in resistance per degree Celsius. For example, with the 2N4948 UJT, this amount of change could be as small as 0.1%/°C or as great as 0.9%/°C.

The UJT is often used as the switching device within a free-running pulse circuit referred to as the **relaxation oscillator.** Figure 7–4(a) shows the schematic diagram of a basic UJT relaxation oscillator, and Figure 7–4(b) illustrates the sawtooth waveform produced by this device. The relaxation oscillator is similar to the astable multivibrator in that it relies on RC timing to control its operating frequency.

At the instant the circuit in Figure 7–4(a) is energized, C_1 is assumed to be totally discharged and, thus, momentarily behaving as a short circuit. At this instant in time, with a reverse-biased condition at the emitter junction of the UJT, this device would be operating in the cutoff region of its characteristic curve. As C_1 charges via R_1, the voltage at the emitter input to the UJT eventually attains the level of V_P. When its *p-n* junction becomes forward biased, the UJT begins to operate in the negative resistance region of its curve. This forward-biased condition allows C_1 to begin discharging toward circuit ground through r_{B1}. This action causes the ohmic value of r_{B1} to decrease toward the level

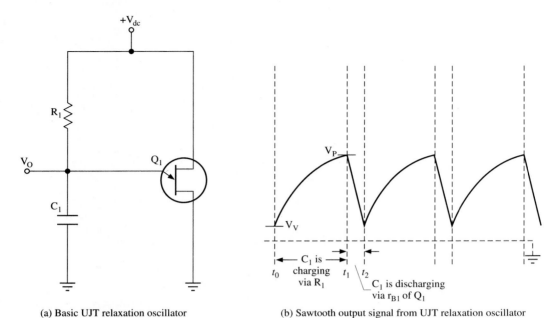

(a) Basic UJT relaxation oscillator

(b) Sawtooth output signal from UJT relaxation oscillator

Figure 7–4

of r_S as I_E rapidly increases. As C_1 rapidly discharges, the voltage between the emitter of the UJT and ground falls from V_P toward V_V. However, as the value of V_{EB1} approaches its saturation level (which is slightly higher than V_V) the UJT enters the cutoff state, again allowing C_1 to begin charging toward the level of the applied voltage. However, once the level of V_P is attained, the UJT again begins to operate in the negative resistance region, as C_1 rapidly discharges through r_{B1}. This relaxation oscillation process continues as long as the dc power source is present.

There are maximum and minimum limits for the value of R_1 (emitter resistance) contained within the relaxation oscillator. These ohmic values are determined using the data sheets for the given UJT. The maximum value of R_1 is calculated as follows:

$$R_{1(max)} = \frac{V_{app} - V_{P(max)}}{I_{P(min)}} \tag{7–7}$$

In this equation, $V_{P(max)}$ is determined by multiplying the maximum intrinsic-standoff ratio (specified on the data sheet) by the value of applied voltage (V_{app}) for the given circuit; $I_{P(min)}$ is simply taken from the data sheet.

If the value of R_1 chosen for a UJT relaxation oscillator exceeds the value calculated by Eq. (7–7), due to an insufficient flow of peak current, the UJT may not be able to trigger into its negative resistance mode of operation. With this condition, since C_1 is unable to discharge via r_{B1}, it charges toward the supply voltage and remains at that level. If too small a value of R_1 is selected for the relaxation oscillator, once C_1 discharges through r_{B1}, the UJT may not be able to return to the cutoff condition. This is because the valley current would be able to flow continuously. To prevent this latch-on condition from occurring, the value of emitter resistance must be no less than the value calculated as follows:

$$R_1(\text{min}) = \frac{V_{app} - V_{EB1}(\text{sat})}{I_{V(\text{min})}} \tag{7-8}$$

where both the minimum value of $V_{EB1}(\text{sat})$ and the minimum value of valley current should be taken directly from the data sheets for the given UJT.

The basic formula for computing RC time, which was introduced in the previous chapter as Eq. (6–1), can be used in the calculation of frequency for the UJT relaxation oscillator. It applies directly to the calculation of the charge time for C_1, equivalent to the period from t_0 to t_1 in Figure 7–4(b):

$$t = (RC) \ln\left(\frac{V - V_0}{V - V_C}\right)$$

For the UJT relaxation oscillator, the value of V_0, which represents the residual charge on the capacitor, is equivalent to the saturation level of V_{EB1}. The value of V_C, which represents the switching threshold of the capacitor, is equivalent to the peak voltage V_P. Thus, substituting into Eq. (6–1), the following expression is derived for the charge time of capacitor C_1 within the relaxation oscillator:

$$t = (R_1 \times C_1) \times \ln\left(\frac{V_{app} - V_{EB1}(\text{sat})}{V_{app} - V_P}\right) \tag{7-9}$$

The discharge time for C_1, represented as the period from t_1 to t_2 in Figure 7–4(b), is much more difficult to determine, because the value of r_S is a dynamic quantity that varies while saturation current is flowing. Since the ohmic value of r_S is likely to be much less than that of R_1, the period from t_1 to t_2, as shown in Figure 7–4(b), is likely to be much less than the charge time for the capacitor. Consequently, the discharge time for the capacitor is often ignored during the approximation of frequency for the UJT relaxation oscillator. Thus, frequency may be estimated as follows:

$$f \cong \frac{1}{t} \tag{7-10}$$

where t is derived from the application of Eq. (7–9).

Example 7–1

For the UJT contained in Figure 7–5, assume r_{BB} equals 11.76 kΩ and r_{B2} equals 3.293 kΩ. Solve for the intrinsic-standoff ratio, the value of V_P, and the approximate frequency of oscillation. Using the data sheet for the given UJT, determine the maximum and minimum values of R_1 that could be used within the relaxation oscillator.

Solution:

Step 1: Solve for the intrinsic ratio:

$$r_{B1} = R_{BB} - r_{B2} = 11.76 \text{ k}\Omega - 3.293 \text{ k}\Omega = 8.467 \text{ k}\Omega$$

$$\eta = \frac{r_{B1}}{R_{BB}} = \frac{8.467 \text{ k}\Omega}{11.76 \text{ k}\Omega} = 0.72$$

Figure 7–5

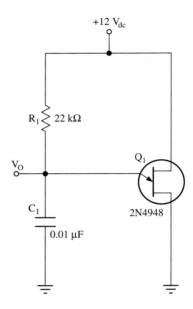

Step 2: Solve for V_P:

$$V_P = V_{p\text{-}n} + (\eta \times V_{app}) = 0.7 + (0.72 \times 12\ V) = 9.34\ V$$

Step 3: Solve for the approximate operating frequency of the relaxation oscillator (using the calculated value of V_P, and the typical value of $V_{EB1}(\text{sat})$ specified in the data sheet for the 2N4948):

$$t = 22\ k\Omega \times 0.01\ \mu F \times \ln\left(\frac{12\ V - 2.5\ V}{12\ V - 9.34\ V}\right) = 280\ \mu sec$$

$$f \cong \frac{1}{t} = \frac{1}{280\ \mu sec} \cong 3.571\ kpps$$

Step 4: Solve for the maximum allowable value of R_1 that could be used in the relaxation oscillator in Figure 7–5. Determine V_P using the maximum value of η listed in the data sheet:

$$V_{P(max)} = V_{p\text{-}n} + [\eta_{(max)} \times V_{app}] = 0.7\ V + (0.82 \times 12\ V) = 10.54\ V$$

Using the calculated value of V_P and the minimum value of I_P listed in the data sheet:

$$R_{1(max)} = \frac{12\ V - 10.54\ V}{2\ \mu A} = 730\ k\Omega$$

Step 5: Solve for the minimum value of R_1 using the typical value of $V_{EB1}(\text{sat})$ and the minimum value of I_V listed in the data sheet:

$$R_1(\text{min}) = \frac{12\ V - 2.5\ V}{2\ mA} = 4.75\ k\Omega$$

In addition to its temperature instability, a problem associated with the UJT is the wide variance in the value of η. For example, as indicated by the data sheet for the

2N4948, η could be as low as 0.55 or as high as 0.82. Thus, if the UJT contained in Figure 7–5 were replaced by another device of the same part number, it is likely that the relaxation oscillator would operate at a different frequency. This problem could, of course, be compensated for by making R_1 a potentiometer. However, if wide variations in frequency are to be achieved through use of a variable emitter resistor, care must be taken that its ohmic value does not go above $R_{1(max)}$ or below $R_{1(min)}$ for the given value of V_{app}.

Discrepancies in UJT circuit performance resulting from wide variations in η may be avoided through use of a more complex device called the **programmable unijunction transistor** (PUT). As is evident in Figure 7–6, this device takes its name from the fact that the effective R_{BB} is established through use of an external voltage divider. The effective intrinsic-standoff ratio for the device may now be precisely controlled through varying the values of R_1 and R_2. The PUT is not actually a UJT. Instead, it could best be described as a form of **silicon-controlled rectifier** (SCR), a device that is studied in detail in the next section. As seen in Figure 7–6, the control input to the PUT is designated the **gate,** and the two main terminals of the device are called the **anode** and **cathode.** Figure 7–6(a) represents the internal structure of the PUT, which consists of two blocks of p-type silicon and two blocks of n-type silicon arranged in a four-layer pattern. The schematic symbol used to represent the PUT is shown in Figure 7–6(b).

The function of the anode terminal of the PUT is similar to that of the emitter input of the conventional UJT. As the anode voltage becomes more positive than the voltage level present at the gate, a relatively small forward **gate current** (I_G) begins to flow. This action, in turn, induces a much larger **anode current** (I_A), which flows through the main terminals of the device. As this occurs, V_{AK} (the difference in potential between the anode and cathode terminals) suddenly decreases to a **forward saturation** level (V_F). When the main terminal current is reduced to a level that no longer sustains forward conduction, V_{AK} returns to its maximum value.

The characteristic curve in Figure 7–7 serves to clarify the operation of the PUT. When the voltage present at the anode is negative with respect to the gate, only a small

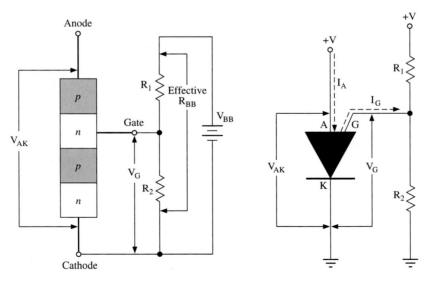

(a) Four-layer structure of the PUT   (b) PUT circuit showing schematic
    connected to external voltage divider   symbol and external voltage divider

Figure 7–6

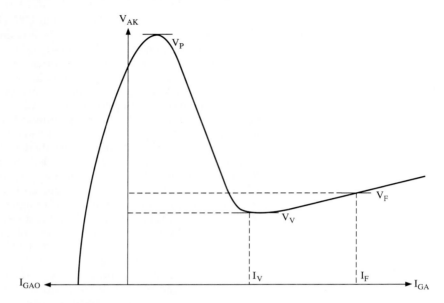

Figure 7–7
Anode characteristic curve for the PUT

gate-to-anode leakage current (I_{GAO}) is able to flow. This reverse gate current is likely to consist of only a few nanoamps. At the point where the anode voltage equals that present at the gate, no bias voltage exists from anode to cathode, resulting in a **forward gate current** (I_{GA}) of 0 A. However, as V_{AK} continues to increase in the positive direction, I_{GA} begins to flow. The peak level of V_{AK} (indicated as V_P on the curve) is attained when V_{AK} is nearly equal to $V_G + 0.7$ V). (Remember that V_G is established by an external source such as the voltage divider in Figure 7–6.) Once this threshold level of V_{AK} is attained, the PUT is triggered on; I_{GA} rapidly increases, causing avalanche breakdown to occur between the main terminals of the device; and V_{AK} rapidly falls toward the valley point. Once triggering has occurred, even as main terminal current is increased, V_{AK} remains near its forward saturation level.

The action of the PUT just described makes it ideal for use within a relaxation oscillator. An example of such an application is shown in Figure 7–8(a). Notice that R_1, within the RC timing combination, is a fixed resistance, whereas R_3, within the external voltage divider, is made variable. When power is applied to the circuit, C_1 begins to charge through R_1, toward the level of the supply voltage. However, at the instant V_{C1} attains the level of V_P, C_1 rapidly discharges into the anode of the PUT. As the charge on the capacitor dissipates, I_A decreases to a level where forward conduction of the PUT ends. At this point, the capacitor again begins to charge through R_1. This action continues as long as the dc source is present, resulting in a sawtooth waveform similar to that shown in Figure 7–4(b). Increasing the ohmic value of R_3 raises the value of V_G. Consequently, the value of V_P increases. Since V_P represents the threshold voltage for the capacitor, which charges via R_1, the period for the output signal of the relaxation oscillator increases, resulting in a lower output frequency. Decreasing the value of V_G lowers this switching threshold, resulting in a shorter period and a higher frequency for the output signal.

As with the relaxation oscillator containing the conventional UJT, the period of the signal produced by the circuit in Figure 7–8 is determined by substituting known values

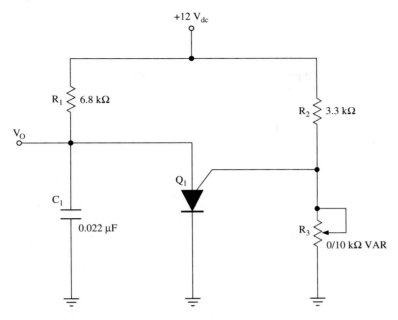

Figure 7–8
PUT relaxation oscillator

into Eq. (6–1). Thus, for the basic PUT relaxation oscillator,

$$t = (R_1 \times C_1) \times \ln\left(\frac{V_{app} - V_F}{V_{app} - V_P}\right) \tag{7–11}$$

where V_{app} represents the applied voltage, V_F represents the forward saturation voltage of the PUT, and V_P is equivalent to ($V_G + 0.7$ V).

Example 7–2

For the relaxation oscillator in Figure 7–8, assume the forward saturation voltage of the PUT equals 1.35 V. If R_3 is adjusted to 6.43 kΩ, what is the approximate frequency of oscillation?

Solution:

Step 1: Solve for V_G:

$$V_G = V_{app} \times \frac{R_3}{R_2 + R_3} = 12\text{ V} \times \frac{6.43\text{ k}\Omega}{3.3\text{ k}\Omega + 6.43\text{ k}\Omega} = 7.93\text{ V}$$

Step 2: Solve for V_P:

$$V_P = (V_G + 0.7\text{ V}) = (7.93\text{ V} + 0.7\text{ V}) = 8.63\text{ V}$$

Step 3: Solve for t:

$$t = 6.8 \text{ k}\Omega \times 0.022 \text{ }\mu\text{F} \times \ln\left(\frac{12 \text{ V} - 1.35 \text{ V}}{12 \text{ V} - 8.63 \text{ V}}\right) = 172 \text{ }\mu\text{sec}$$

Step 4: Solve for approximate frequency of oscillation:

$$f \cong \frac{1}{172 \text{ }\mu\text{sec}} \cong 5.8 \text{ kpps}$$

In Figure 7–9, the concept of the relaxation oscillator is expanded to form a free-running ramp generator. As with the circuit in Figure 7–8, both the peak amplitude and the period of the output signal become a function of the gate voltage of the PUT. With the -5-V source present at point A, C_1 attempts to charge via R_1, positive with respect to point B. Since the inverting input of the op amp functions as a virtual ground point, the charging current for C_1 remains constant, allowing the growth of voltage at point B to be linear. Due to the **Miller effect,** the capacitor attempts to charge toward the positive supply voltage of the op amp. However, with the main terminals of the PUT placed in parallel with C_1, the peak amplitude of the ramp signal developing at point B is limited

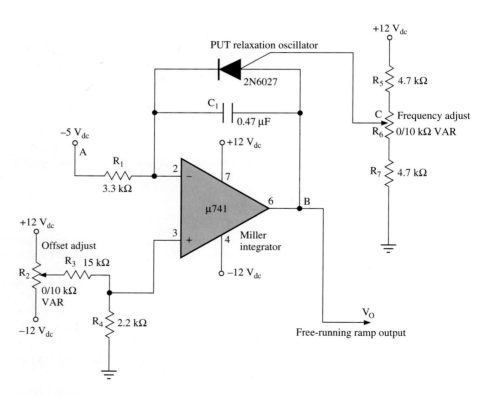

Figure 7–9
VCO with a ramp output consisting of an op amp integrator and a PUT relaxation oscillator

to the level of V_P. Once the voltage at point B attains a level equivalent to (V_G + 0.7 V), triggering occurs, allowing C_1 to discharge rapidly through the main terminals of the PUT. Once the forward current of the PUT dissipates to a level that no longer sustains forward conduction, the PUT switches off, again allowing C_1 to charge via R_1. As with the basic UJT or PUT relaxation oscillator, this cycle continues as long as dc voltage is applied to the circuit. However, as shown in Figure 7–10, the output waveform has a linear slope, as opposed to the exponential or sawtooth shape characteristic of the simple relaxation oscillator. This is due to the function of the op amp as a constant current source. Also, the presence of the resistive network consisting of R_2, R_3, and R_4 allows fine adjustment of the dc reference of the output signal.

Recalling the basic mathematics for the Miller integrator, as developed in Section 2–6 of Chapter 2, the operating frequency of the circuit in Figure 7–9 may be approximated. For the Miller integrator,

$$\Delta V = \frac{I \times \Delta t}{C}$$

Here, ΔV represents the change in voltage over the capacitor for a given period, represented as Δt; and I represents the constant current flowing through R_1, easily determined using Ohm's law. For the circuit in Figure 7–9, ΔV is established by the PUT circuitry. The peak amplitude of the signal developed at point B is equivalent to (V_G + 0.7 V), whereas the minimum amplitude of this signal would be V_F of the PUT. Thus,

$$\Delta V = (V_P - V_F)$$

Since the gate voltage is being varied to produce changes in the period of the ramp signal, it is necessary to develop an expression for time:

$$\Delta t = \frac{(V_P - V_F) \times C}{I} \tag{7–12}$$

Finally, since the discharge time for the capacitor in Figure 7–9 is much less than the ramp time, the frequency of oscillation may be approximated simply as the reciprocal of the result of Eq. (7–12).

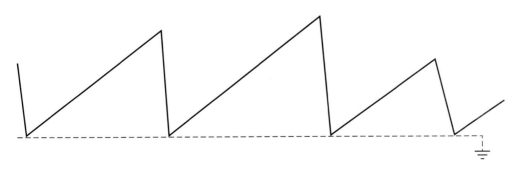

Figure 7–10
Output of the ramp generator in Figure 7–9

Example 7–3

For the circuit in Figure 7–9, assume that the forward saturation voltage of the PUT is 2.2 V and that the potentiometer is adjusted such that V_G equals 6.8 V. Given these values, solve for the frequency of oscillation.

Solution:

Step 1: Solve for the value of constant current:

$$I = \frac{-5 \text{ V}}{3.3 \text{ k}\Omega} = -1.515 \text{ mA}$$

Step 2: Solve for V_P:

$$V_P = 6.8 \text{ V} + 0.7 \text{ V} = 7.5 \text{ V}$$

Step 3: Solve for ΔV:

$$\Delta V = (V_P - V_F) = 7.5 \text{ V} - 2.2 \text{ V} = 5.3 \text{ V}$$

Step 4: Solve for Δt (using the absolute value of current):

$$\Delta t = \frac{(5.3 \text{ V} \times 0.47 \text{ }\mu\text{F})}{1.515 \text{ mA}} = 1.644 \text{ msec}$$

Step 5: Solve for approximate frequency of oscillation:

$$f = \frac{1}{1.644 \text{ msec}} \cong 608 \text{ pps}$$

SECTION REVIEW 7–1

1. For a 2N4948 UJT, assume r_{B1} equals 5.62 kΩ and η is 0.74. What are the ohmic values of R_{BB} and r_{B2}?
2. For the UJT contained in Figure 7–4(a), assume η to be 0.67 and $V_{EB1}(\text{sat})$ to equal 2.35 V. Also assume R_1 equals 12 kΩ and C_1 is 0.047 μF. If the supply voltage equals 15 V_{dc}, what is the operating frequency of the circuit?
3. What is negative resistance? How is this phenomenon exploited in the operation of the UJT relaxation oscillator?
4. Sketch a block diagram of the internal structure of a programmable UJT. Regarding both structure and operation, what control device does the PUT resemble?
5. For the circuit in Figure 7–8, assume R_3 is set to 8.26 kΩ. What are the values of V_G and V_P? Assuming V_F equals 2.1 V, determine the approximate operating frequency of the relaxation oscillator.
6. Explain why the output of the circuit in Figure 7–9 is a linear ramp rather than a sawtooth waveform.
7. For the circuit in Figure 7–9, assume V_F of the PUT to be 1.95 V. What must be the voltage setting at point C to obtain an output signal with a frequency of 575 pps?

7-2   INTRODUCTION TO SILICON-CONTROLLED RECTIFIERS

The idea of a four-layer control device with a gate, or triggering, input has already been introduced in the form of the programmable UJT. The **silicon-controlled rectifier** is also a four-layer (*p-n-p-n*) device, which may be considered as the fusion of an *n-p-n* and a *p-n-p* transistor. Like both the UJT and the PUT, the SCR is an inherently nonlinear switching device.

Figures 7–11(a) and (b) represent the internal structure of the SCR, whereas (c) shows its schematic symbol. In Figure 7–11(a), the four-layer structure of the SCR is clearly shown. Here, the dashed line dividing the n_1 and the p_2 layers clarifies the presence of Q_1 and Q_2 within the four-layer structure. These two transistors comprise a complementary pair, with Q_1 being a *p-n-p* device and Q_2 being *n-p-n*. Represented as n_{1A} and n_{1B}, the base of Q_1 and the collector of Q_2 are formed by a single layer of *n*-type silicon. Shown as p_{2A} and p_{2B}, the collector of Q_1 and the base of Q_2 are formed by a single layer of *p*-type silicon. Note that, for the schematic symbol, the anode and cathode terminals of the SCR are represented as those of a typical diode or PUT. However, to differentiate the device from the PUT, the gate input is shown connected to the cathode rather than the anode side of the diode symbol.

Referring to Figure 7–11(a), consider a condition in which the anode of the SCR is made positive with respect to the cathode, but with 0 V present at the gate. Due to the reverse-biased condition at the junction formed by the n_1 and p_2 materials, the SCR behaves identically to a reverse-biased diode. With this condition, only a slight **leakage current** (I_{FX}) flows through the main terminals of the device. This parameter, which may also be referred to as **forward blocking current** or **forward peak off-state current** (IDR_{MX}), is likely to be only 100 μA. If, however, with the gate still at 0 V, the forward voltage between the anode and cathode ($+V_{AK}$) is steadily increased, a forward potential is eventually attained where the SCR is forced into forward conduction. The maximum forward voltage that the SCR is able to withstand before being forced into this avalanche

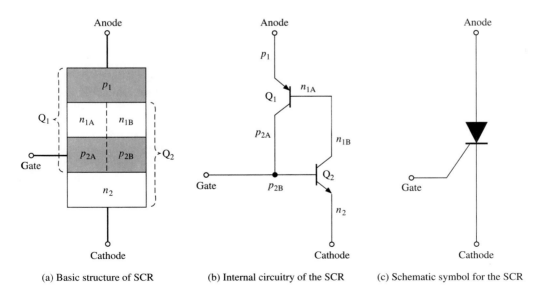

(a) Basic structure of SCR   (b) Internal circuitry of the SCR   (c) Schematic symbol for the SCR

Figure 7–11

breakdown condition is called the **peak-forward blocking voltage** (V_{FOM}). This maximum forward potential may also be referred to as the **forward breakover voltage** [$V_{BR(F)}$ or $V_{F(BO)}$]. Also, manufacturers are likely to differentiate between a repetitive and nonrepetitive peak-forward blocking voltage, which further complicates the use of SCR data sheets. For example, one manufacturer designates the **nonrepetitive peak-forward off-state voltage** as V_{DSMX} and the **repetitive peak-forward off-state voltage** as V_{RSMX}. The nonrepetitive peak potential represents a significantly higher voltage than the repetitive. As an example, for a given SCR, V_{DSMX} could be 250 V, whereas V_{RSMX} is only 200 V.

The actual peak-forward blocking voltage for an SCR may be as low as 25 V or as high as 800 V, depending on the type of device. However, once avalanche breakdown does occur, Q_1 and Q_2 rapidly switch into a forward saturation condition, causing the voltage between the main terminals to fall to a minimal forward conduction level. This forward, or on-state, voltage for an SCR may be designated simply V_F. In some manufacturers' data sheets, this parameter is referred to as **instantaneous on-state voltage** and designated V_T. Typically, V_F (or V_T) is between 1 and 2.5 V, depending on the type of SCR. The operation of an SCR in the forward saturation condition is similar to that of a forward-biased rectifier diode, with current flow being limited only by a load resistance placed in series with the main terminals of the device. Manufacturers' data sheets specify a maximum forward current that may safely flow through the main terminals of the SCR. This **forward** or **on-state current** (I_F or I_T) is likely to be expressed as an rms value and is assumed to be continuous. Manufacturers also specify a **peak one-cycle surge forward current** (I_{FM}) or **peak surge on-state current** (I_{TSM}), which is nonrepetitive and much greater in value than I_F (or I_T). For instance, the data sheets for a given SCR might specify an I_F of only 1.6 A_{rms} and an I_{FM} of 10 A_{rms}. The form of avalanche breakdown just described, induced by exceeding the forward blocking voltage, is often referred to simply as **breakover.** This breakover process, which occurs without the flow of gate current, is to be avoided in the normal operation of an SCR.

Now consider a condition in which a forward voltage much lower than V_{FOM} is applied between the main terminals of the SCR. With no gate voltage present, the SCR functions as an open switch. However, if a sufficient level of positive voltage is applied to the gate terminal, the device is immediately triggered into forward conduction. The threshold level of gate voltage, at which an SCR is triggered into forward conduction, is referred to as the **dc gate trigger voltage** (V_{GT}). This parameter is typically between 0.8 and 1.5 V. The amount of gate current that must be flowing to induce forward conduction of the SCR is referred to as **dc gate trigger current** and designated as I_{GT}. The value of dc gate trigger current may vary greatly, even among SCRs of the same type. For example, according to the manufacturers' data sheets, an RCA S2062 series SCR could have an I_{GT} as low as 100 μA or as high as 2 mA.

Once the SCR is triggered into forward conduction, it remains in this condition, even after the positive gate voltage is removed. This process, which represents the proper triggering technique for the SCR, is referred to as **latching-on.** Once triggered, the device remains latched on until V_{AK} is reduced to nearly 0 V. Not until forward current has fallen below a mimimum holding level does the SCR again behave as an open switch. The minimum amount of main terminal current required for sustaining forward conduction of the SCR is called **instantaneous holding current** (I_H) and usually represents only a few milliamperes.

Referring again to Figure 7–11(b) and to Figure 7–12 should aid in visualizing how the SCR triggering process occurs. Assuming a forward voltage ($+V_{AK}$) is already present between the anode and cathode terminals, a positive voltage at the gate then causes the flow of base current within Q_2, the conduction path for this current being the p_{2B} and the n_2 materials. This initial conduction induces the flow of current through the p_1 and n_1

Figure 7–12
SCR voltages and currents

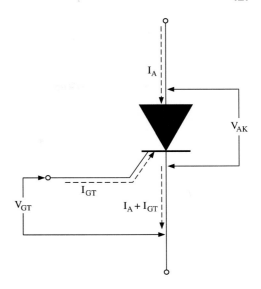

materials. Since the n_{1A} material functions as the base of Q_1, this transistor rapidly switches on, with its collector current now functioning as the base current for Q_2. Also, the base current induced in Q_1 becomes the collector current of Q_2. Thus the supply of current may be removed from the gate, but main terminal current continues to flow until V_{AK} approaches 0 V.

The SCR is *not* designed to be latched on with a negative voltage ($-V_{AK}$) present from anode to cathode. The only method by which a reverse current could be made to flow between the main terminals of the SCR would be to induce a **reverse breakover** condition. This occurs when $-V_{AK}$ exceeds the **peak-reverse blocking voltage** for the device. This maximum reverse potential [V_{ROM} or $V_{BD(R)}$] is likely to have the approximate absolute value of the peak-forward blocking voltage. We also stress that reverse gate voltage should be avoided for an SCR, because it could easily destroy the device. The maximum allowable peak gate reverse voltage (designated V_{GRM}) could be only a few volts.

Figure 7–13 contains the characteristic curve for a typical SCR. Note that in the **forward characteristic** portion of the curve, with a gate current I_G held at 0 A, forward conduction does not occur until V_{AK} attains the level of V_{FOM}. However, for successively larger values of gate current, forward conduction is induced at successively lower values of forward voltage. Since the SCR may not be triggered with a reverse voltage applied over its main terminals, the **reverse characteristic** consists only of the **reverse blocking region,** in which reverse leakage current (I_R) increases only slightly with large increases in negative voltage. However, reverse current increases greatly at the V_{ROM} point, where reverse breakover begins. Note that the distance between the V_{ROM} point and the Y axis is nearly the same as that from the Y axis to the V_{FOM} point, indicating that V_{ROM} and V_{FOM} have nearly the same absolute value.

Figure 7–14 contains a very basic form of SCR control circuit. In this example, the SCR is serving the same function as a rheostat in controlling the intensity of the lamp. However, due to its inherent on/off behavior, the SCR performs this task with much greater efficiency. When constructing a circuit such as that in Figure 7–14, it is imperative that the peak amplitude of the source waveform not exceed the peak-forward and peak-reverse blocking voltages of the SCR. For example, assume an SCR is to be used in

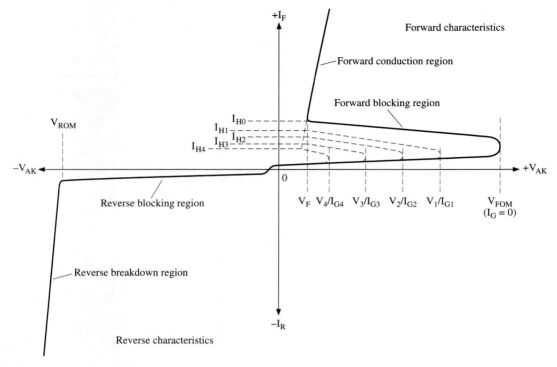

Figure 7–13
SCR forward and reverse characteristic curves

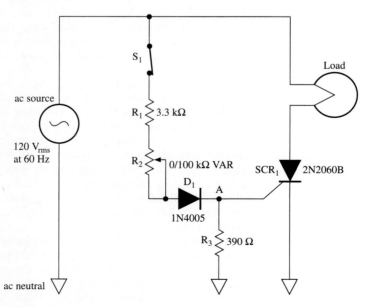

Figure 7–14
Basic SCR lamp-dimmer control circuit

conjunction with a 120-V_{rms} power source. In examining the data for the S2060 series SCR being used, we discover that the suffix B indicates a V_{DRMX} (equivalent to V_{FOM}) of 200 V. Since the 120-V_{rms} source waveform actually attains a peak amplitude of only 170 V, the S2060B device can be used safely within the circuit. (The value of V_{ROM} for the SCR should, of course, be -200 V, which would allow it to withstand the negative cycle of the source waveform.) If an SCR with sufficient V_{FOM} and V_{ROM} is selected, providing S_1 in Figure 7–14 remains open, the condition of the SCR is constantly that of an open switch in series with the lamp, causing it to remain de-energized throughout the source cycle.

Ideally, prior to the instant of triggering, the function of the SCR would be as represented in Figure 7–15(a). Once the SCR is triggered on, it could be compared to a closed switch, as represented in Figure 7–15(b). After triggering has occurred, during the remainder of the positive alternation of the source cycle, virtually all the available power from the source is transferred to the lamp.

R_1, R_2, D_1, and R_3 in Figure 7–14 control the flow of gate current for the SCR: D_1 allows gate current to flow only during the positive half-cycle of the source waveform, thus preventing the development of negative gate voltage and current. Potentiometer R_2 varies the triggering point or **firing-delay angle** of the SCR, which, for the circuit in Figure 7–14, must occur between 0° and 90°. According to the characteristic curve for the SCR, it would appear that a value of V_{AK} at which forward conduction is to be induced could be reliably determined, providing that the required value of I_{GT} has been determined. However, as has already been shown, the actual gate trigger current for an SCR may vary greatly. In Figure 7–14, R_3 serves as **shunt resistor,** the purpose of which is to negate the effect of widely varying values of I_{GT}. This is accomplished by ensuring that the value of I_{R_3} is much greater than 200 μA, the maximum value of I_{GT} specified in the data sheets. Assuming I_{R_3} to be at least 10 times greater than I_{GT}, the value of resistor that satisfies this condition is easily selected as follows:

$$I_{R_3} = 10 \times I_{GT}(max) = 10 \times 200 \ \mu A = 2 \ mA$$

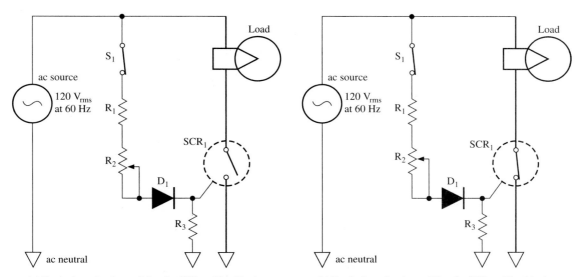

(a) Equivalent circuit condition for SCR and load in the off state: v_i of $SCR_1 = v_i$ of source; v_i of load = 0 V

(b) Equivalent circuit condition for SCR and load in the on state: v_i of $SCR_1 = v_F$; v_i of load $\cong v_i$ of source

Figure 7–15

Using the maximum specified V_{GT} of 0.8 V, a nominal value of R_3 may now be calculated:

$$R_3 = \frac{V_{GT}(max)}{I_{R_3}} = \frac{0.8 \text{ V}}{2 \text{ mA}} = 400 \text{ }\Omega \text{ (Use 390 }\Omega \text{ as standard value.)}$$

Since I_{R_3} and I_{GT} both originate at node A in Figure 7–14, at the instant of triggering, the current flowing through R_1, R_2, and D_1 must equal ($I_{GT} + I_{R_3}$). Thus, at the triggering point:

$$I_{R_3} + I_{GT} = \frac{0.8 \text{ V}}{390 \text{ }\Omega} + 200 \text{ }\mu\text{A} = 2.05 \text{ mA} + 200 \text{ }\mu\text{A} = 2.25 \text{ mA}$$

Also at the instant of triggering, the instantaneous value of source voltage (abbreviated V_i) is determined as follows:

$$V_i = (V_{GT} + V_{D1}) + [(I_{GT} + I_{R_3}) \times (R_2 + R_1)]$$

Thus,

$$V_i = (0.8 \text{ V} + 0.7 \text{ V}) + [2.25 \text{ mA} \times (R_2 + 3.3 \text{ k}\Omega)]$$

From this above equation, it becomes evident that the instantaneous value of source voltage at which triggering occurs is a function of the ohmic setting of potentiometer R_2.

Example 7–4

For the circuit in Figure 7–14, assume R_2 is adjusted to 27.3 kΩ. At what instantaneous value of source voltage may triggering of the SCR be expected to occur? What would be the firing-delay angle?

Solution:

Step 1: Solve for V_i:

$$V_i = 1.5 \text{ V} + [2.25 \text{ mA} \times (27.3 \text{ k}\Omega + 3.3 \text{ k}\Omega)] = 70.35 \text{ V}$$

Step 2: Solve for the firing-delay angle:

$$V_{peak} = 1.414 \times 120 \text{ V}_{rms} = 169.7 \text{ V}$$

$$\sin\theta = \frac{V_i}{V_{peak}} = \frac{70.35 \text{ V}}{169.7 \text{ V}} = 0.4146$$

$$\text{Firing-delay angle} = \theta = \text{nearly } 24.5°$$

Figure 7–16 is a graphic representation of how the SCR is expected to operate with R_2 set as specified in Example 7–4. At t_0, with the source waveform at 0 V, the SCR is in the off condition, equivalent to that shown in Figure 7–15(a). The lamp remains de-energized until t_1, where the instantaneous amplitude of the source waveform approaches that required for triggering. At this point, V_{AK} falls to V_F, allowing virtually all the source potential to develop over the lamp. The lamp remains energized until a few degrees before the 180° point, where I_A falls below the level of I_H. **Commutation** then occurs, with the SCR again behaving as an open switch. After t_2, of course, the SCR is reverse biased, blocking the flow of current to the lamp.

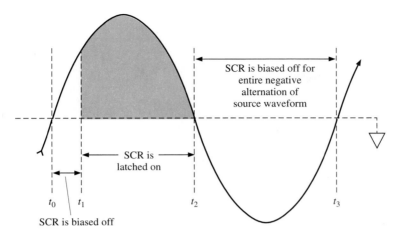

Figure 7–16
Operation of SCR in Figure 7–14

The actual difference in time between t_0 and t_1 in Figure 7–16 may be easily determined, once the firing-delay angle is known. For the 60-Hz input waveform:

$$T = \frac{1}{f} = \frac{1}{60 \text{ Hz}} = 16.667 \text{ msec}$$

$$\Delta t = \frac{\theta}{360°} \times T = \frac{24.5°}{360°} \times 16.667 \text{ msec} \cong 1.13 \text{ msec}$$

For the circuit in Figure 7–14, if R_2 is adjusted to 0 Ω, the firing-delay angle is reduced to only 3°. Thus, power is consumed by the lamp during virtually all the positive half-cycle. However, if R_2 is set at too great an ohmic value, the gate voltage will be held below the level of V_{GT}. If the SCR is not triggered by the 90° point in the source waveform it will not conduct at all.

This problem of limited control range may be remedied through use of **phase-shift control,** as illustrated in Figure 7–17. Before the point at which the SCR is triggered, no current would be flowing through D_1 and R_3. With this condition, R_1, R_2, and C_1 could be considered as a simple series RC circuit. Not until V_{C1} attains a level equal to $[V_{GT} + V_{D1} + (I_{GT}(\text{max}) \times R_3]$ can the SCR be expected to switch on. At this instant, the gate-cathode junction of the SCR would become forward biased, allowing the capacitor to discharge through R_3 and D_1.

Since the source frequency for the circuit in Figure 7–17 is a constant 60 Hz, the capacitive reactance of C_1 is also constant and is calculated as follows:

$$X_{C_1} = \frac{1}{2\pi f C} = \frac{1}{2\pi \times 60 \text{ Hz} \times 2.2 \text{ μF}} = 1.206 \text{ k}\Omega$$

With R_2 adjusted to 0 Ω, the conduction angle for the SCR is at its minimum value. With this condition, X_{C_1} is much greater than R_1, and total impedance for the RC network is only slightly greater than the reactance of C_1:

$$Z_T = \sqrt{R_1^2 + X_{C_1}^2} = \sqrt{330 \ \Omega^2 + 1.206 \text{ k}\Omega^2} = 1.25 \text{ k}\Omega$$

Figure 7–17
SCR lamp-dimmer control circuit with RC phase-shift control

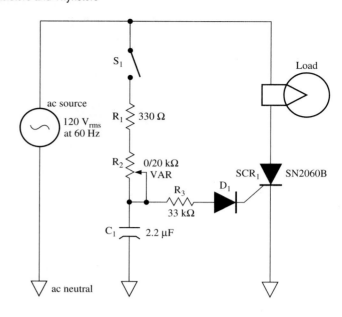

The impedance triangle representing this circuit condition could be graphed as shown in Figure 7–18(a). The phase angle for the RC network would then be determined as follows:

$$\sin\theta = \frac{-X_{C_1}}{Z_T} = \frac{-1.206\ \text{k}\Omega}{1.25\ \text{k}\Omega} = -0.965$$

$$\theta \cong -75°$$

The angle just derived *is not* the firing-delay angle for the SCR. Rather, this is the angle by which resistive voltage and current within the RC network lead the source voltage. Since capacitive current leads capacitive voltage by 90°, the capacitive voltage lags behind that of the source. Where the difference in phase between the source and capacitive voltage waveforms is represented as ϕ:

$$\phi = \theta + 90° = -75° + 90° = 15°$$

This angle ϕ still does not represent the firing-delay angle, but only the point at which the capacitive waveform crosses the zero reference line. Recall that the SCR is not triggered until the capacitor voltage attains the following threshold level:

$$V_{C_1} = V_{GT} + V_{D1} + [I_{GT}(\text{max}) \times R_3] = 0.8\ \text{V} + 0.7\ \text{V} + (200\ \mu\text{A} \times 33\ \text{k}\Omega) = 8.1\ \text{V}$$

If the SCR were not present in the circuit, with R_2 adjusted to 0 Ω, V_{C_1} would attain a peak amplitude calculated as follows:

$$V_{C_1(\text{peak})} = \frac{X_{C_1}}{Z_T} = \frac{1.206\ \text{k}\Omega}{1.25\ \text{k}\Omega} \times 169.7\ \text{V} = 163.7\ \text{V}$$

With R_2 set to 0 Ω, the switching threshold is attained less than 3° beyond the point where the capacitive waveform crosses the reference line. This additional delay angle is determined as follows:

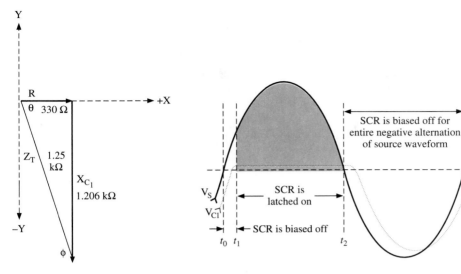

(a) Impedance triangle for minimum conduction angle

(b) Source and capacitive waveforms occurring with conduction angle at minimum

Figure 7–18

$$\sin\theta = \frac{V_{C_1}}{V_{C_1(\text{peak})}} = \frac{8.1 \text{ V}}{163.7 \text{ V}} = 0.0495$$

$$\theta \text{ (for additional delay)} = 2.8°$$

Finally, the minimum conduction angle for the SCR may be determined, representing the sum of ϕ and the additional delay angle:

$$\text{Firing-delay angle} = 15° + 2.8° \cong 18°$$

The source and capacitive waveforms representing this minimum firing-delay condition are contained in Figure 7–18(b). Here, as indicated by the shaded area, the power transfer to the load is maximum. At t_1, where the SCR is triggered into forward conduction, the source waveform is already at an instantaneous amplitude of nearly 50 V, being determined as follows. For the firing-delay angle:

$$\sin\theta = \sin 18° = 0.309$$
$$V_i = 0.309 \times 169.7 \text{ V} = 52.44 \text{ V}$$

The time delay between t_0 and t_1 is also easily calculated:

$$\Delta t = \frac{18°}{360°} \times 16.667 \text{ msec} \cong 833 \text{ μsec}$$

For the circuit in Figure 7–17, when R_2 is adjusted to its maximum ohmic value, the firing-delay angle for the SCR is also at maximum. With R_2 now at 20 kΩ, X_{C_1} is much lower than the series resistance within the RC network. The triangle representing this impedance condition is contained in Figure 7–19(a).

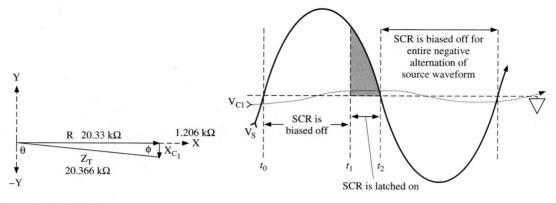

(a) Impedance triangle for maximum conduction angle

(b) Source and capacitive waveforms occurring with conduction angle at maximum

Figure 7–19

Total impedance for the RC network is now determined as follows:

$$Z_T = \sqrt{20.33 \text{ k}\Omega^2 + 1.206 \text{ k}\Omega^2} = 20.366 \text{ k}\Omega$$

The angles ϕ and θ are determined next:

$$\sin\theta = \frac{-1.206 \text{ k}\Omega}{20.366 \text{ k}\Omega} = -0.0592$$

$$\theta = -3.395°$$

$$\phi = -3.4° + 90° = 86.6°$$

With this maximum impedance condition for the RC network, if the SCR were removed from the circuit, the theoretical peak amplitude developed over the capacitor would be much less than with the minimum impedance condition. Solving for $V_{C_1(\text{peak})}$:

$$V_{C_1(\text{peak})} = \frac{1.206 \text{ k}\Omega}{20.366 \text{ k}\Omega} \times 169.7 \text{ V} = 10.05 \text{ V}$$

With the maximum impedance condition, the delay between the point where the capacitive waveform crosses the reference line and the point where the SCR is triggered is considerable, being calculated as follows:

$$\sin\theta = \frac{8.1 \text{ V}}{10.05 \text{ V}} = 0.806$$

$$\theta \cong 54°$$

The maximum firing-delay angle is now determined:

$$\text{firing-delay angle} = 86.6° + 54° \cong 140°$$

The source and capacitive waveforms occurring with this maximum firing-delay condition are contained in Figure 7–19(b). As indicated by the shaded area, power transfer to the load is now at a minimum. At t_1, where the SCR is finally triggered into forward conduction, the instantaneous voltage of the source waveform is determined as follows:

$$\sin\theta = \sin 140° = 0.643$$

Finally, the difference in time between t_0 and t_1 is calculated:

$$\Delta t = \frac{140°}{360°} \times 16.667 \text{ msec} = 6.48 \text{ msec}$$

SECTION REVIEW 7-2

1. For the circuit in Figure 7–20, what is the function of R_3? Explain why such a relatively small ohmic value was selected for this component.
2. What is the maximum value of current that could be expected to flow through R_3 in Figure 7–20?
3. What is the purpose of D_1 in Figure 7–20?
4. If, in Figure 7–20, S_1 is opened, can we safely assume that the load will remain de-energized? Give a specific reason for your answer.
5. If, in Figure 7–20, R_1 is adjusted to 397 Ω, what would be the firing-delay angle? What would be the instantaneous value of the source voltage at the triggering point?
6. True or false?: If, in Figure 7–20, R_1 is set to its maximum ohmic value, it is likely that the load will remain de-energized. (Explain the reason for your answer.)
7. True or false?: For the circuit in Figure 7–20, the 50-Ω load resistor may safely be replaced with a resistor of only 10 Ω. (Explain the reason for your answer.)
8. What is the primary advantage of the circuit in Figure 7–21 over that in Figure 7–20?
9. What type of control is represented by the circuit in Figure 7–21?
10. For the circuit in Figure 7–21, what voltage must be present at point A to trigger the SCR into forward conduction?

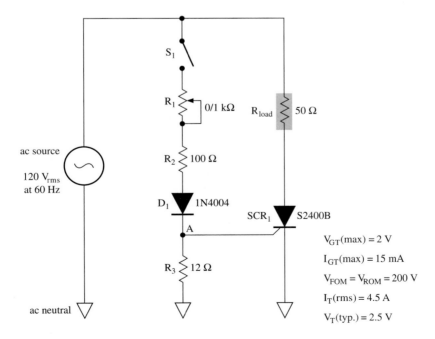

Figure 7–20

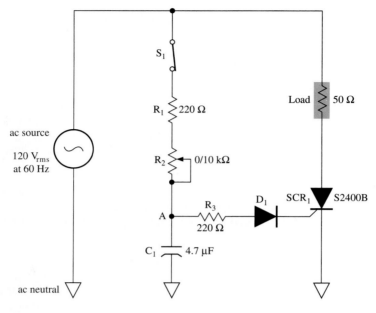

Figure 7–21

11. For the circuit in Figure 7–21, what is the minimum firing-delay angle? What instantaneous value of source voltage corresponds with this delay angle?
12. For the circuit in Figure 7–21, what is the maximum firing-delay angle? What instantaneous value of source voltage corresponds with this delay angle?

7-3   APPLICATION OF UJTs IN SCR CONTROL CIRCUITRY

If sufficient forward voltage is present between the anode and cathode of an SCR, forward conduction may be induced by a quick positive-going current pulse at the gate terminal. Due to the wide variations possible for the value of dc gate trigger current, this method of triggering is the most reliable, since it represents much greater control of the conduction angle than either of the simple control circuits represented thus far. Gate current could be held at virtually 0 A until the desired instant of triggering. At that point, a rapid pulse of current could be provided at the gate terminal of the SCR, inducing the flow of main-terminal current. Even though the gate voltage and current would quickly fall toward zero, the SCR would remain latched on until commutation occurred near the 180° point of the source waveform. The UJT is an ideal device for performing the triggering function just described. A simple SCR control circuit containing a UJT is shown in Figure 7–22.

The operation of the UJT gate control circuitry in Figure 7–22 is similar to that of a relaxation oscillator. As the UJT switches into forward saturation, a trigger pulse develops over R_4, resulting in forward conduction of the SCR. A few degrees into the positive alternation of the source waveform, a constant 12-V level is established at point A. The approximate point in the cycle where this level is attained may be calculated as follows:

$$\sin\theta = \frac{12\ V}{169.7\ V} = 0.07071$$

$$\theta \cong 4°$$

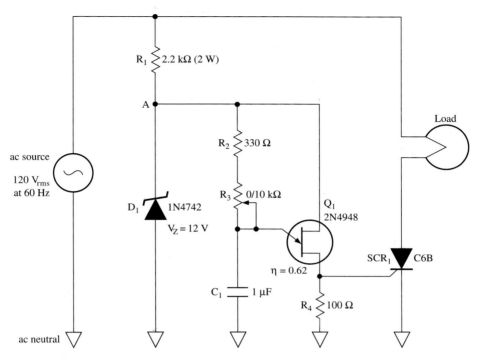

Figure 7–22
Improved SCR control circuit containing a UJT

This 12-V_{dc} level, which is so quickly established at point A, becomes the target voltage for C_1, which attempts to charge to this potential during the positive alternation of the source waveform. The rate of this charging process is determined by the ohmic setting of R_3. Since the intrinsic-standoff ratio of the UJT is nearly equal to 0.632, the switching threshold of V_P is attained in nearly one RC time constant. This voltage may be approximated as follows:

$$V_P \cong \eta \times V_Z \cong 0.62 \times 12 \text{ V} \cong 7.44 \text{ V}$$

Before this level of capacitor voltage is attained, the UJT is operating in the cutoff region of its characteristic curve. With this condition, the 2N4948 UJT would have a minimal R_{BB} of 4 kΩ. Thus the voltage being developed over R_4 would be much lower than the minimum level of V_{GT} for the SCR. This value of V_{R_4} is determined as follows:

$$V_{R_4} = \frac{100 \text{ }\Omega}{4.1 \text{ k}\Omega} \times 12 \text{ V} = \text{less than } 0.3 \text{ V}$$

Once V_{C_1} approaches the threshold level of 7.44 V, the UJT begins to operate in the negative resistance portion of its characteristic curve. It is at this point that the trigger pulse is developed over R_4. Figure 7–23 illustrates the waveforms produced by the control circuit in Figure 7–22. Note that, with R_3 set at a relatively low ohmic value, several trigger pulses may occur during a given positive alternation. However, since the first pulse is the one that latches on the SCR, any subsequent pulses are insignificant. Note that the voltage at point A goes to −0.7 V during the negative alternation of the source waveform.

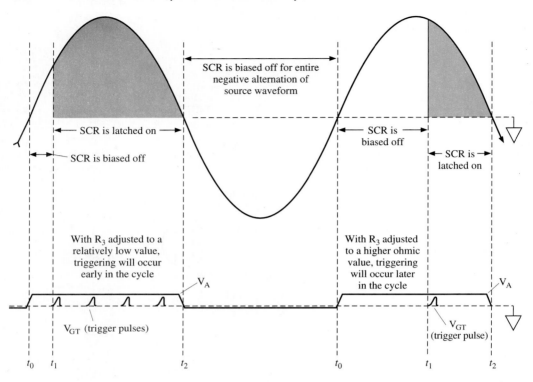

Figure 7–23

This is because the Zener diode is now forward biased. This condition is desirable since it effectively clamps the UJT control circuitry in an off state until the subsequent positive alternation.

Example 7–5

For the circuit in Figure 7–22, assume R_3 has been adjusted to 5.62 kΩ. What firing-delay angle does this ohmic setting represent?

Solution:

Step 1: Solve for the RC time constant:

$$\tau = (R_1 + R_3) \times C_1 = (5.62 \text{ k}\Omega + 330 \ \Omega) \times 1 \ \mu\text{F} = 5.95 \text{ msec}$$

Step 2: Convert the difference in time represented by the time constant to electrical degrees:

$$\frac{\tau}{T} \times 360° = \frac{5.95 \text{ msec}}{16.667 \text{ msec}} \times 360° = 128.5°$$

Step 3: Determine the approximate firing-delay angle by adding the previous result to the offset number of degrees required to attain V_Z:

$$\text{Firing-delay angle} = 128.5° + 4° \cong 133°$$

SECTION REVIEW 7–3

1. Explain the primary advantage of using a UJT in SCR gate control circuitry.
2. Describe the operation of D_1 in Figure 7–22 during both the positive and negative alternations of the source waveform.
3. Assuming R_3 in Figure 7–22 is set to 4.38 kΩ, approximate the firing-delay angle for the SCR. What is the instantaneous amplitude of the source waveform at the instant the SCR is triggered?

7–4   DC SWITCHING APPLICATION OF AN SCR

In many industrial applications, thyristors are substituted for mechanical and electrome-chanical switching devices. The SCR, in particular, is well suited for dc switching functions. As a solid-state device, the SCR has the advantage of being rapidly latched on by a single control pulse at its gate. However, a disadvantage of using an SCR as a dc switch lies in

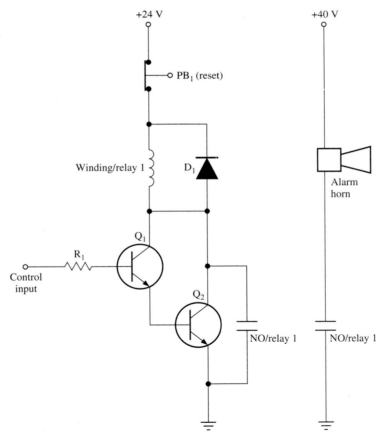

(a) Simple alarm horn switching circuit containing
Darlington pair and electromechanical relay

Figure 7–24

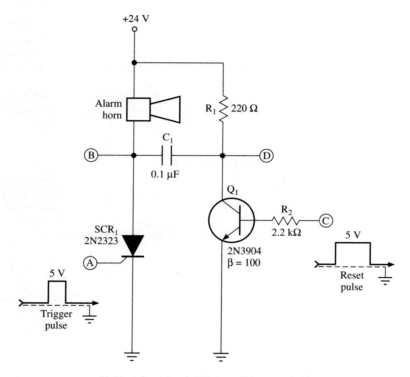

(b) Alarm horn circuit utilizing solid-state switching

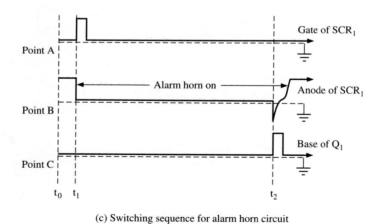

(c) Switching sequence for alarm horn circuit

Figure 7–24 (continued)

the fact that commutation (turn off) must be forced by causing forward current to fall below its holding level. Once this occurs, the SCR switches off and remains off until the next control pulse arrives at its gate.

Figure 7–24(a) illustrates a simple method of activating an alarm horn through use of transistor and electromechanical switching. Here, a strobe pulse arriving at the base of the Darlington pair initiates the flow of current through the relay winding, which serves as the collector load. If PB_1 is in its closed position when a strobe pulse occurs, the coil of the relay is energized. With normally open contacts being placed in parallel with the Darlington pair, the relay coil remains latched on after the strobe pulse has disappeared, allowing the alarm horn to continue sounding. Only when PB_1 is pressed does the relay drop out, turning off the alarm horn. Diode D_1 serves as a kickback diode, protecting the Darlington pair from the transient response of the inductor at the instant PB_1 is pressed. As the electromagnetic magnetic field around the coil collapses, induced current is shunted through the diode.

Figure 7–24(b) illustrates how the switching sequence just described might be accomplished by totally electronic means. The switching sequence for this circuit is shown in Figure 7–24(c). If, at t_0, both the SCR and the transistor switch are biased off, 24 V would be present at both points B and D. At this time, C_1 would be held in a discharged condition. At t_1, when a strobe pulse arrives at point A, the alarm horn is activated as the voltage at the anode of the SCR falls to the level of V_F. With Q_1 still in a cutoff condition, C_1 instantly acquires a potential equivalent to the applied voltage minus V_F of the SCR. This charge over C_1 would be positive with respect to point D.

Since the SCR in Figure 7–24(b) is now being used in a dc switching application, it is able to latch on indefinitely. The SCR does not turn off until **forced commutation** occurs. This action results from a positive-going pulse at point C in Figure 7–24(b). If R_1 is of sufficient ohmic value to ensure saturation of Q_1, point D falls to virtually 0 V at the arrival of the commutating pulse at point C. As represented at t_2 in Figure 7–24(c), the sudden drop in potential at point D momentarily clamps the voltage at point B to nearly $-(V_{source} - V_F) + V_{CE}(sat)$. This momentary $-V_{AK}$ condition instantly stops the flow of forward current through the main terminals of the SCR, thus resulting in commutation. With the SCR switched off and Q_1 again in a cutoff state, V_{B-D} returns to 0 V and the alarm horn remains off until another trigger pulse arrives at point A.

Example 7–6

For the switching circuit in Figure 7–24(b), if the SCR is an S2061Y, what would be the voltage developed over C_1 after the SCR is latched on? What would be the voltage at point B at the instant a high pulse arrives at point C?

Solution:

Step 1: According to the data sheets on the S2061 SCR, maximum V_T (or V_F) could be 2.2 V. Thus, the minimum charge attained by the capacitor would be determined as:

$$V_{C_1} = 24\ V - 2.2\ V = 21.8\ V$$

Step 2: Assuming the saturation voltage of Q_1 to be 0.2 V, the voltage at point B at the instant of forced commutation becomes:

$$V_B = -(24\ V - 2.2\ V) + 0.2\ V = -21.6\ V$$

SECTION REVIEW 7-4

---

1. Explain the advantages of the solid-state switching circuit in Figure 7–24(b) over the equivalent electromechanical circuit in Figure 7–24(a).
2. Define the term *forced commutation*. How is this action accomplished for the circuit in Figure 7–24(b)? Explain the function of capacitor C_1 during this process.

7-5   SILICON-CONTROLLED AND GATE TURN-OFF SWITCHES

The **silicon-controlled switch** (SCS) is a relatively low-power thyristor that could be considered as an SCR with two gates. Figure 7–25(a) shows the equivalent circuit for the SCS, while Figure 7–25(b) contains its schematic symbol. Note that the transistor configuration for the SCS is identical to that of the SCR, with the addition of the anode gate terminal. Because of its flexibility, the SCS is used prevalently in pulse circuits, especially digital control circuitry.

The major advantage of the SCS lies in the fact that forward current may be controlled either by signals at the **cathode gate** or the **anode gate.** As illustrated in Figure 7–26, with a forward voltage present from the anode to the cathode of the SCS, forward current could be induced by a positive-going pulse at the cathode gate. The device would then remain latched on until a negative-going pulse occurred at the same terminal. A complementary switching sequence could also occur at the anode gate, producing the same result. Main terminal current would be induced by a negative-going pulse, whereas a positive-going pulse would terminate this current flow.

The SCS is ideally suited for dc switching applications. As illustrated in Figure 7–27(a), an active-high strobe pulse from a digital circuit could be differentiated at the cathode gate, controlling a load placed in series with the main terminals of the SCS. The quick positive-going gate pulse, which occurs on the rising edge of the control signal, is sufficient for latching on the SCS. The negative-going spike, occurring on the falling edge of the control signal, interrupts the flow of forward current and thus forces commutation. Figure 7–27(b) illustrates an alternative method for controlling the same load. An active-low pulse, now present at the anode gate of the SCS, induces the flow of forward current through the SCS. A positive spike present at the same terminal interrupts the flow of main terminal current, resulting in commutation.

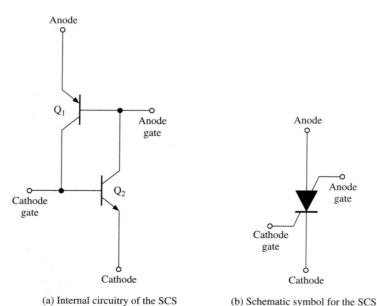

(a) Internal circuitry of the SCS   (b) Schematic symbol for the SCS

Figure 7–25

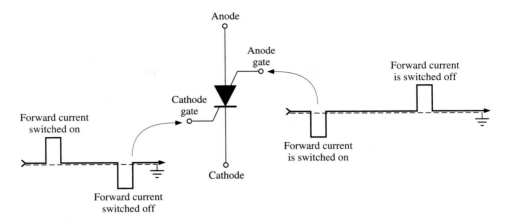

Figure 7–26
SCS switching pulses

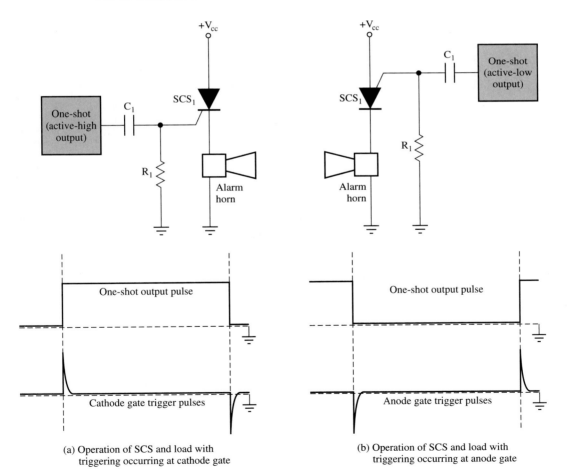

(a) Operation of SCS and load with
triggering occurring at cathode gate

(b) Operation of SCS and load with
triggering occurring at anode gate

Figure 7–27

The **gate turn-off switch** (GTO) is a form of SCR that has the capability of being triggered on and switched off by control pulses at a single gate terminal. As with the conventional SCR, the device is latched on by a positive-going current pulse at its gate. In addition, a negative-going pulse at the gate terminal stops the flow of forward current. Two possible schematic symbols for the GTO are contained in Figure 7–28. (Note that the symbol in Figure 7–28(a) clearly illustrates the capability of bidirectional gate current.) A design compromise within the structure of the GTO is that, compared to a conventional SCR, relatively large values of positive and negative gate current are required to control the device. Therefore, the GTO is limited to low- and medium-power applications. However, the GTO's capability of commutation by means of negative gate current gives the device a distinct advantage over the conventional SCR for high-speed dc switching applications. Remember, as illustrated in Figure 7–24(b), for a conventional SCR used in a dc switching application, commutation must be forced through use of a negative current pulse at its anode terminal. An attempt to interrupt the flow of forward current through a negative pulse at the gate of the conventional SCR would almost certainly destroy that device.

Figure 7–29(a) illustrates an application of solid-state switching within an automotive ignition system. Here, a simple transistor switch and a GTO are being used to control the transfer of a high-voltage pulse to the spark plugs. In earlier model automobiles, this task was performed by mechanical points, which required frequent replacement. The dc power source for this circuit is, of course, the automobile's battery, which is assumed to be providing 12 V_{dc}. As illustrated by the waveforms in Figure 7–29(b), Q_1 operates in either saturation or cutoff, being controlled by the current pulse occurring at its base. These timing pulses at point A are generated by a proximity sensor (possibly in the form of a Hall-effect switch) connected to the rotor shaft. Operating in conjunction with L_1 and C_1, Q_1 controls the development of trigger pulses at the gate of the GTO. When operating properly, the GTO rapidly switches on and off at a rate determined by the angular velocity of the rotor. Consequently, the GTO is able to control the rate at which high-voltage pulses occur over the ignition coil.

The following analysis of the operation of the circuit in Figure 7–29(a) should aid in the understanding of how gate pulses are developed for the GTO. Assume first that Q_1 is in the cutoff state while the GTO is conducting. At this time, point B would be at 12 V while point C, representing the gate voltage of the GTO, would be estimated at nearly 1 V. Thus, the instantaneous voltage acquired by C_1 as a difference in potential

Figure 7–28

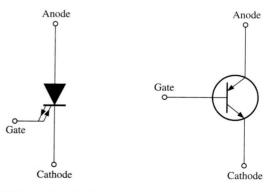

(a) Schematic symbol for
gate turn-off switch

(b) Alternative schematic
symbol for the GTO

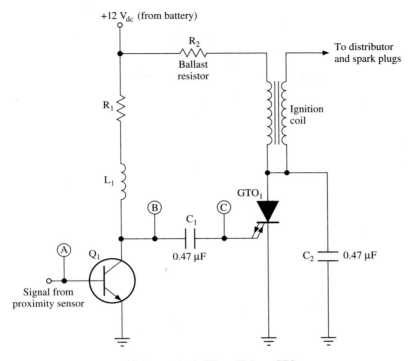

(a) Automotive ignition utilizing a GTO

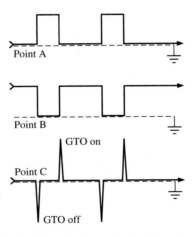

(b) Waveforms for automotive ignition system

Figure 7–29

between points B and C would be determined as follows:

$$V_{C_1} = V_{BB} - V_{GT} = 12 \text{ V} - 1 \text{ V} = 11 \text{ V}$$

Also, on the subsequent positive transition at point A, as Q_1 switches on in saturation, energy is stored by the inductor in the form of an electromagnetic field. Due to the relatively low ohmic value of R_1, the inductor voltage approaches nearly 11 V. When Q_1 is again biased off, the GTO is triggered on and the ignition coil is energized. It is important to note that, at this instant, the electromagnetic field of L_1 collapses and its energy is transferred to the electrostatic field of the capacitor. The counter EMF produced by the inductor aids the charge already present over the capacitor, allowing V_{C_1} to attain nearly 22 V. Thus, at the instant Q_1 goes into saturation, the voltage present at the gate of the GTO is about -22 V. Associated with this negative peak gate voltage is a negative gate current that, while lasting only a few microseconds, could be as high as -3 A. It is this reverse gate current that interrupts the flow of main-terminal current for the GTO. On the next negative transition at point A, as Q_1 goes into cutoff, a positive-going current pulse is again induced at point C allowing forward conduction of the GTO.

SECTION REVIEW 7–5

1. What type of switching device is formed by R_1 and C_1 in Figures 7–27(a) and (b)? Why is this device necessary for the proper operation of the SCS?
2. If an SCS is to be controlled by a signal at its anode gate, what must be the direction and polarity of the trigger signal at the instant of turn on? What must be the direction and polarity of the gate signal at the instant of commutation?
3. Explain the purpose of L_1 and C_1 in Figure 7–29(a). Why are these devices necessary for the reliable operation of the GTO?
4. What is the major advantage of the GTO over the conventional SCR? What limitation of the GTO is associated with this design advantage?

7–6   INTRODUCTION TO TRIACs

During ac operation, the major limitation of the SCR is its inability to conduct during the negative alternation of the source sine wave. The need for control of an ac load throughout the full 360° of the source cycle has led to the development of the TRIAC. The internal structure of the TRIAC is illustrated in Figure 7–30(a), and its equivalent circuitry is shown in Figure 7–30(b). The schematic symbol for the device, contained in Figure 7–30(c), indicates the ability of the TRIAC to conduct main-terminal current in two directions.

Since the structure and operation of the SCR have already been examined in detail, the function of the TRIAC may be readily understood. As represented in Figures 7–30(a) and (b), the TRIAC consists of a complementary pair of SCRs that share a common gate. Note that instead of having a cathode and an anode as its main terminals, the TRIAC has two anodes, designated A_1 and A_2.

Referring to Figure 7–30, assume that the A_2 terminal is made positive with respect to A_1. As with a single SCR, as long as the gate voltage remains at 0 V, only a small reverse leakage current flows between the main terminals of the device. However, if a positive-going trigger pulse occurs at the gate, the internal SCR consisting of Q_3 and Q_4 is latched on. Forward current continues to flow until the difference in potential between the two anodes drops to nearly 0 V. Meanwhile, the SCR consisting of Q_1 and Q_2 would have remained in the off state, functioning as an open switch. Now assume that gate terminal is again at 0 V while the A_1 terminal is made positive with respect to A_2. With this condition, the TRIAC is again in the off state, opposing the flow of main-terminal current in either direction. If, however, another current pulse occurs at the gate terminal, the internal SCR consisting of Q_1 and Q_2 is triggered on, with the opposite SCR remaining in the off state.

The characteristic curve for a TRIAC is shown in Figure 7–31. Note that in either

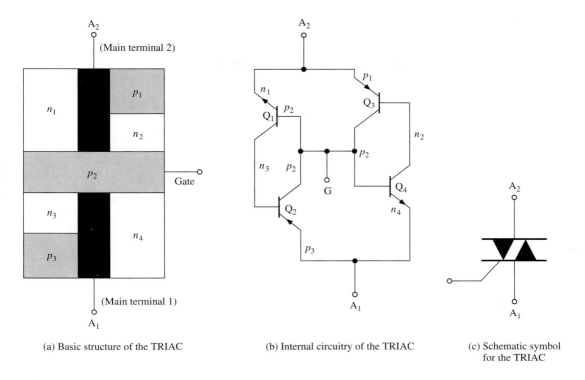

(a) Basic structure of the TRIAC   (b) Internal circuitry of the TRIAC   (c) Schematic symbol for the TRIAC

Figure 7–30

the forward or reverse portion of the characteristic curve, with I_{GT} at 0 A, breakover does not occur until V_A exceeds the level of $+V_{DROM}$ or $-V_A$ falls below the level of $-V_{DROM}$. For the TRIAC, $+V_{DROM}$ and $-V_{DROM}$ represent the peak-forward and peak-reverse blocking voltages. Exceeding either of these voltage levels would not necessarily destroy the TRIAC. However, this action would represent a temporary loss of control during circuit operation. Once proper triggering or breakover does occur, the main-terminal voltage of the TRIAC falls to a **peak-forward conduction** level (designated V_{TM}). This parameter is equivalent to V_F for the SCR. However, for the TRIAC, V_{TM} occurs in either the positive or negative direction. The TRIAC has a maximum rating of forward and reverse main-terminal current (designated I_T), which is usually specified as an rms value. Ideally, the specified values of V_{DROM}, V_{TM}, and I_T are the same for both SCRs within a given TRIAC. Thus, if a data sheet specifies a V_{DROM} of 200 V and a V_{TM} of 2 V, it is implied that these parameters are valid for both positive and negative conduction of the TRIAC.

Example 7–7

A TRIAC is to be used in controlling an ac load that requires 220 V_{rms} and is expected to dissipate 195 W. It is found that T2316A, T2316B, and T2316D TRIACs are available in stock. Which of these devices, if any, can be safely used in this control application?

Solution:

Step 1: Solve for the required V_{DROM}:

$$V_{peak} = 1.414 \times V_{rms} = 1.414 \times 220 \text{ V}_{rms} \cong 311 \text{ V}$$

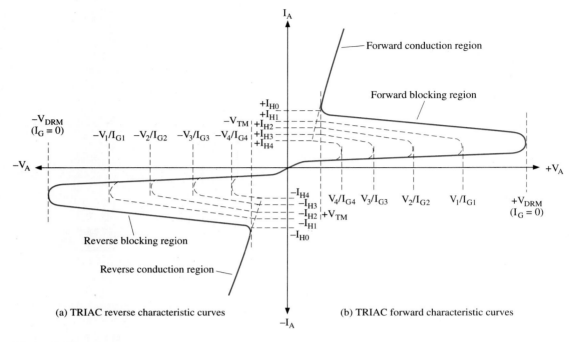

(a) TRIAC reverse characteristic curves   (b) TRIAC forward characteristic curves

Figure 7–31

Only the T2316D, with a V_{DROM} of 400 V, satisfies this requirement.

Step 2: Determine if the T2316D meets the current requirement.

$$I_{rms} = \frac{P}{V_{rms}} = \frac{195 \text{ W}}{220 \text{ V}_{rms}} = 886.4 \text{ mA}_{rms}$$

According to the data sheets, the T2316D TRIAC has an I_T of 2.5 A_{rms} and would, therefore, be suitable for the given application.

Figure 7–32 contains a control circuit that incorporates the switching technique introduced in Figure 7–22 (the use of a UJT to control the firing-delay angle). In Figure 7–32, however, with the TRIAC, control of the load is extended to nearly the full 360° of the source cycle. The circuit in Figure 7–32 also introduces the use of a pulse transformer for development of trigger signals at the gate of the TRIAC. The pulse transformer provides total isolation between the low-power UJT control circuit and the relatively high-power TRIAC and load circuit. Also, the pulse transformer removes the dc component from the current pulse developed at the B_1 terminal of the UJT. Diodes D_1, D_2, D_3, and D_4 form a bridge rectifier. These four components allow a fully rectified pulsating dc waveform to develop at point A. Through the action of R_1 and the Zener diode, the potential developed between point B and the dc ground is limited to the V_Z level of 12 V. This voltage is attained very early in the positive alternation of the source cycle, near the point where D_3, D_5, and D_2 begin to conduct:

$$V_i \cong V_{D_3} + V_{D_5} + V_{D_2} \cong 0.7 \text{ V} + 12 \text{ V} + 0.7 \text{ V} \cong 13.4 \text{ V}$$

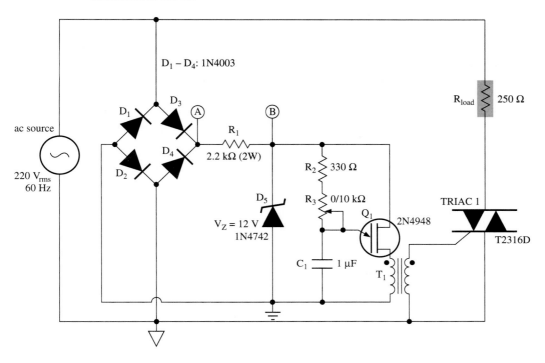

Figure 7–32
TRIAC control circuit with UJT and pulse transformer-type triggering

$$\sin\theta = \frac{13.4 \text{ V}}{311.1 \text{ V}} = 0.043$$

$$\theta \cong 2.5°$$

At nearly the same instant in the negative alternation, where D_4, D_5, and D_1 begin to conduct, 12 V would again be present between point B and the dc ground.

Through adjustment of potentiometer R_3, the firing-delay angle is made variable for both the positive and negative alternations of the source cycle. Thus, if R_3 is adjusted such that triggering of the TRIAC occurs at the 35° point in the positive alternation, triggering should also occur again near the 215° point (35° into the negative alternation). As indicated by the waveforms in Figure 7–33, several trigger pulses could occur during either alternation of the cycle, depending on the setting of R_3. However, similar to the SCR circuit in Figure 7–22, the TRIAC responds only to the first trigger pulse within either alternation. These initial pulses latch the device on until commutation occurs automatically near the 0° and 180° points.

Example 7–8

For the circuit in Figure 7–32, assume that the intrinsic-standoff ratio of the UJT is 0.63. Also assume that R_3 has been adjusted to 6.8 kΩ. Given these values, determine the firing-delay angle for both the positive and negative alternations of the source cycle. Also determine the instantaneous voltages of the source waveform at both of these trigger points.

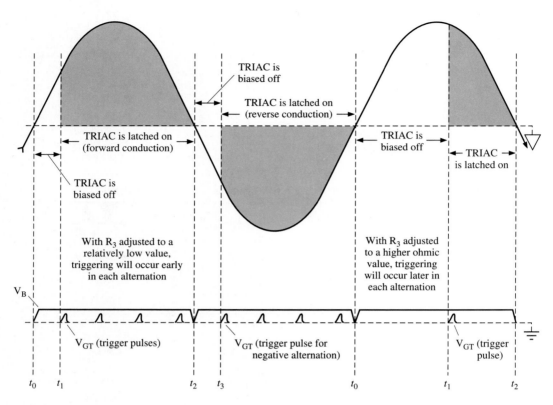

Figure 7–33

Solution:

Step 1: Solve for the firing-delay angles: With an η value nearly equal to 0.632, Δt could be assumed to equal the time constant τ. Thus, during the positive alternation,

$$\Delta t \cong \tau \cong (R_2 + R_3) \times C_1 \cong (330\ \Omega + 6.8\ k\Omega) \times 1\ \mu F \cong 7.13\ msec$$

$$\theta = \frac{\Delta t}{T} \times 360° = \frac{7.13\ msec}{16.667\ msec} \times 360° = 154°$$

For the negative alternation:

$$\text{Firing-delay angle} \cong (180° + \theta) \cong (180° + 154°) \cong 334°$$

Step 2: Solve for instantaneous voltages:

$$+V_i = 311\ V \times \sin 154° = 311\ V \times 0.438 \cong 136\ V$$
$$-V_i = 311\ V \times \sin 334° = 311\ V \times -0.438 \cong -136\ V$$

SECTION REVIEW 7–6

1. Describe the basic structure and operation of the TRIAC. What is the primary advantage of this device over the SCR in controlling an ac load?

2. For the circuit in Figure 7–32, what must be the setting of R_3 to achieve firing-delay angles of 35° and 215°?

3. With regard to Review Problem 2, what would be the instantaneous amplitude of the source waveform at the instant the TRIAC is triggered during the positive alternation? What would be the instantaneous amplitude of the source waveform at the instant of triggering during the negative alternation?

7–7   UNILATERAL BREAKOVER DEVICES

Although untriggered avalanche breakdown is to be avoided in the operation of either the SCR or TRIAC, this breakover effect represents the normal mode of operation of such devices as the **four-layer diode** and the **silicon unilateral switch** (SUS). Both of these thyristors are referred to as **unilateral breakover devices** because they are designed to go safely into avalanche breakdown in only one direction. Unilateral breakover devices do have a reverse breakdown potential, the absolute value of which is much greater than the **forward breakover voltage.** Reverse breakdown is likely to be destructive to unilateral breakover devices and should, therefore, be avoided in the design of the circuitry in which they are contained.

The four-layer diode, also known as the **Shockley diode** (after its inventor William Shockley), has only two terminals, whereas the SUS has two main terminals and a gate. Use of the gate is not required for operation of the SUS. However, the gate terminal does allow for the adjustment of the breakover voltage of the device. The breakover potential of a four-layer diode, sometimes referred to as the **forward-switching voltage** (designated V_S) is not adjustable. However, the fact that four-layer diodes are available in a wide variety of breakover potentials makes up for this limitation. Also, the four-layer diode is better suited than the SUS for high-power applications. The operation of the four-layer diode is essentially that of an SCR without a gate terminal. When a minimum value of V_S is attained and a minimum value of forward-switching current (designated I_S) is flowing, breakover occurs, causing main-terminal voltage to fall to a minimum forward potential. Like an SCR, the four-layer diode remains switched on in saturation until forward current falls below a minimum holding level. Figure 7–34(a) represents the internal structure of the four-layer diode and (b) shows its schematic symbol. Figure 7–34(c) shows the equivalent circuitry of the SUS and (d) shows the schematic symbol for the device.

The SUS behaves as an open switch until a threshold voltage (V_{BO}) is attained between its two main terminals. At that point, the breakover effect occurs and the device begins to conduct rapidly. This action is similar to the negative-resistance characteristic that occurs between the emitter and B_1 terminals of the UJT. This effect is illustrated in Figure 7–35, which contains the characteristic curve for an SUS. Note that, before V_{BO} is attained, only a small leakage current flows between the main terminals of the breakover device. When breakover does occur, the main-terminal voltage suddenly decreases to a saturation level as current rapidly increases. As shown on the characteristic curve, the difference in potential between the breakover voltage and the saturation (on-state) voltage is referred to as the **breakback voltage.** For example, if an SUS has a V_{BO} of 8 V and a $+V_{sat}$ of 1 V, the breakback voltage would be 7 V. Because of their ability to produce a sudden surge of current at a predictable voltage level, breakover devices are ideally suited for use in the gate triggering circuitry of SCRs.

To understand how the SUS operates, both with and without the gate terminal connected, the internal operation of the device should first be analyzed. Referring to the equivalent circuit for this device, contained in Figure 7–34(c), consider a condition in which the gate terminal is left unconnected. For breakover to occur, the anode-to-cathode

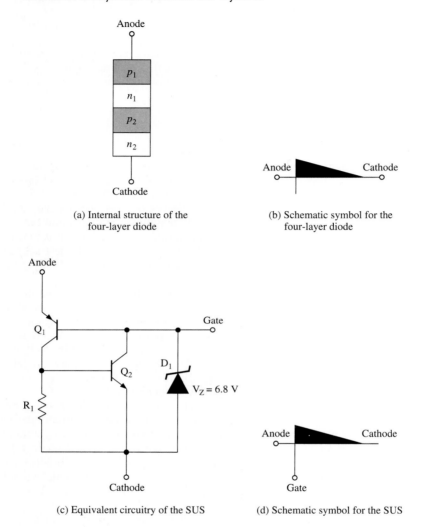

(a) Internal structure of the
four-layer diode

(b) Schematic symbol for the
four-layer diode

(c) Equivalent circuitry of the SUS

(d) Schematic symbol for the SUS

Figure 7–34

voltage must exceed the combined value of V_{BE} for Q_1 and V_Z of D_1. Once this voltage level is exceeded, Q_1 and D_1 begin to conduct. As a result, as Q_2 switches on into saturation, the anode-to-cathode potential of the SUS breaks back to a level equivalent to the combined values of V_{BE} for Q_1 and $V_{CE}(sat)$ for Q_2. With the Zener voltage of D_1 equaling 6.8 V and V_{BE} of Q_1 being nearly 0.7 V, V_{BO} for the SUS may be expected to be less than 8 V. With Q_1 switching on in saturation when breakover does occur, $+V_{sat}$ for the SUS may be expected to be slightly less than 1 V. Thus, the breakback potential for the device is estimated at nearly 7 V.

Now consider a condition in which a 3.3-V Zener diode is placed, as shown in Figure 7–36(a), between the gate and cathode terminals of the SUS. With this external diode effectively in parallel with the internal diode D_1, breakover is expected to occur when the difference in potential between the main terminals of the SUS approaches 4 V (V_Z

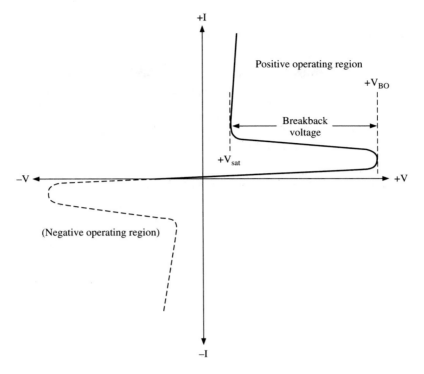

Figure 7–35
Characteristic curve for the SUS

of the external diode plus V_{BE} of Q_1). An alternative method of using the gate terminal of the SUS is shown in Figure 7–36(b). With this example, as anode voltage increases above ground potential, Q_1 becomes forward biased as base current begins to flow through the gate terminal. Breakover is expected to occur when the anode potential (with respect to the ground) approaches 1 V. The operation of the SUS in this example is quite similar to that of an SCR, with breakover voltage becoming a function of the gate current.

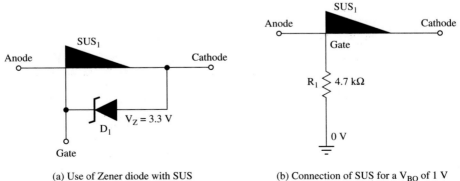

(a) Use of Zener diode with SUS   (b) Connection of SUS for a V_{BO} of 1 V

Figure 7–36

Figure 7–37(a) illustrates a possible application of a unilateral breakover device, serving as the switching element within a relaxation oscillator. The two output waveforms produced by this circuit are shown in Figure 7–37(b). Although this form of circuit is commonly associated with the UJT or PUT, both the four-layer diode and the SUS exhibit the negative-resistance characteristic necessary to function in this capacity. Recall that, with a UJT relaxation oscillator, the threshold voltage V_C is equivalent to V_P, which is, in turn, a function of both the supply voltage and the intrinsic-standoff ratio. For the SUS in Figure 7–37(a), however, since the breakover voltage is a constant value inherent to the device, V_C also becomes constant, regardless of deviations in the supply voltage.

At the instant power is applied to the circuit in Figure 7–37(a), C_1 behaves as a dead short and point A is at 0 V. Prior to the time at which V_A attains the level of V_{BO}, the SUS exhibits a high impedance to current flow, thus allowing C_1 to charge via R_1. Capacitor C_1 attempts to charge toward the level of the supply voltage. However, once V_A approaches V_{BO}, breakover occurs and the SUS begins to operate in the negative-resistance region of its characteristic curve. During this time, the SUS rapidly discharges via the main terminals of the SUS through R_2. The voltage at point A continues to decrease until current flow through the SUS and R_2 falls below a minimum holding level. At this point, the SUS is biased off and again exhibits a high impedance to source current. Thus C_1 again charges via R_1 until V_{BO} is again attained. As with the UJT relaxation oscillator, a sawtooth (exponential) waveform develops between point A and the ground. Also, a pulse develops over R_2 during the time in which the SUS is operating in the negative-resistance region.

As with all other pulse circuits presented thus far in this text, the period of the output signal of the SUS relaxation oscillator may be derived using Eq. (6–1):

$$t = RC \times \ln\left(\frac{V - V_0}{V - V_C}\right)$$

In applying this formula to the circuit in Figure 7–37(a), R would be equivalent to R_1 and C would be equivalent to C_1. Although V still represents the supply voltage, V_C

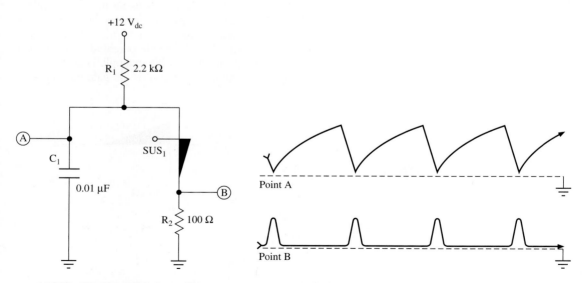

(a) SUS utilized in a relaxation oscillator   (b) Output waveforms for SUS relaxation oscillator

Figure 7–37

becomes V_{BO} of the SUS and V_0 is equivalent to $+V_{sat}$ of the SUS. Thus, Eq. (6–1), when applied to the SUS relaxation oscillator, may be transformed as follows:

$$PW \cong (R_1 \times C_1) \times \ln \left(\frac{V_{source} - V_{sat}}{V_{source} - V_{BO}} \right) \qquad (7-13)$$

Example 7–9

For the circuit in Figure 7–37(a), assume V_{BO} of the SUS is 7.8 V, while $+V_{sat}$ equals 1.2 V. What would be the pulse width of the sawtooth waveform? What would be the approximate frequency of this signal?

Solution:

Step 1: Solve for pulse width:

$$PW = 2.2 \text{ k}\Omega \times 0.01 \text{ } \mu F \times \ln \left(\frac{12 \text{ V} - 1.2 \text{ V}}{12 \text{ V} - 7.8 \text{ V}} \right) = 20.8 \text{ } \mu sec$$

Step 2: Solve for frequency:

$$f \cong \frac{1}{20.8 \text{ } \mu sec} \cong 48 \text{ kpps}$$

SECTION REVIEW 7-7

1. Describe the basic operation of a unilateral breakover device. How does the operation of such a device differ from that of an SCR or TRIAC?
2. What form of circuit is contained in Figure 7–38?
3. For the SUS in Figure 7–38, assume that V_{BO} is 9.3 V and V_{sat} is 1.2 V. Also assume that the 4.7-V Zener diode is disconnected from the circuit at this time. What is the pulse width of the signal taken at point A? What is the approximate frequency of that signal?
4. Using the information given in Review Problem 3, determine the breakback voltage of the SUS.
5. Now assume that the jumper is inserted into the circuit in Figure 7–38. What would be the pulse width and approximate frequency of the signal at point A?

7-8   BILATERAL BREAKOVER DEVICES

Bilateral breakover devices are thyristors that can be triggered safely in both the forward and reverse directions. Such devices include the **silicon bilateral switch** (SBS), the **bilateral four-layer diode,** and the **DIAC.** Because of their ability to produce a triggering pulse in either a positive or negative direction, bilateral breakover devices are ideally suited for use in TRIAC power-control circuitry. As stated earlier, unilateral breakover devices do have a reverse breakover voltage. However, this reverse potential is much greater than the forward V_{BO}. Most bilateral devices, by contrast, are designed symmetrically, in that the absolute values of their forward and reverse breakover voltages are nearly the same.

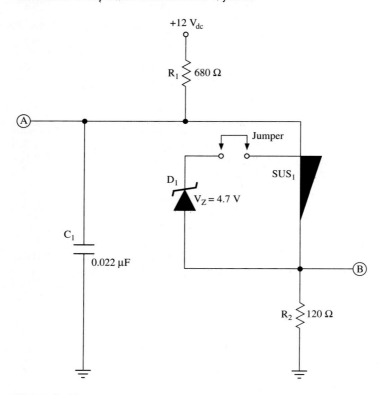

Figure 7–38

Figure 7–39(a) shows the internal circuitry of the SBS and (b) shows its schematic symbol. The SBS, like its unilateral counterpart the SUS, has a gate terminal, which may be used to alter the breakover potential of the device. Ideally, with the gate unconnected, the forward and reverse breakover voltages of the SBS are the same absolute value, typically around 8 V. In actuality, variations of 0.2 to 0.5 V may occur in this **absolute switching voltage.**

The operation of the SBS may be best understood through analysis of the operation of its internal circuitry. Referring to Figure 7–39(a), assume initially that no voltage is present between the two main terminals of the SBS. Next, assume that anode 1 is becoming increasingly positive with respect to anode 2. With this condition, the emitter-base junction of transistor Q_1 and diode D_1 are able to function as forward-biased diodes, while Q_3 and Q_4 remain in a reverse-biased state. Diode D_2 functions as a Zener diode, assumed here to have a V_Z of 6.8 V. Until the voltage at anode 1 approaches the level of $(V_{BE1} + V_{Z2})$ with respect to anode 2, Q_2 and D_2 remain reverse biased, allowing virtually no current to flow through the main terminals of the SBS. As this difference in potential attains a threshold level of nearly 7.5 V, the base-emitter junction of Q_1 and D_1 become forward biased and D_2 begins to operate in its Zener region. However, the flow of base current in Q_1 causes collector current to flow, which results in the flow of base current for Q_2. Thus, the main terminal voltage of the SBS breaks back sharply as Q_1 and Q_2 now conduct in saturation. The saturation voltage now present between the main terminals of the SBS is likely to be much less than 1 V.

Referring again to Figure 7–39(a), assume that the voltage at anode 1 is now becoming negative with respect to anode 2. With this circuit condition, Q_1 and Q_2 remain in a

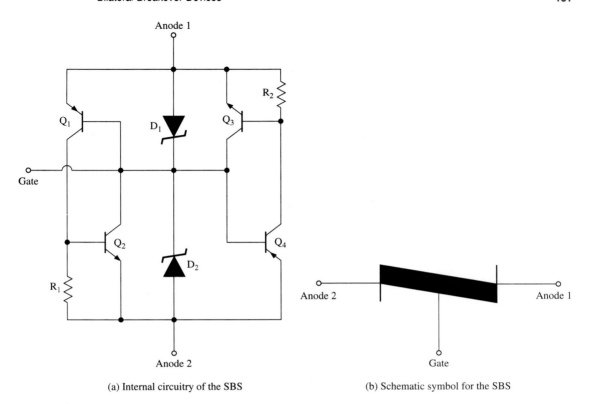

(a) Internal circuitry of the SBS

(b) Schematic symbol for the SBS

Figure 7–39

reverse-biased state. However, the base-emitter junction of Q_4 and D_2 will now be enabled to function as forward-biased diodes. With D_1 now functioning as a 6.8-V Zener diode, the flow of current from anode 2 to anode 1 will be virtually blocked until the difference in potential between the main terminals approaches $-(V_{BE4} + V_{Z1})$. At this instant D_1 begins to operate in its Zener region, allowing base and collector current to flow through Q_4. As a result, base current is able to flow for Q_3. Thus, breakover now occurs in the negative direction, as Q_3 and Q_4 both conduct in saturation.

A possible connection for the gate terminal of the SBS is shown in Figure 7–40(a). The resultant change in behavior of the device is illustrated in Figure 7–40(b). After referring to Figure 7–39(a), it becomes evident that the external Zener diode is in parallel with the internal Zener diode D_2. Since the external diode has a much lower Zener voltage of 3.3 V, forward breakover is forced to occur at around 4 V (equivalent to $V_{BE1} + V_{Zext}$). In reverse operation, however, the external diode has no effect on the SBS. Referring again to Figure 7–39(a) helps visualize why this is so. As the voltage at anode 1 increases negatively with respect to anode 2, both the external Zener diode and D_2 eventually become forward biased. Meanwhile, D_1 still operates in its Zener region, allowing for the full reverse breakover potential of around -7.5 V.

Figure 7–41(a) contains the schematic symbol for the bilateral four-layer diode and (b) shows the schematic symbol for the DIAC. The DIAC, which is also referred to as a **symmetrical trigger diode** or a **bilateral trigger diode,** may be considered simply as an ungated TRIAC. This is true regarding both the internal structure and the operating characteristics of the device. The DIAC behaves identically to other breakover devices

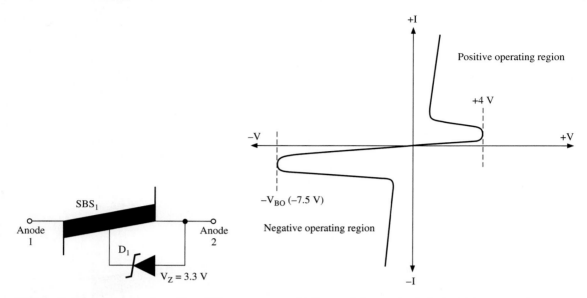

(a) Zener diode used in conjunction with an SBS

(b) Characteristic curve for SBS with 3.3 V Zener diode connected between the gate and anode 2

Figure 7–40

in that its negative resistance characteristic allows a sudden surge of either forward or reverse current, accompanied by a sharp drop in forward or reverse voltage.

The advantages of the DIAC are its excellent temperature stability and its symmetrical triggering characteristic. Thus the DIAC is ideally suited for use within industrial environments where there are likely to be wide ranges of operating temperature. Also, if a DIAC with a designated V_{BO} of 32 V was to be used in a trigger circuit, the absolute value of the reverse breakover voltage would be within 1 V of the forward breakover potential (no more than 33 V but no less than 31 V). DIACs with such high breakover voltages are often found in high-power TRIAC control circuitry. Since such circuits dissipate considerable heat, the excellent temperature stability of the DIAC makes it particularly suitable for such applications.

Figure 7–42 contains an economical form of lamp-dimmer circuit in which the gate-triggering of a TRIAC is controlled by a DIAC. Through the action of the RC network

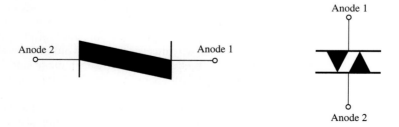

(a) Schematic symbol for the bilateral four-layer diode

(b) Schematic symbol for the DIAC

Figure 7–41

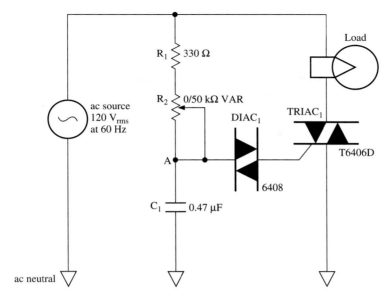

Figure 7–42
Lamp-dimmer control circuit utilizing a DIAC to trigger a TRIAC

consisting of R_1, R_2, and C_1, the voltage at point A lags behind the source sine wave. The degree of phase shift is controlled by R_2. An increase in its ohmic value lengthens the time delay between the source waveform and the capacitive waveform developed at point A. Decreasing the ohmic value of R_2 shortens this time delay.

Example 7–10

For the circuit in Figure 7–42, assume that R_2 is adjusted to its maximum ohmic value. What would be the difference in phase between the source waveform and the waveform at point A? Assume the absolute switching voltage of the DIAC is 32 V. What would be the condition of the load at this time?

Solution:

Step 1: Solve for delay angle:

$$R_1 + R_2 = 330\ \Omega + 50\ \text{k}\Omega = 50.33\ \text{k}\Omega$$

$$X_{C_1} = \frac{1}{2\pi \times f \times C_1} = \frac{1}{2\pi \times 60\ \text{Hz} \times 0.47\ \mu\text{F}} = 5.644\ \text{k}\Omega$$

$$\sin\theta = \frac{X_{C_1}}{Z_T} = \frac{-5.644\ \text{k}\Omega}{\sqrt{5.644\ \text{k}\Omega^2 + 50.33\ \text{k}\Omega^2}} = -0.1114$$

$$\theta = -6.4°$$

$$\phi = \theta + 90° = -6.4° + 90° = 83.6°$$

Step 2: Solve for V_A(peak):

$$V_A(\text{peak}) = \sin\theta \times V_{\text{source}}(\text{peak}) = 0.1114 \times 169.7\ \text{V} = 18.91\ \text{V}$$

Since V_A must attain nearly 33 V for triggering to occur (V_{BO} of DIAC + V_{GT} of TRIAC), the load will remain de-energized for the entire source cycle.

The two waveforms in Figure 7–43(a) represent the circuit condition analyzed in Example 7–10, where $\pm V_A$ is much less than $\pm V_{BO}$ of the DIAC. An uninterrupted sine wave appears at point A, trailing the source waveform by nearly 84°. If the ohmic value of R_2 is slowly decreased, the amplitude of the waveform at point A would gradually increase, eventually allowing the peak amplitude of the waveform at A to approach the level of $V_{BO} + V_{GT}$. At this instant, breakover of the DIAC occurs, allowing the TRIAC to be triggered and the load to be energized.

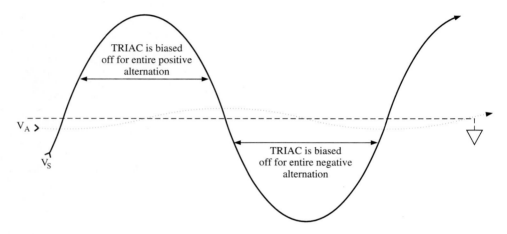

(a) Waveforms representing a circuit condition where V_A is lower than V_{BO} of the DIAC

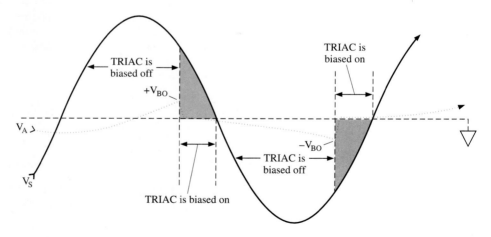

(b) Waveforms representing a circuit condition where V_A begins to attain a high enough level to switch on the DIAC

Figure 7–43

Example 7–12

Assume the ohmic value of R_2 in Figure 7–42 is gradually decreased from its maximum value. At what firing-delay angle would the TRIAC first be able to conduct?

Solution:

Step 1: Solve for ϕ, which is the angle by which V_A lags behind the source waveform:

$$\sin\theta = \frac{-(V_{BO} + V_{GT})}{V_{S(peak)}} \cong \frac{-33 \text{ V}}{169.7 \text{ V}} \cong -0.19446$$

$$\theta = -11.2°$$

$$\phi = \theta + 90° \cong -11.2° + 90° \cong 78.8°$$

Step 2: Solve for the actual firing-delay angle of the TRIAC: With the maximum firing-delay angle, the breakover voltage of the DIAC is assumed to be the plus or minus peak amplitude of the waveform at point A. Thus, conduction of the TRIAC would not begin until 90° beyond the point where V_A crosses the zero reference line.

$$\text{Firing-delay angle} = \phi + 90° = 78.8° + 90° \cong 169°$$

During the negative alternation, when reverse breakover of the DIAC occurs:

$$\text{Firing-delay angle} \cong 169° + 180° \cong 349°$$

The circuit condition analyzed in the Example 7–11 is represented by the waveforms of Figure 7–43(b). As the ohmic value of R_2 is continually decreased, the DIAC is able to break over earlier in each alternation of the source waveform. As a result, the lamp glows with greater intensity.

A problem commonly associated with the simple form of the lamp-dimmer circuit shown in Figure 7–42 is that of **flash-on,** the technical term for which is **hysteresis.** As was just explained, when the ohmic value of R_2 is slowly decreased from its maximum value, a point is eventually reached where the DIAC is enabled to break over and trigger the TRIAC into conduction. Assume that the TRIAC is first triggered on during the positive alternation of the source cycle. Due to the negative-resistance characteristic of the DIAC, the capacitor is able to discharge rapidly through the DIAC into the gate terminal of the TRIAC and, thus, have a head start charging negatively. As a result of this action, the reverse breakover potential for the DIAC is reached much earlier in the negative alternation, allowing the TRIAC to be triggered much sooner than in the previous positive alternation. Consequently, the capacitor now has a head start at charging in the positive direction, allowing the TRIAC to be triggered sooner during the next positive alternation. Because of the effect just described, no matter how slowly the wiper of R_2 is turned, the lamp flashes on. To dim the lamp, the ohmic value of the potentiometer must be increased **after** the initial flash-on occurs. Thus, a given ohmic setting for R_2 may produce two distinctly different intensities of light, depending on the direction by which this level of resistance was approached. Improved types of lamp dimmers, referred to as **hysteresis-free,** are commercially available. Due to its much lower value of breakover voltage, an SBS is likely to replace the DIAC in such a circuit, greatly reducing the flash-on effect. A hysteresis-free lamp dimmer is analyzed in detail in the following chapter.

So common is the application of the DIAC as a gate-triggering device for the TRIAC that the two components are made available within a single package. The internal structure

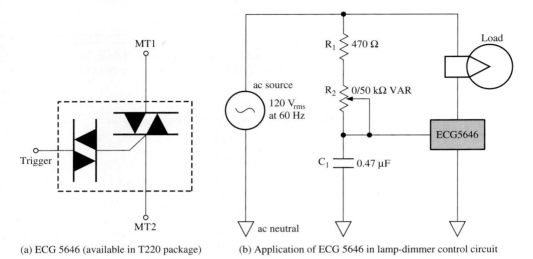

(a) ECG 5646 (available in T220 package)   (b) Application of ECG 5646 in lamp-dimmer control circuit

Figure 7–44

of such a device, the ECG 5646 is shown in Figure 7–44(a), and a possible application is shown in (b). One terminal of the internal DIAC is connected directly to the gate terminal of the TRIAC; the other input serves as the trigger input to the device. The ECG 5646 is designed for control of medium- to high-power loads. Since the V_{DROM} of the internal TRIAC is 600 V, it is compatible with high-voltage ac sources. Also, the main-terminal current of the ECG 5646 may be as high as 10 A_{rms}. While the absolute switching voltage of the DIAC within the ECG 5646 may be as high as 35 V, the gate current required for triggering of the internal TRIAC may be as low as 200 μA. Such high sensitivity for the trigger input of the device is extremely beneficial, since it allows for the use of compact, low-power RC circuitry in controlling firing delay.

SECTION REVIEW 7–8

1. Assume D_1 in Figure 7–40(a) is now a 4.7 V Zener diode. Using Figure 7–40(b) as a guide, sketch the new characteristic curve for the device.
2. True or false?: With a Zener diode connected as shown in Figure 7–40(a), the reverse breakover voltage (measured from anode 2 to anode 1) is no more than 0.7 V.
3. Define the term *absolute switching voltage* as it applies to the operation of a bilateral breakover device.
4. Identify two operating advantages of the DIAC.
5. List two alternative names for the DIAC.
6. For the circuit in Figure 7–44(b), assume the absolute switching voltage of the internal DIAC to equal 33 V and V_{GT} of the internal TRIAC to be 1 V. If R_2 is set to 7.3 kΩ, what would be the firing-delay angles for the positive and negative alternations of the source sine wave. At what instantaneous values of source voltage could conduction of the TRIAC occur?
7. Explain in detail the problem of flash-on as it pertains to the operation of the circuit in Figure 7–44(b). What is another term used to identify this problem?

7–9   LIGHT-ACTIVATED SCRs AND PHOTOTHYRISTORS

In the final section of this chapter, the types of thyristors that incorporate optoelectronic switching techniques into their basic operation are briefly surveyed. A logical starting point in this exercise is the **light-activated SCR** (LASCR). As indicated in Figure 7–45(a), which represents the physical structure of the LASCR, main-terminal current flow through the LASCR may be induced through a rise in light intensity at the gate-to-cathode *p-n* junction of the device. Light is able to strike this *p-n* junction through a transparent window, similar to that of the photodiode or phototransistor. Like a conventional SCR, this LASCR may also be triggered into forward conduction by a positive-going current pulse at its gate terminal. Analysis of the equivalent circuitry of the LASCR, contained in Figure 7–45(b), should help visualize how this dual operation is possible. The schematic symbol for the LASCR is shown in Figure 7–45(c).

At this juncture, we stress that the operation of the LASCR is distinctly different from that of the phototransistor. Remember that the phototransistor is a linear device, for which the flow of forward current is directly proportional to the intensity of light at the base input. However, since the LASCR is a thyristor and, therefore, nonlinear, its applications are limited to switching circuitry. The primary advantage of the LASCR is the fact that it may be triggered into forward conduction by a sudden pulse of light, remaining latched on in saturation until commutation occurs. A problem associated with the LASCR is that of temperature instability: The sensitivity of the device to light increases significantly with operating temperature.

Figure 7–46 serves to illustrate the operation of the LASCR. Connection of R_1 between the gate of the LASCR and circuit ground allows adjustment of the sensitivity of the device. Reduction of gate sensitivity, accomplished by reducing the ohmic value of R_1, may be necessary to inhibit spurious triggering of the load. A single pulse of light penetrating the window of the LASCR is capable of latching on the load. After this light pulse disappears, the load remains energized until the flow of forward current is interrupted through actuation of S_2.

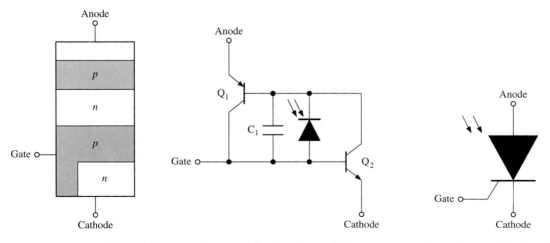

(a) Internal structure of the LASCR   (b) Internal circuitry of the LASCR   (c) Schematic symbol for the LASCR

Figure 7–45

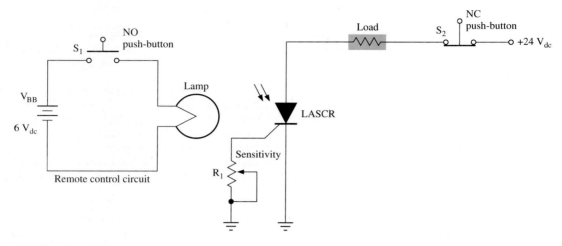

Figure 7–46
Simple on/off control utilizing a LASCR

Photothyristors are specialized forms of optoisolators, two basic forms of which are shown in Figure 7–47. The **MOC3003 photo-SCR optoisolator,** shown in Figure 7–47(a), contains a gallium-arsenide infrared LED as its emitter element, and it has an LASCR as the detector. Note that a gate terminal is provided with this device, to allow adjustment of sensitivity. The **MOC3009 photo-TRIAC driver** of Figure 7–47(b), like its SCR counterpart, contains a gallium-arsenide infrared LED as its emitter. The device is sometimes referred to as a **photo-TRIAC optoisolator.** However, this description, as well as the schematic representation in Figure 7–47(b), might be misleading. The thyristor functioning as the detector within the MOC3009 is not a TRIAC; nor is it a DIAC, as implied by the schematic symbol. Rather, the detector element is actually a silicon bilateral switch. When no current is flowing through the emitter LED of the MOC3009, the device is able to withstand nearly ±250 V between its main terminals without being forced into breakover. (The MOC3020, which is structurally identical to the MOC3009, is able to withstand a main-terminal difference in potential of nearly ±400 V.)

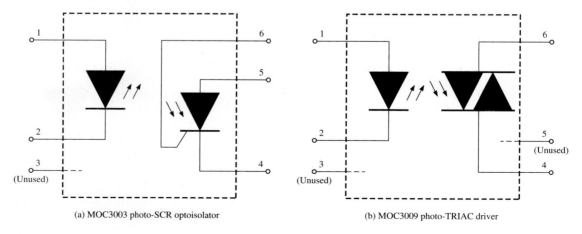

(a) MOC3003 photo-SCR optoisolator   (b) MOC3009 photo-TRIAC driver

Figure 7–47

Figure 7–48
Application of MOC3009 photo-TRIAC
driver

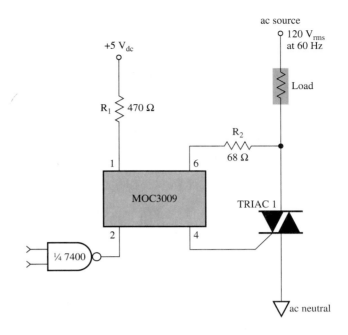

Figure 7–48 illustrates a simple application of the photo-TRIAC driver. Here, the primary advantage of this form of optoisolator is clearly shown. Since the **isolation voltage** (designated V_{iso}) between the emitter and detector of the device is nearly 7.5 kV, the low-power digital control circuitry is well protected from surges and noise spikes that are likely to occur in the high-power ac circuitry.

With a logic high present at the output of NAND gate in Figure 7–48 reverse-biased state, the flow of current to the infrared LED within the MOC3009 is inhibited. Consequently, the SBS within the device is unable to break over and no trigger current flows through R_2. With this condition, the TRIAC functions as an open switch, inhibiting the flow of current to the load. When the output of the NAND gate switches to a logic low, nearly 7.5 mA flows through the infrared emitter of the MOC3009. This current is sufficient for inducing breakover within the SBS on the detector side of the device, providing at least ±8 V is present between the main terminals of the TRIAC. Thus, in a simple on/off control application, as long as a logic low is present at the output of the NAND gate, the load is triggered on virtually at the beginning of each alternation of the source waveform.

SECTION REVIEW 7–9

1. Describe the internal structure and operation of the LASCR. Describe two ways by which the device might be triggered.

2. What is the purpose of R_1 in Figure 7–46? Why would this component be necessary when using a LASCR in an industrial environment?

3. True or false?: The detector element within the MOC3003 is a LASCR.

4. True or false?: The MOC3009 photo-TRIAC driver contains a light-sensitive TRIAC as its detector element.

5. Define the term *isolation voltage* as it applies to the operation of either the MOC3003 or MOC3009.

6. What is the primary advantage of using a MOC3009 to trigger a TRIAC, as opposed to using a UJT or an SBS?

SUMMARY

Unijunction transistors (UJTs) are nonlinear switching devices found in various forms of industrial control circuitry. Their negative-resistance characteristic makes them ideally suited for use in pulse circuits, a common example of which is the relaxation oscillator. Also, due to their ability to produce quick current pulses, they are often used to control the triggering of SCRs and TRIACs. A major problem associated with UJTs is that of temperature stability. Also, significant variation in the intrinsic-standoff ratio η for a given device further complicates design problems involving this device. The programmable UJT (PUT) offers greater performance stability than the conventional UJT, due primarily to its gate terminal, which allows the switching threshold of the device to be established by an external voltage source.

SCRs and TRIACs, along with various forms of unilateral and bilateral breakover devices, comprise a family of devices known as *thyristors*. SCRs may be used to control the transfer of power from either ac or dc sources to various industrial loads. SCRs are limited in that they perform half-wave rectification of the waveform from an ac power source. Therefore, actual control of a load by a single SCR is limited to only the positive alternation of the source cycle. TRIACs, which may be considered to be complementary pairs of SCRs within a single package, are better suited for the control of ac loads. Their design allows control to take place throughout the full 360° of a given source cycle.

Unilateral and bilateral breakover devices are thyristors that may be used to control the operation of SCRs and TRIACs. Unilateral breakover devices include the four-layer diode and the silicon unilateral switch (SUS), whereas bilateral breakover devices include the bidirectional four-layer diode, the silicon bilateral switch (SBS), and the DIAC.

The light-activated SCR (LASCR) is a versatile switching device that is triggered by either a positive-going current pulse at its gate or by an increase in light intensity. LASCRs serve as the detector elements within photo-SCR optoisolators. Photo-TRIAC drivers are optoisolators especially designed for triggering TRIACs. Due to their high isolation voltage, photo-TRIAC drivers allow direct control of high-power ac loads by low-power digital control circuitry.

SELF-TEST

1. Identify and briefly describe the operation of each of the devices represented in Figure 7–49.

2. For the circuit in Figure 7–9 assume R_6 is adjusted such that 8.3 V occurs between point C and the ground. Given this condition, what would be the peak amplitude of the output signal? What would be the approximate frequency of the output signal?

3. For the circuit in Figure 7–50, assume the actual value of gate trigger current for the SCR is 180 μA. Also assume that V_{GT} for the SCR and the forward voltage of the diode are both 0.7 V. If R_2 is adjusted to 12.5 kΩ, what would be the instantaneous value of source voltage at the point of triggering?

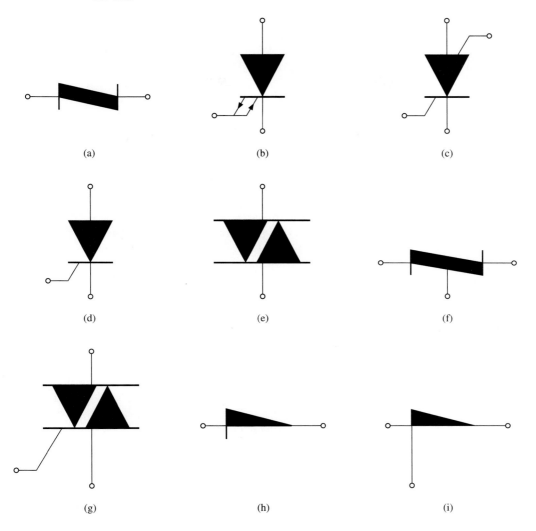

(a)   (b)   (c)

(d)   (e)   (f)

(g)   (h)   (i)

Figure 7–49

4. Assume the ohmic setting of R_2 in Figure 7–50 is still 12.5 kΩ. What would be the firing-delay angle for the SCR?
5. The purpose of the diode in Figure 7–50 is
   a. to provide temperature compensation for the SCR
   b. to protect the gate of the SCR from negative current flow
   c. to minimize the effects of deviations in the value of I_{GT}
   d. described correctly by choices b and c
6. The purpose of D_1 in Figure 7–51 is
   a. to provide a source of dc voltage for Q_1 during the positive alternation of the source waveform
   b. to provide temperature stabilization for Q_1
   c. to protect Q_1 during the negative alternation of the source waveform
   d. described correctly by choices a and c

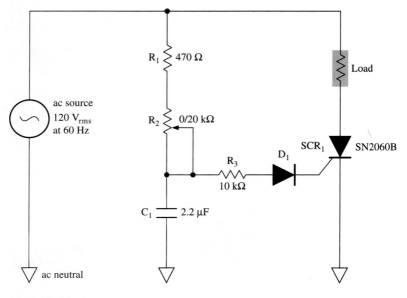

Figure 7–50

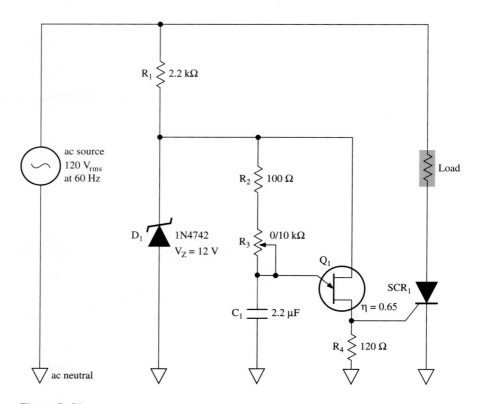

Figure 7–51

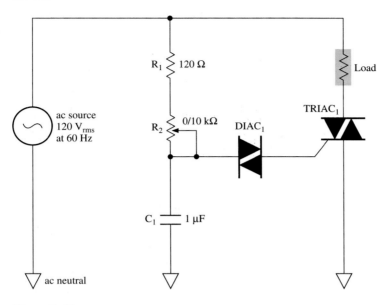

Figure 7–52

7. If r_{B2} for Q_1 in Figure 7–51 equals 4.83 kΩ, what are the ohmic values of R_{BB} and r_{B1}?

8. For the circuit in Figure 7–51, assume that R_3 is set to its maximum ohmic value. What would be the condition of the load?

9. For the circuit in Figure 7–51, assume that R_3 is adjusted to 2.4 kΩ. What would be the approximate firing-delay angle?

10. For the circuit in Figure 7–51, assume that R_3 is still set at 2.4 kΩ. What would be the instantaneous value of source voltage at the point of triggering?

11. For the circuit in Figure 7–51, the range of control
   a. is limited to the first 90° of the source cycle
   b. lies within the positive alternation of the source cycle
   c. extends the entire 360° of the source cycle
   d. is difficult to determine, due to the hysteresis problem associated with Q_1

12. Describe the operation of the GTO. What is the primary advantage of this device over a conventional SCR? What is an inherent weakness of this device?

13. Define flash-on. Why is this problem inherent to the circuit in Figure 7–52?

14. What is the advantage of using a DIAC in Figure 7–52, especially if the circuitry is subjected to wide deviations in temperature?

15. For the circuit in Figure 7–52, assume the breakover voltage of the DIAC is ±33 V and the gate trigger voltage of the TRIAC is ±1 V. Also assume that R_1 is set to 3.7 kΩ. Given these circuit conditions, what are the firing-delay angles for the TRIAC?

8 INTRODUCTION TO THYRISTOR CONTROL CIRCUITS

CHAPTER OUTLINE

LEARNING OBJECTIVES

On completion of this chapter, the student should be able to:

- Explain how a bridge rectifier can be used with an SCR for control of dc loads.
- Describe how an SCR and bridge rectifier operate within a battery charger.
- Describe the operation of a hysteresis-free lamp dimmer containing a TRIAC and an SBS.
- Explain the need for LC noise suppression in thyristor control circuitry.
- Explain the operation of a snubber network in thyristor control circuitry.
- Describe the basic operating concept of zero-point switching circuitry.
- Identify the advantages of zero-point switching circuitry.
- Describe the operation of a precision oven temperature controller using an SCR in an on/off zero-point switching configuration.
- Describe the operation of a temperature controller using a TRIAC and a pulse width modulator.
- Describe the internal structure and operation of a monolithic zero-point switching controller.
- Explain the operation of a digitally controlled firing-delay circuit containing a TRIAC.
- Identify the advantages of using digitally controlled firing-delay circuitry.

471

INTRODUCTION

In this chapter, several forms of thyristor control circuitry are analyzed in detail. Included in this survey are a silicon-controlled rectifier battery charger, a hysteresis-free lamp dimmer control circuit, two forms of zero-point switching circuitry, and a digitally controlled firing-delay circuit. Although the examples in this chapter have simple lamps and heating elements as their loads, they demonstrate control principles found in a wide variety of industrial systems. The problem of RF noise inherent to phase-shift control circuitry is discussed and techniques used to suppress such noise are analyzed.

8-1   DC CONTROL APPLICATIONS OF SILICON-CONTROLLED RECTIFIERS

Figure 8–1 illustrates how a silicon-controlled rectifier (SCR) might be used in conjunction with a bridge rectifier to provide proportional control for a load operating on pulsating dc. The bridge rectifier allows unidirectional current flow through the SCR and the load during both alternations of the source sine wave. This form of circuitry enhances the control capability of the SCR, since the device is now able to trigger twice during the source cycle. Capacitor C_1 allows control of the firing-delay angle to be extended throughout most of each alternation; stability of triggering is enhanced through use of the silicon unilateral switch (SUS) and the Zener diode.

Utilizing the procedures introduced in the previous chapter, minimum and maximum firing-delay angles for SCR_1 in Figure 8–1 may be estimated. With a source voltage of

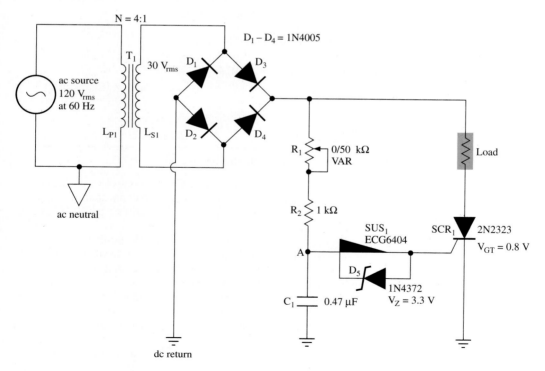

Figure 8–1
SCR used in conjunction with a bridge rectifier to control a dc load

30 V_{rms}, the absolute value of peak amplitude of each alternation is determined as follows:

$$V_{peak} = 1.414 \times 30 \text{ V}_{rms} = 42.42 \text{ V}$$

Allowing for the forward voltage drop of the diodes within the bridge rectifier, the peak amplitude of the pulsating dc voltage available to the SCR and load is no more than 41 V. The 3 V Zener diode connected between the gate and cathode of the SUS allows the device to break over at a forward potential of nearly 4 V. Thus, the instantaneous value of voltage that must be attained at point A for triggering to occur would be determined as follows:

$$V_i = V_{GT} + V_{BO} = 0.8 \text{ V} + 4 \text{ V} = 4.8 \text{ V}$$

The minimum firing-delay angle, representing a maximum power condition for the load, can now be determined as follows:

$$R = R_1 + R_2 = 0 \text{ }\Omega + 1 \text{ k}\Omega = 1 \text{ k}\Omega$$

$$X_C = \frac{1}{2\pi f C} = \frac{1}{2\pi \times 60 \text{ Hz} \times 0.47 \text{ }\mu\text{F}} = 5.644 \text{ k}\Omega$$

$$Z_T = \sqrt{R^2 + X_C^2} = \sqrt{1 \text{ k}\Omega^2 + 5.644 \text{ k}\Omega^2} = 5.732 \text{ k}\Omega$$

$$\sin\theta = \frac{-5.644 \text{ k}\Omega}{5.732 \text{ k}\Omega} = -0.98466$$

$$\theta \cong -80°$$

$$\phi = \theta + 90° \cong 10°$$

Since the charge over the capacitor must equal 4.8 V for breakover of the SUS, an additional delay angle must be determined. If the SCR were removed from the circuit, the peak amplitude of the capacitive waveform would be determined as follows:

$$V_{C_1}(\text{peak}) = \sin\theta \times 41 \text{ V} = 0.98466 \times 41 \text{ V} = 40.37 \text{ V}$$

The additional delay angle is then determined as follows:

$$\sin\theta = \frac{4.8 \text{ V}}{40.37 \text{ V}} = 0.1189$$

$$\theta = 6.8°$$

The minimum value of firing-delay angle for the circuit in Figure 8–1 is then determined:

$$\text{Firing-delay angle} = 10° + 6.8° = 16.8°$$

The pulsating waveforms developed over the SCR and the load during the minimum firing-delay condition are illustrated in Figure 8–2. With R_2 adjusted to its maximum ohmic value, no power is made available to the load. This may be proven as follows:

$$R = 50 \text{ k}\Omega + 1 \text{ k}\Omega = 51 \text{ k}\Omega$$

$$Z_T = \sqrt{51 \text{ k}\Omega^2 + 5.644 \text{ k}\Omega^2} = 51.311 \text{ k}\Omega$$

$$\sin\theta = \frac{-5.644 \text{ k}\Omega}{51.311 \text{ k}\Omega} = 0.11$$

$$\theta \cong -6.3°$$

$$\phi = -6.3° + 90° = 83.7°$$

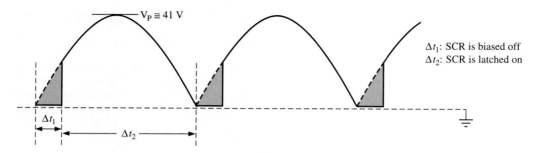

Δt_1: SCR is biased off
Δt_2: SCR is latched on

Minimum firing-delay condition for SCR and load

Figure 8–2

If the SCR is removed from the circuit, the maximum amplitude of the waveform developed over the capacitor would equal nearly 4.5 V, being determined as follows:

$$0.11 \times 41 \text{ V} = 4.51 \text{ V}$$

Since this peak value of V_{C_1} is less than the required 4.8 V threshold at point A, the SCR remains off and the load remains de-energized.

Figure 8–3 shows a battery charger that uses two SCRs. The operating principle of this circuit is similar to that introduced in Figure 8–1. As in Figure 8–1, a bridge rectifier is used to convert the ac source voltage to pulsating dc. This circuit could function without full-wave rectification, relying on the reverse-blocking characteristic of the SCRs to prevent negative voltage from developing over the battery. However, with full-wave rectification, the battery is charged more rapidly and more efficiently. An optional branch of the circuit, containing D_5 and the indicator lamp, can be used to indicate that the battery is approaching the fully charged condition. Diode D_5 would be necessary to prevent the battery from discharging as the source voltage drops to a level lower than the charge on the battery. If the indicator lamp is used, the transformer secondary voltage must be of sufficient magnitude to allow the lamp to glow brightly when the battery approaches its full charge. (During this condition, SCR_1 would be biased off.) If the circuit in Figure 8–4 is used to charge a 12-V (six-cell) battery, the minimum value of secondary voltage should be 20 V_{rms}. This would allow the battery to charge to a level of 13.8 V, with 6.2 V still available for the lamp and D_5. Charging a battery is a slow process, often taking several hours. Attempting to charge a battery too quickly, through use of a higher secondary voltage, is likely to destroy the battery.

The path for charging the battery is through SCR_1, which is triggered on by the flow of current through R_4 and D_6. When the battery voltage measured at point A (designated V_{BB}) is much lower than the target level of 13.8 V, SCR_1 switches on early in each alternation of the source waveform. SCR_2, however, remains off until the voltage at point B attains a level equal to the sum of V_{GT} of SCR_2 and V_{BO} of D_7. Figure 8–4(a) contains the equivalent circuit for the battery charger during the time when SCR_1 is active. Here, the internal resistance of the battery is represented simply as a load resistor.

The firing-delay angle for SCR_1 can be easily calculated. Assume the required I_{GT} of the 2N4444 to be 10 mA. The instantaneous amplitude of the rectified source waveform that must be attained before SCR_1 is able to trigger can then be determined as follows:

$$v_i = (I_{GT} \times R_4) + V_{D_6}$$
$$v_i = (10 \text{ mA} \times 680 \text{ } \Omega) + 0.7 \text{ V} = 7.5 \text{ V}$$

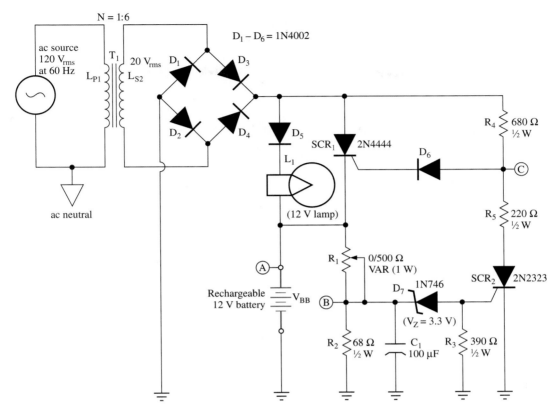

Figure 8–3
SCR-controlled battery charger

The peak voltage attainable over the SCR is estimated as:

$$V_{peak} = (1.414 \times 20 \text{ V}) - 1.4 \text{ V} \cong 26.9 \text{ V}$$

$$\sin\theta = \frac{7.5 \text{ V}}{26.9 \text{ V}} = 0.279$$

Firing-delay angle = 16.2°

Thus, as shown by the shaded area of the waveform in Figure 8–4b, during the early portion of the charging cycle, most of the source power is made available to the battery.

The function of R_1, C_1, and SCR_2 in Figure 8–3 must now be considered. Potentiometer R_1 allows the adjustment of a target voltage for charging the battery, and it should be adjusted to an ohmic value that allows a 12 V battery to charge toward a level of 13.8 V. The presence of C_1 in the circuit is important for two reasons. First, early in the charging process, it creates a phase shift, thus delaying the triggering of SCR_2. Because a dc voltage is slowly developing over the battery, a residual dc voltage is also developing over C_1. At any time, V_{C_1} could be determined as follows:

$$V_{C_1} = (V_{BB} - V_{R_1})$$

In order for SCR_2 to be triggered, the following circuit conditions must exist:

$$V_{C_1} = (V_{BB} - V_{R_1}) = (V_{GT} + V_{DZ})$$

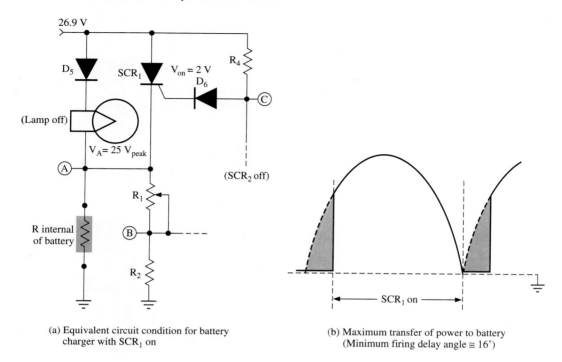

(a) Equivalent circuit condition for battery
charger with SCR_1 on

(b) Maximum transfer of power to battery
(Minimum firing delay angle $\cong 16°$)

Figure 8–4

If R_1 is adjusted properly, the triggering of SCR_2 will occur automatically. This happens whenever the dc charge on the battery attains 13.8 V. The second purpose of C_1 now becomes obvious. The growth of voltage over C_1 can be considered analogous to that developed over the battery. As the level of dc voltage increases over C_1, the firing-delay angle for SCR_2 decreases. Eventually, SCR_2 begins to switch on before SCR_1. Finally, the voltage over C_1 stabilizes at a level equal to $(V_{GT} + V_{DZ})$. This condition for C_1 allows SCR_2 to remain latched on, forcing SCR_1 to remain off. The conditions causing the commutation of SCR_1 are clearly shown in Figure 8–5, which illustrates the circuit condition during the time V_{BB} equals 13.8 V. With SCR_2 latched on, point C is held at a level of around 2 V. With point A remaining at around 13.8 V, D_6 is reverse biased, protecting the gate of SCR_1 from the flow of reverse current.

SECTION REVIEW 8–1

1. Describe the advantages of using a bridge rectifier in conjunction with an SCR when controlling a dc load.
2. What is the purpose of the SUS in Figure 8–1?
3. For the circuit in Figure 8–1, assuming R_1 is adjusted to 27 kΩ, what would be the firing-delay angle for the SCR?
4. Describe the basic operation of the battery charger in Figure 8–3.
5. Explain the purpose of capacitor C_1 in Figure 8–3.
6. What is the purpose of D_6 in Figure 8–3?

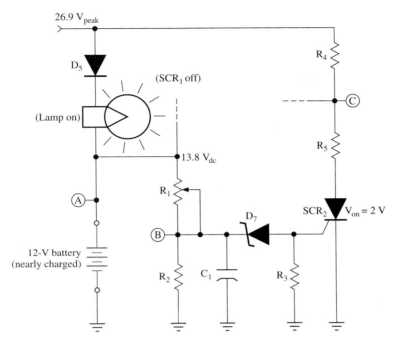

Figure 8–5
Equivalent circuit condition for battery charger with 12-V battery nearly charged

8–2   A HYSTERESIS-FREE LAMP DIMMER

The circuit in Figure 8–6 contains a high-quality commercial lamp dimmer in which the problem of flash-on is virtually eliminated. That a silicon bilateral switch (SBS) is being used instead of a DIAC to trigger the TRIAC helps alleviate the hysteresis problem so often associated with lamp-dimmer circuitry. This is because the breakover potential of the SBS is only around 8 V, whereas that of the DIAC is typically more than 30 V. Diodes D_1 and D_2 are also essential to the hysteresis-free operation of the circuit in Figure 8–6. Diode D_1, which is connected directly to the gate of the SBS, allows the flow of current from the gate of the SBS whenever the voltage at point A is more positive than the voltage at point C. When SBS gate current starts to flow, forward breakover of the SBS takes place, as long as point A is at least 1 V more positive than point B. With D_1 connected as shown, the only possible path for SBS gate current is toward point C. Therefore, without the flow of gate current, reverse breakover of the SBS may not occur until the difference in potential from point A to point B reaches nearly -8 V.

As the ac source waveform approaches the 180° point, commutation of the TRIAC occurs automatically. During the negative alternation, the TRIAC is not able to switch on until the SBS is able to break over in the negative direction. Without D_2 in the circuit, C_1 would continue to discharge via the SBS, D_1, and R_3. Therefore, the capacitor would be unable to charge to the -8 V necessary for breakover during the negative alternation. However, with D_2 present in the circuit, an alternative path for reverse current exists through D_2 and R_3. As the source voltage goes negative, D_2 becomes forward biased. While D_2 is conducting, the voltage present at its cathode is held at -0.7 V. Since, at

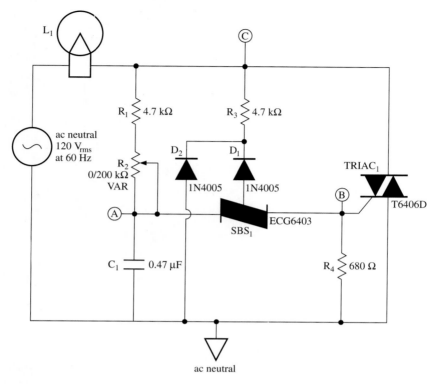

Figure 8–6
Hysteresis-free commercial lamp dimmer control circuit

this time, D_1 is effectively in series with a forward-biased Zener diode within the SBS, -1.4 V would be required at its cathode to continue the flow of reverse gate current. Thus, D_1 is able to shut off the flow of reverse gate current, allowing C_1 to charge negatively. Thus, the presence of D_2 is essential for triggering of the TRIAC during the negative alternation of the source cycle.

At this point in the analysis of the lamp-dimmer circuit in Figure 8–6, the reason for placing the load directly between point C and the ac source should become evident. Once the TRIAC has been triggered during the positive alternation, the voltage at point C is held at the forward saturation level of the TRIAC (around 2 V) until commutation occurs near the 180° point. Not until the beginning of the negative alternation does C_1 begin to charge negatively. After breakover has occurred in the positive alternation, the voltage at point A remains at a positive level, slightly lower than that present at point C. This voltage level is present until virtually the end of the positive alternation. Therefore, with the circuit in Figure 8–6, for a given setting of potentiometer R_2, triggering of the TRIAC is nearly symmetrical, with the firing-delay angle for the negative alternation nearly corresponding to that of the positive alternation. Once triggering of the TRIAC has occurred during the negative alternation, the voltage at point C remains at the reverse-saturation level, preventing C_1 from charging in the positive direction. Thus, by placing the load between point C and ground, the problem of flash-on is greatly reduced.

The condition that exists in Figure 8–6 with R_2 set to its maximum ohmic value is represented in Figure 8–7. In this maximum resistance condition, since the voltage at point A reaches a peak amplitude of no greater than 4.5 V, breakover of the SBS does

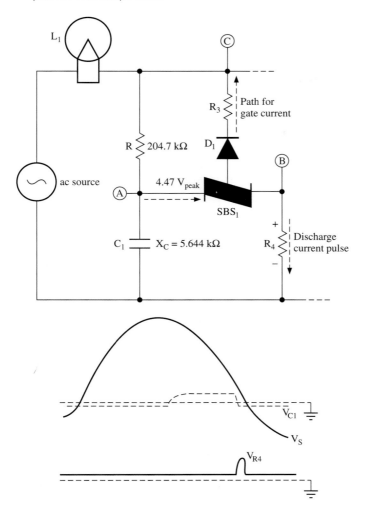

Figure 8–7
Circuit condition for lamp-dimmer control with R_2 set to maximum

not occur until current flows from the gate of the SBS toward point C. As shown in Figure 8–7, the capacitive waveform at point A does not cross the 0 reference line in the positive direction until nearly 90° into the positive alternation. Thus, the voltage at point A does not attain its positive peak level until virtually 180° into the positive alternation, when point C is approaching 0 V. At this time, D_1 becomes forward biased, allowing the flow of gate current from the SBS. Once this occurs, since point A is still above 0 V, the SBS breaks over, allowing C_1 to discharge rapidly through R_4. It is important to note that, since the source waveform is at nearly 0 V at the instant the SBS breaks over, the TRIAC is unable to switch on. (A minimum forward or reverse potential of around 2 V must be present between the main terminals of the TRIAC in order for that device to be latched on.) Regardless of whether the TRIAC is triggered on during any given alternation, the charge on C_1 is nearly 0 V at the instant the source waveform approaches the 0° or 180° point. Thus, for the circuit in Figure 8–6, the problem of hysteresis is virtually eliminated, allowing a smooth transition in light intensity to take place as the shaft of potentiometer R_2 is turned in either direction.

SECTION REVIEW 8-2

1. True or false?: The SBS in Figure 8–6 breaks over at least twice during each source cycle, regardless of the setting of R_2.
2. What is the purpose of D_2 in Figure 8–6? Explain its function during the negative alternation of the source waveform.
3. Explain the function of R_4 in Figure 8–6.
4. Explain why, in Figure 8–6, the load is placed between the ac source and point C.

8-3   NOISE SUPPRESSION IN THYRISTOR CONTROL CIRCUITRY

One of the problems inherent to thyristor control circuits operating on the principle of phase shift is that of high-frequency noise generation. As an example, consider the lamp-dimmer circuit in Figure 8–6. Assume that potentiometer R_2 has been adjusted such that the TRIAC triggers 90° (halfway) into each alternation. The waveforms developed over the TRIAC and load would then appear as illustrated in Figures 8–8(a) and (b).

With the conditions shown in Figure 8–8, due to the regenerative switching action of the TRIAC, load current is able to increase from a few microamperes to nearly 1 A in a few microseconds. Although this high-speed switching action is a desirable quality for thyristor control circuits, it is also a major source of noise. The nearly instantaneous rise in main-terminal current that occurs when the TRIAC is triggered generates high-frequency harmonics. This noise is capable of causing spurious triggering within industrial control circuits. Also, since this interference extends into the RF range, it is even capable of traveling through the air as radio waves, further compounding the noise suppression problem. In low- to medium-power applications, this RF noise problem may be remedied by using an inductor and a capacitor to form a simple low-pass filter. In Figure 8–9, this form of passive filter is incorporated into the design of the hysteresis-free lamp dimmer.

The purpose of inductor L_1 in Figure 8–9 is to inhibit the rapid growth of load current that could occur if an SCR or TRIAC is triggered on when the source waveform has already reached a high amplitude. Since the inductor opposes a change in current, the nearly instantaneous rise in load current (as was represented by the waveforms in Figure 8–8) does not occur. Instead, load current increases gradually at first, as shown in Figure 8–10. The action performed by L_1 may be compared to that of an RF choke, attenuating the high-frequency harmonics generated by the rapid rise in either forward or reverse load current. Thus the propogation of noise along power lines or through the air is greatly inhibited.

For the circuit in Figure 8–9, an inductor must be selected that offers substantial impedance to RF noise. At the same time, this inductor should offer minimal impedance to the 60-cycle source waveform. For example, assume L_1 equals 300 μH. The value of inductive reactance offered to the ac source, as determined here, is much less than an ohm. Thus the only significant opposition that L_1 offers to the source is a few ohms of resistance.

$$X_L = 2\pi fL = 2\pi \times 60 \text{ Hz} \times 300 \text{ μH} = 0.113 \text{ }\Omega$$

However, using 1 MHz as an example of an RF noise frequency, inductive reactance would approach 2 kΩ:

$$X_L = 2\pi fL = 2\pi \times 1 \text{ MHz} \times 300 \text{ μH} = 1.885 \text{ k}\Omega$$

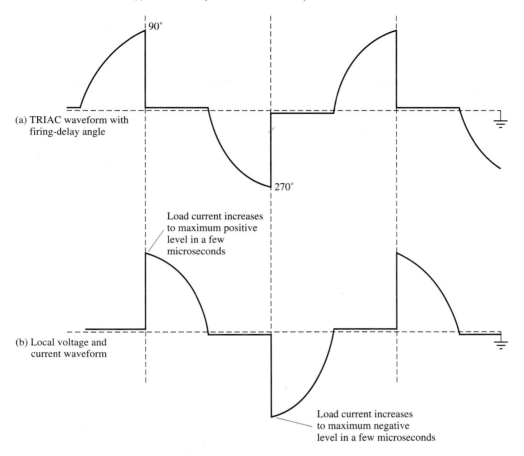

(a) TRIAC waveform with firing-delay angle

90°

270°

Load current increases to maximum positive level in a few microseconds

(b) Local voltage and current waveform

Load current increases to maximum negative level in a few microseconds

Figure 8–8

If higher frequency RF noise attempts to enter the circuit, L_1 would develop greater inductive reactance and, consequently, operate even more effectively as a filter.

In Figure 8–9, additional noise elimination is accomplished by capacitor C_2, which shunts any RF noise still present in the circuit directly toward circuit ground. With C_2 equaling 0.1 μF, the capacitive reactance offered to the 60-cycle source approaches 27 kΩ.

$$X_C = \frac{1}{2\pi f C} = \frac{1}{2\pi \times 60 \text{ Hz} \times 0.1 \text{ μF}} = 26.53 \text{ k}\Omega$$

This response of the capacitor to the source frequency is desirable, causing it to function as a high-impedance in parallel with the control circuitry. However, in response to the 1-MHz frequency, the capacitive reactance of C_2 is less than 2 Ω, being determined as follows:

$$X_C = \frac{1}{2\pi \times 1 \text{ MHz} \times 0.1 \text{ μF}} = 1.6 \text{ }\Omega$$

Thus, RF noise bypasses the TRIAC control circuitry. Since capacitive reactance is inversely proportional to frequency, higher frequency RF noise would be shunted even more easily toward circuit ground.

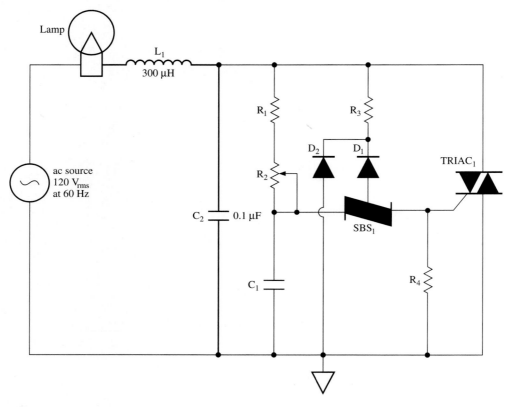

Figure 8–9
TRIAC control circuit containing passive LC filter for noise suppression

Figure 8–11 contains the same lamp-dimmer circuit as was presented in Figure 8–6. However, this circuit now contains an additional feature known as a **snubber network.** This snubber network may be used in SCR and TRIAC circuits where a rapid rise in main-terminal voltage could result in spurious triggering of the device. Even without a trigger pulse present at the gate, such spurious triggering may occur when a noise spike across the main terminals of a TRIAC or an SCR exceeds the maximum forward- or reverse-blocking voltage of the given device. Also, if the rise in main-terminal voltage exceeds a rate specified by the manufacturer, this undesired breakover may even occur at a voltage much lower than the specified blocking potential. To understand why this form of spurious triggering may occur, the internal structure of the SCR must be reexamined. Such analysis is also applicable to the TRIAC, since the device essentially consists of two complementary SCRs.

When an SCR is forward biased, as shown in Figure 8–12(a), the junction formed by the n_1 and p_2 material becomes reverse biased. This condition results in a small parasitic capacitance at the junction. Assume at first that the SCR in Figure 8–12(a) is in the off state, with no gate current flowing. If the source voltage is a 60 Hz sine wave with a peak amplitude below the peak forward voltage of the SCR, the device may be expected to behave as an open switch. However, if a sudden surge in voltage occurs between the main terminals of the SCR, the junction capacitance will, for a brief instant, behave as a dead short, allowing a charge current to flow within the device. This action might be

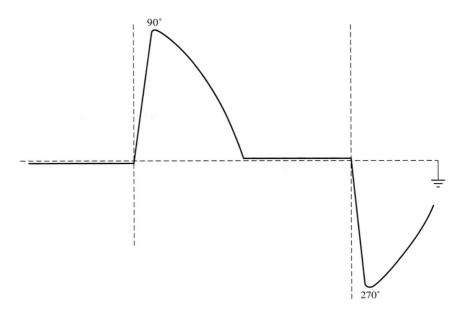

Figure 8–10
Dampened rise in load current due to action of the LC filter

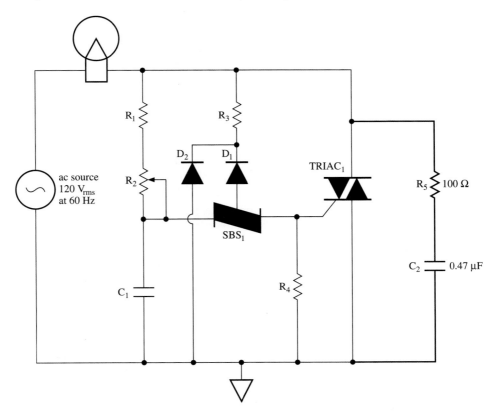

Figure 8–11
Use of a snubber network in conjunction with a TRIAC control circuit

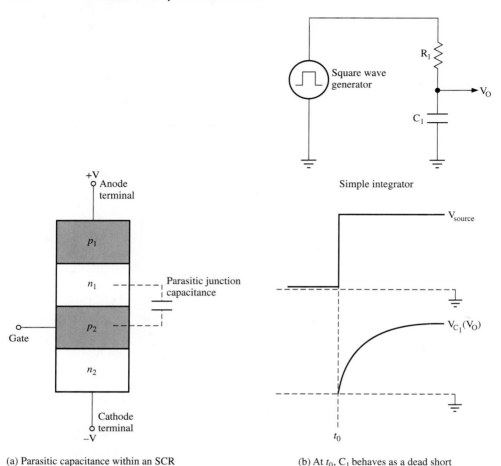

(a) Parasitic capacitance within an SCR

(b) At t_0, C_1 behaves as a dead short and circuit current is at maximum

Figure 8–12

compared to the initial response of a capacitor to a rectangular wave, as illustrated by the simple RC integrator in Figure 8–12(b).

The important factor in determining the probability of spurious triggering is not so much the voltage level that a noise spike might attain as the *rate* at which its amplitude increases. The following equation is used to calculate the amount of charge current that would be flowing within the SCR or TRIAC as a result of a rapid surge in main-terminal voltage. If this calculation yields a value of charge current for the junction capacitance that equals or exceeds the minimum gate trigger current for the device, spurious triggering is likely to occur.

$$i_C = C \times \frac{V_C}{\Delta t} \qquad (8\text{–}1)$$

where C represents the parasitic capacitance at the junction, i_C is the current flowing into the junction capacitance, and $V_C/\Delta t$ is the slope or rate of voltage increase across the main terminals of the SCR or TRIAC.

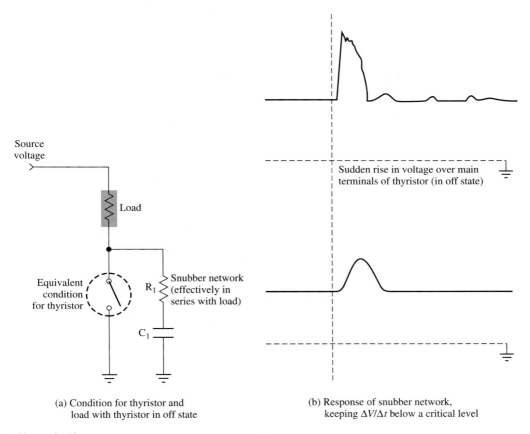

(a) Condition for thyristor and
load with thyristor in off state

(b) Response of snubber network,
keeping $\Delta V/\Delta t$ below a critical level

Figure 8–13

Since an SCR or TRIAC behaves as an open switch before triggering, the load may be considered as effectively in series with the snubber network when the thyristor is in the off state. This circuit condition is illustrated in Figure 8–13(a). If a sudden increase in voltage does occur at the source, as represented in Figure 8–13(b), through the action of the snubber network, the rate of voltage increase over the main terminals of the SCR is substantially reduced. This dampening in the growth of main-terminal voltage greatly decreases the amount of charging current for the junction capacitance. Thus, an SCR or TRIAC may be protected from voltage surges and noise spikes present at the power source. At the same time, spurious triggering of the device is avoided. Considering that an SCR or TRIAC may be functioning as a main power switch for a powerful piece of industrial equipment, the importance of the simple RC snubber network becomes important indeed!

SECTION REVIEW 8-3

1. Explain the response of the inductor in Figure 8–9 to RF noise. What is the response of this component to the 60-Hz source waveform?
2. Explain the response of C_2 in Figure 8–9 to RF noise. What is the response of this component to the 60-Hz source waveform?

3. Explain the operation of a snubber network. How does this simple RC network reduce the possibility of spurious triggering of a thyristor?

8–4   ZERO-POINT SWITCHING CIRCUITS

Such LC noise-suppression networks as that introduced in Figure 8–9 operate effectively in low- or medium-power applications. In high-power thyristor circuitry, however, where load current is likely to be several amperes, an inordinately large value of inductance would be required to attenuate RF interference effectively. For this reason, in the design of high-power thyristor control circuitry where RF noise is a major concern, the idea of phase-shift control may be abandoned entirely, being replaced by a form of load control called **integral-cycle, synchronous switching** or, simply, **zero-point switching.**

This form of thyristor control may also be referred to as **soft-start switching,** implying that a load may be turned on at the instant the source waveform passes the 0-V point, allowing load current to begin flowing gradually. In a phase-shift control system, the amount of power available to a load is dependent on the amount of time in which current is allowed to flow from the source to the load during a given cycle. However, with zero-point switching, load power is dependent on the number of entire cycles permitted to pass from the source to the load in a given amount of time.

Although zero-point switching circuits are generally more complex than phase-shift circuits, they do have two major advantages. First, compared to phase-shift control circuitry, zero-point switching circuits are virtually noise free. Whereas noise can be suppressed in phase-shift control circuits, in zero-point switching circuitry it is less likely to occur, because rapid changes in load current are avoided. In high-power circuits, sudden surges in power could cause damage to a load, especially during a startup condition. Rotating machinery is particularly susceptible to this form of **thermal shock.** Within a zero-point switching circuit, however, since load current always increases gradually, such damaging current surges are avoided.

Thus far in the study of thyristors, the load has been represented simply as a lamp or a resistor. Also, control of the load has been accomplished through the adjustment of a potentiometer. In Figure 8–14, however, control of load power becomes automatic. Here, potentiometer R_3 is used only for initial calibration of the **set-point** temperature. (The set-point represents the desired level of temperature for the environment being controlled.) Once the circuit begins to operate, thermistor R_2 becomes the actual controlling device. As explained in Chapter 3, the thermistor is a temperature-sensing transducer. The device in Figure 8–14 has a **negative temperature coefficient** (NTC). Thus, the thermistor's resistance decreases as it becomes warmer. The purpose of the comparator, SCR, and TRIAC in Figure 8–14 is to turn the load on or off, as the temperature fluctuates around the set point. For optimal effectiveness, the thermistor is placed in a central location within the environment being controlled.

The load within this circuit is a heating element that is designed to dissipate a large amount of power as it converts electrical current into thermal energy. Since the load is rated at 2000 W with an applied voltage of 240 V_{rms}, a TRIAC must be selected that is capable of withstanding the large quantity of main-terminal current that flows as the load is operating. Assuming the load power dissipation to be 2000 W_{rms}, the amplitude of TRIAC main-terminal current could be estimated at around 8.33 A_{rms}. A TRIAC suitable for this application would be a T6406D, a device especially designed for zero-point switching applications. Its gate characteristics assure reliable operation over a wide range of temperatures. Also, the device is capable of supplying a load current of as much as 40 A_{rms}. This TRIAC has a V_{DROM} of 400 V, which allows it to commutate reliably when operating in conjunction with a 240-V_{rms} source. Remember that the off-state voltage of

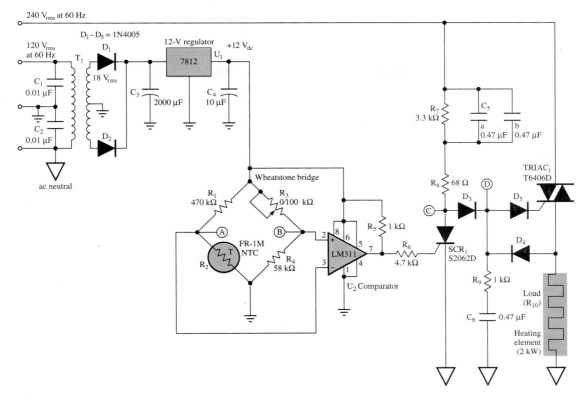

Figure 8–14
Oven-control circuit utilizing zero-point switching

a TRIAC is rated as a peak value, but the line voltage is specified as rms:

$$240 \text{ V}_{rms} \times 1.414 = 339.4 \text{ V}_{peak}$$

SCR_1 functions as an intermediate switching device between the low-power dc control portion of the circuit and the high-power ac section. The S2062D is a **sensitive-gate SCR.** Although capable of withstanding a main-terminal current of 4 A, this device may be triggered by less than 2 mA of gate current. Through the action of the 7812 IC, a fully regulated 12-V_{dc} source is provided for stable operation of the Wheatstone bridge and the LM311 high-speed precision comparator. Because of its high output current capability, the LM311 easily drives the gate of the S2062D.

Before analyzing the operation of the two thyristors within Figure 8–14, we must first understand the function of the Wheatstone bridge and comparator. Figure 8–15 contains a schematic diagram of the Wheatstone bridge, showing points A and B simply as open terminals. (These points serve as the inverting and noninverting inputs, respectively, to the comparator.) The Wheatstone bridge circuit allows adjustment of the set-point temperature for the environment being controlled. For instance, assume that the typical operating range of the FR-1M thermistor is from around 1 MΩ (cold) to 600 kΩ (hot). Then R_3 could be adjusted such that, with R_2 equaling 600 kΩ, a balanced condition exists across the bridge. The difference in potential from point A to point B would then equal 0 V. The mathematical relationship for the resistances within the Wheatstone bridge

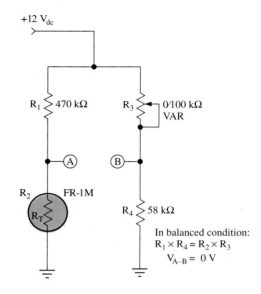

Figure 8–15
Detail of Wheatstone bridge

during a balanced condition is expressed as follows:

$$(R_1 \times R_4) = (R_2 \times R_3) \tag{8–2}$$

From this relationship, a formula may be derived for the setting of R_3 that would allow 0 V to occur between points A and B at the set-point temperature:

$$R_3 = \frac{(R_1 \times R_4)}{R_2} \tag{8–3}$$

Substituting the known values of resistance for the Wheatstone bridge, the setpoint value of R_3 is calculated as follows:

$$R_3 = \frac{(470 \text{ k}\Omega \times 58 \text{ k}\Omega)}{600 \text{ k}\Omega} = 45.433 \text{ k}\Omega$$

If the preceding solution for R_3 is correct, at the set-point temperature, the voltage measured from point A to dc ground would equal that measured between point B and dc ground:

$$12 \text{ V} \times \frac{600 \text{ k}\Omega}{(600 \text{ k}\Omega + 470 \text{ k}\Omega)} = 12 \text{ V} \times \frac{58 \text{ k}\Omega}{(58 \text{ k}\Omega + 45.433 \text{ k}\Omega)} = 6.73 \text{ V}$$

Thus, when thermistor R_3 is at the set-point temperature, $V_{A–B}$ equals 0 V. If the oven temperature increases above the set-point level, thermistor resistance decreases. This, in turn, causes the voltage at point A to decrease while the voltage at point B remains constant. As a result, the voltage at the inverting input of the comparator becomes more negative than the voltage at the noninverting input. This difference in potential from point A to point B causes the output of the comparator to switch immediately to a high level. As long as thermistor resistance is lower than its set-point value, point A remains more negative than point B and the output of the comparator remains high. As the oven begins to cool, the ohmic value of the thermistor increases, causing the voltage level at point A to become more positive. Once the voltage level at point A exceeds the voltage level at point B, the output of the comparator immediately falls low and remains low until the oven temperature increases above the set-point level.

Now consider a circuit condition where the oven has just been turned on. At this time, the ohmic value of the thermistor would be well above the established set-point level of 600 kΩ, causing the voltage at point A to be more positive than that present at point B. As a result, the inverting input of the comparator is now more positive than the noninverting input, causing the output of that device to remain at virtually 0 V. Thus, no gate current is able to flow to SCR_1, which now functions as an open switch. Assume, with this startup condition, that the source waveform is approaching the 0-V reference line from the negative direction. During this time, the TRIAC is biased off. Not until point C begins to exceed a threshold level established by the forward voltage of D_3, the forward voltage of D_5, and the gate voltage of the TRIAC does the load become energized. Assuming a V_{GT} of 1 V for the TRIAC, this minimum switching voltage is determined as follows:

$$v_i = V_{D_3} + V_{D_5} + V_{GT} = 0.7 \text{ V} + 0.7 \text{ V} + 1 \text{ V} = 2.4 \text{ V}$$

Until the TRIAC is triggered on, because no gate current is flowing through R_7 and R_8, capacitors C_5 and C_6 are unable to charge. Instead, the source voltage develops over the main terminals of the TRIAC. Considering that R_7 and R_8 have very low ohmic values and that capacitor C_5 functions as a dead short at the instant gate current begins to flow, triggering of the TRIAC may be expected to occur near the calculated threshold level of 2.4 V. Since this switching threshold may be considered to be an instantaneous value of source voltage, the expected firing-delay angle may be approximated as follows:

$$\sin\theta = \frac{v_i}{V_{peak}} = \frac{2.4 \text{ V}}{339 \text{ V}} = 0.00708$$
$$\theta = 0.4° \cong \text{less than } 1°$$

As indicated by this calculation, if SCR_1 is in an off state, triggering of the TRIAC occurs virtually at the beginning of the positive alternation of the source waveform. This rapid switching action allows virtually all the source voltage to be developed over the load during the positive alternation. If the SCR remains off for the duration of the positive alternation, current continues to flow from point C into the gate of the TRIAC, via D_3 and D_5. This current allows C_5 to charge during the positive alternation. The waveform developed over C_5 will, of course, lag behind the source waveform. Thus, when the source waveform approaches 180°, the voltage level at the junction of R_7 and R_8 is negative with respect to ac neutral. Then, since D_3 is reverse biased, no trigger current flows to the TRIAC.

However, shortly after the TRIAC is triggered at the beginning of the same positive alternation, C_6 is able to charge in the positive direction, through D_4 and R_9. After the source waveform attains a 90° point and the load voltage begins to fall below its peak positive amplitude, D_4 becomes reverse biased. In this circuit condition, the only path for discharge of C_6 is then through R_9, D_5, and the gate of the TRIAC. As the source waveform crosses the 180° point, the TRIAC momentarily turns off. However, as the source waveform approaches an instantaneous amplitude of around −2 V, the TRIAC immediately switches on again, allowing load current to flow in the negative direction. This switching action is possible because a large enough positive voltage is still present at point D. As long as the SCR is still in the off-state, C_5 is able to charge during the negative alternation, positive with respect to point C. This charge is present over C_5 as the source approaches the 0° point. Thus, as the waveform attains around 2 V, the TRIAC again switches on.

Now consider the condition that would exist when the SCR is able to latch on. This would occur as the oven temperature begins to exceed the set-point level. As the thermistor resistance falls below 600 kΩ, the inverting input to the comparator becomes slightly more negative than the fixed voltage level at the noninverting input. Once this occurs, the output of the comparator toggles to a logic high, providing a source of current for the gate of the SCR. If the source waveform is already in its positive alternation (at or above 2 V), the SCR immediately triggers on. If triggering of the SCR occurs during the positive alternation of the source waveform, the voltage at point C falls to the saturation level of the device, which should be no more than 2 V.

Commutation of the TRIAC, however, does not occur immediately. This device remains latched on for the duration of the positive alternation. As the source waveform progresses into the negative alternation, both the SCR and the TRIAC are forced into commutation. However, enough positive voltage is still present at point D to trigger the TRIAC back on during the negative alternation. If the output of the comparator is still high at the beginning of the next positive alternation, the SCR is triggered on again virtually at the 0° point, holding the voltage at point C at less than 2 V. Thus, D_3 and D_5 become reverse biased, causing the TRIAC to remain off for the duration of the positive alternation. Since the TRIAC is not triggered on at the beginning of the positive alternation, C_6 is unable to charge. Thus, point D remains at 0 V. As a result, the TRIAC remains untriggered during the negative alternation.

It should now be evident that, if the comparator toggles to a high condition at any point during the positive alternation, the SCR latches on for the duration of that alternation. However, the load does not turn off until just after the next 0° point. As long as the oven temperature is above the set-point level, SCR_1 latches on virtually at the 0° point of each cycle, holding the heating element in the off condition.

Finally, we consider the circuit condition where the oven has been cooled and the thermistor resistance has increased beyond the set-point value of 600 kΩ. The output of the comparator is now low and no current flows to the gate of the SCR. With the SCR now functioning as an open switch, the TRIAC is able to switch on at the beginning of each alternation of the source waveform. The load now remains energized until the oven temperature again exceeds the set-point value. It is important to note that, although the output of the comparator may toggle at any instant during a given cycle of the source waveform, the load itself switches off or on only at the 0° point. Through this action, precise temperature control is achieved while RF noise is kept to a minimum.

A more energy-efficient temperature control circuit is shown in Figure 8–16. Although this second example of zero-point switching is more complex than that in Figure 8–14, it is also more efficient, because it requires the use of only one thyristor within the ac portion of the circuit. Recall that in the previous example, SCR_1 had to be triggered on at the beginning of a positive alternation to hold the load in the off condition. Current from the source would, therefore, be shunted past the TRIAC and load, flowing through R_7, R_8, and the main terminals of the SCR. This current would continue to flow until the 180° point of the source cycle. The energy dissipated by R_7 and R_8 in Figure 8–14 during the positive alternation would then be estimated as follows:

$$I_{R_7} = I_{R_8} \cong \frac{V_{\text{source rms}}}{(R_7 + R_8)} \cong \frac{240 \text{ V}}{(3.3 \text{ k}\Omega + 68 \text{ }\Omega)} \cong 71.3 \text{ mA}_{\text{rms}}$$

$$P_{R_7} = 71.3 \text{ mA}^2 \times 3.3 \text{ k}\Omega = 16.7 \text{ W}_{\text{rms}}$$

$$P_{R_8} = 71.3 \text{ mA}^2 \times 68 \text{ }\Omega = 346 \text{ mW}$$

Although this resistive heat loss is minimal in comparison to the energy consumed by the load itself, this design requires the use of a large, high-wattage resistor as R_7 in Figure 8–14. In control applications where component heat dissipation is crucial, a more efficient design approach must be used.

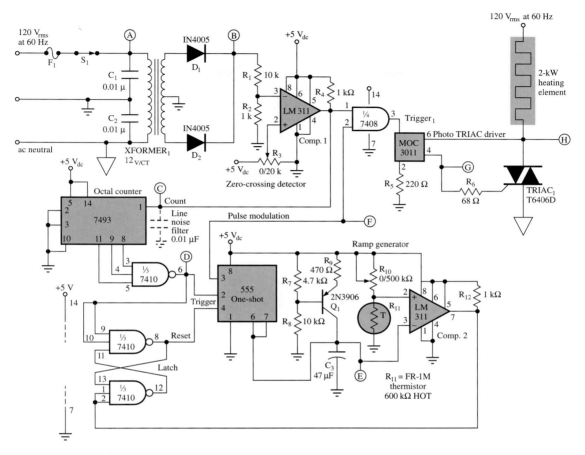

Figure 8–16
Improved oven-control circuit utilizing zero-point switching and pulse width modulation

The circuit example in Figure 8–16 demonstrates how TTL logic may be used to control the timing within a zero-point switching system. As was introduced in Chapter 6, the 7493 is a typical IC ripple counter, configured here to perform a divide-by-eight operation. The counter receives its clocking pulses from the output of the first 311 comparator, which is functioning as a zero-crossing detector. To understand the intricate timing of this system, it is necessary to trace the signal conditioning process that begins at the secondary of transformer 1. This component, in conjunction with D_1 and D_2, provides a fully rectified, but unfiltered, pulsating dc signal. Note that this signal is fed to the inverting input of the comparator, which is represented as waveform B in Figure 8–17. Since the source waveform crosses the zero-point refrence line twice during a given cycle, the clock signal taken at the output of the first comparator, represented as waveform C in Figure 8–17, occurs at twice the frequency of the ac source, or nearly 120 pulses per second. The duty cycle of this clock pulse is varied through adjustment of potentiometer R_3. The pulse width of this signal should be made narrow enough to allow triggering of the TRIAC to occur within one or two degrees of the zero-crossing point.

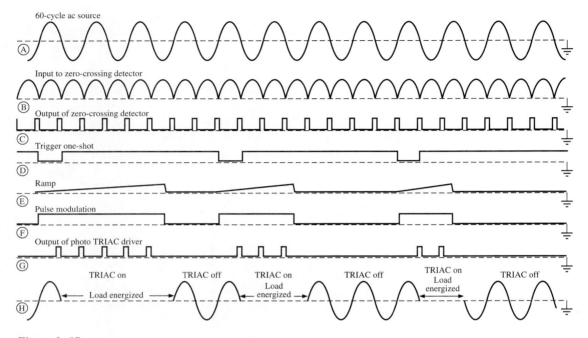

Figure 8–17
Pulse train representing action of circuit as oven gradually warms toward set-point temperature

The 7493 octal counter receives a clock signal from the zero-crossing detector at a rate of 120 pps and is thus able to increment from 000_2 to 111_2 fifteen times per second. With the three Q outputs of the counter connected to the inputs of the first NAND gate, an active-low pulse occurs at the output of that gate every eight clock states (at each instant the counter attains 111). This signal, which serves as a trigger pulse to the 555 one-shot, is represented as waveform D in Figure 8–17. The 555 one-shot is connected to transistor Q_1 in such a way that the collector of that device functions as a constant current source, used for charging capacitor C_3. Thus, when the one-shot is triggered, a linear ramp signal develops over C_3. The slope of this signal is easily determined as follows:

$$V_{R_7} = \frac{R_7}{(R_7 + R_8)} \times 5 \text{ V} = \frac{4.7 \text{ k}\Omega}{14.7 \text{ k}\Omega} \times 5 \text{ V} = 1.6 \text{ V}$$

$$V_{R_9} \cong 1.6 \text{ V} - 0.6 \text{ V} \cong 1 \text{ V}$$

$$I_{R_9} = \frac{1 \text{ V}}{470 \text{ }\Omega} = 2.13 \text{ mA}$$

$$\text{Slope} = \frac{I}{C} = \frac{\Delta V}{\Delta t}$$

$$\Delta t = \frac{C \times \Delta V}{I} = \frac{47 \text{ }\mu\text{F} \times 3.3 \text{ V}}{2.13 \text{ mA}} \cong 73 \text{ ms}$$

where I equals I_{R_9}, C equals C_3, and ΔV equals 2/3 V_{CC} or 3.3 V.

The ramp signal, which is represented as waveform E in Figure 8–17, is fed to the inverting input of the second comparator. When the instantaneous amplitude of the ramp signal exceeds either the reference level established at the non-inverting input of the comparator or the 2/3 V_{CC} threshold of the 555 timer, the output of the second comparator falls low. If this action occurs before the 7493 reaches count 111, the NAND latch resets. An active-low pulse then occurs at pin 4 of the 555 timer, causing C_3 to discharge quickly through pin 7 of that device. The output of the comparator is low for only a few nanoseconds. Once the instantaneous voltage at the inverting input falls below the threshold level at the non-inverting input, the output of comparator 2 immediately switches to a logic high. The NAND latch then returns to a hold condition, allowing the one-shot to be triggered at the instant the octal counter again attains 111.

The 555 timer in Figure 8–16 functions as a pulse width modulator. Although the output signal taken at pin 3 occurs at a steady frequency, the pulse width of that signal is variable, being a function of the threshold voltage at the noninverting input of comparator 2. Since the 555 timer is retriggered every eight clock states, the operating frequency of the pulse width modulator is determined as follows:

$$T = 8 \times 8.833 \text{ ms} = 66.67 \text{ ms}$$
$$f = \frac{1}{T} = \frac{1}{66.67 \text{ ms}} = 15 \text{ pps}$$

During the time in which C_3 is charging, a logic high is present at pin 3 of the 555 timer. The duration of this pulse is controlled by the threshold voltage present at the noninverting input of the second comparator. If this voltage is made higher, the ramp signal is extended, causing the pulse width of the signal at pin 3 of the 555 timer to increase by an equal amount of time. If the comparator threshold is decreased, the duration of the ramp signal and the pulse width of the signal at pin 3 of the 555 timer also decrease. Since the slope of the ramp signal is highly linear, the relationship of the comparator threshold voltage and the output pulse width of the 555 timer is also linear. Thus, an increase in pulse width is directly proportional to an increase in the level of reference voltage. The mathematical expression for this relationship is shown in the following equation:

$$\Delta PW \propto \Delta V_{ref} \tag{8–4}$$

The output pulse from the 555 timer, which is represented as waveform F in Figure 8–17, serves as the enable input for the AND gate. Thus, when pin 3 of the 555 is at a logic high, the clock signal at pin 3 of the 555 timer is at a logic high and the clock signal from the zero-crossing detector is enabled to pass to the **MOC3011 photo TRIAC driver.** Each time this device is triggered, the T6406D TRIAC is also triggered, allowing complete alternations to pass from the ac source to the heating element. If the thermistor temperature increases, the threshold voltage for the second comparator s lowered. This causes the ramp signal to decrease and the number of trigger pulses that pass to the photo TRIAC driver within an eight-clock-state period to be reduced. As the temperature of the thermistor decreases, the threshold voltage of the second comparator increases, allowing more trigger pulses to pass to the MOC 3011 within the same clocking period. The control system in Figure 8–17 is able to control the transfer of energy to the load in half-cycle increments. Thus, within a given clocking period, there are eight possible gradations of control. The relationship of the photo TRIAC driver output and the signal developed over the T6406D is shown in Figure 8–17 as waveforms G and H.

The need for zero-point switching systems within industry is great enough that the gate-control circuitry for such devices is made available in monolithic form. An example of such a zero-point switch would be the **CA3059.** With the addition of only a limiting resistor and a filter capacitor, this IC could be placed directly into a TRIAC control circuit where the line voltage could be as high as ± 220 V_{ac}. Figure 8–18(a) contains a functional diagram of this IC, and Figure 8–18(b) shows its actual circuitry.

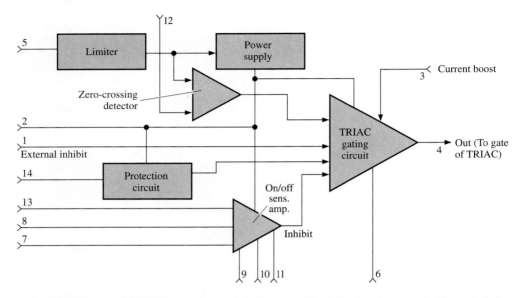

(a) Functional block diagram of CA3059 zero-voltage switch (Courtesy of Harris Semiconductor, used with permission)

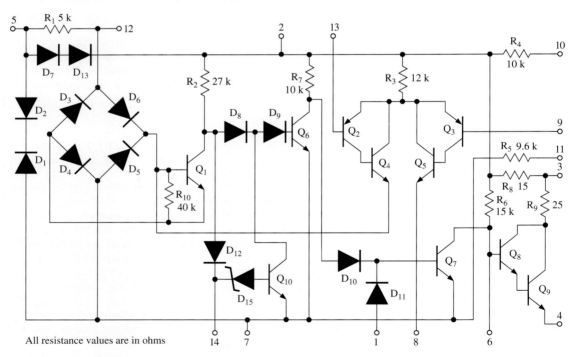

(b) Internal circuitry of CA3059 (Courtesy of Harris Semiconductor, used with permission)

Figure 8–18

Figure 8–19 illustrates how the CA3059 may be incorporated into the design of an oven controller. The function of this circuit is similar to that of the circuit in Figure 8–14. It provides precise, high-speed on/off control of oven temperature, the set-point value of which is established through the adjustment of R_1. External resistor R_5 functions as the limiting resistor for the IC, keeping the amplitude of the ac signal at pin 5 within a safe margin. Inside the CA3059 IC (referring momentarily to Figure 8–18[b]), D_7 and D_{13} provide half-wave rectification of the ac line voltage. With filter capacitor C_1 connected between pin 2 and ac neutral, approximately 6 V_{dc} is provided as V_{CC} for the 10 internal switching transistors. The zero-crossing detector portion of the IC consists of D_3, D_4, D_5, D_6, and transistor Q_1. The on/off sensing amplifier consists of Q_2, Q_3, and Q_4 and Q_5, which form a differential comparator. Transistors Q_8 and Q_9 form a Darlington pair, allowing the gate of the TRIAC to be driven directly from pin 4. With pins 9, 10, and 11 of the CA3059 tied together, a voltage divider, consisting of internal resistors R_4 and R_5, is created between pin 2 and ac neutral. If the dc level at pin 2 is 6 V, the threshold voltage present at the base of Q_3 (pin 9) may be calculated as follows:

$$V_{TH} = 6\ V \times \frac{R_5}{(R_4 + R_5)} = 6\ V \times \frac{9.6\ k\Omega}{19.6\ k\Omega} = 2.94\ V$$

In Figure 8–19, potentiometer R_1 and thermistor R_2 form an external voltage divider. Because thermistor R_2 has a negative temperature coefficient, as oven temperature increases, the voltage at pins 13 and 14 of the CA3059 decreases. External resistors R_3 and

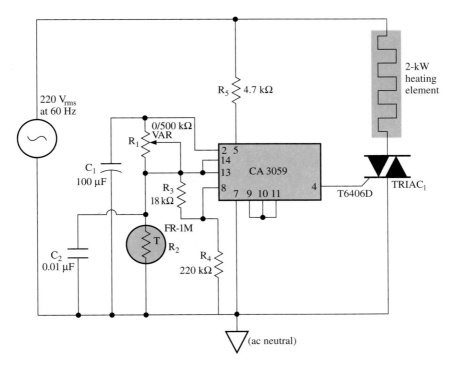

Figure 8–19
CA3059 zero-voltage switch used in conjunction with TRIAC in heating element control circuit
(Courtesy of Harris Semiconductor, used with permission)

R_4 allow switching hyteresis. With the values shown, oven temperature may vary within ±10% of the set-point value. Such hysteresis may be necessary for stable operation within some types of control systems.

As an example of circuit operation, assume the oven is cold and that the dc level at pin 13 is higher than the threshold voltage established at pin 9 of the CA3059. With this condition, transistors Q_2 and Q_4 are biased off, holding Q_1 in the off state as the source voltage approaches the zero-crossing point. With Q_1 off, Q_6 is on, Q_7 is off, and the Darlington pair is biased on, providing current flow to the gate of the TRIAC. As long as the dc level at pin 13 remains higher than that at pin 9, this switching action occurs as the source waveform approaches 0 V.

Next, after the oven has been on for a while, assume that the voltage present at pin 13 has fallen below the reference level established at pin 9. This input condition causes Q_2 and Q_4 to switch on. As a result, Q_1 is biased on and remains on as the source waveform passes 0 V. Therefore, Q_6 switches off and Q_7 switches on, causing the Darlington pair to be biased off. As long as the voltage at pin 13 is lower than the reference level at pin 9, no gate current is provided to the TRIAC and the load remains de-energized.

The advantage of using a zero-voltage switch such as the CA3059 should now be evident. Because the device contains its own power supply components, the need for costly transformers, rectifiers, and voltage regulators is eliminated. Also, zero-voltage switches are easily incorporated into more complex forms of thyristor control circuitry, an example of which is shown in Figure 8–20.

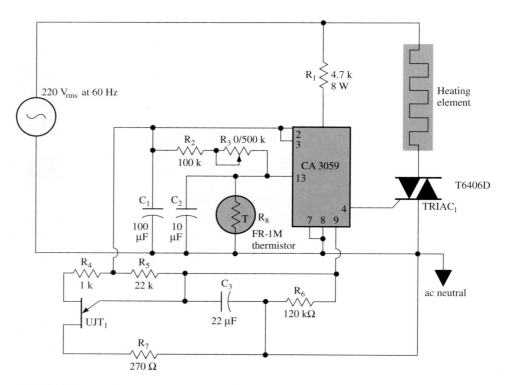

Figure 8–20
Proportional zero-voltage switching control (Courtesy of Harris Semiconductor, used with permission)

SECTION REVIEW 8–4

1. What are two alternative terms used to describe zero-point switching?
2. What are the advantages of zero-point switching over phase control switching? Why is zero-point switching essential in the control of high-power loads?
3. Explain the purpose of the Wheatstone bridge in Figure 8–14.
4. Explain the purpose of the comparator and SCR in Figure 8–14.
5. Identify the major advantages of the circuit in Figure 8–16 over that in Figure 8–14.
6. Describe the purpose of the zero-crossing detector in Figure 8–16. What is the frequency of its output signal?
7. Describe the function of the octal counter in Figure 8–16.
8. What is the function of the 555 timer in Figure 8–16? Describe the signals present at pins 3 and 6 of this device.
9. What is the function of Q_1 in Figure 8–16?
10. What type of circuit is formed by Q_2, Q_3, Q_4, and Q_5 within the CA3059 zero voltage switch?
11. True or false?: The basic function of the circuit in Figure 8–20 is similar to that of the circuit in Figure 8–16.
12. In Figure 8–20, what type of circuit is formed by the UJT, R_4, R_5, R_7, and C_3?
13. Describe the signal present at pin 9 of the CA3059 in Figure 8–20.
14. In Figure 8–20, what effect does an increase in the temperature of R_8 have on the pulse width of the signal at pin 4 of the CA3059?

8–5 DIGITALLY CONTROLLED FIRING DELAY OF A TRIAC

Figure 8–21 is a block diagram of a complete digital logic-based system for controlling the firing delay of a TRIAC. Rather than use an RC phase-shift network for controlling the trigger pulse at the gate of the TRIAC, this system contains an 8-bit binary up-counter as the controlling element in producing firing delay. The system is much more complex than the thyristor control circuits examined thus far. Such circuitry is necessary to allow a digital device such as a microcomputer or microcontroller to control an analog operation. Figures 8–22 through 8–26 contain the components found within each of the major blocks of Figure 8–21. After the operation of each of these major sections is examined in detail, the overall function of the circuit is analyzed as various binary operands are entered at the control input. If this circuit is constructed for laboratory investigation, control variables may be entered easily with an 8-bit DIP switch. This process simulates the input of control data from either a keypad or a microcomputer port. Like the proportional temperature controller examined previously, this digital-based controller could be part of a larger, closed-loop system in which a microcomputer is the controlling element.

In Figure 8–22, D_1, D_2, D_3, and D_4 form a full-wave bridge rectifier, which is connected across the secondary of a 12-V_{ac} step-down transformer. The unfiltered output of the full-wave bridge rectifier occurs at point A, as shown in the waveform diagram below the circuit. This signal serves as the active input to U_1, which is a high-speed comparator functioning as a zero-crossing detector. The pulse width of the output signal at point B depends on the threshold level set at the noninverting input of comparator U_1. The closer this threshold level is to 0 V, the narrower the pulse width of this signal. The pulse at point B occurs at each instant the source waveform crosses the zero-reference line. Since the pulses at point B occur at the half-cycle points of the source waveform, these signals appear as shown at the bottom of Figure 8–22. Since the frequency of these trigger pulses is twice that of the ac source, they occur at intervals of 8.33 ms.

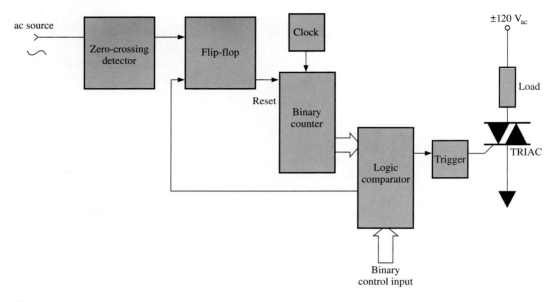

Figure 8–21
Functional block diagram of digitally controlled firing-delay circuit for a TRIAC

The presence of D_5 in Figure 8–22 allows full filtering of the source waveform to occur at points C and D. Also, D_5 allows the desired unfiltered signal to occur at point A. As point A begins to fall below its peak level of nearly 17 V, D_5 becomes reverse biased, preventing the growth of a steady dc potential at point A. This action is necessary for the proper function of the zero-crossing detector. Through the filtering action of C_1, a steady dc level of about 12 V is made available at point D, which serves as V_{CC} for the transistor switch in Figure 8–26. Further regulation is provided at point C by the Zener diode D_6. Here 5 V_{dc} is made available as V_{CC} for the comparator, the TTL ICs and the two 555 timers.

The two 7493s (U_2 and U_3 in Figure 8–23) together form an 8-bit binary upcounter. The clock source for this ripple counter is the 555 timer U_8, which is configured to function as an astable multivibrator. The output of this clock source is taken at point E. With potentiometer R_7 set to 10 kΩ, the operating frequency of this clock circuit is determined as follows:

$$PW = 0.693 \times (R_7 + R_8) \times C_3 = 0.693 \times 230 \text{ k}\Omega \times 100 \text{ pF} = 15.94 \text{ }\mu\text{sec}$$
$$SW = 0.693 \times R_8 \times C_3 = 0.693 \times 220 \text{ k}\Omega \times 100 \text{ pF} = 15.25 \text{ }\mu\text{sec}$$
$$T = 15.94 \text{ }\mu\text{sec} + 15.25 \text{ }\mu\text{sec} = 31.2 \text{ }\mu\text{sec}$$
$$f = \frac{1}{T} = 32 \text{ kpps}$$

8-bit up-count = 31.2 μsec × 256 ≅ 8 msec

The pulse repetition frequency for the astable multivibrator would allow 256 clock states to occur within a period of 8 msec, which is slightly less than the time for one alternation of the source waveform. This relationship between the signal at point A and the clock pulse occurring at point E allows the counter to begin incrementing from a reset condition (00000000), near the zero-crossing point of the source signal. The time required to attain

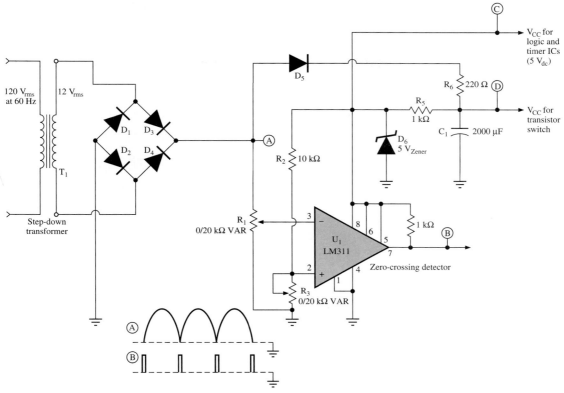

Figure 8–22
Zero-crossing detector section of digitally controlled firing-delay circuitry

the maximum count (11111111) is 8 msec, slightly less than the time required for one alternation of the source waveform. To achieve optimal control of the load, it is essential that this relationship exist between the source sine wave and the astable clock source. Otherwise, if the maximum counting time were to exceed the 8.33-msec period of an alternation, **cycle-skipping** would occur and control of the load would be lost. Fine adjustment of potentiometer R_7 allows for calibration of this clock signal. The importance of this crucial timing relationship of the clock and ac source signals becomes increasingly evident as this control circuit is examined in greater detail.

The two 7485's (U_4 and U_5) together form an 8-bit magnitude comparator. The pins of each IC are connected in such a way that a logic high occurs at output pin 6 at the instant a nibble from the control input equals the input nibble from one of the 7493s. Magnitude comparator U_4 compares the lower nibbles of the counter and the control variable, while U_5 performs the same function with the upper two nibbles. Only when the 8-bit number from the counter equals the 8-bit control variable do both 7485s produce a logic high (at pin 6) at the same instant. At that instant, the output of NAND gate U_{6A} falls low. This action triggers the 555 one-shot U_7, causing a logic high to occur momentarily at the output of that device. The action of this device is examined in detail later in this analysis.

The same active-low pulse from U_{6A} that triggers the one-shot also occurs at pin 4 of U_{6B}. If the counter had, until this instant, been incrementing, pin 6 of NAND gate U_{6B} would have been at a logic low. (Recall that if either pin 2 or 3 of a 7493 is held low,

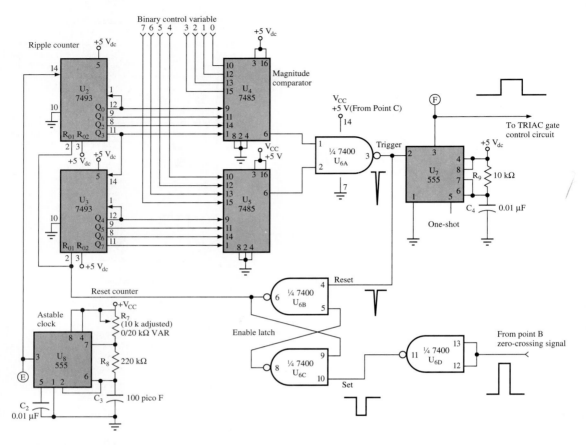

Figure 8–23
Digital timing circuitry

the device is able to increment on the negative transition of each clock pulse.) The logic low present at the output of NAND gate U_{6B} is fed back to pin 9 of NAND gate U_{6C}. Since U_{6B} and U_{6C} are functioning as a latch, the device may be considered to have already been placed in the set condition by an input pulse from point B, which is the output of the zero-crossing detector. Note that this signal is inverted by NAND gate U_{6D}, the output of which is the set input to the NAND latch. At this instant in the operation of the circuit, it is assumed that the set input to the NAND latch (pin 10 of U_{6C}) has returned to a logic 1.

With the NAND latch in a set condition, before the arrival of a trigger pulse from pin 3 of U_{6A}, the condition of the latch would be as shown in Figure 8–24(a). At the instant the count from the 7493s equals the 8-bit control variable, the falling edge of the trigger pulse at the input of the one-shot causes the NAND latch to toggle into the reset condition, as represented in Figure 8–24(b). The trigger pulse occurring at pin 3 of U_7 is extremely narrow, lasting only as long as the propagation time of the latch, the counter, the magnitude comparators, and NAND gate U_{6A}. Once the NAND latch resets, causing the output of U_{6B} to go high, the counter immediately resets to 00000000. Assuming the input control variable is a number other than 00000000, at least one of the inputs to U_{6A} falls to a logic low, causing its output to return to a logic high. Assuming pin 10 of U_{6C}

Figure 8–24
Operation of NAND latch

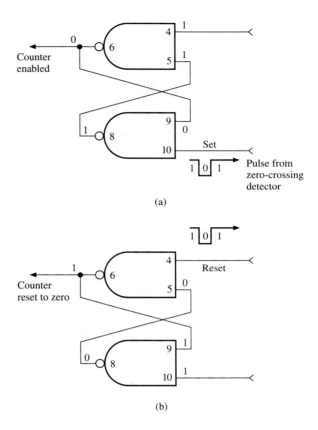

(a)

(b)

remains at a logic 1, as represented in Figure 8–24(b), the latch remains in a reset condition after the trigger pulse disappears. Thus, the counter remains in a reset condition until the arrival of the next pulse from the zero-crossing detector. The logic inverse of this signal, occurring at pin 10 of U_{6C}, toggles the NAND latch back into the set condition. Once this occurs, the counter is able to increment. The timing sequence analyzed thus far may be summarized as follows:

1. A start signal from the zero-crossing detector sets the NAND latch, allowing the ripple counter to increment from the reset condition of 00000000.
2. Once the number being formed by the counter equals the input control word, the one-shot is triggered and the NAND latch is reset.
3. The counter remains in a reset condition until the next start pulse arrives from the zero-crossing detector.

An illustration of the timing sequence for the entire control circuit is shown in Figure 8–25.

The purpose of the one-shot (555 timer U_7) is to provide an output pulse that activates the gate-triggering circuitry of the TRIAC. The pulse width of the one-shot output is determined as follows:

$$PW = 1.1 \times R_9 \times C_4$$
$$PW = 1.1 \times 10 \text{ k}\Omega \times 0.01 \text{ }\mu\text{F} = 110 \text{ }\mu\text{sec}$$

Figure 8–25
Waveforms for digitally controlled firing-delay circuitry

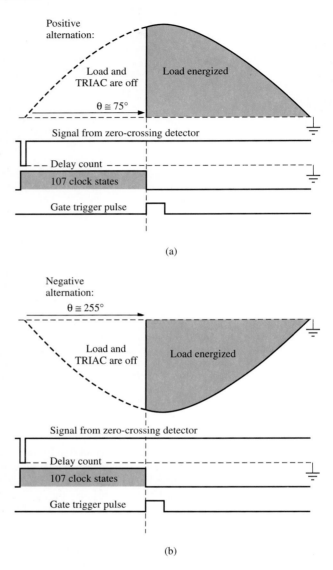

(a)

(b)

In Figure 8–23, when a strobe pulse occurs at point F, transistor Q_1 in Figure 8–26 switches on. The collector load for this transistor consists of the primary winding of **pulse transformer** T_2. As Q_1 switches on in saturation, nearly 12 V_{dc} develops over the primary winding of T_2 as its forward current increases to the maximum value in only a few microseconds. This rapid growth of current in the primary of the pulse transformer induces current flow in the secondary. As is indicated by the schematic symbol for T_2 in Figure 8–26, the pulse transformer is wound in such a way that that secondary current flows in the same direction as primary current. The induced current pulse in the gate circuit is sufficient magnitude and duration to cause triggering of the TRIAC. When the output of the one-shot falls low, Q_1 rapidly switches off. Diode D_7 functions as a kickback diode. (The function of this device was explained in detail in Chapter 1.) Once triggered, the

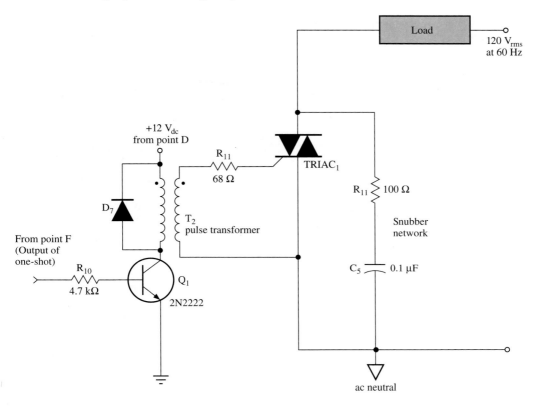

Figure 8–26
Output section of digitally controlled firing-delay circuitry

TRIAC remains on for the remainder of the alternation. As the source waveform approaches the 0-V point, commutation of the TRIAC occurs automatically. The TRIAC remains off until the next trigger pulse at pulse transformer T_2.

The operation of the digital control circuitry is now investigated, as the effects of changes in the value of the 8-bit control word on the firing-delay angle of the TRIAC are examined in detail. Ideally, having an 8-bit control input would allow the TRIAC to be triggered at any of 256 different points within a 180° alternation of the source waveform. The digital control source would then be capable of controlling the load power with a resolution of around 0.7°, which is calculated as follows:

$$\text{Resolution} = \frac{180°}{256} = 0.7° \text{ (per count)} \qquad (8\text{–}5)$$

The 7493s begin to count from 00000000 at the instant the output from the zero-crossing detector is high. This occurs slightly ahead of the 0° and 180° points of the source waveform. Also, if the astable clock source U_7 has been adjusted for an output frequency of 32 kpps, the counter arrives at its maximum count of 11111111 8 msec after the output of the zero-crossing detector goes high. Therefore, even if the control input has been set to its maximum value of 11111111, the TRIAC is still triggered on briefly for the last few degrees of each alternation of the source waveform. Since the time of one-half of the source cycle is 8.33 msec, the maximum firing-delay angle could be estimated as follows:

$$\frac{8 \text{ msec}}{8.33 \text{ msec}} = \frac{N}{180°}$$

$$N = \frac{8 \text{ msec} \times 180°}{8.33 \text{ msec}} = 173°$$

This result would indicate that, with the maximum control word of 11111111, the load could be energized only during the last 7° of an alternation. Also, consider that the TRIAC itself turns off as its main-terminal current falls below a minimum holding value. This is likely to occur very near the estimated 173° and 353° commutation points in the source cycle. Thus, with the maximum control word of 11111111, the load remains virtually off. At the other extreme, with an input word of 00000000, the load is likely to be turned on for virtually all the source cycle.

Between these two extremes of control range, which represent a fully on or off state for the load, gradations of load control may be achieved, as illustrated in the following examples.

Example 8–1

Assume the binary input to the control system represented in Figure 8–21 is 01101011. At what approximate points during the positive and negative alternation of the source waveform will the TRIAC be triggered?

Solution:
Step 1: Convert the input variable to a decimal value:

$$01101011_2 = 2^0 + 2^1 + 2^3 + 2^5 + 2^6 = 1 + 2 + 8 + 32 + 64 = 107_{10}$$

Step 2: Convert the count to a firing-delay angle:

$$0.7° \times 107 \cong 74.9° \cong 75°$$

Step 3: Since the TRIAC may be expected to trigger at nearly the same point in the negative alternation:

$$\text{Firing-delay angle} \cong 180° + 75° \cong 255°$$

These conditions are represented in Figure 8–25.

Example 8–2

Assume the firing-delay angles in Figure 8–21 are to be for the TRIAC 38° and 218°. What binary operand at the control input would allow this phase angle to occur?

Solution:
Step 1: Solve for the number of clock states between 0° and 38°:

$$N \text{ (clock states)} = \frac{38°}{0.7°} = 54.3 \cong 54$$

Step 2: Solve for the binary value of the control word:

$$54_{10} = 2^1 + 2^2 + 2^4 + 2^5 = 00110110_2$$

In these two examples, two conditions were randomly selected for demonstrating the method by which firing delay of the TRIAC might be controlled. Under normal operating conditions, however, digital control would occur through gradually incrementing or decrementing the input variable, depending on the power requirements of the load. Slowly decreasing the binary value of the control word would result in a gradually decreasing firing-delay angle. This, in turn, would result in a gradual increase in the amount of load power. Increasing the value of the binary control word would have the opposite effect, increasing the firing-delay angle and reducing the amount of power available to the load.

In this final circuit example, the versatility of thyristors and their associated control circuitry should have become quite evident. Through laboratory analysis and field testing of such a system in a temperature control application, a lookup table could be derived in which each of the 256 possible binary control words could be cross-referenced to both a phase angle and an oven temperature. Thus, a control code, representing the desired operating condition for the load, could be entered directly into the system by way of a keypad or DIP switch. Also, control of the load could be automatic, through operation of an embedded microcomputer, microcontroller, or programmable logic controller. Such digital control devices will be thoroughly investigated in later chapters.

SECTION REVIEW 8–5

1. Describe the operation of the two 7485 ICs in Figure 8–23.
2. Explain the operation of the NAND latch in Figure 8–23.
3. Explain the operation of the pulse transformer shown in Figure 8–26. Could this device be replaced by a photo TRIAC driver?
4. Define cycle skipping. Why could this be a serious problem within the control circuit just analyzed?
5. Assuming the control variable entering the circuitry in Figure 8–23 equals 10110100_2, what would be the firing-delay angles for both the positive and negative alternations of the source waveform?
6. If triggering of the TRIAC in Figure 8–26 is to occur 2.5 msec into each alternation of the source cycle, what binary number should be present at the control inputs of the circuitry in Figure 8–23?

SUMMARY

Dynamic control of dc loads may be accomplished by SCRs used in conjunction with bridge rectifiers. Common applications of such circuitry include dc motor controllers and battery chargers.

Although the simplest form of dynamic control for an ac load would consist of a TRIAC being triggered by a DIAC, more complex circuitry is required to avoid the problem of hysteresis. A hysteresis-free lamp dimmer is likely to contain an SBS rather than a DIAC as its gate triggering device. Since this device breaks over at a much lower potential than the DIAC, the phase-shift capacitor develops significantly less voltage than with the DIAC. Consequently, the problem of flash-on is significantly reduced. The problem of hysteresis is even further reduced by the fact that the SBS is forced to break over at the end of every alternation of the source waveform, thus preventing the capacitor from having a head start charging in either direction. The example of hysteresis-free phase-shift control presented in this chapter is a high-quality lamp dimmer. However, the basic design idea of this circuit may be incorporated into all forms of phase-shift control where flash-on might be a problem.

A problem inherent to thyristor phase-shift control circuitry is that of RF noise. Such high-frequency interference is often generated by SCRs and TRIACs as a result of the rapid rise in main-terminal current that takes place at the instant of triggering. Not only does this noise interfere with the operation of the circuitry containing the thyristor, it also enters ac power lines and even travels through the air in the same manner as radio signals. For most low- and medium-power applications of thyristor phase-shift circuitry, a simple LC filter is sufficient for noise suppression. The inductor serves to block RF noise leaving the circuit, while the capacitor shunts such noise to ground. Snubber networks are also important in suppressing noise within thyristor circuitry, especially those containing inductive loads. These RC networks, when placed in parallel with the main terminals of an SCR or TRIAC, dampen noise spikes that may be present on the ac line. This action reduces the possibility of accidental triggering of these devices by such noise spikes.

In high-power thyristor applications, where passive LC filters would prove inadequate in suppressing noise, phase-shift control is often abandoned in favor of zero-point switching. Zero-point switching circuits are often more complex than those involving phase shift. However, they have a major advantage in that they produce virtually no noise. This is because only complete alternations of a source signal are developed over a given load. Zero-point switching may be used in either on/off or proportional control. Where load control is proportional, pulse width modulation is often incorporated into gate-control circuitry. Gate-control circuits especially designed for zero-point switching applications of thyristors are available in monolithic form. These devices allow reduction in both size and complexity of zero-point switching circuitry.

Digital timing devices may be incorporated into thyristor phase-shift control circuitry. For example, the use of an 8-bit counter within the gate-control circuitry of a TRIAC would allow nearly 256 gradations of control within a given alternation of the source sine wave. While digital control of thyristors allows for precise control of firing delay, it also requires precise timing adjustment to avoid cycle skipping.

SELF-TEST

1. If R_1 in Figure 8–1 is set to 5.75 kΩ, what would be the firing delay-angle for the SCR?
2. For the circuit in Figure 8–1, what would be the minimum and maximum firing-delay angles if the 3.3-V Zener diode is replaced by a 4.7-V device?
3. Explain the purpose of R_1 in Figure 8–3.
4. What is the function of D_6 in Figure 8–3?
5. For the circuit in Figure 8–3, explain how the current at point A is reduced to a trickle-charge value as the voltage on the battery approaches its maximum level.
6. For the circuit in Figure 8–6, what happens at point B near the 0° and the 180° points of the source waveform?
7. For the circuit in Figure 8–6, what design features prevent C_1 from having a head start charging in the negative direction?
8. Explain the purpose of L_1 and C_2 in Figure 8–9.
9. For the circuit in Figure 8–9, assume RF noise with a frequency of 350 kHz is present on the ac line. How much inductive reactance does L_1 offer to this noise component? How much capacitive reactance does C_2 offer to this noise component?
10. What is the primary advantage of zero-point switching over phase-shift control circuitry? Why is zero-point switching preferred for controlling high-power inductive loads?

11. Define the term *set point* as it applies to the circuit in Figure 8–14. How is the set point in this circuit made variable?

12. In Figure 8–14, what happens to the voltage at point A as oven temperature increases?

13. While oven temperature is above the set-point level, what is the condition of the comparator in Figure 8–14? What is the condition of the SCR? Would the TRIAC and load be conducting at this time?

14. What are three advantages of the circuit in Figure 8–16 over that in Figure 8–14?

15. For the circuit in Figure 8–16, what form of pulse circuit allows proportional control of the load to take place?

16. What component in Figure 8–16 allows adjustment of the set-point temperature?

17. For the circuit in Figure 8–16, explain how a constant flow of current is made available at point E.

18. For the circuit in Figure 8–16, if oven temperature increases, what happens to the pulse width of the signal at point F?

19. For the circuit in Figure 8–16, what logic condition is present at point F while pin 4 of the 555 timer is at a logic low?

20. What two components in Figure 8–19 provide switching hysteresis?

21. What form of pulse circuit is contained in Figure 8–20?

22. What is the purpose of R_3 in Figure 8–22?

23. What is the purpose of the two 7485 ICs in Figure 8–23?

24. For the circuit in Figure 8–23, what adjustment should be made to ensure that cycle skipping does not occur?

25. For the circuit in Figure 8–23, assume the binary control variable equals 11010111. What would be the firing-delay angles for the positive and negative alternations of the source signal? (Assume R_7 is set to 10 kΩ).

9 INDUSTRIAL POWER SOURCES

CHAPTER OUTLINE

LEARNING OBJECTIVES

On completion of this chapter, the student should be able to:

- Identify the two basic configurations for three-phase power sources.
- Determine line voltage for a three-phase power source.
- Determine load voltage, current, and power consumption within a three-phase power distribution system.
- Explain the need for EMI and RFI filtering within industrial power distribution systems.
- Describe the operation of EMI and RFI filters.
- Explain the need for such protective devices as varistors, circuit breakers, and line fuses within industrial power distribution systems.
- Describe the operation of circuit breakers, line fuses, and varistors.
- Describe the basic structure and operation of the type of ac power distribution system commonly found in an industrial complex.
- Explain the function of varistors, circuit breakers, fuses, line filters, isolation transformers, and ac relays within ac power distribution systems.
- Explain the need for emergency-stopping and interlock circuitry within industrial systems.
- Describe how emergency stopping and interlock circuitry may be integrated into the ac power distribution system and motor control circuitry of a complex industrial machine.
- Identify the basic parts of a conventional linear power supply. Describe the function of each of these sections.

- Explain how three-phase transformers may be connected to brute force dc power supplies.
- Explain the need for voltage regulators within industrial circuitry. Describe the internal structure and operation of various IC voltage regulators.
- Identify problems associated with conventional power supplies.
- Explain the major advantages of high-frequency switching power supplies over conventional power supplies.
- Identify the basic parts of a typical high-frequency switching power supply. Describe the function of each of these sections.
- Explain how a multiple-output high-frequency switching power supply might be employed within an industrial system.
- Explain methods by which control circuitry is able to monitor the status of ac and dc power within an industrial system.

INTRODUCTION

In this chapter, methods by which ac and dc power are distributed within a modern industrial system are investigated. This survey begins with an introduction to the two basic configurations of three-phase power sources. Methods by which these configurations are used to power various ac loads are then briefly examined. After the various components common to industrial ac power distribution systems are introduced, a typical system is presented. Methods by which such a system is activated and controlled are examined in detail. Included in this study is the function of such protective devices as circuit breakers, fuses, and varistors. Also, emergency-stopping and interlock circuitry, which are safety requirements for most rotating machinery, are thoroughly examined.

Next, methods by which single-phase power and three-phase power are converted to dc are investigated. After the operation of conventional dc power supplies is investigated, high-frequency switching power supplies are examined in detail. The advantages of these innovative power sources over conventional dc supplies are stressed. Also, the application of a typical multiple-output switching power supply within an industrial system is investigated.

Finally, methods by which the system control electronics is able to detect failures within both the ac and dc power distribution systems of a complex industrial machine are investigated. The integrated circuits (ICs) commonly used in this fault-detection process are introduced. Diagnostic techniques used to isolate and identify failure of ac and dc voltage sources are also thoroughly examined.

9–1   INTRODUCTION TO THREE-PHASE POWER

The ideas of induced voltage and current should already be familiar from the study of basic electricity. Hence, it is assumed during this chapter that generator and transformer action, along with the fundamentals of the sine wave, are already understood. The most common form of sine wave is the 60-Hz/120-V rms line voltage shown in Figure 9–1(a). As seen in Figure 9–1(b), this voltage can be represented as a single vector, rotating counterclockwise on the reference axis.

If the peak amplitude of the ac source voltage is known, the instantaneous amplitude at any point on the sine wave may be determined using the basic formula for the sine wave:

$$v_i = V_{peak} \times \sin\theta \tag{9-1}$$

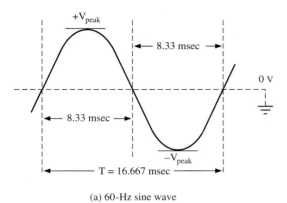

(a) 60-Hz sine wave

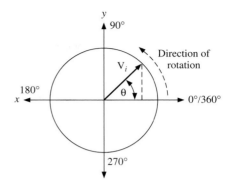

(b) Sine wave as rotating vector on reference axis

Figure 9–1

Often in the analysis of ac power sources and control circuitry, we need to equate an instantaneous value of voltage with a phase angle and/or a point in time. This is easily accomplished, as long as you remember the following basic relationship:

$$\frac{t}{T} = \frac{\theta}{360°} \qquad (9-2)$$

Example 9–1

For the sine wave represented in Figure 9–1(a), assume 2.7 msec have passed since the vector crossed the 0° reference point. Solve for the angle θ and the instantaneous voltage:

Solution:

Step 1: Solve for angle θ:

$$\frac{2.7 \text{ msec}}{16.667 \text{ msec}} = \frac{\theta}{360°}$$

$$0.162 = \frac{\theta}{360°}$$

$$\theta = 0.162 \times 360° \cong 58.3°$$

Step 2: Solve for v_i:

$$v_i = \sin 58.3° \times 169.7 \text{ V} = 144.4 \text{ V}$$

Although a single ac power line is adequate in many low- to medium-power industrial applications, in high-power systems it is often abandoned in favor of a **polyphase** power source. The most common polyphase power source is **three-phase power** (abbreviated 3-ϕ). As represented in Figure 9–2(a), three-phase voltage may be generated by placing three stator windings at equal distances around a rotating field winding. The three resultant output waveforms are spaced 120° apart, as shown in Figure 9–2(b). As seen in Figure 9–2(c), these ac voltages can be represented as three rotating vectors spaced at equal distances on the reference axis. With the **wye configuration** represented here, the common point for the three stator windings serves as ac neutral. With respect to this neutral point,

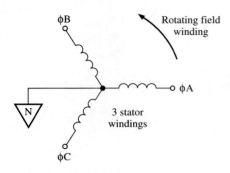

(a) 3-φ generator in wye configuration

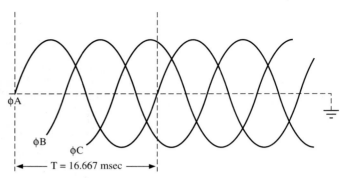

(b) Sine waves produced by 3-φ generator spaced 120° apart

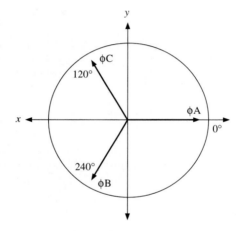

(c) 3-φ voltage represented as three
rotating vectors on a reference axis

Figure 9–2

each of the output waveforms is equivalent to the single ac line voltage represented in Figure 9–1, having a peak amplitude of nearly 169.7 V and an rms value of 120 V.

Note that, at any instant in time, the algebraic sum of the instantaneous voltages of the three waveforms is zero. For instance, when φA is at the 90° point, its instantaneous amplitude is equivalent to 169.7 V. At the same time, the voltage of φB would be determined as follows. Since φB lags behind φA by 120°:

$$\theta = 90° - 120° = -30°$$
$$v_b = 169.7 \text{ V} \times \sin{-30°} = -84.85 \text{ V}$$

Since φC lags behind φB by 120°:

$$\theta = -30° - 120° = -150°$$
$$v_c = 169.7 \text{ V} \times \sin{-150°} = -84.85 \text{ V}$$

Thus,

$$v_a + v_b + v_c = 169.7 \text{ V} + -84.85 \text{ V} + -84.85 \text{ V} = 0 \text{ V}$$

Although each phase voltage (V_ϕ) within the wye configuration is 120 V_{rms}, the actual **line voltage** (V_L), which is measured between any pair of stator windings, is 208 V_{rms}. This may be proven through phaser analysis.

As illustrated in Figure 9–3(a), assume the line voltage is being measured between ϕA and ϕC. Remember that, in order to determine a difference in voltage between two points, the voltage at the second point must be *subtracted* from the voltage at the first point. Thus, as shown in Figure 9–3(b), the vector for ϕC is flipped 180°. The vector representing a $-\phi C$ is now placed 60° clockwise with respect to the vector for ϕA. To determine the length of the vector representing the difference in voltage between ϕA and ϕC, a parallelogram is completed as shown in Figure 9–3(c). Next, a vector is extended from the neutral point to the V_L point. It is this vector that represents the magnitude of the line voltage. As illustrated in Figure 9–3(c), the vector representing line voltage divides the parallelogram into two equal isosceles triangles. Finally, a straight line drawn from the ϕA point to the $-\phi C$ point divides each isosceles triangle into two right triangles. For each right triangle, the base line is equal to half the V_L vector, the hypotenuse is equal to the voltage of a single phase, and the angle θ becomes 30°. Thus, for each right

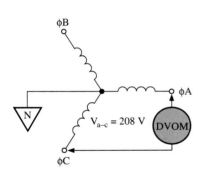

(a) DVOM measuring line voltage

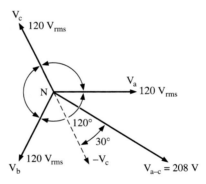

(b) Vector analysis of 3-ϕ line voltage

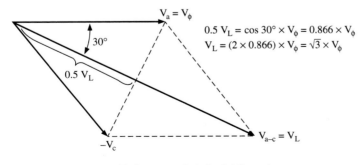

$$0.5 \, V_L = \cos 30° \times V_\phi = 0.866 \times V_\phi$$
$$V_L = (2 \times 0.866) \times V_\phi = \sqrt{3} \times V_\phi$$

(c) Detail of vector analysis for 3-ϕ line voltage

Figure 9–3

triangle, the ratio of the base line to the ϕA vector is equivalent to $\cos 30°$. Once this important mathematical relationship is realized, an expression showing the relationship of a given phase voltage to line voltage can be developed:

$$0.5 \times V_L = \cos 30° \times V_\phi = 0.866 \times V_\phi$$

Thus,

$$2 \times 0.5 \times V_L = 2 \times 0.866 \times V_\phi$$
$$V_L = 1.732 \times V_\phi = \sqrt{3} \times V_\phi$$

(9–3)

Thus, for a generator with stator windings in a wye configuration and a phase voltage of 120 V_{rms}:

$$V_L = 120 \text{ V}_{rms} \times 1.732 = 207.8 \text{ V} \cong 208 \text{ V}$$

For all 3-ϕ generators in a wye configuration, the factor 1.732 remains constant, regardless of the value of phase voltage. Thus, if the phase voltage is known, the line voltage may be easily determined.

Example 9–2

Assume the peak amplitude of each phase voltage of a 3-ϕ generator in a wye configuration is 311 V. Determine the line voltage.

Solution:
Step 1: Solve for the rms value of phase voltage:

$$V_{rms} = 0.707 \times 311 \text{ V} \cong 220 \text{ V}_{rms}$$

Step 2: Solve for line voltage:

$$V_L = 1.732 \times 220 \text{ V}_{rms} \cong 381 \text{ V}$$

Figure 9–4 illustrates two methods by which loads may be connected to a 3-ϕ power source in a wye configuration. If, as seen in Figure 9–4(a), a resistive load is connected between any pair of phases, the load voltage becomes the 208 V line voltage. However, if the load is connected between a single phase and neutral, as shown in Figure 9–4(b), its voltage is limited to the 120 V_{rms} phase voltage. For a load connected between two phases, current is easily determined using Ohm's law:

$$I_L = \frac{208 \text{ V}}{R_L}$$

(9–4)

Here, according to Kirchhoff's current law, the current entering the load must equal the current leaving the load. Thus, in Figure 9–4(a), the current entering and leaving the winding of ϕA, the load, and the winding of ϕC must be equal. For a load connected between any phase and neutral, the current is again determined using Ohm's law:

$$I_\phi = \frac{120 \text{ V}_{rms}}{R_\phi}$$

(9–5)

Here, Kirchhoff's current law would dictate that the current leaving the ϕA winding must equal the current returning to the power source through the neutral line.

Now assume, as shown in Figure 9–5, that three resistive loads are connected between each of the three phases and the neutral line. According to Kirchhoff's current law, the

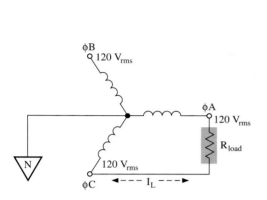

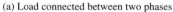

(a) Load connected between two phases

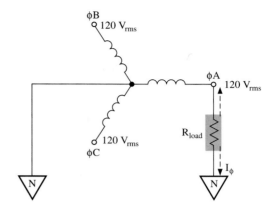

(b) Load connected between a phase and neutral

Figure 9–4

algebraic sum of the currents entering and leaving the neutral node equals zero. Thus, at any instant:

$$I_N = I_{\phi 1} + I_{\phi 2} + I_{\phi 3} \tag{9–6}$$

However, since these phase currents are 120° out of phase, they cannot be added together by simple algebraic means. To determine the value of current flowing through the neutral line, it is necessary to convert the three phase currents from polar to rectangular form.

As an example of this operation, assume that each of the resistive loads in Figure 9–5 equals 47 Ω, causing the phase currents to be equal. It can be proven that, with this

Figure 9–5
Three loads connected from each phase to neutral

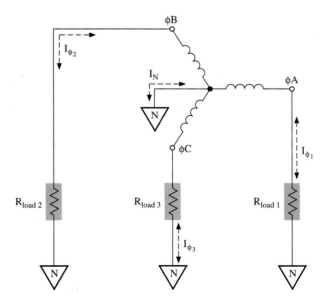

condition, the load currents cancel each other, causing the neutral current to be 0 A:

$$I_{\phi 1} = I_{\phi 2} = I_{\phi 3} = \frac{120 \text{ V}_{rms}}{47 \text{ }\Omega} = 2.553 \text{ A}_{rms}$$

Thus, in polar form:

$$I_N = 2.553 \text{ A} \angle 0° + 2.553 \text{ A} \angle 120° + 2.553 \text{ A} \angle -120°$$

To add the currents of ϕB and ϕC to that of ϕA, these currents must first be converted to rectangular form. Beginning with current from ϕC (Figure 9–6):

$$\sin\theta = \sin 120° = 0.866$$
$$0.866 \times 2.553\text{A} = 2.21 \text{ A}$$
$$\cos\theta = \cos 120° = -0.5$$
$$-0.5 \times 2.553 \text{ A} = -1.277 \text{ A}$$

Thus,

$$2.553 \text{ A} \angle 120° = (-1.277 \text{ A} + j2.211 \text{ A})$$

For ϕB current (Figure 9–7):

$$\sin\theta = \sin-120° = -0.866$$
$$-0.866 \times 2.553 \text{ A} = -2.211 \text{ A}$$
$$\cos\theta = \cos-120° = -0.5$$
$$-0.5 \times 2.553 \text{ A} = -1.277 \text{ A}$$

Figure 9–6
Polar and rectangular representations of current for ϕC

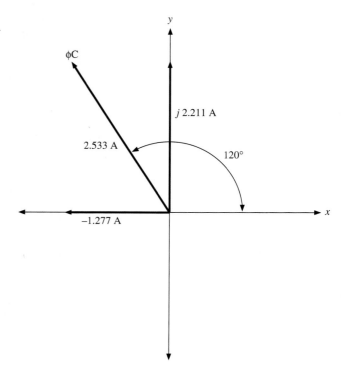

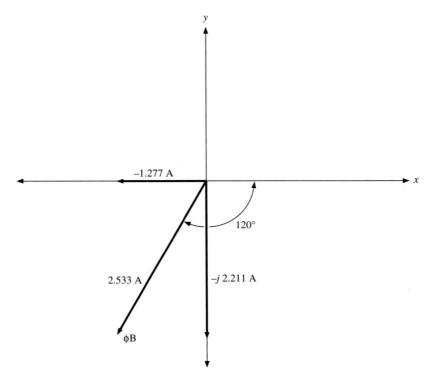

Figure 9–7
Polar and rectangular representations of current for ϕB

Thus,

$$2.553 \text{ A} \angle -120° = (-1.277 \text{ A} + -j2.211 \text{ A})$$

Solving for current in the neutral line:

$$I_N = 2.553 \text{ A} + (-1.277 \text{ A} + j2.211 \text{ A}) + (-1.277 \text{ A} + -j2.211 \text{ A})$$
$$= 2.553 \text{ A} - 1.277 \text{ A} - 1.277 \text{ A} + j2.211 \text{ A} - j2.211 \text{ A} \cong 0 \text{ A}$$

In a well-designed 3-ϕ power distribution system, the loading for the three phases is kept as uniform as possible. However, in most industrial 3-ϕ phase systems, phase currents vary significantly, resulting in a substantial amount of current flow in the neutral line.

Example 9–3

Assume for the circuit in Figure 9–5 that R_{Load1} equals 33 Ω, R_{Load3} equals 68 Ω, and R_{Load2} equals 100 Ω. What are the values of phase current? What is the current flow in the neutral line? Indicate the neutral current in both rectangular and polar forms.

Solution:

Step 1: Solve for the phase currents:

$$I_{\phi 1} = \frac{120 \text{ V}_{rms} \angle 0°}{33 \Omega} = 3.636 \text{ A}_{rms} \angle 0°$$

$$I_{\phi 3} = \frac{120\ \text{V}_{\text{rms}} \angle\ 120°}{68\ \Omega} = 1.765\ \text{A}_{\text{rms}} \angle\ 120°$$

$$I_{\phi 2} = \frac{120\ \text{V}_{\text{rms}} \angle\ -120°}{100\ \Omega} = 1.2\ \text{A}_{\text{rms}} \angle\ -120°$$

Step 2: Convert I_{L_2} and I_{L_3} to rectangular form:

$$\sin 120° \times I_{\phi 3} = 0.866 \times 1.765\ \text{A} = 1.529\ \text{A}$$
$$\cos 120° \times I_{\phi 3} = -0.5 \times 1.765\ \text{A} = -882.5\ \text{mA}$$
$$I_{\phi 3} = -882.5\ \text{mA} + j1.529\ \text{A}$$
$$\sin -120° \times I_{\phi 2} = -0.866 \times 1.2\ \text{A} = -1.039\ \text{A}$$
$$\cos -120° \times I_{\phi 2} = -0.5 \times 1.2\ \text{A} = -600\ \text{mA}$$
$$I_{\phi 2} = -600\ \text{mA} + -j1.039\ \text{A}$$

Step 3: Solve for current flow in the neutral line:

$$I_N = 3.636\ \text{A} + (-882.5\ \text{mA} + j1.529\ \text{A}) + (-600\ \text{mA} + -j1.039\ \text{A})$$
$$= 3.636\ \text{A} - 882.5\ \text{mA} - 600\ \text{mA} + j1.529\ \text{A} - j1.039\ \text{A}$$

Step 4: Convert I_N to polar form:

$$I_N = 2.154\ \text{A} + j490\ \text{mA}$$
$$\tan\theta = \frac{0.49}{2.154} = 0.2275$$
$$\theta = 12.82°$$
$$I_N = \sqrt{2.154\ \text{A}^2 + 490\ \text{mA}^2} = 2.209\ \text{A}$$
$$\cong 2.2\ \text{A} \angle\ 12.8°$$

SECTION REVIEW 9–1

1. For the circuit in Figure 9–8, what voltages would be indicated by DVOM 1 and DVOM 2?
2. True or false?: DVOM 1 in Figure 9–8 is reading phase voltage and DVOM 2 is reading line voltage.

Figure 9–8

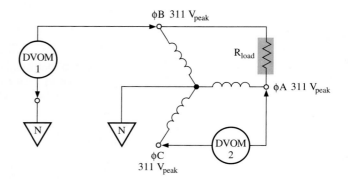

Figure 9–9

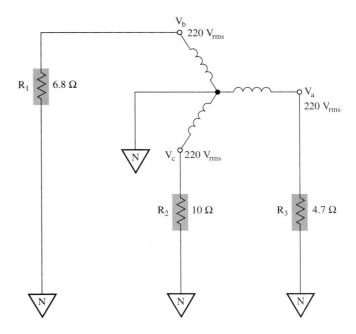

V_b 220 V_{rms}

R_1 6.8 Ω

V_a
220 V_{rms}

N

V_c 220 V_{rms}

R_2 10 Ω

R_3 4.7 Ω

N

N

N

3. True or false?: Ideally, for the circuit in Figure 9–8, the rms value of current flowing through the ϕA winding would be the same as that for ϕB.
4. For the circuit in Figure 9–9, determine the current flow for each of the loads. What is the amplitude of the current flowing through the neutral line? Express your answers in both rectangular and polar forms.
5. For the circuit configuration in Figure 9–9, what condition must exist for the current through the neutral line to be 0 A?

9–2 LINE FILTERS AND PROTECTIVE DEVICES

Within an industrial facility, there is likely to be a considerable amount of noise riding on the ac power lines. Such noise could exist in the form of **electromagnetic interference** (EMI) and **radio-frequency interference** (RFI). EMI results from the presence of such inductive devices as motors, generators, clutch/brakes, solenoids, and relay coils. RFI, however, is likely to be generated by electromechanical and electromagnetic switching devices. Likely sources of RFI include SCR and TRIAC-based control circuits (other than those using zero-point switching) and high-frequency switching power supplies. To control such electromagnetic and RF noise problems, a wide variety of line filters are commercially available. A large 3-ϕ EMI filter with a high current rating, such as that shown in Figure 9–10(a), is usually found at the main ac power input of an industrial system. The purpose of this filter is to attenuate noise that might be entering or leaving a system through the ac power bus. Smaller, single-phase RFI filters, such as that shown in Figure 9–10(b), are placed at the inputs to switching power supplies. Although these filters do attenuate line noise entering a system, their primary function, when used with switching power supplies, is to inhibit noise generated by the power supply itself.

Important specifications for line filters include the **current rating** and the **insertion loss.** The current rating represents the maximum rms line current that may be drawn continuously by a load through a given filter. Insertion loss is the ability of the line filter

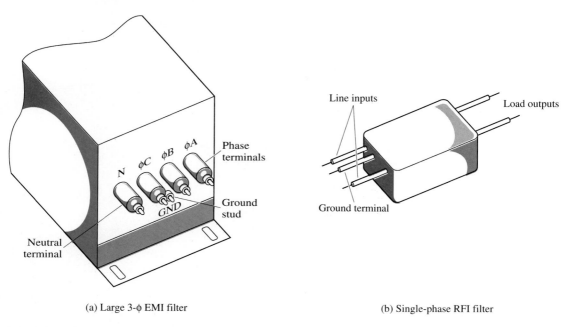

(a) Large 3-ϕ EMI filter

(b) Single-phase RFI filter

Figure 9–10

to block interference frequencies, and it is specified in decibels of attenuation. As an example, the data sheet for a certain filter gives a current rating of 30 A per phase. Insertion loss is usually specified in tabular form, being listed in the data sheets for both differential-mode (L/N) and common-mode (L/G) interference. Data sheets for EMI and RFI filters also specify **leakage current.** This parameter, which represents the minute amount of current lost from line to line or line to ground within the filter, is, at worst case, only a few milliamperes.

Figure 9–11 contains the schematic diagram for a typical 3-ϕ line filter. This passive LC device contains three identical τ **filters.** It is designed to attenuate both **differential-mode** noise and **common-mode** noise that may be present on the input power bus. For a 3-ϕ line filter, differential-mode interference would be present simultaneously between a given phase line and the neutral line or any pair of phase lines. Common-mode noise, however, would be present between a phase line and ground. For a single-phase line filter, differential-mode noise would occur simultaneously between the two lines, whereas common-mode noise would occur between either line and ground.

For the filter in Figure 9–11, the nine chokes offer minimal impedance to the 60-Hz ac voltage passing through the filter. At the same time, they exhibit extremely high impedance to any noise spikes or high-frequency interference riding on this 60-Hz waveform. The **X capacitors,** placed between each of the three phase lines and the neutral line, offer a high impedance to the 60-Hz waveform. At the same time, these X capacitors offer minimal impedance to noise spikes and high-frequency interference. Thus, these higher frequency components of the source waveform are effectively shorted between a given phase line and neutral. The **Y capacitors** perform the same low-pass filtering action as the X capacitors. As part of the common-mode filtering process, they serve to shunt noise spikes and high-frequency interference directly toward ac ground. Resistors R_1, R_2, and R_3 serve as discharge (bleeder) resistors for capacitors C_{X4}, C_{X5}, and C_{X6}.

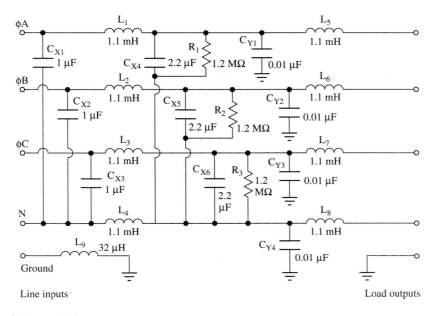

Figure 9–11
Schematic diagram of a 3-ф EMI line filter

Example 9–4

For the line filter represented in Figure 9–11, what impedance would C_{X1}, C_{X2}, or C_{X3} offer to the 60-Hz line voltage? What impedance would one of these components offer to interference occurring at 50 kHz? What impedance would one of the 1.1-mH inductors offer at each of these frequencies? Finally, assume the 50-kHz interference is entering the line filter as a component of all three phase voltages. Assume this noise is also present on the neutral line. If this source interference has a peak amplitude of 450 mV and a frequency of 50 kHz, what would be its amplitude measured from line to neutral at the filter output? What would be its amplitude measured from line to ground?

Solution:

Step 1: At 60 Hz:

$$X_C = \frac{1}{2\pi \times 60 \text{ Hz} \times 1 \text{ } \mu\text{F}} = 2.653 \text{ k}\Omega$$

$$X_L = 2\pi \times 60 \text{ Hz} \times 1.1 \text{ mH} = 0.415 \text{ } \Omega$$

Step 2: At 50 kHz:

$$X_C = \frac{1}{2\pi \times 50 \text{ kHz} \times 1 \text{ } \mu\text{F}} = 3.18 \text{ } \Omega$$

$$X_L = 2\pi \times 50 \text{ kHz} \times 1.1 \text{ mH} = 345.6 \text{ } \Omega$$

Step 3: For the 3-ф line filter in Figure 9–11, at 50 kHz, L/N insertion loss equals 15 dB and L/G insertion loss is 10 dB. Therefore, between a given output phase and neutral:

$$-15\,\mathrm{dB} = 20\log_{10}\left(\frac{V_{\mathrm{out}}}{450\,\mathrm{mV}}\right)$$

$$-0.75 = \log_{10}\left(\frac{V_{\mathrm{out}}}{450\,\mathrm{mV}}\right)$$

$$\mathrm{Antilog}\,{-0.75} = \frac{V_{\mathrm{out}}}{450\,\mathrm{mV}}$$

$$\frac{V_{\mathrm{out}}}{450\,\mathrm{mV}} = 0.17783$$

$$V_{\mathrm{out}} = 0.17783 \times 450\,\mathrm{mV} = 80\,\mathrm{mV}$$

Between a given phase and ac ground:

$$-10\,\mathrm{dB} = 20\log_{10}\left(\frac{V_{\mathrm{out}}}{450\,\mathrm{mV}}\right)$$

$$V_{\mathrm{out}} = 450\,\mathrm{mV} \times 0.31623 = 142.3\,\mathrm{mV}$$

Figure 9–12 contains the schematic diagram for a **dual τ section** RFI filter. The design of this device is essentially the same as that of its EMI counterpart in that it contains high-frequency chokes, X capacitors, and Y capacitors. For effective operation in the RFI range, X capacitors range from 0.1 to 2 μF; Y capacitors range from 2200 pF to 0.033 μF; and inductor values range from 1.8 to 47 mH. If the value of X capacitance within the RFI line filter exceeds 1 μF, it is recommended that a discharge resistor, such as R_1 in Figure 9–12, be incorporated into the design. To meet industrial safety standards, a value of discharge resistance must be selected according to the following formula:

$$R = \frac{1}{2.21 \times C_X} \tag{9–7}$$

where C_X represents the sum of the X capacitors.

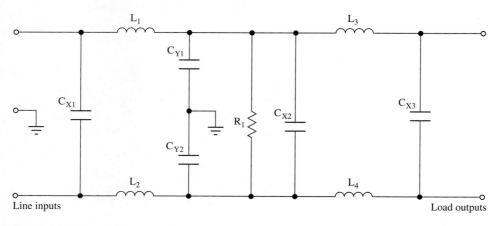

Figure 9–12
Schematic diagram of a dual τ section RFI line filter

Example 9–5

For the RFI line filter represented in Figure 9–12, assume that each of the X capacitors equals 0.1 μF. What value of R_1 must be selected to comply with the accepted industrial safety standards?

Solution:

$$R = \frac{1}{2.21 \times C_X} = \frac{1}{2.21 \times 0.3 \ \mu F} = 1.508 \ M\Omega$$

Choose 1.5 MΩ as a standard value.

Besides electromagnetic and radio-frequency interference, another serious problem associated with the 60-Hz power line is that of transient voltage spikes. Special devices called **varistors** are often used to suppress such sudden increases in line potential, which may occur during thunderstorms or as a result of power surges. A typical varistor is shown in Figure 9–13(a) and its internal composition is shown in 9–13(b). Figure 9–13(c) shows two possible schematic symbols for the varistor.

Although, as seen in Figure 9–13(a), a varistor is likely to resemble a large mica dip capacitor, its operation is similar to that of a thermistor with a **negative temperature coefficient** (NTC). As shown in Figure 9–14, varistors are usually placed between the individual phase lines and ac neutral, ahead of the EMI filters on the main power buses. At the instant a voltage spike begins to develop on an ac line, current through the varistors begins to increase. As this current increases, the temperature of the varistor rises. Consequently, due to its NTC, the resistance of the device rapidly decreases, causing still more current to be shunted past the line filter. Thus, potentially dangerous voltage spikes are dampened and costly system circuitry is protected through the action of the varistor. However, while performing its protective function, especially during an electrical storm, a varistor is likely to be damaged or destroyed. For this reason, varistors placed within main power distribution systems should be examined frequently and replaced when necessary as part of the preventive maintenance program for a given piece of machinery.

(a) Typical varistor

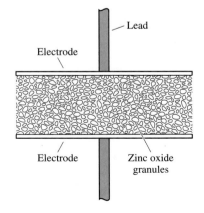

(b) Composition of varistor

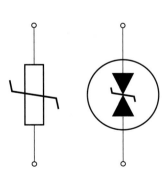

(c) Schematic symbols for varistor

Figure 9–13

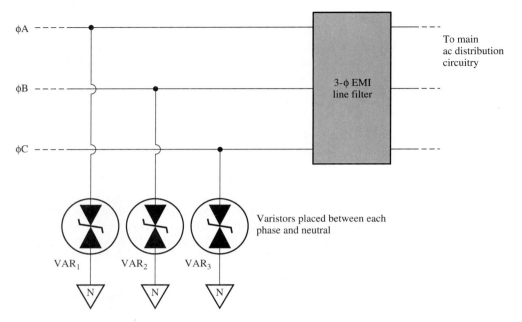

φA

φB

3-φ EMI
line filter

φC

To main
ac distribution
circuitry

Varistors placed between each
phase and neutral

VAR₁   VAR₂   VAR₃

N   N   N

Figure 9–14
Varistors placed ahead of main line filter

    Before discussing ac power distribution in detail, it is necessary to analyze the operation of circuit breakers and fuses. In the control of ac power, circuit breakers provide a dual function. First, they serve as manually operated switches. Large 3-φ circuit breakers, such as that represented in Figure 9–15, are used to connect or disconnect ac voltage to the main power buses of an industrial system. Smaller circuit breakers are used to control the distribution of power to various branch lines. To perform their second purpose, which is to prevent an overload condition, circuit breakers are designed to trip automatically when their rated current limit is exceeded, thus preventing damage to costly system motors and control circuitry.

    Figure 9–16 illustrates the operating principle of a **thermal circuit breaker.** Current flows through this device along a bimetallic strip, which is constructed of two different metal alloys. The alloy that comprises the lower metallic strip has a greater coefficient of expansion than the upper strip. During a condition of normal current flow, as represented in Figure 9–16(a), the contacts of the circuit breaker remain closed. However, if the bimetallic strip is heated by excessive current flow, the bottom metal expands faster than the metal on the top. Consequently, the bimetallic strip bends upward, opening the contacts of the circuit breaker.

    As shown in Figure 9–16(b), the bending action of the bimetallic strip allows the spring-loaded **holding lever** to slip into position between the bimetallic strip and the lower contact of the circuit breaker. Thus, closure of the contacts is inhibited as the bimetallic strip begins to cool due to the absence of current flow. Only when the circuit breaker is mechanically reset are the contacts again able to close.

    Figure 9–17(a) contains the schematic symbol for a single-phase circuit breaker, and Figure 9–17(b) contains the symbol for a 3-φ device. Many of the circuit breakers found in industrial control systems contain **sensing contacts.** This is especially true for those smaller, single-phase devices designed for use within branch circuitry. The operation of

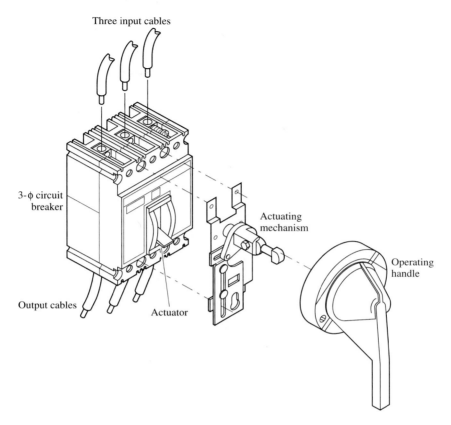

Figure 9–15
Structure and operation of a typical 3-φ main circuit breaker

the sensing contact is synchronized with that of the main contacts of the circuit breaker. In a large industrial system, the presence of these sensing contacts allows control circuitry to monitor the status of the branch circuit breakers. A possible method of using a sensing contact is illustrated in Figure 9–18.

When the sensing contact in Figure 9–18 is closed, current flows through the LED within the optoisolator. As a result, the phototransistor is switched on in saturation and a logic low is maintained at point B. Thus, the system control circuitry is able to sense that the circuit breaker is closed by detecting a logic high at point C. If the circuit breaker is manually opened, or tripped automatically as the result of an overcurrent condition, point C falls to a logic low. When the control circuitry detects this transition at point C, it quickly responds by removing power from all the machinery. As a part of its diagnostic routine, the system control could, through use of a display module or even a single LED, indicate the location of the tripped circuit breaker. At this juncture, we should stress that the frequent tripping of a motor circuit breaker is likely to indicate a problem with the motor rather than the breaker itself. For instance, the motor could be rotating slowly, possibly as a result of worn bearings. Consequently, it draws excessive current, causing the circuit breaker to be tripped. Therefore, the frequent tripping of any circuit breaker within an industrial system should not be ignored. The proper response to this problem would be a thorough examination of the circuitry and components in line with that circuit breaker.

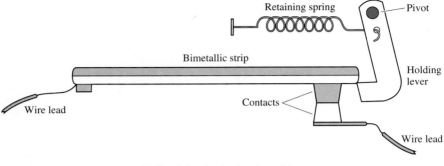

(a) Circuit breaker in closed condition

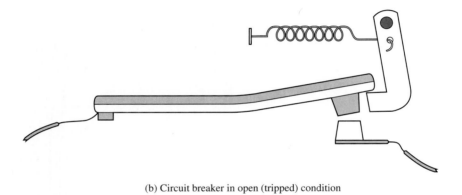

(b) Circuit breaker in open (tripped) condition

Figure 9–16

Fuses are the oldest form of electrical protective device. Like circuit breakers, they are used to protect electrical components from an overcurrent condition. Whereas circuit breakers are placed in either single-phase or 3-φ power distribution lines, fuses are usually placed in line with ac or dc components within individual branch circuits. Fuses are classified according to their **fusing characteristic,** which may be either **standard (normal-blo)** or **slow-blo** (also referred to as **time delay** or **dual-element**).

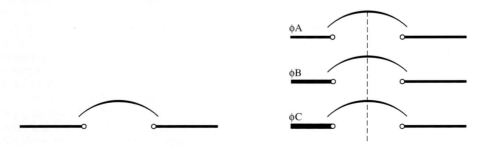

(a) Schematic symbol for single-phase circuit breaker

(b) Schematic symbol for 3-φ circuit breaker

Figure 9–17

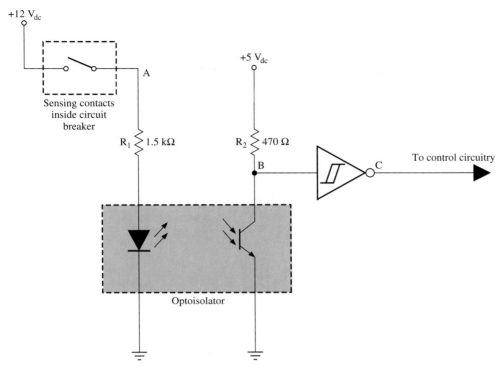

Figure 9–18
Conditioning of signal from sensing contacts inside a circuit breaker

The internal structures of both a standard and a slow-blo fuse are illustrated in Figure 9–19. As shown in Figure 9–19(a), the standard fuse contains a **metal fuse link,** either end of which is soldered to a metal cap. This fuse link is placed within either glass or ceramic tubing. This tubing could contain a chemical fill, a vacuum, or simply air. As seen in Figure 9–19(b), slow-blo fuses, like their standard counterparts, contain a metal fuse link. These devices also contain a spring, which is connected to the fuse link by a solder joint.

Important specifications for line fuses are their current and voltage ratings. The current rating for a standard fuse is that amount of rms current that causes the fusible link within the fuse to melt. The ambient temperature at which this melting effect occurs is usually designated as 25°C. The actual melting current for a fusible link is, however, affected by environmental temperature. Thus, a fuse placed in a hot environment may be expected to blow at a significantly lower value of current than one placed in a relatively cool location. The voltage rating for a line fuse becomes important after the fuse has blown. Providing that a fuse remains intact while power is applied to a circuit, it offers virtually no impedance to current flow and thus drops nearly 0 V. However, if the fuse opens due to an overcurrent condition, it acts like an open switch. Thus, the voltage normally developed over the load protected by the fuse is now developed over the fuse itself. To prevent arcing and a possible electrical fire, the fuse must be able to withstand the applied voltage. A line fuse must, therefore, be rated at a voltage significantly higher than that normally applied to the load.

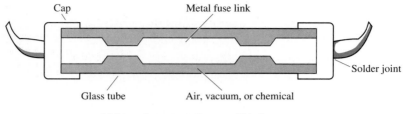

(a) Internal structure of a normal-blo fuse

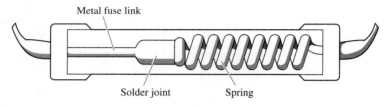

(b) Internal structure of a slo-blo fuse

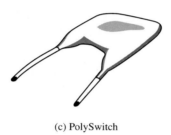

(c) PolySwitch

Figure 9–19

Standard fuses are designed for fast response to a sudden increase in current. They are usually placed in series with loads that are primarily resistive. Since such loads are expected to exhibit a minimal surge effect, either at turn on or during their operation, a large increase in current should be considered to be the result of a short circuit. Thus, the standard fuse should quickly blow at the instant a current surge begins. When selecting a standard fuse, the current rating for this fuse must be greater than the current that is normally drawn by the load. However, this rated current must still be less than that which would occur during an overcurrent condition for the load.

Slow-blo fuses are designed for use in loads that are likely to undergo current surges, especially at the time of turn on. Such loads would include motors and circuitry containing large capacitors. If an abnormally large surge in current were to occur (well above the rated value of the slow-blo fuse), the metal fuse link would quickly melt, breaking the conduction path. If the surge of current were only slightly above the rated current value for the fuse, the fuse would remain intact. However, a prolonged overcurrent condition would eventually cause the solder connection within the fuse to melt. The spring mechanism within the fuse would then quickly pull away from the fuse link, interrupting the flow of current.

If a component within a system frequently blows fuses, that component should be carefully examined for a possible short circuit or partial short circuit condition. When a line fuse blows, it should never be replaced by a fuse of a higher current rating or a lower voltage rating. A standard (normal-blo) fuse should never be replaced by a slow-blo fuse.

A weakness of conventional line fuses is that they are destroyed as a result of protecting a circuit during an overcurrent condition. This problem is overcome by specially designed **positive temperature coefficient (PTC)** devices. An example of this special type of fuse is the **PolySwitch,** which is shown in Figure 9–19(c). This device is similar to disk capacitors and thermistors in size and appearance. As long as branch current is at a safe level, the conductive polymer compounds within the PolySwitch offer minimal resistance. However, an overcurrent condition causes the PolySwitch to heat rapidly and become nonconductive, functioning identically to an open fuse. Only when circuit current is reduced to a safe level does the PolySwitch again become conductive.

SECTION REVIEW 9–2

1. What is an EMI filter? Describe its purpose within a power distribution system.
2. What is an RFI filter? Describe its purpose within a power distribution system.
3. A single-phase RFI filter exhibits an insertion loss of 65 dB to noise signals occurring at a frequency of 1 MHz. If a noise component of the output of the filter had a peak amplitude of 235 μV at that frequency, what would be the peak amplitude of that noise component at the input to the filter?
4. Describe the operation of a varistor. What purpose does this device serve in an ac power distribution system?
5. Describe the operation of a thermal circuit breaker. How does this device react to an overcurrent condition? After the tripping of a thermal circuit breaker, what prevents its contacts from closing as the device cools?
6. What is the reason for placing sensing contacts within a circuit breaker?
7. Describe the internal structure and operation of a slow-blo fuse.
8. What type of load is likely to be protected by a slow-blo fuse?

9–3   AC POWER DISTRIBUTION SYSTEMS

Figure 9–20 illustrates how a 3-ϕ power source (in a wye configuration) could be connected to the ac power distribution buses within a large industrial system. Circuit breaker CB_1, which is a manually operated contactor, serves as the main on/off switch to the power distribution system. Although closing CB_1 connects facility power to the machinery, it does **not** cause the system to begin to operate. Immediately beyond this main contactor are three neon lamps that indicate the status of each of the three phase voltages. These phase lamps are placed on the front panel of the main power distribution cabinet and should be observed whenever power is first applied to the system. While failure of one of the lamps to illuminate could be the result of a burned out bulb, it could also be indicating the absence of a phase. Therefore, the system should not be allowed to run until it has been verified through voltage measurement that all three phases are present. Three varistors, V_{R1}, V_{R2}, and V_{R3}, are placed in parallel with the phase lamps. As explained in the previous section, these devices, due to their negative temperature coefficient, protect the system from power surges.

Beyond the main circuit breaker, the 3-ϕ power line divides into two branches called the **motor bus** and the **electronics bus.** Each of these branches has its own 3-ϕ line filter. The purpose of these filters is twofold. Not only do they attenuate noise originating from outside the system, they also inhibit noise produced by either the motors or control

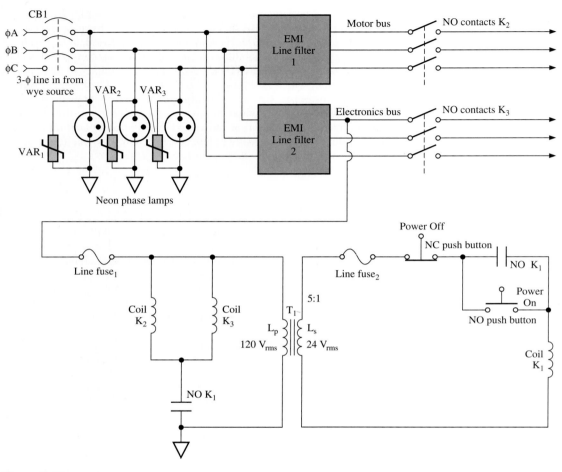

Figure 9–20
AC power distribution system

electronics from feeding back into the facility power lines. Although motors become immediately suspect as likely sources of line noise, it is difficult at first to imagine that this problem could originate in relatively low-power electronic control circuitry. However, switching power supplies, which are rapidly replacing conventional linear power supplies within dc control circuits, often produce a significant high-frequency ripple. If not controlled, this ripple can enter the main power lines, causing noise problems in other circuitry.

For the system in Figure 9–20, we assume that the operator control station is to be located several feet away from the main power distribution cabinet. For this reason transformer T_1 is placed in the circuit, providing isolation between the main distribution circuitry and the operator controls. As an additional safety feature, this transformer, which has a turns ratio of $5:1$, steps the 120-V_{rms} ϕC voltage down to 24 V_{rms}. Although the closure of CB_1 brings 3-ϕ voltage as far as the outputs of the line filters in Figure 9–20, the dc power supplies within the system are not energized until the Power On switch S_1 is pressed. When this normally open push-button switch is actuated, 24 V_{rms} develops over the coil of relay K_1. This causes both of the normally open contacts of K_1 to close.

Consequently, the coils of both 3-φ contactors, K_2 and K_3, become energized. As the contacts of K_2 and K_3 close, 3-φ power is made available on the electronics and motor buses. With the holding contacts of relay K_1 placed in parallel with the contacts of S_1, K_1 is latched on when the Power On switch is pressed. Thus the normally open contacts of K_1 remain closed and the contactors remain energized until the Power Off switch is actuated. This action breaks the flow of current to the coil of K_1, causing this relay, along with contactors K_2 and K_3, to drop out.

It is important to realize that the holding relay and the contactors involved in the power-on sequence for the circuit in Figure 9–20 are ac devices. This is a design requirement, because ac power must first be made available on the electronics bus before any dc components within the system are enabled to operate. Although closure of K_1 allows the control electronics and other dc circuitry within the system to become operative, it should be stressed that the motors within an industrial system are not directly started by this action, even though ac voltage is now available to the motor bus. In a microcomputer-based industrial system, there is likely to be a wait state, occurring between the instant of power-on and the point at which the control circuitry enables the actual startup sequence to begin. The length of such a wait state may vary from a few seconds to a few minutes, depending on the complexity of the system in which the control circuitry is operating. During this wait state, the system control circuitry checks the status of various parts of the system. Before allowing the ac motors within a system to rotate, the control microprocessor must check the status of the motor circuit breakers by reading the input signals from their sensing contacts. The control microprocessor must also verify that all dc power supplies are operative and that all line fuses are intact. If the system contains embedded microcomputers, the control microcomputer must check the operation of these subordinate devices. Finally, if the system contains photosensors, light barriers, and magnetic proximity sensors, the control microcomputer must also check their status before enabling the motors to operate.

As illustrated in Figure 9–21, only after verifying that all parts of the system are operative does the control microcomputer indicate that it is able to respond to the actuation of the Start switch. In many industrial systems, the control microcomputer sends a System ready signal to the operator station. This signal could be in the form of an actual message, sent to an alphanumeric display module on the control panel of the operator station. However, in a simpler system, the ready indication could consist of only a single illuminator, possibly contained in the Start switch itself.

Due to the extreme hazards presented by rotating machinery, industrial safety codes require that systems containing motors be equipped with **emergency-stopping** and **interlock circuitry.** Large industrial systems are likely to contain several emergency-stop switches, the purpose of which is to stop all motors instantly. Such an action could be necessary to prevent serious injury to personnel or costly damage to equipment. Interlock switches are often attached to doors and covers within industrial systems. As a safety precaution, these switches serve to remove power from rotating machinery and/or hazardous circuitry contained within cabinets and enclosures at the instant their lids or doors are opened.

Figure 9–22 illustrates the internal structure and operation of a typical emergency-stop switch, which is essentially a double-pole, double-throw device. Note that one set of contacts is normally open, while the other set is normally closed. For ease of operation, the emergency-stop switch is likely to have a plunger as its actuator. When the plunger is pressed, the circuit in series with contacts 3 and 4 is broken. At the same instant, continuity is provided between contacts 1 and 2. In a large industrial system, as shown Figure 9–23, the normally closed contacts of several emergency-stop switches are connected in series. Providing these contacts remain closed, the 12 V_{dc} originating at point A will be present at point B. However, if one of the emergency-stop switches is actuated,

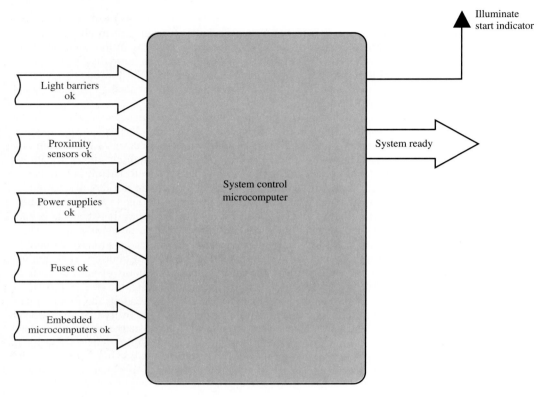

Figure 9–21
System control microprocessor in system start function

the circuit loop is broken, removing 12 V_{dc} from point B. Through interaction with additional circuitry, which is discussed later, this loss of voltage at point B causes power to be removed from all system motors. When one of the emergency-stop switches is pressed, a path is created between a small lamp, contained below the cap of the plunger mechanism, and the buffered output of a low-frequency astable multivibrator. When the output of the astable multivibrator falls to a logic low, 12 V is developed over the lamp, causing it to illuminate. When the multivibrator output toggles to a logic high, the lamp turns off. The rate of flashing for an emergency-stop lamp is typically 1 or 2 pps. (The design and operation of this form of astable flasher was covered in detail during the first section of Chapter 6.)

The structure and operation of an interlock switch is shown in Figure 9–24(a), and the standard symbol for an interlock switch is shown in Figure 9–24(b). The device shown here, which contains a normally open microswitch, is typically used to sense the closure of a heavy cover on a piece of machinery. When the cover is closed, its lateral pressure causes the plunger to be forced inward. This, in turn, causes the plunger to exert pressure on the leaf spring actuator of the microswitch. As a result, continuity is provided between the common and normally open contacts of the microswitch. When the cover of the machinery is opened, the spring loaded plunger is released, causing the leaf spring actuator of the microswitch to move upward. This action breaks continuity between the common

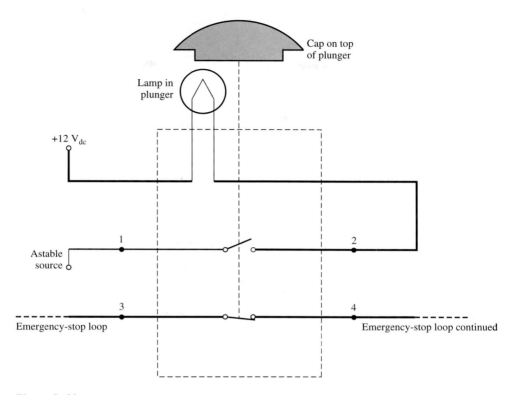

Figure 9–22
Internal structure and operation of an emergency-stop switch

and normally open contacts of the microswitch. Normally this open condition for an interlock is interpreted by the control electronics as an unsafe condition, causing the control electronics to remove power from the equipment. However, because it is often necessary to perform maintenance on a piece of equipment while it is operating, interlocks must have the capability of being defeated. For the interlock shown in Figure 9–24, this is accomplished by pulling outward on the plunger. Due to the shape of the plunger, when it is pulled to this full outward position, it presses on the leaf spring. Since continuity is again established between the common and normally open contacts of the microswitch, the control electronics assumes the cover to be closed and allows the machinery to operate. In many industrial systems, the common and normally closed contacts of the interlock switch are used as sensing contacts. As a cover or door on a piece of machinery is opened, the establishment of continuity at the normally closed terminal may be quickly sensed by the system control circuitry. The signal-conditioning circuitry involved in this process could be identical to that for the circuit breaker sensing contacts shown in Figure 9–18.

Like emergency-stop switches, interlocks are normally connected in series. A simple example of an interlock loop is illustrated in Figure 9–25. As long as all of the interlock switches are either closed or defeated, 12 V_{dc} is present at point B. However, if one of the interlock switches in the series loop is opened, continuity is broken and point B falls to nearly 0 V. This action also interrupts the flow of current through the emitter portion

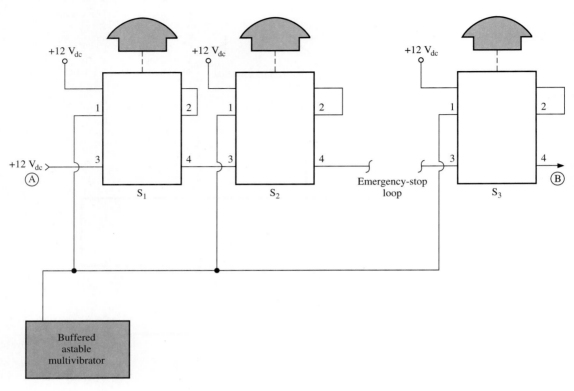

Figure 9–23
Structure of an emergency-stop loop

of an optoisolator located at the end of the interlock loop, causing the collector output of the phototransistor to toggle from 0 to 5 V. This logic high tells the control electronics that a lid or cover within the system has been opened.

A possible method by which the emergency stop and interlock loops could be incorporated into the control of the system motors is now examined. First, we consider the prerequisite conditions for energizing the system motors. For any given motor to operate:

1. The Emergency Stop Loop must be intact.
2. The Interlock Loop must be intact.
3. A System Ready signal must be generated by the System Control Microcomputer.
4. The Main Motors Circuit Breaker must be closed.
5. A motors enable signal (Motors On) must be generated by the System Control Microcomputer.
6. The Start signal for the given motor must be present.

From the standpoint of Boolean logic, since *all* the preceding requirements must be met for a given motor to operate, the start condition for that motor may be represented as a six-input AND function. This logic condition is shown in Figure 9–26. However, the actual motor on/off control circuitry within this hypothetical power distribution system consists of more than just a six-input AND gate. As seen in Figure 9–27, the logic ANDing of the Interlocks OK and the System Ready signals is accomplished through use of relay logic. The electromechanical devices involved in this process are two small SPDT PC

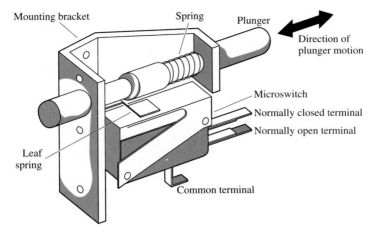

(a) Internal structure of typical interlock switch

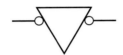

(b) Schematic representation of interlock switch

Figure 9–24

board relays, designated as K_1 and K_2 in Figure 9–27. If all of the interlocks within the system are in a closed or a defeated condition, 12-V_{dc} is present at point B in Figure 9–25, allowing the coil of K_1 to be energized. Also, the presence of 12 V at point B in Figure 9–25 produces a logic low at point C. This active-low Interlocks OK signal is monitored continuously by the System Control Microcomputer. During system start-up, if the System Control Microcomputer detects a logic low at point C, it generates the

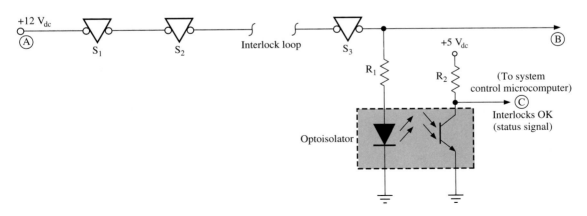

Figure 9–25
Typical interlock loop

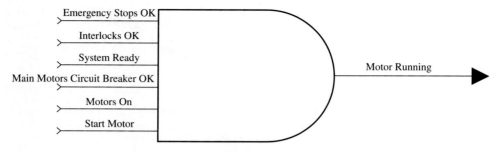

Figure 9–26
Conditions that must exist for a given ac motor to operate

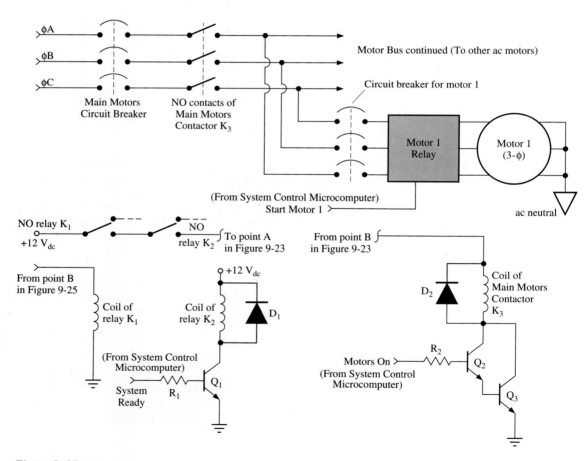

Figure 9–27
Incorporation of interlock and emergency-stop loops into motor on/off circuitry

System Ready signal, which arrives at the base of Q_1 in Figure 9–27. As Q_1 switches on in saturation, the coil of relay K_2 is energized. Thus, as long as the interlock loop is intact and the System Control Microcomputer senses this condition and responds with the System Ready signal, the NO contacts of K_1 and K_2 are both closed. This allows 12 V_{dc} to be present at point A in Figure 9–23, which is the beginning of the emergency stop loop.

As long as all the emergency stop switches are closed, 12 V_{dc} is present at point B in Figure 9–23. As indicated in Figure 9–27, Point B in Figure 9–23, serves as the voltage source for the coil of the Main Motors Contactor K_3. The opposite terminal of the contactor coil is connected to the collector terminal of a Darlington pair, consisting of Q_2 and Q_3. Thus, for the contactor coil to be energized, the emergency-stop loop must be intact and an active-high Motors On signal generated by the System Control Microcomputer must be present at the base of the Darlington pair. Only with these conditions do the NO contacts of K_3 close. Assuming the Main Motors Circuit Breaker and the Motor 1 Circuit Breaker are both closed, 3-ϕ Motor 1 begins to rotate with the arrival of an active-low Start Motor 1 signal at the control input to the Motor 1 Relay. This commercially manufactured **solid-state relay** is likely to consist of three identical zero point switches (one for each phase), similar in structure to the CA3059 IC introduced in the Chapter 8. These three switches would be fabricated into a monolithic package and share a common control input. As was explained, zero point switches are preferred for activating high-power AC loads because they eliminate the problems of thermal shock and RF noise propagation. For Motor 1 in Figure 9–27, regardless of when the Start Motor 1 signal falls low, the phase voltages do not begin to develop over the motor's stator windings until the points at which these waveforms cross the 0° reference in the positive direction. Thus, Motor 1 always begins its rotation smoothly.

Under normal machine operating conditions, the Start Motor 1 signal remains low as long as operation of Motor 1 is required. At points during normal operation, where the System Control Microcomputer senses that Motor 1 must be temporarily halted, it brings the Start Motor 1 signal to a logic high, switching off the solid-state relay. Thus, Motor 1 turns off smoothly while other motors in the system are allowed to continue rotating. Assume, however, that an alert machine operator sensed a dangerous condition and pressed the nearest emergency stop switch. Due to the resultant loss of 12 V_{dc} at point B in Figure 9–23, Main Motors Contactor K_3 drops out, immediately removing power from *all* motors. Now assume that a less careful operator attempts to open a door to the rotating machinery while the motors are turning. Since this action would open one of the interlock switches in Figure 9–25, 12 V_{dc} would no longer be present at point B and relay K_1 in Figure 9–27 would drop out. This action would cause the Emergency Stop loop to open and force the Main Motors Contactor K_3 to drop out. Thus, with a well-designed power distribution system, opening a door to hazardous machinery produces the same result as pressing an emergency stop switch!

SECTION REVIEW 9–3

1. What is the purpose of the three neon lamps in Figure 9–20? What safety precautions should be observed if one or more of these lamps fail to illuminate when CB_1 is closed?
2. Explain the purposes of transformer T_1 in Figure 9–20.
3. Explain why an ac relay must be used for K_1 in Figure 9–20.
4. Describe the operation of relay K_1 in Figure 9–20 at the instant the Power On switch in Figure 9–20 is pressed. How does relay K_1 function when the Power Off switch is pressed?

5. True or false?: A typical emergency-stop switch is likely to contain two sets of normally open contacts.

6. True or false?: Within high-power machine control systems, the normally closed contacts of emergency-stop switches are usually connected in a series loop.

7. Describe the structure of a typical interlock. Explain the process for defeating an interlock.

8. For the circuitry in Figures 9–23 and 9–27, explain how it is possible for the actuation of any emergency-stop switch to de-energize instantly all of the motors within the system.

9. Explain how opening an interlock switch in Figure 9–25 is able to remove 12 V_{dc} from the coil of K_3 in Figure 9–27.

10. Assume that the bearings of Motor 1 in Figure 9–27 are worn, causing that motor to draw excessive current. What switching device in the motor control circuitry should react automatically to this condition?

9–4   DC VOLTAGE SOURCES

Within a typical industrial control system, a wide variety of dc voltages is required. For example, printed circuit boards within a given system could require 5 V_{dc} for operation of TTL, as well as ± 12 V_{dc} for op amp circuitry requiring two supplies. In addition, photosensors, magnetic proximity sensors, and low-power solenoids might operate on 24 V_{dc}. High-power solenoids and clutch/brakes are likely to require a dc source of at least 40 V.

Circuits designed to convert ac line voltage to a steady dc level are referred to as **power supplies.** Figure 9–28 contains a block diagram of a **series-pass linear regulated power supply.** With this conventional form of power supply, the first stage is an isolation power transformer, which represents most of the circuit's weight and bulk. Following the transformer is a diode rectifier, which converts the ac voltage to pulsating dc. The filter stage then converts the pulsating dc voltage to a nearly constant dc level. If a highly regulated dc output voltage is required, the final stage of the linear power supply could consist of a series-pass element as well as feedback and control circuitry.

Figure 9–29 illustrates how full-wave rectification is accomplished through use of a single-phase, center-tapped transformer and two diodes. Note in this example that the

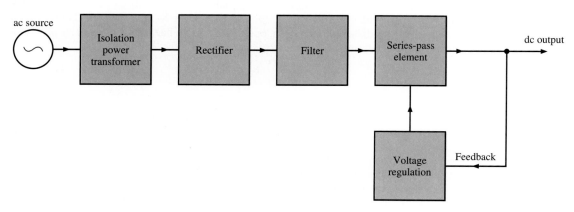

Figure 9–28
Functional block diagram of typical series-pass linear regulated power supply

Figure 9–29
Full-wave rectifier with two diodes and cen-
ter-tapped transformer

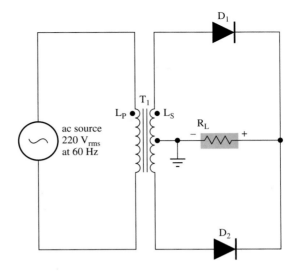

center-tap is established as circuit ground. During the positive alternation of the source waveform, as represented in Figure 9–30(a), D_1 is forward biased and D_2 is reverse biased. During the negative alternation, as shown in Figure 9–30(b), D_2 becomes forward biased and D_1 is reverse biased. In either alternation of the source waveform, electron flow is from the ground reference point through the load resistor. Thus, as shown in Figure 9–30(c), resistor voltage and current constantly vary in amplitude without changing polarity.

Example 9–6

Assume that the transformer in Figure 9–29 has a turns ratio of 6 : 1 and that both diodes are made of silicon. Solve for the peak, rms, and average voltages developed over R_1.

Solution:

Step 1: Determine the peak primary voltage:

$$V_{peak} = 1.414 \times 220 \text{ V}_{rms} = 311 \text{ V}$$

Step 2: Solve for the peak voltage developed over the secondary winding:

$$V_{peak} = \frac{311 \text{ V}}{6} = 51.8 \text{ V}$$

Step 3: Solve for the peak voltage developed over the load, accounting for the center-tap and the forward voltage of the diodes:

$$V_{R_L(peak)} = \frac{51.8 \text{ V}}{2} - 0.7 \text{ V} = 25.2 \text{ V}$$

Step 4: Solve for the average and rms values:

$$V_{R_L(avg)} = 0.636 \times 25.2 \text{ V} = 16 \text{ V}$$

$$V_{R_L(rms)} = 0.707 \times 25.2 \text{ V} = 17.83 \text{ V}$$

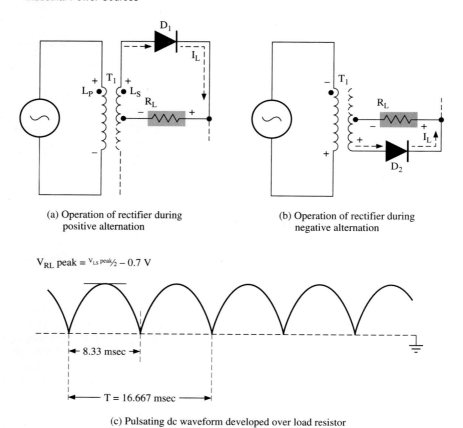

(a) Operation of rectifier during
positive alternation

(b) Operation of rectifier during
negative alternation

V_{RL} peak $= {}^{V_{LS}\,peak}\!/_2 - 0.7$ V

(c) Pulsating dc waveform developed over load resistor

Figure 9–30

The most common form of diode rectifier is the full-wave bridge, an example of which is shown in Figure 9–31. A major advantage of this diode configuration is that full-wave rectification may be obtained without a center-tapped transformer. As shown in Figure 9–32(a), during the positive alternation of the source waveform, D_3 and D_2 are forward biased, while D_1 and D_4 are in a reverse-biased condition. Thus, electron flow occurs from point B to point A. During the negative alternation of the source, as seen in Figure 9–32(b), D_1 and D_4 are forward biased, while D_3 and D_2 are in a reverse-biased condition. Thus, electron flow is from point A to point B. Note, however, through the

Figure 9–31
*Full-wave bridge rectifier with transformer
and load resistance*

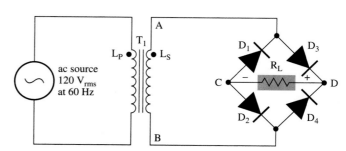

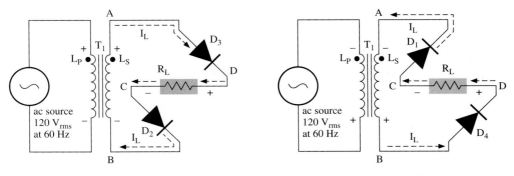

(a) Operation of full-wave bridge rectifier during positive alternation

(b) Operation of full-wave bridge rectifier during negative alternation

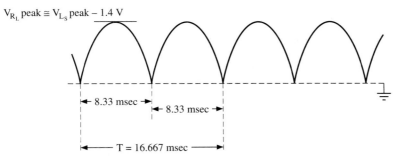

V_{R_L} peak $\cong V_{L_S}$ peak $-$ 1.4 V

8.33 msec

8.33 msec

T = 16.667 msec

(c) Pulsating dc waveform developed over R_L

Figure 9–32

action of the full-wave bridge, electron flow occurs from point C to point D during both alternations of the source. Thus, the resultant load voltage appears as shown in Figure 9–32(c).

Example 9–7

Assume for the circuit in Figure 9–31 that the primary winding consists of 600 turns, and the secondary winding has 200 turns. Also assume the four diodes are silicon and that R_L equals 220 Ω. Given these conditions, determine the peak, rms, and average values of V_{R_L} and I_{R_L}.

Solution:

Step 1: Determine the peak values of load voltage and current:

$$V_{peak} = 1.414 \times 120 \; V_{rms} = 169.7 \; V$$

$$\frac{200}{600} \times 169.7 \; V = 56.57 \; V$$

Step 2: Determine the rms and average values of load voltage and current:

$$V_{R_L(peak)} = 56.57 \; V - 1.4 \; V = 55.17 \; V$$

$$I_{L\,(\text{peak})} = \frac{55.17\text{ V}}{220\ \Omega} \cong 251\text{ mA}$$

$$V_{R_L(\text{rms})} = 55.17\text{ V} \times 0.707 = 39\text{ V}$$

$$I_{R_L(\text{rms})} = \frac{39\text{ V}}{220\ \Omega} = 177.3\text{ mA}$$

$$V_{R_L(\text{avg})} = 0.636 \times 55.17\text{ V} = 35.09\text{ V}$$

$$I_{R_L(\text{avg})} = \frac{35.09\text{ V}}{220\ \Omega} = 159.5\text{ mA}$$

In a linear power supply, the filtering process involves the conversion of pulsating dc voltage to a nearly constant level. This process requires at least one **filter capacitor.** Such components are usually large electrolytic capacitors, ranging in value from as low as 40 μF to as high as 3000 μF. In Figure 9–33, a filter capacitor is placed in parallel with the load resistance. Through the low-pass filtering process, the output of the bridge rectifier is converted to a smoother form of dc. However, the output voltage still contains some degree of slope, referred to as **ripple.** The action of the filter capacitor in improving the quality of the rectified output is illustrated in Figure 9–34.

Assume that at the instant the circuit in Figure 9–33 is energized the source waveform is in the first 90° of its positive alternation. Since the filter capacitor is totally discharged at this time, it behaves as a virtual short. With D_2 and D_3 now forward biased, C_1 charges rapidly, with electron flow occurring from point C into the capacitor via D_2. During the next 90° of the positive alternation (between the 90° and 180° points) the source amplitude begins to decrease toward 0 V. As the voltage at point A becomes more negative than the voltage over the capacitor (between point B and ground), both D_2 and D_3 become reverse biased. Thus, the only path for discharge of the capacitor is through the load resistance. For a well-designed filter, the RC time constant for the filter capacitor and resistive load should be much greater than the period of the pulsating dc waveform. If the ac source has a frequency of 60 Hz, the period of this waveform is 16.667 msec. Since the rectified waveform occurs at twice the frequency of the source, its period is only 8.333 msec. If the RC time constant τ is much greater than the period T, then the capacitor would lose only a small portion of its charge. This loss in charge occurs during

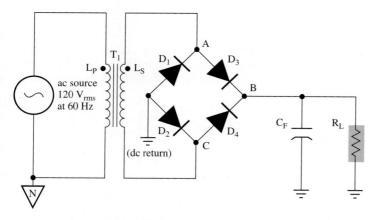

Figure 9–33
Filter capacitor placed in parallel with the load resistance

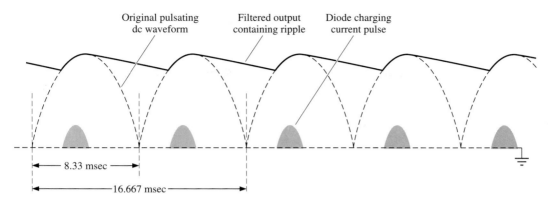

Original pulsating
dc waveform

Filtered output
containing ripple

Diode charging
current pulse

8.33 msec

16.667 msec

Figure 9–34
Waveforms for circuit in Figure 9–33

the time in which the capacitor is not receiving current from the rectifier. The current pulses from the bridge rectifier, which replenish the charge on the capacitor, are shown as the darkened areas in Figure 9–34.

Because of the storage capability of the filter capacitor, it is sometimes referred to as a **reservoir capacitor.** Ideally, if the load resistance were not present in Figure 9–33, the filter capacitor would not lose any of its charge. Therefore, as soon as the circuit is energized, the capacitor would quickly charge to the peak level of the rectified dc waveform and remain at that potential. The ideal filtered output would be a pure dc level, devoid of any ripple.

In choosing the proper value of filter capacitor, the acceptable amount of ripple voltage (represented as V_r) must first be determined. Figure 9–35 serves to clarify the key concepts involving ripple voltage. The **maximum output voltage** from the filter [represented as $V_O(max)$] is the peak amplitude of the pulsating dc waveform. The **minimum output voltage** [represented as $V_O(min)$] is the lowest voltage level developed over the capacitor. The ripple voltage, representing the difference between the maximum and minimum output voltages, is determined as follows:

$$V_r = V_O(max) - V_O(min) \qquad (9\text{–}8)$$

Ripple is often specified as a percentage of deviation above and below the average output voltage of the filter. The average output voltage [represented as $V_O(avg)$] is determined as follows:

$$V_O(avg) = \frac{V_O(max) + V_O(min)}{2} \qquad (9\text{–}9)$$

Figure 9–35(b) represents the important time relationships involved in the design of filter circuits. Here t_1 represents the time during which the filter capacitor is being discharged, supplying current to the load. The time during which the capacitor is being charged via the bridge rectifier is indicated as t_2. Note that t_1 consists of the last 90° of the given alternation of the ac source plus that portion of the subsequent alternation required for the voltage at either point A or point C in Figure 9–33 to become 0.7 V more positive than point B. If the frequency of the ac source is 60 Hz, the sum of t_1 and t_2 [represented as T in Figure 9–35(b)] always equals 8.333 msec. Also, as clarified in Figure 9–31(b), the first portion of t_1 must always be equivalent to one-quarter of a source cycle. With

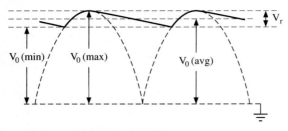

(a) Voltage levels for pulsating dc waveform

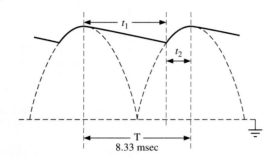

(b) Time periods for pulsating dc waveform

Figure 9–35

the standard 60-Hz signal, this time is easily calculated as follows:

$$\Delta t = 0.25 \times 16.667 \text{ msec} = 4.167 \text{ msec}$$

The second portion of t_1 is determined through a more complex process. Since $V_0(\text{max})$ is the peak point on a sine curve and $V_0(\text{min})$ is an instantaneous point on that curve, the number of degrees represented by the second portion of t_1 is determined from the following:

$$\sin\theta = \frac{V_0(\text{min})}{V_0(\text{max})}$$

After the angle θ is derived, the second portion of t_1 may be determined:

$$\Delta t = \frac{\theta}{360°} \times 16.667 \text{ msec} \qquad (9–10)$$

The total duration of t_1 is then calculated as follows:

$$t_1 = 4.167 \text{ msec} + \Delta t \text{ for } \theta \qquad (9–11)$$

Once t_1 is determined, t_2 is easily calculated:

$$t_2 = 8.333 \text{ ms} - t_1 \qquad (9–12)$$

All of the above concepts become important in determining the minimum value of filter capacitor. The maximum allowable percentage of ripple would be decided by circuit designers before selection of this component. Once the value of load current is known, the minimum value of required filter capacitance is then determined:

$$C_F = \frac{I_L \times t_1}{V_r} \tag{9-13}$$

Example 9-8

For the circuit in Figure 9–33, assume the ripple voltage must represent no more than 5% of the average output voltage. Assume the source voltage is 120 V_{rms}, the transformer has a turns ratio of 5 : 1, and the load resistance is 470 Ω. Determine the minimum value of filter capacitance (standard value).

Solution:

Step 1: Solve for the peak value of voltage at point B:

$$V_{L_p}(\text{peak}) = 1.414 \times 120 \ V_{rms} = 169.7 \ V$$
$$V_{L_s} = \frac{169.7 \ V}{5} = 33.94 \ V$$
$$V_B(\text{peak}) = 33.94 \ V - 1.4 \ V = 32.54 \ V$$

Step 2: Given the specified ripple factor of 5%, determine the average and minimum values of output voltage. Also determine the value of ripple voltage:

$$V_O(\text{max}) = 102.5\% \text{ of } V_O(\text{avg}) = 1.025 \times V_O(\text{avg})$$
$$V_O(\text{avg}) = \frac{32.54 \ V}{1.025} = 31.746 \ V$$
$$V_O(\text{min}) = 97.5\% \text{ of } V_O(\text{avg}) = 0.975 \times 31.746 \ V = 30.952 \ V$$
$$V_r = V_O(\text{max}) - V_O(\text{min}) = 32.54 \ V - 30.952 \ V = 1.588 \ V$$

Proof: This difference in voltage should represent 5% of the average output voltage:

$$\frac{1.588 \ V}{31.746 \ V} \times 100\% = 5\%$$

Step 3: Solve for t_1 and t_2:

$$\sin\theta = \frac{30.952 \ V}{32.54 \ V} = 0.9512$$
$$\Delta t = \frac{72°}{360°} \times 16.667 \ \text{msec} = 3.333 \ \text{msec}$$
$$t_1 = 4.167 \ \text{msec} + 3.333 \ \text{msec} = 7.5 \ \text{msec}$$
$$t_2 = 8.333 \ \text{msec} - 7.5 \ \text{msec} \cong 833 \ \mu\text{sec}$$

Step 4: Solve for the nominal value of capacitance based on a ripple factor of 5%.

$$I_L = \frac{31.746 \ V}{470 \ \Omega} = 67.545 \ \text{mA}$$
$$C_F = \frac{I_L \times t_1}{V_r} = \frac{67.545 \ \text{mA} \times 7.5 \ \text{msec}}{1.588 \ V} = 319 \ \mu\text{F}$$

Choose 330 μF as a standard value with a breakdown potential of at least 50 V.

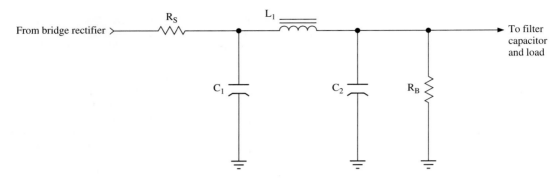

Figure 9–36
PI (π) filter with surge and bleeder resistors

Figure 9–36 contains an improved form of filter that could be connected to the output of the bridge rectifier. Resistor R_S is a **surge resistor,** the purpose of which is to protect the bridge rectifier from the large flow of current that could occur at the instant the circuit is energized. This surge in current is likely because, at the instant the circuit is energized, the filter capacitors behave as dead shorts. The maximum value of surge current for a rectifier diode is typically specified as I_{FSM} in the manufacturer's data sheets. For the 1N4001 rectifier diode this **nonrepetitive surge current** could be as high as 30 A_{peak}. Thus, the ohmic value of R_S could be relatively low compared to the impedance of the load. However, the wattage rating of the surge resistor must be very high.

Together, C_1, L_1, and C_2 in Figure 9–36 form a **PI filter.** This form of filter derives its name from the fact that its appearance in the schematic vaguely resembles the Greek letter π. The inductor L_1 is often referred to as a **filter choke,** because it blocks or chokes the higher frequency components of the rectified ac signal. Not only does L_1 assist in smoothing out the ripple from the rectified voltage, it also attenuates any high-frequency noise that may be present on the ac line. Resistor R_B is a **bleeder resistor,** the purpose of which is to allow the quick discharge of the two capacitors and subsequent filter capacitor once the circuit is energized. It also provides a minimum amount of loading for the power supply, stabilizing its operation while the actual load condition is changing.

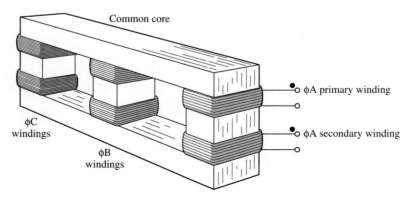

Figure 9–37
Representation of structure of 3-φ transformer

In industrial power applications where large amounts of dc current are required, single-phase isolation transformers are often abandoned in favor of three-phase transformers. The basic structure of such a device is shown in Figure 9–37. Note that on each of the three sections of the core a primary phase winding and its equivalent secondary winding are coiled together. Since the currents in each of the windings are 120° out of phase, at any time, flux in a given winding is aided by flux in another winding. Three-phase transformers may be configured in four possible ways, since either the primary or the secondary may be wired in a delta or wye configuration.

In Figure 9–38, a three-phase transformer with a delta primary and a wye secondary is connected to a **full-wave line-to-line three-phase bridge rectifier.** While this form of rectifier is more complex than the single-phase full-wave bridge, it has two important advantages. First, it is capable of providing large amounts of current to high voltage dc loads and, therefore, is often used as a power source for such electromechanical devices as dc solenoids and clutch/brakes. This type of unregulated power source is sometimes referred to as a **"brute force"** power supply.

The second advantage of the full-wave line-to-line three-phase bridge rectifier is its ability to minimize ripple. As shown in Figure 9–39, the frequency of its pulsating dc output is six times greater than that of the individual phases. Thus, if the line voltage occurs at 60 Hz, the ripple frequency becomes 360 Hz. Since the time between peak points on the wave is then only 2.78 msec, the filtering process is greatly alleviated.

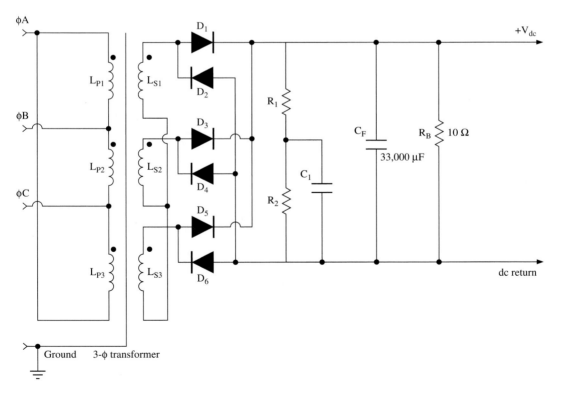

Figure 9–38
Three-phase transformer with delta primary and wye secondary used in conjunction with a full-wave line-to-line three-phase bridge rectifier

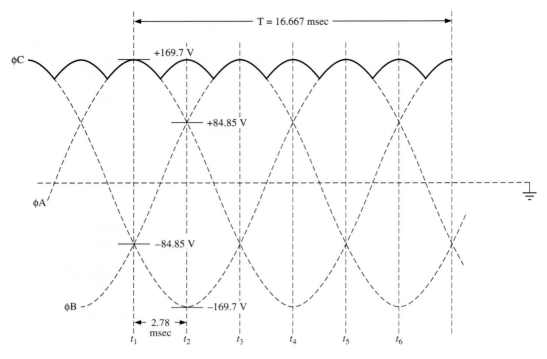

Figure 9–39
Pulsating dc output of full-wave line-to-line three-phase bridge rectifier

In Figure 9–38, R_1, R_2, and C_1 form a noise elimination (snubber) network, while the actual filter capacitor and bleeder resistor consist of C_F and R_B. If the load consists of several solenoids and clutch/breaks, the extremely large capacitance of C_F is necessary. In its reservoir function, this component must maintain a nearly constant dc level while the load current varies greatly.

The operation of the power supply in Figure 9–38 might best be understood by comparing the six circuit conditions represented in Figure 9–40 with the six points in time designated in Figure 9–39. At t_1, ϕA is at its positive peak amplitude of 169.7 V while ϕB and ϕC are both at −84.85 V. With this voltage condition, as shown in Figure 9–40(a), the voltages induced over L_{P1} and L_{P3} are at maximum. Since both ϕB and ϕC are at −84.85 V, no difference in potential occurs across L_{P2} and L_{S2}. However, the voltages induced over L_{S1} and L_{S3} are series aiding, allowing a path for load current through D_1 and D_6. At t_2, with ϕC at its negative peak potential, the voltages induced over L_{P2} and L_{P3} are at maximum and the voltage over L_{P1} is at minimum. Thus, as represented in Figure 9–40(b), secondary windings L_{S2} and L_{S3} are series aiding, whereas L_{S1} has no induced voltage. The path for load current is now through D_3 and D_6. As shown by the four remaining examples in Figure 9–40, this pattern continues throughout the three-phase cycle. Different pairs of diodes allow current pulses to flow to the capacitor every 2.77 msec. The amplitude and duration of these current pulses are, of course, a function of load current.

Connection of high-power dc devices such as solenoids and clutch/brakes directly to the output of the power supply in Figure 9–40 is possible, since these electromechanical devices operate quite satisfactorily on pulsating dc. However, for most dc circuit applications, connection of a load directly in parallel with the filter stage of a power supply is

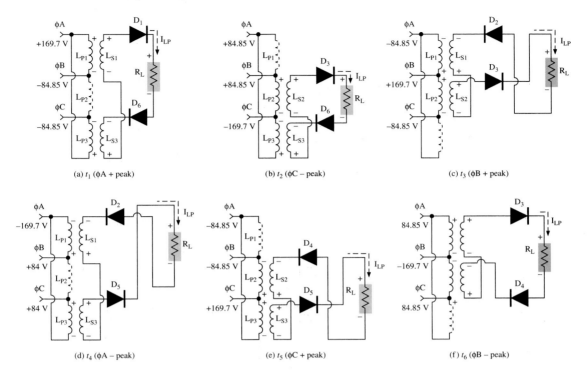

(a) t_1 (ϕA + peak)   (b) t_2 (ϕC – peak)   (c) t_3 (ϕB + peak)

(d) t_4 (ϕA – peak)   (e) t_5 (ϕC + peak)   (f) t_6 (ϕB – peak)

Figure 9–40

unacceptable. As represented by the block diagram in Figure 9–28, a high-quality linear power supply is likely to contain a series-pass element with feedback and control. This output stage of the power supply provides the voltage regulation and current limitation necessary for controlling such dc loads as electronic control and sensing circuitry.

Voltage regulation is the process by which a power supply is able to maintain a constant output voltage, despite deviations in both its input voltage and output current. The concept of voltage regulation is introduced by the simple **Zener regulator** in Figure 9–41(a), which could be placed at the output of the filter stage of a power supply. The circuit in Figure 9–41(a) is essentially a step-down device, since the regulated output at point B must be lower in amplitude than the unregulated source. Since the load is placed in parallel with the Zener diode, the regulated load voltage is equivalent to the Zener voltage (designated V_Z). Series resistance R_S develops the difference in potential between the source and point A. Since total current must flow through R_S, its power dissipation becomes an important concern in circuit design. According to the data sheets for the 1N755 7.5 V Zener diode in Figure 9-41(a), **Zener test current** (designated I_{ZT}) is 20 mA and **maximum Zener current** (designated I_{ZM}) is 50 mA. With S_1 opened, the circuit is in a no-load condition. At this time, current through the Zener diode is at maximum. Thus, a value of R_S must be selected such that, in a no-load condition, I_{RS} nearly equals but does not exceed I_{ZM}.

Example 9–9

For the circuit in Figure 9–41(a), assume that the input voltage is 12 V_{dc}. Choose a suitable value of R_S. Also determine the necessary wattage value for this series resistance.

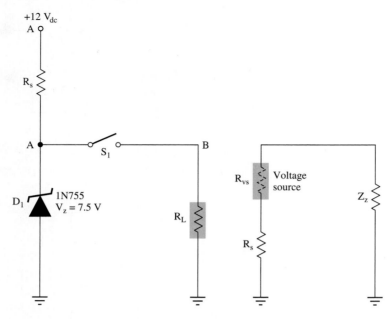

(a) Basic Zener voltage regulator

(b) AC equivalent circuit for basic
Zener voltage regulator

Figure 9–41
Basic Zener voltage regulator

Solution:

Step 1: Solve for resistance:

$$V_{RS} = V_S - V_Z = 12\text{ V} - 7.5\text{ V} = 4.5\text{ V}$$

$$R_S = \frac{V_{RS}}{I_{ZM}} = \frac{4.5\text{ V}}{50\text{ mA}} = 90\ \Omega$$

Choose the next highest standard value of 100 Ω.

Step 2: Solve for power dissipation:

$$P = I_{RS}^2 \times R_S = 50\text{ mA}^2 \times 100\ \Omega = 250\text{ mW}$$

Select the next highest rating of 1/2 W.

When S_1 is closed, the current I_{RS} branches to become I_Z and I_L. According to Kirchhoff's current law:

$$I_{RS} = I_Z + I_L \tag{9–14}$$

Assuming the Zener diode is functioning properly, the voltage at point A remains virtually the same as the switch closes. Thus, I_{RS} is nearly unchanged. For stable circuit operation, with S_1 closed, I_Z should nearly equal I_{ZT}. Thus, maximum load current may be easily determined as follows:

$$I_L(\text{max}) = I_{RS} - I_{ZT} \tag{9–15}$$

Minimum load resistance may now be determined using a transposition of Ohm's law:

$$R_L(\text{min}) = \frac{V_Z}{I_L(\text{max})} \tag{9–16}$$

Example 9–10

For the circuit in Figure 9–41(a), determine the maximum load current and minimum load resistance.

Solution:

Step 1: Solve for maximum value of load current. Since, for the 1N755, I_{ZT} equals 20 mA:

$$I_L = 50\,\text{mA} - 20\,\text{mA} = 30\,\text{mA}$$

Step 2: Solving for minimum load resistance:

$$R_L(\text{min}) = \frac{7.5\,\text{V}}{30\,\text{mA}} = 250\,\Omega$$

An important parameter for any voltage regulator is its **stabilization ratio** (designated S_V). This parameter is a measurement of deviations in output voltage (designated ΔV_0) with respect to changes in the source voltage (designated ΔV_S).

$$S_V = \frac{\Delta V_0}{\Delta V_S} \tag{9–17}$$

As a figure of merit, the ideal stabilization ratio is zero, since, for the ideal regulator, there would be no change in output voltage. In determining the stabilization ratio, it is first necessary to calculate the **output impedance** of the regulator (designated Z_O). This parameter is the ac or dynamic impedance of the series resistance (R_S) and the Zener impedance (designated Z_Z). The ac equivalent circuit for the Zener regulator in Figure 9–41(a) is shown in Figure 9–41(b). Note that the source resistance R_{VS} is in series with R_S. These two resistances, in turn, are in parallel with Z_Z. Since the ohmic value of R_S is likely to be much greater than R_{VS}, the output impedance of the voltage regulator may be approximated as follows:

$$Z_O = \frac{Z_Z \times R_S}{Z_Z + R_S} \tag{9–18}$$

As a transposition of Ohm's law, output impedance defines the change in output voltage for a given change in load current:

$$\Delta V_O = \Delta I_L \times Z_O \tag{9–19}$$

Looking from the voltage source, with no load present, the series resistance and the Zener impedance form a simple series voltage divider. Therefore,

$$\Delta V_O = \frac{Z_Z}{R_S + Z_Z} \times \Delta V_S \tag{9–20}$$

Thus, the stabilization ratio may also be expressed as follows:

$$S_V = \frac{Z_Z}{R_S + Z_Z} \tag{9–21}$$

Example 9–11

For the Zener voltage regulator in Figure 9–41(a), determine the output impedance and stabilization ratio. What would be the amount of change in output voltage if load current varies by 24 mA?

Solution:

Step 1: Solve for output impedance:
According to the data sheets for the 1N755, $Z_Z(max)$ is 6 Ω. Therefore,

$$Z_O = \frac{100\,\Omega \times 6\,\Omega}{6\,\Omega + 100\,\Omega} = 5.66\,\Omega$$

Step 2: Solve for the stabilization ratio:

$$S_V = \frac{6\,\Omega}{100\,\Omega + 6\,\Omega} = 0.0566$$

Step 3: Solve for the deviation in output voltage, given a change in load current of 24 mA:

$$\Delta V_O = 5.66\,\Omega \times 24\,mA = 138.5\,mV$$

The output stability of a voltage regulator may also be defined in terms of **line regulation.** With this parameter, deviation in output voltage is expressed as the change in output voltage per a given change in source voltage divided by the nominal output voltage. As shown in Eq. (9–22), line regulation is usually expressed as a percentage:

$$\text{Line regulation} = \frac{(S_V \times \Delta V_S)}{V_O} \times 100\% \qquad (9\text{–}22)$$

Load regulation is another important figure of merit for a voltage regulator, representing the ability of the device to maintain a nearly constant output voltage as the output current changes from a no load to a full load condition. As shown in Eq. (9–23), load regulation is also expressed as a percentage:

$$\text{Load regulation} = \frac{\Delta V_O(max)}{V_O} \times 100\% = \frac{[\Delta I_L(max) \times Z_O]}{V_O} \times 100\% \qquad (9\text{–}23)$$

Example 9–12

For the Zener voltage regulator in Figure 9–41, determine the line regulation, assuming a 25% change in source voltage. Also determine the load regulation, assuming a change from a no-load to full-load condition.

Solution:

Step 1: Solve for line regulation:

$$S_V \times 25\% \times V_S = 0.0566 \times 0.25 \times 12\,V = 169.8\,mV$$

$$\text{Line regulation} = \frac{169.8\,mV}{7.5\,V} \times 100\% = 2.26\%$$

Step 2: Solve for load regulation:

$$\text{Load regulation} = \frac{30 \text{ mA} \times 5.66 \ \Omega}{7.5 \text{ V}} \times 100\% = 2.26\%$$

The Zener regulator introduced in Figure 9–41 is limited to low-power applications, where load current is only a few milliamperes. In medium- to high-power applications, more complex voltage regulators are required. Such devices fall into two major classes: **linear-mode** or **switch-mode** regulators. Linear voltage regulators, which are classified as either **series** or **shunt** devices, are covered first. Switch-mode voltage regulators and power supplies are covered in detail in the final section of this chapter.

Figure 9–42 contains a basic form of series regulator. With this design, a Zener diode is still required for stable circuit operation, providing a constant reference voltage at the noninverting input to an op amp. Resistors R_2 and R_3 form a voltage divider, the purpose of which is to sense any deviation in the amplitude of V_O. Virtually all the current from the unregulated source leaves the emitter of pass transistor Q_1, flowing to the load and the sensing resistors. Although V_A remains nearly constant as V_B attempts to vary, the function of the op amp is best understood by considering it to be in a noninverting configuration with the Zener reference voltage as its active input. Resistor R_3 then becomes the input resistor, and R_2 provides feedback. Thus, ignoring V_{BE} of the pass transistor, the output voltage of the regulator could be approximated as follows:

$$V_O \cong V_{ref} \times 1 + \left(\frac{R_2}{R_3}\right) \tag{9-24}$$

Note that, in the preceding formula, the source voltage is not a factor in determining V_O. If source voltage does change, a proportional change in amplitude occurs at point B, causing the error voltage V_{A-B} to either increase or decrease. For example, if V_S decreases, the voltage at point B falls below the reference voltage, causing the output voltage of the

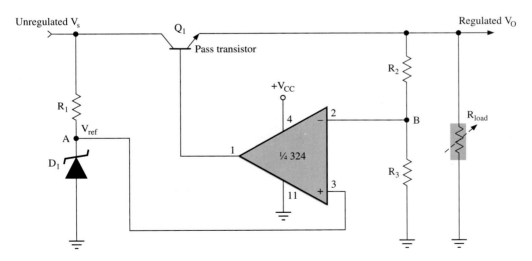

Figure 9–42
Basic series regulator

op amp to increase. This action causes an increase in emitter voltage for the pass transistor, thus compensating for the drop in V_S. The same corrective action would occur if the output voltage attempted to drop due to a decrease in load resistance. In a practical series regulator, due to the large current flow it must sustain, Q_1 must be a power transistor, possibly requiring a large heat sink.

Example 9–13

For the series regulator in Figure 9–42, assume V_S equals 24 V and that D_1 is a 1N750 4.7-V Zener diode. Assuming R_2 equals 10 kΩ and R_3 equals 4.7 kΩ, solve for the value of V_O.

Solution:

$$V_O = 4.7 \text{ V} \times 1 + \left(\frac{10 \text{ k}\Omega}{4.7 \text{ k}\Omega}\right) = 14.7 \text{ V}$$

Most voltage regulators used in industry require some form of short circuit or overload protection. As illustrated in Figure 9–43, this may be accomplished by adding a **current limiter** stage to the basic series regulator. The current limiter in Figure 9–43 consists of Q_2 and R_2. The ohmic value of R_2 is extremely small compared to the impedance of the load. Thus, prior to an overload condition, transistor Q_2 is biased off, since V_{R2} is less than 0.7 V. During an overload condition, load current increases greatly as load resistance approaches 0 Ω. Without the current-limiting components in the circuit, the regulator would attempt to maintain the output voltage at the desired level, until either the load or

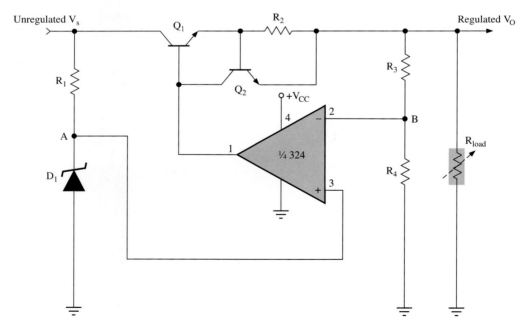

Figure 9–43
Series regulator with current limiter

the regulator itself was destroyed. However, in Figure 9–43, as V_{R_2} approaches 0.7 V, Q_2 switches on, holding V_{R_2} at the level of V_{BE}. Thus, even with a shorted load, regulator output current is held at a maximum level, which is approximated as follows:

$$I_L(\max) = \frac{V_{BE}}{R_2} = \frac{0.7 \text{ V}}{R_2} \tag{9–25}$$

Figure 9–44 contains a basic shunt regulator. This configuration takes its name from the fact that the transistor now functions as a voltage-variable resistor, effectively in parallel with the load. As with the series regulator, the sensing voltage divider, consisting of R_3 and R_4, detects any deviations in either V_S or V_O. However, it is important to note that the voltage at point B is now fed to the noninverting input of the op amp, whereas the reference voltage from point A is present at the inverting input. Assume, for example, that V_O attempts to decrease as a result of an increase in load current. Since the voltage at point B also begins to decrease, the error voltage between points A and B becomes smaller, lowering the output voltage of the op amp. Consequently, base current for Q_1 is reduced, increasing the resistivity between its collector and emitter. This action tends to raise V_O, compensating for the increase in load current.

A variety of voltage regulators is available in monolithic form. They are widely used in linear power supplies, being placed between the filter stage and the final output. **Fixed-voltage regulators** are designed to serve as either positive or negative voltage sources, with voltage ratings of ±5 and ±12 V being typical. Also, adjustable voltage regulators are available, designed to produce positive or negative dc levels varying from less than 2 V to more than 30 V.

One of the most difficult problems in voltage regulator design is maintaining a constant reference voltage over a wide range of operating temperatures. If the reference voltage

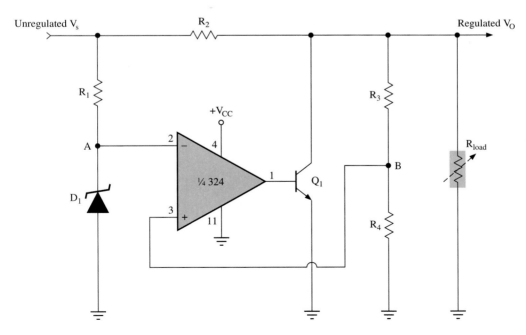

Figure 9–44
Basic shunt regulator

established within an IC regulator is allowed to deviate with changing temperature, control of the output voltage is lost. Therefore, IC voltage regulators must contain their own temperature compensation circuitry. The **Widlar bandgap reference** (named for its inventor Joseph Widlar) is a reference voltage circuit that provides its own temperature compensation. This type of reference circuitry is incorporated into the design of the 7800 series voltage regulators, which are used widely in industrial control systems.

The basic Widlar bandgap reference is contained in Figure 9–45. The 1.2-V output of this circuit is equivalent to the bandgap voltage of pure (undoped) silicon at 0 K. The **bandgap voltage** (designated V_{g0}) is the potential required to produce a free electron in pure silicon. Ideally, when the reference voltage in Figure 9–45 is at 1.2 V, the temperature coefficient is considered to be at zero. To understand how the Widlar bandgap circuit is able to maintain a constant reference level, the following voltage and current relationships must be established:

$$V_{ref} = V_{R_2} + V_{BE_3}$$

$$V_{R_2} \cong 10 \times V_{R_3}$$

For the current mirror consisting of Q_1 and Q_2:

$$V_{BE_1} = V_{BE_2} + V_{R_3}$$

$$I_{R_1} \cong 10 \times I_{R_2} \quad \text{Thus, } I_{E_1} \cong 10 \times I_{E_2}$$

The base-emitter junction of Q_3 has a negative temperature coefficient of nearly 2 mV/°C. Thus, as temperature increases, the base voltage for Q_3 decreases.

However, because I_{E_1} is 10 times greater than I_{E_2} the difference in voltage between V_{BE_1} and V_{BE_2} is allowed to have a positive temperature coefficient. Since V_{R_3} must equal the difference between V_{BE_1} and V_{BE_2}, V_{R_3} increases as V_{BE_2} decreases. Since I_{R_2} virtually equals I_{R_3}, V_{R_2} also increases. Since the ohmic value of R_2 is 10 times greater than R_3, ΔV_{R_2} is 10 times greater than ΔV_{R_3}. Thus, since V_{R_2} varies in proportion to V_{R_3}

$$\Delta V_{R_2} \propto \Delta(V_{BE_1} - V_{BE_2}) \tag{9–26}$$

Figure 9–45
Basic Widlar bandgap voltage reference circuit

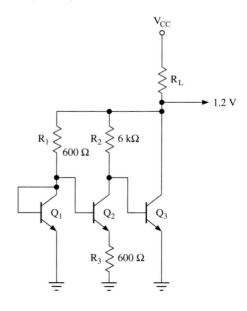

In a well-designed Widlar bandgap circuit, the increase in V_{R_2} is equal and opposite to the decrease in V_{BE_3}, thus maintaining V_{ref} at a constant 1.2-V level.

Figure 9–46 contains the internal circuitry of a 7805 three-terminal 5-V regulator. A more complex form of Widlar bandgap reference is found here than was shown in Figure 9–45. Transistors Q_1 and Q_2 comprise the current mirror, with R_3 developing the difference in voltage between V_{BE_1} and V_{BE_2} and R_2 amplifying the positive temperature coefficient of ($V_{BE_1} - V_{BE_2}$), increasing in voltage at a rate of $+8$ mV/°C. This action compensates for the combined negative temperature coefficient of Q_3, Q_4, Q_5, and Q_6. Since each transistor has a temperature coefficient of -2 mV/°C, the combined temperature coefficient is -8 mV/°C. Thus, throughout the operating temperature range of the 7805:

$$V_{BE_4} + V_{BE_3} + V_{R_2} + V_{BE_5} + V_{BE_6} = 5 \text{ V}$$

Figure 9–47 illustrates how the 7805 5-V regulator may be used within a linear power supply. Along with the filter capacitor C_1, input capacitor C_2 and output capacitor C_3 are also recommended for optimal performance. Capacitor C_2 is necessary if the regulator is located at a considerable distance from the filter capacitor, and C_3 improves the transient response of the regulator. Capacitor C_4 provides additional filtering, with R_1 serving as a bleeder resistor.

Figure 9–48(b) illustrates an interesting application of the **LM317T adjustable voltage regulator.** Besides being capable of withstanding 15 W of power dissipation, this

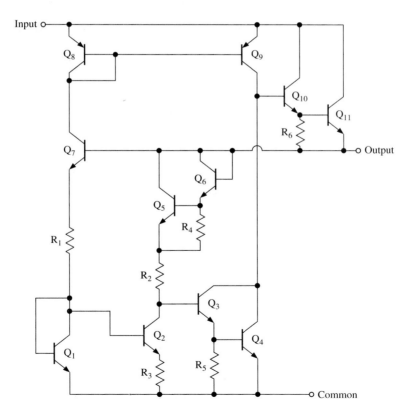

Figure 9–46
Internal circuitry of 7805 5-V regulator

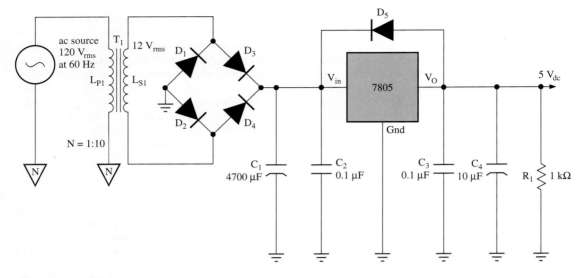

Figure 9–47
Basic 5-V regulated power supply containing 7805

device exhibits a line regulation of 0.01%, a load regulation of 0.1%, and a ripple rejection of 80 dB. It also has a wide output voltage range (1.2 to 37 V). Since dc power supplies in industrial control systems are usually committed to produce a few standard output voltages such as +5, +12, −12, and +24 V, adjustable regulators are often used to establish other voltage levels as necessary.

Assume that, for proper rotational speed, a dc motor must operate at 7.4 V_{dc}. In Figure 9–48(a), a rheostat is used to establish this voltage level. Three problems are associated

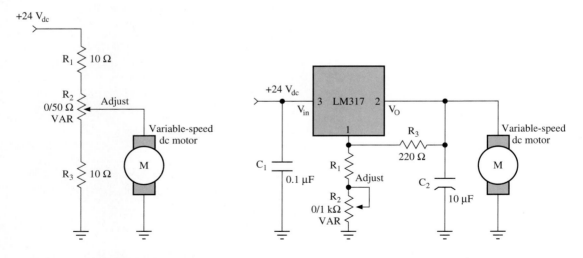

(a) Rheostat used to adjust speed of a dc motor   (b) Adjustable regulator used to control speed of a dc motor

Figure 9–48

with this method of control. First, a rheostat capable of withstanding the current flowing through the motor is likely to be quite bulky and dissipate considerable heat. Also, the wiper voltage is likely to drift slightly with deviations in temperature, causing motor speed to fluctuate. Any changes in the source potential would also cause deviations in motor speed. All of these problems are overcome by the design alternative shown in Figure 9–48(b). Even with required heat sinking, the TO-220 package is much smaller than a typical rheostat. Also, since the motor voltage is now fully regulated, deviations in motor speed as a result of drifting source voltage are virtually eliminated. A one-time speed adjustment could easily be made by placing a tachometer on the shaft of the motor and varying the setting of R_2 until the desired rpm is achieved.

SECTION REVIEW 9–4

1. Draw the block diagram of a series-pass linear power supply. Describe the function of each major section.
2. What is an alternative word used to describe the filter capacitor within a dc power supply?
3. Describe the operation of a full-wave line-to-line three-phase bridge rectifier. What are two advantages of this form of rectification over a single-phase bridge rectifier?
4. Define the terms *stabilization ratio, line regulation,* and *load regulation* as they apply to the operation of a linear voltage regulator.
5. Identify the two basic forms of linear voltage regulator. Describe the basic operation of each.
6. Describe the structure and operation of the current limiter stage within a linear voltage regulator.
7. Describe the operation of a Widlar bandgap reference circuit.

9–5   SWITCHING REGULATORS AND SWITCH-MODE POWER SUPPLIES

Switching regulators, like their linear counterparts, are capable of converting unregulated dc voltages to stable dc outputs. Switching regulators are far more complex than linear regulators and are prone to producing more RF noise and ripple. However, a primary advantage of switching regulators is power efficiency. The operating principle of the switching power supply is derived from the concept that an ideal switch, being either fully off or fully on, dissipates no power. Since bipolar transistors within switching regulators operate at saturation and cutoff, they approximate the action of an ideal switch. Thus, when compared to the pass transistors found within series regulators, their power expenditure is minimal.

As indicated by the block diagram in Figure 9–49, the control technique used in switching regulators is **pulse width modulation** (PWM). The process of PWM should be familiar, since it was introduced during the fourth section of Chapter 6 and incorporated into the design of an oven controller in Chapter 8. The feature of a pulse width modulator salient to the operation of switching regulators is that changes in its average output voltage are proportional to changes in its duty cycle:

$$\Delta\delta \propto \Delta V_{avg} \qquad\qquad (9\text{–}27)$$

where δ represents duty cycle.

When frequency is held constant, both PW and SW change as a result of pulse width modulation. Pulse width modulation may also result from holding either PW or SW constant while modulating the operating frequency.

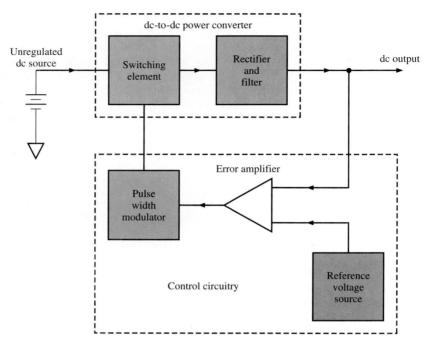

Figure 9–49
Block diagram of a switching voltage regulator

Assume, as shown in Figure 9–50(a), that a digital VOM, adjusted to measure dc voltage, is placed in parallel with a resistive load. When S_1 is closed, V_{R_L} equals 12 V_{dc}, which would be displayed by the meter. When S_1 is opened, V_{R_L} falls to 0 V, which is displayed by the meter after a slight delay. Next, as seen in Figure 9–50(b), we assume that the mechanical switch and battery are replaced by a pulse width modulator. As with the mechanical switch, the PWM is able to change the load voltage instantly from the level of V_{BB} to nearly 0 V. However, due to the rapid switching action of the PWM, the meter is no longer able to display both 12 and 0 V. Rather, it indicates the average output voltage. Thus, if the duty cycle of the PWM is increased, the value of V_{avg} indicated by the meter increases proportionally.

The pulsating dc voltage produced by the PWM could be used directly to control some loads, such as dc solenoids and motors. However, in the development of switch-mode power supplies suitable for use with electronic control circuitry, the pulsating dc voltage must undergo further processing. A **dc-to-dc power converter** must first be created, the purpose of which is to transform the output of the PWM to a dc voltage with an acceptable degree of ripple. A logical step in this conversion process might, at first, appear to be the placement of a capacitor in parallel with the load. However, due to the rapid voltage changes of the PWM output, capacitor current peaks at extremely high values, being limited only by source resistance. This transient response of the capacitor could be counteracted by an inductor, which would oppose these rapid changes in current. However, a design problem would still exist, since the inductive response to changes in source current would be high-amplitude voltage spikes. Through precise switching action, the operation of both the capacitor and the inductor may be synchronized to produce a dc output voltage. The three basic dc-to-dc power converter configurations used in switching

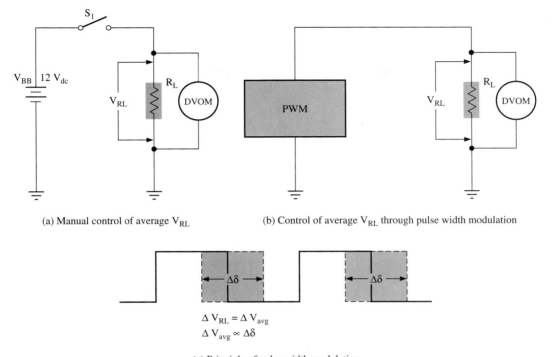

(a) Manual control of average V_{RL}

(b) Control of average V_{RL} through pulse width modulation

$$\Delta V_{RL} = \Delta V_{avg}$$
$$\Delta V_{avg} \propto \Delta \delta$$

(c) Principle of pulse width modulation

Figure 9–50

regulators are **boost, forward** (also referred to as **buck**), and **flyback** (sometimes referred to as **buck-boost**).

Figure 9–51 illustrates the operation of the basic **boost power converter.** Assume, as represented in Figure 9–51(a), that S_1 closes at the instant S_2 opens. Current flows through the inductor developing a counter emf with the polarity shown. While this current is flowing, energy is stored by the inductor in the form of an electromagnetic field. When, as shown in Figure 9–51(b), S_1 opens and S_2 closes, inductive current flow attempts to decrease, because of the residual charge developed over the capacitor. However, as the electromagnetic field of the inductor begins to collapse, the polarity of the induced voltage reverses, allowing V_{BB} and V_{L_1} to be series-aiding with respect to C_1 and the load. Thus, with the boost configuration in Figure 9–51, the output potential becomes greater than that of the dc source.

Figure 9–52 illustrates the operation of a **forward power converter.** As S_1 closes and S_2 opens, current flows to both the capacitor and the load, developing voltages with the polarities shown. As S_1 opens and S_2 closes, current flow from the source is interrupted. However, as the electromagnetic field of the inductor begins to collapse, the polarity of the inductor voltage reverses. Thus, the inductor is able to maintain the flow of current to C_1 and the load. Since the inductor is now the series component in a series/parallel configuration, the output voltage for the forward configuration in Figure 9–52 must be less than that of the source.

Figure 9–53 illustrates the operation of the **flyback power converter.** Assume, as represented in Figure 9–53(a), that S_1 closes as S_2 opens. Current flows through L_1

Figure 9–51
Operation of boost power converter

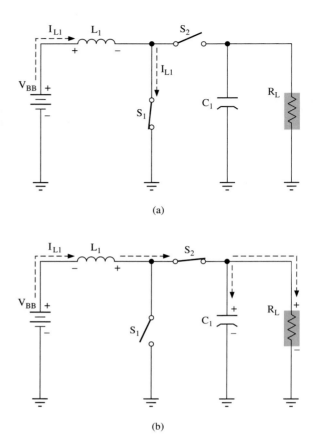

(a)

(b)

producing a counter emf with the polarity shown. While this current is flowing, energy is stored by the inductor in the form of an electromagnetic field. Next assume, as shown in Figure 9–53(b), that S_1 opens at the instant S_2 closes. The polarity of the inductive voltage reverses as the electromagnetic field collapses. Since the polarity of the inductive current is maintained, it flows to C_1 and the load in the direction shown. Note that, with the flyback configuration of Figure 9–53, the output voltage becomes negative with respect to the ground. With the flyback power converter, the amplitude of the output becomes a function of duty cycle. If S_1 is closed longer than S_2, effectively creating a duty cycle greater than 50%, V_O may be greater than V_{BB}. If S_2 is closed longer than S_1, resulting in a duty cycle of less than 50%, V_O is less than V_{BB}.

Each of the switching conditions represented in Figures 9–51, 9–52, and 9–53 can be produced using a transistor switch and a diode. In the next three circuit examples, the operation of switching transistor Q_1 is equivalent to S_1, while the function of D_1 is equivalent to that of S_2.

Consider first the boost power converter shown in Figure 9–54, the operation of which is analogous to the circuit in Figure 9–51. While the dc source is still shown as a battery, Q_1 is controlled by a pulse width modulator. During the pulse time of the PWM, Q_1 is switched on in saturation. With this condition, D_1 is reverse biased, allowing L_1 to develop a counter emf equivalent to $[V_{BB} - V_{CE}(sat)]$. At the beginning of the space time of the PWM, as the output of the PWM falls low, Q_1 becomes reverse biased. At that

Figure 9–52
Operation of forward power converter

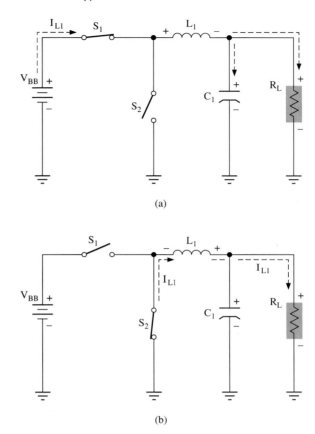

(a)

(b)

instant, the polarity of V_{L_1} reverses, attempting to oppose the change in current flow. Thus, D_1 becomes forward biased, allowing current from the inductor to flow to C_1 and the load. When the space time of the PWM begins, V_O is equivalent to $(V_{BB} + V_{L_1} - 0.7 \text{ V})$.

Figure 9–55 illustrates the operation of the forward power converter. During the pulse time of the PWM, Q_1 is switched on in saturation, allowing current to flow through L_1 to C_1 and the load. At this time, due to the polarity of V_{L_1}, D_1 is reverse biased. At the beginning of the space time, however, as Q_1 becomes reverse biased, current flow from the source is interrupted. Since the polarity of V_{L_1} is now reversed, D_1 becomes forward biased, providing a path for inductor current. The operation of this circuit is analogous to that in Figure 9–52, with load voltage being less than that of the source.

Figure 9–56 contains a flyback power converter, the operation of which is equivalent to the circuit in Figure 9–53. During the pulse time of the PWM, Q_1 is switched on in saturation. Emitter current now flows through the inductor, which develops a counter emf opposing the source. The polarity of V_{L_1} causes D_1 to be reverse biased, preventing current from flowing to C_1 and the load. At the beginning of the space time of the PWM, Q_1 switches off. The resulting interruption in current flow causes the polarity of V_{L_1} to be reversed. As a result, D_1 becomes forward biased. The inductor now serves as a current source for C_1 and the load. As in Figure 9–53, the voltage over the capacitor and load develops negatively with respect to ground.

Figure 9–53
Operation of flyback power converter

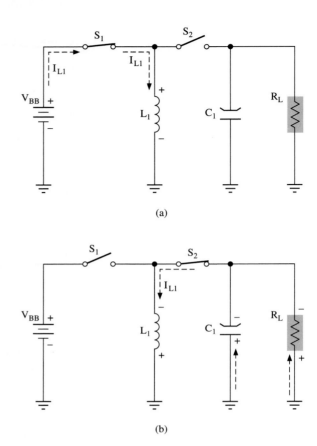

(a)

(b)

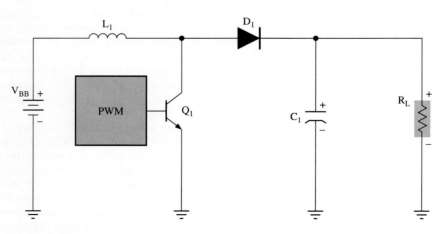

Figure 9–54
Basic boost power converter

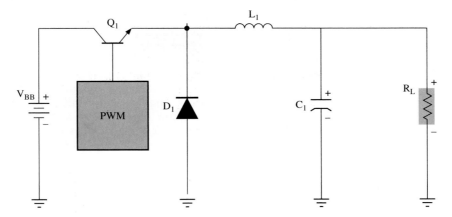

Figure 9–55
Basic forward power converter

Figure 9–57 contains a complete switching regulator, designed to convert a nonregulated 24 to a fully regulated 5 V. In Figure 9–57(a), Q_1, D_1, L_1, and C_3 comprise a forward power converter. The ECG2314 is a high-speed Darlington driver, which is capable of passing up to 6 A of load current. The flow of base current to the Darlington driver is controlled by switching transistor Q_3, which buffers the output of the LM311 comparator. This comparator serves as the output stage of the pulse width modulator, the triggering and ramp signals of which are controlled by two 555 timers. The first timer functions as an astable multivibrator, the active-low output of which serves as the trigger pulse to the ramp generator. The ramp generator consists of a 555 timer in a monostable configuration and a constant current source. The voltage divider, consisting of R_4 and R_5, holds the value V_{R6} at a constant level. Thus, the charging current for C_5 is held constant, allowing the growth in voltage over C_5 to be nearly linear. With the component values used in the pulse width modulator, the trigger and ramp signals should appear as the waveforms in Figure 9–58(a) and (b).

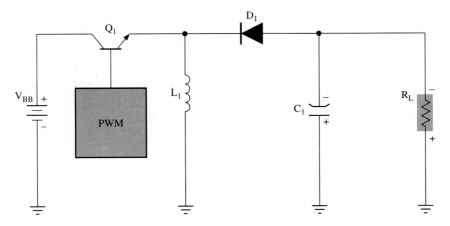

Figure 9–56
Basic flyback power converter

Figure 9–57(b) contains the reference voltage and error amplifier sections of the switching regulator. While a 7812 provides a regulated 12-V_{dc} source for the pulse width modulator and error amplifier, the LM317T provides an adjustable **reference voltage** V_{ref}. This reference voltage is fed to the noninverting input of the error amplifier, which consists of an OP-22 op amp in an inverting amplifier configuration. This device amplifies the difference in potential between V_{ref} and V_O, which is fed back to the error amplifier. For stable operation, this amplifier is **frequency compensated,** being designed to offer relatively high voltage gain to low-frequency error signals and to attenuate high-frequency ripple and switching noise. The error signal (designated V_e) is fed to the noninverting input of the comparator. The output of the comparator is high while V_e is greater than the instantaneous value of the ramp signal and falls low at the instant the ramp signal exceeds V_e. While the output of the comparator is high, Q_3 is switched on in saturation, allowing current to pass to the load via the Darlington driver. When the comparator output falls low, Q_3 switches off, reverse biasing the Darlington driver and interrupting current flow to the load.

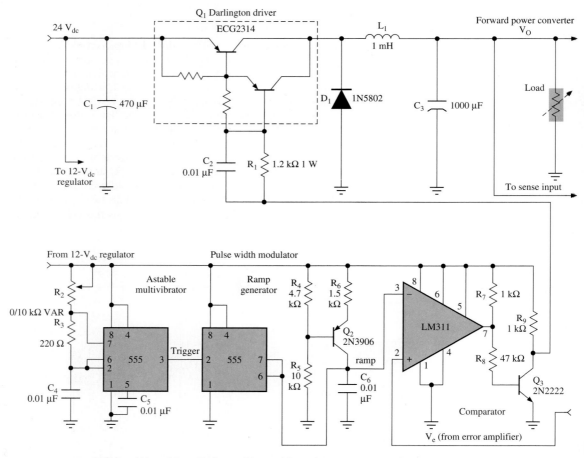

(a) Pulse width modulator, Darlington driver, and forward power converter sections of a switching regulator

Figure 9–57

output having a peak amplitude of nearly 340 V. This high dc voltage is chopped into rectangular pulses by switching circuitry and fed to the power conversion portion of the switching power supply.

During the positive alternation of the source sine wave, with S_1 closed, the path for current is from point A through D_1. As this current flows through R_2, C_1 charges with the polarity shown. At this time, the return path is through S_1 toward point D. During the negative alternation, the path for current is from point D through S_1 and R_3, charging C_2 in the direction shown. With D_4 now forward biased, current returns to the source at point A. Although, with S_1 closed, only two of the four diodes in the bridge actually conduct, both capacitors are able to charge to a peak level equivalent to $[V_S(peak) - 0.7$ V]. Thus, the dc output available between points B and C is determined as follows:

$$V_{B-D}(peak) = (2 \times 120 \text{ V}_{rms} \times 1.414) - 1.4 \text{ V} \cong 338 \text{ V}$$

If the source is 240 V_{ac}, S_1 is opened, allowing the bridge rectifier to function in the conventional manner. During the positive alternation, current flows from point A through D_1, R_2 and R_3 and D_3, allowing C_1 and C_2 to charge simultaneously. During the negative alternation, current flow is maintained in the same direction, with D_2 and D_4 now forward biased. Since R_2 equals R_3 and C_1 equals C_2, the peak voltage developed over each of the filter capacitors equals half the peak voltage developed between points B and C. Thus,

$$V_{B-C}(peak) = (240 \text{ V}_{rms} \times 1.414) - 2.8 \text{ V} = 336.6 \text{ V}$$

Also,

$$V_{C_1} = V_{C_2} = \frac{(240 \text{ V} \times 1.414) - 2.8 \text{ V}}{2} = 168.3 \text{ V}$$

With the ac line being taken directly to the bridge rectifier portion of the switching power supply, special precautions must be taken to avoid damage from high-voltage spikes and in-rush current. As shown in Figure 9–60, the noise spike problem is easily remedied by placing a varistor in parallel with the ac source. A possible solution to the problem of in-rush current would be the placement of the TRIAC and current limiting resistor R_1 directly in line with the ac source. The gate control circuitry for the TRIAC would be designed such that, at the instant power is applied to the circuit, the device is biased off. Thus, the in-rush current that attempts to flow to the discharged filter capacitors is limited by R_1. After a few seconds, with a residual charge having developed over C_1 and C_2, the TRIAC is allowed to trigger. Since the gate control for the TRIAC is a zero-point switching circuit, the TRIAC now turns on at the 0° and 180° points of every source cycle, causing current to be shunted past the limiting resistor.

A disadvantage of the TRIAC form of current limiting would, of course, be the need for precise gate control signals, possibly generated by ancillary circuitry contained within the switching power supply. An easier solution to the problem of in-rush current could be the use of NTC thermistors. As shown in Figure 9–61, one thermistor could be placed between the D_1, D_2 node of the full-wave bridge and the positive terminal of C_1, and a second thermistor is located between the D_4, D_3 node and the negative terminal of C_2. This allows a thermistor to be in series with a capacitor, regardless of the position of S_1. When power is first applied to the switching power supply, the thermistors are relatively cold, allowing their resistance to be relatively high. Thus, they function as current-limiting resistors in series with C_1 and C_2. However, through the self-heating action of the thermistors, their resistance drops greatly as current begins to flow. After a few seconds, in which the filter capacitors are able to acquire their residual charge, the resistance of the thermistors decreases to an insignificant level.

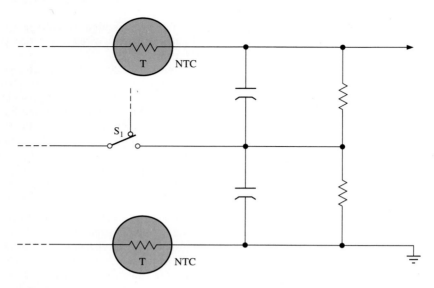

Figure 9–61
Optional use of thermistors for prevention of in-rush current

    In an off-of-the-line switch-mode power supply, isolation is usually provided between the filter stage, the output of which is a raw high-voltage dc, and the power converter stage. As stated earlier, this isolation is accomplished by using a transformer-choke, which serves as the inductive component within the power converter. Figure 9–62 illustrates how a transformer-choke may be incorporated into the design of a basic flyback converter.

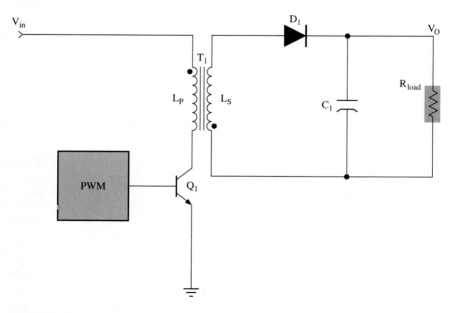

Figure 9–62
Flyback converter with isolation transformer-choke

While Q_1 is biased on, current flows through the primary winding of the transformer-choke. While this current is flowing, L_P develops a counter emf opposing the source, storing energy in the form of an electromagnetic field. Due to the opposite phasing of the secondary winding, D_1 is now reverse biased, blocking the flow of secondary current to C_1 and the load. When Q_1 becomes reverse biased, the magnetic field of the primary winding begins to collapse, causing the primary voltage to reverse polarity, becoming series-aiding with respect to the source. The resultant reversal of polarity in the secondary winding forward biases D_1, allowing induced current to flow to the load and replenish the charge on C_1.

A design problem associated with the flyback converter of Figure 9–62 is that the switching transistor must withstand an extremely high peak collector voltage. As indicated by the following equation, this voltage is determined by the dc input voltage and the duty cycle of the pulse width modulator:

$$V_{CEpeak} = \frac{V_{input}}{1 - \delta_{max}} \tag{9–29}$$

From this equation, it becomes evident that, as pulse width increases, the peak voltage developed over the collector of the switching transistor also increases. For this reason, the duty cycle of the pulse width modulator is usually kept below 40%. Even so, the peak amplitude developed over the collector of Q_1 may still attain a level of nearly $(2.2 \times V_{in})$.

A solution to the problem of high collector voltages in flyback converters is offered by the circuit in Figure 9–63. Here two transistors, which are biased on and off simultaneously, control the flow of current through the primary winding of the transformer-choke. Since the main terminals of Q_1 and Q_2 are effectively in series with L_P, the maximum values of collector-emitter voltage (abbreviated V_{CEO}) for the two transistors become additive. Thus, if two power transistors, each with a V_{CEO} of 600 V are used in Figure 9–63, their effective V_{CEO} becomes 1200 V. To further protect the two switching transistors, D_1 and D_2 are connected as shown. These diodes function as negative clippers. At the instant Q_1 and Q_2 are biased off, D_1 and D_2 become forward biased, providing a path for the reverse current that results from the collapse of the magnetic field around L_P. Thus, the peak value of V_{CE} for each of the switching transistors is limited to the approximate value of V_{in}.

Figure 9–64 contains a circuit commonly referred to as a **push-pull converter.** The term *push-pull* is misleading because the action of the circuit, which effectively functions as two isolated forward converters, could better be described as *push-push.* In contrast to the flyback converter in Figure 9–63, the two transistors within the push-pull converter must never be biased on at the same time. Thus, within a given clocking period, the duty cycle for either transistor must be less than 50%.

During the time Q_1 is on and Q_2 is off, current flows through L_{P1}, developing a counter emf that is positive with respect to point B. Since no phase reversal exists between L_{P1} and L_{S1}, D_1 is reverse biased. However, as Q_1 switches off, the polarity of the primary and secondary voltages of transformer-choke T_1 instantly reverses. This action forward biases D_1, allowing secondary current to flow to the load and C_1 via inductor L_1. While Q_1 is off and current is flowing to the load through D_1, Q_2 switches on, allowing the flow of current through L_{P2}. Thus, a counter emf develops over the primary winding of T_2, positive with respect to point B. Because of the lack of phase reversal in T_2, D_2 is now reverse biased. However, as Q_2 switches off, the polarities of the primary and secondary voltages developed over T_2 immediately reverse. This action forward biases D_2, thus maintaining the flow of current to L_1, C_1, and the load. The output voltage of the push-

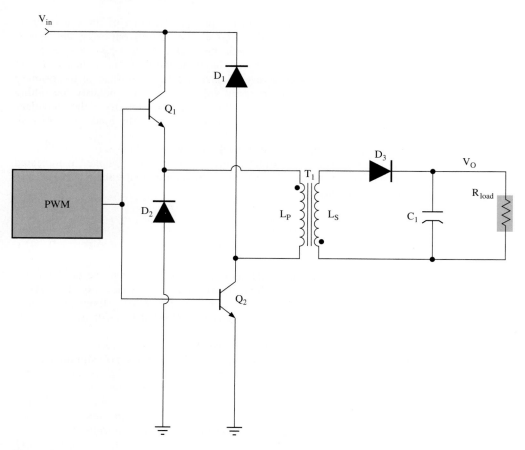

Figure 9–63
Flyback converter that uses two transistor switches and two clipping diodes

pull converter may be determined as follows:

$$V_O = \frac{2\delta_{max} \times V_{in}}{n} \qquad (9\text{–}30)$$

where δ_{max} represents the maximum duty cycle and n represents the turns ratio of the two transformer-chokes.

The push-pull converter design has two major advantages. First, since each of the transistors is biased on for less than 50% of the clocking period, the likelihood of thermal runaway is greatly reduced. Also, since the two isolation transformer-chokes are utilized, each section is active for only half of the clocking period. This greatly reduces the possibility of core saturation, which tends to occur when transformer primary current is pulsating dc.

Because the switching transistors within the push-pull converter require intricate timing, a more complex form of PWM control would be required in Figure 9–64 than has been introduced thus far. Such precise timing is provided by various forms of monolithic PWM controllers. An example of such an IC is the **TL494 PWM controller,** the internal circuitry of which is shown in Figure 9–65. The oscillator portion of this IC is

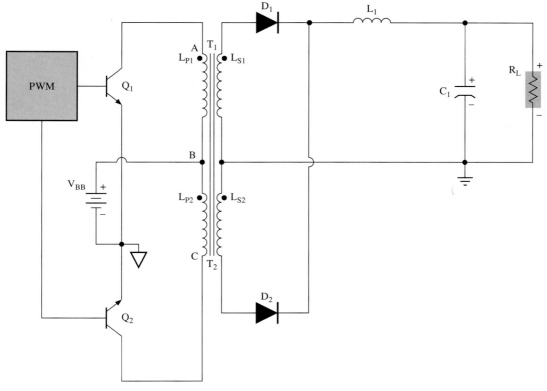

Figure 9–64
Basic push-pull power converter

a linear sawtooth (ramp) generator, similar in operation to the dual 555 timer circuitry in Figure 9–57(a). The frequency of the oscillator within the TL494 is determined by an external resistor and capacitor, which are connected from the R_T and C_T terminals to ground. For the TL494, the frequency of oscillation is determined as follows:

$$f = \frac{\ln 3}{R_T \times C_T} \cong \frac{1.1}{R_T \times C_T} \qquad \textbf{(9–31)}$$

The internal comparators of the TL494 allow the instantaneous amplitude of the ramp signal present over C_T to be compared to a **dead-time control** signal at pin 4, as well as the outputs of the internal error amplifiers. If pin 4 is held at ground potential, an offset potential of 0.12 V is present at the noninverting input of the **dead-time comparator.** Thus, the output of the OR gate is at a logic high at the beginning of each ramp cycle, at least until the instantaneous value of V_{CT} exceeds 0.12 V. During this dead time, a logic high is present at the lower input of each of the NOR gates, holding both internal transistors Q_1 and Q_2 in a reverse-biased condition. These transistors can function as either current sinks or current sources for external power transistors contained within the power converter stage of a switch-mode power supply. Thus, during the dead time, these external transistors would also be reverse biased.

    If Q_1 and Q_2 of the TL494 are to switch on and off simultaneously, as would be necessary for controlling the transistors within the flyback converter in Figure 9–63, then

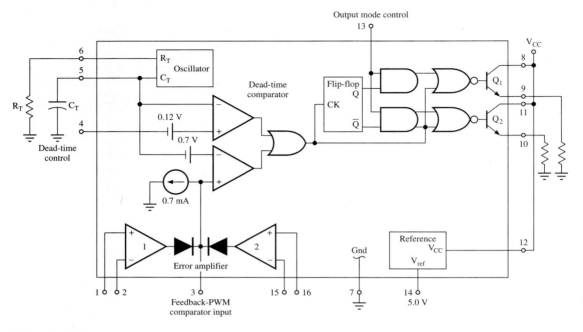

Figure 9–65
Internal circuitry of TL494 PWM controller. (Courtesy of Motorola.)

the **output mode control** input (pin 13) would be tied to circuit ground. Since the outputs of the two AND gates would then remain at a logic low, the outputs of the NOR gates would be at a logic high whenever the output of the OR gate is at a logic low. With pin 13 held low, the output switching frequency of the TL494 is equivalent to that of the internal oscillator. Although the flip-flop is toggling at this time, it is effectively removed from the circuit due to the disabled condition of the AND gates.

If pin 13 of the TL494 is tied back to pin 14, 5 V_{dc} is applied to the enabling inputs of both AND gates. With this logic high condition for the **output mode control,** the switching conditions for Q_1 and Q_2 are controlled by the Q and $\overline{Q}$ outputs of the flip-flop. Because of the divide-by-two action of the flip-flop, the switching frequency of Q_1 and Q_2 becomes only half that of the internal oscillator. Whenever the output of the OR gate is high, both Q_1 and Q_2 are biased off. However, with a high at pin 13, Q_1 and Q_2 are in complementary conditions while the output of the OR gate is low. Thus, with the output mode control tied high, the TL494 may be used to control a push-pull converter. With pin 4 still held at 0 V, the duty cycle for either transistor is held to a maximum duration of 48%.

Figure 9–66 serves to clarify the operation of the TL494 PWM controller. Note that during the first 11 cycles of the ramp signal, while the output mode control input is high, transistor Q_1 may be biased on only while the Q output of the flip-flop is high. When the flip-flop is reset, with $\overline{Q}$ high, transistor Q_2 is enabled to conduct. During the last four ramp cycles, while the output control mode input is low, transistors Q_1 and Q_2 switch simultaneously, at the same frequency as the ramp signal. Also note that the pulse widths of the emitter signals for transistors Q_1 and Q_2 are inversely proportional to the amplitude of the feedback-PWM comparator input, which is represented as a dashed line juxtaposed over the sawtooth signal.

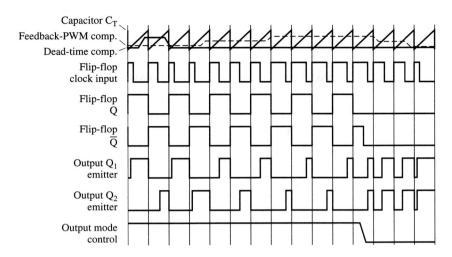

Figure 9–66
Waveforms produced by TL494 IC. (Courtesy of Motorola.)

In Figure 9–67, a TL494 IC is used to control the operation of a push-pull converter stage within a switching power supply. With pin 13 tied to pin 14, the internal flip-flop controls the operation of the two *p-n-p* power transistors within the converter. To determine the switching frequency of the push-pull converter, the oscillating frequency of the TL494 must first be determined:

$$f_{ramp} \cong \frac{1.1}{15 \text{ k}\Omega \times 0.001 \text{ }\mu\text{F}} \cong 73.333 \text{ kpps}$$

$$f_{out} = \frac{f_{ramp}}{2} = \frac{73.333 \text{ kpps}}{2} = 36.667 \text{ kpps}$$

The TL494 contains two differential amplifiers, either or both of which may be used as error amplifiers within the control loop of a switching power supply. For the power supply in Figure 9–67, only amplifier 1 is being incorporated within the feedback loop. With amplifier 2 being left in an open-loop condition, its inverting input (pin 15) is held at a level slightly above ground potential, while its noninverting input (pin 16) is connected directly to dc ground. This ensures that the output of amplifier 2 is held low. When incorporating the TL494 into a switching power supply, it is the responsibility of the engineer to design the error amplifier. Factors affecting the design of the error amplifier include the switching frequency and output amplitude of the power supply. Since the circuit configuration used for the error amplifier in Figure 9–67 is difficult to ascertain from the figure, an equivalent amplifier circuit is provided in Figure 9–68. The 28-V quiescent output of the power supply is divided down to nearly 5 V by R_1 and R_2. This sense voltage is fed back to the non-inverting input of the error amplifier:

$$V_{sense} = V_{out} \times \frac{R_2}{R_1 + R_2} = 28 \text{ V} \times \frac{4.7 \text{ k}\Omega}{22 \text{ k}\Omega + 4.7 \text{ k}\Omega} \cong 4.93 \text{ V}$$

The regulated 5-V level available at pin 14 of the TL494 serves as the reference input to the error amplifier. When the level of the sense input decreases, as a result of increased

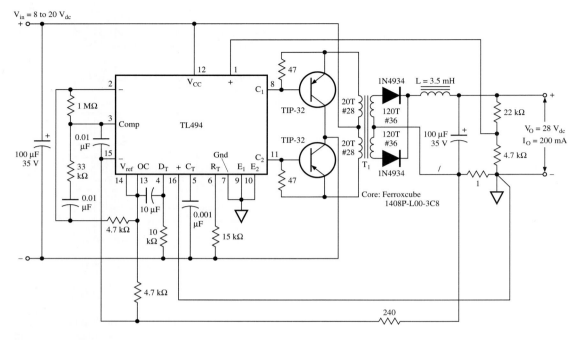

Figure 9-67
Application of TL494 PWM controller within a push-pull dc-to-dc power converter. (Courtesy of Motorola.)

loading of the power supply, the output voltage of the error amplifier is also reduced. This increases the space width of the signal at the output of the OR gate of the dead-time comparator, thus increasing the duty cycle of the push-pull transistors. Consequently, the output voltage increases toward the desired level.

As with the error amplifier design in Figure 9–57(b), frequency compensation is provided in Figure 9–68 to ensure stability of the feedback loop. This design provides high amplification for deviations between the dc levels of V_{sense} and V_{ref}, while attenuating the ripple and high-frequency noise that could be present at the amplifier inputs. With this design, the input impedance (Z_{in}) consists only of R_3 and, therefore, is not sensitive to deviations in frequency. However, the feedback impedance (Z_{fb}), consisting of R_5, C_1, and R_6, is frequency sensitive. Since C_1 and C_2 offer virtually infinite reactance to gradual variations in dc, the amplifier responds rapidly to deviations in V_{sense}. However, C_1 offers only a few ohms of reactance to high-frequency noise, resulting in a greatly reduced loop gain. Also, C_2 performs a shunting action with respect to the high-frequency component of the amplifier output, further attenuating high-frequency noise and increasing the stability of the control loop.

In concluding the study of switching regulators and power supplies, **ancillary** (or supporting) devices used to monitor and control their operation are introduced. These devices perform such functions as soft-starting, current-limiting, and overvoltage protection.

A **soft-start switching circuit,** in its simplest form, is an RC network, usually placed between the output of the error amplifier and its input to the PWM comparator. A typical application of this form of circuit is shown in Figure 9–69. The soft-start action of this RC network begins at the instant ac voltage is applied to the power supply. At this point,

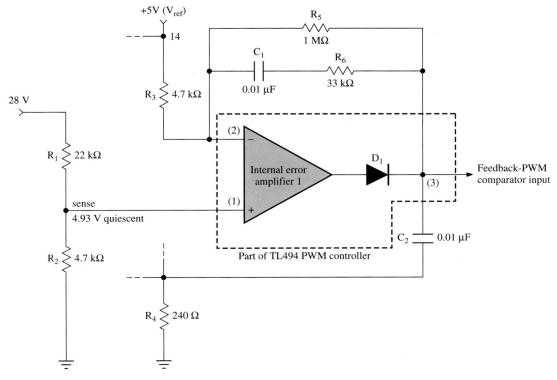

Figure 9–68
Equivalent circuit for error amplifier stage in Figure 9–67

with C_1 in a discharged state, point A is at virtually 0 V. Then C_1 begins to charge through R_1 toward the level of V_{ref}. The rate at which C_1 charges is a function of the RC time constant τ, determined simply as $(R_1 \times C_1)$. Early in the charging cycle, while D_2 is forward biased, the output of the error amplifier is limited to $(V_{C_1} + 0.7\text{ V})$. As a result, the pulse width of the PWM output signal is quite narrow when the power supply is first turned on. The pulse width increases slowly, at a rate determined by τ. This soft-start action is desirable because it prevents thermal shock and core saturation of the transformer-choke(s) within the power converter circuitry of a switch-mode power supply. As the voltage over C_1 continues to increase, it eventually attains a level greater than $(V_{error} - 0.7\text{ V})$. At this point, D_1 becomes reverse biased. Although C_1 continues charging toward V_{ref}, it is effectively removed from the control loop, since D_2 now functions as an open switch. When ac voltage is removed from the power supply, D_1 becomes operative, allowing for a rapid discharge of C_1. This action becomes important if source voltage is only momentarily removed from the power supply. With the capacitor quickly discharged, a complete soft-start cycle may still occur as source voltage is restored.

A commonly used form of **current-sensing circuit** is shown in Figure 9–70. The term **hiccup** is appropriate in describing the operation of this circuit because it abruptly and, if necessary, repeatedly interrupts the flow of current. To sense an overcurrent condition, the primary winding of a **toroid current transformer,** which may consist of only one turn, could be placed in series with the primary windings of the transformer-choke within the power converter. The function of the current transformer is to convert

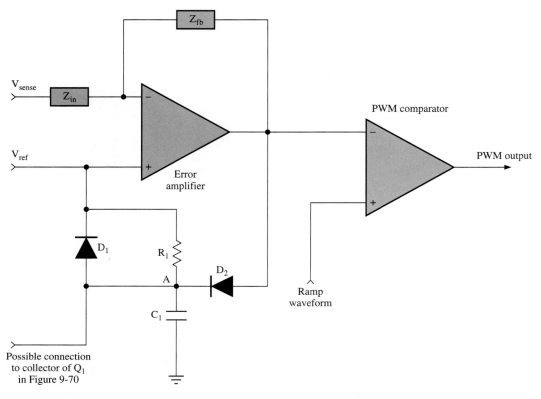

Figure 9–69
RC soft-start circuitry incorporated into the control loop of a switching power supply

a change in primary current to a proportional change in secondary voltage. This secondary voltage is converted to pulsating dc by a bridge rectifier and filtered by capacitor C_1. Potentiometer R_1 allows the adjustment of a threshold voltage at the inverting input to the LM311 comparator. This threshold voltage represents the maximum allowable level of load current. Thus, at the instant the threshold level exceeds a reference voltage present at the noninverting input, the output of the comparator falls low. This negative transition triggers the 555 one-shot, causing its output to go high for a period approximately equal to ($1.1 \times R_5 \times C_2$). The logic high pulse present at the output of the 555 timer forward biases switching transistor Q_1, the collector of which could be tied to the positive terminal of capacitor C_1 of the soft-start circuit in Figure 9–69. Thus, during the pulse time of the one-shot, the soft-start capacitor is held in a nearly discharged state, momentarily interrupting the operation of the power supply. If, during the pulse time of the one-shot, the load condition that caused the excess current is corrected, the output of the comparator remains high, allowing the power supply to return to operation. If, however, an overcurrent condition occurs, the one-shot is triggered again. This ''hiccup'' process continues until the excess load condition is remedied.

Besides overcurrent protection, ancillary circuitry is also provided to protect switching power supplies from an **overvoltage condition.** An IC commonly used for overvoltage protection (abbreviated OVP) is the MC3423, the internal circuitry of which is represented in Figure 9–71. This device contains a reference voltage source, which produces a constant

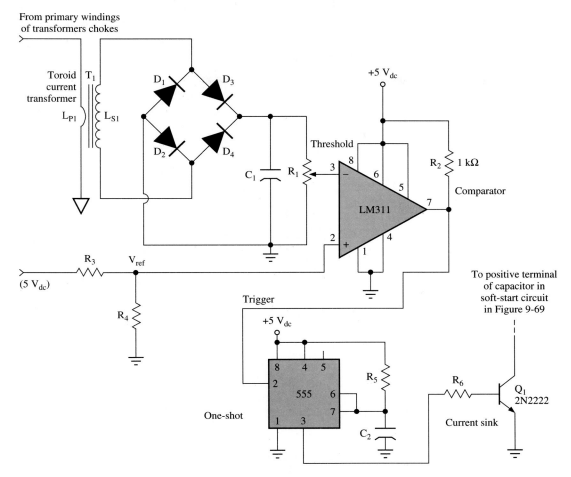

Figure 9–70
A 555 one-shot incorporated into a current-limiting "hiccup" circuit

2.6 V_{dc}. This internal V_{ref} is applied to the noninverting input of one comparator and the inverting input of a second comparator. This allows the first comparator to detect if the $V_{sense\,1}$ input (pin 2) exceeds 2.6 V. The second comparator detects if the $V_{sense\,2}$ input (pin 3) remains below the reference level. The use of two comparators within the MC3423 allows for an adjustable time delay between the instant the IC detects an overvoltage condition at pin 2 and the time it responds with a logic high at pin 8. Such dampening of the sensitivity of an OVP device is necessary to prevent spurious triggering by noise spikes.

Figure 9–72 illustrates a typical application of the MC3423 IC, which is being used in conjunction with a **crowbar SCR** to detect to an overvoltage condition. This OVP circuit is connected directly between the switching power supply and the dc load. Resistors R_1 and R_2, which are connected between the output of the power supply and the dc ground, form a voltage divider. The ohmic value of R_1 is adjusted such that, when the output of the power supply reaches a maximum allowable level (designated V_{trip}), 2.6 V is present

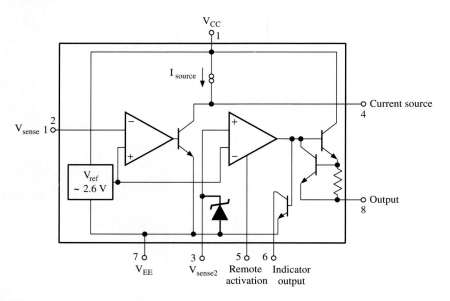

Figure 9–71
Internal circuitry of MC3423 overvoltage protection IC. (Courtesy of Motorola.)

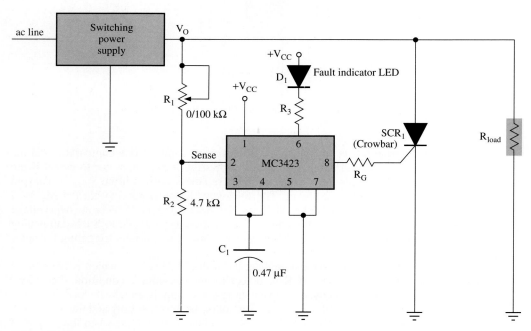

Figure 9–72
MC3423 overvoltage protection IC used in conjunction with a crowbar SCR

at pin 2 of the MC3423. The relationship of V_{trip}, V_{ref}, and the resistors within the voltage divider is expressed as follows:

$$V_{trip} = 2.6 \text{ V} \times \left(1 + \frac{R_1}{R_2}\right) \qquad (9\text{--}32)$$

Providing the voltage at pin 2 is less than 2.6 V, the transistor at the output of the first comparator is forward biased and is thus able to shunt current from the internal constant current source past C_1. Thus, the capacitor remains in a nearly discharged state. However, at the instant the voltage at pin 2 exceeds 2.6 V, the output of the first comparator falls low, reverse biasing its output transistor. With this transistor now functioning as an open switch, current from the internal source is diverted to C_1 via pin 4. Since C_1 is charging via a constant current source, the voltage over C_1 increases as a linear ramp. The delay period is that time required for the instantaneous value of capacitor voltage to attain the reference level of 2.6 V. For the MC3423, this delay time, which is primarily a function of the capacitance of C_1, is estimated as follows:

$$t_d = \frac{V_{ref}}{I_{source}} \times C \cong (12 \times 10^3) \times C \qquad (9\text{--}33)$$

where V_{ref} is the 2.6-V reference, I_{source} is the output of the internal constant current source, and C is the capacitance connected to pins 3 and 4 of the MC3423.

If an overvoltage condition sensed at pin 2 of the MC3423 persists through the time-delay period, the capacitor voltage attempts to exceed the 2.6-V reference. This causes the output of the second comparator to toggle to a high condition, forward biasing the two output transistors. A current pulse then occurs at pin 8, latching on the crowbar SCR. Thus the SCR shunts current past the load, limiting the output voltage of the power supply to a forward saturation level. It should be stressed that, although the output voltage is now being held low, a large current is now flowing through the SCR. Once this crowbar condition has occurred, commutation of the SCR must be forced by temporarily removing ac voltage from the power supply. In most industrial systems, this is easily accomplished by switching off and on the circuit breaker in line with the power supply. An overvoltage condition could be the result of a power surge or, in the case of an adjustable power supply, it could result from a technician inadvertently adjusting an output beyond the trip point. In the latter instance, it is advisable to adjust the power supply output to its lowest level, remove and restore ac power, then slowly adjust the output to its proper level.

SECTION REVIEW 9-5

1. Identify the three major forms of dc-to-dc power converters found within switching regulators and power supplies. Sketch each converter configuration correctly showing the placement of the transistor, diode, inductor, capacitor, and load resistor.
2. Draw the block diagram of a typical switching regulator. Briefly explain the function of each section.
3. What is the purpose of S_1 in Figure 9-60? Describe the operation of the circuit with S_1 closed. Describe circuit operation with S_1 open.
4. What is the purpose of R_1 in Figure 9-60? What is the purpose of the TRIAC placed in parallel with this resistor?
5. What is the major advantage of the power converter shown in Figure 9-63 over that contained in Figure 9-62?
6. Explain the operation of the power converter contained in Figure 9-64. Identify two advantages of this converter design over that represented in Figure 9-63.

7. Draw the block diagram of a typical high-frequency switch-mode power supply. Explain the function of each major section of the power supply.

8. Define the term *off-of-the-line* as it applies to the operation of a switch-mode power supply.

9. True or false?: High-frequency switching power supplies produce much less ripple and RFI than their linear counterparts.

10. List two major advantages of switch-mode power supplies over linear power supplies.

11. What power converter configuration is utilized in Figure 9–67?

12. True or false?: For the circuit in Figure 9–67, the output switching frequency is one-half that of the internal sawtooth waveform generator.

SUMMARY

AC power is usually delivered to industrial machinery in polyphase form, with three-phase power being the most common. A three-phase power source may be wound in either a delta or a wye configuration. Because of their versatility, wye configurations are often used as the main power sources for industrial machinery. Powerful three-phase motors may operate directly from the line voltage (approximately 208 V from phase-to-phase), whereas 120-V_{ac} loads, such as dc power supplies, may be connected between individual phases and ac neutral.

Important devices contained within ac power distribution systems include line filters, varistors, circuit breakers, line fuses, emergency-stop switches, and interlocks. The two basic forms of line filters are EMI and RFI. EMI filters control the propagation of electromagnetic noise, which is prevalent in environments containing motors, relays, and solenoids. RFI filters attenuate radio-frequency interference, which is likely to be produced by high-frequency power supplies. Varistors are expendable NTC devices that protect circuitry from the high-voltage spikes that occasionally occur across ac lines. Besides being safety devices, protecting loads from overcurrent conditions, circuit breakers also serve as on/off switches for various subassemblies within large industrial systems. Circuit breakers often have sensing contacts, allowing control circuitry to detect their status.

Besides the familiar single-phase bridge rectifier, three-phase bridge rectifiers are commonly used in industrial systems. An example of such a device is the full-wave line-to-line three-phase bridge rectifier. Such devices are often used within brute force power supplies, which provide unregulated dc voltage to such loads as solenoids and clutch/brakes. This form of rectifier is capable of sourcing a large amount of current with a relatively low degree of ripple.

DC power supplies used in conjunction with electronic control circuitry require a high degree of filtering and voltage regulation. The simplest form of voltage regulation would be the basic Zener regulator. These devices are limited in their current sourcing capability and are temperature sensitive. Better load regulation is accomplished by series and shunt voltage regulators containing active feedback and control circuitry. Although Zener diodes may be used as reference voltage sources within series and shunt voltage regulators, Widlar bandgap references are used more frequently within IC voltage regulators, especially those which are subject to wide variations in operating temperature.

Problems with linear power supplies include their relatively low power efficiency and large size. Both of these problems result from the use of large isolation transformers, designed to operate directly from the ac line. Switch-mode power supplies are more power efficient and compact than linear power supplies. Since they operate at higher frequencies (20 kHZ to 1 MHZ), switch-mode power supplies may contain relatively small and lightweight isolation transformer-chokes within their dc-to-dc power converter stages. The basic control element within a switch-mode power supply is a pulse width modulator, which controls the amplitude of the dc output through variations in duty cycle. Switch-mode power supplies are far more complex than their linear counterparts. Care must be taken in their design to limit RF noise. Most commercially manufactured switch-mode power supplies are usually equipped with sophisticated ancillary circuits that provide current limiting, overvoltage protection, and soft-start switching.

SELF-TEST

1. The peak phase voltage produced by a three-phase generator in a wye configuration is 339 V. If a digital VOM is adjusted to measure ac voltage and placed between any pair of phases, what value would it indicate?
2. For the circuit in Figure 9–5, assume $R_{Load\ 1}$ equals 2.4 Ω, $R_{Load\ 2}$ equals 3.7 Ω, and $R_{Load\ 3}$ equals 2.1 Ω. Solve for the phase currents and the value of I_N. Express the value of I_N in both rectangular and polar form.
3. Explain the operation of a thermal circuit breaker. What prevents this device from closing once the flow of current has been interrupted?
4. Identify the two basic types of passive filters commonly used in power distribution systems.
5. For the circuit in Figure 9–33, assume the turns ratio for the transformer is 4 : 1. Also assume that the minimum allowable ripple is 4% and the load resistance is 100 Ω. Give these values, choose the minimum standard value of electrolytic capacitor that could serve satisfactorily as C_F. What would be the recommended voltage rating for this capacitor?
6. What is the approximate frequency range for switch-mode power supplies?
7. For the Zener regulator in Figure 9–41, assume the input voltage drops from 12 to 9.96 V. Assuming the stabilization ratio is still 0.0566, solve for the resultant change in output voltage and the percentage of line regulation.
8. For the circuit in Figure 9–43, what would be the maximum value of load current if R_2 equals 2.2 Ω?
9. For the circuit in Figure 9–62, assume the input voltage is 14 V and the duty cycle of the PWM is 40%. Given these conditions, solve for the peak value of V_{CE} developed over transistor Q_1.
10. For the circuit in Figure 9–64, assume the maximum duty cycle for each transistor is 46%, the turns ratio is 6 : 1 and the input is 26 V. What is the approximate output voltage?
11. For the TL494 PWM controller in Figure 9–67, assume C_T now equals 0.022 μF and R_T equals 4.7 kΩ. What is the operating frequency of the internal oscillator? What is the switching frequency for the output transistors?
12. What is the purpose of the 10-kΩ resistor and the 10-μF capacitor connected to the D_T input of the TL494 in Figure 9–67? How do they control the operation of the IC and power converter stage at the instant power is applied to the circuit?
13. Why is some dead time necessary for the reliable operation of the circuit in Figure 9–67? How does the internal circuitry of the TL494 ensure that a minimal dead time does occur?
14. How could the operation of the TL494 IC in Figure 9–65 be made to function as the PWM controller for the power converter in Figure 9–63? Sketch a circuit showing how this could be accomplished.

15. For the circuit sketched in response to Question 14, show how a soft-start switching network and current-limiting circuit could be incorporated in the overall design.

16. Explain the purpose and operation of the soft-start switching circuit in Figure 9–69. Why is this circuit connected between the error amplifier and the PWM comparator of the switching power supply? What is the purpose of D_1 in this circuit? What is the purpose of D_2?

17. Explain the operation of the current-limiting circuit in Figure 9–70. Why is the term "hiccup" appropriate in describing its operation? Why would the collector output of this circuit be connected between the soft-start timing capacitor and resistor in Figure 9–69?

18. For the OVP circuit in Figure 9–72, what must be the ohmic setting of R_1 to establish a trip potential of 31 V?

19. What is the delay time for the OVP circuit in Figure 9–72? Explain the purpose of this time delay.

20. What is the purpose of the crowbar SCR in Figure 9–72? What action must be taken once this SCR is latched on? What additional precautions should be taken if the crowbar SCR latches on as the result of a technician accidentally adjusting an output of a switch-mode power supply above the maximum allowable voltage level?

10 INTRODUCTION TO MOTORS AND MOTOR CONTROL CIRCUITRY

CHAPTER OUTLINE

LEARNING OBJECTIVES

On completion of this chapter, the student should be able to:

- Describe the structure and operation of a single-phase ac induction motor.
- Describe the structure and operation of a 3-ϕ ac induction motor.
- Explain the advantages of a 3-ϕ ac motor over a single-phase device.
- Describe the structure and operation of a basic dc motor.
- Identify the various configurations used to connect dc motors.
- Explain the operation of control circuitry used to govern the rotational speed and direction of dc motors.
- Describe the basic structure and operation of a synchro. Explain the applications of synchros within industrial controllers.
- Describe the structure and operation of a resolver.
- Describe the structure and operation of a 3-ϕ brushless dc servomotor.
- Describe the basic structure and operation of the control circuitry used to govern a 3-ϕ brushless dc servomotor.
- Describe the structure and operation of stepper motors.

INTRODUCTION

This chapter consists of a detailed survey of the structure and operation of ac and dc motors most commonly used in modern industry. The major ac and dc motor configurations are compared to one another in terms of power, speed, torque, and control capability.

Standard methods of controlling conventional dc motors are thoroughly examined. Also, the basic circuitry used to control the direction and speed of motor rotation is introduced. Special types of motors and their associated control circuitry are then investigated. The internal structure and operation of three-phase brushless dc servomotors, along with circuitry used to control these devices, are examined in detail. Finally, the three basic forms of industrial stepper motors are introduced.

10–1   INTRODUCTION TO AC AND DC MOTORS

Single-phase and three-phase (3-φ) ac motors are widely used within automated industrial systems. Single-phase ac motors are limited to low-power applications, whereas various forms of 3-φ ac motors are used in medium- and high-power functions. For example, single-phase ac motors are used to operate cooling fans, while 3-φ ac motors, often working in tandem, are capable of driving large conveyance systems. As explained later, 3-φ motors are less complicated than their single-phase counterparts because they require no starting circuitry.

Figure 10–1 shows a cutaway view of an **ac induction motor.** The two principal parts of the motor are the **rotor** and the **stator.** (The term *rotor* generally identifies any rotating machine part; the term *stator* is used to indicate a stationary machine part within which a rotor revolves.) The type of rotor contained in Figure 10–1 is sometimes referred to as a **squirrel cage,** because it is composed of several aluminum bars that function as conductors. Through the process of die-casting, aluminum is poured into holes in the ferromagnetic core material of the rotor, forming the bars and end rings. During this process, the rotor is temporarily mounted on a dowel. After partial cooling, the dowel is removed and the rotor is pressed onto the motor shaft. As the rotor continues to cool, it contracts, becoming tightly bonded to the shaft. Like the rotor of an ac induction motor,

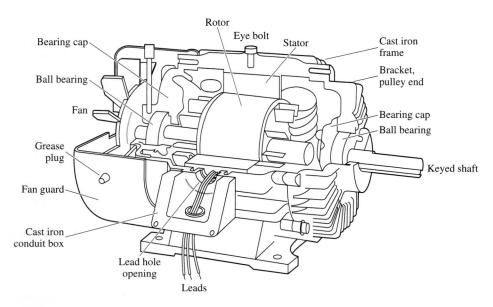

Figure 10–1
Cutaway view of an ac induction motor

the stator is composed of several laminations. After being fastened to an outer shell, these laminations are wrapped with insulated wiring. It is the stator windings of an induction motor that are connected to the ac source.

To initiate rotation within induction motors that operate from a single-phase ac source, some form of phase-shift circuitry must be used. For this reason, such devices are often referred to as **split-phase induction motors.** An example of such a motor is shown in Figure 10–2. A common form of split-phase motor is the **permanent-split capacitor (PSC) motor.** As seen in Figure 10–2, the electrolytic capacitor used within the starting circuitry of larger induction motors is often contained within the motor housing. (Note that the PSC form of a motor would not contain either the stationary or centrifugal switch.) The purpose of the capacitor is to create a phase shift between the current of the main stator windings and the current flowing in the start windings. As implied by its name, a constant phase shift (or phase split) exists between the start windings and the main windings of the PSC motor. The function of the PSC motor may best be understood through analysis of the equivalent circuit contained in Figure 10–3(a).

The schematic diagram in Figure 10–3 clearly shows that the starting circuitry, consisting of a series parallel network, is placed in parallel with the two main motor windings. Remember that, with a pure inductance, induced voltage tends to lead induced current by 90°. Thus, within both main motor windings, induced current tends to lag behind the source voltage by nearly 90°. (This phase difference is never quite 90° due to the dc resistance of the motor windings.) In the starting circuitry, however, due to the presence of the series capacitance, branch current tends to be more nearly in phase with the applied voltage. A large enough value of start capacitor allows the current within the start windings (designated I_S) to lead current within the main windings (designated I_N) by nearly 90°. This phase-shift condition is represented by the waveforms in Figure 10–3(b).

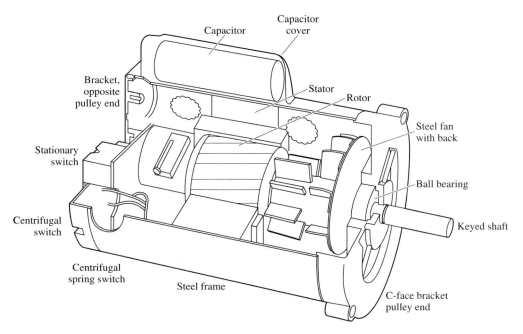

Figure 10–2
Cutaway view of a split-phase induction motor

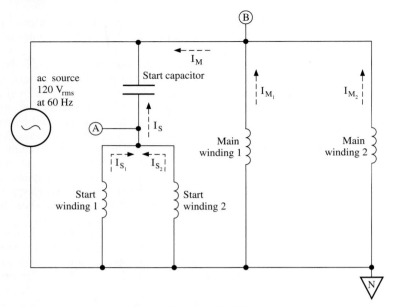

(a) Equivalent circuit for PSC motor

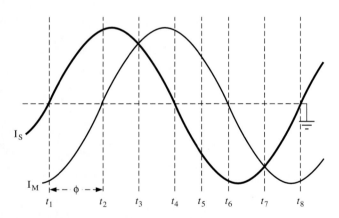

(b) Phase relationship of currents in start and main windings of PSC motor

Figure 10–3

How this disparity in phase between the start and main windings induces rotation of the rotor can now be explained. Remember that the strength of the magnetic field surrounding an inductance is directly proportional to the amplitude of induced current. Thus, as seen at t_2 in Figure 10–3(b), I_S is approaching its peak level, and the magnetic fields associated with the start windings are at their maximum strength. With no current flowing through the main stator windings at this time, the condition of the motor is that represented in Figure 10–4(a). As the magnetic flux lines of the start windings cut through the rotor, a magnetic field is induced. The polarity of this induced rotor field opposes that of the start windings. Induced current flows through the aluminum bars of the rotor as a result of the expanding and contracting stator fields. In Figure 10–4, small circles are used to show the direction of rotor current. Dots within the circles indicate conventional current flowing out of the page, and an X within a circle indicates conventional current entering the page.

One explanation for rotor movement within the ac induction motor involves the physical force that results from the interaction of the induced rotor current and the stator field. With rotor current flowing in the direction shown in Figure 10–4(a), counterclockwise rotation results. The physical force that produces this rotation is exerted by the aluminum bars as current is flowing. The direction of this force may be determined using the right-hand rule for magnetomotive force. As shown by the detail in Figure 10–5, where the direction in which the fingers are pointing is equivalent to the direction of conventional current flow, the right thumb points in the direction the aluminum bars tend to move.

Movement of the rotor results when, as shown in Figure 10–4(b), the magnetic field begins to develop in the main windings. This condition corresponds with t_3 in Figure 10–3(b). As the induced rotor field attempts to align itself with the resultant stator field, the interaction of the rotor currents and stator fields sustains the physical force exerted by the aluminum bars. At t_4, with I_S at 0 A, the stator field is entirely the result of current

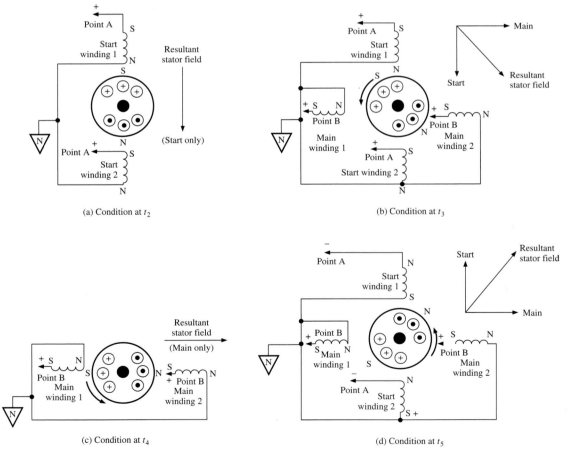

(a) Condition at t_2

(b) Condition at t_3

(c) Condition at t_4

(d) Condition at t_5

Figure 10–4

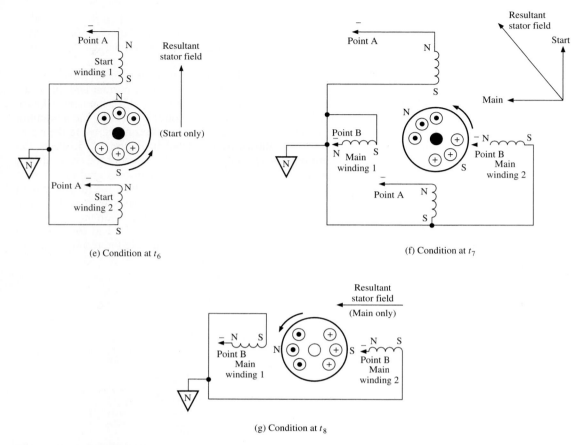

(e) Condition at t_6

(f) Condition at t_7

(g) Condition at t_8

Figure 10–4 ***(continued)***

flow in the main windings. As shown in Figure 10–4(c), the rotor continues to turn counterclockwise, attempting to align itself with the rotating stator field. By t_5, the currents flowing through the main and stator winding are of nearly equal amplitude but opposing polarity. At this time, as shown in Figure 10–4(d), the resultant magnetic field has shifted still further, causing continued rotor movement. Figures 10–4(e), (f), and (g), corresponding with t_6 through t_8 in Figure 10–3, illustrate how the resultant stator field and the induced rotor field complete their 360° of rotation.

As the rotational velocity of the rotor increases, due to the physical force being exerted by the aluminum bars, it is possible for the motor to continue operating without the presence of the start capacitor and windings. A **capacitor-start motor,** represented schematically in Figure 10–6, exploits this property of the ac induction motor. A **centrifugal switch** within the motor housing is designed to open automatically as motor speed approaches its maximum velocity. As seen in Figure 10–2, the actuating mechanism for the centrifugal switch is a bobbin mounted on the rotor shaft. While the rotor is stationary or turning at less than approximately 80% of full speed, the bobbin is held against the switch contacts by a retaining spring, allowing current to flow to the start capacitor and windings. At this time, the operation of the motor is identical to that illustrated in Figure 10–4. However, as motor speed approaches around 80% of full velocity, centrifugal force

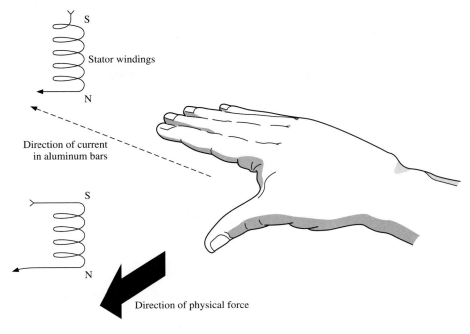

Figure 10–5
Physical force resulting from interaction of stator field and rotor current

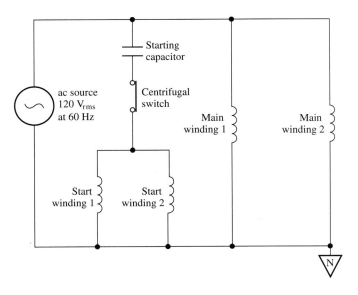

Figure 10–6
Schematic diagram of capacitor-start motor containing centrifugal switch

allows the bobbin to overcome the force of the retaining spring. As the bobbin pulls away from the switch contacts, the start circuit drops out, allowing the motor to continue to rotate through the interaction of only the main windings and the rotor. If motor speed falls below the 80% point, the bobbin engages the switch contacts again, bringing the start components back into the circuit.

Although 3-ϕ motors are usually larger and more powerful than single-phase motors, they are likely to contain fewer components. The phase shift necessary for initiating rotation of an induction motor is inherent to the 3-ϕ power source. Thus, no start capacitor or centrifugal switch is necessary within the 3-ϕ motor. As shown in Figure 10–7(a), the electrical arrangement of the field windings of the 3-ϕ induction motor is a delta (Δ) configuration. The physical placement of these field windings within the stator of the 3-ϕ induction motor is represented in simplified form by Figure 10–7(b).

Remember from the previous chapter that a phase voltage is the rms voltage measured between an individual phase and ac neutral, whereas a line voltage is the rms voltage existing between a pair of phases. With the delta configuration, the three possible combinations of line voltage are V_{A-B}, V_{B-C}, and V_{A-C}. According to Eq. (9–3), providing the phase voltage is known, line voltage may be easily determined as follows:

$$V_L = V_\phi \times \sqrt{3}$$

Thus, with a phase voltage of 120 V_{rms}, the line voltage for the stator windings in Figure 10–7 is determined as:

$$V_{A-B} = V_{B-C} = V_{A-C} \cong 120\ V_{rms} \times \sqrt{3} = 208\ V_{rms}$$

As illustrated in Figure 10–8, with a line voltage of 120 V_{rms}, a digital voltmeter would indicate 208 V_{rms} if placed between any pair of phases. The reason for this is made clear in Figure 10–9. As shown in Figure 10–9(a), the maximum difference in voltage between ϕA and ϕC occurs when ϕA is at its 120° point and phase C is at its 240° point.

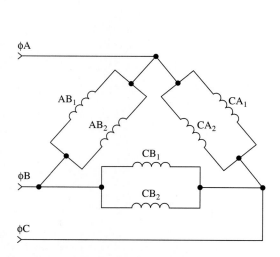

(a) Field windings of a 3-ϕ motor in a delta configuration

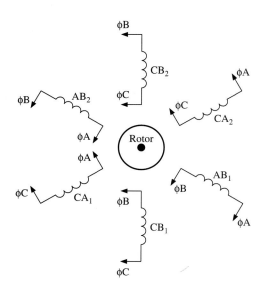

(b) Possible placement of field windings in a 3-ϕ motor

Figure 10–7

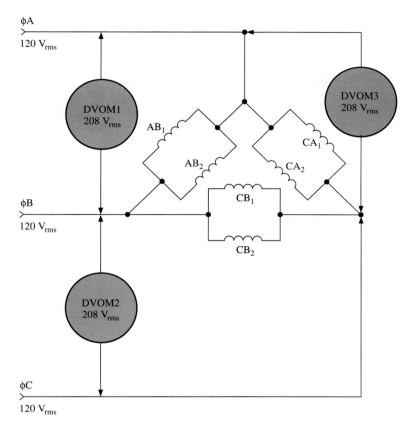

Figure 10–8
Stator voltages for a 3-ϕ ac motor

At this time, ϕB is at the 0° and thus has an instantaneous value of 0 V. For this to occur, according to Kirchhoff's law, the instantaneous voltages of ϕA and ϕC must be equal and opposite. This is easily proven as follows:

$$V_A = \sin 120° \times 169 \; V_{peak} \cong 147 \; V$$
$$V_B = \sin 240° \times 169 \; V_{peak} \cong -147 \; V$$

Thus,

$$V_{A-B} \cong 147 \; V_{peak} - -147 \; V_{peak} \cong 294 \; V_{peak}$$

Finally, the rms voltage indicated by the digital voltmeter is determined as:

$$V_{A-B}(rms) \cong 0.707 \times 294 \; V_{peak} \cong 208 \; V_{rms}$$

At nearly 90° beyond the peak voltage point, as shown in Figure 10–9(b), the current flow between the two phases (designated I_{CA}) reaches its peak amplitude, with conventional current flowing from ϕA to ϕC.

At a point 180° beyond that shown in Figure 10–9(a), the voltage condition would be equivalent to that shown in Figure 10–10(a). Thus, nearly 90° beyond this point, as

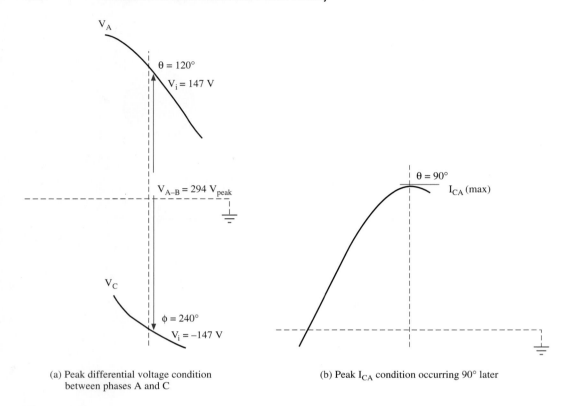

(a) Peak differential voltage condition between phases A and C

(b) Peak I_{CA} condition occurring 90° later

Figure 10–9

shown in Figure 10–10(b), I_{CA} would attain a peak amplitude equal to and opposite that shown in Figure 10–9(b).

The voltage and current relationships just explained for ϕA and ϕC also exist between ϕB and ϕC, as well as ϕA and ϕB. Thus, as illustrated in Figure 10–11, three sine waves may be drawn to represent currents I_{AB}, I_{CB}, and I_{CA}.

The operation of the 3-ϕ induction motor may now be explained with reference to the 12 points in time shown in Figure 10–11. At t_1, current flow is as shown in Figure 10–12(a). At this instant, the north poles produced by the CA windings are contiguous to the south poles of the AB windings. With this condition, two highly concentrated magnetic fields develop between these two pairs of windings, with virtually none of their flux cutting into the rotor. Thus, as far as operation of the motor is concerned, the AB and CA fields effectively cancel. At the same time, however, the currents and magnetic fields of the CB windings are at a peak level. The orientation of the fields developed over these windings causes their flux to cut directly across the rotor, producing the resultant field shown. In the same manner as was described for the single-phase induction motor, currents are induced in the aluminum bars of the rotor, resulting in angular force.

At t_2, 30° further into the cycle, current flow is as shown in Figure 10–12(b). At this instant, the CA windings have no current and, thus, produce no flux. However, with current flowing through the AB and CA windings in the direction shown, a resultant magnetic field is produced. The orientation of this field is 30° counterclockwise with respect to that produced by the CB fields in Figure 10–12(a). Thus, the angular motion of the rotor is counterclockwise as it attempts to align itself with this field.

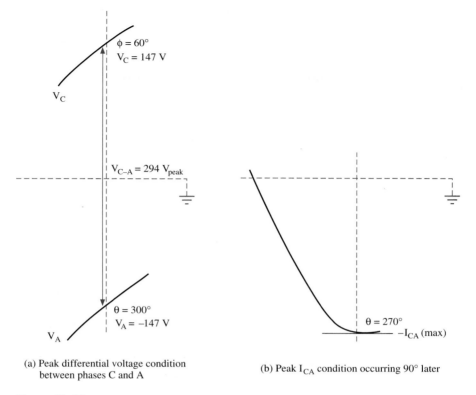

(a) Peak differential voltage condition
between phases C and A

(b) Peak I_{CA} condition occurring 90° later

Figure 10–10

By t_3 in Figure 10–11, the condition for the stator windings is as shown in Figure 10–12(c). As in Figure 10–12(a), one pair of windings is dominant, and the other two pairs effectively cancel each other. While virtually none of the flux produced by the CB and CA windings cuts into the rotor, the AB fields are at their maximum strength, cutting directly across the rotor as shown. The orientation of these fields is another 30° counterclockwise with respect to the fields in Figure 10–12(b). Thus, angular motion continues in a counterclockwise direction as the rotor attempts to align itself as shown.

Figure 10–12(d) corresponds with t_4 in Figure 10–11. Here, the motor condition is similar to that in Figure 10–12(b). With I_{CB} now at 0 A, the CB windings produce no flux. However, the flow of current through the other two pairs of windings produces a resultant field, the alignment of which is 30° counterclockwise with respect to the fields produced at t_3. Thus, the angular motion of the rotor continues, as it attempts to align itself with this resultant field.

By t_5, the current and magnetic fields of the CA windings are at their peak levels. Current also flows in the other four windings. However, as illustrated in Figure 10–12(e), the fields produced by the AB and CB windings at this time effectively cancel each other. Consequently, the rotor responds only to the fields produced by the CA windings. Since these fields are aligned an additional 30° counterclockwise with respect to the fields at t_4, the angular movement of the rotor continues.

Finally, at t_6, the condition of the motor currents and fields is as shown in Figure 10–12(f). Current I_{AB} is momentarily at 0 A and, thus, no flux is produced by the AB windings. However, the currents present in the CA and CB windings produce a resultant

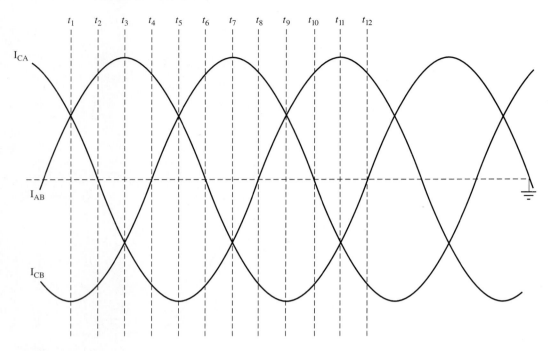

Figure 10–11
Current waveforms for field windings of 3-ϕ induction motor

field, the alignment of which is 30° counterclockwise with respect to the fields produced by the CA windings at t_5. Thus, rotor movement continues another 30°. By t_7, the condition of the motor would be complementary to that shown in Figure 10–12(a). This analysis of changing motor currents and fields could continue from t_7 through t_{12}, illustrating how the rotor makes one complete rotation. However, the preceding analysis should be sufficient for illustrating the principles involved.

Like the ac induction motor, a dc motor is composed essentially of a rotor and a stator. However, since a dc motor is connected to a voltage source that maintains a constant polarity, its operation is markedly different from that of an ac device. The simplest form of dc motor is the **permanent magnet** (PM) motor. This form of motor contains no stator windings. Instead, the required stator field is provided by a strong permanent magnet. Since the field strength of a permanent magnet is limited, it is often replaced by field windings in more powerful motors. Figure 10–13 contains a cutaway view of this form of dc motor, which contains at least one group of **field windings.** These windings are wrapped around ferromagnetic cores called **field poles,** thus forming powerful electromagnets. Due to the relatively thin wire and large number of turns used to form these field windings, they offer fairly high resistance to a dc voltage source.

Remember that, with an ac induction motor, current flows through the aluminum conductors of the squirrel cage rotor as a result of the continuously changing flux of the stator windings. It is the interaction of the rotor and stator currents and fields that produces angular motion. With a dc motor, however, source voltage is likely to be nearly constant. Since current flows through the field windings in only one direction, the polarity of the stator field does not change. Thus, with this static field condition, an alternative method must be used to induce angular motion of the rotor. With a dc motor, rotor movement is

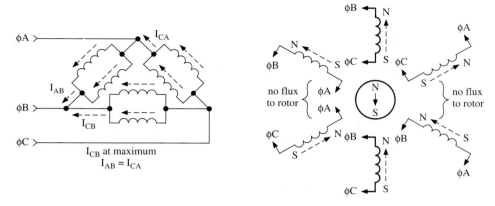

(a) Motor condition at t_1: ϕB–ϕC fields dominant; ϕA–ϕB and ϕC–ϕA fields effectively cancel each other

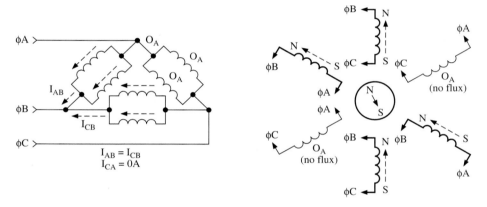

(b) Motor condition at t_2: ϕC–ϕA fields at minimum; ϕA–ϕB and ϕC–ϕB fields produce a resultant field

Figure 10–12

achieved through interaction of an **armature** and **commutator,** which are mounted on the rotor shaft.

As seen in Figure 10–13, the armature contains several coils that are inserted into slots within the cylindrical rotor. In contrast to the field windings, the armature windings are composed of fewer turns of relatively thick wire, thus offering minimal resistance (often less than 1 Ω) to the dc source. Each of these coils is connected to several segments of the commutator. Current flows through a given armature winding when one of its commutator segments contacts one of two carbon **brushes** mounted on opposite sides of the rotor shaft. These brushes are connected to an external dc voltage source.

The interaction of the stator and rotor fields of the dc motor may be best understood through analysis of the simplified illustrations in Figure 10–14. Here the armature, which may consist of several sets of windings within an actual motor, is represented by a single turn of wire. Also note that, for the sake of clarity, half of this armature is darkened. The stator of this simplified motor is represented as a permanent magnet. When power is applied to the motor, as represented in Figure 10–14(b), the direction of current through

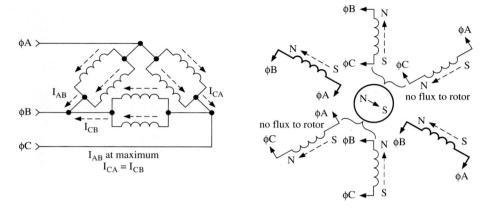

(c) Motor condition at t_3: ϕA–ϕB fields dominant; ϕC–ϕA and ϕC–ϕB fields effectively cancel each other

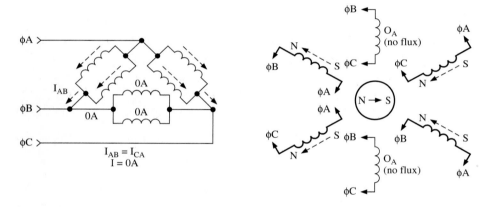

(d) Motor condition at t_4: ϕC–ϕB fields at minimum; ϕA–ϕB and ϕC–ϕA fields produce a resultant field

Figure 10–12 (continued)

the armature is such that the north pole of the rotor field is next to the north pole of the stator field, and the south pole of the rotor field is next to the south pole of the stator field. Note that the darkened half of the armature is assumed to align at the south pole of the rotor field. The magnetomotive force resulting from the interaction of the rotor and stator fields produces angular motion of the rotor. It is at this juncture that the action of the commutator and brushes becomes clear. As the rotor turns, the ends of the armature momentarily break contact with the commutator segments. When the armature again touches the commutator segments, as shown in Figure 10–14(c), its darkened half is approaching the north pole of the stator field. Since the direction of current flow through the two sides of the armature is now opposite to that shown in Figure 10–14(b), the darkened half develops a north pole and the other half becomes the south pole. As a result, due to the repulsion of the like poles, the rotor continues to turn.

At the instant power is applied to the dc motor, the armature condition is as shown in Figure 10–15(a). Since the change in current is at maximum, the armature winding develops a self-induced counter electromotive force (emf) that equals and opposes the

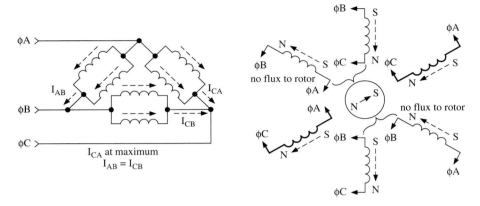

(e) Motor condition at t_5: ϕC–ϕA fields dominant; ϕA–ϕB and ϕC–ϕB fields effectively cancel each other

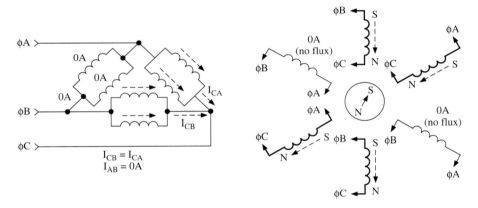

(f) Motor condition at t_6: ϕA–ϕB fields at minimum; ϕC–ϕB and ϕC–ϕA fields produce a resultant field

Figure 10–12 (continued)

dc source. Thus, for a brief instant, armature current (L) is at nearly 0 A. However, after only a few milliseconds, armature current rapidly increases. With the armature still stationary, as shown in Figure 10–15(b), current flow is limited only by the series resistance (r_S) of the winding. At this time, armature current is at maximum. Due to the low series resistance of the armature windings, this initial value of current is likely to equal several amperes. With the entire armature voltage (V_A) being developed by the series resistance, the value of armature current being drawn during the stationary armature condition is easily determined using Ohm's law:

$$I_A = \frac{V_A}{r_S} \tag{10–1}$$

As the rotor shaft begins to turn, the armature windings cut across the magnetic field of the stator. As a result of this action, a strong counter emf is induced in the armature windings. When the rotor is turning at full speed, this counter emf (V_{cemf}) can equal nearly 90% of the source voltage. Since this induced voltage opposes the dc source, it can be

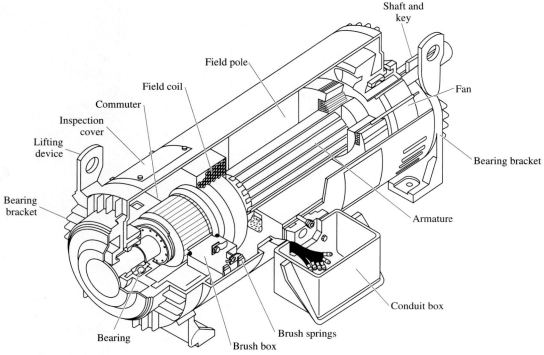

Figure 10–13
Cutaway view of a dc motor

represented as the series-opposing battery contained in Figure 10–15(c). While the rotor is turning, the current drawn by the armature is greatly reduced. According to Kirchhoff's voltage law, the voltage available to the series resistance of an armature winding is limited to the difference between V_A and V_{cemf}. Thus, with the armature turning,

$$I_A = \frac{V_A - V_{cemf}}{r_S} \qquad (10–2)$$

Example 10–1

Assume that, with 12.2 V_{dc} applied to the armature winding of a small dc motor, it draws 2.58 A when first energized. What is the value of r_S for that armature winding? Assume that the counter emf induced in the armature is 90% of the source voltage when the motor is running at full speed. What would be the reduced value of I_A?

Solution:

Step 1: Solve for r_S:

$$r_S = \frac{V_S}{I_A} = \frac{12.2 \text{ V}}{2.58 \text{ A}} \cong 4.73 \ \Omega$$

Step 2: Solve for current with the motor running at full speed:

$$12.2 \text{ V} \times 0.9 = 10.98 \text{ V}$$
$$12.2 \text{ V} - 10.98 \text{ V} = 1.22 \text{ V}$$
$$I_A \cong \frac{1.22 \text{ V}}{4.73 \ \Omega} \cong 258 \text{ mA}$$

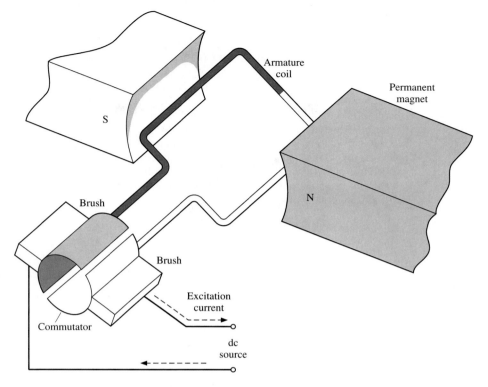

(a) Simplified representation of dc motor

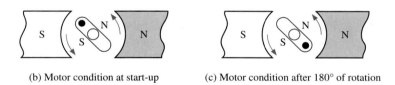

(b) Motor condition at start-up

(c) Motor condition after 180° of rotation

Figure 10–14

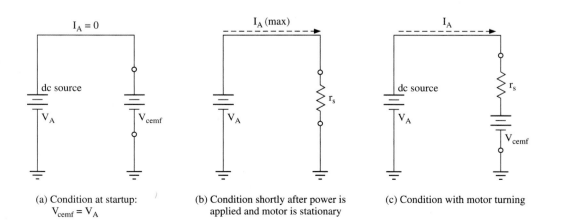

(a) Condition at startup:
$V_{cemf} = V_A$

(b) Condition shortly after power is applied and motor is stationary

(c) Condition with motor turning

Figure 10–15

It should now be evident that current flow through the armature winding of a dc motor is dependent on the angular velocity of the rotor. By contrast, field current remains nearly constant, as long as field voltage remains at a steady dc level. Assume, with field current and voltage remaining constant, that rotor speed decreases, due to an increase in mechanical load. This condition causes a reduction in the counter emf being induced in the armature windings. Consequently, according to Eq. (10–2), armature current must increase. However, if the angular velocity of the rotor increases due to a reduction in mechanical loading, a greater part of the armature voltage occurs in the form of induced counter emf. Consequently, I_A decreases. Although I_A varies as a function of changing load conditions, it is not in itself a control variable for motor speed. As explained in detail later in this chapter, speed control for dc motors involves control of the voltage over the field and/or armature windings.

SECTION REVIEW 10–1

1. What type of ac induction motor is represented in Figure 10–3? Explain the purpose of the start capacitor and windings contained within this motor.
2. What type of ac induction motor is represented in Figure 10–6? Explain the purpose and operation of the centrifugal switch contained within this form of motor. At approximately what points during the operation of this form of motor does this switch open and close?
3. For the motor circuits in Figures 10–3 and 10–6, what is the approximate phase relationship between the currents of the main winding and start winding?
4. Describe the structure and operation of a squirrel cage rotor. In what types of motor is this device found?
5. Explain the right-hand rule as it applies to the interaction of the rotor current and stator fields within an ac or dc motor.
6. Identify at least two major advantages of a 3-ϕ induction motor over a single-phase device.
7. What electrical configuration is typically used for the stator windings of a 3-ϕ induction motor? Carefully sketch this wiring configuration. Also illustrate how the stator windings are likely to be arranged within the motor.
8. Following the method shown in Figure 10–12, sketch the circuit conditions that would exist for t_7 through t_{12} in Figure 10–11. Clearly show the direction of conventional current flow and the polarity of the magnetic fields. Identify which stator fields aid each other and which fields effectively cancel each other.
9. Identify the principal components of a dc motor.
10. What type of dc motor requires no windings in its stator? What is a limitation of this form of motor?
11. Describe in detail the structure and operation of the armature of a dc motor. Why is it necessary for the armature to be connected to an outside dc current source?
12. Carefully describe the operation of the simplified dc motor represented in Figure 10–14. Be sure to explain how angular motion of the rotor is maintained within this device.
13. Assume a dc motor has 15 V_{dc} applied to its armature windings. While the rotor is stationary, the armature windings draw 1.35 A of current. When the rotor is turning at full speed, the armature develops a counter emf equivalent to 87% of the source potential. How much current does the armature draw at this time?

10-2   MECHANICS OF MOTOR CONTROL

Before investigating the electronic control of either dc or ac motors, it is necessary to gain a fundamental knowledge of the mechanics of motor control. A grasp of such key concepts as torque, inertia, velocity, and acceleration is an essential basis for understanding the circuitry used in the control of industrial machinery. In this section, after the basic terminology of motor mechanics is introduced, basic drive systems used in operating motors are investigated.

The term **torque,** which is derived from the Latin word for ''twist,'' is frequently used with regard to the operation of drive systems. Torque may be simply defined as the tendency of a force to produce a rotation around an axis. An automobile engine is able to deliver torque to the fly wheel through the force exerted on the crankshaft by the pistons. Similarly, an ac or dc motor delivers axial force as the result of the interaction of stator fields and rotor currents. The term torque is also used to describe the effectiveness of an axial force. In this case, torque is considered to be the product of the axial force and the perpendicular distance of that force from the rotational axis. This concept is illustrated in Figure 10–16. For electric motors, torque is usually quantified in **pounds per inch** (lb-in.) or **ounces per inch** (oz-in.). Torque may also be specified in **grams per centimeter** (gm-cm).

Data sheets usually specify both the **running torque** and **stall torque** for a motor. The stall torque is that torque which is available when the motor shaft is at 0 rpm (stalled). The stall torque of a motor is significantly higher than its running torque. Another important motor parameter involving torque is the **torque constant.** This number, which represents the relationship between the motor input current and the resultant output torque, is usually expressed in lb-in./A or oz-in./A.

The angular velocity or, simply, the speed at which the rotor shaft of a motor turns is specified as **revolutions per minute** (rpm). For ac induction motors, such as the split-phase and 3-ϕ phase devices analyzed in the previous section, the concept of **synchronous speed** becomes important. Synchronous speed may be defined as the rate at which the rotating magnetic field revolves around the motor stator. Assuming a source frequency

Figure 10–16
Concept of torque

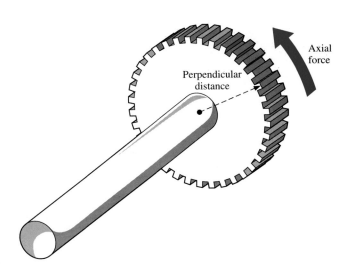

of 60 Hz, for either the split-phase or 3-ϕ motor, this field rotation occurs once each 16.67 ms. Ideally, the resultant field in the rotor would exactly follow the stator field, allowing the rpm of the motor shaft to be calculated as follows:

$$\text{Rotational speed} = 60\,\text{Hz} \times 60\,\text{sec} = 3600\,\text{rpm}$$

In actuality the rotor field and, hence, the rotor turning speed lags slightly behind that of the stator field. This **slip** may vary, depending on the amount of torque the motor must deliver to a load. Slip, which is usually expressed as a percentage, is determined as follows:

$$\text{Slip} = \frac{\text{Synchronous speed} - \text{Rotational speed}}{\text{Synchronous speed}} \times 100\% \qquad \textbf{(10–3)}$$

We must also stress that, with many of the ac motors used in industry, the stator configuration is such that rotational velocity is only 1800 rpm (half the ideal velocity of 3600 rpm).

Example 10–2

Assume that a 3-ϕ phase motor with a rotational speed of 1800 rpm has a slip of about 3.8%. What is the rpm of the rotor shaft?

Solution:

Step 1: Convert the percentage of slip to rpm:

$$\text{Slip} = 0.038 \times 1800\,\text{rpm} = 68.4\,\text{rpm}$$

Step 2: Solve for rpm of the rotor shaft:

$$\text{Rotational speed} = \text{Synchronous speed} - \text{Slip} = 1800\,\text{rpm} - 68.4\,\text{rpm} \cong 1732\,\text{rpm}$$

Motors are also rated by **wattage** and **horsepower.** Horsepower is a term commonly used to specify the amount of work that can be performed by an engine or motor. As a unit of measure, one horsepower (hp) represents the amount of force a horse exerts while pulling a load. One horsepower is equivalent to 756 watts. For smaller dc motors, which may expend only a few watts, horsepower units are not practical. However, horsepower units are commonly used to specify the load-handling capability of larger 3-ϕ motors.

Inertia is an important concept in motor mechanics since it represents the degree of torque that must be applied to start, stop, or change the rotational velocity of a load. Inertia may be defined as the resistance that a body exhibits toward being accelerated or decelerated. While a motor shaft must apply considerable torque to a stationary load to achieve angular motion, additional torque is also required to increase or decrease rotational speed. Thus, to determine the torque requirements of a motor drive system, it is first necessary to determine the load inertia (J_L). Inertia is specified in **pound-inch second**^{2} (lb-in-s^2).

For the solid cylinder represented in Figure 10–17(a), if weight and radius have been determined, inertia may be determined as follows:

$$J_L = 0.5 \times \frac{W}{g} \times r^2 = 0.0013 \times W \times r^2 \qquad \textbf{(10–4)}$$

where J_L represents inertia in lb-in.-s^2, W represents weight in pounds, r represents the radius of the solid cylinder, and g is the gravitational constant (equivalent to 386 in./s^2).

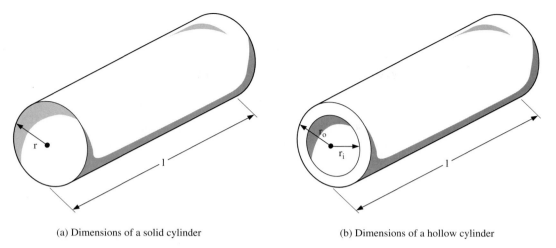

(a) Dimensions of a solid cylinder   (b) Dimensions of a hollow cylinder

Figure 10–17

Providing density (Table 10–1), radius, and length of the cylinder have been determined, inertia for a solid cylinder is determined as:

$$J_L = 0.5 \times \pi \times p \times \frac{r^4}{g} = 0.00407 \times l \times p \times r^4 \qquad \textbf{(10–5)}$$

where l represents the length of the solid cylinder and p represents the density of the material in lb/in.³

If a cylinder is hollow, as represented in Figure 10–17(b), different formulas must be applied to determine inertia. Providing the weight and radii of the hollow cylinder are known, inertia is calculated as follows:

$$J_L = 0.5 \times \frac{W \times (r_o^2 + r_i^2)}{g} = 0.0013 \times W \times (r_o^2 + r_i^2) \qquad \textbf{(10–6)}$$

where r_i represents the inside radius and r_o represents the outside radius of the cylinder in inches.

Table 10–1
Densities of common materials

Type of Material	Density (lb/in.³)
Aluminum	0.096
Brass	0.300
Bronze	0.295
Copper	0.322
Cold-rolled steel	0.280
Plastic	0.040
Hard wood	0.029
Soft wood	0.018

Where density, radii, and length have been determined:

$$J_L = \frac{0.5 \times \pi \times l \times p \times (r_o^4 - r_i^4)}{g} = 0.00407 \times l \times p \times (r_o^4 - r_i^4) \qquad \textbf{(10–7)}$$

Example 10–3

Assume a hollow brass cylinder has an inner radius of 0.875 inches, an outer radius of 1.22 inches, and a length of 4.96 inches. What is the inertia of the cylinder? What is the weight of the cylinder?

Solution:

Step 1: Solve for inertia using Eq. (10–7):

$$J_L = 0.00407 \times 4.96 \times 0.3 \times (1.22^4 - 0.875^4) = 9.866\ 10^{-3}\ \text{lb/in.}^3$$

Step 2: Transpose Eq. (10–6) to solve for weight:

$$W = \frac{9.866 \times 10^{-3}}{0.0013 \times (1.22^2 + 0.875^2)} = 3.367\ \text{lb}$$

The basic mechanical drive systems used in the connection of motors to industrial machinery include **direct drive, gear drive, leadscrew drive,** and **tangential drive.** A direct drive system, which is the simplest form of mechanical drive, is represented in Figure 10–18. With direct drive, the rpm of the load (W_L) is always equivalent to the rpm of the motor shaft (W_M). Also, the load torque reflected to the motor shaft (T_L) must equal the **unreflected torque** of the load (T'). The total inertia (J_T) for the direct drive system is the sum of the motor inertia (J_M) and load inertia (J_L):

$$J_T = J_M + J_L \qquad \textbf{(10–8)}$$

Figure 10–19 illlustrates a gear drive system. With this form of mechanical drive, both the rpm and the torque of the load are controlled by the **gear ratio.** Where N_L represents the number of load gear teeth and N_M represents the number of motor gear teeth, the gear ratio is determined as follows:

$$\text{Gear ratio} = \frac{N_L}{N_M} \qquad \textbf{(10–9)}$$

Figure 10–18
Direct drive motor connection

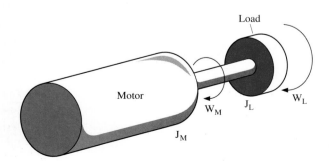

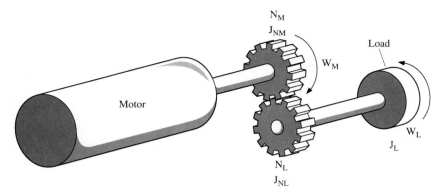

Figure 10–19
Gear drive motor connection

Thus, if the motor gear has 24 teeth and the load gear has 72 teeth, the gear ratio is determined and expressed as follows:

$$\text{Gear ratio} = \frac{N_L}{N_M} = \frac{72}{24} = 3:1$$

Equations (10–10) and (10–11) clarify how load rpm and load torque become the function of the gear ratio of the drive system:

$$W_L = \frac{W_M \times N_M}{N_L} \tag{10–10}$$

$$T' = \frac{T_L \times N_L}{N_M} \tag{10–11}$$

According to Eq. (10–10), if the load gear has fewer teeth than the motor gear, the rpm of the load becomes greater than motor rpm. However, as made evident by Eq. (10–11), the **unreflected torque** of the load is decreased. If the load gear has more teeth than the motor gear, load rpm is less than motor rpm. However, the unreflected torque of the load is increased.

The operating principles of the gear drive system are exploited in the manufacture of mechanical devices commonly referred to as **gearheads** or **gear reducers.** Such devices are available in a wide variety of sizes, torque capabilities, and gear ratios. As shown in Figure 10–20, for use in larger drive systems, gear reducers may be manufactured as separate items. For such applications, gear reducers are usually fabricated to fit on the keyed shaft of the motor. As shown in Figure 10–20(a), the output of the gear reducer could be a splined shaft, fabricated to fit within a pulley, sprocket, or gear within a larger mechanical drive system. Figure 10–20(b) illustrates a smaller ac motor that contains a built-in (permanently attached) gearhead. In this example, the output shaft of the gearhead is shown as being eccentric with respect to the motor shaft. Gearhead motors are also available for which the output shaft is concentric to that of the motor.

Example 10–4

The rotor shaft of a small permanent-split capacitor motor operating from a 120 V$_{rms}$/60 Hz source is connected to a gear reducer with a ratio of 24:1. The synchronous speed of the motor is 3600 rpm. The motor shaft under load has a slip of 4.7% and a load torque

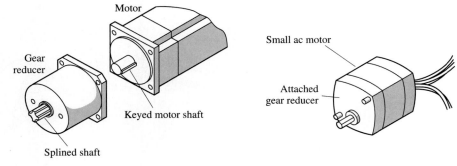

Motor

Gear
reducer

Small ac motor

Attached
gear reducer

Keyed motor shaft

Splined shaft

(a) Gear reducer with splined shaft being   (b) Small ac motor with a permanently
     attached to a motor with a keyed shaft   attached gear reducer

Figure 10–20

reflected to the motor shaft of 127 oz-in. What are the rpm and unreflected torque of the
load?

Solution:

Step 1: Solve for the rpm of the motor shaft: Accounting for slip:

$$W_M = 3600 \text{ rpm} - (0.047 \times 3600 \text{ rpm}) = 3430.8 \text{ rpm}$$

Step 2: Solve for the rpm of the load:

$$W_L = \frac{W_M \times N_M}{N_L} = \frac{3430.8 \text{ rpm} \times 1}{24} \cong 143 \text{ rpm}$$

Step 3: Solve for the unreflected torque of the load:

$$T' = \frac{T_L \times N_L}{N_M} = \frac{127 \times 24}{1} = 3048 \text{ oz-in.}$$

Figure 10–21 represents a leadscrew mechanical drive. With this form of motor drive,
a leadscrew, or worm, is connected to the shaft of the motor. The load is then moved
linearly along the length of the leadscrew through rotation of the motor. For such a system,
the linear velocity of the load is determined as follows:

$$V_L = \frac{W_M}{P}$$

where V_L represents the rate of linear movement (linear velocity) of the load in inches
per minute (in./min) and P represents the pitch of the leadscrew in revolutions per inch
(revs/in.).

The total inertia of the leadscrew drive system is calculated as follows:

$$J_T = \left[\frac{W}{g} \times \left(\frac{1}{2\pi \times p} \right)^2 \times \frac{1}{e} \right] + J_{LS} + J_M$$

which simplifies to:

$$J_T = \left[(6.56 \times 10^{-5}) \times \frac{W}{e \times p^2} \right] + J_{LS} + J_M \qquad (10\text{–}12)$$

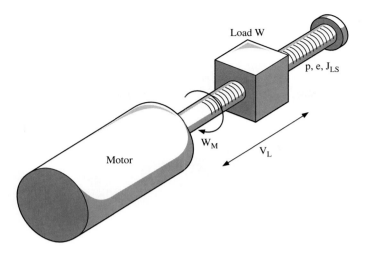

Figure 10–21
Leadscrew drive motor configuration

where J_{LS} represents the inertia of the leadscrew and e represents leadscrew efficiency, which may vary from 0.4 to 0.9.

A major consideration within the leadscrew drive system is that of friction. As shown in Eq. (10–13), **frictional force** (F_F) is a product of the **coefficient of friction** (u) and the weight of the load (W). (Both weight and frictional force are expressed in pounds.) Table 10–2 lists a few common materials used in leadscrew systems and specifies their coefficients of friction.

$$F_F = u \times W \qquad (10\text{–}13)$$

Friction torque (T_F) is also an important concept in the analysis of a leadscrew drive system. Friction torque is defined as the torque required to overcome the force that retards or prevents the relative motion of mechanical components while they are in contact with each other. Although friction torque may be calculated with Eq. (10–14), it is readily measured by placing a torque wrench at the drive shaft of the motor and determining the amount of torque required to initiate rotation of the leadscrew.

$$T_F = \frac{F_F}{2\pi \times p \times e} = \frac{0.159 \times F_F}{p \times e} \qquad (10\text{–}14)$$

Figure 10–22 represents a tangential drive system. This form of mechanical drive serves as the basis for conveyor belt, rack and pinion, and chain and sprocket systems. For a

Table 10–2
Coefficients of friction

Material	Coefficient
Steel on steel (unlubricated)	0.580
Steel on steel (lubricated)	0.150
Teflon on steel	0.040
Ball bushing	0.003

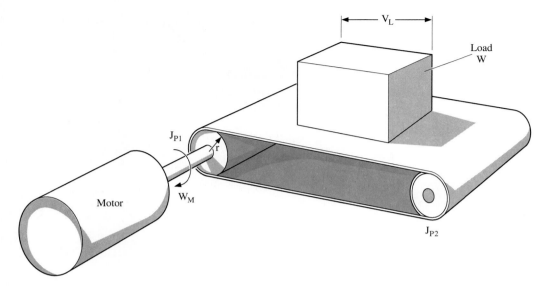

Figure 10–22
Tangential motor drive configuration

tangential drive system, linear velocity of the load is calculated as follows:

$$V_L = W_M \times 2\pi \times r = \frac{W_M \times r}{0.159} \tag{10–15}$$

where r represents the radius of the motor pulley (in inches).

The load torque reflected back to the motor shaft becomes the product of load force and the motor pulley radius. Thus:

$$T_L = F_L \times r \tag{10–16}$$

The total system inertia is determined as follows:

$$J_T = \frac{W \times r^2}{g} + J_{P1} + J_{P2} + J_M \tag{10–17}$$

where J_{P1} represents the inertia of the motor pulley and J_{P2} represents the inertia of the second pulley.

Within a tangential drive system, the required friction torque is a product of the frictional force of the load and the radius of the motor pulley. Thus:

$$T_F = F_F \times r \tag{10–18}$$

SECTION REVIEW 10–2

1. Define the terms *inertia* and *torque* as they apply to the mechanics of motor drive systems. In what units is each of these parameters specified?
2. What is the torque constant of a motor? In what units is this parameter expressed?

3. A 3-φ motor operating from a 60-Hz source has a rotational velocity of 1783 rpm. What is the percentage of slip for the motor if its synchronous speed is 1800 rpm?

4. A 3-φ motor expends 189 W. What would be the horsepower rating of the motor?

5. Assume a solid cylinder, composed of cold-rolled steel, has a radius of 1.5 inches and is 5.72 inches in length. What is the inertia of the cylinder?

6. Identify and briefly describe the four basic mechanical drive systems used in connecting motors to industrial loads.

7. The load gear within a gearhead contains 108 teeth, while a motor gear contains only 24 teeth. What is the gear ratio of the gearhead?

8. A single-phase ac motor, operating from a 60-Hz source, is connected to a gearhead. Assume the motor gear has 24 teeth and the load gear has 96 teeth. If the motor has a slip of 2.7%, what is the rpm of the load?

9. A leadscrew drive system, consisting of lubricated steel on steel, has a load weight of 87 pounds. What is the frictional force of the system?

10. Identify two common industrial applications of the tangential form of drive system.

10-3   BASIC DC MOTOR CONTROL CIRCUITRY

A major advantage of ac motors is their ability to run at a nearly constant speed in spite of widely varying load conditions. The running speeds of ac motors can be varied; however, this effort often involves complex control circuitry. By comparison, dc motors are relatively easy to control. Because of their high torque capabilities and precise response to velocity control circuitry, dc motors are preferred in industrial applications involving rapid and accurate control of moving machinery.

Figure 10–23 illustrates three basic methods of connecting a dc motor to a power source. In Figure 10–23(a), the field windings are placed in series with armature windings. The main advantage of this **series-field** configuration is a large amount of starting torque. However, the series-field configuration is difficult to use in velocity-control applications. Figure 10–23(b) illustrates a basic **shunt-field** configuration for a dc motor. This configuration derives its name from the fact that the field windings are placed in parallel, or shunted, with respect to the armature. The advantage of the shunt-field configuration is accuracy of velocity control. A **compound** motor connection scheme is represented in Figure 10–23(c). In this configuration, one field winding, referred to as the series winding, is placed in series with the dc source. A second field winding is placed in the shunt position with respect to the armature. Because this last motor configuration is a combination of the series-field and shunt-field configurations, it exhibits properties of both schemes. Thus, a dc motor in a compound connection is likely to offer better torque capability but less accurate velocity control than a motor in a shunt-field connection.

The rotational velocity of a **shunt-field** (sometimes referred to as shunt-wound) dc motor may be controlled by varying the armature voltage, the armature resistance, or **field strength** (ϕf). (As was stated earlier, armature current varies as a function of changing load conditions but is not in itself a control variable.) The mathematical relationship of these variables is as follows:

$$W_M = \frac{V_A - (I_A \times R_A)}{\phi f} \tag{10-19}$$

where R_A represents the combined ohmic value of the series resistance of the armature windings (r_s) and any external resistance placed in series with the armature windings.

As is made evident in Eq. (10–19), the rotational velocity of the shunt-wound motor is inversely proportional to the strength of the electromagnetic field produced by the field

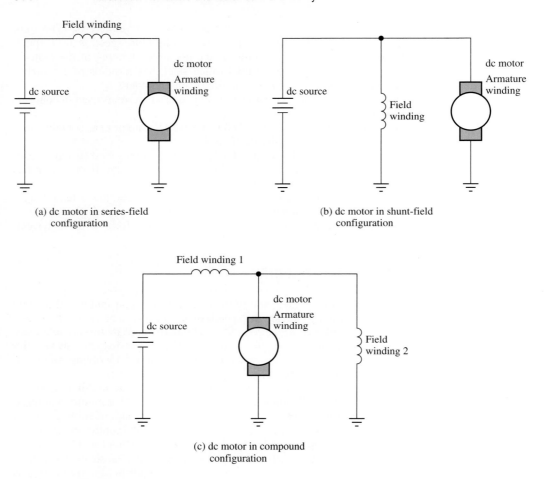

(a) dc motor in series-field
configuration

(b) dc motor in shunt-field
configuration

(c) dc motor in compound
configuration

Figure 10–23

windings. This idea might, at first, be difficult to grasp. However, it should become plausible by considering that, as the field strength weakens, less counter emf is induced within the armature. Because of the resultant increase in armature current, both the velocity and the torque of the motor shaft are increased. The relationship of motor torque, field strength, and armature current is defined as follows:

$$T_M = \kappa \times I_A \times \phi f \tag{10–20}$$

where κ represents the torque constant of the motor. From Eq. (10–20), it becomes evident that, for motor torque to remain constant under a given load condition, armature current must increase as the stator field is weakened. This increase in armature current accompanies a decrease in counter emf within the armature windings that occurs as a result of the weakening stator field.

Figure 10–24 contains the characteristic curve for a typical shunt-wound motor. Such performance curves are frequently included in motor manufacturers' data sheets. The shaded area of this curve is referred to as the continuous duty zone. The **continuous rated torque** (T_{CR}) is determined by the maximum operating temperature of the motor. The rated speed of the motor (W_R) is a function of the applied voltage. The motor may

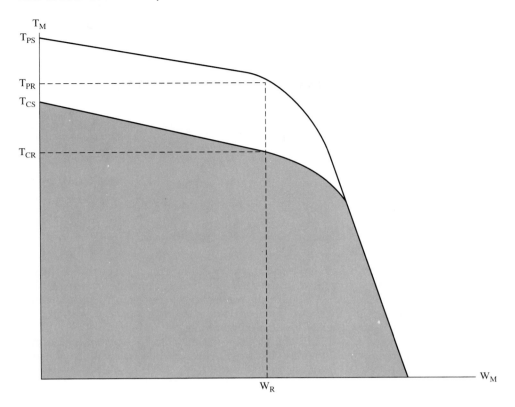

Figure 10–24
Characteristic curve for a typical shunt-wound motor

operate continuously within the shaded area of the characteristic curve. It may, however, operate only for brief periods (a couple of seconds at a time) within the intermittent duty zone. Thus, with shaft speed approaching 0 rpm, motor torque may remain indefinitely at the **continuous stall torque** (T_{CS}) level. However, the motor torque may be at the **peak stall torque** (T_{PS}) level for only a couple of seconds as rpm approaches zero. Also, during intermittent heavy load conditions, the motor torque may briefly increase to the **peak rated motor torque level** (T_{PR}).

Figure 10–25 illustrates how control of dc motor speed may be achieved by placing a rheostat in series with the shunted field windings. This **field-weakening speed control** technique is both simple and power efficient. The rheostat dissipates relatively little power in comparison to the armature windings, due to the relatively high resistivity of the field windings. Increasing the ohmic value of R_1 reduces both the voltage drop and the current flow for the field windings. The resultant reduction in stator field strength causes an increase in both armature current and motor rpm. Reducing the ohmic value of the rheostat in Figure 10–25 causes the stator field to strengthen. The resultant increase in the counter emf of the armature windings causes a reduction in armature current. Consequently, the rpm of the rotor shaft decreases. A major design consideration for the field-weakening method of motor speed control is the maximum value of series resistance. In Figure 10–25, too high a value of rheostat resistance could result in excessive weakening of the stator field. An overcurrent/overtemperature condition could then result for the motor armature, accompanied by excessive rotor speed.

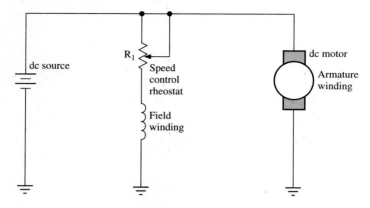

Figure 10–25
Field-weakening control configuration for a shunt-wound dc motor

Figure 10–26 represents the set of characteristic curves produced by the shunt-wound motor in Figure 10–25. Although only three performance curves are shown, a separate performance curve would exist for each setting of the rheostat. Note that, as the ohmic setting of the rheostat (R) is increased, each curve represents a slightly higher maximum torque and a slightly lower maximum rpm.

Figure 10–27 illustrates a shunt-wound motor with **armature voltage speed control.** In Figure 10–27(a), a single rheostat is placed in series with the armature windings of a

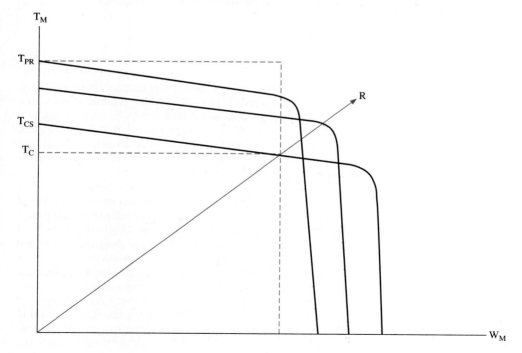

Figure 10–26
Characteristic curve for a shunt-wound motor with rheostat control

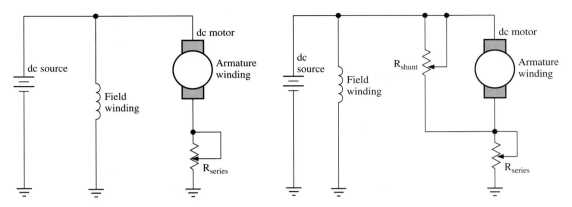

(a) Armature voltage control configuration
using a single series rheostat

(b) Armature voltage control configuration
using both a series and a shunt rheostat

Figure 10–27

dc motor while the field windings are placed directly in parallel with the dc voltage source. Since the field voltage is now at a fixed value, ϕf in both equations (10–19) and (10–20) remains constant. However, an increase in motor rpm can now be accomplished through a reduction in the ohmic value of the rheostat, which would result in an increase in the value of V_A. According to Eq. (10–19), this increase in armature voltage would produce a proportional increase in rotational velocity. As long as the load remains constant, motor torque also increases as the ohmic value of the rheostat is decreased. Since I_A increases with V_A, as becomes evident in Eq. (10–20), motor torque must increase proportionally. An increase in the ohmic value of the rheostat, of course, produces opposite effects than those just described, with motor speed and torque both decreasing.

If, with armature voltage remaining constant, the load on the dc motor in Figure 10–27(a) is increased, the speed of the motor decreases. As the armature current increases, V_A must decrease, allowing more voltage to develop over the rheostat. This effect may be dampened significantly by the placement of a shunt resistor directly in parallel with the armature windings, as shown in Figure 10–27(b). The presence of the shunt resistor allows for improved stability of motor speed with varying load conditions. The series and shunt rheostats are ganged, allowing them to be controlled in tandem. A disadvantage of the motor configurations shown in Figure 10–27 is that of I^2R loss. Since the armature current is significantly greater than that of the field winding, power loss in the rheostats of an armature voltage-controlled motor is much greater than the power loss of the rheostat used in field weakening.

For a dc motor with armature voltage speed control, varying the ohmic values of the rheostats produces the set of characteristic curves shown in Figure 10–28. It is important to note the major difference between these performance curves and those for the field-weakening form of speed control in Figure 10–26. With armature voltage control, an increase in armature voltage results in an increase in both torque and rpm. With field-weakening control, however, an increase in either one of these two variables is associated with a decrease in the other.

Within industrial systems, dc motors are often powered and controlled from pulsating dc sources. An example of such a motor control circuit is shown in Figure 10–29. The operation of this circuit is similar to that of the circuit in Figure 7–22. However, with the presence of the bridge rectifier, the circuit in Figure 10–29 is capable of controlling armature voltage and current during both alternations of the source cycle. [The operation

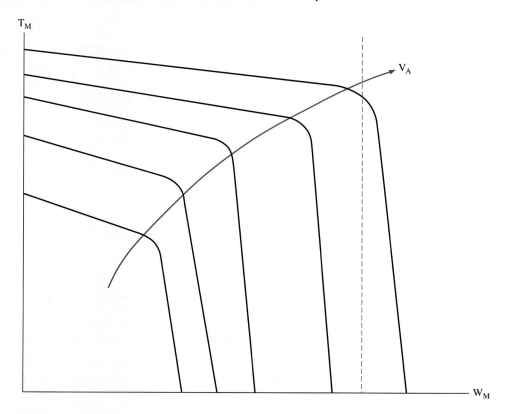

Figure 10-28
Characteristic curve for armature voltage in armature voltage speed control

of a bridge rectifier in conjunction with a silicon-controlled rectifier (SCR) was covered in detail in the first section of Chapter 8.]

With the turns ratio shown for the transformer in Figure 10-29, the peak amplitude of the voltage present at point C is limited to around 27 V, which is determined as follows:

$$V_A(\text{peak}) = \left(\frac{169.7\ \text{V}}{6}\right) - 1.4\ \text{V} = 26.9\ \text{V} \cong 27\ \text{V}$$

The pulsating dc waveform developed at point A, which is represented as waveform A in Figure 10-30, is developed over the field winding of the dc motor. With the field winding being connected directly between point A and ground, its average voltage and current remain virtually constant. Thus, the average field strength (ϕf) for the field winding remains constant. With Zener diode D_5 connected between point B and ground, the dc supply voltage for the unijunction transistor (UJT) trigger circuit is limited to 12 V_{dc}, as shown by waveform B in Figure 10-30. The point at which this peak value of V_B is attained during each source alternation is determined as follows:

$$\sin \theta \cong \left(\frac{12\ \text{V}}{27\ \text{V}}\right) \cong 0.4444$$

$$\theta \cong 26°$$

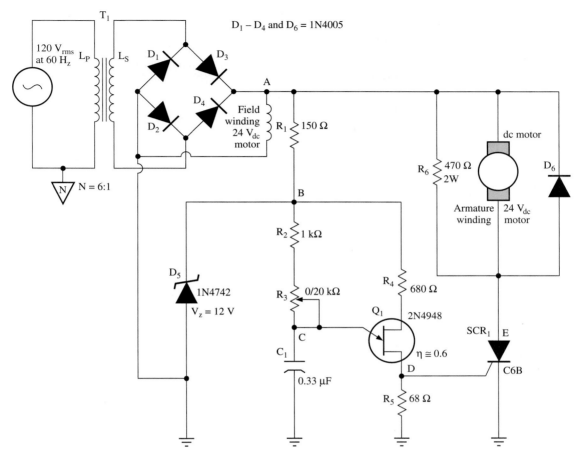

Figure 10–29
Pulse-width-modulated armature voltage speed control of a dc motor using a UJT and SCR

During the on time of the UJT trigger circuit, C_1 charges toward the 12-V level via the combined resistance of R_2 and R_3. As the voltage at point C approaches the level of V_P for the UJT, that device rapidly switches into forward saturation. The resultant current pulse, which develops over R_5, switches the SCR into forward saturation. The minimum and maximum time constants for the UJT trigger circuit may be estimated as follows:

$$\tau_{min} = 1 \text{ k}\Omega \times 0.33 \text{ } \mu\text{F} = 330 \text{ } \mu\text{sec}$$
$$\tau_{max} = 21 \text{ k}\Omega \times 0.33 \text{ } \mu\text{F} \cong 6.93 \text{ mS}$$

The first of these results indicates that, with R_3 set at 0 Ω, the SCR switches on virtually at the beginning of each alternation. With this condition, the armature waveform (equivalent to V_{R6}), V_D, and V_E, are as shown from t_0 through t_2 in Figure 10–30. The second of the above results indicates that, with R_3 set at its maximum ohmic value, the SCR is not triggered, if at all, until the end of each alternation. The waveforms representing this condition are shown from t_2 through t_4 in Figure 10–30.

It should now be obvious that the circuit in Figure 10–29 represents a pulse-width-modulated approach to armature voltage speed control. While the average voltage and

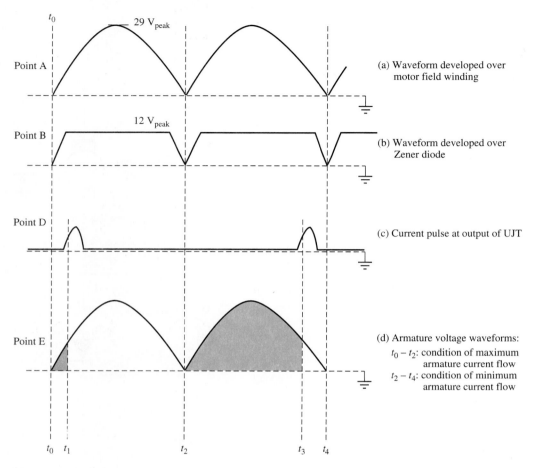

Point A — 29 V$_{peak}$ — (a) Waveform developed over motor field winding

Point B — 12 V$_{peak}$ — (b) Waveform developed over Zener diode

Point D — (c) Current pulse at output of UJT

Point E — (d) Armature voltage waveforms:
$t_0 - t_2$: condition of maximum armature current flow
$t_2 - t_4$: condition of minimum armature current flow

t_0 t_1 t_2 t_3 t_4

Figure 10–30
Waveforms for the circuit in Figure 10–29

current for the field windings are held at fixed values, the average voltage and current for the armature windings are made variable through the adjustment of R_3. Thus, both equations (10–19) and (10–20) and the characteristic curves of Figure 10–28 are still applicable for defining circuit operation. Like the shunt resistor in Figure 10–27(b), R_6 in Figure 10–29, although not variable, helps to stabilize motor speed as the load varies. Diode D_6 serves as a kickback diode, protecting the SCR from the large counter emf that is induced within the armature windings of the motor as power is removed.

DC motors can be made to rotate in two directions. One method of providing this capability is to make field current reversible while armature current remains unidirectional. An alternative method of directional control is to keep field current unidirectional while allowing armature current to be bidirectional. Figure 10–31 contains an example of this second form of control circuit. As with the control circuitry in Figure 10–29, the motor field winding is connected directly across the bridge rectifier, again allowing for virtually constant average values of field voltage, current, and magnetic field intensity. The operation of the RC timing components within this circuit is similar to that of the **silicon unilateral switch** (SUS) relaxation oscillator in Figure 7–37. The firing-delay angle of the SCRs is

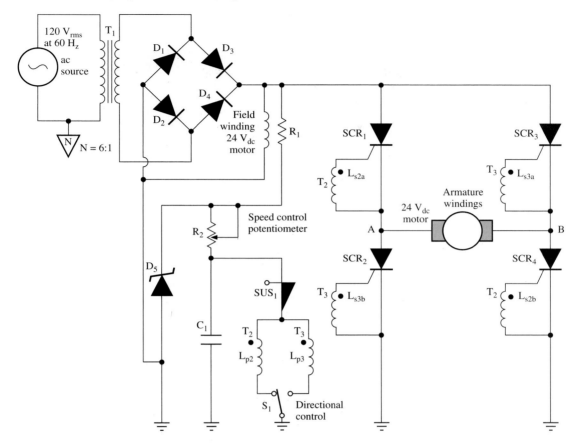

Figure 10–31
Pulse-width-modulated armature voltage speed control with direction-reversing capability

controlled by potentiometer R_2, which governs the flow of charging current to C_1. As the SUS breaks over, a current pulse is developed in the primary of one of the two pulse transformers, T_2 and T_3. Each of the two pulse transformers has two secondaries, which are connected to the gates of two SCRs. Directional control is provided by S_1, which determines which pair of SCRs is to be enabled.

The operation of the circuit in Figure 10–31 is as follows. Assume first that S_1 is positioned such that the primary of T_2 (designated L_{P2}) is in series with the main terminals of the SUS. When the voltage over C_1 attains the level of V_{BO} for the SUS, that device breaks over. This action produces a current pulse in the primary of T_2 that induces simultaneous current pulses in the secondary windings, L_{S2a} and L_{S2b}. These current pulses cause SCR_1 and SCR_4 to switch on, allowing armature current to flow from point A to point B. If S_1 is placed in the opposite position, the primary of T_3 is now in series with the SUS. With this condition, the current pulses developed in T_3 cause SCR_2 and SCR_3 to switch on simultaneously. This action allows the flow of armature current from point B to point A. This reversal in current flow causes a reversal in the direction of motor rotation.

The basic forms of dc motor control circuitry covered thus far share a common limitation in that they are not **self-regulating.** Since they lack a path for feedback from the motor armature to the controlling device, such control circuits are referred to as **open-loop** circuits. Referring back to the circuit in Figure 10–29, assume that R_3 is adjusted for a motor speed of 1800 rpm under a given load condition. If the load on the motor is increased, the motor slows down. This results in less counter emf, accompanied by an increase in armature current.

The motor control problem just described would be unacceptable in most industrial motor applications. This problem is very similar to that of a poorly regulated power supply, for which a large increase in load current could result in an unacceptable drop in output voltage. This concept of **load regulation** applies to motors as well as power supplies. Remember, with dc power sources, load regulation is a figure of merit describing the ability of a power supply to maintain a nearly constant output voltage during variations in load current. As a figure of merit for a motor control circuit, load regulation refers to the ability of the control system to maintain nearly constant motor speed while variations in load torque occur. As a percentage, load regulation for a motor control circuit is determined as follows:

$$\text{Load regulation} = \frac{W_{\text{MNL}} - W_{\text{MFL}}}{W_{\text{MNL}}} \times 100\% \qquad (10\text{–}21)$$

where W_{MNL} represents motor speed (in rpm) during a no-load condition and W_{MFL} represents motor speed under full load.

Example 10–5

Under a no-load condition, assume a dc motor turns at a speed of 1785 rpm. If the motor control circuitry has a load regulation of 2.3%, what is the motor speed under full load?

Solution:
Step 1: Solve for the difference in speed:

$$\Delta \text{ rpm} = 0.023 \times 1785 \text{ rpm} = 41 \text{ rpm}$$

Step 2: Solve for the full load speed:

$$W_{\text{MFL}} = 1785 \text{ rpm} - 41 \text{ rpm} \cong 1744 \text{ rpm}$$

The circuit in Figure 10–32 represents a design improvement over the open-loop form of controller in Figure 10–29. Here, a path for feedback from the motor is provided, allowing for a self-regulating action within the motor control circuitry. Thus, motor speed remains nearly constant under changing load conditions. Like the circuit in Figure 10–29, the circuit in Figure 10–32 uses a full-wave bridge rectifier to provide pulsating dc for the operation of the motor and control circuit. Also, as in Figure 10–29, the field winding is placed directly in parallel with the diode bridge, resulting in virtually constant values of average current, voltage, and field intensity. However, in Figure 10–32, the UJT is replaced by an SUS as the triggering device for the SCR.

For the circuit in Figure 10–32, during a given pulse from the bridge rectifier, the path for charging capacitor C_1 is via the motor armature windings, R_3, potentiometer R_2, and D_6. While the source is approaching 0 V, a path for discharging C_1 toward point B is provided, via D_5, R_1, and the motor field winding. The relatively short time constant for these components allows C_1 to be nearly discharged at the beginning of each source alternation.

The self-correcting ability of the circuit in Figure 10–32 becomes evident by first considering that the voltage present between point C and dc ground must always equal

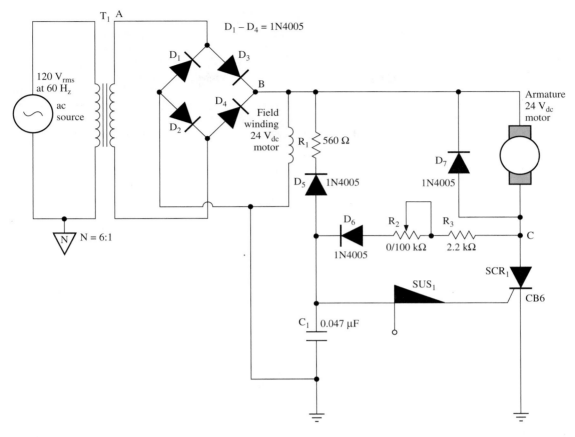

Figure 10–32
A dc motor control circuit with feedback path for self-regulation

the difference between the instantaneous voltage at point B and the voltage developed over the motor armature:

$$V_C = V_B - V_{cemf}$$

While the SCR is conducting, the armature voltage equals the instantaneous voltage at point B minus the small value of V_F developed over the SCR. However, while the SCR is still biased off, the voltage present at point C is determined as:

$$V_C = V_{C_1} + [I_A \times (R_2 + R_3)] + 0.7 \text{ V}$$

It should be stressed that, when the SCR is not conducting, the armature current is quite small, being limited by the high ohmic values of R_2 and R_3.

During an increasing load condition, the difference in voltage between points B and C begins to decrease as a consequence of reduced armature counter emf. Thus, while the SCR is biased off, the voltage at point C begins to increase. This action results in an increase in charging current for C_1, allowing V_{C_1} to attain the triggering threshold earlier in a given alternation. This triggering voltage is equal to the combined breakover voltage of the SUS and gate trigger voltage of the SCR. At the instant of triggering:

$$V_{C_1} = V_{BO} + V_{GT}$$

With an increase in load, as I_A begins to increase, the threshold level of V_{C_1} is attained earlier in the subsequent source alternations. Consequently the firing-delay angle of the SCR is reduced. This action, in turn, increases the average armature voltage, allowing it to compensate quickly for the increased load condition. Thus, motor speed remains nearly constant with the increase in load. If the load is decreased, the motor attempts to turn faster. Since this action results in an increase in counter emf, the voltage at point C begins to fall, causing a reduction in charging current for C_1 and an increase in the firing-delay angle for the SCR. Thus, the control circuit is also able to compensate for a reduction in loading by decreasing the average values of armature voltage and current.

In industrial applications where dc motors are required to drive heavy loads, 3-ϕ power is often utilized. As with ac 3-ϕ motors, dc motors operating from 3-ϕ sources are often rated at several horsepower. The advantage of dc 3-ϕ motors over their ac counterparts is their precise speed control capability. The reason for the high torque and power capability of dc 3-ϕ motors becomes evident through examination of the simplified control circuitry in Figure 10–33. The associated waveforms for the dc 3-ϕ motor are contained in Figure 10–34.

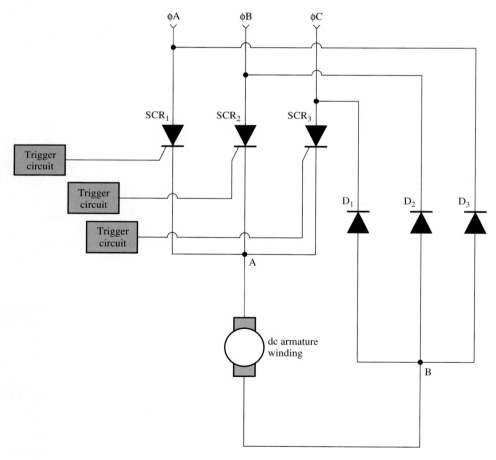

Figure 10–33
A 3-ϕ dc motor with phase-shift control

As shown in Figure 10–33, each phase input to the motor armature has its own SCR and phase-control circuitry. Because the cathode terminals of these three SCRs are common to point A, a positive voltage is always present at the top side of the motor armature. Also, at point B, on the bottom side of the armature windings, the anodes of three diodes are connected in common. Since the cathode of each of these diodes is connected to one of the phase lines, a negative voltage is always present at the lower terminal of the armature. Thus, throughout the 3-ϕ cycle, a dc potential is continuously present across the armature, positive with respect to point A.

To understand how armature current is maintained in the dc 3-ϕ motor, consider two circuit conditions represented in Figure 10–34. From t_0 to t_1, SCR_1 and SCR_3 are enabled to switch on, since phases A and C are in their positive alternations. Also during this time, D_2 is biased on because phase B is in its negative alternation. From t_1 to t_2, only SCR_1 is enabled to switch on. However, since phases B and C are now in their negative alternations, both D_1 and D_2 are biased on. Each of the three phase-shift control circuits, represented as blocks in Figure 10–33, could be similar to the circuit in Figure 10–32, with counter emf feedback being used to maintain constant motor speed as the load varies. The operation of these three control circuits could be enabled via a 3-ϕ transformer. Also, the three potentiometers used in the phase-shift control circuits could be ganged together, allowing close control of average armature current and voltage.

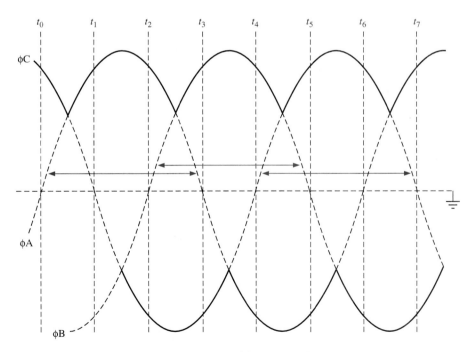

Note: Arrows represent variable portions of waveforms

Figure 10–34
Armature voltage for a 3-ϕ dc motor

A problem associated with the control circuitry just described is that of noise. As explained in Chapter 7, the use of phase-shift control in high-power control applications is likely to generate RF noise that can easily disrupt the operation of electronic control circuitry. Also, the arcing that occurs between the brushes and commutators of conventional dc motors is a further source of RF noise. For these reasons, in high-power applications within modern industrial facilities, conventional dc motors and their 3-φ controllers are likely to be abandoned in favor of brushless dc motors and their associated controllers. The operation of such state-of-the-art motors and control circuitry is covered in detail in Section 10–5.

SECTION REVIEW 10–3

1. Why are dc motors preferred over ac motors in many precision machinery applications?
2. Identify and briefly describe the three basic configurations for connecting a dc motor to a power source.
3. With the field-weakening form of motor speed control represented in Figure 10–25, what happens to rotor speed as rheostat resistance is increased? What could happen to the motor if rheostat resistance is set too high?
4. What is the purpose of the shunt rheostat in Figure 10–27(b)? How does it improve the stability of motor operation?
5. Explain how average armature voltage is controlled in Figure 10–29.
6. Describe two methods by which dc motors can be made to reverse direction of rotation. Which technique is used in Figure 10–31?
7. Under a full-load condition, a dc motor turns at 1534 rpm. If the load regulation equals 2.9%, what is the speed of the motor under a no-load condition?
8. Explain how the motor control circuit in Figure 10–32 represents a design improvement over that in Figure 10–29. Describe how the circuit in Figure 10–32 is able to maintain a nearly constant motor rpm in spite of changing load conditions.
9. What is an advantage of using a 3-φ power source with a dc motor as shown in Figure 10–33?
10. What is a problem associated with the type of motor control circuitry contained in Figure 10–33?

10–4   INTRODUCTION TO SYNCHROS AND RESOLVERS

Synchros and **resolvers** are electromechanical devices used within industrial systems to detect the positions of a rotating shafts. Synchros are similar in appearance to motors and, like motors, are composed of rotors and stators. Synchros are similar in operation to transformers. The function of the rotor, which is connected to the ac power source, is similar to the transformer primary, whereas the stator windings are equivalent to the transformer secondary. The stator windings are not connected directly to any power source. Rather, they rely on the magnetic flux produced by the current in the rotor windings to induce voltage and current.

Figure 10–35 illustrates the basic structure of a synchro. The rotor of the synchro may consist of either one or three windings that are connected to **slip rings** mounted on the synchro shaft. These slip rings provide continuity of current flow to the rotor windings throughout the 360° of shaft rotation. The stator windings are placed in a wye configuration and spaced 120° apart.

Figure 10–36 illustrates the operation of a synchro. Assume at first that the rotor is positioned as shown in Figure 10–36(a). With this condition, the flux lines produced by

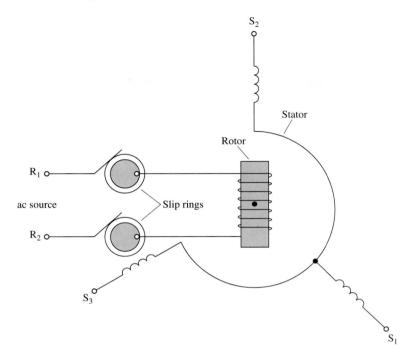

Figure 10–35
Basic structure of a synchro

the rotor winding are parallel with the stator windings. As a result, induced voltage and current in the stator are at a minimum level. Next, consider the rotor to be turned 90° counterclockwise to the position shown in Figure 10–36(b). With the flux lines now cutting across the stator windings at a 90° angle, the induced voltage and current in the stator are at their maximum amplitudes. When the rotor is turned 90° further, as seen in Figure 10–36(c), flux linkage between the rotor and stator is again at minimum, causing induced voltage and current to be at minimum. When the rotor is turned an additional 90°, as seen in Figure 10–36(d), flux linkage is again at maximum, with the magnetic flux lines again cutting across the stator windings at a 90° angle. However, with the position of the rotor now opposite to that in Figure 10–36(b), the induced stator voltage is 180° out of phase with respect to that in Figure 10–36(b).

If the peak amplitudes developed over the stator windings are known, the instantaneous voltages for each angular position of the rotor may be predicted. Assume, for example, that the rotor turns counterclockwise from the position shown in Figure 10–36(a) toward the position shown in Figure 10–36(c). This would cause flux linkage to increase from minimum to maximum then return to minimum. The corresponding changes in amplitude for the stator voltage would be as illustrated in Figure 10–37(a).

As the rotor is turned from the position in Figure 10–36(c) through the starting position in 10–36(a), the signal in Figure 10–37(b) is produced. While the transitions between minimum and maximum amplitude are the same, the waveforms in Figures 10–37(a) and (b) are 180° out of phase, representing the reversal in rotor position. For synchros containing three stator windings, the neutral terminal (S_N) is normally not accessible. With these devices, three differential outputs are normally taken: between the S_1 and S_2 terminals, the S_2 and S_3 terminals, and the S_3 and S_1 terminals, as shown in

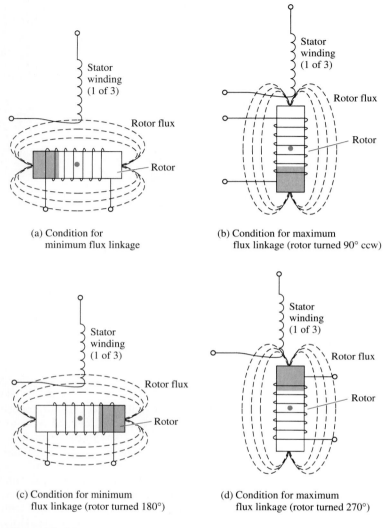

(a) Condition for
minimum flux linkage

(b) Condition for maximum
flux linkage (rotor turned 90° ccw)

(c) Condition for minimum
flux linkage (rotor turned 180°)

(d) Condition for maximum
flux linkage (rotor turned 270°)

Figure 10–36
Operation of a synchro

Figure 10–38. Since the three stator windings of the synchro in Figure 10–38 are connected in a wye configuration, the peak amplitude between a pair of output terminals is determined as follows:

$$V_{S1}(\text{peak}) - V_{S2}(\text{peak}) = V_S(\text{peak}) \times \sqrt{3} \qquad (10\text{–}22)$$

While the rotor of a synchro system may be connected directly to the 60-Hz line voltage, in many applications the rotor is excited by an ac voltage occurring at a much higher frequency, with 400 Hz being typical. The advantage of operating at a higher frequency is greater noise immunity. At 400 Hz, the synchro system would be less susceptible to the noise associated with the 60-Hz power line. The constant excitation voltage developed

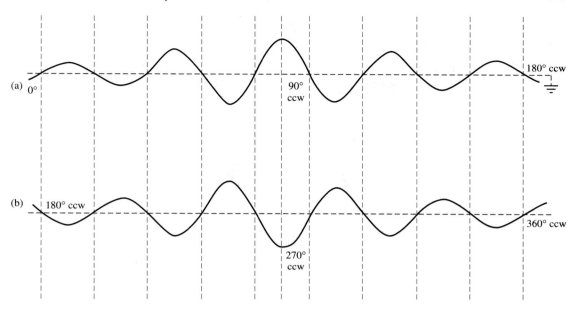

Figure 10–37
Changes in synchro stator voltage corresponding to Figure 10–36

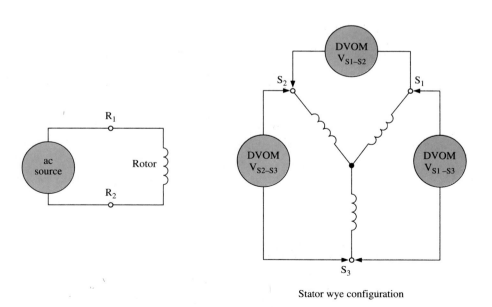

Figure 10–38
Electrical connection of synchro, showing three differential outputs

across the rotor of the synchro is called the **reference voltage.** As the rotor turns, changes in the peak voltage developed over a given stator are proportional to the cosine of an angle representing the difference in axes between the rotor and that stator. With the outputs of the synchro taken between each of the three pairs of stators, the three differential voltages are developed as follows:

$$V_{S1-S3}(\text{rms}) = 0.707 \times V_{\text{peak}} \times \sin\theta \qquad (10\text{--}23)$$

where θ represents the rotor angle with respect to the S_1 axis and V_{peak} represents the peak differential voltage developed between each pair of stators.

$$V_{S1-S2}(\text{rms}) = 0.707 \times V_{\text{peak}} \times \sin(\theta + 120°) \qquad (10\text{--}24)$$
$$V_{S2-S3}(\text{rms}) = 0.707 \times V_{\text{peak}} \times \sin(\theta + 240°) \qquad (10\text{--}25)$$

Example 10–6

For the synchro in Figure 10–38, assume that the rotor is aligned on an axis 47° positive with respect to the axis of stator 1. Also assume that the peak differential voltage developed between each pair of stator windings is 87 V. Given these conditions, solve for the rms values of V_{S1-S3}, V_{S1-S2}, V_{S2-S3}.

Solution:

Step 1: Solve for the rms value of V_{S1-S3}:

$$V_{S1-S2}(\text{rms}) = 0.707 \times 87\ V_{\text{peak}} \times \sin 47° \cong 44.98\ V$$

Step 2: Solve for the rms value of V_{S1-S2}:

$$V_{S1-S2}(\text{rms}) = 0.707 \times 87\ V_{\text{peak}} \times \sin(47° + 120°) \cong 13.84\ V$$

Step 3: Solve for the rms value of V_{S2-S3}:

$$V_{S2-S3}(\text{rms}) = 0.707 \times 87\ V_{\text{peak}} \times \sin(47° + 240°) \cong -58.82\ V$$

Proof: According to Kirchhoff's voltage law, since the three differential voltages of the wye stator configuration form a closed loop, their algebraic sum must always equal 0 V. Thus:

$$V_{S1-S3} + V_{S1-S2} + V_{S2-S3} = 44.98\ V + 13.84\ V - 58.82\ V \cong 0\ V$$

The synchros used within industrial control systems fall into two basic classifications, commonly referred to as **torque synchros** and **control synchros.** Torque synchros are capable of transmitting angular displacement information directly to the shaft of another synchro. With these devices, which are confined to relatively low power applications, no intermediate amplification or gearing is required. Control synchros, which are used in higher power applications that require greater accuracy of angular displacement, are not capable of driving a load directly. Control synchro outputs are, therefore, connected to a servo amplifier rather than another synchro.

Figure 10–39 illustrates the connection of two torque synchros in a basic repeater system. Here, the synchro transmitting displacement information is referred to as the **torque transmitter** (TX), while the synchro receiving this signal is called the **torque receiver** (TR). In this basic repeater system, an ac signal is applied to both rotors simultaneously. With both the TX and the TR synchros in the home position, the **format voltages** are present at the S_1, S_2, and S_3 terminals of both synchros. In this balance condition, no

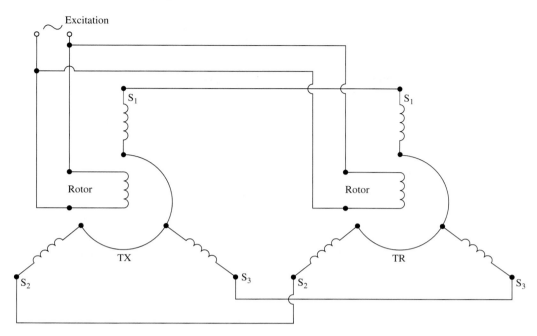

Figure 10–39
Basic synchro repeater system containing a torque transmitter and a torque receiver

current flows between the TX and the TR stator windings. If the rotor of the TX synchro is moved either manually or mechanically, an unbalanced condition momentarily occurs, resulting in stator current flow between the two synchros. This current continues to flow until the TR rotor is aligned at the same angle as the TX rotor. Once this alignment is achieved, a balanced condition again exists and stator current flow between the two synchros again approaches 0 A.

In some industrial applications involving synchros, it is necessary to add or subtract two angular movements to produce a resultant displacement. A device used to perform this task is the **differential synchro transmitter** (TDX), the schematic symbol for which is shown in Figure 10–40. As indicated here, the stator windings for the TDX are nearly identical to those of the basic TX device. The rotor of the TDX, however, consists of three windings spaced 120° apart.

A simple application of the TDX is illustrated in Figure 10–41. As seen here, the rotor of the TDX is not directly connected to an excitation voltage source. Rather, it must rely on its induced stator voltages for the generation of its own rotor signals, which serve as outputs to the TR. As an example of the operation of the synchro system in Figure 10–41, consider a condition where the mechanical displacement of the TX rotor is 68° and the displacement of the rotor of the TDX is 27°. With the TDX connected to perform a subtract function, the resultant angular displacement of the TR is 41°. If the TDX were connected to perform an additive function, then the displacement of the TR would become 95°.

Figure 10–42 briefly illustrates the operation of a control synchro system. Here, a **control transmitter** (CX) is chained to a **control transformer** (CT). An excitation voltage is fed to the rotor of the CX. The orientation of the rotor of the CX, as indicated in the schematic diagram, is shifted 90° with respect to the rotor orientation of the TX. The operation of the synchro system in Figure 10–42 is such that, as the rotor of the CX is

Figure 10–40
Schematic symbol for differential synchro
transmitter (TDX)

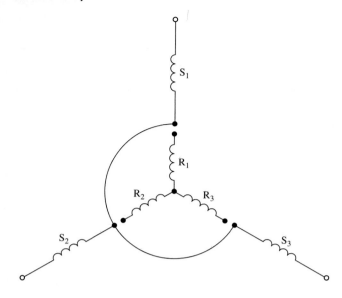

turned, the output of the CT is directly proportional to the difference in angular position between the CX and the CT rotors.

Figure 10–43 contains the schematic symbol for an electromechanical position-sensing device called a **resolver.** The resolver is similar in appearance to the synchro and, like the synchro, contains rotor and stator windings. However, the function of the resolver is distinctly different from that of the synchro. The resolver's primary purpose is to allow the solution for an angle θ, which may be accurately determined from the relative amplitude of two output signals. This angle may represent the shaft position for a rotating piece of machinery.

In a typical application of the resolver, a high-frequency excitation voltage is applied to one of the stator windings while the second, unused, stator winding is shorted. The

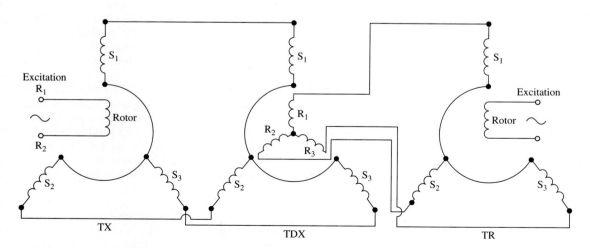

Figure 10–41
Use of differential synchro transmitter (TDX) in a repeater system

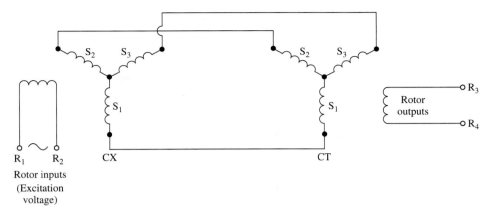

Figure 10–42
Basic control synchro system containing a control transmitter and a control transformer

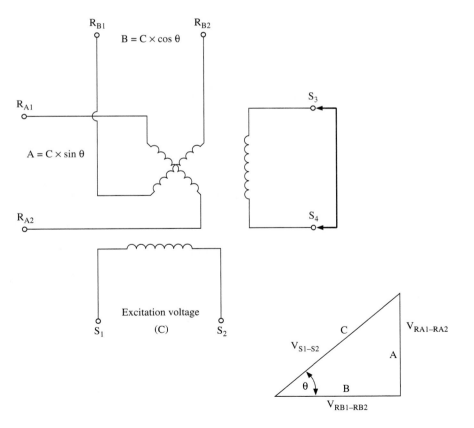

Figure 10–43
Schematic representation of a resolver

amplitude of the excitation voltage may be considered to represent the hypotenuse of a right triangle. The two output signals are derived from the rotors, which are electrically isolated and oriented at a 90° angle from each other. The amplitudes of the two rotor outputs, as indicated in Figure 10–43, are considered to represent the opposite and adjacent sides of a right triangle. Thus, the ratio of the amplitude of the A rotor winding signal to the stator excitation signal may be considered as **sinθ** for the right triangle. The ratio of the B rotor output to the excitation signal would thus be **cosθ**. Also the ratio of the A output to the B output would represent **tanθ**. An application of a resolver as a feedback device is covered in detail in the next section of this chapter.

SECTION REVIEW 10–4

1. What are slip rings? Describe their purpose within synchro systems.
2. Describe the basic structure and operation of a synchro.
3. Why are synchros often operated at around 400 Hz rather than at the line frequency of 60 Hz?
4. Define the term *format voltage* as it applies to the operation of a synchro.
5. For the synchro system in Figure 10–38, assume the rotor is aligned on the 63° axis with respect to the axis of stator 1. Also assume that the peak differential voltage between each pair of stator windings is 92 V. Given these conditions, solve for V_{S1-S3}, V_{S1-S2}, and V_{S2-S3}.
6. What are the two basic categories of industrial synchro?
7. Describe the structure and operation of a basic synchro repeater system.
8. Describe the basic structure and operation of a differential synchro transmitter.
9. Describe the basic structure and operation of a resolver. Explain how its output signals may be used to determine the position of a rotating shaft.
10. Define the term *excitation voltage* as it applies to the operation of a resolver.

10–5   INTRODUCTION TO BRUSHLESS DC SERVOMOTORS

In industrial applications where both high torque and precise motion control are required, conventional ac and dc motors are often abandoned in favor of **brushless dc servomotors.** In modern industrial systems, such motors are frequently driven by state-of-the-art controllers. These commercially manufactured control systems, which are often microcomputer based, may be easily programmed to perform a variety of tasks. The internal circuitry and programming techniques for this form of motor control device, along with the underlying control principles of the **servo system,** are covered in detail later in this text. At this juncture, analysis is limited to the brushless dc servomotor and its immediate drive circuitry.

Figure 10–44 contains a cutaway view of a brushless dc servomotor. It is evident with this form of dc motor that no armature windings are present. In fact, the internal structure of the brushless dc servomotor is essentially that of a 3-ϕ ac induction motor. However, the stator windings of the motor in Figure 10–44 are placed in a wye rather than a delta configuration. Also the rotor is likely to consist of a permanent magnet. This rotor magnet may consist of a ferrite material or, in more powerful motors, a rare-earth material such as **samarium cobalt.** Though more expensive than a ferrite rotor, the samarium cobalt rotor is preferred due to its low inertia and its ability to focus magnetic flux. Like the ac induction motor, the brushless dc servomotor relies on the interaction of the expanding and contracting stator fields and the induced currents within the rotor

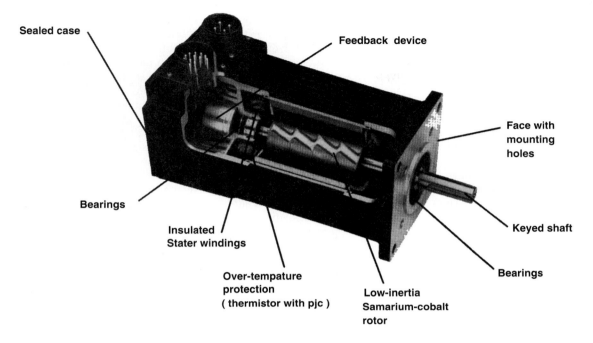

Sealed case

Feedback device

Face with
mounting
holes

Bearings

Insulated
Stater windings

Over-tempature
protection
(thermistor with pjc)

Low-inertia
Samarium-cobalt
rotor

Keyed shaft

Bearings

Figure 10–44
Internal structure of a typical brushless dc servomotor (Courtesy of Pacific Scientific)

to produce angular motion of the motor shaft. As with the ac induction motor, the field windings of the brushless dc servomotor are arranged to produce a rotating magnetic field when being driven by the controller's output circuitry. Thus, the brushes and commutators of conventional dc motors may be abandoned, allowing for low-noise operation.

As with any **servomechanism** used in the automated control of machinery, the dc brushless servomotor must be provided with at least one form of **feedback** device. Feedback signals representing the rpm of the motor shaft and/or the angular position of the shaft must be continuously provided to the motor control circuitry if precise control of a moving piece of machinery is to be maintained. As indicated in Figure 10–44, a manufacturer may include one or two feedback devices within the housing of a dc brushless servomotor. Feedback devices commonly used within servomotors include **brush-type dc tachometers, brushless tachometers, Hall-effect sensors, synchros, resolvers,** and **shaft encoders.**

The brush-type dc tachometer contains brushes and a commutator that is mounted on the motor shaft. The function of this feedback device is equivalent to that of a dc generator. As the motor shaft turns, a dc voltage is produced at the output of the tachometer. The amplitude of this voltage is directly proportional to the rpm of the motor shaft. The brush-type tachometer is also able to detect a change in direction of motor rotation, since a reversal in direction results in a reversal of output signal polarity. The operation of the brushless tachometer is essentially the same as that of the brush type. Although brushless tachometers are virtually noise free, they contain electronic components that are subject to damage by the high-temperature conditions that could occur within the motor housing. The main advantage of the brush-type tachometer is its ability to withstand high temperatures. However, as with conventional dc motors, the brushes and commutators are prone to noise generation.

Hall-effect sensors were introduced during the fourth section of Chapter 1 as part of an introductory survey of on/off devices. Within a 3-φ brushless dc servomotor, three Hall-effect sensors, placed at 120° intervals around the path of shaft rotation, can be used for determining course shaft position. To develop signals from the Hall-effect sensors, a magnetized wheel may be mounted at the end of the motor shaft as shown in Figure 10–45(a). When a notch in the magnetized wheel passes under one of the three Hall-effect sensors, the momentary absence of magnetic flux results in the momentary absence of Hall voltage (designated V_H). Signal conditioning circuitry within the motor controller, usually containing Schmitt triggers, converts the outputs of the Hall-effect sensors to the cascaded rectangular waves shown in Figure 10–45(b). These signals may be used to control the commutation of the power transistors at the output of the motor controller. Although Hall-effect sensors, in themselves, are highly linear, their function within motor circuitry is essentially that of switching. They are frequently used for detecting approximate motor shaft position and for commutation control; however, they are less suitable for measurement of rpm or exact shaft position.

Synchros and resolvers are often used as feedback elements within brushless dc servomotors. As was explained in the previous section of this chapter, synchros and servos are capable of generating output voltages that precisely represent the angular position of a rotating piece of machinery. Resolvers are particularly useful as feedback devices. The outputs of the resolver require considerable signal conditioning within the servomotor controller, as is evident in Figure 10–46. However, the resolver offers a distinct advantage as a feedback device in that it effectively performs the functions of both a Hall-effect sensor and a brushless tachometer.

The block diagram in Figure 10–46 illustrates how a resolver might be used as the primary feedback device within a servomotor. In this example, as compared to the schematic representation in Figure 10–43, the functions of the rotors and stators are reversed. One rotor now serves as the input, while the windings of the second one are shorted. Both stator windings are now active, being used to generate the sine and cosine functions.

As shown in Figure 10–46, the rotor excitation voltage is produced by a relatively high frequency sinusoidal oscillator located on the servomotor controller. The function of this excitation voltage is similar to that of the carrier frequency for an AM radio signal. As the rotor shaft of the servomotor turns, amplitude-modulated signals are developed

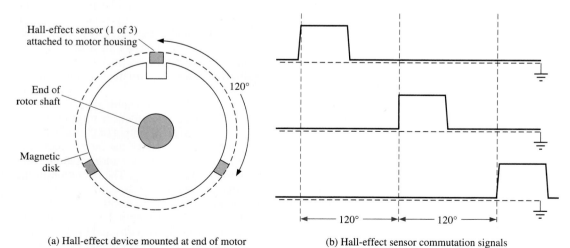

(a) Hall-effect device mounted at end of motor

(b) Hall-effect sensor commutation signals produced during one motor revolution

Figure 10–45

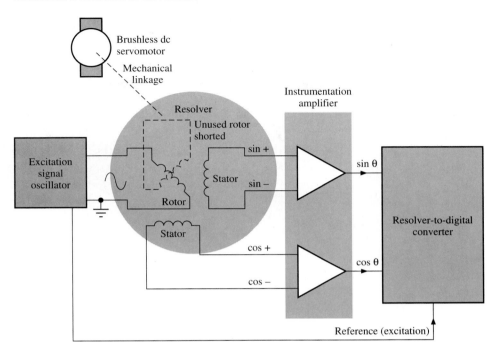

Figure 10–46
Use of a resolver as the primary feedback device for a brushless dc servomotor

over the two stator windings. These sine and cosine outputs of the resolver, which represent motor shaft position, are sent to buffer amplifiers within the servomotor controller. The amplified sine and cosine signals, as well as the excitation signal, are sent to a **resolver-to-digital converter** (RDC), which is part of the servomotor controller's signal conditioning circuitry.

Figure 10–47 contains an example of a sinusoidal generator that could produce the excitation voltage required for the resolver circuitry in Figure 10–46. This **precision Wien bridge oscillator** may be used to generate a sine wave exhibiting virtually constant amplitude and frequency, as well as minimal harmonic distortion. However, this form of oscillator is limited to applications where a frequency of less than 10 kHz is adequate. The actual Wien bridge consists of R_1, R_2, C_1, and C_2. With R_1 equaling R_2 and C_1 equaling C_2, the center (or resonant) frequency of the Wien bridge oscillator is determined as follows:

$$f_c = \frac{1}{2\pi \times R \times C} \tag{10–26}$$

where $R = R_1 = R_2$ and $C = C_1 = C_2$.

The basic operating principle for the oscillator is that, at its resonant frequency, the phase shift across the Wien bridge is 0°. Thus, only a sine wave occurring at the resonant frequency of the Wien bridge oscillator is developed and sustained via the feedback loop. Although a Wien bridge oscillator may be constructed using only a single op amp, the more complex form of circuit shown in Figure 10–47 is necessary for maintaining a stable output signal with minimal harmonic distortion. This is essential for accurate operation of the RDC, since the excitation voltage produced by the resolver must function as the

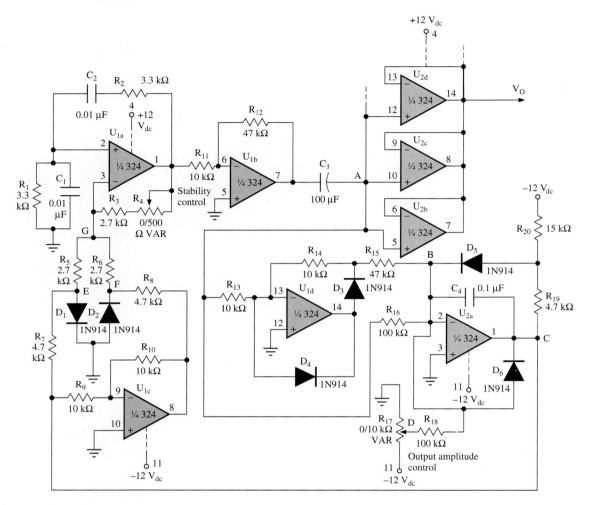

Figure 10–47
Precision Wien bridge oscillator to be used as sinusoidal generator for resolver excitation

reference voltage in the analog-to-digital conversion process. The circuit in Figure 10–47 is capable of producing a sine wave with a harmonic distortion factor of less than 0.01%!

The basic Wien bridge oscillator portion of the circuitry in Figure 10–47 consists of U_{1a} and U_{1b}. The gain control for U_{1a} is accomplished in two ways. Manual stability adjustment is accomplished by varying the ohmic value of R_4. Automatic gain control is provided by an **absolute value circuit** and an **error integrator.** Op amp U_{1d}, D_3, D_4, R_{13}, and R_{14} form the absolute value circuit, and U_{2a}, D_6, C_4, and R_{16} comprise the error integrator. Potentiometer R_{17}, considered to be part of the error integrator, allows for an adjustable reference voltage at point D. The purpose of the error integrator is to sense deviations in the absolute value of the signal at point A. This is accomplished by continually comparing the absolute value of this signal to that of the reference voltage. If the voltage at point A attempts to increase, the negative feedback at the output of the error integrator increases. This error signal is sent back to a **diode bridge** containing D_1 and D_2 and unity-gain inverting amplifier U_{1C}. The increased negative feedback temporarily decreases the gain of the first stage of the basic Wien bridge oscillator, thus reducing the amplitude of the signal at point A.

Since a single op amp lacks the current sourcing capability for driving the rotor windings of a resolver, several op amps may be connected in a voltage-follower configuration and connected in parallel, as are U_{2b}, U_{2c}, and U_{2d} in Figure 10–47. Although this may at first appear as an excessive use of op amps, it is, in fact, very economical with regard to both cost and space. This should become apparent when considering that no extra components, such as resistors and coupling capacitors, are required. Two quad op amp packages could be used to create as many as eight buffer amplifiers. Impedance matching with the relatively low ohmic value of the resolver rotor could thus be achieved in a minimal amount of space on a printed circuit board.

Figure 10–48 illustrates the instrumentation amplifier circuitry used to buffer the outputs of the resolver and shows how this circuitry could be connected to the RDC. Because the resistor values used in these instrumentation amplifiers are all 12 kΩ, both amplifier stages have a gain of approximately -1. While the necessary voltage and current amplification of the excitation signal is provided within the Wien bridge circuitry in Figure

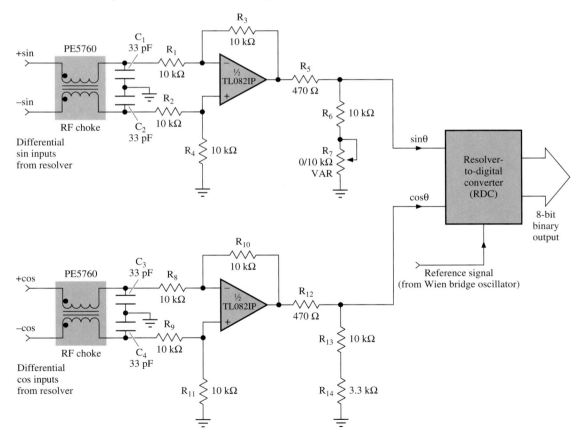

Figure 10–48
Instrumentation amplifier circuitry used to condition differential output signals from a resolver

10–47, the instrumentation amplifiers in Figure 10–48 provide common-mode noise rejection and buffering for the differential inputs to the RDC. Additional high-frequency noise attenuation is provided by RF chokes.

As shown in Figure 10–48, the buffered outputs of the resolver are fed to the sine and cosine inputs of the RDC. The excitation signal is also fed to the RDC, where it serves as a reference voltage. Through low-pass filtering action the sine, cosine, and excitation signals are demodulated. **Analog-to-digital conversion** (ADC) circuitry within the RDC samples the instantaneous values of the demodulated sine and cosine signals, converting them to binary numbers. Special arithmetic circuitry within the RDC rapidly compares these binary values and determines an 8-bit variable representing an instantaneous motor shaft position. The RDC also produces an analog voltage representing motor shaft rpm. These two variables are sent to the motor speed and commutation control circuitry within the servomotor controller.

As an example of resolver operation, consider the condition shown at t_0 in Figure 10–49, with the motor shaft at its 0° reference position. At this time, the cosine of the right triangle equals unity and the sine equals zero. The resolver accurately reflects this 0° condition, with flux linkage from the rotor to the cosine stator winding now at maximum. At the same instant, flux linkage to the sine stator winding is at minimum. As shown at

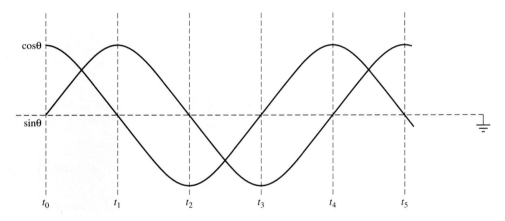

(a) Demodulated $\sin\theta$ and $\cos\theta$ as seen by RDC

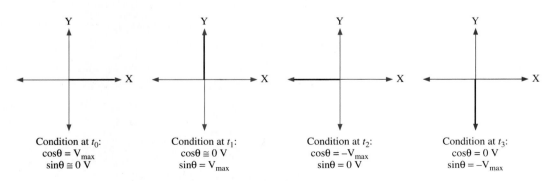

(b) Vector conditions for $\sin\theta$ and $\cos\theta$ signals

Figure 10–49

t_1 in Figure 10–48, by the 90° point in the shaft rotation, the conditions for the sine and cosine windings are reversed. With the sine of the right triangle now at unity, maximum flux linkage occurs between the rotor and sine stator winding. Since the cosine of the triangle now equals zero, the flux linkage between the rotor and cosine stator winding is at minimum. At t_2, 180° into the shaft rotation, the sine of the right triangle again equals zero and the cosine equals -1. Thus, the cosine waveform is at its negative peak potential. At t_3, 270° into the shaft rotation, the sine of the right triangle equals -1 and the cosine again equals zero. Thus, the sine waveform is at its negative peak potential, while the cosine waveform is at 0 V.

From the waveforms in Figure 10–49, it should become evident that the period of either the sine or the cosine waveform represents exactly one revolution of the motor shaft. This allows easy determination of motor rpm. For example, assume that T in Figure 10–49 equals 39 msec:

$$f = \frac{1}{39 \text{ msec}} = 25.641 \text{ Hz}$$

$$\text{rpm} = f \times 60 \text{ sec} = 25.641 \times 60 \cong 1538.5 \text{ rpm}$$

Example 10–7

For the circuit in Figure 10–47, what is the frequency of the excitation signal? Assuming the output of the precision Wien bridge oscillator is 4.2 V_{rms} and the coefficient of coupling within the resolver is 0.62, what are the peak amplitudes of the sine and cosine signals entering the RDC in Figure 10–48? If the motor is turning at 967 rpm, what is the frequency of the demodulated sine and cosine signals? At a point 5.3 mS into the demodulated sine waveform, what would be the instantaneous values of the demodulated sine and cosine signals? What would be the instantaneous position of the motor shaft?

Solution:

Step 1: Solve for the frequency of the excitation signal. For the Wien bridge oscillator in Figure 10–47:

$$f = \frac{1}{2\pi \times R \times C} = \frac{1}{2\pi \times 3.3 \text{ k}\Omega \times 0.01 \text{ }\mu\text{F}} \cong 4.823 \text{ kHz}$$

Step 2: Solve for the peak amplitudes of the sine and cosine signals. For the Wien bridge oscillator:

$$V_O(\text{peak}) = 4.2 \text{ V}_{rms} \times 1.414 \cong 5.94 \text{ V}_{peak}$$

For the sine and cosine signals:

$$V_{peak_{sin}} = V_{peak_{cos}} = 0.62 \times 5.94 \text{ V} = 3.682 \text{ V}$$

Step 3: Solve for the frequency of the demodulated sine and cosine signals:

$$f = \frac{967 \text{ rpm}}{60 \text{ sec}} = 16.117 \text{ Hz}$$

Step 4: Solve for the instantaneous values of sine and cosine signals:

$$T = \frac{1}{16.11 \text{ Hz}} = 62.05 \text{ msec}$$

$$\theta = \frac{5.3 \text{ msec}}{62.05 \text{ msec}} \times 360° = 30.75°$$

$$\sin 30.75° = 0.5113$$

For the demodulated sine signal:

$$V_i = 0.5113 \times 3.682 \text{ V} = 1.88 \text{ V}$$

For the demodulated cosine signal:

$$\cos 30.75° = 0.8594$$

$$V_i = 0.8594 \times 3.682 \text{ V} = 3.164 \text{ V}$$

Optical shaft encoders, similar in operation to those introduced in Section 5–4, may also be used as feedback devices within servomotors. The two basic forms of shaft encoders are **incremental** and **absolute.** An advantage of incremental shaft encoders is their ability to indicate rotational direction. For an incremental encoder, directional sensing may be accomplished by using two Mylar disks, such as that illustrated in Figure 10–50(a). Each of these disks is composed of alternating opaque and translucent stripes. A light source, such as an LED or a small lamp, is placed on one side of each disk. On the

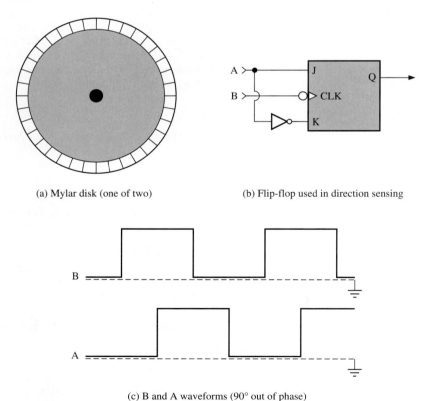

(a) Mylar disk (one of two)   (b) Flip-flop used in direction sensing

(c) B and A waveforms (90° out of phase)

Figure 10–50

opposite side of the disk, directly in line with the light source, is placed a detector element, possibly a photodiode or phototransistor. As the motor shaft turns, light pulses are generated by the alternating stripes passing in front of the detector element. For the purpose of sensing direction, the two disks are arranged such that their output pulses are out of phase with each other by about 90°, as shown in Figure 10–50(c). Figure 10–50(b) illustrates how a JK flip-flop, effectively functioning as a D flip-flop, may be used to detect the direction of shaft rotation. With the B signal leading the A signal by around 90°, as shown in Figure 10–50(c), the Q output of the JK flip-flop remains high. This is because, with each negative transition of the B signal, the high condition present at the J input is updated at the Q output. However, if the direction of shaft rotation is reversed, the A signal begins to lead the B signal by 90°, thus causing the Q output of the flip-flop to be held low. Absolute encoders are more complex than incremental encoders, consisting of several tracks. However, their advantage is the ability to indicate exact shaft position. For most shaft encoders used as feedback devices, the signal conditioning circuitry required for converting light pulses to digital signals is contained within the motor housing.

The commutation signals for the brushless dc servomotor are sent to power transistors that source current to and sink current from the stator windings. As shown by the simplified diagram in Figure 10–51, each output phase of the motor controller may be provided with a current-sourcing and a current-sinking Darlington power transistor. Using a design technique similar to that shown in Figure 10–33, a positive pulsating dc source is provided by three SCRs. However, in Figure 10–51, the phase-shifting technique of Figure 10–33 is replaced by zero-point switching. (Zero-point switching is covered in detail in Section 8–4.) Thus, the SCRs remain on during the entire positive alternation of their associated phase voltages. Three diodes allow current flow to continue during the negative alternations of each of the phase voltages.

As seen in Figure 10–51, the controller for a powerful 3-ϕ brushless dc servomotor may contain as many as six PWM-controlled power converters! The purpose of these converters is to govern the flow of current to each of the three sets of stator windings, designated L_X, L_Y, and L_Z. Since current-sourcing and current-sinking power Darlingtons are connected to each winding, bidirectional current flow may occur through L_X and L_Y, L_X and L_Z, or L_Y and L_Z. For example, if the ϕX sourcing transistor and the ϕY sinking transistor are biased on simultaneously, current flows from point E into point F, energizing stator windings L_X and L_Y. For the same two windings, if the ϕY sourcing and the ϕX sinking transistors are on simultaneously, current flows in the opposite direction.

Figure 10–52 contains a detail of the power converter and output circuitry represented in Figure 10–51. Each of the six power converters is in a dual push-pull configuration. (The push-pull power converter was examined in detail in Section 9–5.) The switching frequency of the first power converter, which may be as high as 5 MHz, is established by the CD4078B, which is a high-speed astable multivibrator. The buffered Q and $\overline{Q}$ outputs of this device serve as the gate inputs to the two n-channel power MOSFETs. The second power converter is driven by a relatively low-frequency pulse width modulator, contained within the system control circuitry. The operating frequency of the PWM, typically around 3 kpps, is established by a triangle wave generator. (The operation of this circuitry is covered in detail in Section 14–1.)

The operation of the astable multivibrator and the first push-pull converter is as follows. When the Q output of the CD4078B is high, as long as an active-low enable signal is present at point A, current flows through L_{P1} of T_1 toward point C. At this time, with no phase inversion across the transformer choke, Schottky diode D_3 is reverse biased. As the Q output of the CD4078B falls low, current through L_{P1} is interrupted, causing the polarity of the voltage over L_{S1} to reverse. As a result, D_3 is forward biased, allowing current to flow toward point E, charging C_1. At the same time, with the $\overline{Q}$ output high, current flows through L_{P2} toward point D. As the $\overline{Q}$ output of the CD4078B falls low,

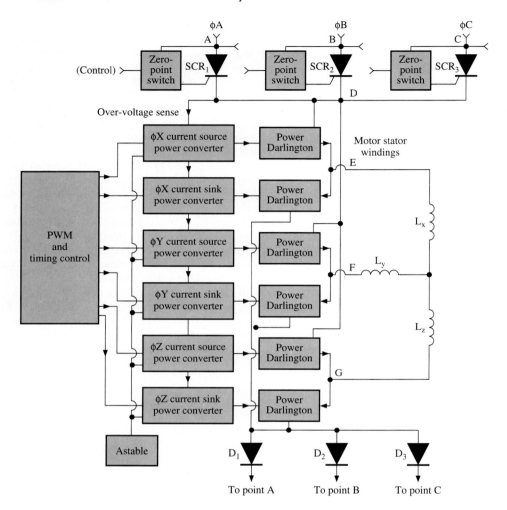

Figure 10–51
Simplified diagram of commutation control circuitry for 3-ϕ brushless dc servomotor

the phase of the voltage over L_{S2} reverses, forward biasing D_4 and allowing charging current to flow to C_1. Thus, with the drive enable signal low, C_1 rapidly charges to a positive dc level. Since the current pulses at points C and D have a virtually constant frequency and duty cycle, the dc voltage at point E remains constant. This steady potential at point E establishes the anode voltage for SCR_1. The voltage at point F, which controls the amplitude of the base current for the power Darlington, is a function of the Commutation signal at point B. During the space width of this signal, D_1 becomes forward biased, allowing current to flow from point G to point C when the Q output of the CD4047B pulses high. When the $\overline{Q}$ output of the CD4047B pulses high, providing a logic low is still present at point B, D_2 becomes forward biased and current flows from point G to point D. As the Q output of the CD4047B pulses low, interrupting the flow of current in L_{P1} of T_2, Schottky diode D_5 becomes forward biased, resulting in the flow of current toward point F. When the $\overline{Q}$ output of the CD4047B pulses low and current flow in L_{P2}

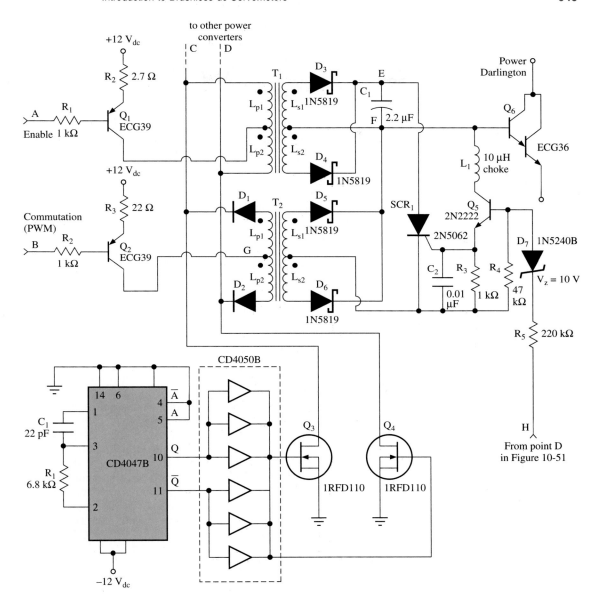

Figure 10–52
Detail of one of six power converter stages within brushless dc servomotor controller

of T_2 is interrupted, D_6 is forward biased, thus allowing the continuation of current flow toward point F.

The amplitude of the voltage at point F is directly proportional to the space width of the rectangular waveform at point B. During the time in which the power Darlington in Figure 10–52 is to be biased off, the system control circuitry holds the base of Q_2 at a logic high. With Q_2 biased off, the second push-pull converter is inactive and no current flows to point F. Thus, Q_6 is deprived of base current and remains in the off state. When

Q_6 is to be switched on, a logic low appears at point B, forward biasing Q_2. If the base current to Q_6 is to be increased, the space width of the signal at point B increases. A reduction in the space width of the signal at point B decreases the base current for Q_6. Thus, collector and emitter currents for the power Darlingtons become functions of the space width of the Commutation signal. The waveforms represented in Figure 10–53 should aid the reader in understanding the operation of the power converter in Figure 10–52.

SCR_1, switching transistor Q_5, and zener diode D_7 together protect the servomotor and its control circuitry from an overvoltage condition. To understand how this is accomplished, it is necessary to note that point H in Figure 10–52 is connected to point D in Figure 10–51. Thus, a surge in the 3-ϕ line voltages is immediately detected at point H in Figure 10–52. Under normal operating conditions, the ohmic values of R_4 and R_5 are high enough to hold Q_5 in an off condition. Thus, SCR_1 remains biased off. However, if the overvoltage condition at point D is great enough, Q_2 switches on, inducing the flow of gate current in the crowbar SCR. As this SCR switches on in forward saturation, points E and F are clamped at nearly 0 V and the power Darlington is biased off. As indicated by the block diagram in Figure 10–51, the overvoltage sensing circuits are cascaded, causing all six power converters to shut down if any one of them detects an overvoltage condition. Thus, both the motor controller and the servomotor are protected from severe damage or destruction.

For a high-power brushless dc servomotor, as shown in Figure 10–54(a), there are likely to be at least two sets of X, Y, and Z windings. As seen in Figure 10–54(b), these windings are arranged for maximum flux linkage between the stator and the PM rotor.

Figure 10–55 illustrates the conditions of the power Darlingtons and stator windings during one rotation of the motor shaft. (Note, for simplicity, only single transistors and stator windings are represented here.) At t_0, with the ϕY current-sourcing and ϕX current-sinking transistors forward biased, the stator current and magnetic field conditions are as shown in Figure 10–55(a). At t_1, commutation occurs with the ϕZ current-sourcing and ϕX current-sinking transistors active. Thus, the magnetic field shifts 60° clockwise, causing the PM rotor to align with the shifting stator field. At t_3, with the ϕZ current-sourcing and ϕY current-sinking transistors biased on, the rotor turns an additional 60°.

Two major design advantages of the motor control system just analyzed should now be apparent. Motor rpm may be controlled by the rate at which the commutation sequence illustrated in Figure 10–55 is made to occur. By continually monitoring data from one, or possibly two, feedback devices, system control circuitry may precisely govern this commutation sequence and, thus, exercise exact control of motor rpm. As was explained for the individual power converter in Figure 10–52, stator current becomes a function of the space width of the signals arriving from the pulse width modulator (PWM). For the

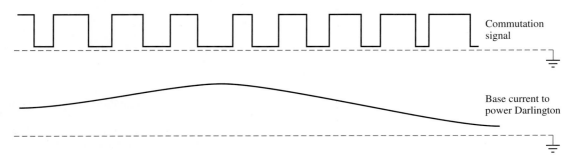

Commutation signal

Base current to power Darlington

Figure 10–53
Waveforms for power converter

Figure 10–54

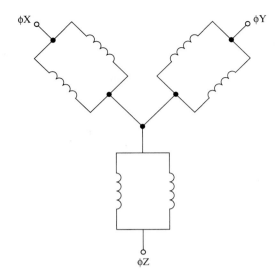

(a) Wye configuration of stator windings for brushless dc servomotor

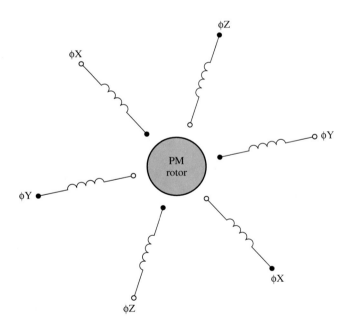

(b) Orientation of stator windings for brushless dc servomotor

3-ɸ brushless dc servomotor containing a PM rotor, torque becomes a function of the stator current. Thus, system control circuitry may also precisely govern motor torque as necessary, based on changing load conditions. The operation of this servomotor control circuitry is covered in detail in Section 14–1.

    In some industrial applications, it is necessary for a motor to run continuously at an exact speed. For such motor operation, control circuitry must be provided that continuously

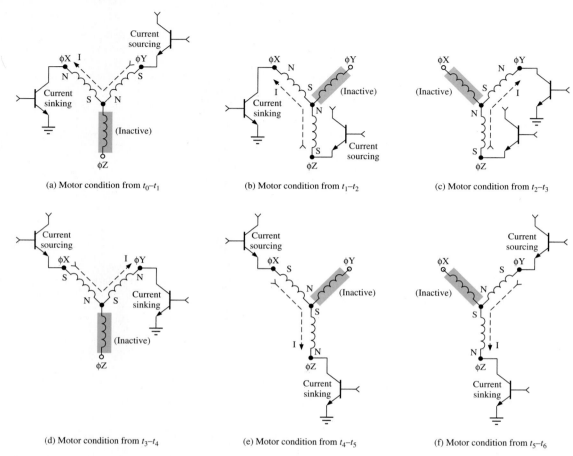

(a) Motor condition from t_0–t_1

(b) Motor condition from t_1–t_2

(c) Motor condition from t_2–t_3

(d) Motor condition from t_3–t_4

(e) Motor condition from t_4–t_5

(f) Motor condition from t_5–t_6

Figure 10–55
Condition of power Darlingtons and stator windings during one revolution of the
servomotor shaft

monitors motor rpm and instantaneously responds to any deviations in speed. As an
example of such a motor application, consider the circuit in Figure 10–56. Here a **phase-
locked loop** (PLL), working in conjunction with a Hall sensor commutation control device
and two JK flip-flops, maintains a virtually constant speed for a brushless dc motor. This
low-power motor is used to spin a mirror wheel within a bar code reader system. The
rotating mirror is used in the development of a laser beam into an X pattern. Such X
patterns are used in scanning bar code labels on large containers. A typical rotational
speed for the mirror motor within such a system is 3000 rpm.

The placement and operation of the three Hall sensors for the mirror motor would
be identical to that illustrated in Figure 10–45, with a slotted wheel being placed on the
rotor shaft. These feedback signals directly enter the UDN2936W commutation control
integrated circuit (IC) as shown in Figure 10–56. However, for increased noise immunity
and TTL compatibility, the Hall sensor signals pass through optoisolators and Schmitt-
triggered inverters before arriving at the PLL circuitry. Here, the Schmitt-triggered Hall
sensor signals, which now appear as the rectangular waveforms in Figure 10–57, enter a
pair of cascaded exclusive OR gates. These three logic gates effectively form a three-

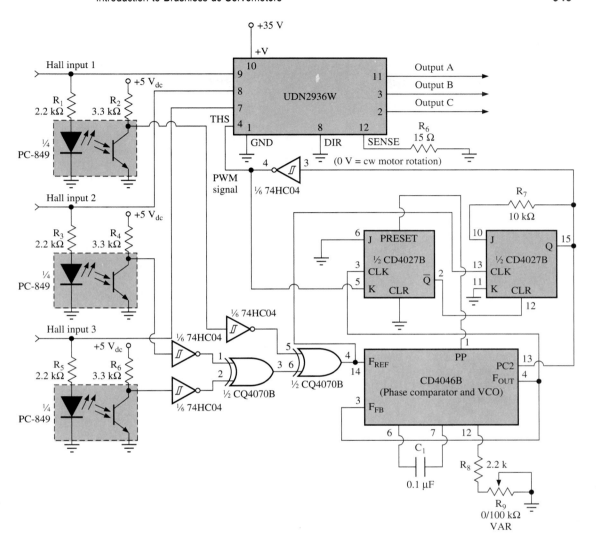

Figure 10–56
Mirror motor control circuitry for a bar code scanner

input XOR function. The output of this XOR function is plotted below the three Hall sensor signals in Figure 10–57.

Since the XOR function responds to odd parity, it produces a logic high only when a single Hall sensor signal is high. Thus, as indicated in Figure 10–57, the output of the XOR function is low during the overlap time of the Hall sensor signals. The resultant output of the XOR function is a clocking signal occurring at three times the frequency of motor rotation. The output of the XOR function serves as the reference frequency input (F_{REF}) for the CD4046B PLL IC and also serves as the clock input to the second JK flip-flop.

The CD4046B is a monolithic device containing the phase comparator, low-pass filter and voltage control oscillator (VCO) portions of the PLL. The center frequency for the CD4046B is determined by the values of C_1 and ($R_8 + R_9$). The adjustment of potentiometer

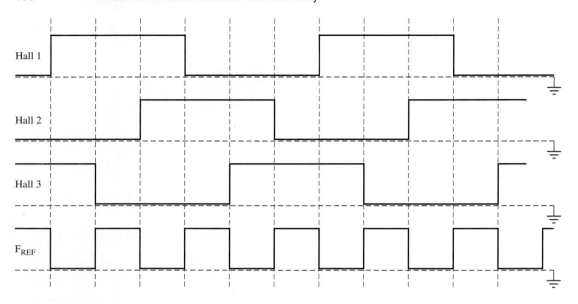

Figure 10–57
Waveforms for mirror motor control circuitry

R_9 determines the center frequency of the VCO and, therefore, controls the steady-state rpm of the mirror motor. Assuming the constant operating frequency of the mirror motor is to be 3000 rpm, the number of rotations occurring per second would then be determined as follows:

$$\frac{3000 \text{ rpm}}{60 \text{ sec}} = 50 \text{ rotations per second}$$

A timing pulse at the THS input to the UDN2936W is required to trigger each of the three commutation signals that occur during one shaft rotation. Thus, to obtain a motor speed of 3000 rpm, the required frequency of the signal at the THS input is determined as follows:

$$f_{THS} = 3 \times 50 \text{ rotations} = 150 \text{ pulses per second}$$

Since the two JK flip-flops together perform a divide-by-four function, the required center frequency of the VCO would be determined as follows:

$$f_c = F_{OUT} = 150 \text{ pps} \times 4 = 600 \text{ pps}$$

The operation of the PLL in this example is distinctly different from that which controlled the video scanning circuitry in Section 5–7. For the PLL in Figure 10–56, there is no active low-pass filter. Therefore, the output frequency of the VCO is not slaved to the reference frequency. Rather, in this example, the pulse width of the THS input to the UDN2936W is modified as necessary to hold the mirror motor speed at 3000 rpm.

While the F_{FB} signal is occurring at the adjusted frequency of 600 pps, the F_{REF} signal is occurring at nearly 150 pps. The 600-pps signal clocks the first JK flip-flop, while the F_{REF} signal clocks the second flip-flop. Since the Q output of the second JK flip-flop is fed to the second phase comparator input of the CD4046B, the phase detector portion of the CD4046B is able to monitor the parity condition between the Q output of the second

JK flip-flop and the F_{REF} signal. During an odd-parity condition, the PP output of the CD4046B is high, presetting the first JK flip-flop. With this condition, the CLR input to the second JK is disabled, allowing the logic condition at the J input to the second flip-flop be latched at Q as the F_{REF} signal switches to a logic high. During an even-parity condition, the logic condition at the THS input to the UDN2936W, which is present at the K input to the first JK flip-flop, is updated at the $\overline{Q}$ output on each positive pulse of the F_{OUT} signal. Thus, when $\overline{Q}$ of the first JK is high, the CLR input to the second JK is high, causing its Q output to be low, regardless of the condition of the F_{REF} signal. Thus, through the action of the fourth inverter, the THS input to the UDN2936W remains high. When the PP output of the phase detector again pulses high, the $\overline{Q}$ output of the first JK goes low, again disabling the clear function of the second flip-flop. Thus, the second JK is able to toggle on the positive transition of the F_{REF} signal. Through this action, the PLL is able to compensate for any drift in the frequency of the F_{REF} and hold motor rpm at a steady speed.

SECTION REVIEW 10–5

1. Describe the structure and operation of a 3-ϕ brushless dc servomotor. Explain how this device is able to operate without the brushes, commutator, and armature of the conventional dc motor. What materials could comprise the rotor of the 3-ϕ brushless dc servomotor?
2. List those feedback devices commonly found within 3-ϕ brushless dc servomotors. Briefly describe the operation of each device. Which feedback devices provide commutation control? Which devices are able to represent shaft position accurately?
3. Explain how a shaft encoder system within a 3-ϕ brushless dc servomotor is able to indicate direction of shaft rotation.
4. Explain the operation of the resolver circuitry diagrammed in Figure 10–46. Explain in detail the purpose and operation of the RDC. What types of output signal does this device provide?
5. Why is it necessary for the Wien bridge oscillator in Figure 10–47 to have a stable output amplitude and minimal harmonic distortion?
6. For the Wien bridge oscillator in Figure 10–47, what ohmic value for R_1 and R_2 would result in an operating frequency of around 8.84 kHz?
7. Identify which components comprise the error integrator in Figure 10–47. Explain the operation of this portion of the circuit. How is this circuitry able to maintain a virtually constant output amplitude for the Wien bridge oscillator?
8. Explain how the power converter circuitry in Figure 10–52 is able to shut down automatically in the event of an overvoltage condition.
9. For the power converter in Figure 10–52, what must be the condition of the enable signal for base current to flow to the power Darlington?
10. For the power converter in Figure 10–52, what controls the amplitude of the base current flowing to the power Darlington?

10–6   INTRODUCTION TO STEPPER MOTORS

As their name implies, **stepper motors** are devices that rotate by discrete incremental steps, as opposed to the relatively smooth and continuous shaft rotation characteristic of conventional ac and dc motors. The primary advantage of stepper motors is that they can be controlled almost directly by digital circuitry. Although current amplifiers must be placed between the digital control source and the stator windings of a stepper motor, no complex digital-to-analog conversion circuitry is required for controlling motor speed. Another major advantage of stepper motors is their ability to operate reliably in an open-loop control mode, eliminating the need for complex feedback circuitry. A major structural advantage of stepper motors is that they contain no brushes or commutators. As with brushless dc servomotors, no part of the stator of a stepper motor physically contacts the rotor. Thus, stepper motors exhibit relatively high durability and produce minimal RF noise.

Although all forms of stepper motors operate on the same electromagnetic principles, they may be separated into three basic groups. These classifications are **permanent magnet** (PM), **variable reluctance** (VR) and **hybrid stepper motor** (HSM). The number of phase windings present in the stator determines the degree by which the motor steps. For the most commonly used industrial stepper motors, this **step angle** (designated θ_S) may be as low as 7.5° or as high as 30°. A stepper motor with a small step angle is capable of a higher resolution of motion than a motor with a larger step angle. The number of steps required for a given stepper motor to complete one revolution is its **stepping rate** (designated R_S). The relationship of the step angle and stepping rate may be expressed as follows:

$$\theta_S = \frac{360°}{R_S} \qquad (10\text{--}27)$$

Example 10–8

For one stepper motor, the step angle is specified as 7.5°. What is its stepping rate? If, for another motor, the stepping rate is 12, what is the step angle?
Solution:

For the first motor:

$$R_S = \frac{360°}{7.5°} = 48$$

For the second motor:

$$\theta_S = \frac{360°}{12} = 30°$$

Figure 10–58 illustrates the basic structure and operation of a stepper motor. In this example, the rotor contains a permanent magnet, and the motor stator consists of pairs of **phase windings.** A stepper motor may contain several sets of phase windings. However, for the sake of simplicity, only two pairs are shown in Figure 10–58. Mounted opposite each other on the stator, L_1 and L_3 form one pair of phase windings, and L_2 and L_4 form the second pair.

To understand the fundamental operation of the stepper motor, consider first the condition represented in Figure 10–58(a). Prior to the closure of S_1 and S_3, no current flows through windings L_1 and L_3. However, as S_1 and S_3 close simultaneously, L_1 and L_3 develop magnetic fields with the polarity shown in Figure 10–58(b). Due to the resultant magnetomotive force, the rotor moves in the direction shown, aligning itself with the stator field created by L_1 and L_3. With the PM rotor now aligned as shown in Figure 10–58(b), an equilibrium condition exists. Thus, with L_1 and L_3 still energized, the rotor remains fixed in the position shown.

Next, consider the conditions represented in Figure 10–58(c). With switches S_1 and S_3 now in the open position, S_2 and S_4 are closed simultaneously. As phase windings L_2

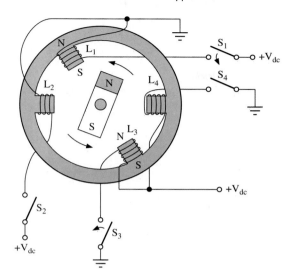

(a) Stepper motor condition with S_1 and S_3 closing

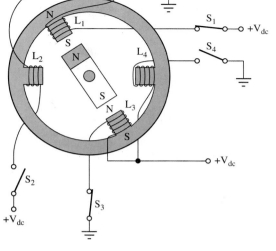

(b) Stepper motor condition after S_1 and S_3 close

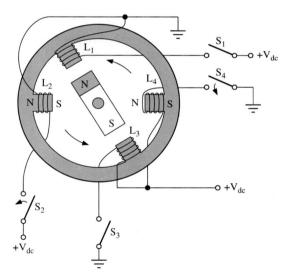

(c) Stepper motor condition with S_2 and S_4 closing

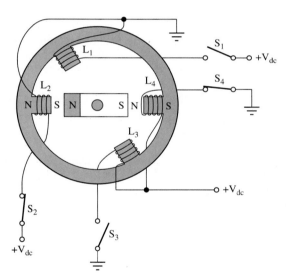

(d) Stepper motor condition after S_2 and S_4 close

Figure 10–58

and L_4 are energized, they develop magnetic fields with the polarity shown, causing the rotor to move in the direction indicated. Equilibrium is again achieved when, as shown in Figure 10–58(d), the rotor aligns itself with this new stator field. While current continues to flow through L_2 and L_4, the rotor remains fixed in this position.

Figure 10–59(a) illustrates the internal structure of a complete PM stepper motor. Since this device contains four pairs of phase windings it is referred to as a **four-phase motor.** For the sake of clarity, a schematic representation of this four-phase stepper motor is contained in Figure 10–59(b).

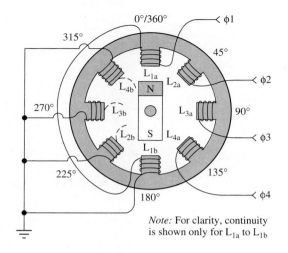

(a) Basic structure of four-phase PM stepper motor

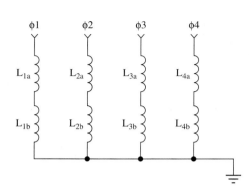

(b) Schematic representation of four-phase PM stepper motor

Figure 10–59

    With the design format shown in Figure 10–59, a bidirectional current source is provided for each of the four pairs of phase windings. As shown in Figure 10–59(b), the two windings within a given pair are placed in series with their current source. Since the PM stepper motor shown here has eight stator windings, the stepping rate is also 8. This is because there are eight possible positions for the rotor within a single revolution. Thus, the stepping angle is determined as:

$$\theta_S = \frac{360°}{8} = 45°$$

Figure 10–60 represents the sequence of current pulses that must occur at the phase inputs of the PM stepper motor in Figure 10–59 to produce one revolution.

    At t_0 in Figure 10–60, assume that the PM rotor is aligned with the L_{1a} and L_{1b} windings and that positive current is flowing from the $\phi 1$ source toward ground. Between t_0 and t_1, an equilibrium condition exists, causing the PM rotor to remain aligned with the $\phi 1$ windings. At t_1, current flow through the $\phi 1$ windings terminates while a positive current pulse occurs at the $\phi 2$ input. According to the principles demonstrated in Figure 10–58, the rotor in Figure 10–59 aligns itself with the magnetic field produced by the L_{2a} and L_{2b} windings and remains in that position until t_2. This stepping process continues through t_3, with positive current pulses occurring at the $\phi 3$ and $\phi 4$ inputs.

    At t_4 in Figure 10–60, to cause the rotor to increment to the 180° point in its rotation, the magnetic poles of the stator must be reversed with respect to their orientation at t_0. This is accomplished, as shown in Figure 10–60, by reversing the polarity of the $\phi 1$ current pulse. With the current through the L_{1a} and L_{1b} windings now flowing in the negative direction, the south pole of the rotor magnet is attracted toward the L_{1a} winding while the north pole is attracted toward L_{1b}. At t_5, with a negative current pulse arriving at the $\phi 2$ input, the rotor increments to the 225° position. With negative current pulses arriving at the $\phi 3$ and $\phi 4$ inputs at t_6 and t_7, the stepping action continues. By t_8, the rotor is again at the 0° position.

    Figure 10–61(a) illustrates the structure of a single-stack variable reluctance (VR) stepper motor. (The term single-stack indicates that the VR motor contains only one rotor/

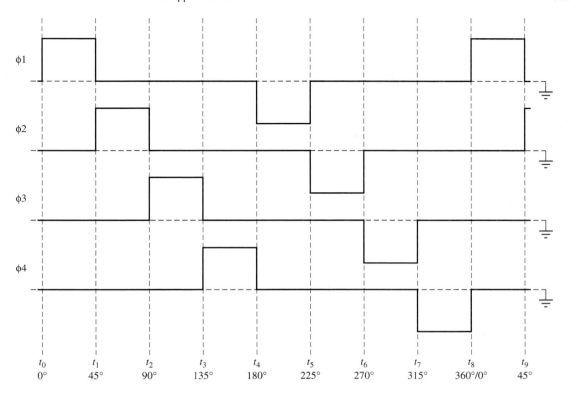

Figure 10–60
Pulse train required for operation of four-phase PM stepper motor

stator assembly.) The rotor of this device is composed of a ferrous material but is not a permanent magnet. For the single-stack VR stepper motor shown here, there are three phases, each of which contains four stator windings. The electrical arrangement of these windings is represented in Figure 10–61(b).

As seen in Figure 10–61(a), both the rotor and the stator of the single-stack VR stepper motor contain **teeth.** For the stator, these teeth function as poles, around which the phase windings are wrapped. The four windings of a given phase are wrapped on every third stator tooth. Thus, for $\phi 1$, L_{1a}, L_{1b}, L_{1c}, and L_{1d} are wound in series and placed on teeth 1, 4, 7, and 10, respectively. The $\phi 2$ windings are wrapped on poles 2, 5, 8, and 11, and the $\phi 3$ windings are placed on poles 3, 6, 9, and 12.

Important design parameters for the single-stack VR stepper motor include the number of stator teeth (designated N_S) and the number of rotor teeth (designated N_R). The pitch, or degree of slant, of the stator and rotor teeth is also important. The pitch of the stator teeth is designated P_S, and the pitch of the rotor teeth is designated P_R. The parameters of P_S and P_R are specified in degrees, as illustrated in Figure 10–62. The angle formed by two contiguous stator teeth, P_S, is determined as follows:

$$P_S = \frac{360°}{N_S} \qquad (10–28)$$

Thus, for the stepper motor in Figure 10–61, since there are 12 stator teeth, P_S becomes 30°.

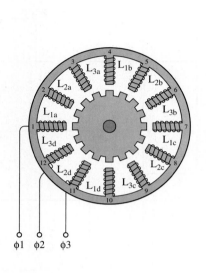

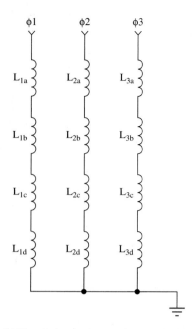

(a) Structure of a single-stack variable reluctance stepper motor

(b) Electrical arrangement of stator windings for variable reluctance stepper motor

Figure 10–61

The angle formed by two contiguous rotor teeth, P_R, is determined as follows:

$$P_R = \frac{360°}{N_R} \qquad (10-29)$$

Thus, for the motor in Figure 10–61, since there are 16 rotor teeth, P_R becomes 22.5°.

For the single-stack VR stepper motor, the step angle may be determined in two ways. If P_R is known, then θ_S is derived as follows:

$$\theta_S = \frac{P_R}{N_P} \qquad (10-30)$$

Figure 10–62
Parameters for P_S and P_R

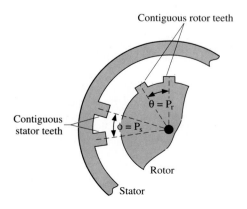

where N_P represents the number of phases for the single-stack V_R stepper motor. Thus, for the stepper motor in Figure 10–61, since there are three phases, the step angle becomes:

$$\theta_S = \frac{22.5°}{3} = 7.5°$$

According to Eq. (10–27), for any stepper motor, the step angle may also be determined as follows:

$$\theta_S = \frac{360°}{R_S}$$

Thus, for the single-stack VR stepper motor represented in Figure 10–61:

$$\theta_S = \frac{360°}{R_S} = \frac{P_R}{N_P}$$

If, according to Eq. (10–29),

$$P_R = \frac{360°}{N_R}$$

then,

$$P_R \times N_R = 360°$$

Formula (10–30) may now be developed as follows:

$$\theta_S = \frac{P_R}{N_P} = \frac{(P_R \times N_R)}{(N_P \times N_R)}$$

Substituting:

$$\theta_S = \frac{360°}{(N_P \times N_R)} \qquad (10\text{–}31)$$

Thus, as may be proven for the motor in Figure 10–61:

$$\theta_S = \frac{360°}{(3 \times 16)} = 7.5°$$

Again referring to formula (10–27), if

$$\theta_S = \frac{360°}{R_S}$$

then,

$$\frac{360°}{R_S} = \frac{360°}{(N_P \times N_R)}$$

Thus, an alternative expression for the stepping rate of a single-stack VR stepper motor would be:

$$R_S = (N_P \times N_R) \qquad (10\text{–}32)$$

As may be proven for the motor in Figure 10–61:

$$R_S = \frac{360°}{7.5°} = (3 \times 16) = 48$$

The formation of the magnetic fields between the rotor and stator of the VR stepper motor is distinctly different from the process just analyzed for the PM device. Referring back to the PM stepper motor in Figure 10–59 and the pulse train in Figure 10–60, recall that the internal poles of the stator windings, which are 180° apart, always have opposing magnetic polarities. This is necessary to attract the opposite poles of the PM rotor. For the VR stepper motor in Figure 10–61, however, where the rotor consists of a ferrous material rather than a permanent magnet, the internal poles of the opposing stator teeth have the same magnetic polarity. Thus, as shown in Figure 10–63(a), when the $\phi 1$ windings are energized, L_{1a} and L_{1c} both develop their magnetic north poles closest to the rotor. However, L_{1b} and L_{1d} both develop their south poles closest to the rotor. Consequently, following paths of least reluctance, four closed loops of magnetic flux are formed. The flux leaving the north pole of L_{1a} enters tooth 1 of the rotor, then diverges to leave the north poles developed at rotor teeth 5 and 13. After entering the stator at the south poles of L_{1b} and L_{1d}, this flux then travels via the stator, converging at the north pole of L_{1a}. Also, the flux leaving the stator at the north pole of L_{1c} enters tooth 9 of the rotor, then diverges to leave through the north poles present at rotor teeth 5 and 13. This flux then travels the path of least reluctance through the stator, entering the south poles of L_{1b} and L_{1d} and converging at the north pole of L_{1c}. With current flowing through the phase 1 windings and magnetic flux flowing along four paths of minimum reluctance, the motor is in an equilibrium condition and the rotor shaft remains stationary.

To advance the rotor one step clockwise, the $\phi 2$ windings are energized. This causes the rotor to move just to the point where the reluctance to magnetic flux is again at minimum. As shown in Figure 10–63(b), this occurs when rotor teeth 2, 6, 10, and 14 are aligned with the stator teeth containing windings L_{2a}, L_{2b}, L_{2c}, and L_{2d}, respectively. The degree of clockwise rotation required to achieve this new equilibrium condition is only 7.5°, equivalent to the calculated step angle of the motor. At this time, the formation of the four magnetic flux loops along minimum reluctance paths is identical to that

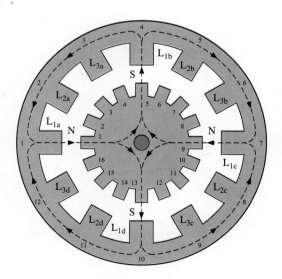

(a) Stepper motor condition with $\phi 1$
windings energized

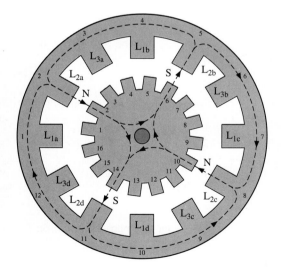

(b) Stepper motor condition with $\phi 2$
windings energized (the rotor
has moved 7.5° clockwise)

Figure 10–63

described for the φ1 windings and rotor teeth 1, 5, 9, and 13. This switching sequence continues with the activation of the φ3 windings, advancing the rotor an additional 7.5°. Rotor teeth 3, 7, 11, and 15 are then aligned with the stator teeth containing the φ3 windings. Since there are 16 rotor teeth, this switching sequence involving each of the three sets of phase windings must repeat 16 times to complete one cycle of the rotor. Hence, there are 48 steps within a given cycle.

Example 10–9

A 4-φ single-stack VR stepper motor has a step angle of 1.5°. Also, the pitch of the stator teeth for this motor is 8°. Given this information, solve for the pitch of the rotor teeth, the number of rotor teeth, the number of stator teeth, and the stepping rate.

Solution:

Step 1: Solve for the pitch of the stator teeth. Transposing formula (10–30):

$$P_R = \theta_S \times N_P = 1.5° \times 4 = 6°$$

Step 2: Solve for the number of rotor teeth. Transposing formula (10–29):

$$N_R = \frac{360°}{P_R} = \frac{360°}{6°} = 60$$

Step 3: Solve for the number of stator teeth. Transposing formula (10–28):

$$N_S = \frac{360°}{P_S} = \frac{360°}{8°} = 45$$

Step 4: Solve for the stepping rate. Transposing formula (10–27):

$$R_S = \frac{360°}{\theta_S} = \frac{360°}{1.5°} = 240$$

The multiple-stack VR stepper motor represents a modification of the single-stack design concept just introduced. With the multiple-stack format, a separate stator/rotor assembly is provided for each phase. However, the rotors are still mounted on a common shaft. Figure 10–64(a) contains a cutaway view of a multiple-stack VR motor containing three phases or stacks. As illustrated in Figure 10–64(a), each of the three stacks contains four poles. Although these stator poles may be considered as large teeth, they each, in turn, have three smaller teeth. It is important to note that each rotor has 12 teeth that may be aligned with the 12 teeth of a given stator. While the teeth of the three rotors are perfectly aligned with each other, the stator teeth are deliberately misaligned. To allow the motor to operate, the φ2 stator poles are skewed by 10° with respect to the φ1 poles, while the φ3 poles are skewed an additional 10° with respect to the φ2 poles. This relationship is clearly represented in Figure 10–64(b).

The step angle for a multiple-stack VR stepper motor may be determined by using Eq. (10–30), with the understanding that N_P now represents the number of stacks within the device. Thus, for the motor represented in Figure 10–64, since there are 12 rotor teeth and three phases:

$$\theta_S = \frac{P_R}{N_P} = \frac{30°}{3} = 10°$$

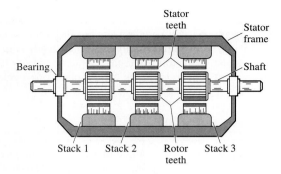

(a) Multiple-stack VR stepper motor (cutaway view)

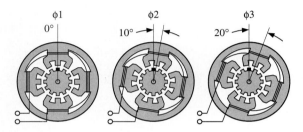

(b) Placement of stacks in a VR stepper motor

Figure 10–64

Figure 10–65 represents the switching sequence for the multiple-stack VR stepper motor in Figure 10–64. Here, the first column represents the advance of the rotor within the φ1 stack through two switching states, resulting in a 20° movement of the rotor shaft. The second column represents the advance of the rotor within the φ2 stack during this same period, while the third column represents the advance of the φ3 rotor.

At t_0, the teeth of the φ1 rotor are in alignment with the teeth of the φ1 stator poles. With current flowing through the φ1 windings, four magnetic loops are formed in the same manner as was shown in Figure 10–63. At t_1, with the excitation of the φ2 stator windings, the rotor shaft advances 10°, to the point where the teeth of the φ2 rotor teeth are aligned with the teeth of the stator poles. Here, equilibrium is again attained with the development of the four magnetic loops between the rotor and stator. When, at t_2, the φ3 windings are energized, the rotor shaft advances an additional 10°, just to the point where the φ3 rotor teeth align with the teeth of the φ3 stator poles. Again equilibrium occurs and the rotor shaft remains stationary until the φ1 windings are activated.

Although the rotors of the multiple-stack VR stepper motor are mounted on a common shaft, they are well isolated from one another magnetically. This allows for various switching options that are impractical for the single-stack motor design. Figure 10–66(a) represents the waveforms produced by a digital control source for conventional **single-phase excitation** of the VR stepper motor.

Figure 10–66(b) introduces a control technique called **dual-phase excitation.** The simultaneous excitation of the φ1 and φ2 windings between t_0 and t_1 causes the rotor teeth of these phases to align in a position one-half step (5°) short of the equilibrium point for single-phase excitation. Next, the simultaneous excitation of the φ2 and φ3 windings between t_1 and t_2 causes the rotor teeth of these phases to align in another one-

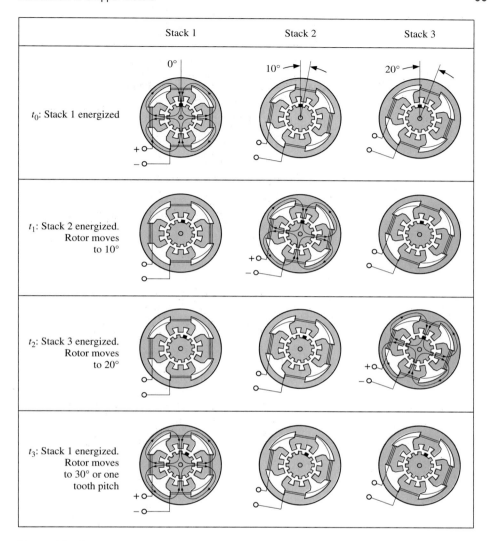

Figure 10–65
Commutation sequence for a three-stack VR stepper motor

half step position. Thus, the effective step angle remains 10°. The primary advantage of dual-phase excitation is the greater torque achieved through the greater values of stator current flow.

The stepping index for a multiple-stack VR stepping motor may effectively be doubled by the use of a combination of single- and dual-phase excitation. As shown by the control waveforms in Figure 10–66(c), from t_0 through t_1, both $\phi 1$ and $\phi 2$ are active, causing the rotor teeth of those two phases to align in a one-half step position. At t_1, with only the $\phi 2$ windings active, the rotor shaft turns only 5°, as the $\phi 2$ rotor teeth align directly with the $\phi 2$ stator teeth. From t_2 to t_3, excitation current flows through both the $\phi 2$ and $\phi 3$ windings, causing the rotor teeth of these phases to advance only 5° to another one-

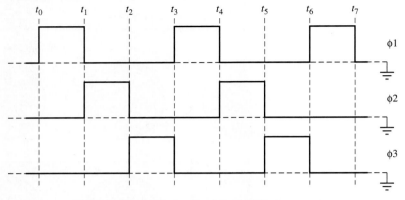

(a) Single-phase excitation signals for 3-φ VR stepper motor

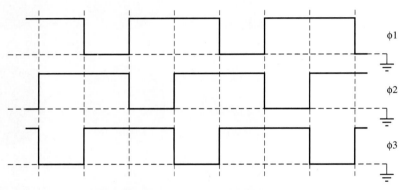

(b) Dual-phase excitation signals for 3-φ VR stepper motor

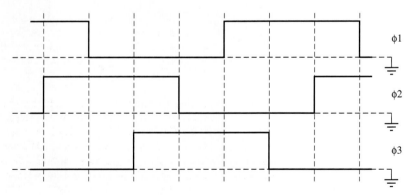

(c) Half-stepping excitation signals for 3-φ VR stepper motor

Figure 10–66

half step position. At t_3, with only the $\phi3$ windings active, the rotor shaft increments another 5°, as the $\phi3$ rotor teeth align directly with their stator poles. Thus, for the motor in Figure 10–64, with combined single- and dual-phase clocking. θ_S becomes only 5° and R_S doubles to 60.

The hybrid stepping motor has structural characteristics of both the PM and the multiple-stack VR stepping motor (VRSM). As shown by the cutaway view in Figure 10–67(a), the hybrid stepper motor is, like the multiple-stack VR device, divided into two or more stator/rotor assemblies. However, there are three important structural differences between the hybrid and the multiple-stack VRSM designs. First, unlike the VR device, the hybrid stepper motor contains a permanent magnet, which is mounted axially between the rotor assemblies. Also, for the multiple-stack VRSM, the number of rotor teeth for a given stack equals the number of stator teeth. However, for the hybrid stepper motor in Figure 10–67(b), the rotors contain more teeth than their stators. Finally, remember that, with the multiple-stack VRSM, the rotor teeth are perfectly aligned from stack to stack, while the stator poles are skewed. With the hybrid stepper motor, just the opposite occurs, with the stator teeth being perfectly aligned and the rotor teeth being skewed from stack to stack.

The method by which the phase windings are wrapped on the stator poles also differs greatly between the hybrid and the multiple-stack VR stepper motors. With the multiple-stack VRSM, recall that the stator within a given stack contains the windings of only one

Figure 10–67

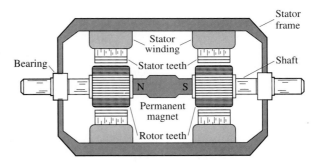

(a) Hybrid stepper motor (cutaway side view)

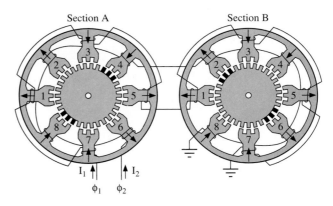

(b) Stack orientation in a hybrid stepper motor

phase. Thus, with this form of motor, the terms *stack* and *phase* are often used interchangeably. With the hybrid stepper motor, however, as shown in Figure 10–67(b), the windings of two different phases are placed on a single stator assembly. In this example, the windings of φ1 are mounted on the odd-numbered poles of both stator assemblies, and the φ2 windings are wrapped on the even-numbered poles.

As shown in Figure 10–68, the permanent magnet mounted on the shaft of the hybrid stepper motor establishes a constant north pole at the rotor of the first stack and a constant south pole at the rotor of the second stack. With current applied to the φ1 windings, magnetic flux passes from the stack 1 rotor to the stack 1 stator at the points of minimum reluctance. With the two rotors positioned as seen in Figure 10–68, these minimum reluctance points are at stator poles 1 and 5, where the stator and rotor teeth are in their closest alignment. The path for flux is then via the stator housing toward the stator of stack 2. Flux then enters the rotor at the points where the stator teeth are aligned with the rotor teeth. These minimum reluctance points are at stack 2 stator poles 3 and 7. The two magnetic loops then converge into the stator 2 rotor to pass through its axial magnet toward their origin point at rotor 1.

For the rotor shaft to increment by one step, excitation of the φ2 windings must occur. At this time, the rotor advances just to the point where the stack 1 rotor teeth are aligned with the teeth of stacker poles 4 and 8. At the same time, the stack 2 rotor teeth are most closely aligned with the teeth of stator poles 2 and 6. Once this degree of motion is achieved, an equlibrium condition exists, with two magnetic loops being formed as was just described for φ1 excitation. For the hybrid motor represented in Figure 10–67, four steps of the motor shaft are required to advance the rotor a distance equivalent to

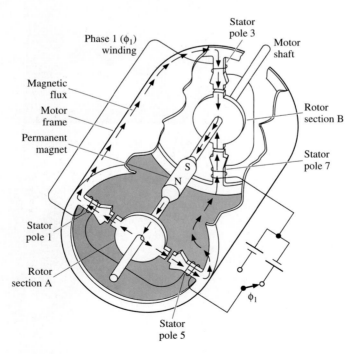

Figure 10–68
Magnetic flux conditions for a hybrid stepper motor

the pitch of the rotor teeth. Thus, for the motor in Figure 10–67:

$$\theta_S = \frac{P_R}{4}$$

Since each stack has 30 rotor teeth:

$$P_R = \frac{360°}{30} = 12°$$

$$\theta_S = \frac{12°}{4} = 3°$$

SECTION REVIEW 10–6

1. What are the three basic forms of stepper motor? Briefly describe the structure of each type of motor.
2. Define the term *stack* as it applies to the structure of the variable reluctance stepper motor.
3. A single-stack VR stepper motor has a P_S of 15°, an N_R of 32, and an N_P of 3. How many stator teeth does the motor have? What is the pitch of the rotor teeth? What is the step angle? What is the stepping rate?
4. What are the three forms of excitation that can be used in the operation of a multiple-stack VRSM? Briefly describe each operating method.
5. Explain how the stepping index of a multiple-stack VRSM may be effectively doubled.

SUMMARY

AC induction motors are the workhorses of industry. Single-phase ac induction motors are used in low- to medium-power applications, and 3-ϕ ac induction motors are used where large amounts of power are required. Single-phase ac induction motors require some form of phase-shift circuitry to initiate and maintain rotation. Examples of split-phase induction motors include the permanent-split capacitor and the capacitor-start designs. In addition to increased power capability, an advantage of 3-ϕ motors is their ability to operate without phase-shift circuitry.

DC motors are often used in industrial applications where precise control of machinery is required. The simplest form of conventional dc motor is the PM type, which uses a permanent magnet to produce its stator field. More powerful dc motors use electromagnets to produce a stator field. Conventional dc motors contain brushes that directly contact commutators mounted on the armature shaft. Arcing between brushes and commutators produces noise and is also a cause of wear in conventional dc motors. For these reasons, brushless dc motors are replacing conventional dc motors in many industrial applications.

One of the most important concepts in the mechanics of motor control is that of torque. Torque may be generally defined as the twisting force that a motor is able to produce. Motor data sheets are likely to specify values of running and stall torque. The torque constant, which represents the relationship between input current and resultant output torque, is also an important motor parameter. The rotational velocity of a motor is specified in rpm or revolutions per minute. For an ac induction motor operating directly from a 60-Hz ac line, rotational speed is typically around 1800 rpm (equivalent to ½ 60 Hz × 60 sec). This value, which represents the speed of rotation for the stator field, is referred to as the *synchronous speed.* The rpm of the induction motor lags slightly behind the synchronous speed. The disparity between synchronous speed and rpm is called *slip.*

Basic mechanical drive systems used in connecting motors to industrial machinery include direct drive, gear drive, leadscrew drive, and tangential drive. Direct drive represents the simplest form of mechanical linkage, with the load being connected directly to the motor shaft. Gear drive is often used as a means of varying rpm or amplifying torque. Commercially manufactured gearheads, or gear reducers, are commonly used in such operations. Also, a variety of ac motors are readily available with gearheads already attached. In a leadscrew system, a leadscrew, or worm, is connected to the shaft of a motor and used to move the load linearly. Conveyor belt, rack and pinion, and chain and sprocket systems are common examples of tangential drive.

The three basic dc motor configurations are series-field, shunt-field, and compound. Although the series-field configuration provides the most starting torque, the shunt-field configuration allows greater accuracy of velocity control. The compound configuration exhibits characteristics of both the shunt-field and the series-field configurations. Two basic types of dc motor speed control are field-weakening and armature-voltage. While field-weakening speed control is more power efficient, armature-voltage speed control is more easily implemented, especially with solid-state control devices such as SCRs. Pulse width modulation is often used to control the average armature voltage and, thus, govern motor speed. Direction of rotation may be controlled in two ways. Field current may be made bidirectional with armature current being kept unidirectional. Also, field current could be made unidirectional with armature current becoming bidirectional. Feedback may be used to improve the load regulation of a motor control circuit, allowing speed to remain constant with varying load conditions. In high-power applications, dc motors may be connected to 3-ϕ sources, with SCR control circuitry being connected to each phase input.

Synchros and resolvers are electromechanical devices capable of sensing the position of a rotating shaft. Industrial synchros may be classified as either *torque synchros* or *control synchros.* Torque synchros are commonly used in repeater systems for the purpose of transmitting angular displacement information. Control synchros are used in conjunction with control transformers and amplifier circuitry in higher power servo applications.

Three-phase brushless dc servomotors and their associated controllers are becoming widely used in industrial machinery. Such servomotors may contain a variety of feedback devices, including dc tachometers, Hall-effect sensors, synchros, resolvers, and shaft encoders. Hall-effect devices are commonly used to control motor commutation, whereas resolvers and their associated signal-conditioning circuitry are used to indicate motor shaft position. The rotor inside a 3-ϕ brushless dc servomotor is likely to contain a permanent magnet; its field windings are typically placed in a wye configuration. Current flow to these windings may be governed by power converters within a commercially manufactured servomotor controller.

Stepper motors are devices that rotate by discrete incremental steps rather than continuous angular motion. The main advantage of stepper motors is their ability to operate almost directly from digital control circuitry. Due to their predictable performance, stepper motors may be operated reliably within open-loop control systems. The three classifications of stepper motors are permanent magnet, variable reluctance, and hybrid.

SELF-TEST

1. What is the purpose of the centrifugal switch in Figure 10–6? What is the condition of this switch at the instant the motor is energized? What motor conditions causes this switch to open?

2. Identify two types of split-phase induction motor. Explain the basic differences between the two forms of motor.

3. Why is the basic structure of the 3-ɸ induction motor likely to be simpler than that of its single-phase counterpart?

4. What wiring configuration is likely to be used for the stator windings of a 3-ɸ induction motor? Describe how these windings could be arranged within the stator.

5. Which dc motor configuration allows the most reliable velocity control?

6. Which dc motor configuration provides the most starting torque?

7. Explain why, with field-weakening control of a dc motor, a reduction in current flow through the field winding produces an increase in rpm.

8. With armature-voltage speed control of a dc motor, an increase in armature voltage results in
   a. an increase in rpm
   b. an increase in torque
   c. both choices a and b
   d. a decrease in torque

9. True or false?: With field-weakening speed control of a dc motor, an increase in rpm is accompanied by a decrease in motor torque.

10. For the circuit in Figure 10–32, explain how an increase in the mechanical loading of the motor causes the SCR to switch on earlier in a given cycle.

11. How does the circuit in Figure 10–32 represent a design improvement over that in Figure 10–29?

12. In a gear drive system, if the number of load gear teeth is increased while the number of motor gear teeth remains the same,
   a. load rpm is increased
   b. load rpm is decreased
   c. unreflected load torque is decreased
   d. both choices b and c occur

13. True or false?: Connection of a motor to a chain and sprocket system would represent tangential drive.

14. A 3-ɸ induction motor operating from a 60-Hz source has a synchronous speed of 1800 rpm. If the actual rotational velocity of the motor is 1726 rpm, what is the percentage of slip?

15. Under a full-load condition, a dc motor rotates at 1237 rpm. If load regulation for the motor is 3.7%, what would be the motor rpm under no load?

16. For the synchro in Figure 10–38, assume the peak differential voltage developed between each pair of stator windings is 92 V. If the rotor is aligned at 72° with respect to the stator 1 axis, what are the rms values of V_{S1-S3}, V_{S1-S2}, and V_{S2-S3}?

17. Define TX, TR, TDX, CX, and CT as they apply to the operation of synchro systems.

18. Assume the resolver represented in Figure 10–43 is connected to the shaft of a servomotor. If the motor shaft is aligned at 27° with respect to its home (0°) position and $V_{R_{B1}-R_{B2}}$ has an instantaneous value of 235-mV, what would be the instantaneous amplitude of $V_{R_{A1}-R_{A2}}$?

19. What are the two basic forms of shaft encoder used as feedback devices within brushless dc servomotors? Explain the differences in structure and operation between these two types of shaft encoder.

20. Explain the advantages of using a resolver as a feedback device within a servomotor. What is a disadvantage of using the resolver?
21. Referring to the block diagram contained in Figure 10–51, identify an IC that could comprise each of the three zero-point switches.
22. Assume a multiple-stack VRSM has three stacks with 12 rotor teeth per stack. If the motor is being controlled by half-stepping excitation signals, what are the effective step angle and stepping rate?
23. True or false?: In a multiple-stack VRSM, each stack contains the windings of only one phase.
24. True or false?: Within a hybrid stepper motor, the rotor teeth are perfectly aligned from stack to stack, while the stator poles are skewed.
25. True or false?: The rotor of a hybrid stepper motor contains an axially mounted permanent magnet.

11 INDUSTRIAL APPLICATIONS OF DIGITAL DEVICES

CHAPTER OUTLINE

LEARNING OBJECTIVES

On completion of this chapter, the student should be able to:

- Identify the families of digital logic devices commonly used within industrial control circuitry.
- List the Boolean postulates.
- Describe DeMorgan's theorems and explain their importance to an industrial electronics technician.
- Optimize a combinatorial logic circuit using Karnaugh mapping.
- Describe how D flip-flops may be used within industrial control circuitry.
- Explain how JK flip-flops may be used to form state counters and other sequential control circuitry.
- Identify IC counters and registers commonly used in industrial control circuitry.
- Describe the various types of on/off interface modules that allow digital control circuitry to send control signals to and receive signals from industrial machinery.
- Describe the structure and operation of a basic digital-to-analog converter consisting of a resistive network and an op amp.
- Describe the structure and operation of monolithic digital-to-analog converters.
- Describe the structure and operation of the four basic types of analog-to-digital converters. Describe the advantages and disadvantages of each type.
- Describe the structure and operations of various monolithic analog-to-digital converters.

INTRODUCTION

The purpose of this chapter is to survey the industrial applications of digital logic devices, beginning with simple combinatorial logic circuits. During this chapter, it is assumed that the student already has a solid foundation in logic gates, flip-flops, counters, and registers. Thus, emphasis is placed on actual industrial applications of digital devices rather than the fundamentals of their operation.

Also in this chapter, various interfacing devices are introduced. The study begins with the signal conditioning requirements of basic on/off, or discrete, devices. Next, various forms of **digital-to-analog** (D/A) and **analog-to-digital** (A/D) converters are covered.

Figure 11–1 contains a block diagram of a typical closed-loop digital control system. Here, the **reference** signal could be a binary control variable encoded by an input device such as a dip switch. At a summing point (Σ in Figure 11–1), this value is compared to the output of an A/D converter. The output of the summing point is an **error** signal, representing the deviation between the desired reference-point condition and the instantaneous condition of the **plant.** The plant, which is the industrial device being controlled, could be a servomotor, clutch/brake, or some other piece of machinery. The error signal is fed to the digital controller, which could be a microcomputer, microcontroller, or programmable logic controller. In a simpler control application, the digital control circuitry might consist of only a few flip-flops and logic gates. In older control circuitry, designed before the appearance of the bus-organized microprocessor, the digital controller might consist of numerous logic gates, data registers, counters, and arithmetic circuits.

The digital control circuitry processes the error signal, outputting a control variable to a D/A converter. The output of the D/A converter controls power devices that, in turn, govern the plant. The plant usually contains some form of sensor, examples of which are resolvers, tachometers, Hall-effect devices, or shaft encoders. The sensor provides an analog feedback signal to the A/D converter. The A/D converter completes the control loop by providing digital information to the summing point.

It should be evident at this point that digital control circuitry is not capable of directly controlling an industrial process. In proportional control applications, such as that represented in Figure 11–1, extensive interface circuitry is required, both to convert the binary output of the controller to a usable analog signal and to translate the analog output of a sensory device to a binary code. Even in relatively simple discrete digital control applications, special input and output modules are required. Input modules convert ac

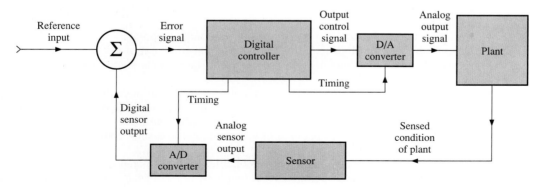

Figure 11–1
Block diagram of a digital control system

line voltages and high dc voltages to low dc levels that may be safely processed by digital control circuitry. Output modules perform an opposite function, allowing digital control circuitry to switch on devices that operate from high dc voltage sources or the ac line.

11–1   INTRODUCTION TO DIGITAL LOGIC FAMILIES

Various digital logic families are represented within modern industrial control systems. Examples of earlier forms of digital logic, which may still be found in some older control circuitry, include **diode-transistor logic** (DTL), **resistor-transistor logic** (RTL), and **high-threshold logic** (HTL). These older forms of logic are still commercially available on a limited scale.

Figure 11–2(a) contains the internal circuitry of a DTL expandable (or extendable) four-input NAND gate. The schematic symbol for this device is shown in Figure 11–2(b). Diodes D_1 through D_4, along with R_1, form a simple diode AND function. The remaining components in Figure 11–2(a) comprise the inverting switch. To fulfill the truth table for the NAND function, the output of the gate must be at a logic high when one or more of its inputs is at a logic low. Thus, if the cathode of one of the input diodes is at 0 V, Q_1 must be in a cutoff condition, with V_0 virtually equaling V_{CC}. As an example, if input 1 is at 0 V, D_1 becomes forward biased, holding point A at 0.7 V. Due to the presence of series diodes D_5 and D_6, the base voltage of Q_1 becomes -0.7 V. Thus, V_0 equals nearly 5 V. However, if all four inputs are at a logic high level (approximately 5 V), virtually no current flows through the four input diodes. The resultant increase in voltage at point A causes the base-emitter junction of Q_1 to become forward biased, allowing that transistor to switch into saturation. Thus V_0 is at a logic low level of less than 0.2 V.

An advantage of DTL logic gates is that they may be easily expanded to accommodate additional inputs. For example, if the gate in Figure 11–2 were contained within an integrated circuit (IC) package, a pin would be provided for the expandable input. Switching diodes could then be connected with their anodes common to this pin, in the same format as D_1 through D_4. Note in Figure 11–2(b) that the expandable input is shown with an arrow bending toward the front of the gate. This representation of the expandable input looks similar to and, therefore, should not be confused with the enabling input to a tri-state logic gate. (The operation of tri-state logic gates is covered later in this section.)

Figure 11–3 contains the internal circuitry and schematic symbol for an RTL three-input NOR gate. With this device, in conformance with the truth table for the NOR function, a logic high at the base input of any of the three switching transistors causes the output to fall to a logic low. Only when all three inputs are at a logic low, placing the three transistors in a reverse-biased condition, does the output switch to a logic high. Since each of the three inputs to the RTL gate in Figure 11–3 goes directly to the base of a transistor switch, the device is not expandable. However, an advantage of commercially manufactured RTL devices is that they operate reliably at low values of V_{CC}, with 3.6 V being typical.

Figure 11–4 contains the internal circuitry and schematic symbol for an HTL expandable three-input NAND gate. High-threshold logic evolved from the basic DTL family out of the necessity to improve noise immunity. Noise immunity is especially important within an industrial environment, where noise may be produced by motors, generators, relays, and switching power supplies. The use of two transistors within the logic gate in Figure 11–4, in itself, increases input impedance and, thus, attenuates noise. However, it is the Zener diode that is primarily responsible for decreasing susceptibility to spurious triggering. As indicated for the circuit in Figure 11–4(a), HTL operates at relatively high values of V_{CC}, with 15 V_{dc} being typical. This allows the establishment of a switching threshold well above the level of noise that is likely to exist in an industrial environment.

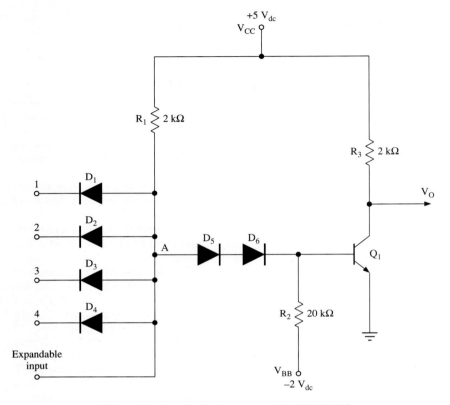

(a) Internal circuitry of a four-input expandable DTL NAND gate

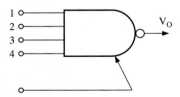

(b) Schematic symbol for a four-input expandable NAND gate

Figure 11–2

This threshold voltage is equivalent to the combined values of V_{BE} for the two transistors and V_Z of the Zener diode minus the forward voltage of the input diodes. Thus, for the logic gate in Figure 11–4(a):

$$V_{TH} = 0.7 \text{ V} + 0.7 \text{ V} + 6.8 \text{ V} - 0.7 \text{ V} = 7.5 \text{ V} \cong 1/2 \, V_{CC}$$

Figure 11–5 clarifies the operating parameters of an HLT gate. Any noise occurring between ground potential and the threshold level of 7.5 has no effect on the operation of the HTL gate. Thus, any input voltage below the 7.5-V level is seen by the gate as a logic low, whereas any pulse exceeding the threshold potential is read as a logic high.

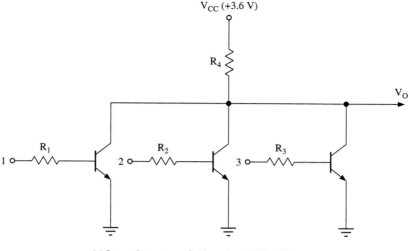

(a) Internal structure of a three-input RTL NOR gate

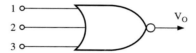

(b) Schematic symbol for three-input NOR gate

Figure 11–3

Although HTL circuitry is able to operate reliably in noisy industrial environments, its requirement for high levels of V_{CC} represents a disadvantage. In recent years, HTL devices have been rapidly replaced by faster, more power efficient logic circuitry.

   Transistor-to-transistor logic (TTL) is, at present, the most widely used form of digital logic within industrial control systems. In the pursuit of ever greater switching speed and power efficiency, standard **7400 series TTL** has been superseded by **low-power Schottky** (LS) TTL in many industrial control circuits. The internal design of a given LS TTL logic gate is essentially the same as that of its standard TTL counterpart. However, the *n-p-n* transistors of the standard device are replaced by Schottky switching transistors, an example of which is shown in Figure 11–6. Here, a high-speed Schottky clamping diode is connected between the base and collector of the *n-p-n* transistor. This greatly improves the switching speed of the transistor, since the problems of saturation and its associated storage time are virtually eliminated. The low-power operation of this improved form of TTL is achieved by increasing the ohmic values of the internal resistors.

   Figure 11–7(a) shows the internal circuitry of a standard two-input TTL NAND gate. The switching circuitry shown here comprises one of the four NAND gates contained within the 7400 quad two-input NAND IC contained in Figure 11–7(b). If the device were LS, the 16 switching transistors contained within the IC would be as shown in Figure 11–6(a) and the IC would be designated as 74LS00.

   The significant structural features of the TTL NAND gate include multiple-emitter inputs and totem-pole outputs. When either of the emitter inputs is brought near ground potential by a logic low, that base-emitter junction becomes forward biased. When this occurs, the base-collector junction of Q_1 becomes reverse biased, interrupting the flow of

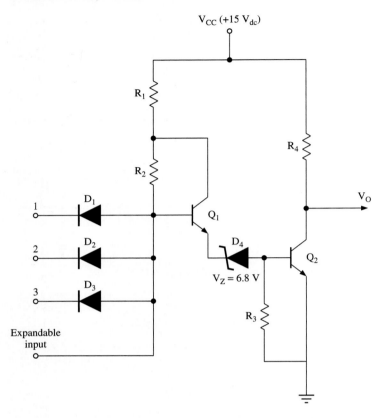

(a) Internal circuitry of a three-input expandable HTL NAND gate

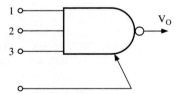

(b) Schematic symbol for a three-input expandable NAND gate

Figure 11–4

base current to Q_2. Since Q_2 is now in a cutoff condition, no base current flows to Q_4, which is also in a cutoff state. However, with R_2 functioning as the base resistance for Q_3, this transistor is biased on, functioning as a current source to an external load. Only when both emitter inputs to the NAND gate are at a logic high is the current flowing through R_1 diverted toward the base of Q_2. As Q_2 switches on, the base-emitter junction of Q_3 becomes reverse biased, resulting in a cutoff condition for Q_3. However, with Q_2 now biased on, a path for base current exists for Q_4, allowing that transistor to function as a current sink for an external load.

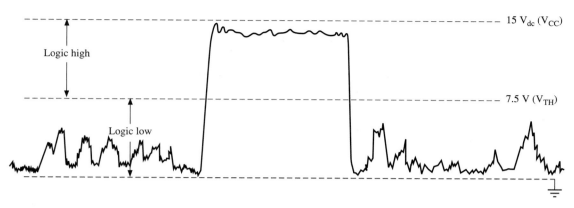

Figure 11–5
Operating parameters of an HTL gate

The totem-pole outputs of TTL gates should never be directly wired together. If the current-sourcing transistor of one of the outputs is forward biased at the instant the current-sinking transistor of another gate is biased on, a virtual short-circuit condition exists that could easily result in the destruction of both logic gates. With **open-collector TTL,** however, such wire ORing of logic outputs is possible. As the term *open-collector* implies, this form of TTL has the collector terminal of an *n-p-n* transistor as its output. To develop an output signal, as shown in Figure 11–8(a), this output must be connected to a load, or **pull-up** resistor. An open-collector logic gate is inherently active-low, due to the current-sinking action of its output transistor. Figure 11–8(b) shows how two open-collector outputs may be wire ORed, sharing a common pull-up resistor. If either transistor is biased on, the output falls low. Only when both base inputs are low does the output become high.

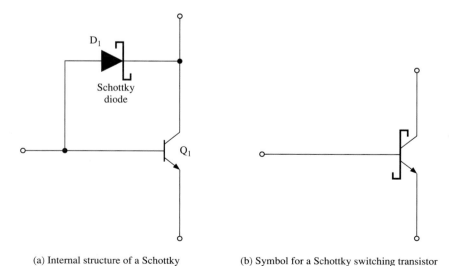

(a) Internal structure of a Schottky
    switching transistor

(b) Symbol for a Schottky switching transistor

Figure 11–6

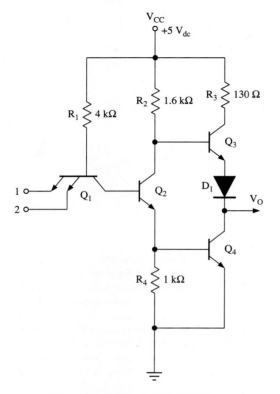

(a) Internal circuitry of a TTL NAND gate

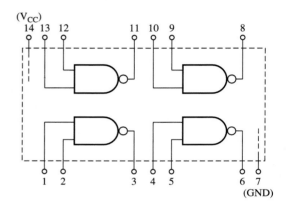

(b) 7400 Quad two-input NAND IC (J- or N-type package)

Figure 11–7

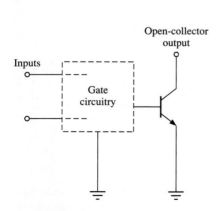

(a) Logic gate with open-collector output

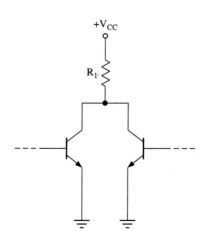

(b) Wire ORing of open-collector outputs

Figure 11–8

Figure 11–9(a) shows how four open-collector inverters could be connected to a common pull-up resistor to form the NOR function represented in 11–9(b). With open-collector logic, the ability to direct-wire outputs together reduces the need for multiple-input gates. However, open-collector logic is susceptible to spurious triggering by high-frequency noise. Thus, care must be taken to keep such noise to a minimum when incorporating open-collector TTL into digital control circuitry. Open-collector TTL is frequently used when interfacing with such switching devices as optocouplers and miniature electromechanical relays.

In large industrial control systems, digital pulses must often travel considerable distances along large wire lengths, often being buffered by optoisolators when passing from one unit of the machinery to another. It is, therefore, difficult to maintain the integrity of rectangular waveforms, since these signals become sluggish due to the stray capacitances within a system. The output pulses of optoisolators are often regenerated by **Schmitt-triggered TTL** gates. (The operation of the basic op amp Schmitt trigger is covered in detail in Section 2–7.) Figure 11–10(a) contains the schematic symbol for a Schmitt-triggered inverter, and (b) and (c) represent the input and output waveforms for this device. During the rise time of the input signal, at the instant its amplitude exceeds the **upper trigger point** (UTP) of the Schmitt-triggered inverter, the output rapidly switches to a logic low. At the point where the instantaneous amplitude of the input signal falls below the **lower trigger point** (LTP) of the Schmitt trigger, the output rapidly switches to a logic high.

Complementary metallic-oxide semiconductor (CMOS) is the most prevalent of the three major divisions of MOS logic devices. The other two divisions include **positive-channel metallic oxide semiconductor** (PMOS) and **negative-channel metallic-oxide semiconductor** (NMOS). Since CMOS logic gates are composed of both PMOS and NMOS switches, these devices are briefly introduced prior to the analysis of a CMOS device.

Figure 11–11 illustrates the operation of a PMOS inverter. Since the device is operating in the enhancement mode, no channel exists as long as the input voltage equals the potential at the source terminal, in this instance 0 V. With this condition, the device is

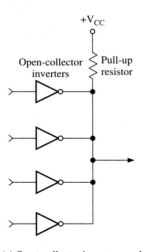

(a) Open-collector inverters used
   to form a NOR gate

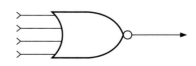

(b) Equivalent logic function

Figure 11–9

Figure 11–10

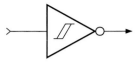

(a) Schematic symbol for a Schmitt-triggered inverter

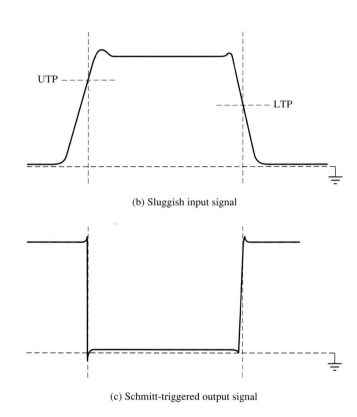

UTP

LTP

(b) Sluggish input signal

(c) Schmitt-triggered output signal

equivalent to an open switch, allowing the output voltage to equal the value of $-V_{DD}$. As the input falls to its negative peak level, the device switches on, allowing the output to rise to nearly ground potential.

Figure 11–12 shows the operation of an NMOS inverting switch. Like its PMOS counterpart, this device operates in the enhancement mode. Thus, while the input is at ground potential, the output nearly equals $+V_{DD}$. As the input voltage switches to its positive peak level, the output switches to nearly 0 V.

Figure 11–13 contains the internal circuitry of a CMOS logic inverter. Note that, with this device, the gates of the PMOS and NMOS switches are connected to a common input terminal. When the input switches to a logic high level of nearly 5 V, the *p*-channel switch (Q_1) is biased off as the *n*-channel device (Q_2) switches on. This causes the output voltage to fall to nearly ground potential. As the input falls to 0 V, Q_1 is biased on while Q_2 functions as an open switch. This causes the output to switch to a logic high. At this point, it should be evident that the operation of the CMOS inverter in Figure 11–13 is

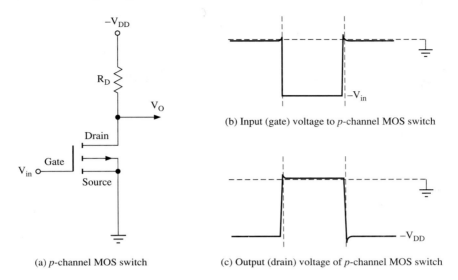

(a) *p*-channel MOS switch

(b) Input (gate) voltage to *p*-channel MOS switch

(c) Output (drain) voltage of *p*-channel MOS switch

Figure 11–11

analogous to that of the bipolar transistor totem pole, with Q_1 functioning as a current source and Q_2 functioning as a current sink.

Compared to TTL, the advantages of the CMOS logic family are as follows. Whereas all forms of TTL must be operated with a V_{CC} of 5 V_{dc}, CMOS operates reliably with a V_{DD} as low as 3 V_{dc} or as high as 15 V_{dc}. Due to their high input impedance, CMOS logic gates are capable of much greater **fan-out** than TTL. As represented in Figure 11–14, a CMOS gate is capable of a fan-out of 50, providing it is outputting to other CMOS

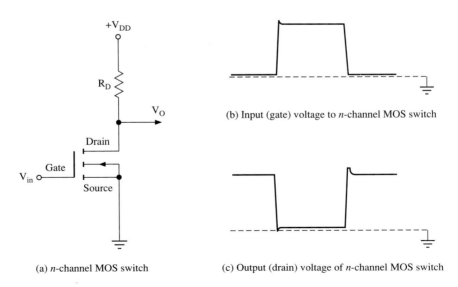

(a) *n*-channel MOS switch

(b) Input (gate) voltage to *n*-channel MOS switch

(c) Output (drain) voltage of *n*-channel MOS switch

Figure 11–12

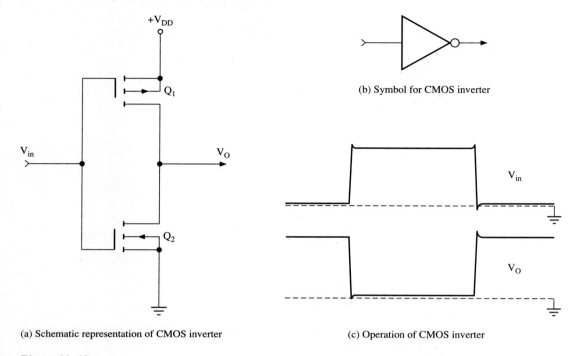

(a) Schematic representation of CMOS inverter

(b) Symbol for CMOS inverter

(c) Operation of CMOS inverter

Figure 11–13

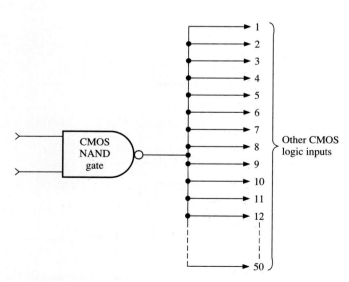

Figure 11–14
Fan-out capability of CMOS logic

devices. Again due to their high input impedance, CMOS devices provide better noise immunity than TTL. Because CMOS devices are voltage operated rather than current operated, they are much more power efficient than even low-power TTL. For this reason, CMOS logic is often preferred for the manufacture of portable, battery-operated equipment.

While the high input impedance of CMOS contributes to its excellent power efficiency and noise immunity, it also has an adverse effect—it is susceptible to damage by **electro-static discharge** (ESD). In recent years, increasing attention has been paid to this problem. Improper or careless handling of circuit boards containing MOS devices can easily lead to impaired performance or total failure of electronic systems. Therefore, many companies involved in the assembly and maintenance of electronic systems now train their technicians in the proper handling and storage of CMOS ICs and circuit boards containing MOS devices.

The basic precautions to be followed when handling CMOSs are summarized as follows. The working environment where CMOS is handled should be made as static free as possible. The chassis ground for all test equipment should be connected to the earth ground, and soldering irons should also be equipped with an earth ground prong. Metal workbenches, as well, should be connected to the earth ground. As a further precaution, in a very dry, static-prone environment, a length of wire should be wrapped around the wrist and connected to an earth-grounded workbench via a 1-MΩ resistor. To prevent damage by ESD, CMOS ICs should be stored in **conductive foam.** If this material is not available, aluminum foil may be used. Styrofoam has become a popular medium for organizing resistors and TTL ICs. However, since this material easily develops static electricity, it should **never** be used for storing CMOS!

Although TTL and CMOS ICs are often found on the same circuit boards, certain design precautions must be adhered to when interfacing between the two logic families. TTL and CMOS devices should never be directly connected within the same circuit. Due to an impedance-matching problem, oscillations may occur at points in a circuit where TTL and CMOS are connected. Also, as illustrated in Figure 11–15, the two logic families have significantly different **noise margins.** Note that TTL has a maximum logic low level of 0.8 V and a minimum logic high level of 2 V. However, a CMOS device operating

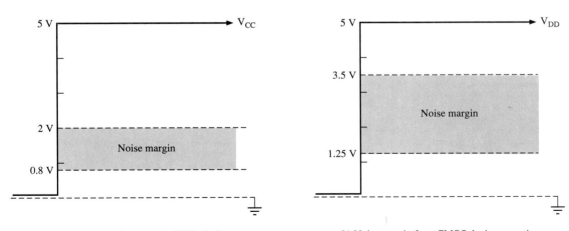

(a) Noise margin for a standard TTL device

(b) Noise margin for a CMOS device operating at the same source voltage as TTL

Figure 11–15

with the same supply voltage has a maximum logic low of 1.25 V and a minimum logic high of 3.5 V.

Figure 11–16(a) shows how, through the use of a transistor switch, the output of a TTL NAND gate may be successfully coupled to CMOS circuitry operating at a higher supply voltage. When both inputs to the TTL NAND gate are at a logic high, the output of the NAND gate falls low. This reverse biases the transistor switch, causing 12 V to be present at the input to the CMOS NAND gate. Since the inputs to this device are tied together, it behaves as an inverter. Thus the input to the CMOS logic circuit is a logic low, equivalent to the TTL output. If either of the inputs to the TTL NAND gate falls to a logic low, a logic high is present at the output of that gate, causing transistor Q_1 to switch on in saturation. With the collector output near 0 V, the output of the CMOS NAND gate becomes nearly 12 V, equivalent to the logic high output of the TTL NAND gate.

If a TTL device is driving CMOS logic with a V_{DD} of 5 V_{dc}, as seen in Figure 11–16(b), the only extra component required is a 10-kΩ pull-up resistor. A separate pull-

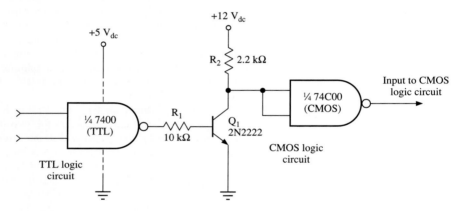

(a) Interfacing between TTL logic circuit and CMOS logic circuit operating with a higher supply voltage

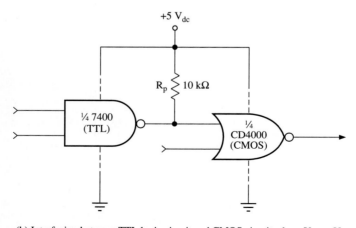

(b) Interfacing between TTL logic circuit and CMOS circuit where $V_{CC} = V_{DD}$

Figure 11–16

up resistor is required for each junction point between the two logic families. Thus, for the sake of conserving space on crowded circuit boards, monolithic pull-up resistor packages are often used at interfacing points between TTL and CMOS.

When inputting from CMOS to TTL, due to the low current-sourcing capability of CMOS, it is necessary to use a current amplifying buffer. Examples of such interfacing circuitry are shown in Figure 11–17. In Figure 11–17(a), a noninverting CMOS buffer provides the boost in current necessary for TTL compatibility. As shown in Figure 11–17(b), an inverting CMOS buffer may also be used for current amplification. In this example, since the CMOS circuit has a V_{DD} of 12 V_{dc}, the CMOS inverting buffer must be connected to the TTL power source.

The evolution of modern microprocessor-based control systems is largely the result of the development of **tri-state logic** (TSL). This form of logic, which is readily available in both TTL and CMOS packages, has allowed the development of **common-bus** organized

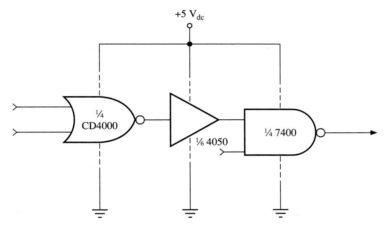

(a) Use of noninverting converter/buffer to drive TTL from a CMOS logic circuit

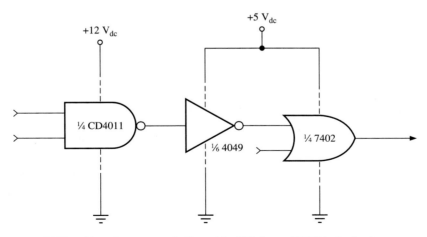

(b) Use of inverting converter/buffer to drive TTL from a CMOS logic circuit

Figure 11–17

digital control systems. Because several tri-state devices may be wire ORed to a common bus line, the need for extremely large OR functions within such circuits is eliminated. Thus efficiency is increased due to the reduced number of ICs required to perform a given task.

Figure 11–18(a) contains the internal circuitry of a TTL NAND gate with a tri-state output. Since the enable, or control input to this device is active low, it may be represented as shown in Figure 11–18(b). The symbol for a tri-state NAND gate with an active-high enable input is represented in Figure 11–18(c). When the enable input in Figure 11–18(a) is low, the output of the internal inverter is at a logic high. Since this condition reverse biases D_1, the device is enabled to function as an ordinary three-input TTL NAND gate. However, if the enable input switches to a logic high, the low output of the inverter holds Q_2 in the off condition while forward biasing D_1. Since Q_2 is reverse biased, Q_4 is also off. If D_1 were not in the circuit, a logic high would occur at the output as Q_3, allowing this transistor to function as a current source for an outside load. However, as D_1 conducts,

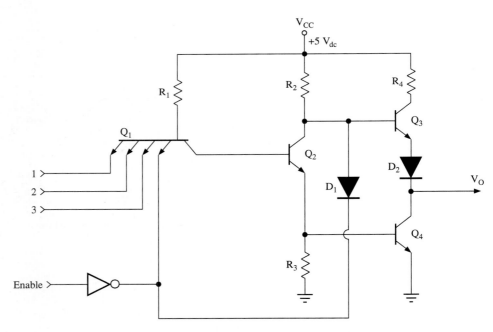

(a) Internal circuitry of three-input TTL NAND gate with tri-state output

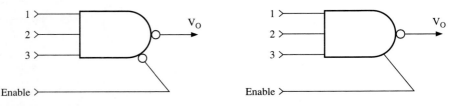

(b) Schematic symbol for three-input NAND gate with active-low enable

(c) NAND gate with active-high enable

Figure 11–18

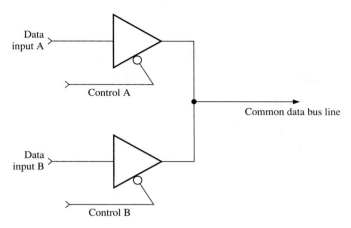

Figure 11–19
Direct coupling of tri-state buffers to a common-bus line

it holds Q_3 in the off state. Thus, neither transistor of the totem pole is on and the gate offers extremely high impedance to any outside circuitry. It should now be evident that the outputs of several gates such as that in Figure 11–18(a) could be directly wired to a common bus line, providing *only one* of the gates is enabled at a time.

Figure 11–19 illustrates how two noninverting tri-state buffers can be connected to a common data bus. If the A control line is low while the B control line is high, the logic level present at data input A passes to the common bus. Because the B buffer is now in a high-impedance condition, it has virtually no effect on the A data. Next, assume the logic conditions of the two enable inputs are reversed. This allows the B data to pass to the bus while the A buffer remains in the high-impedance state. This concept of tri-state controlled data transmission is explained further in the next chapter, where the common-bus microcomputer is introduced.

Two recently developed forms of digital logic are **emitter-coupled logic** (ECL) and **integrated injection logic** (I²L). With logic devices containing bipolar transistors, one of the greatest limitations to switching speed is storage time. As explained in Section 1–3, storage time is that time required for a transistor to recover from a saturated condition. With ECL devices, as with Schottky TTL, transistors are prevented from entering a deep saturated condition. Thus, switching speed is greatly enhanced. The I²L logic family approaches the ideal form of digital switching circuitry. Since it contains no internal resistors, its power dissipation is minimal, being comparable to that of CMOS. It is easily fabricated, with extremely high levels of integration being readily achieved. Like ECL and Schottky TTL, I²L is also capable of high switching speeds.

SECTION REVIEW 11-1

1. Describe the internal operation of the logic gate shown in Figure 11–2. Explain the purpose of the expandable input.
2. Explain the internal operation of the logic gate contained in Figure 11–3. What input conditions must exist to produce a logic high at the output?
3. What is an advantage of the form of logic represented in Figure 11–4? What internal component is responsible for its high switching threshold?
4. Identify two advantages a 74LS00 IC would have over a conventional 7400 package.

5. Describe the structure and operation of a Schottky switching transistor.
6. What logic family is characterized by multiple-emitter inputs and totem-pole outputs?
7. What is an advantage of open-collector TTL?
8. What is the purpose of Schmitt-triggered TTL?
9. Identify two major advantages of the CMOS logic family. What problem is associated with this form of logic?
10. Explain the operation of the tri-state logic gate in Figure 11–18. Describe the condition of its internal switching transistors during each of the three possible conditions for the device.
11. What is the primary advantage of tri-state logic?
12. Identify the main design advantage of ECL. Identify a design advantage of I²L.

11-2   INDUSTRIAL APPLICATIONS OF COMBINATORIAL LOGIC DEVICES

Within industrial control systems, combinations of logic gates are often used to control the transmission of signals from control sources to actuating devices. The specialized mathematics used in the design and optimization of such circuits was first presented by the British scholar George Boole at the middle of the nineteenth century. The important **Boolean laws,** or **postulates,** are identified and briefly explained here. Also included are two important theorems developed by DeMorgan, who was a student of Boole.

Figure 11–20 illustrates the **commutative** and **associative** laws. The first law states that the arrangement of the inputs to an AND function has no effect on its logic output; the second law allows similar flexibility with an OR function. These laws allow the development of multiple-input AND gates and OR gates through **cascading** of smaller gates.

Figure 11–21 demonstrates the **distributive** law, which, like its counterpart in conventional algebra, involves the process of factoring. A simple arithmetic example of factoring would be as follows:

$$3 \times (6 + 2) = (3 \times 6) + (3 \times 2)$$

The equivalent operation in Boolean algebra would be as follows:

$$A \cdot (B + C) = (A \cdot B) + (A \cdot C)$$

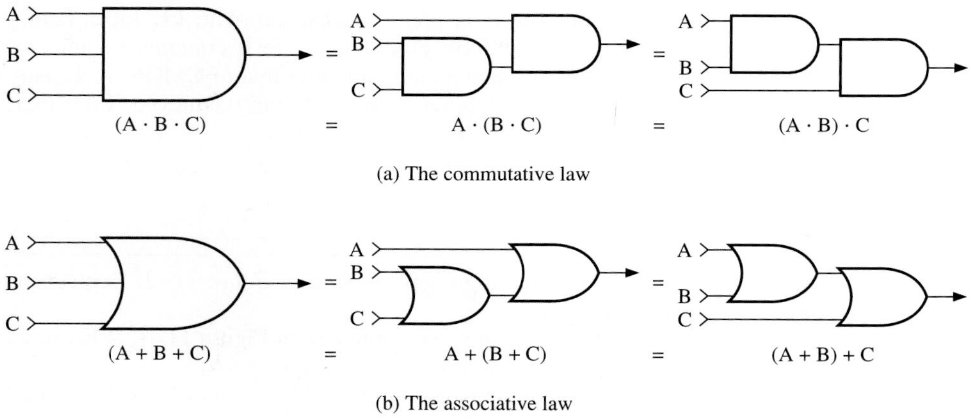

(a) The commutative law

(b) The associative law

Figure 11–20

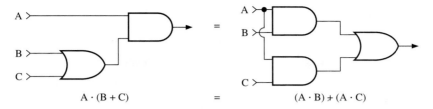

$$A \cdot (B + C) \qquad = \qquad (A \cdot B) + (A \cdot C)$$

Figure 11–21
The distributive law

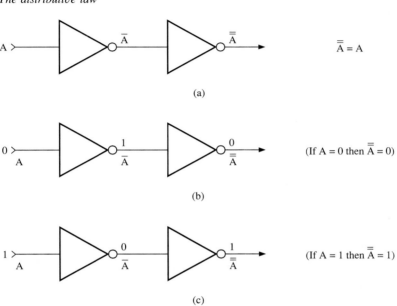

Figure 11–22

In Figure 11–22, the **complementation** laws are illustrated. Although these laws may, at first, appear obvious, they are among the most important rules in the design of combinatorial logic circuits. The complementation laws are as follows:

$$\text{If } A = 1 \text{ then } \overline{A} = 0$$
$$\text{If } A = 0 \text{ then } \overline{A} = 1$$
$$\overline{1} = 0$$
$$\overline{0} = 1$$
$$\overline{\overline{A}} = A$$

The AND and OR laws, like the complementation laws, appear simple. However, they become important in the design of data multiplexing and data transfer circuitry. These laws are listed next and illustrated in Figure 11–23.

$$A \cdot 1 = A$$
$$A \cdot \overline{A} = 0$$

$$A \cdot A = A$$
$$A + 0 = A$$
$$A + 1 = 1$$
$$A + A = A$$
$$A + \overline{A} = 1$$

DeMorgan's theorems are constantly used in the design and implementation of logic circuits. They allow the designer to construct circuits containing noninverting functions (AND and OR gates) from inverting devices. This flexibility results in design simplicity, since it significantly reduces the number of ICs required in a combinatorial logic circuit. DeMorgan's first and second theorems are listed next and illustrated in Figure 11–24:

DeMorgan's first theorem: $\overline{A \cdot B} = \overline{A} + \overline{B}$

DeMorgan's second theorem: $\overline{A + B} = \overline{A} \cdot \overline{B}$

A common application of DeMorgan's first theorem is illustrated in Figure 11–25. Here, three of the four two-input NAND functions available on a single TTL 7400 IC are connected to form a sum of products function. If actual AND and OR gates were used in this application, two ICs would be required, adding to the size and complexity of the circuit.

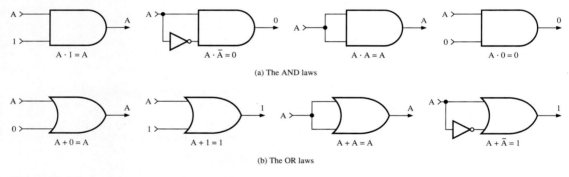

(a) The AND laws

(b) The OR laws

Figure 11–23

Figure 11–24

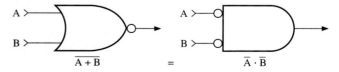

(a) DeMorgan's first theorem

(b) DeMorgan's second theorem

Figure 11–25

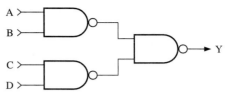

(a) Sum of products function
constructed from 7400 IC

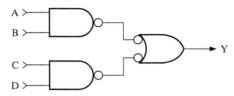

(b) Same logic function with DeMorgan's
first theorem applied to output gate

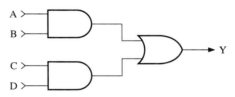

(c) Equivalent logic function after applying
law of double complementation

Where Y represents the output condition, the Boolean expression for the basic NAND circuit in Figure 11–25(a) is as follows:

$$Y = \overline{(A \cdot B) \cdot (C \cdot D)}$$

By applying DeMorgan's first theorem to the output NAND function, the circuit in Figure 11–25(a) may be represented as shown in Figure 11–25(b). The Boolean expression for this altered circuit would be as follows:

$$Y = \overline{(\overline{A \cdot B})} + \overline{(\overline{C \cdot D})}$$

Applying the law of double complementation, the circuit may be simplified to the sum of products statement in Figure 11–25(c), and its Boolean expression may be simplified as follows:

$$Y = (A \cdot B) + (C \cdot D)$$

The following Boolean laws of simplification allow the digital designer to arrive at the simplest possible combinatorial logic circuit for a given control application. These simplification laws are stated next and illustrated in Figure 11–26:

$$(A \cdot B) + (A \cdot \overline{B}) = A$$
$$A + (A \cdot B) = A$$
$$A \cdot (A + B) = A$$

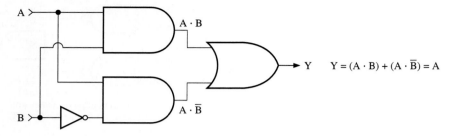

(a) Equivalent circuit for first simplification law

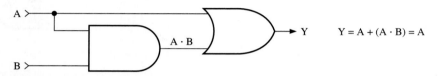

(b) Equivalent circuit for second simplification law

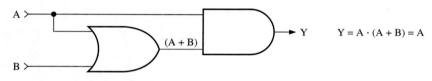

(c) Equivalent circuit for third simplification law

Figure 11–26

Note that the third simplification law is equivalent to the second since

$$A \cdot (A + B) = (A \cdot A) + (A \cdot B) = A + (A \cdot B) = A$$

    A common characteristic of the three simplification laws is that the *B* variable becomes irrelevant. Since *B* has no effect on the logic condition of the output *Y*, all three logic functions in Figure 11–26 serve no purpose. This is clearly indicated by Table 11–1, which is the **truth table** for all three of the preceding logic functions. Note that the output condition is totally dependent on the condition of the *A* variable.

    Once a control application for a combinatorial logic circuit has been defined, digital design engineers use Boolean algebra and DeMorgan's theorems, along with a circuit optimization process such as Karnaugh mapping, to obtain the simplest possible circuit for a given control application.

Example 11–1

As a technician working with an engineering team, you are assigned the task of designing a small combinatorial logic circuit for controlling a solenoid. This solenoid controller is to operate either in the normal mode or in the maintenance mode. When in the normal mode of operation, the solenoid responds to a trigger signal during the time an enable signal is present. During the maintenance mode, the solenoid responds only to a test signal, disregarding the enable and trigger inputs. In the normal mode, however, the

Table 11–1

A	B	Y
0	0	0
0	1	0
1	0	1
1	1	1

solenoid *must not* respond to the test signal. Your task requires the design of the simplest possible solenoid control circuit using the fewest possible inputs.

Solution:

Step 1: After assigning letter designations to the input signals, write the truth table for the circuit as shown by Table 11–2.

Let *A* represent the *normal* signal.
Let *B* represent the signal for *enable* operation.
Let *C* represent the *test* signal.
Let *D* represent the *trigger* signal.

Step 2: Create a Karnaugh map from the truth table of Table 11–2:

	$\overline{A}\,\overline{B}$	$\overline{A}B$	AB	$A\overline{B}$
$\overline{C}\,\overline{D}$				
$\overline{C}D$				1
CD	1	1	1	
$C\overline{D}$	1	1		

Step 3: Derive a Boolean expression from the Karnaugh map. For the group of four products:

$$(\overline{A} \cdot \overline{B} \cdot C \cdot D) + (\overline{A} \cdot \overline{B} \cdot C \cdot \overline{D}) + (\overline{A} \cdot B \cdot C \cdot D) + (\overline{A} \cdot B \cdot C \cdot \overline{D}) = (\overline{A} \cdot C)$$

Table 11–2

A	B	C	D	Y
0	0	0	0	0
0	0	0	1	0
0	0	1	0	1
0	0	1	1	1
0	1	0	0	0
0	1	0	1	0
0	1	1	0	1
0	1	1	1	1
1	0	0	0	0
1	0	0	1	0
1	0	1	0	0
1	0	1	1	0
1	1	0	0	0
1	1	0	1	1
1	1	1	0	0
1	1	1	1	1

For the pair of products:

$$(A \cdot B \cdot \overline{C} \cdot D) + (A \cdot B \cdot C \cdot D) = (A \cdot B \cdot D)$$

Thus, the final Boolean expression becomes:

$$Y = (A \cdot B \cdot D) + (\overline{A} \cdot C)$$

Step 4: Sketch the logic circuit representing the above simplified Boolean expression. [This logic circuit is shown in Figure 11–27(a).]

Step 5: Draw the actual logic circuit using the fewest possible inputs and TTL ICs. [This circuit is shown in Figure 11–27(b).]

Step 6: Use Boolean algebra to prove that the circuit in Figure 11–27(b) performs the equivalent function of the circuit in 11–27(a). Again using variables *A*, *B*, *C*, and *D* to represent the control signals:

$$Y = \overline{\overline{(A \cdot B \cdot D)} \cdot \overline{(A \cdot B \cdot D)} \cdot \overline{(\overline{A} \cdot \overline{A} \cdot C)}}$$

Applying DeMorgan's first theorem:

$$Y = \overline{\overline{(A \cdot B \cdot D)}} + \overline{\overline{(A \cdot B \cdot D)}} + \overline{\overline{(\overline{A} \cdot \overline{A} \cdot C)}}$$

Applying the law of double complementation:

$$Y = (A \cdot B \cdot D) + (A \cdot B \cdot D) + (\overline{A} \cdot \overline{A} \cdot C)$$

Applying the AND and OR laws involving redundancy:

$$Y = (A \cdot B \cdot D) + (\overline{A} \cdot C)$$

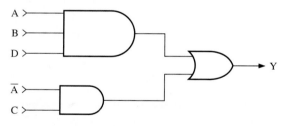

(a) Circuit for simplified Boolean expression

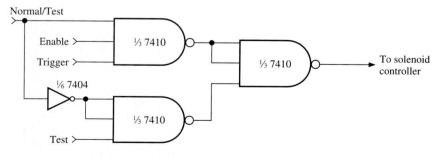

(b) Actual circuit using one 7410 IC and part of an inverter

Figure 11–27

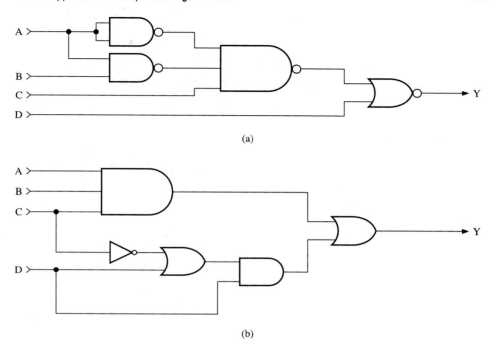

(a)

(b)

Figure 11–28

SECTION REVIEW 11–2

1. Simplify the logic circuit in Figure 11–28(a).
2. Simplify the logic circuit in Figure 11–28(b). Show how the simplified logic function may be implemented if 7400s, 7402s, and 7410s are the only available ICs. If possible, use only one IC.
3. Show how an eight-input AND function could be created using a complete 7420 TTL IC package and one-quarter of a 7402.
4. Show how an eight-input OR function could be created using one-half of a 7420 IC package and a complete 7402.
5. Assume that, under normal operation, the ac motors within a conveyance system may be energized by an active-high *motors on* signal, providing that active-high outputs from four light barriers are all high and an active-low *enable* signal is also present. The motors may also be started by a *motors test* signal, providing that the *enable* signal is low. During the test mode, the condition of the light barrier signals is irrelevant. Design the simplest possible combinatorial logic circuit that will fulfill these conditions. Assume available TTL ICs are 7400, 7402, 7410, and 7420.

11–3   CONTROL APPLICATIONS OF SEQUENTIAL LOGIC DEVICES

Frequency-division and time-delay applications of bistable multivibrators within industrial control circuits were introduced as part of Chapters 6 and 8. In this section of the current chapter, the ability of counters and shift registers to control directly such devices as solenoids and motors is investigated. Figure 11–29 illustrates how a single D flip-flop

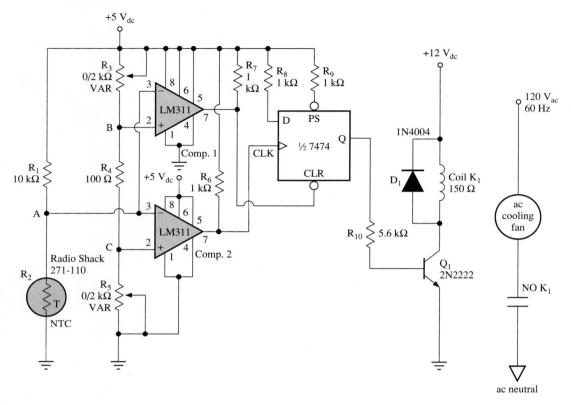

Figure 11–29
Two-state deadband fan control circuit

could be used within a simple two-state (on/off) control system. The Q output of this device controls a transistor switch that governs the operation of a 12-V_{dc} PC board relay. As the relay coil energizes, an ac fan, powered by a small single-phase induction motor, begins to rotate. The purpose of the control circuit is to switch on the fan automatically as the temperature within a piece of industrial equipment attempts to exceed a maximum allowable limit. The fan is allowed to operate until the temperature within the equipment attempts to fall below a level substantially lower than the upper limit. This is done to provide operating stability. With two-state control, it is virtually impossible to maintain a constant environmental temperature. If the fan was turned on at the instant temperature attained a maximum level and turned off at the instant temperature fell below that level, oscillation would result as the fan rapidly switched off and on. Thus, a **hysteresis** or **deadband** is deliberately introduced into the system. As an example, the fan might be turned on as temperature begins to exceed 95°F. The fan could then be allowed to run until temperature decreased toward 85°F.

In Figure 11–29, the deadband is established by the **window detector,** consisting of R_3, R_4, R_5, and the two comparators. To understand the operation of the window detector, which governs the operation of the D flip-flop, it is first necessary to analyze the operation of the two voltage dividers. The first voltage divider consists of R_1 and R_2. Since R_2 is a thermistor with a negative temperature coefficient, located within the machinery being monitored, the voltage at point A decreases as machine temperature increases. The variable voltage at point A serves as the active input to both comparators. The second voltage divider, consisting of R_3, R_4, and R_5, establishes the reference voltages for the comparators. The output of comparator 1 switches to a logic low as the voltage at point A exceeds the reference level at point B. The output of comparator 2 switches to a logic high as the voltage at point A falls below the reference level at point C.

For the D flip-flop in Figure 11–29, only the CLK and CLR inputs are active. Pull-up resistor R_8 holds the D input at a constant logic high and pull-up resistor R_9 disables the PS input. Thus, if a logic high is already present at CLR as a positive transition occurs at CLK, the Q output of the flip-flop switches to a logic high. This action turns on Q_1, energizing the relay coil and switching on the cooling fan. A negative transition at CLR switches the Q output to a low level. As a result, Q_1 turns off and K_1 drops out, switching off the fan. Since a negative-going pulse at the output of comparator 1 turns off the fan, the voltage at point B must represent the lower temperature limit of the deadband. Since a positive going pulse at the output of comparator 2 turns on the fan, the voltage at point C must represent the upper temperature limit of the deadband.

Example 11–2

For the circuit in Figure 11–29, determine the ohmic settings of R_3 and R_5 that are required to establish an upper limit of 95°F and a lower limit of 85°F. Use Table 3–5 (the data sheet for the 271-110 thermistor) as a guide.

Solution:

Step 1: For compatibility with the data sheet, convert the deadband temperatures to degrees Celsius. Using a derivation of formula (3–12):

$$°C_{max} = \frac{(°F - 32°)}{1.8} = \frac{(95° - 32°)}{1.8} = 35°C$$

$$°C_{min} = \frac{85° - 32°}{1.8} = 29.4°C$$

Step 2: Match the corresponding thermistor ohmic values with the deadband temperatures: At 35°C, R_2 equals 6.941 kΩ. Interpolating for 29.4°C, R_2 equals approximately 8.5 kΩ.

Step 3: Determine the upper and lower threshold voltages for the window detector: Since V_B must equal the maximum value of V_A:

$$V_B = \frac{8.5\ k\Omega}{(8.5\ k\Omega + 10\ k\Omega)} \times 5\ V \cong 2.297\ V$$

Since V_C must equal the minimum value of V_A:

$$V_C = \frac{6.941\ k\Omega}{(6.941\ k\Omega + 10\ k\Omega)} \times 5\ V \cong 2.049\ V$$

Step 4: Solve for the current flow through the second voltage divider. Due to the high input impedance of the comparators, assume $I_{R_3} = I_{R_4} = I_{R_5}$.

$$V_{R_4} = V_{R_3} - V_{R_5} = 2.297\ V - 2.049\ V = 248\ mV$$

$$I_{R_4} = \frac{V_{R_4}}{R_4} = \frac{248\ mV}{100\ \Omega} = 2.48\ mA$$

Step 5: Solve for the ohmic setting of R_5:

$$R_5 = \frac{2.049\ V}{2.48\ mA} \cong 826\ \Omega$$

Step 6: Solve for the ohmic setting of R_3.

$$R_3 = \frac{(5 \text{ V} - V_B)}{I_{R_3}} = \frac{(5 \text{ V} - 2.297 \text{ V})}{2.48 \text{ mA}} \cong 1.09 \text{ k}\Omega$$

In more complex industrial control applications, several flip-flops may be required. To understand the operation of such circuits, it is first necessary to review the operation of data registers. Figure 11–30(a) contains a basic **shift register**; part (b) shows a register capable of **parallel** (or **broadside**) loading. With the circuit in Figure 11–30(a), data bits are shifted one position to the right on each clock pulse. This form of device is used extensively in serial data acquisition systems. During four consecutive positive clock transitions, data bits arriving at the serial data input form a 4-bit data word. This information could either be accessed immediately at the Q outputs of the flip-flops or, during four more clock pulses, shifted out of the register through the serial data output. The major advantage of the register in Figure 11–30(b) is its ability to acquire a data word present at the D inputs in a single positive clock transition. A disadvantage of this register is that it requires four data input lines rather than one.

Although the data registers in Figure 11–30 serve to demonstrate the principles of serial and parallel loading, both devices are too simple for direct use in an industrial control system. As shown by the TTL device in Figure 11–31, data registers used in the control of industrial machinery are likely to contain a considerable number of logic gates.

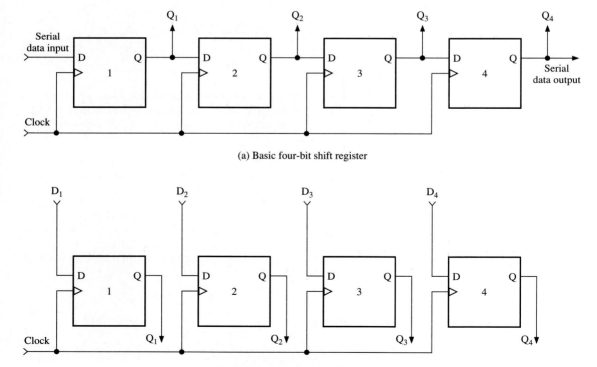

(a) Basic four-bit shift register

(b) Basic four-bit parallel-loading register

Figure 11–30

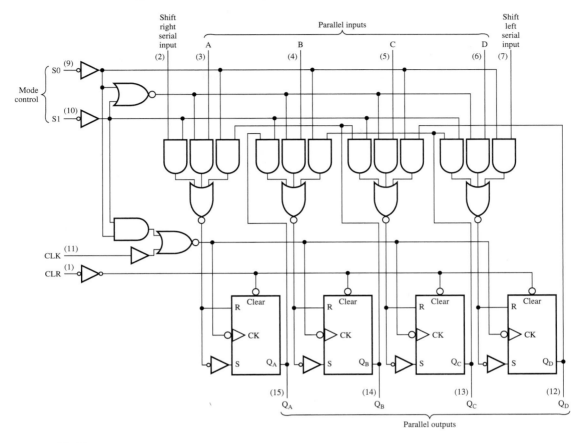

Figure 11–31
Internal circuitry of 74194 4-bit bidirectional universal shift register

The 74194 IC is a TTL **universal shift register.** As the term *universal* implies, the device in Figure 11–31 is capable of loading four data bits in parallel, as well as shifting data to the right or left. This versatility is provided by four data selectors that control the flow of data to each of the four flip-flops. Each data selector consists of three two-input AND functions and a three-input NOR gate. The operation of the data selectors is governed by the control inputs S0 and S1. Careful analysis of the internal design of the 74194 would yield the results in Table 11–3 for the operation of these control inputs.

The operation of the controlling inputs to the 74194 may be summarized as follows. When both the S0 and S1 inputs are low the CLK input is effectively inhibited, allowing the Q outputs to hold their present data. With S0 low and S1 high, the device is enabled to shift data to the left one position on each positive clock transition. If the logic conditions of S0 and S1 are reversed, the device is enabled to shift data right one position per clock pulse. Finally, with both control inputs at a logic high, the device is enabled to load four data bits in parallel on the next positive clock transition. Additional operating flexibility is provided by an asynchronous CLR input that has priority over all other inputs. Whenever CLR is at a logic low, all four Q outputs of the 74194 are low.

The Shift Right Serial Input and Shift Left Serial Input may be used to cascade 74194s, effectively forming an 8-bit universal shift register. In fact, the two ICs may be

Table 11–3

S0	S1	Resultant Operation
0	0	Hold current data
0	1	Shift data left
1	0	Shift data right
1	1	Load data in parallel

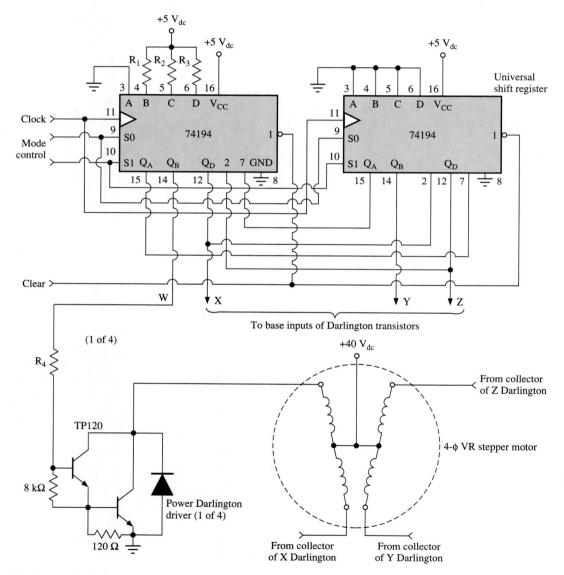

Figure 11–32
Universal shift register controlling a 4-φ VR stepper motor

connected to form a bidirectional **ring counter,** as demonstrated by the stepper motor control circuit in Figure 11–32. A path for left-hand rotation is created by connecting pin 7 of the first IC to pin 15 of the second IC and pin 7 of the second IC to pin 15 of the first IC. A path for right-hand rotation is created by connecting pin 12 of the first IC to pin 2 of the second IC and pin 12 of the second IC to pin 2 of the first IC. Also, the CLK inputs of both ICs are connected to a common clocking source, and the S0 and S1 control lines of both ICs are connected to form two common control lines. Connecting the CLR inputs to a common active-low control line allows all eight Q outputs to be brought instantly to a logic low, regardless of the conditions of S0, S1, and CLK.

The configuration of the two 74194 ICs in Figure 11–32 allows the drive signals at the outputs of the four power Darlingtons to produce a half-wave stepping pattern for controlling a 4-ϕ variable reluctance (VR) stepper motor. Immediately after power is applied to the circuit, system control circuitry must ensure that both mode control inputs are momentarily at a logic high, allowing the registers to load the data present at the parallel inputs on the arrival of the first positive clock transition. For the first 74194, the A data input is tied to a ground, whereas inputs B, C, and D are tied high via pull-up resistors. For the second 74194, all four data inputs are tied to ground. Thus, the register always loads the 01110000 as its initial control variable. If the S0 input is brought low while S1 remains high, this control word is shifted in the left-hand direction through the register, producing the sequence of output conditions shown in Table 11–4. The resultant half-stepping excitation signals are shown in Figure 11–33.

If the logic conditions of S0 and S1 in Figure 11–32 are reversed, the control variable rotates through the register in the right-hand direction. As a result, the direction of rotation for the stepper motor reverses. If the S0 and S1 inputs are both low, since clock pulses to the register are inhibited, the motor stops, holding firmly in its current position. If the register is cleared, power is effectively removed from the motor since all four power Darlingtons are switched off. It should be mentioned that, while the shift register is capable of starting and stopping the stepper motor, as well as controlling its direction of rotation, it does not control rotational speed. For the circuit in Figure 11–32, this may be accomplished only through increasing or decreasing the frequency of the clock source.

Binary up/down counters may also be used in the control of stepper motors. A monolithic device suitable for use in such an application is the **74193 binary up/down counter.** The internal logic circuitry of this device is shown in Figure 11–34. The asynchronous inputs to this device are CLR, which is active-high, and Load, which is active-low. A logic high at the CLR input instantly brings all four Q outputs to a logic low. A logic low at Load allows the logic conditions present at the four data inputs to be instantly latched at the four Q outputs. The 74193 has two clocking inputs, one for counting up

Table 11–4
Stepper motor half-stepping excitation signals

	Q_{A1}	Q_{B1}	Q_{C1}	Q_{D1}	Q_{A2}	Q_{B2}	Q_{C2}	Q_{D2}	W	X	Y	Z
t_0	0	1	1	1	0	0	0	0	1	1	0	0
t_1	0	0	1	1	1	0	0	0	0	1	0	0
t_2	0	0	0	1	1	1	0	0	0	1	1	0
t_3	0	0	0	0	1	1	1	0	0	0	1	0
t_4	0	0	0	0	0	1	1	1	0	0	1	1
t_5	1	0	0	0	0	0	1	1	0	0	0	1
t_6	1	1	0	0	0	0	0	1	1	0	0	1
t_7	1	1	1	0	0	0	0	0	1	0	0	0
t_8	0	1	1	1	0	0	0	0	1	1	0	0
t_9	0	0	1	1	1	0	0	0	0	1	0	0

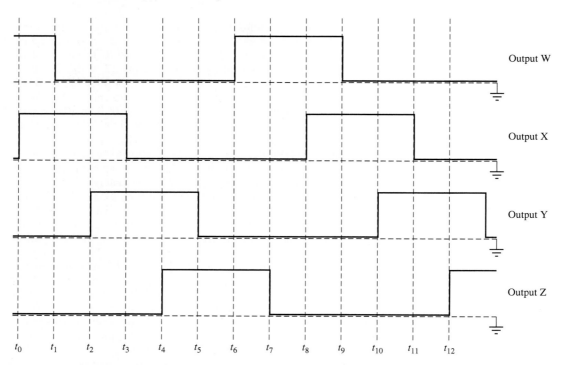

Output W

Output X

Output Y

Output Z

t_0 t_1 t_2 t_3 t_4 t_5 t_6 t_7 t_8 t_9 t_{10} t_{11} t_{12}

Figure 11–33
Half-step excitation waveforms produced by the universal shift register of Figure 11–32

and one for counting down. If pin 4 of the device is held at a logic high, the device is enabled to increment (count up) on each positive transition of a clocking signal at pin 5. With a logic high present at pin 5, the device is enabled to decrement (count down) on each positive transition of a clocking signal at pin 4.

Figure 11–35 illustrates how a 74193 IC, operating in conjunction with a 7400 quad NAND and a 7486 quad XOR, could create dual-phase excitation signals for a 4-ϕ VR stepper motor. Here, a simple combinatorial logic circuit, consisting of four two-input NAND gates, allows the up/down counter to operate from a single, free-running clock source. Providing a logic high is present at the $\overline{\text{Hold}}$ input, a logic high at the direction control input allows the counter to increment on each positive transition at CLK. If the direction control input falls low, with $\overline{\text{Hold}}$ still at a logic high, the counter begins to decrement on each positive clock transition. This capability allows bidirectional rotation of the stepper motor. If the $\overline{\text{Hold}}$ input falls low, the counter holds its present count, effectively inhibiting the clock signal and allowing the stepper motor to remain stationary. A logic high at the CLR input instantly returns the Q outputs of the counter to 0000. Tying pin 11 to a logic high via a pull-up resistor disables the Load capability of the counter, and tying the four DATA input lines to a logic high reduces the possibility of noise interference. With only the Q_A and Q_B outputs of the 74193 being accessed, the counting range is limited to 00_2 through 11_2.

With pin 1 of the 7486 tied low and pins 6 and 8 tied high, the resultant control outputs of the XOR function are as shown in Table 11–5. The equivalent waveforms are contained in Figure 11–36. If the counter is made to decrement, the control sequence for the XOR outputs also reverses, causing the stepper motor to rotate in the opposite direction.

'193, 'L193, 'LS193

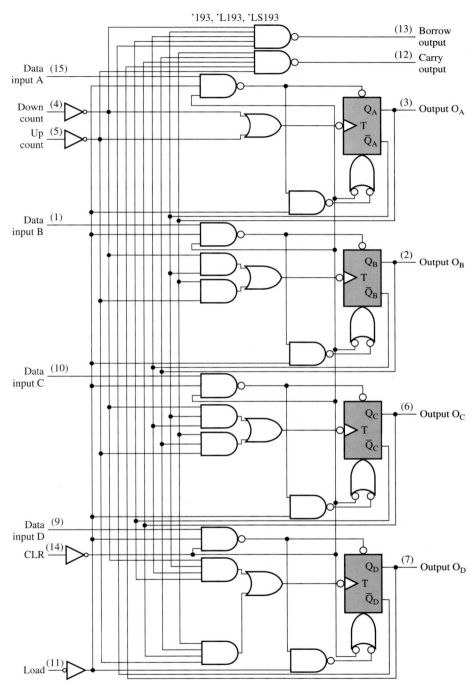

Figure 11–34
Internal circuitry of 74193 synchronous 4-bit up/down counter

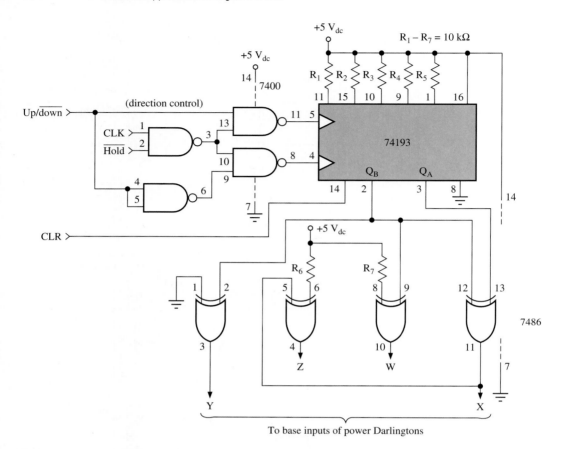

Figure 11–35
74193 and 7486 used to create dual-phase excitation signals for a 4-ϕ VR stepper motor

Table 11–5

	Q_B	Q_A	W	X	Y	Z
t_0	0	0	1	0	0	1
t_1	0	1	1	1	0	0
t_2	1	0	0	1	1	0
t_3	1	1	0	0	1	1
t_4	0	0	1	0	0	1
t_5	0	1	1	1	0	0
t_6	1	0	0	1	1	0
t_7	1	1	0	0	1	1

SECTION REVIEW 11–3

1. For the circuit in Figure 11–29, what must be the ohmic settings of R_3 and R_5 if the deadband for the control circuit is to be 105°F to 90°F?
2. What is an advantage of the basic shift register shown in Figure 11–30(a)?

3. What is the primary advantage of the register shown in Figure 11–30(b)?
4. The 74194 may be enabled to
    a. shift data to the left
    b. shift data to the right
    c. load data in parallel
    d. do all the above
5. What is the condition of the 74194 while both the S0 and the S1 control inputs are at a logic low?
6. For the 74194, what conditions must occur for data present at the parallel inputs to be registered at the Q outputs?
7. For the circuit in Figure 11–32, assume the A data input of the first flip-flop is tied to V_{CC} via a pull-up resistor rather than to dc ground as shown. Using the format shown in Figure 11–33, sketch the resultant waveforms as they would appear at the W, X, Y, and Z outputs.
8. For the 74193, what conditions of the control inputs would allow the device to increment?
9. True or false?: A logic low at pin 14 of the 74193 immediately brings the four Q outputs of that device to a logic low.
10. Assume for the circuit in Figure 11–35 that, with the direction control at a logic low, the stepper motor is turning clockwise. What must be the sequence of control signals that would cause the motor to stop turning for two clock states then start turning in the opposite direction?

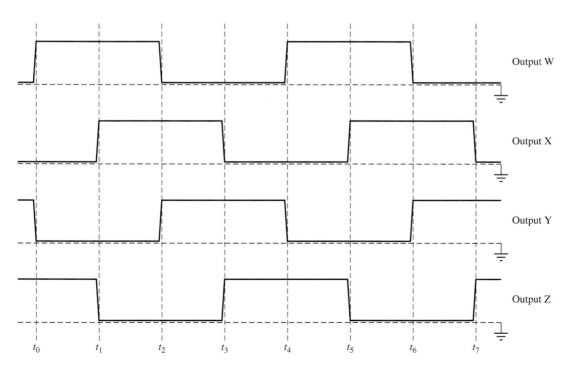

Figure 11–36
Dual-phase excitation waveforms produced by counter and XOR gates in Figure 11–35

11-4   BASIC ON/OFF INTERFACE DEVICES

In this section of the chapter, four basic types of on/off interface devices are covered. These devices, which are readily available in modular form, fall into two basic categories. Input modules are commonly used to inform digital control circuitry of the on/off status

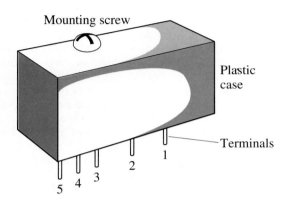

(a) Typical input/output module

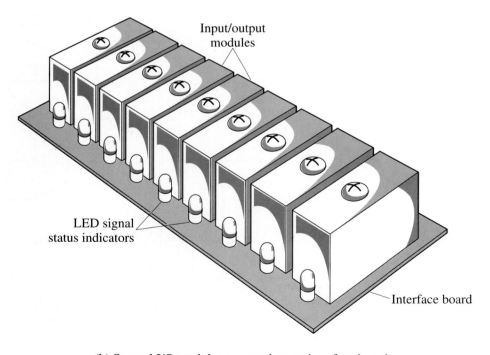

(b) Several I/O modules mounted on an interface board

Figure 11–37

of either ac or dc loads within a piece of industrial equipment. Output modules allow digital control circuitry to energize or de-energize these ac or dc loads. The **input/output modules** (often referred to simply as **I/O modules**) introduced in this section are limited to two-state control applications. They neither provide proportional control to a load or sense proportional changes in a load. In applications where a digital system must provide proportional control, digital-to-analog (D/A) and analog-to-digital (A/D) conversion is necessary. D/A and A/D converters are covered in detail in the last section of this chapter. A representative I/O module is shown in Figure 11–37(a); a typical arrangement of these devices on an interface circuit board is shown in Figure 11–37(b).

Figure 11–38 illustrates the internal circuitry of a **dc input module.** Since no voltage rectification is required, the circuitry for this device is relatively simple. The use of an optoisolator allows a digital controller to receive a status signal from a load that might be several feet away, operating from a separate dc power source. Resistor R_X serves to limit the flow of current to the emitter side of the optoisolator, which consists of an LED. The output of the phototransistor on the detector side is fed to an input voltage hysteresis circuit, which contains a Schmitt trigger. The output of the Schmitt trigger feeds a current-regulating circuit, which controls the flow of base current to a transistor switch that can be either current-sinking or current sourcing. The dc input module is connected such that, when a dc load is energized, current flows through the LED via R_X, thus allowing base current to flow to the output transistor.

Figure 11–39 illustrates how a dc input module could be used to inform a digital control source of the status of a magnetic sensor such as the proxistor illustrated in Figures 4–15, 4–16, and 4–17. (Magnetic sensors are covered in detail in Section 4–3.) As part of a large piece of machinery, such a device is likely to be located several feet away from the digital control circuitry, possibly operating from a separate power source. By monitoring the input signal from the magnetic sensor, a digital control source is able to check indirectly the status of a moving part on a machine.

In Figure 11–39, the output of the magnetic sensor is fed to pin 1 of the dc input module. Thus, as the magnetic sensor detects the proximity of a metallic surface, its

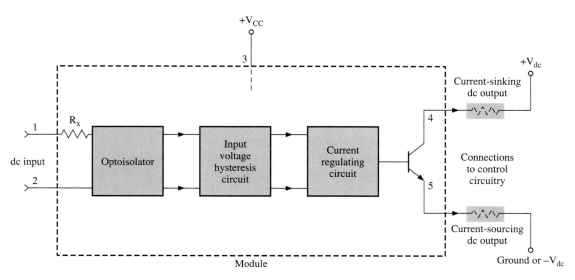

Figure 11–38
Internal structure of a dc input module

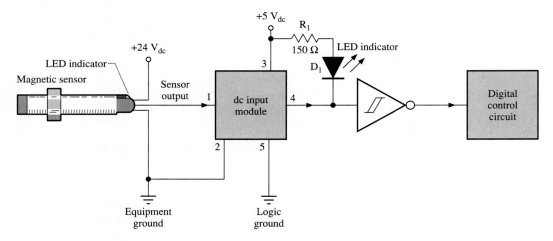

Figure 11–39
Application of dc input module

output toggles to nearly 24 V_{dc}, initiating current flow through R_X and the LED. The module's output transistor is then able to conduct, producing a logic low at pin 4 and forward biasing D_1. Through the action of the Schmitt-triggered inverting buffer, a logic high is sent to the digital control circuit, signaling that a metallic object has approached the magnetic sensor. As the object moves away, the output of the sensor falls low, interrupting the flow of current to the LED inside the input module. As a result, pin 4 of the module toggles to a logic high, causing D_1 to turn off and a logic low to occur at the output of the inverting buffer. Both the miniature LED attached to the magnetic sensor and D_1 are valuable troubleshooting aids. The sensor LED allows the technician to verify that the sensor is operative, while D_1 indicates that the dc module is producing an output signal. If, for example, the sensor LED is on while D_1 is off, the technician should check the voltage at pin 1 of the module. If 24 V is present at pin 1, the technician should replace the module. If 0 V is measured at pin 1, the technician should look for a continuity problem in the cabling between the module and sensor.

Figure 11–40 contains the internal circuitry of an **ac input module.** These devices are used when a digital control circuit must check the status of a load operating from a single-phase ac source. A bridge rectifier within the module converts the incoming ac sine wave to pulsating dc. At the detector side of the optoisolator, this pulsating signal is fed to signal-conditioning circuitry that converts it to a constant dc level for compatibility with the digital control circuitry. The functions of the input voltage hysteresis and current-regulating circuitry are identical to those of the dc input module.

Figure 11–41 illustrates an application of an ac input module. Here, the device is used to inform digital control circuitry of the status of a solenoid within an electronic disk brake assembly. This type of brake is often mounted at the end of a rotating shaft within a piece of machinery, preventing the shaft from turning while power is removed. When the solenoid is de-energized, due to the force exerted by a pressure spring, friction disks within the brake housing press firmly against stationary disks, thus inhibiting shaft motion. When the brake is to be released, the system control circuitry initiates an active low signal that energizes the coil of K_1. As the normally open contacts of K_1 close, current flows through the coil of the brake solenoid, allowing the armature of the solenoid to defeat the pressure spring. Thus the shaft is able to turn freely when driven by a motor.

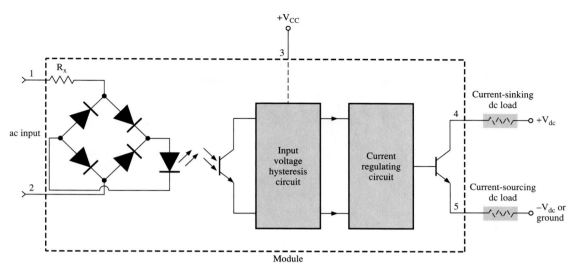

Figure 11–40
Internal structure of an ac input module

While the contacts of K_1 are closed, the ac supply voltage is present at pin 1 of the input module, initiating current flow through its internal bridge rectifier and LED. With the internal phototransistor now conducting, a logic low occurs at pin 4 of the input module. As with the dc input module, this condition forward biases the LED indicator and produces a logic high at the input to the digital control circuit.

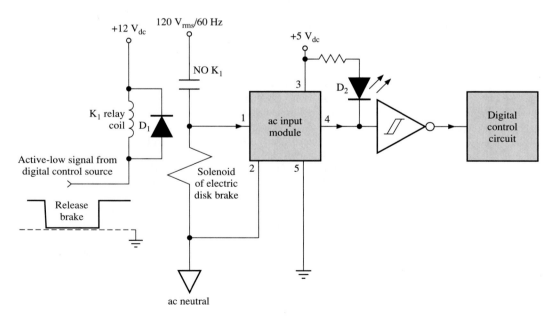

Figure 11–41
Application of ac input module

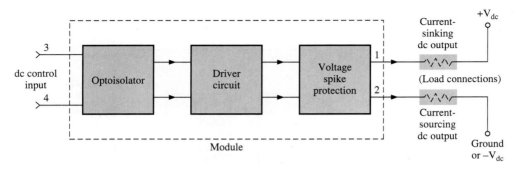

Figure 11–42
Internal structure of a dc output module

When the control circuitry receives a logic high from the output of the Schmitt trigger in Figure 11–41, it assumes the brake has been released. The control circuitry may then safely initiate a signal turning on the drive motor for the rotating shaft. If relay K_1 fails, 0 V is present at pin 1 of the ac input module, causing a logic low to be present at the input to the control circuitry. With this condition, the control circuit withholds the start signal for the motor and possibly initiates a fault message.

Figure 11–42 contains the internal circuitry of a **dc output module.** In this device, the digital control source initiates the flow of current through the emitter side of the internal optoisolator. The phototransistor at the detector side serves as the input to a driver circuit equipped with overvoltage protection at its output. If the input to the module is to be driven by an active-low signal from the control source, pin 3 would be connected to logic V_{CC}, while the control signal is sent to pin 4. If the control signal is active-high, pin 4 is tied to dc ground while the control signal is fed to pin 3.

Figure 11–43 illustrates an application of the dc output module. With reference to Figure 11–41, the circuitry in Figure 11–43 could be serving as the interface between the digital control source and the relay coil. When a logic high occurs at pin 3 of the dc output module in Figure 11–43, the module's output drive transistor is forward biased, serving as a current sink for the relay coil. As pin 1 falls toward ground potential, D_2 becomes forward biased, indicating that a start signal has been initiated for energizing the load.

Figure 11–44 contains the internal circuitry of an **ac output module.** As with the dc output module, the first stage of this device is an optoisolator. The subsequent trigger circuit stage is likely to be a **zero-voltage switch,** similar to the CA3059 illustrated in Figure 8–19. Recall that, with **zero-point switching,** also referred to as **integral-cycle, synchronous switching,** a thyristor is only enabled to turn on at the beginning of an ac source cycle. As explained in Section 8–4, such a switching technique is preferred since it reduces the likelihood of noise propagation within a system and minimizes the possibility of thermal shock to an electromagnetic load. To avoid the possibility of accidental triggering of the output TRIAC by high-voltage noise spikes on the ac line, a **snubber network** is placed in parallel with the output terminals of the module. (The operation of snubber networks is covered in detail in Section 8–3.)

A common application of an ac output module containing a zero-point switch is illustrated in Figure 11–45. Here, a medium power single-phase ac induction motor serves as the load for the module's output TRIAC. Regardless of when the active-high signal from the control source occurs, the TRIAC is not turned on until the beginning of the next ac source cycle. Also, regardless of when the high signal at pin 3 of the module is

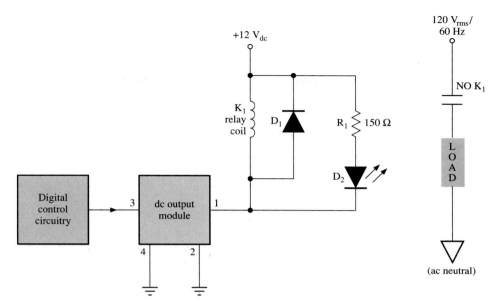

Figure 11–43
Application of dc output module

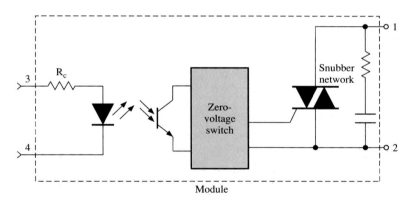

Figure 11–44
Internal structure of an ac output module

removed, the TRIAC does not turn off until the end of the current ac cycle. This action ensures smooth turn on and turn off of the motor and inhibits the propagation of RF noise. While a logic high is present at the output of the digital control source, D_1 is forward biased, indicating the presence of a drive signal for the motor.

SECTION REVIEW 11–4

1. What is the common element contained within the four types of interface module just described? Why is this device so prevalent within interface circuitry?

Figure 11–45
Application of ac output module

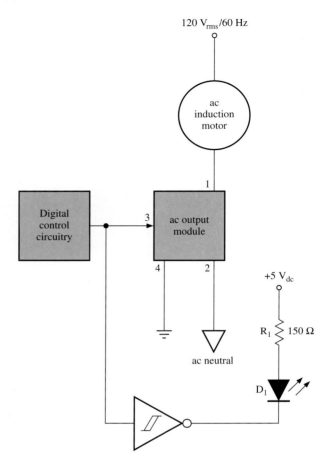

2. Could the device contained in Figure 11–42 be used to energize the armature winding of a conventional dc motor? If so, how could motor rpm be controlled?

3. Why are devices such as that shown in Figure 11–44 likely to contain zero-voltage switches?

4. If necessary, could the device in Figure 11–40 receive a signal from a dc load? Explain the reason for your answer.

5. For the interface circuit in Figure 11–39, assume that, while the LED indicator on the proxistor is lit, D_1 remains off. If 0 V is measured at pin 1 of the input module,
   a. the DC input module is probably defective
   b. the proxistor is almost certainly defective
   c. a continuity problem exists in the signal cable from the proxistor
   d. none of the above

6. For the circuit in Figure 11–41, assume that the solenoid of the disk break is energized by an output module and relay connected as shown in Figure 11–43. When the break is to be released, the LED's in both figures light, indicating that the control circuitry is initiating a signal to energize the solenoid and receiving a signal that the solenoid is connected to the ac line voltage. However, once the drive motor attempts to turn the shaft to which the brake is connected, the motor circuit breaker trips almost immediately. Identify a possible cause of this problem.

11-5   DIGITAL-TO-ANALOG AND ANALOG-TO-DIGITAL CONVERTERS

The simple two-state control devices covered in the previous section are limited to on/off control applications. For example, an ac output module would be suitable for energizing a servomotor controller and a dc input module could be used to inform system control circuitry of the on/off status of that controller. However, since they cannot produce or sense proportional changes in a load, these devices are incapable of controlling or monitoring the operation of the servomotor itself. A digitally based control device such as the modern servomotor controller must contain both **digital-to-analog** and **analog-to-digital conversion** circuitry. In fact, a block diagram of the servomotor controller could be identical to that contained in Figure 11–1. At the output side of the device, the process of digital-to-analog (D/A) conversion could be used in governing motor speed. At the input side, analog-to-digital (A/D) conversion could be used in conditioning signals from such feedback devices as tachometers or resolvers. In this final section of the chapter, D/A converters are investigated first, because D/A converters are contained in some types of A/D converter. The chapter concludes with a survey of the basic forms of A/D converter.

Figure 11–46 contains a simple D/A converter, consisting of a **binary-weighted resistor network,** an op amp summer amplifier, and an op amp level-shifting output

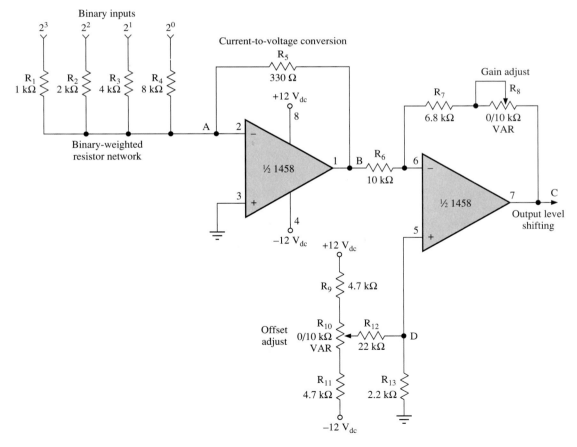

Figure 11–46
Binary-weighted resistor network and dual op amp IC used to form a basic D/A converter

stage. Resistors R_1 through R_4 comprise the binary-weighted resistor network, the purpose of which is to convert the 16 possible binary input numbers (0000 through 1111) to 16 corresponding values of current. This output current is measurable at the summing point A. The function of the first op amp stage is to convert changes in current at point A to voltage changes at point B. Since the voltage at point B develops below ground potential, a second inverting stage is necessary. Besides bringing the output signal to a positive level, this second op amp stage provides the capabilities of offset nulling and gain adjustment.

As the term *binary-weighted* implies, the ohmic values of R_1 through R_4 in Figure 11–46 change by a factor of two. Since R_1 has the lowest ohmic value and thus allows the most current flow, it corresponds with 2^3, the most significant binary input. Since R_4 has the greatest ohmic value, it corresponds with 2^0, the least significant input. Thus, assuming a logic high at the 2^3 input is the same amplitude as a logic high at the 2^0 input,

$$I_{R_1} = 2^3 \times I_{R_4}$$

Considering the ohmic values of R_2 and R_3, this concept may be expanded as follows:

$$I_{R_1} = 2 \times I_{R_2} = 4 \times I_{R_3} = 8 \times I_{R_4}$$

If the voltage level of a logic high at the binary inputs and the voltage level at the virtual ground point are known, the currents within the resistive network may be approximated using Ohm's law. In Figure 11–46, assume that a logic high at the binary inputs equals 3.8 V, and a logic low equals virtually 0 V. Also assume the virtual ground potential at point A is 0.2 V. Given these values, for any of the four resistors:

$$I_N = \frac{(3.8 \text{ V} - 0.2 \text{ V})}{R_N}$$

Once the individual resistor currents have been calculated, the summing current at point A is easily calculated as follows:

$$I_A \cong I_{FB} \cong (I_{R_1} + I_{R_2} + I_{R_3} + I_{R_4})$$

In Figure 11–46, potentiometer R_{10} allows adjustment of the minimum level of output voltage. Normally, with the binary inputs at 0000, R_{10} is adjusted such that the output at point C is at 0 V. With the binary inputs at 1111, R_8 is then adjusted to achieve a desired maximum output potential.

Example 11–3

For the D/A conversion circuit in Figure 11–46, assume that, with the binary input variable at 1111, the voltage at point C is to equal 3 V. What must be the ohmic setting of R_8? (Assume an offset nulling adjustment has already been made.)

Solution:

Step 1: Solve for the individual input currents:

$$I_{R_1} = \frac{(3.8 \text{ V} - 0.2 \text{ V})}{1 \text{ k}\Omega} = 3.6 \text{ mA}$$

$$I_{R_2} = \frac{3.6 \text{ V}}{2 \text{ k}\Omega} = 1.8 \text{ mA}$$

$$I_{R_3} = \frac{3.6\ \text{V}}{4\ \text{k}\Omega} = 900\ \mu\text{A}$$

$$I_{R_4} = \frac{3.6\ \text{V}}{8\ \text{k}\Omega} = 450\ \mu\text{A}$$

Step 2: Solve for the value of summing (feedback) current:

$$I_A \cong I_{FB} \cong 3.6\ \text{mA} + 1.8\ \text{mA} + 900\ \mu\text{A} + 450\ \mu\text{A} \cong 6.75\ \text{mA}$$

Step 3: Solve for the voltage at point B:

$$V_B = -(I_{FB} \times R_5) = -(6.75\ \text{mA} \times 330)\ \Omega = -2.2275\ \text{V}$$

Step 4: Calculate the current flow through R_6, assuming the voltage at pin 6 of the second op amp is 0.2 V:

$$I_{R_6} = \frac{-(-2.2275\ \text{V} - 0.2\ \text{V})}{10\ \text{k}\Omega} = 242.75\ \mu\text{A}$$

Step 5: Calculate the combined ohmic value of R_7 and R_8:

$$(R_7 + R_8) = \frac{(3\ \text{V} - 0.2\ \text{V})}{242.75\ \mu\text{A}} \cong 11.535\ \text{k}\Omega$$

Step 6: Solve for the ohmic setting of R_8:

$$R_8 = 11.535\ \text{k}\Omega - 6.8\ \text{k}\Omega = 4.735\ \text{k}\Omega$$

If a binary up-count occurs at the inputs to the D/A converter in Figure 11–46, assuming the circuit is adjusted in accordance with Example 11–3, its ideal output would appear as shown in Figure 11–47(a). (This pattern may be viewed on the display of an oscilloscope if the inputs to the circuit in Figure 11–46 are connected to the Q outputs of a 4-bit binary up-counter.)

Three important parameters for a D/A converter are **monotonicity, linearity,** and **resolution.** A D/A converter is said to have monotonicity if all the steps in the ramp signal at its output progress in the same direction, as illustrated in Figure 11–47(a). An output signal displaying a lack of monotonicity is illustrated by the detail in Figure 11–47(b). If a D/A converter has perfect linearity, the tips of the steps in its output ramp signal would form a straight line, indicating that all the steps are of equal amplitude. A detail of an output signal with a nonlinear step progression is shown in Figure 11–47(c). The sensitivity and accuracy of a D/A converter are functions of its resolution. The resolution of a D/A converter is determined by its number of binary inputs. Since the highest count attainable at the input to the D/A in Figure 11–46 is 1111, the resolution of the circuit is only 15. Thus, accuracy is equivalent to only 6.67% or 1 part per 15. This degree of resolution is unsatisfactory for most industrial control applications. However, resolution improves significantly as the number of binary inputs is increased.

Example 11–4

A D/A converter has 8 binary inputs. What is its resolution expressed as a percentage? If the maximum output voltage of the D/A converter is to be 12 V, what is the minimum difference in voltage that the device could produce?

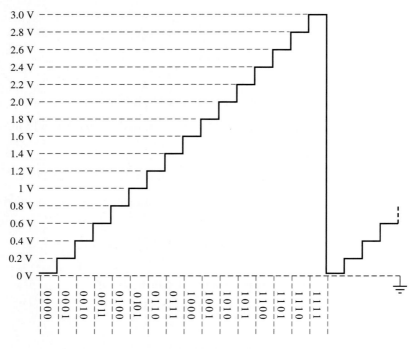

(a) Ideal output of four-bit D/A converter (response to binary up-count)

(b) Detail of D/A output showing
lack of monotonicity

(c) Detail of D/A output showing
nonlinearity

Figure 11–47

Solution:

Step 1: Determine the highest input count. Since the highest input variable equals 11111111_2, the maximum count is 255_{10}. Thus, resolution equals one part per 255.
Step 2: Determine percentage of resolution:

$$\text{Resolution} = \frac{1}{255} \times 100\% = 0.39\%$$

Step 3: Determine the minimum difference in output voltage:

$$\Delta V_{min} = \frac{12 \text{ V}}{255} = 47 \text{ mV}$$

In fabricating D/A converters with higher resolutions, especially those devices manufactured in monolithic form, the concept of the binary-weighted resistor network becomes unwieldy. This is due to the large differences in resistive values that must be created within the IC. As an example, for a 10-bit system to have a 1-kΩ resistor at its most significant input, it must provide a 512-kΩ resistor at its least significant input! A more practical solution to this fabrication problem has been found in a device called the **R/2R ladder,** which is utilized within many monolithic D/A converters. As its name implies, this resistive network requires only two values of resistance, regardless of the number of binary inputs to the D/A converter.

A 4-bit model of an R/2R ladder is shown in Figure 11–48(a). As becomes evident in Figures 11–48(b) and (c), the operation of this device is far more complex than that of the binary-weighted resistor network. For each of the 16 possible input combinations to the network, there is a unique equivalent circuit condition. As an example, consider the condition represented in Figure 11–48(b), which occurs when the binary inputs form 0110 (equivalent to decimal 6). As represented in (b), the logic highs present at inputs 2^1 and 2^2 may be represented as a common positive voltage source, whereas the logic lows at the 2^0 and 2^3 inputs may be shown as equivalent ground points. In this example, it is assumed that R equals 50 kΩ, 2R equals 100 kΩ, and a logic high is 5 V. Since the output of the R/2R ladder serves as a current source to the virtual ground point of an op amp, the I_{out} point is considered to be pulled toward ground potential. Also note that, with both sides of R_8 clamped at virtual ground potential, this component is effectively removed from the circuit.

With R_5 now functioning as a bridge between nodes B and C, the network condition in Figure 11–48(b) may not be approached directly as a conventional compound (series/ parallel) circuit. Using a network theorem, the delta (Δ) configuration consisting of R_4, R_5, and R_6 may first be converted to its equivalent wye (Y) configuration. First, the sum of these three resistors (designated as Σ) is determined:

$$\Sigma = R_4 + R_5 + R_6 = 100 \text{ k}\Omega + 50 \text{ k}\Omega + 100 \text{ k}\Omega = 250 \text{ k}\Omega$$

The R_a, R_b, and R_c of the equivalent wye configuration of Figure 11–48(b) are determined as follows:

$$R_a = \frac{(R_4 \times R_6)}{\Sigma} = 40 \text{ k}\Omega$$

$$R_b = \frac{(R_4 \times R_5)}{\Sigma} = 20 \text{ k}\Omega$$

$$R_c = \frac{(R_5 \times R_6)}{\Sigma} = 20 \text{ k}\Omega$$

We can now calculate total resistance for the network condition. After calculating the resistance from the imaginary node X of the wye configuration toward the equivalent ground points, total network resistance is determined as follows:

$$\frac{(R_1 \times R_2)}{(R_1 + R_2)} + R_3 + R_b = 120 \text{ k}\Omega$$

$$R_c + R_7 = 70 \text{ k}\Omega$$

$$R_T = \frac{(120 \text{ k}\Omega \times 70 \text{ k}\Omega)}{(120 \text{ k}\Omega + 70 \text{ k}\Omega)} + 40 \text{ k}\Omega = 84.2105 \text{ k}\Omega$$

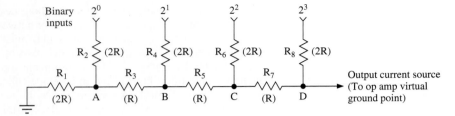

(a) Four-bit R/2R ladder network

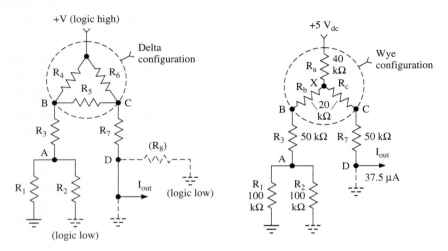

(b) Equivalent network condition for binary input of 0110

(c) Reduction of condition at 0110 to compound circuit using network theorem

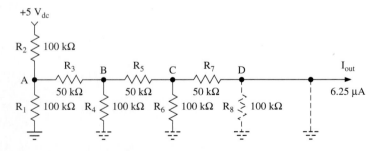

(d) Equivalent network condition for binary 0001

Figure 11–48

Total current for the network can now be calculated:

$$I_T = \frac{5 \text{ V}}{84.2105 \text{ k}\Omega} = 59.375 \text{ μA}$$

The voltage present at imaginary node X must then be determined:

$$I_T \times R_a = 59.375 \text{ μA} \times 40 \text{ k}\Omega = 2.375 \text{ V}$$
$$V_X = 5 \text{ V} - 2.375 \text{ V} = 2.625 \text{ V}$$

Finally, the network output current is calculated:

$$I_{out} = \frac{2.625 \text{ V}}{70 \text{ k}\Omega} = 37.5 \text{ μA}$$

Figure 11–48(d) shows the equivalent circuit condition for the R/2R ladder when binary 0001 is present at the inputs. Analysis of the circuit would prove the output current to be 6.25 μA, exactly one-sixth the current calculated in the preceding example, where the input variable was assumed to be 0110. Analysis of the 13 remaining network conditions would prove that, with each increment of the binary inputs, current increases by 6.25 μA.

A monolithic D/A converter suitable for use in industrial control systems contains components that are not contained in the introductory example shown in Figure 11–46. For most control applications, the D/A converter must accommodate at least six binary inputs. Therefore, the resistive network within the device is likely to consist of a diffused R/2R ladder network. As shown in Figures 11–48(c) and (d), the ohmic values within the resistive network are usually made large enough to avoid errors resulting from temperature effects. The binary inputs are not connected directly to the resistive network. Instead, these inputs are fed to either current or voltage switches. These switches, consisting of either FET or bipolar technology, must offer high impedance to the digital source. They are usually designed for compatibility with major logic families, including TTL and CMOS. Functioning effectively as single-pole, double-throw devices, in response to active logic levels at the binary inputs, the switches connect the inputs of the resistive network to a precision reference voltage source. This reference voltage source, along with the output op amp, can be contained within the monolithic D/A converter package.

A monolithic D/A converter identical to that just described is the **DAC-01 6-bit voltage-output D/A converter** (manufactured by PMI). Because of its accuracy, high speed, and low power consumption, this device is widely used in digitally programmed power supplies, digital filters, pulse generators, and servo-positioning control systems. The internal circuitry of the DAC-01 is represented in Figure 11–49.

The DAC-01 accommodates six active-low (complementary binary) inputs. Thus, when operating in the unipolar summing mode, if all six of the binary inputs are at a logic high, the analog output becomes 0 V. Adjustment of the 500-Ω potentiometer connected between the −15-V supply and pin 14, as shown in Figure 11–50(a), allows for a full-scale output voltage of 10 V with all six binary inputs at a logic low. Thus, 64 gradations of output voltage could be achieved between 0 and 10 V. Actual resolution would be 1 part per 63, or 1.587%. In the bipolar summing mode, 64 gradations in output voltage may be achieved within a ±5 or ±10-V range. Shorting pin 10 to pin 11 results in the 0- to 10-V range for unipolar operation or the ±5-V range for bipolar mode. Leaving pin 10 unconnected allows a ±10-V range for bipolar operation. Refer to the diagram of the DAC-01 in Figure 11–49 to understand why this is so. The feedback path for the internal op amp consists of two 5.4-kΩ resistors, with a node between these two resistors being accessible at pin 10. Thus, since the first feedback resistor is shorted, connecting pin 10 to pin 11 reduces the op amp closed-loop gain by half. To achieve

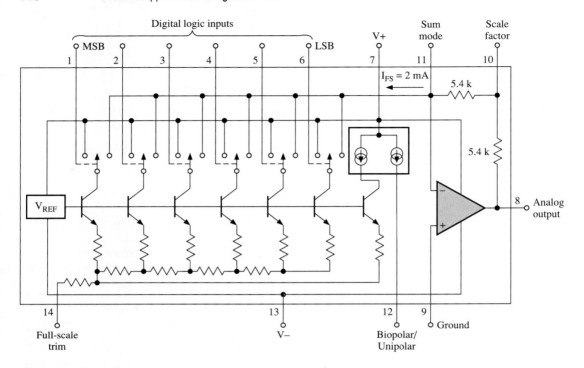

Figure 11–49
Simplified internal circuitry of DAC-01 D/A converter (Courtesy of Analog Devices, Inc., used
with permission.)

bipolar operation (in complementary offset binary), pin 12 is bridged to pin 11, and a
100-kΩ potentiometer and 470-kΩ resistor are connected as shown in Figure 10–50(b).

In the remainder of this section, the main types of A/D converter are introduced.
The simplest and fastest form of A/D converter is the **flash-comparator network.** This
device was introduced in Section 4–4, where it functioned as part of the signal conditioning
circuitry for a capacitive level-sensing transducer. The flash-comparator circuitry is found
within the larger control circuit in Figure 4–27. The actual flash comparator consists of
R_1 through R_5, which form a voltage divider, and the four comparators. Since the operation
of this device is thoroughly explained in Chapter Four, we will only briefly review it at
this juncture. The voltage divider provides reference voltages to the four comparators,
representing 25%, 50%, 75%, and 100% fluctuation of the analog input. Since the input
voltage is fed directly to the noninverting inputs of the four comparators, each comparator
is able to detect when the input fluctuates above or below its reference voltage level.
Thus, as the analog input increases from 0 V toward its maximum level, the comparators
produce the four outputs 0000, 1000, 1100, 1110, and 1111. (As illustrated in Figure 4–
28, for display purposes, the three XOR gates translate the comparator outputs to 0000,
1000, 0100, 0010, and 0001.)

Limitations of the flash-comparator network in Figure 4–28 should now be obvious.
First, although its output is a 4-bit variable compatible with digital circuitry, it is not a
true binary code. While there are 16 possible conditions of a 4-bit binary number, there
are only 5 possible output conditions for the 4-bit flash-comparator circuit. To produce
16 possible output conditions, the flash-comparator network would require 16 voltage-
divider resistors, 15 comparators, and 15 output lines! Thus, while suitable for simple

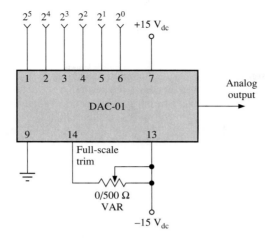

(a) Connection of DAC-01 for 0- to 10-V unipolar output deflection

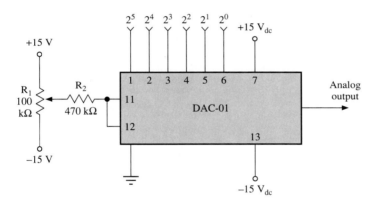

(b) Connection of DAC-01 for −10- to +10-V bipolar output deflection

Figure 11–50

display applications such as that shown in Figure 4–27, the flash-comparator A/D converter concept is usually abandoned for actual A/D conversion involving true binary numbers.

The advantage of the flash-comparator A/D converter is its speed. For the circuitry in Figure 4–27, the response time of the comparators and logic gates to fluctuations in the analog input voltage is likely to be only a few nanoseconds. However, for A/D conversion systems that derive actual binary or BCD numbers from an analog input voltage, significant conversion time is required. Typically with such systems, a sample of the analog input is taken by a device called a **sample-and-hold gate** or simply a **sampling gate.** This device holds the sample of the analog signal at a constant level while the conversion process takes place. At the end of the conversion cycle, the A/D converter usually feeds the digital result to a data latch, where it is held until being received by the control circuitry. After sampling gates are introduced, the **single-slope** and **dual-slope** models of A/D converter, devices that require significant processing time, are introduced. The chapter concludes with a study of **successive-approximation** and **continuous-tracking** A/D converters, devices that operate at relatively high speeds.

Figure 11–51(a) illustrates the basic concept of the sample-and-hold module. Such devices are necessary when a rapidly changing analog input signal is being sampled by an A/D converter with a relatively long processing time. The ideal operation of the sample-and-hold device is best understood by considering the condition where S_1 is momentarily closed, allowing C_1 to charge immediately to the instantaneous amplitude of the input. After S_1 opens, the sample of analog voltage, obtained while S_1 was closed, is held by the capacitor indefinitely, regardless of rapid changes in amplitude at the analog input. This ideal operation of the sample-and-hold module is based on the assumption that, with S_1 closed, virtually no resistance exists between the analog input and the capacitor. Also, the input impedance to the A/D conversion circuitry is assumed to approach infinity.

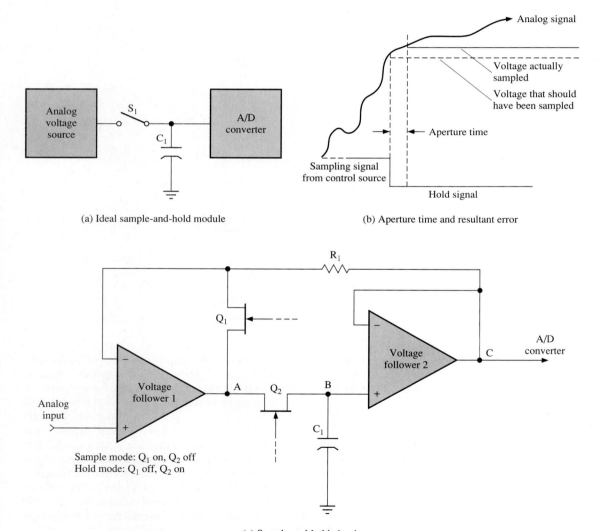

(a) Ideal sample-and-hold module

(b) Aperture time and resultant error

(c) Sample-and-hold circuit

Figure 11–51

Considerations involving the operation of the real-world sample-and-hold module are as follows. First, there is a slight time delay between the instant a digital control source initiates a command to sample the analog input and the time at which this action takes place. This delay is referred to as **aperture time.** The concepts of aperture time and its resultant error are represented graphically in Figure 11–51(b). After the sampling switch is closed, a minimum time is required for the module to obtain a valid sample of the analog input. This period is referred to as the **acquisition time.** A major factor in determining acquisition time is the resistance of the analog source. To achieve a high degree of accuracy (within 0.01%), the acquisition time for the sample-and-hold module must be at least $(9 \times \tau)$, where τ is the time constant or product of the source resistance R_S and the value of the sampling capacitor C_1. A source of error lies in the fact that, during the acquisition process, the analog input voltage may be changing. Thus, for the greatest possible accuracy, input impedance to the sample-and-hold switch must be kept as low as possible, while aperture and acquisition time are kept at a minimum.

Figure 11–51(c) illustrates a sample-and-hold gate that approaches nearly ideal operation. In this circuit, the first voltage follower provides impedance matching with the analog source. During the sampling time, Q_1 is biased on while Q_2 is off. This allows the voltage present at point A to virtually equal that present at the analog input. Thus, the time required for the development of a valid input voltage at point A is determined by the slew rate of the op amp. At the instant a sample of the analog input voltage is to be taken, Q_1 switches off as Q_2 turns on. This action effectively removes the analog source from point A while allowing the sampled input voltage to be present at points B and C. The discharge path for C_1 becomes the noninverting input of the second voltage follower. Since this point represents a few megohms of impedance, the voltage at point B remains virtually constant during the sampling cycle of the A/D converter.

Figure 11–52(a) contains a block diagram of a single-slope A/D converter. Along with its timing and control circuitry, a single-slope A/D converter contains a linear ramp source and counter. The ramp source within the single slope could be purely analog, involving a constant current source and capacitor. (An example of this form of pulse circuit is the linear ramp generator contained in the pulse modulation section of Figure 8–17.) In a monolithic single-slope A/D converter, however, the ramp source is likely to be a D/A converter. Remember that, as shown by the step waveform in Figure 11–47(a), a D/A converter connected to a binary counter is capable of producing a signal approximating a linear ramp. The greater the number of binary inputs to the D/A converter, the smoother the ramp output. For example, if an 8-bit counter is connected to a D/A converter with eight inputs, the output ramp signal has 255 steps.

Regardless of the internal structure of the single-slope A/D converter, the basic operation of this device is similar to that shown in Figure 11–52(b). At t_0, the control circuitry simultaneously resets the ramp source and the counter. Thus, the initial voltage for the ramp source is always 0 V and the starting count is always 000 . . . . 0. When the sample-and-hold module switches to the Hold mode, a sample of the analog input is held at the noninverting input of the comparator in Figure 11–52(a). This steady dc level serves as a reference voltage to which the increasing ramp amplitude is compared. Between t_1 and t_2, while the sample voltage is higher than the ramp potential, the output of the comparator, which functions as the Clock Enable input to the AND gate, is high. Thus, pulses from the clock source located within the control circuitry are able to pass through the AND gate, incrementing the counter. At t_2, as the ramp potential begins to exceed the sample voltage, the output of the comparator falls low, interrupting the flow of clock pulses to the counter. Thus, the counter holds its highest count. On the first positive clock transition after the strobe pulse at t_3, the binary number being held by the counter is fed to the data latch. After the data latch receives the contents of the counter, the ramp generator and counter are reset until another conversion cycle begins.

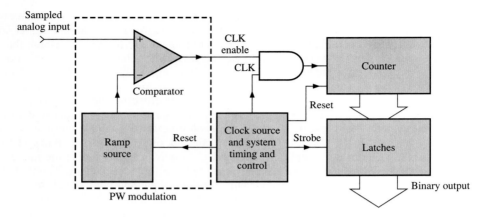

(a) Block diagram of a single-slope A/D converter

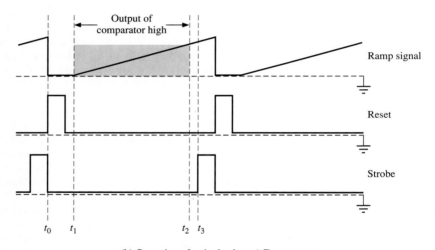

(b) Operation of a single-slope A/D converter

Figure 11–52

At this point, it should be evident that the comparator and ramp source in Figure 11–52(a) together comprise a pulse width modulator. As the analog input voltage increases, the pulse width of the Clock Enable signal increases, allowing the counter to increment to a higher binary number. As the analog input decreases in amplitude, the pulse width of the Clock Enable signal is reduced, causing the counter to increment to a lower binary number. As illustrated in Figure 11–52(b), due to the linearity of the ramp waveform, the following mathematical relationships exist:

$$\Delta \text{ Binary count} \propto \Delta\text{PW}_{\text{Clock Enable}} \propto \Delta V_{\text{analog input}}$$

Example 11–5

Assume for an 8-bit A/D converter that an input voltage of 2.36 V produces a binary output of 10101011. If the analog input voltage decreases to 1.87 V, what is the new binary output of the device?

Solution:

Step 1: Determine the amount of change in the analog input:

$$\Delta V = 2.36 \text{ V} - 1.87 \text{ V} = 0.49 \text{ V}$$

Step 2: Determine the degree (or percentage) of change in analog input:

$$\frac{0.49 \text{ V}}{2.36 \text{ V}} = 0.20763 = 20.763\%$$

Step 3: Convert the original binary value to decimal:

$$10101011_{(2)} = 171_{(10)}$$

Step 4: Determine the degree of change in the output variable:

$$\Delta V = 0.20763 \times 171 = 35.5$$

Step 5: Determine the decimal value of the new output.

$$171 - 35.5 \cong 135.5$$

Step 6: Convert the decimal value to binary.

$$135.5_{(10)} \cong 10000111_{(2)}$$

Figure 11–53 illustrates how the design concept of the single-slope A/D converter might be implemented. Here, a 6-bit R/2R ladder network, driven by the binary counter, functions as the D/A converter/ramp generator portion of the device. The counter that drives the R/2R ladder also functions as the principal component of the A/D converter. This dual role of a single counter results in design efficiency, especially for monolithic devices. Also, since ramp generation is accomplished through D/A conversion rather than integration, no external RC components are required.

The key to understanding the operation of the circuit in Figure 11–53 is remembering that, according to Kirchhoff's current law, the algebraic sum of the currents entering and leaving node A must equal 0 A. Since, prior to the point where the sample-and-hold module switches into the Hold mode, the counter is reset, all six of the solid-state switches are biased off. With this condition, the flow of current to the R/2 ladder is nearly 0 A. Thus, as the sample-and-hold module switches to the Hold mode, the op amp comparator feedback current (I_{FB}) virtually equals the input current (I_{in}). At this time I_{FB} flows from point A to point B through the forward-biased Zener diode. Because the sampled analog voltage remains at a steady amplitude, the flow of I_{in} remains constant. Since the output of the comparator has slewed in the negative direction, transistor switch Q_1 becomes reverse biased, causing a logic high to be present at the Enable input to the NAND gate. This condition allows inverted clock pulses to pass to the binary counter. As the counter begins to increment, I_{RN} increases in a linear step progression. Again complying with Kirchhoff's current law, since I_{in} does not vary, I_{FB} must decrease in amplitude as I_{RN} increases. Eventually a point is attained where I_{RN} begins to exceed I_{in}. Still adhering to Kirchhoff's current law, I_{FB} must change polarity, now flowing from point B into point A. This action pulls the output of the comparator in the positive direction, switching on Q_1, and interrupting the flow of clock pulses to the counter.

As with the single-slope A/D converter presented in Figure 11–52(a), the highest binary number attained by the counter in Figure 11–53 may be stored in a data latch before the counter resets. The operation of the circuit in Figure 11–53 is represented in

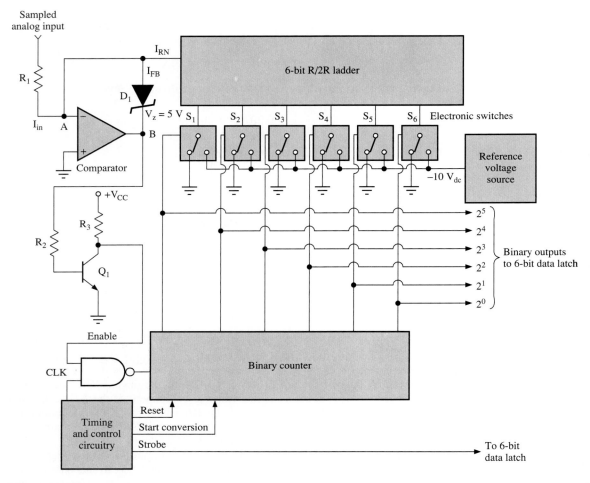

Figure 11–53
Six-bit R/2R ladder and binary counter used to form the D/A converter ramp generator portion
of a single-slope A/D conversion system

Figure 11–54. As shown in the second conversion cycle, if the sample of input voltage is higher than the first, the amplitude of the constant input current increases. As a result, the initial value of feedback current is higher. Thus, more time is required for the feedback current to decrease toward the zero point and the counter is able to attain a higher binary value. As long as the output of the R/2R ladder exhibits monotonicity and linearity, changes in the value of the binary output are approximately proportional changes in input amplitude. However, since the output of the R/2R ladder consists of 65 steps rather than a continuous ramp signal, the output resolution is limited to 1 part in 65 or nearly 1.54%. Increasing the number of bits for the R/2R ladder and counter would, of course, improve the resolution and, thus, the accuracy of the system. However, this improved system would require a greater number of clock pulses for a given conversion cycle. Thus, a limitation of single-slope A/D converters is that of conversion speed.

Dual-slope A/D converters are the most accurate form of analog-to-digital conversion device. At the same time, they represent the slowest form of A/D conversion. Thus, their

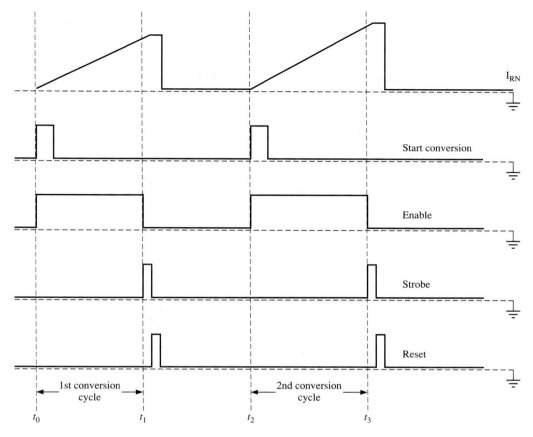

Figure 11–54
Waveforms showing operation of circuitry in Figure 11–53

use is limited to applications such as digital voltmeters, where conversion time is not a major consideration. A block diagram of a dual-slope A/D converter is given in Figure 11–55.

The operation of the dual-slope A/D converter is as follows. Through the action of an electronic switch, the control source holds a sample of the analog input at the current source of an integrator for a fixed time. This initial phase of the conversion cycle is represented as t_0 to t_1 in Figure 11–55(b). During this fixed sampling time, assuming the input voltage is positive, current flows from the input to point A and since the voltage at point A remains at a virtual ground potential, the flow of feedback current is constant. Thus, as seen in (b), a negative-going linear ramp develops at point B.

At t_1, the control circuitry switches the current source of the integrator from the sample input voltage to a fixed value of negative reference voltage. This action causes the output of the integrator to ramp in the positive direction, as seen from t_1 to t_2 in Figure 11–55(b). The second conversion cycle in (b) represents a condition where the sampled input voltage has increased in amplitude, causing the input current to increase. As a result, the ramp signal is able to integrate to a greater negative potential by t_1. It is important to note that, during the second conversion cycle, due to the higher input amplitude, the angle of the negative input ramp is steeper. Thus, for the first phase of any conversion cycle,

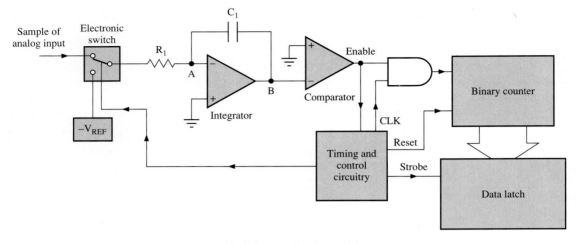

(a) Function block diagram of dual-slope A/D converter

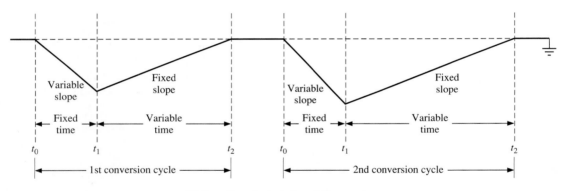

(b) Operation of a dual-slope A/D converter

Figure 11–55

time is fixed but the angle of negative slope is variable. However, during the second phase of the conversion process, due to the fixed negative reference potential, the angle of positive slope is fixed. Since the starting point of this fixed ramp varies with input voltage, the time of the positive slope (t_1 to t_2) becomes variable.

In Figure 11–55(a), the output of the comparator, which functions as a zero-crossing detector, is high during the entire time the integrator output is below ground potential. However, the counter is held in a reset condition until t_1. As with the single-slope device represented in Figure 11–52(a), the output of the comparator serves as the Enable signal, allowing the counter to increment only between t_1 and t_2. Thus, the magnitude of the highest count attained by the counter becomes a function of the amplitude of the input voltage. As with the single-slope device, the operation of the dual-slope A/D converter may be explained mathematically:

$$\Delta \text{ Binary count} \propto \Delta(t_2 - t_1) \propto \Delta V_{\text{analog input}}$$

The reason for the high degree of accuracy of the dual-slope A/D converter may also be explained mathematically as shown below:

$$\frac{V_{input}}{R_1 \times C_1} \times (t_1 - t_0) = \frac{V_{REF}}{R_1 \times C_1} \times t_2 - t_1$$

Since the values of R_1 and C_1 may be canceled from the equation:

$$V_{input} \times (t_1 - t_0) = V_{REF} \times (t_2 - t_1)$$

Thus,

$$(t_2 - t_1) = V_{input} \times \frac{(t_1 - t_2)}{V_{REF}}$$

Thus, since they are canceled from the equation, the values of resistance and capacitance have no effect in determining the time during which the counter in Figure 11–55(a) is enabled to increment. Relating this fact to actual circuit operation, any deviations in the values of R and C affect both the negative and positive slopes of the integrator signal. Thus, such component discrepancies are prevented from introducing errors in the A/D conversion process.

In A/D conversion applications requiring high conversion speeds, the single- and dual-slope designs are abandoned in favor of faster systems. A device capable of relatively high-speed sampling and digitizing of analog signals is the **successive-approximation A/D converter.** Prior to an explanation of the internal structure and operation of this device, the conversion method of successive approximation must be explained. As an analogy to A/D conversion, consider the measurement process illustrated in Figure 11–56. Here, imaginary blocks having specified binary weights are placed on the left-hand side of a scale to determine the unknown binary value of an object on the right-hand side.

Figure 11–56(a) represents the ramping-and-counting operation of either the single- or dual-slope A/D converter with an 8-bit output. Here, as many as 255 blocks must be individually placed on the scale until the most balanced condition is achieved. Relating this process to actual circuit operation, the placement of a block equaling 2^0 on the left-hand side of the scale represents the incrementing of the counter contained within the A/D converter by one step. Thus, if the binary weight of the object being measured is 11111111, 255 blocks must be placed on the scale, representing the expenditure of 255 clock pulses.

With successive approximation involving an 8-bit number, only eight blocks are required. As illustrated in Figure 11–56(b), the binary weights of these blocks are 2^0 through 2^7. In the successive-approximation process, the heaviest block, representing 2^7, is placed on the scale first. If the scale tilts, indicating the unknown weight being measured is less than 2^7, the 2^7 block is immediately removed and set aside. If the scale doesn't tilt, indicating an unknown weight greater than 2^7, the 2^7 block is left on the scale. This trial-and-error measurement system is repeated for the 2^6 through the 2^0 blocks. The binary weights that remain on the scale after the nearest possible balanced condition is achieved represent logic highs within the final result. Those weights that are set aside represent logic lows. Since, for the successive-approximation A/D converter, each trial placement of a block on the scale represents a single clock cycle, the advantage of this conversion technique should be obvious. In achieving an 8-bit result, the successive-approximation method of conversion may be as much as 32 times faster than the single-slope process. When compared to dual-slope operation, successive approximation is faster still.

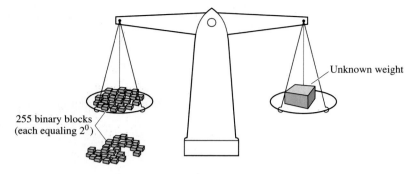

(a) Representation of single- or dual-slope A/D conversion

255 binary blocks
(each equaling 2^0)

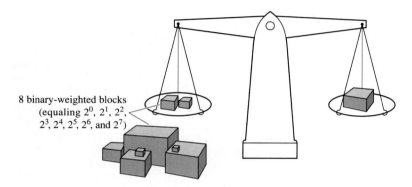

(b) Representation of successive-approximation conversion method

8 binary-weighted blocks
(equaling 2^0, 2^1, 2^2,
2^3, 2^4, 2^5, 2^6, and 2^7)

Unknown weight

Figure 11–56

A schematic diagram of a successive-approximation A/D converter is shown in Figure 11–57. The primary components of the system are an MC14549 successive-approximation register (SAR), an MC1408 D/A converter, and two LM319N op amps. The output of the D/A converter is connected to the inverting input of an op amp that functions as a current-to-voltage converter. Potentiometer R_8, which provides offset adjustment, allows the output of the current-to-voltage converter to be made compatible with either an ac or dc input signal, whereas potentiometer R_3 allows for gain adjustment. For example, assume the full swing of the analog input signal is ± 5 V (10 V peak to peak). With all eight binary inputs to the MC1408 at a logic high, R_3 and R_8 must be adjusted such that the output of the current-to-voltage converter equals $+5$ V_{dc}. With all eight binary inputs to the D/A converter at a logic low, the output of the current-to-voltage converter must equal -5 V. Such a preliminary adjustment assures maximum resolution for the A/D conversion process. The output of the first LM319N current-to-voltage converter feeds the inverting input of the second LM319N, which functions as a comparator. This second op amp compares the amplitude of the sampled analog input to the voltage at the output of the current-to-voltage converter. The comparator output serves as the data input to the SAR. If the output voltage of the current-to-voltage converter is greater than the amplitude of the sampled input, the input to the SAR is a logic low. If the sampled input potential is greater than the output amplitude of the first op amp, the input to the SAR is a logic high.

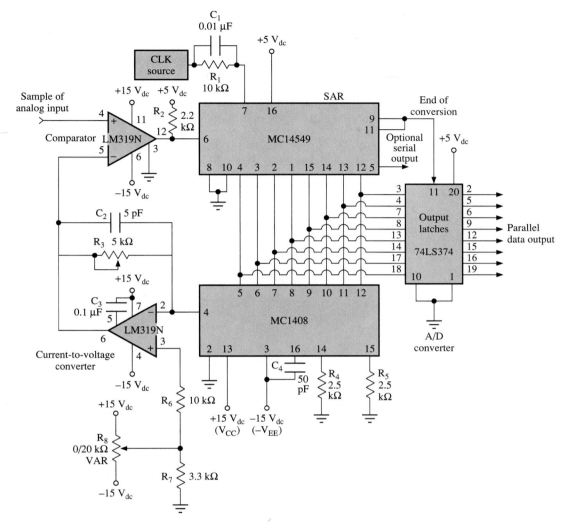

Figure 11–57
Successive-approximation A/D converter

The operation of the circuit in Figure 11–57 is as follows. After a hold signal (not shown) arrives from the control source and a sample of the analog input has settled at the noninverting input of the comparator, the first clock pulse of the conversion cycle occurs at pin 7 of the MC14549 SAR. This action causes a temporary logic high at pin 4, which is connected to the most significant input of the MC1408 A/D converter. As a result, current flows from the output of the MC1408, causing the output of the current-to-voltage converter to slew in the positive direction. If the voltage that settles at the inverting input of the comparator exceeds that of the sampled input, the output of the comparator is low. This logic low informs the SAR that the analog equivalent of the most significant bit (MSB) is greater than the sampled analog input. Consequently, the SAR toggles its MSB output back to a logic low. (This action would be analogous to removing the weight from the left-hand side of the scale.) If, however, the output of the current-to-voltage converter does not exceed the analog input sample, the SAR would allow its MSB to remain at a logic high for the duration of the 8-bit conversion process (analogous to keeping the weight on the scale). This process would continue at the remaining seven outputs of the SAR. After testing of the least significant bit (LSB), the 8-bit result is stored in a data latch. The SAR is then cleared in preparation for the next conversion cycle. Also note that, since the MC14549 is provided with a serial output, the resultant binary number could be fed into a shift register or possibly the serial port of a microcomputer.

Except for the flash-comparator device, the fastest form of A/D converter is the **continuous-counter ramp A/D converter** or, simply, **tracking A/D converter,** an example of which is shown in Figure 11–58. This device is ideally suited for applications requiring high-speed conversion, where the input voltage changes gradually through time. The main components of the tracking A/D converter are a binary up/down counter and a D/A converter with a current-to-voltage converter at its output. At this juncture, the need for separate Up-Count and Down-Count inputs to a monolithic counter such as the 74193 becomes evident. With the tracking A/D converter, the analog input is fed directly to the noninverting input of a high-speed comparator, and the output of the current-to-voltage converter is fed to the inverting input. When the analog input attempts to exceed the voltage present at the comparator's inverting input, the output of the comparator goes high, allowing clock pulses to pass to the Up-Count input of the 74193. As a result, the counter begins to increment, causing the output of the current-to-voltage converter to slew

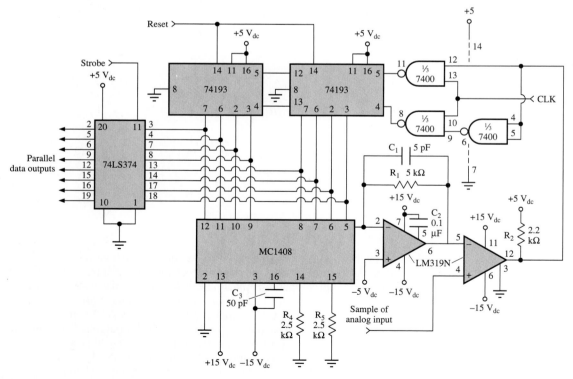

Figure 11–58
Continuous-counter ramp A/D converter

in the positive direction. At the instant the output of the current-to-voltage converter begins to exceed the voltage present at the analog input, the Down-Count input to the 74193 is enabled. Thus the counter begins to decrement, causing the output of the current-to-voltage converter to swing positively. When the analog input attempts to fall below the voltage present at the comparator's inverting input, the output of the comparator goes low, allowing clock pulses to pass to the Down-Count input of the 74193. During a typical, high-speed operation of the circuit in Figure 11–58, the counting mode of the 74193 is changed rapidly, allowing the changes in binary output to reflect closely the changes in the analog input voltage. The accuracy of the circuit in Figure 11–58 becomes a function of the rate at which samples of the output of the counter are captured in the data latches.

SECTION REVIEW 11–5

1. Explain why the R/2R ladder is preferred over the binary-weighted resistor network in the fabrication of high-resolution D/A converters.
2. For the R/2R ladder in Figure 11–48(a), assume R equals 50 kΩ, and a logic high equals 3.8 V. What is the value of output current if the binary inputs are forming the number 1011?
3. Define the terms *monotonicity, linearity,* and *resolution* as they apply to the operation of a D/A converter.
4. A D/A converter with an output range of ±5 V accommodates 12 binary inputs. What is the minimum change in voltage that should occur at the output of this device? What is its resolution expressed as a percentage?
5. What is the main advantage of the flash-comparator A/D converter? What is a disadvantage of this device?
6. What is a sample-and-hold module? Why are these devices necessary in the analog-to-digital conversion process?
7. Define the terms *aperture time* and *acquisition time* as they apply to the operation of a sample-and-hold module.
8. What is the slowest form of A/D converter? Why is this device also the most accurate form of A/D converter?
9. Explain the operation of a successive-approximation A/D converter. What specialized device must this form of A/D converter contain? Why is the conversion rate of this device likely to be much faster than that of a single- or dual-slope device?
10. Explain the operation of a tracking A/D converter.

SUMMARY

While low-power Schottky TTL and CMOS are currently the most common forms of digital logic found within industrial control systems, older control circuitry could contain examples of DTL, RTL, and HTL. Newer forms of logic, including ECL and I²L, are also found in more modern control devices. Low-power Schottky TTL is characterized by fast switching speeds, whereas CMOS is characterized by extremely low power consumption and high noise immunity. Both TTL and CMOS are available with tri-state outputs, allowing them to be used in common-bus organized digital control systems.

Combinatorial logic circuits are widely used in industrial control systems. Through the implementation of Boolean algebra and Karnaugh mapping, conditional logic circuits may be designed for the purpose of controlling the on/off conditions of such common devices as relays, motors, and solenoids.

Individual flip-flops and monolithic shift-registers and counters may be used in the control of industrial devices. While a single flip-flop could serve as a simple on/off controller for a conventional dc or ac motor, a universal shift register could be configured for more precise control involving a stepper motor. A binary up/down counter, along with supporting logic gates, could also be used in the control of a stepper motor.

Interfacing modules are used in industrial applications where digital circuits control the on/off conditions of high-power ac and dc loads. AC and dc input modules are often used to inform the digital control circuitry of the status of various loads, whereas ac and dc output modules provide actual control signals to such loads. All four types of interface module are likely to contain optoisolators. AC output modules often contain zero-voltage switches, allowing inductive loads such as motors to be switched on with minimal thermal shock and RF noise.

D/A and A/D converters are used in industrial applications where proportional control of a load is to be provided by a digital source. The basic components of a D/A converter are a resistive network, either binary-weighted or R/2R ladder, and a current-to-voltage converter. These components may be easily contained in a single monolithic package. Analog-to-digital converters are more complex than D/A converters. In fact, many forms of A/D converter contain D/A converters. Examples of such devices are successive-approximation and single-slope A/D converters. While the dual-slope device represents the slowest form of A/D converter, it is also the most accurate. The fastest forms of A/D converter are the flash-comparator and tracking A/D converters.

SELF-TEST

1. From which logic family did HTL evolve? What was the reason for developing this specialized form of digital logic?
2. What earlier forms of digital logic provided gates with expandable inputs?
3. What form of TTL allows direct coupling of logic outputs?
4. Why can't the outputs of standard TTL logic gates (containing totem-pole switching transistors) be directly wired together?
5. Explain why an LS7493 is able to operate reliably at higher switching speeds than a standard 7493.
6. Draw the schematic symbol for a three-input Schmitt-triggered NAND gate. What is the purpose of this type of gate?
7. What form of logic is likely to be contained inside a portable, battery-powered piece of equipment?
8. Write the unsimplified Boolean expression for the circuit in Figure 11–59(a). Using DeMorgan's theorems and Boolean algebra, simplify this circuit as much as possible. Write the simplified Boolean expression for the circuit.
9. Write the unsimplified Boolean expression for the logic circuit contained in Figure 11–59(b). Write a simplified logic expression for the same circuit. Using DeMorgan's theorems, draw a circuit equivalent to that in Figure 11–59(b) containing noninverting gates. Include inverters where necessary.
10. For the circuit in Figure 11–29, assume the deadband is to be 100°F to 80°F. What must be the ohmic settings of R_3 and R_5?
11. For the 74193, which input has priority over all other inputs? What is the active condition of this input?
12. For the circuit in Figure 11–32, what connections ensure the internal flip-flops of the 74194 are always placed in the proper logic conditions prior to circuit operation?
13. Identify the major components likely to be found in an ac input module.

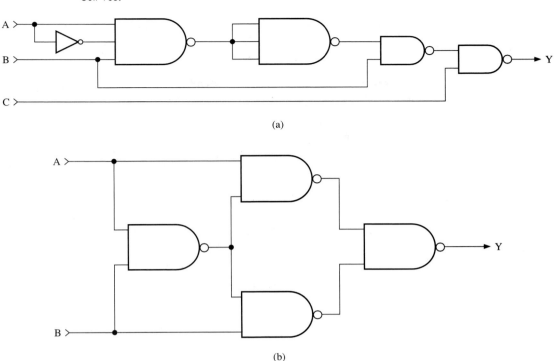

(a)

(b)

Figure 11–59

14. Identify the major components likely to be found in an ac output module.
15. Identify the major components contained in a D/A converter.
16. For the DAC-01 D/A converter, what external connections must be made to have a ±10-V output swing? What connections should be made to have a 0- to 10-V output swing?
17. What is the percentage of resolution for an MC1408 D/A converter?
18. Explain the operation of a dual-slope A/D converter. Identify the main components contained in the device.
19. Explain the operation of a single-slope A/D converter. What are two possible methods of creating a ramp source for this device?
20. Explain the operation of a successive-approximation A/D converter. Identify the main components contained within the device.

12 MICROPROCESSORS AND MICROCONTROLLERS

CHAPTER OUTLINE

LEARNING OBJECTIVES

On completion of this chapter, the student should be able to:

- Describe the function of an arithmetic logic unit.
- Describe the function of a central processing unit.
- Briefly explain the evolution of the microprocessor.
- Describe the internal architecture of the microprocessor.
- Identify the types of memory devices likely to be used in conjunction with a modern microprocessor.
- Describe the overall structure of a modern, bus-organized microcomputer.
- Define the term *microprogramming* as it pertains to the operation of a microcomputer.
- Define the term *algorithm* as it pertains to the operation of a microcomputer.
- Write a simple computer program in assembly language.
- Write a simple computer program in machine language.
- Briefly explain the evolution of the microcontroller.
- Describe the internal architecture of a typical microcontroller.
- Identify the advantages of using a state-of-the-art microcontroller rather than a microprocessor in an industrial process control application.

INTRODUCTION

In this chapter, the internal structure and operation of the microprocessor are introduced. Next, the various forms of memory used in industrial computer systems are surveyed. The structure of a typical microcomputer is then introduced at a block level, illustrating how a microprocessor interfaces with ports and memory. Programming is introduced at a very limited level, primarily as a means of understanding the internal operation of the microcomputer. The chapter concludes with an introduction to microcontrollers, with a state-of-the-art example of this device being examined in detail.

Figure 12–1 is representative of the operator control stations in many modern industrial facilities. The personal computer (PC), with its familiar keyboard, monitor, and printer, is gaining increasing importance in the industrial workplace. In some industrial applications, such computers may be serving in a supportive function, with duties being limited to data collection or assistance in running system diagnostics. If the external PC is limited to a supporting role, a microcomputer, contained on a circuit board within the machinery's system electronics, usually assumes the active control function. This embedded microcomputer communicates with the supporting PC, sending it machine operating statistics and allowing the PC to run diagnostics if there is a system failure. In some control applications, a PC may assume a more active role, directly governing the entire operation of a machine. In such a hierarchy, the role of the PC is analogous to that of a chief executive, ultimately responsible for the overall operation of a complex system. The function of each embedded microcomputer is then similar to that of a specialized, lower level manager, who is responsible for a particular phase of machine operation. These subordinate, or slave, microcomputers communicate with the PC via a common data bus, receiving commands from and sending information to the PC.

In this chapter, emphasis is placed on the structure and operation of the embedded, machine-control computer rather than the PC. (It is assumed that, through an advanced digital electronics course, the student has already gained, or soon will acquire, a basic knowledge of PC architecture and operation.) The justification for this approach to the

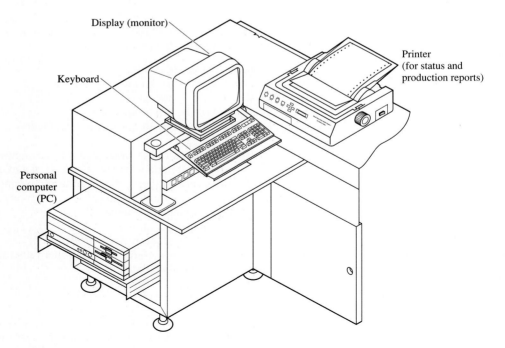

Figure 12–1
Typical PC-based control station for an industrial system

role of the computer in machine control is that a technician is more likely to encounter faults within embedded control devices rather than supporting PCs. Thus, it is the goal of this and the following chapter to present a thorough background in those microprocessors and microcontrollers commonly used in machine control.

12–1   BASIC MICROPROCESSOR ARCHITECTURE AND OPERATION

Contained within the integrated circuitry of the modern microprocessor are circuits that, in the recent past, could only be contained within several cabinets. One such circuit, the focal point of the microprocessor, is the **arithmetic logic unit** (ALU). Suggestive of the earlier days of computer development, separate ALUs are still commercially available, an example of such an IC being the transistor-transistor logic (TTL) 74181. The 74181 ALU (and even the 7483 full adder) may still be found in earlier forms of industrial control circuitry. While these TTL integrated circuits (ICs) accommodate only 8 input bits, producing a 4-bit result, ALUs within 8-bit microprocessors must process 2 bytes of information. Thus, such devices accommodate 16 input bits, producing an 8-bit result.

An ALU must be capable of performing addition and subtraction, as well as logic ANDing, ORing, and XORing of 2 bytes of data. Closely associated with the ALU is a register called the **accumulator,** which is *always* involved in the microprocessor's arithmetic and logic functions. During arithmetic operations, data bytes are either added to or subtracted from the contents of the accumulator. After such an operation, the result is stored in the accumulator. During logic functions, data bytes are ANDed, ORed, or XORed with the contents of the accumulator.

The microprocessor ALU is likely to contain a **two's-complement full adder/subtracter,** a block diagram of which is contained in Figure 12–2. The logic condition at the Mode control input determines whether the device adds or subtracts. A logic low is equivalent to an add command, while a logic high results in a subtract operation. During an add operation, the operand present at the data input is added to the contents of the accumulator. The result of this operation is placed into the accumulator, and the original accumulator data byte is destroyed. A carry bit produced by the add function is stored in a single flip-flop called the **carry flag.** If an addition produces no carry condition, the carry flag is reset. If the addition produces a carry condition, the carry flag is set.

During a two's-complement subtract operation, the ALU first complements the subtrahend (number being subtracted). It then adds this complemented number to the accumulator data while automatically adding a carry-in bit at the least significant bit position. As with an add operation, the original data byte in the accumulator is destroyed. If the two's-complement subtract operation produces a carry bit, the carry flag, which may now be considered as the "borrow" flag, is *not* set. This low condition for the carry bit indicates that the value of the subtrahend is less than the value of the number in the accumulator. If, however, the result of the two's-complement subtract operation does not produce a carry, the subtrahend must be larger than the accumulator number. With this condition, the carry flag is set, indicating a borrow condition.

Example 12–1

Assume that, while the Mode control line to the two's-complement full adder/subtracter in Figure 12–2 is low, the accumulator contains 01101101 and the data input equals 10110101. What would be the result stored in the accumulator? What would be the condition of the carry flag? With the result of this operation still in the accumulator, assume the Mode control input goes high after the data input changes to 00011101. What

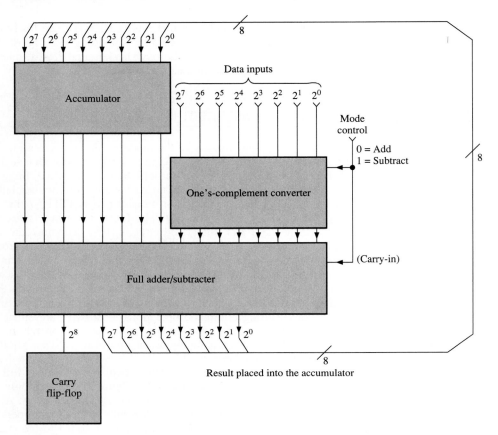

Figure 12–2
Block diagram of a Central Processing Unit

is the next result stored in the accumulator? What is the condition of the carry flag? What does this flag condition indicate about the input data?

Solution:

Step 1: Determine the operation of the two's-complement full adder/subtracter when the Mode control input is low. Since binary addition is being performed:

$$01101101_{(2)} + 10110101_{(2)} = 100100010_{(2)}$$

Thus, the accumulator would hold 00100010 and the carry flag would be set. This operation is equivalent to $[109_{(10)} + 181_{(10)} = 290_{(10)}]$.

Step 2: With the result from step 1 stored in the accumulator, and the mode control line high, two's-complement subtraction is performed, equivalent to $[00100010_{(2)} - 00011101_{(2)}]$ or $[34_{(10)} - 29_{(10)}]$. Adding the contents of the accumulator to the one's complement of the subtrahend and adding a carry in bit:

$$00100010_{(2)} + 11100010_{(2)} + 00000001_{(2)} = 100000101_{(2)}$$

Complementing the carry bit, the result becomes $000000101_{(2)}$.

The reset condition of the carry bit indicates that the result stored in the accumulator is positive. Thus, the original number contained in the accumulator was greater than the subtrahend.

Most modern microprocessors have the capability of performing **binary-coded decimal** (BCD) addition. Recall that, with the BCD number system, the highest value for a byte of data is 10011001, representing $99_{(10)}$. While actual conversion of a binary number to BCD requires a program routine consisting of several steps of machine language, maintaining an addition operation in a BCD format requires only a single machine language instruction. During BCD addition, this operation, usually referred to as **decimal adjusting,** is placed after each add instruction.

The decimal adjustment function is best explained with a simple example. Assume that the ALU is performing a BCD addition equivalent to $[79_{(10)} + 23_{(10)}]$. Assume the accumulator data holds the BCD number equivalent to $79_{(10)}$, while the addend at the data input represents $23_{(10)}$. Converting these numbers to BCD, the input data (addend) would equal 00100011, while the contents of the accumulator would be 01111001. The adder/subtracter would first add these two numbers as if they were true binary values, yielding an interim result as follows:

$$01111001_{(BCD)} + 00100011_{(BCD)} = 10011100$$

Thus, the accumulator would hold 10011100 and the carry flag would be reset. Note that the result stored in the accumulator is not in BCD format. To convert the interim result to a valid BCD number, the decimal adjust circuit first tests the value of the lower nibble of accumulator data. If this lower nibble is greater than 1001, the decimal adjust circuitry immediately adds 0110 to this nibble. The number 0110 is an offset, representing the difference between the highest 4-bit binary value, 1111, and the highest 4-bit BCD number, 1001. Since the lower nibble of accumulator data currently equals 1100, the decimal adjustment of the lower nibble takes place as follows:

$$1100 + 0110 = 10010$$

In adding 0110 to the lower nibble, a carry condition occurs. This carry condition between the lower and upper nibbles is referred to as the **auxiliary carry flag.** If this flag is set, as in the preceding example, the decimal adjust circuitry adds 0001 to the upper nibble of the accumulator data and tests the result. Since the upper nibble now equals 1010, 0110 is added to it:

$$1010 + 0110 = 10000$$

Thus, the upper nibble of accumulator data becomes 0000 and the carry flag is set. Considering the set carry flag condition as the most significant bit of the final result, the BCD number 1 0000 0010 is produced, equivalent to decimal 102. [This is the result of $79_{(10)} + 23_{(10)}$.]

The presence of the carry and auxiliary carry flags allows the microprocessor to perform 16-bit binary and BCD addition, as well as 16-bit binary subtraction. Most instruction sets for 8-bit microprocessors contain commands that allow add-with-carry or subtract-with-borrow operations. For example, assume that two registers within a microprocessor, referred to as B and C, contain 2 bytes of data that, together, form a 16-bit number. Assume the B data comprise the upper byte of the number and the C data form the lower byte. A second pair of registers, D and E, contains another 16-bit BCD number, with D holding the upper byte and E holding the lower. As an example of a 16-bit math operation, assume the BCD number in B and C is to be added to the BCD contents of D and E. The result of this operation is to be stored in a third register pair that is identified as H and L. A definite routine for accomplishing this add function is first outlined. In the following outline, for the sake of brevity, the accumulator register is referred to as A.

1. Move E data into A.
2. Add C data to A data.
3. Decimal adjust the result in A.
4. Save the lower byte of the result by moving the A data into L.
5. Move D data into A.
6. Add with carry B data to A data. (In this step the carry-out condition produced by the result in either step 2 or step 3 is treated as a carry-in, effectively functioning as the carry from the 2^7 to the 2^8 position.
7. Decimal adjust the result in A.
8. Save the upper byte of the result in H.

Example 12–2

Following the procedure outlined, demonstrate how a full adder/subtracter would add two 16-bit BCD numbers. Assume B and C form the BCD equivalent of $3792_{(10)}$, and D and E contain the equivalent of $4653_{(10)}$.

Solution:

Step 1: Determine the contents of the registers:

B and C equal 0011 0111 1001 0010.
D and E equal 0100 0110 0101 0011.

Step 2: Perform addition steps for the two lower bytes:

$$A = E = 01010011$$
$$A + C = 01010011 + 10010010 = 11100101$$

Decimal adjust the accumulator: The lower nibble of the result is left alone since it is less than 1001.
     The upper nibble is decimal adjusted:

$$1110 + 0110 = 10100$$

A now equals 01000101 and the carry flag is set: L = A = 01000101.

Step 3: Perform addition steps for the 2 upper bytes:

$$A = D = 01000110$$
$$A + B + \text{carry bit} = 01000110 + 00110111 + 00000001 = 01111110$$

Decimal adjust the accumulator: The lower nibble is decimal adjusted.

$$1110 + 0110 = 10100$$

The lower nibble of A data is now 0100 and the auxiliary carry flag is set. The auxiliary carry bit is added to the upper nibble of the A data.

$$0111 + 0001 = 1000$$

Since the upper nibble is less than 1001, no further decimal adjustment is necessary.

The final result contained in the H and L register pair becomes 10000100 01000101$_{(BCD)}$, equivalent to 8445$_{(10)}$. *Proof:*

$$3792_{(10)} + 4653_{(10)} = 8445_{(10)}$$

To expedite the processes of multiplication and division, the accumulator register of a microprocessor must be able to shift its contents to either the right or left. As explained later in detail, with the introduction to the instruction set of an 8-bit microprocessor, these shift operations are usually data rotations involving the carry flip-flop.

Along with arithmetic operations, a microprocessor must be able to perform Boolean logic functions. For this reason, the ALU contains gate array circuitry, equivalent to that represented in Figure 12–3. As with addition and subtraction, Boolean logic operations

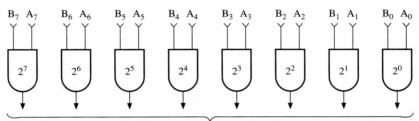

(a) Logic ANDing of accumulator data and data from register B

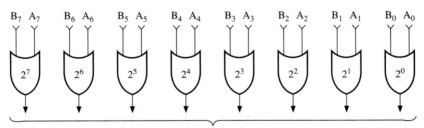

(b) Logic ORing of accumulator data and data from register B

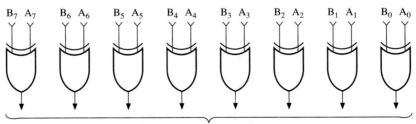

(c) XORing of accumulator data and data from register B

Figure 12–3

within the ALU *always* involve the accumulator. Also, the result of any Boolean logic function is placed in the accumulator, with the original contents of the accumulator being destroyed.

Assume, for example, that the data byte in register B is to be ANDed with the accumulator data. Assume that the accumulator contains 00001011, and the B register holds 10111101. During the AND operation, as shown in Figure 12–13(a), the 2^0 through the 2^7 bits of the accumulator are ANDed with the 2^0 through the 2^7 bits of the B register. Thus, the result stored in the accumulator becomes 00001001, since only for the 2^0 and 2^3 bits of the two numbers do two high conditions occur. As represented in Figure 12–3(b), the ALU provides an ORing function. Thus, if 00001011 and 10111101 are ORed, the accumulator data becomes 10111111. Note that only the 2^6 bit position is low, since this is the only place value for which there were two logic lows. As shown in Figure 12–3(c), the ALU also provides the capability of XORing data with the contents of the accumulator. Thus, a bit-per-bit parity check may be made between the accumulator data and the contents of another register. If an odd-parity condition exists between the two bit positions, a logic high occurs in the result, whereas an even-parity condition produces a logic low. Using the same pair of numbers as in the other examples, XORing 00001011 with 10111101 produces 10110110.

A common application of the ANDing capability of the ALU is for **masking.** Although a microprocessor does not process data at a level lower than the byte, it is often necessary to obtain less than a "byte-size" piece of data. For example, assume that a data word 01101011, stored in the B register, is to be divided into two separate data words. The upper nibble of the B data becomes the lower nibble of a second word stored in register D, and the lower nibble of the B word becomes the lower nibble of a third word stored in register E. Due to the rotating ability of the accumulator, this process is easily accomplished in the following sequence.

1. Move 11110000 into A.
2. AND B data with A data.
3. Rotate the accumulator right four bit positions.
4. Store A data in register D.
5. Move 00001111 into A.
6. AND B data with A data.
7. Store the A data in register E.

As a result of this operation, the D register holds 00000110 and the E register contains 00001011.

Besides the carry and auxiliary carry flags, the microprocessor is likely to have a **zero flag, parity flag,** and **sign flag.** The zero flag is high if, after an arithmetic or logic operation, the accumulator contains 00000000. The parity flag is set if, after an arithmetic or logic operation, the result placed in the accumulator contains an even number of high bits. If an arithmetic or logic function produces an odd-parity condition, the parity flag is reset. The sign flag was originally intended for use with the **sign binary** method of indicating polarity. With this system, the 2^7 bit is reserved to indicate the polarity of the number formed by the remaining seven bits. A logic high in the 2^7 bit position indicates a negative condition for the 7-bit number, while a low in this bit position assigns it a positive value. Thus, if the accumulator contains 01110101, the sign flag would be low, signifying that the number formed by the lower 7 bits is positive. If the number in the accumulator has a high in the 2^7 position, as with 10010101, the sign flag is high, signifying that the number formed by the lower 7 bits is negative. These negative and positive polarities assigned to the accumulator data are irrelevant unless the sign binary system is actually being implemented. However, as demonstrated later, the sign flag may be used for purposes other than indicating polarity.

Example 12–3

Assume that register B holds binary 01001011, register C holds 01010101, and the accumulator holds 11101010. If the number in register B is ORed with the contents of the accumulator and the result of this operation is added to the number in register C, what would be the contents of the accumulator and the status of the carry, zero, parity, and sign flags?

Solution:

Step 1: OR B data with A data:

$$01001011 \text{ ORed with } 11101010 = 11101011$$

Step 2: Add A data to C data:

$$11101011 + 01010101 = 101000000$$

Step 3: Determine the contents of the accumulator and the flag flip-flops:

$$A = 01000000, \quad \text{Carry flag} = 1, \quad \text{Zero flag} = 0, \quad \text{Parity flag} = 0, \quad \text{Sign flag} = 0$$

The ALU is part of a larger section of a microprocessor called the **central processing unit** (CPU). The CPU controls all facets of microprocessor operation. Such CPU operations include fetching and executing instructions that are entered into the microcomputer's memory and performing arithmetic and logic operations such as those just described. To perform these tasks, the CPU must have, along with the ALU, those additional items shown in the block diagram of Figure 12–4.

The **control unit** governs the operation of all of the other components of the CPU. The control unit relies on a 16-bit binary up-counter called the **program counter** to point to the memory address of the next instruction to be executed by the CPU. The 8-bit data word residing at this memory location is latched into the **instruction register** of the CPU. Here, the **instruction decoder** circuitry reads the 8-bit instruction words, enabling portions of the ALU or other parts of the CPU as necessary to execute the instruction. The instruction read from memory is sometimes referred to as a **macroinstruction.** Examples of macroinstructions include such operations as ANDing the accumulator data with data from another source or rotating the accumulator data to the left or right. Those instructions that the control unit must execute to allow the performance of a macroinstruction are referred to as **microinstructions.** As an example of microinstruction execution, consider a simple addition operation. On the arrival of the ADD instruction at the instruction register, the control unit initiates a sequence of control pulses that cause the addend to be placed in a buffer register (sometimes referred to as a **scratch pad register**) of the ALU. Further microinstructions affect execution of the ADD instruction and storage of the result in the accumulator.

Figure 12–5 contains a block diagram of the internal circuitry of the 8085 microprocessor (manufactured by Intel Corporation). This *N*-channel MOS device is designed for processing 8-bit (single-byte) data words and is capable of interfacing with 65,536 memory locations. The 8085 represents the culmination of the first generation of microprocessors. Besides the accumulator and the temporary registers associated with the ALU, the 8085 has six work registers. These registers may be addressed separately or treated as register pairs. As explained later, the macroinstruction set of the 8085 is designed with the assumption that the H and L register pair is to be used as the **memory pointer.** The B and C register pair, as well as D and E, may function in this capacity, though not as easily as the H and L pair. Sometimes during the execution of a program, the six status

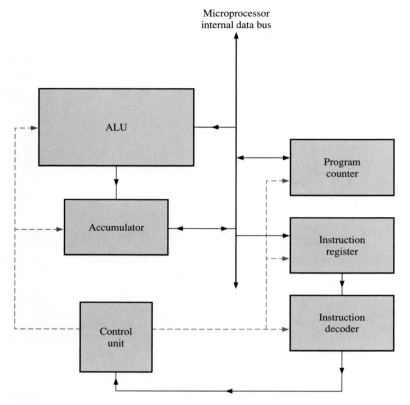

Figure 12–4
Basic structure of a microprocessor CPU

flags and the accumulator are treated as a register pair. (The need for addressing the accumulator and flags as a double byte of data becomes apparent as the branching and interrupt-servicing capabilities of the 8085 are investigated.) As shown in Figure 12–6, the high byte of this register pair is referred to as the **F register,** whereas the accumulator comprises the lower byte. The F register and accumulator together form the **program status word** (PSW). Note in Figure 12–6 that the three undefined bits within the F register, 2^{13}, 2^{11}, and 2^9, are designated by the letter X.

Example 12–4

A hexadecimal read-out of the PSW of the 8085 consists of AEA9. What would be the condition of the flags and the contents of the accumulator? Do the conditions of the sign, zero, and parity flags represent correctly the contents of the accumulator?

Solution:

Step 1: Expand the hexadecimal values into binary:

$$AEA9_{(16)} = 1010111010101001_{(2)}$$

Step 2: Determine the conditions of the flags. Since the high byte represents the F register:

$$Flags = 10101110$$

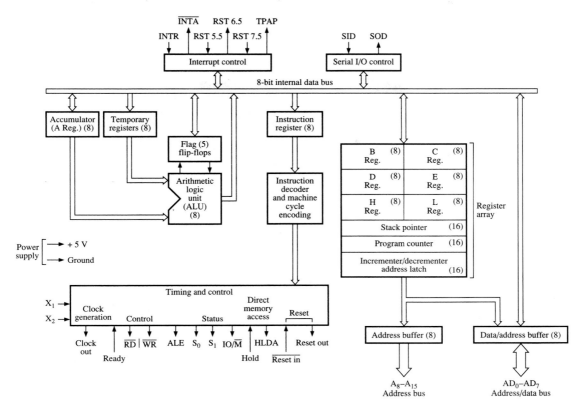

Figure 12–5
Internal block diagram of the 8085 microprocessor

Referring to Figure 12–6,

Sign flag = 1, Zero flag = 0, Auxiliary carry flag = 0, Parity flag = 1, Carry flag = 0

Step 3: Verify that the conditions of the sign, zero, and parity flags concur with the accumulator data.

Since the accumulator data equals 10101001, the 2^7 bit is high. The sign flag correctly indicates this condition.
Since the accumulator data contain 4 high bits, the zero flag must be low and the parity flag should be high. These conditions are correctly represented within the F register.

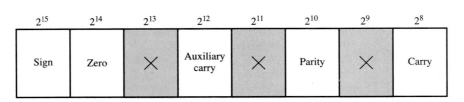

Figure 12–6
Bit configuration for the F register (high byte of PSW)

The **stack pointer** of the 8085 effectively functions as a 16-bit up/down counter. The purpose of the stack pointer is to define the top (lowest address) of the **stack,** which is a temporary storage area within the computer's random access memory (RAM). During the execution of a program involving considerable branching, vital information is sent to, and later taken from, the stack. The top of the stack decrements downward into memory. Thus, as information is placed on the stack, the stack pointer decrements. As information is removed from the stack, the stack pointer increments. (The operation of the stack pointer is clarified when the branch control and machine control instructions of the 8085 are introduced.)

Appendix A contains the instruction set of the 8085 microprocessor. Each instruction consists of an 8-bit **operation code** (op code) and a **mnemonic,** which aids the programmer in remembering the function of the op code. (As seen in Appendix A, the op codes are conveniently written in hexadecimal rather than binary.) The instruction set of the 8085 is divided into four basic groups: the **data transfer group,** the **arithmetic and logical group,** the **branch control group,** and the **I/O and machine control group.**

Op codes may be further classified as 1-byte, 2-byte, or 3-byte instructions. Within the data transfer group, the op codes that begin with the mnemonic **MOV** are single-byte instructions and among the easiest instructions to understand. Within the mnemonic, the letter immediately following MOV represents the destination register, while the letter following the comma represents the source register. Thus, the complete mnemonic *MOV A, B* would be understood as "move the contents of the B register into the accumulator." It is important to realize that, after the execution of this instruction, the original B data would still be present in the B register, while the original contents of the accumulator would be overwritten by the B data.

In examining the mnemonics for the data transfer group, the letter M appears frequently, although there is no M register! Recall that the H and L register pair was described as the memory pointer. In the 8085 instruction set, the letter **M** stands for **memory.** Memory data is the data byte present at a memory location being defined by the H and L register pair. For example, assume register H holds $23_{(16)}$ and register L held $AF_{(16)}$. The instruction *MOV A, M* would cause the data at memory location 23AF to be placed in the accumulator. As explained later, this action involves a **read** function. The instruction *MOV M, B* involves a **write** function, since data from register B is being sent to memory location 23AF.

Those move instructions beginning with **MVI** are 2-byte instructions. In the mnemonics of the 8085 instruction set, the letter **I** stands for **immediate data.** For a 2-byte instruction, immediate data consist of a byte that immediately follows the op code in memory. For example, assume the data word A7 is to be moved into the D register. This instruction could be written mnemonically as *MVI D, A7.* If this instruction begins at memory location 304A, then 304A contains 16, the op code for MVI D, and 304B contains the immediate data byte A7. At the beginning of the **fetch-and-execute cycle** for this instruction, the program counter points to address 304A. Thus, the microinstructions generated by the timing and control circuitry of the 8085 place the op code 16 in the instruction register. As this instruction is decoded, the timing and control circuitry senses that a 2-byte instruction is to be executed. It therefore increments the program counter to 304C, which is the address containing the next instruction.

Those move instructions beginning with **LXI** are three-byte instructions. In this mnemonic, **L** stands for **load** and **X** stands for **extended.** Within the 8085 instruction set, the term *load* signifies the placement of data into a single register or register pair of the 8085. The term *extended* signifies a double-byte, and I denotes immediate data. Thus LXI indicates that a 16-bit number is to be loaded into a register pair or the stack pointer. For example, the mnemonic statement *LXI SP, 30FC* is interpreted as "Load immediate data 30 into the stack pointer high byte and immediate data FC into the stack pointer low byte.'' Thus, the initial stack address is defined as 30FC. For LXI instructions within the 8085 instruction set, it is important to remember that, when listing the immediate data after the op code, the low byte precedes the high byte. For example, assume the command to initialize the stack pointer is the first instruction of a program routine that begins at memory location 0438. The op code and double byte would then be listed as:

0438	31
0439	FC
043A	30

In fetching and executing a load immediate instruction, the timing and control circuitry sense that a 3-byte op code is present in the instruction register. Thus, in response to the instruction in the preceding example, the program counter would have been incremented to address 043B.

The **load/store** commands of the 8085 instruction set are slightly more difficult to understand and remember than the instructions covered thus far. Although the LDAX command involves a sixteen-bit data word, contained in either the B and C or D and E register pair, it is only a single-byte instruction. For example, if the program counter points to a memory location containing the op code 0A, the 8085 executes an *LDAX B* instruction. With this command, the contents of the B and C register pair are placed in the address buffer of the 8085. A read function is then executed, causing the contents of the memory location pointed to by the B and C register pair to be loaded into the accumulator. An *LDAX D* instruction performs a similar function to that just described, with the D and E register pair temporarily functioning as the memory pointer. Like the LDAX commands, the two **STAX** instructions utilize the B and C or D and E register pair as a memory pointer. However, as the letters STA imply, a store function is executed by this instruction. Thus the command *STAX B* performs a write function, saving the contents of the accumulator at a memory location pointed to by the B and C register pair.

The mnemonic **LHLD** is interpreted as ''Load the register pair H and L direct.'' The letter **D** in the mnemonic denotes that the instruction uses the **direct addressing** mode. With direct addressing, the instruction itself forms an address low byte and high byte with the immediate data following the op code. Thus, LHLD is a 3-byte instruction. Assume the LHLD command begins at memory location 3F9B. As shown in the next example, the op code 2A is followed by a low byte and a high byte of a memory address holding data to be placed in the L register:

3F9B	2A
3F9C	4C
3F9D	5E

When the op code 2A is placed into the instruction register, the timing and control circuitry places 5E4C in the incrementer/decrementer address latch. A read function then occurs, causing the contents of address 5E4C to be placed in register L. The timing and control circuitry then increments the address latch, causing it to point to address 5E4D. The contents of this address is then loaded into register H. An **SHLD** instruction performs an opposite function to LHLD. For example, assume the SHLD command begins at memory location 4A37, as shown here:

4A37	22
4A38	4C
4A39	5E

After 5E4C is placed in the incrementing/decrementing address latch, the data in register L are stored at memory location 5E4C. Next, the address latch increments, pointing to address 5E4D. The contents of register H is then stored at that location. The SHLD and LHLD commands allow the status of the H and L memory pointer to be saved in, and later retrieved from, a specified area in memory. In complex programs, this allows the use of multiple memory pointers.

The **LDA** and **STA** commands are 3-byte instructions that allow the contents of the accumulator to be retrieved from or saved in directly addressed locations in memory. For example, consider the following LDA example:

203B	3A
203C	C2
203D	5A

When the op code 3A is fetched into the instruction register, the timing and control circuitry places 5AC2 in the address latch. The data byte at that memory location is then loaded into the accumulator. With the STA instruction demonstrated as follows, the data byte in the accumulator is stored at memory location 5AC2.

203B	32
203C	C2
203D	5A

The **XCHG** instruction is a single-byte command that switches the contents of the D and E and H and L register pairs. Assume, for example, that D holds 48, E holds 3B, H holds 5C, and L holds 2F. After execution of the XCHG instruction, H contains 48 and L contains 3B, while D contains 5C and E holds 2F. The exchange command allows the use of two different memory pointers. Also, as explained later, the XCHG command is often used in operations involving the stack.

The data transfer group of the 8085 instruction set having been introduced, the arithmetic and logical commands are now surveyed. The basic 1-byte **ADD** instructions involve the addition of the accumulator to itself (*ADD A*), one of the six work registers, or the data in the memory location pointed to by the H and L register pair (*ADD M*). The need for an add-with-carry function, which is used in 16-bit binary or BCD arithmetic, has already been explained. The mnemonic **ADC** denotes an add-with-carry function involving the accumulator and itself (*ADC A*), the accumulator and one of the six work registers, or the accumulator and memory (*ADC M*).

The **SUB** command, like ADD, is a 1-byte instruction involving the accumulator and itself, the six work registers, or memory. This instruction causes the ALU of the 8085 to perform two's-complement subtraction, with the data in the specified register or memory being subtracted from the contents of the accumulator. The **SBB** command is a subtract-with-borrow function, which is used in 16-bit subtraction. With this instruction, the data in the specified register or memory are subtracted from the contents of the accumulator, with the borrow (carry) flag condition being included in the operation.

Program Example 12–1 serves to demonstrate how machine language instructions and immediate data are entered into memory. After immediate data bytes are written into the B and C and D and E register pairs, the 16-bit number contained in B and C is subtracted from the data word formed by the D and E register pair. The result of this operation is then stored in the H and L register pair. With the D and E pair containing $4A37_{(16)}$ and the B and C pair containing $1A5B_{(16)}$, the following subtract operation is to be performed:

$$4A37 - 1A5B = 2FDC$$

This hexadecimal expression may be expanded into binary as shown:

$$0100101000110111 - 0001101001011011 = 0010111111011100$$

Program Example 12-1

Address	Op Code	Mnemonic	Comment
3000	11	LXI D, 4A37	Define subtracter as $4A37_{(16)}$.
3001	37		
3002	4A		
3003	01	LXI B, 1A5B	Define subtrahend as $1A5B_{(16)}$.
3004	5B		
3005	1A		
3006	7B	MOV A, E	Perform low-byte subtraction.
3007	91	SUB C	
3008	6F	MOV L, A	Save low-byte result in L.
3009	7A	MOV A, D	Perform high-byte subtraction.
300A	98	SBB B	
300B	67	MOV H, A	Save high-byte result in H.

At address location 3007 in the program example, with the **SUB C** command, the low byte subtraction is performed. As shown next, the data in register C are complemented and added, along with a carry-in bit, to the data in the accumulator. As a result of the move command in the previous step, the data from register E are copied into the accumulator:

$$00110111 + 00000001 + 10100100 = 11011100 = DC_{16}$$

Note that, as a result of this operation, an actual carry condition is *not* produced. Since the SUB C command causes the carry condition to be complemented, the borrow (carry) flag is set. This high condition of the carry bit indicates that a borrow condition exists.

When the high byte subtraction takes place, the **SBB B** command tests the condition of the borrow flag. The high condition of the borrow flag produced during the low byte subtraction at 3007 is tested during the subtract-with-borrow operation at 300A. When the ALU detects the high condition of the borrow flag, it *does not* add a carry-in bit during the high byte subtraction. Thus, at 300A, the ALU complements the contents register B but withholds the carry-in bit. Thus:

$$01001010 + 11100101 = 100101111 = 00101111 \text{ (final borrow (carry) complemented)}$$

As with any subtract instruction, the high carry bit is complemented, indicating that the final result is positive. Thus, after the instruction at 300B, register H holds 00101111, register L holds 11011100, and the borrow (carry) flag is reset.

Commands are also provided in the 8085 instruction set for addition and subtraction involving immediate data. The **ADI** (add immediate) and **SUI** (subtract immediate) instructions are both 2-byte instructions. Assume, for example, that the op code for the ADI instruction, C6, is fetched from memory location 3029 and latched in the instruction register. The data byte stored at 302A is then treated as immediate data, being added to the contents of the accumulator. As with other 8-bit math functions, the result of this addition is placed in the accumulator. If op code D6 were placed at 3029, then an SUI function would take place. The immediate data at location 302A would be subtracted from the contents of the accumulator with the result becoming accumulator data.

Commands also exist for adding and subtracting immediate data with the carry and borrow flag conditions. The **ACI** instruction, like ADI, is a 2-byte instruction. When the op code for this command, CE, is fetched from memory, the data byte at the next location is added along with the carry bit to the contents of the accumulator. The **SBI** command is fetched and executed in an identical manner to the ADI instruction, with the data in the address following that of the op code being subtracted from the contents of the accumulator. As with the SBB in Program Example 12–1, a high condition of the borrow flag causes the SBI function to withhold the carry-in bit. If the borrow flag is low, a normal two's-complement operation takes place, with a carry-in bit being added to the one's complement of the immediate data.

A convenient feature of the 8085 instruction set is the **double add** function, the mnemonic notation for which is **DAD**. With this 1-byte instruction, the contents of a specified register pair, or the stack pointer, are added to the data contained in H and L, with the result being placed in H and L. For example, assume the D and E register pair hold 4A3C, while the H and L pair contain 7AB2. With a *DAD D* command, the contents of D and E are added to the H and L data. Thus, as a result of this operation, H and L hold C4EE.

Increment and decrement instructions are important functions of the ALU. The **INR** and **DCR** functions are 1-byte instructions involving the accumulator, one of the six work registers, or data at a memory location M being pointed to by the H and L register pair.

As an example, assume that, before the execution of an *INR D* instruction, the D register holds 10110101. With the execution of the increment command, the ALU adds 00000001 to the contents of register D. Thus, after the execution of this instruction, the D register holds 10110110. If a *DCR D* instruction follows the increment command, 00000001 is subtracted from the contents of D. Thus, register D again holds 10110101. Assuming a register holds 11111111, an INR instruction toggles the register contents to 00000000. If the register holds 00000000, a DCR function toggles the register contents to 11111111. It is important to remember with the 8085, these INR and DCR instructions affect all flag conditions, regardless of the register with which these commands are performed. As demonstrated later, this versatility allows the work registers to be used effectively as counters, with branch decisions contained in a program being based on one of the flag conditions affected by an increment or a decrement instruction.

The INX and DCX commands are 1-byte instructions involving one of the three register pairs or the stack pointer. Since they involve 16-bit words, they have no effect on the flags within the ALU. Assume, before the execution of an *INX D* command that the D and E pair hold E4FF$_{(16)}$, equivalent to 1110010011111111$_{(2)}$. As a result of the *INX D* command, the carry condition produced by incrementing the data in register E causes the data in register D to increment. Thus, the D and E pair now hold the 16-bit word E500$_{(16)}$, equivalent to 1110010100000000$_{(2)}$. Assume that, with E500$_{(16)}$ now contained in the D and E pair, a *DCX D* command is performed. As a result of this instruction, the register pair again holds E4FF$_{(16)}$. Note that the borrow condition resulting from decrementing the E data causes the D data to decrement. As shown later, the extended increment and decrement comands are often used to control the operation of memory pointers, especially the H and L register pair.

The need for instructions involving logic functions was explained earlier. The instruction set of the 8085 microprocessor provides such instructions, which may be divided into two basic categories. One-byte AND, OR, XOR, and compare functions are performed between data in the accumulator and data in one of the work registers, or data in a memory location pointed to by the H and L pair. As an example of such a command, consider the instruction *ANA D*. Assume register D holds 10110101 and the accumulator contains 11110101. After execution of the *ANA D* command, the accumulator holds 10110101. If, given the same data, an *XRA D* function is performed, the accumulator data becomes 01000000. An *ORA D* function would produce 11110101.

Two-byte logic functions involve the accumulator and the immediate data following the op code in memory. For example, assume the accumulator holds 01001011. If the op code for *XRI (EE)* occurs at memory location 30AD, while 30AE holds 10110101, then the operation produces a result in the accumulator of 11111110. A similar operation would occur with either the ANI and ORI instructions, with data following the op code in memory being ANDed or ORed with the contents of the accumulator.

An important feature of the 8085 microprocessor is the **compare** function. This operation is mathematically equivalent to subtraction. However, with a compare, the original accumulator data *is not* destroyed. One-byte compare operations, designated by the mnemonic **CMP,** involve the contents of the accumulator, the six work registers, or the memory location pointed to by H and L. Assume, for example, prior to a *CMP M* command, the accumulator holds 10110110 and the H and L pair is pointing to memory location 40A3. If this memory location contains 01110101, this data byte is subtracted from the contents of the accumulator. The result of this operation, 01000001, although not actually placed in the accumulator, affects all flags. (Remember the accumulator still holds 10110110.) The **CPI** instruction is a two-byte command involving immediate data that follows its op code, **FE,** in memory. Assume, for example, that the op code FE is placed at memory address 308B and that 308C holds data 10110100$_{(2)}$. If the accumulator also holds 10110100$_{(2)}$, then, after the compare operation, the zero flag is set, indicating that the immediate data at location 308C matches the contents of the accumulator. (Again remember that, after this operation, the accumulator still holds 10110100.)

As was shown by the last example, the compare instructions are frequently used to determine if data are less than, greater than, or equal to the contents of the accumulator. It is important to note that, during a compare operation, the zero flag is set only after an equality condition is detected. If the number being compared is greater than that in the accumulator, the borrow (carry) flag is set while the zero flag is low. If the number being compared is less than the number in the accumulator, neither the carry nor zero flag is high. During the execution of industrial control programs, these flag conditions are often tested immediately after a compare operation.

The four **rotate** instructions of the 8085 instruction set are used in binary multiplication and division, as well as serial data input and output operations. All four of these functions involve the accumulator and the carry flag. The **RAL** and **RAR** instructions involve rotation of accumulator data *through* the carry flip-flop, effectively treating the carry flag condition as a ninth bit of the accumulator. For example, assume that the accumulator and carry flag conditions are as shown:

Carry Flag	Accumulator							
CF	2^7	2^6	2^5	2^4	2^3	2^2	2^1	2^0
0	1	1	0	1	0	1	1	0

After the execution of an RAL instruction (rotate the accumulator left through the carry flag), the contents of the carry flag and accumulator would shift one bit position to the left, thus appearing as shown in the following table. Note that the high bit contained in the 2^7 position of the accumulator is now in the carry flag position, whereas the low bit previously contained in the carry flag is now in the 2^0 position.

Carry Flag	Accumulator							
CF	2^7	2^6	2^5	2^4	2^3	2^2	2^1	2^0
1	1	0	1	0	1	1	0	0

If an RAR instruction (rotate the accumulator right through the carry flag) is now executed, data bits are shifted to the right one bit position through the carry flip-flop. Thus, the data in the accumulator and the carry flag would appear as in the previous example, with the carry flag being reset and the accumulator containing 11010110.

With the **RLC** and **RRC** instructions, the accumulator is treated as an 8-bit shift register, because the accumulator data bits are rotated *with* rather than through the carry flip-flop. For example, consider the accumulator and flag conditions represented here:

Carry Flag	Accumulator							
CF	2^7	2^6	2^5	2^4	2^3	2^2	2^1	2^0
0	1	0	1	1	1	1	1	0

With the execution of an RLC instruction (rotate the accumulator left), data in the accumulator are shifted one bit position to the left, with the high bit in the 2^7 position shifting to the 2^0 position. At the same time this occurs, a copy of the high bit that shifts from 2^7 to 2^0 is placed into the carry flag. Thus, after the RLC instruction is executed, the accumulator and carry flag appear as shown here:

Carry Flag	Accumulator							
CF	2^7	2^6	2^5	2^4	2^3	2^2	2^1	2^0
1	0	1	1	1	1	1	0	1

If an RRC instruction (rotate the accumulator right) is executed next, the accumulator data shifts one bit position to the right, with the high bit now present in the 2^0 position shifted to the 2^7 position. As the high bit at 2^0 moves to 2^7, it is also copied into the carry flag. Thus, the accumulator appears as shown here, with the carry flag remaining high:

Carry Flag	Accumulator							
CF	2^7	2^6	2^5	2^4	2^3	2^2	2^1	2^0
1	1	0	1	1	1	1	1	0

Special instructions within the arithmetic and logical group include the **DAA** (decimal adjust the accumulator), **CMA** (complement the accumulator), **STC** (set the carry flag), and **CMC** (complement the carry flag). The purpose and operation of the decimal adjust instruction has already been explained. However, Program Example 12–2 demonstrates how decimal adjustment is performed using the 8085 instruction set. Before beginning a

Program Example 12-2

Address	Op Code	Mnemonic	Comment
3000	AF	XRA A	Clear the accumulator. Reset the carry and auxiliary carry flags.
3001	3E	MVI A, 45	Move 45$_{(BCD)}$ into the accumulator.
3002	45		
3003	06	MVI B, 27	Move 27$_{(BCD)}$ into B.
3004	27		
3005	80	ADD B	Add B data to A data.
3006	27	DAA	Maintain BCD format.
3007	4F	MOV C, A	Store BCD result in C.

BCD add operation involving the DAA function, it is necessary to clear both the carry and auxiliary carry flags. For this reason, the *XRA A* command at address 3000 is recommended. Not only does it clear the accumulator, it resets both the carry and auxiliary carry flags.

The CMA instruction is quite simple to understand. Before the execution of this 1-byte instruction, assume the accumulator holds 01001101. After the execution of the complement command, the accumulator holds 10110010. The STC function is a 1-byte instruction that directly sets the carry flag to a high condition. The CMC function directly complements the carry condition. If the carry flag is set, the complement instruction resets it to a logic low. If the carry flag is already reset, the CMC command sets the flag to a logic high.

If all industrial control programs could run directly through memory, from start to finish, there would be no need for the **branch control group** of instructions. However, during complex industrial control applications, microcomputers must respond quickly to changes in the process being controlled. For this reason, the branch control group of instructions becomes essential. The branch commands of the 8085 instruction set are classified as either **jumps** or **calls.** Also, these types of branch instruction may be further classified as **conditional** or **unconditional.** The term *jump* is an abbreviation for the instruction ''jump the program counter.'' All jump commands are 3-byte instructions. The op code for a jump instruction is followed by 2 bytes of data that represent the low byte and high byte of an address to be loaded into the program counter. Once this new address is placed in the program counter, the 8085 resumes execution of the program at that location. As shown in the following example, assume the op code for the unconditional jump instruction **JMP** is entered at memory location 3007. The data at memory location 3008 is then considered to represent the low byte of an address word, while address 3009 is considered to contain the high byte.

3007	C3
3008	7C
3009	40

After execution of the preceding 3-byte instruction, rather than being incremented to address 300A, the program counter is jumped to address 407C. Thus, the contents of address 407C, assumed to be the op code for the next instruction of the program, is latched into the instruction register.

Conditional jumps are 3-byte instructions that depend on the status flags to determine whether or not they are executed. Two jump instructions that depend on the condition of the carry flag are **JC** (jump if the carry flag is high) and **JNC** (jump if the carry flag is not high). Consider Program Example 12–3. In this example, for the 8085 to be able to jump the program counter to address 405B, the contents of register B must be greater than $94_{(16)}$. As an example, if $95_{(16)}$ is added to the immediate data $6B_{(16)}$, $100_{(16)}$ is produced, thus setting the carry flag. With this condition, the *XRA A* instruction at address 300C is not executed, since the 8085 fetches its next instruction from address 405B. If, however, the data byte in register B is $94_{(16)}$ or less, the carry flag is not set. With this condition, the 8085 would ignore the jump command at address 3009 and continue execution at location 300C. If op code D2, for the JNC instruction, is placed at address 3009, opposite results to those just described would occur. If the add function did *not* produce a carry, the jump to address 405B would take place. If a carry was produced, program execution would continue at 300C.

Program Example 12-3

Address	Op Code	Mnemonic	Comment
3006	3E	MVI A, 6B	Move immediate data into the accumulator.
3007	6B		
3008	80	ADD B	Add B data to accumulator data.
3009	DA	JC, 405B	Go to address 405B if carry is produced.
300A	5B		
300B	40		
300C	AF	XRA A	Clear the accumulator.

Two jump instructions that depend on the condition of the zero flag are **JZ** (jump if result equals zero) and JNZ (jump if result does not equal zero). Program Example 12–4 serves to demonstrate the operation of the JZ and the JNZ commands. In this program, the 8085 scans a maximum of 256 locations in memory, searching for data that match the contents of the accumulator. During this process, register C serves as a counter. After a location is checked, if a matching data byte is not found, the memory pointer is incremented and the count in C is decremented. If a matching data byte is found, the program counter is jumped to address 5082, which could be the starting address of another program routine. Register C defines the limit of the area in memory being searched. If the down-count in register C decrements to zero, then the unconditional jump instruction at address 3010 is executed. Since the program counter is being jumped to the address of the JMP instruction, the microprocessor is locked in a loop, requiring a reset or restart instruction to allow it to continue operation at another location. This application of the unconditional jump is equivalent to a **halt** command.

Two jump commands that rely on the condition of the parity flag are **JPO** (jump if parity is odd) and **JPE** (jump if parity is even). As with the other 3-byte jump instructions, the op codes are followed by 2 bytes of data representing the branch destination. If an arithmetic or logical operation produces an odd-parity condition in the accumulator, causing the parity flag to reset, a JPO instruction would be executed and a JPE command would be ignored. If the operation produced an even-parity condition, causing the parity flag to set, a JPE command would be executed, and the JPO instruction would be ignored. Two instructions involving the sign flag are **JP** (jump if positive) and **JM** (jump if minus). Remember that, based on the sign binary system, if the accumulator data byte contains a high bit in the 2^7 position, a negative condition is assumed to exist. If the 2^7 bit is low, a positive condition is assumed to exist. Thus, if an arithmetic or logical operation produces 01001111, an ensuing JM command would be ignored and a JP command would be executed. If 10110100 was produced, the JP command would be ignored and the JM instruction would be executed.

Many machine language programs used in industrial process control are primarily compilations of **subroutines.** A subroutine is a miniature program embedded in the memory of the microcomputer that is accessible from more than one point in a given program or, in some cases, from more than one program. To expedite the process of

Program Example 12-4

Address	Op Code	Mnemonic	Comment
3000	21	LXI H, 40C3	Set memory pointer to address 40CB.
3001	CB		
3002	40		
3003	0E	MVI C, FF	Define count in C.
3004	FF		
3005	3E	MVI A, 8B	Place data in accumulator.
3006	8B		
3007	BE	CMP M	Compare memory data to accumulator.
3008	CA	JZ, 5082	Goto address 5082 if memory matches accumulator.
3009	82		
300A	50		
300B	23	INX H	Point to next memory location.
300C	15	DCR C	Count minus 1.
300D	C2	JNZ, 3007	Go back to compare data if count is not complete.
300E	07		
300F	30		
3010	C3	JMP, 3010	Wait in loop.
3011	10		
3012	30		

entering and exiting subroutines, **call** and **return** instructions were developed. Like a jump, a call instruction consists of 3 bytes (an op code followed by a low byte and a high byte of an address word). Unlike a jump, however, the call instruction involves the stack. The involvement of the stack in a call instruction might best be explained through use of an example. Assume that during the execution of a main program, as shown in the following short program segment, a call instruction occurs, accessing a subroutine beginning at memory location 2024:

3002	CD
3003	24
3004	20

Assume that, at the beginning of the main program, the stack pointer was set to address 5002. With the execution of the call instruction, the program counter is jumped to address 2024, while the address to be returned to after the subroutine is placed on the stack. Thus, the stack pointer decrements to address 5000, with the return address being placed on the stack as shown:

5000	05
5001	30
5002	XX

All subroutines end with a return command. Like jumps and calls, returns may be unconditional or conditional, relying on the conditions of the status flags. Unlike jumps and calls, however, returns are 1-byte rather than 3-byte instructions. In programming for the 8085, a return is always the last instruction of a subroutine. As the 8085 fetches and executes a return command, the lowest 2 bytes of data on the stack are placed in the program counter and the stack pointer is incremented by two address locations. Thus, referring to the preceding example, address 3005 would be placed in the program counter, while the stack pointer is incremented back to 5002.

The conditions for 3-byte call instructions are identical to those for conditional jumps. Those call instructions involving the carry flag are **CC** (call if the carry flag is high) and **CNC** (call if the carry flag is not high). Responding to the zero flag are the **CZ** (call if result equals zero) and the **CNZ** (call if result does not equal zero) instructions. The call instructions involving the parity flag are **CPO** (call if parity is odd) and **CPE** (call if parity is even). Two more call instructions, involving the sign flag, are **CP** (call if positive) and **CM** (call if minus). One-byte return instructions involving the carry flag are **RC** (return if the carry flag is high) and **RNC** (return if the carry flag is not high). Involving the zero flag are **RZ** (return if result equals zero) and **RNZ** (return if result does not equal zero). Involving the parity flag are **RPO** (return if parity is odd) and **RPE** (return if parity is even). Finally, responding to the sign flag are **RP** (return if positive) and **RM** (return if minus).

The next group of 8085 instructions to be covered is the **I/O and machine control** group. This group of instructions may be further divided into those involving the stack, those involving the input and output ports, and those involving the maskable interrupts. **PUSH** and **POP** are 1-byte instructions involving the stack. The PUSH instruction allows the storage of register pair data on the stack; the POP instruction allows the retrieval of that data. PUSH commands for the 8085 include *PUSH B, PUSH D, PUSH H,* and *PUSH PSW*. POP commands allow 2 bytes of data stored on the stack to be loaded into a register pair. POP commands include *POP B, POP D, POP H,* and *POP PSW*.

As an introduction to the PUSH instruction, assume that register B holds AF and register C holds B7. Also assume the stack pointer is at memory location 50B7. When the op code for the *PUSH B* instruction, C5, is placed in the instruction register and the high byte and low byte of the B and C pair are placed on the stack as shown here, the stack pointer decrements to 50B5.

50B5	B7
50B6	AF
50B7	XX

During a *POP B* operation, two bytes of data present on the stack are loaded into the B and C register pair. Assume, for example, that the stack pointer currently contains 6AB3 and that the contents of memory are as shown next. When the op code C1 is placed in the instruction register, the contents of address 6AB3 is placed in register C and the data at 6AB4 is placed in register B, while the stack pointer increments by two address locations. Thus, after the *POP B* operation, register B holds 63, register C holds A7, and the stack pointer contains address word 6AB5.

6AB3	A7
6AB4	63
6AB5	XX

The mnemonic **XTHL** may be interpreted as "exchange the data at the top of the stack with the contents of the H and L register pair." This command can, at first, be confusing, since the top of the stack is actually the lowest extent of the stack in memory. (Remember that, as the stack grows, the stack pointer decrements.) As an example of the XTHL command, assume register H holds 35, register L holds A7, and the stack pointer contains the address word 50B3. Also assume that, before the execution of the XTHL command, the condition of the stack is as shown:

50B3	C2
50B4	9A
50B5	XX

After the op code for XTHL, E3, is fetched and executed, register H holds 9A and register L holds C2, while the stack condition changes as shown next. It is important to note that the stack pointer still contains 50B3.

50B3	A7
50B4	35
50B5	XX

SPHL is a 1-byte instruction that places the contents of the H and L register pair into the stack pointer. Assume, for example, that the H and L register pair holds the extended data word 45A3. As the op code F9 is fetched and executed by the 8085, 45A3 is loaded into the stack pointer, causing address 45A3 to become the top of the stack. While 45A3 is still contained in registers H and L, the original contents of the stack pointer are destroyed.

Program Example 12–5 comprises a short subroutine, the purpose of which is to save the contents of the accumulator, flags, work registers, stack pointer, and program counter of the 8085. Such a process is often performed before entering into another routine that, during its own execution, destroys the contents of the registers and flags of the main program. Before returning to the main program, the contents of the flags and registers from the main program would be restored.

Program Example 12-5

Address	Op Code	Mnemonic	Comment
041B	22	SHLD, 4000	Save H and L at 4000 and 4001.
041C	00		
041D	40		
041E	E1	POP H	Retrieve program count into H and L.
041F	22	SHLD, 4002	Store program count at 4002 and 4003.
0420	02		
0421	40		
0422	F5	PUSH PSW	Place accumulator and flags on stack.
0423	E1	POP H	Retrieve accumulator and flags into H and L.
0424	22	SHLD 3FFE	Store accumulator and flags at 3FFE and 3FFF.
0425	FE		
0426	3F		
0427	21	LXI H, 0000	Clear registers H and L
0428	00		
0429	00		
042A	39	DAD SP	Retrieve stack pointer into H and L.

Program Example 12-5 (*continued*)

042B	22	SHLD, 4004	Save stack pointer at 4004 and 4005.
042C	04		
042D	40		
042E	21	LXI H, 3FFE	Define top of stack as 3FFE.
042F	FE		
0430	3F		
0431	F9	SPHL	Set stack pointer to 3FFE.
0432	C5	PUSH B	Save B and C register pair.
0433	D5	PUSH D	Save D and E register pair.
0434	C9	RET	

While most of the commands in Program Example 12–5 are sufficiently clarified in the comments column, two of them require some additional explanation. Remember, in response to a call instruction, the program counter is incremented to the address where the program is to resume *after* a subroutine. This address word is then pushed on the stack. Since the contents of the program counter is the last information placed on the stack, the *POP H* command at address 041E places the contents of the program counter in the H and L pair. The subsequent SHLD command saves this return address in memory. At address 042A, since the H and L registers now contain 0000, the DAD command effectively places the contents of the stack pointer in the H and L pair. As with the program count, a subsequent SHLD command stores the contents of the stack pointer at two locations in memory.

The **input/output instructions** of the 8085 microprocessor involve the **ports** of the microcomputer system in which it is contained. These ports, as explained later, can be located on the system's RAM and ROM ICs. Ports are used to send data to, or receive data from, peripheral devices such as D/A and A/D converters. At the beginning of a program involving ports within a memory device, it is usually necessary to initialize these ports as being either **inputs** or **outputs.** For example, consider the 8155 RAM IC, which is often used in conjunction with the 8085. Along with 256 bytes of memory, this IC contains two ports, addressed as 00 and 01, which must be defined as either inputs or outputs. This is accomplished by outputting to the **data direction registers** of the 8155, which are addressed as ports 02 and 03. Port 02 defines the operation of port 00, and port 03 defines the operation of port 01. By accessing the data registers, ports 00 and 01 may be configured as inputs or outputs at the individual bit level. Placing a logic high in a given bit position of a data direction register causes the corresponding bit position of its associated port to function as an output. Placing a logic low in a given bit position of the data direction register defines its corresponding port bit position as an input.

The op code for an input or output is followed by a port designation. Thus, input and output commands are 2-byte instructions. With an output command, the data in the

accumulator is latched into the port designated by the second byte. With an input instruction, the data contained in the port designated by the second byte is loaded into the
accumulator. For example, consider Program Example 12–6, where we assume that an
8155 is being used in conjunction with the 8085 microprocessor. Thus, we must define
the port being used. This is accomplished through clearing the accumulator and outputting
to the data direction register of port 00. Since all 8 bits of the accumulator are low, all
8 bits of port 00 are defined as inputs. With port 00 now defined as an input, the *IN, 00*
instruction at address 300D causes data at port 00 to be placed in the accumulator. These
data are then saved at address 4050.

Program Example 12–7 illustrates how the same port of the 8155 RAM could be
configured to operate as an output. Outputting FF_{16} to the data direction register of port
00 defines that port as an 8-bit output. Thus, the output instruction at address 300F latches
data from memory location 4050 at port 00.

All the 8085 operations covered thus far have involved the uninterrupted fetching
and execution of instructions originating in memory. These instructions allow the 8085
to perform virtually all forms of mathematics and data transfer necessary for industrial
control. However, to allow the microprocessor to communicate effectively with input
devices, additional branch control instructions involving **restarts** are necessary. There are
eight restarts within the branch control group, RST0 through RST7, that are similar to
call instructions. Like call instructions, these restarts place the return address of the main
program on the stack. Whereas a call is a 3-byte instruction consisting of an op code and
an address word, RST0 through RST7 are only single-byte instructions. As shown in
Table 12–1, each restart has a **vector address.** If one of the eight RST commands is read
by the 8085, after the return address is saved on the stack, the program counter is jumped
to the corresponding vector address. Because of the limited amount of memory between
these vector addresses, the instruction at the vector address usually consists of an unconditional jump (JMP) instruction. This jump instruction vectors the program count to a
location in memory where there is sufficient room to contain the subroutine called by the
RST instruction. For example, assume a RST6 instruction, occurring at address 307D, is
used to call a subroutine that begins at address 507B. Since the RST is a single-byte

Program Example 12–6

3007	21	LXI H, 4050	Set memory pointer at 4050.
3008	40		
3009	50		
300A	AF	XRA A	Clear the accumulator.
300B	D3	OUT, 02	Define port 00 as an 8-bit input.
300C	02		
300D	DB	IN, 00	Input data at port 00.
300E	00		
300F	77	MOV M, A	Store input data in memory.

Program Example 12-7

3007	21	LXI H, 4050	Set memory pointer at 4050.
3008	40		
3009	50		
300A	3E	MVI A, FF	Define port 00 as an 8-bit output.
300B	FF		
300C	D3	OUT, 02	
300D	02		
300E	7E	MOV M, A	Retrieve data from memory.
300F	D3	OUT, 00	Output data at port 00.
3010	00		

instruction, address 307E (the return address) is pushed on the stack. Next, the vector address for RST6 (0030) is loaded in the program counter. Thus, the 8085 fetches an instruction from that address. To access the subroutine beginning at 507B, a jump must be entered in memory as shown next. At the end of the subroutine accessed by the RST6 instruction, a return instruction places 307E back into the program counter, allowing the main program to resume.

0030	C3
0031	7B
0032	50

Although, as just demonstrated, the RST instructions may be used as conventional calls, they may also be used in interrupt service routines. As shown in Figure 12-7, the 8085

Table 12-1
Restart instructions for 8085 microprocessor

Mnemonic	Op Code	Vector Address
RST0	C7	0000_{16}
RST1	CF	0008_{16}
RST2	D7	0010_{16}
RST3	DF	0018_{16}
RST4	E7	0020_{16}
RST5	EF	0028_{16}
RST6	F7	0030_{16}
RST7	FF	0038_{16}

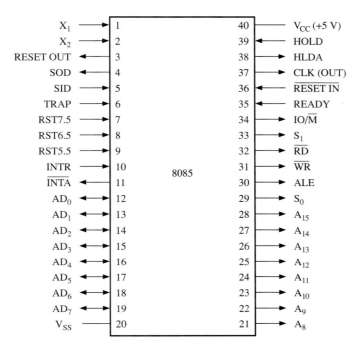

Figure 12–7
Pin assignments for 8085 microprocessor

has an **interrupt request** input (INTR) at pin 10. A peripheral device may initiate a restart routine independently of the program in memory by placing a logic high at the INTR input of the 8085. At the same time, this peripheral device must place the op code for the RST instruction on the AD_0 through the AD_7 lines of the microcomputer. The 8085 then acknowledges receipt of the interrupt service request by outputting an active-low **interrupt acknowledge** signal ($\overline{INTA}$) at pin 11.

Besides the eight restart instructions just described, the 8085 has four **hardware interrupts.** As the term *hardware* implies, these four active-high interrupt signals occur directly at four input pins of the microprocessor. These four inputs are often referred to as **priority interrupts,** since some interrupts take precedence over others in the order they are serviced. The interrupt line with the highest priority is **TRAP.** Because it cannot be disabled through the programming of the 8085, the TRAP is referred to as an **unmaskable** interrupt. Whenever a high occurs at the TRAP input (pin 6) of the 8085, the ongoing program is interrupted and the program counter is loaded with the vector address 0024_{16}. As with the RST inputs, this location is usually loaded with an unconditional jump instruction that jumps the program counter to the location of the TRAP interrupt service routine. Not only is the TRAP input capable of interrupting the main program in memory, it may also temporarily break into any other interrupt routine that might be occurring at the instant a high pulse occurs at pin 6 of the 8085.

Further examination of the pins of the 8085 microprocessor shows there are three restart inputs, designated **RST5.5, RST6.5,** and **RST7.5.** These hardware interrupts are referred to as **maskable** because, as the program in memory is running, they may be enabled and disabled. If the maskable interrupts are to be used during a program, the **interrupt mask** must first be set. This is achieved by entering a control word into the accumulator as immediate data and sending this word to the control section of the 8085.

The **set the interrupt mask instruction,** identified by the mnemonic **SIM,** is a 1-byte command that sends the contents of the accumulator to the control circuitry of the 8085. With the execution of this command, the control circuitry of the 8085 interprets the contents of the accumulator as defined in Figure 12–8. The SIM instruction serves a dual role since it controls both the serial output operation and the status of the maskable interrupts. To enable the serial output function, a logic high must be present in the 2^6 bit position of the accumulator prior to execution of the serial output function. Execution of the SIM command then causes the bit condition at the 2^7 position of the accumulator to be present at the SOD line (pin 4) of the 8085. Definition of the maskable interrupts to be used within a program requires that a logic high be present in the 2^3 position of the accumulator, prior to the execution of the SIM instruction. If the control circuitry senses this bit condition as being high, it checks the conditions of the 2^0, 2^1, and 2^2 bits within the accumulator, which represent RST5.5, RST6.5, and RST7.5, respectively. If a logic low is present in any of these three bit positions, the control circuitry unmasks the corresponding RST input.

Thus, if the immediate data byte 08 is loaded into the accumulator prior to a SIM instruction, RST5.5, RST6.5, and RST7.5 are unmasked simultaneously, allowing the 8085 to respond to interrupt requests from three different sources. Expanding the accumulator data to binary clarifies why this is so. Since the control word 00001000 contains a high in the 2^3 position, a mask set enable condition occurs, allowing the logic lows in the three least significant bit positions to unmask the three interrupt inputs.

The RST7.5 function of the 8085 may be directly reset by placing a logic high at the 2^4 bit position of the accumulator and executing a SIM instruction. Masking of the other two interrupts would require that a logic high be present at the 2^2, 2^1, and 2^0 bit positions before the execution of the SIM instruction.

It must be stressed that unmasking RST5.5, RST6.5, or RST7.5 does not, in itself, cause the 8085 to respond to an active-low transition at pin 9, 8, or 7. After the interrupt mask is set, for the microprocessor to be able to respond to an interrupt request, an instruction to **enable the 8085 interrupt capability,** the mnemonic for which is **EI,** must be fetched and executed. This 1-byte instruction usually does not occur within a program until a point at which the 8085 has already completed critical tasks and may be safely interrupted. In fact, to ensure reliable operation of the 8085, one of the first instructions to occur in a program involving the hardware interrupts is an instruction to **disable the 8085 interrupt capability,** the mnemonic for which is **DI.** Not until the necessary initialization tasks of the program are completed does the EI instruction occur.

Figure 12–9, which illustrates the interrupt control logic of the 8085, should aid in clarifying the operation of the microprocessor's interrupt system. Note that the TRAP line is represented as going directly to its vector address, whereas the maskable interrupts are controlled by AND gates and flip-flops. As indicated in Figure 12–9, the RST7.5 input, which has priority over all interrupts but trap, is controlled by a D flip-flop. When a positive-going transition occurs at the RST7.5 input, the Q output of the D flip-flop switches to a logic high. However, for the RST7.5 service request to be initiated, the

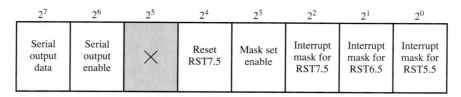

2^7	2^6	2^5	2^4	2^3	2^2	2^1	2^0
Serial output data	Serial output enable	✕	Reset RST7.5	Mask set enable	Interrupt mask for RST7.5	Interrupt mask for RST6.5	Interrupt mask for RST5.5

Figure 12–8
Contents of the accumulator as defined after setting of the interrupt mask

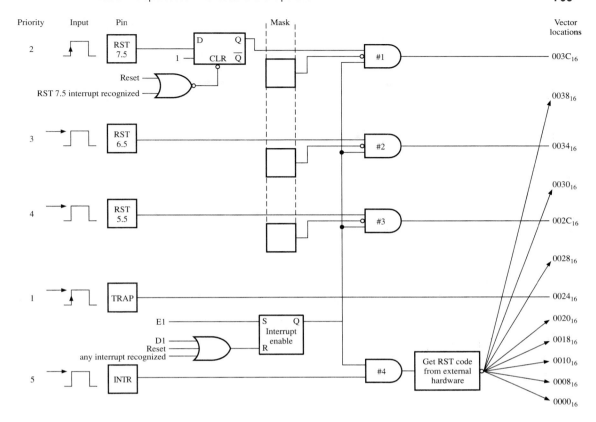

Figure 12–9
Structure of 8085 interrupt circuitry

mask bit for RST7.5 (2^2 within the accumulator) must have been brought low prior to execution of the SIM instruction. Also, an EI instruction must occur. This action sets the Q output of the S/R flip-flop to a logic high, allowing the output of AND gate #1 to go to a logic high. Note that the CLR input of the D flip-flop associated with the RST7.5 interrupt is controlled by a NOR gate. If a high has been placed at the 2^4 bit position of the accumulator, after execution of the SIM instruction, this high appears at the reset input to the NOR gate. This brings the Q output of the D flip-flop to a logic low, disabling the RST7.5 function.

As indicated at the second input to the NOR gate, a high pulse automatically occurs at that point at the instant a RST7.5 service routine is initiated. This high pulse also occurs at the third input to an OR gate that controls the reset input to the S/R flip-flop. Such action is necessary to disable temporarily the maskable interrupt hardware functions while an interrupt service routine is being executed. It is, therefore, necessary, before the return from an interrupt service routine, to re-enable the interrupt capability with an EI instruction. RST5.5, RST6.5, and RST7.5, like TRAP and the RST instructions, have vector addresses. These addresses are shown in Figure 12–9.

The status of the maskable hardware interrupts of the 8085 may be read through use of a **read the interrupt mask** instruction, the mnemonic for which is **RIM.** With the execution of this 1-byte instruction, the status of the hardware interrupts, as shown in Figure 12–10, is loaded into the accumulator. Note that the bit positions for the three maskable hardware interrupts correspond to those associated with the SIM instruction. Immediately after execution of the RIM instruction, a logic low in the 2^0, 2^1, or 2^2 bit position indicates that RST5.5, RST6.5, or RST7.5 has been enabled. A high in one of these three bit positions indicates a disabled interrupt condition. A logic high in the 2^3 position indicates that the interrupt enable flag (as controlled by the SR flip-flop in Figure 12–9) is set. A low in this bit position indicates the interrupt enable flag is reset. As indicated in Figure 12–10, the 2^4, 2^5, and 2^6 positions are used to reveal the status of pending interrupts. A high in one of these bit positions indicates that an interrupt request is occurring at the corresponding hardware input. A logic low in one of these bit positions indicates that the corresponding interrupt line is inactive. Note that, after execution of the RIM instruction, the 2^7 bit position represents the status of the serial input data (SID) line of the 8085.

With the interrupt system of the 8085 having been introduced, the need for the **halt** instruction (the mnemonic for which is **HLT**) should become evident. Often, after the interrupt capability of the microprocessor is enabled, the main program is halted, awaiting a high pulse at an unmasked interrupt line. When the op code for the HLT instruction, 76, is latched in the instruction register, further sequential fetching and execution of program instructions is stopped. Only through intervention of a restart is program execution resumed.

SECTION REVIEW 12–1

1. What is an ALU? What operations is it capable of performing? In what part of a microprocessor is the ALU contained?
2. Identify the main sections of a microprocessor. Identify the essential elements of a microcomputer.
3. With the 8085 microprocessor, when is it advantageous to use a compare instruction rather than a subtract command? Explain the reason for your answer.
4. Assume the accumulator of the 8085 holds 3A. After a *CPI, C4* instruction, what would be the status of the carry, parity, sign, and zero flags? What would be the contents of the accumulator?
5. Define the term *program status word* as it applies to the 8085 microprocessor.
6. Describe the purpose of the stack in an 8085-based microcomputer. What branch commands of the 8085 instruction set automatically place data on the stack?
7. True or false?: For the 8085 microprocessor, the stack pointer increments as data are placed on the stack and decrements as data are removed from the stack.
8. True or false?: The last instruction of a subroutine accessed by a call instruction should always be a return.
9. What is a vectored interrupt? What instruction is usually placed at the vector address?
10. Which hardware interrupt of the 8085 is unmaskable? Does this input have priority over the other hardware interrupts?

2^7	2^6	2^5	2^4	2^3	2^2	2^1	2^0
Serial input data	Interrupt pending flag for RST7.5	Interrupt pending flag for RST6.5	Interrupt pending flag for RST5.5	Interrupt enable flag	Interrupt mask for RST7.5	Interrupt mask for RST6.5	Interrupt mask for RST5.5

Figure 12–10
Contents of the accumulator as defined after reading of the interrupt mask

12–2   INTRODUCTION TO MEMORY DEVICES

For a microprocessor to operate, it must have a storage area or **memory** in which programs routines and immediate data are held. First-generation microprocessors such as the 8085 have no on-board memory; rather, they rely on peripheral ICs to form the memory of the microcomputer. The microcomputer memory is divided into two basic sections called **read-only memory** (ROM) and **random-access memory** (RAM). While several chapters of a microcomputer text could easily be devoted to the operation of microprocessors and their associated memory devices, this section of the chapter serves to introduce RAM and ROM ICs briefly and describe their function within the embedded (or dedicated) microcomputers found in industrial control systems.

As the term *read-only* implies, the microprocessor is able to read instructions and data from a ROM but is not able to send or **write** data to it. However, the microprocessor is able to both read data from and write data to a RAM. A major distinction between a ROM and a RAM is that a ROM is **nonvolatile.** This means that a ROM is able to contain its data indefinitely after power is removed from a microcomputer. With a RAM, however, unless it has battery backup, data are lost almost immediately as power is removed from the IC. Equally important, each time power is applied to a ROM, the data that appear at each of its memory locations are always the same. However, at the instant power is applied to a RAM, the logic conditions that appear at its data registers are unpredictable.

The ROM within an embedded microcomputer of an industrial control system contains those program routines that are necessary for initializing the microprocessor IC and allowing that device to communicate with the main computer system, as well as peripheral devices. Usually consisting of several interconnected subroutines, the main program in ROM is often referred to as the **monitor program,** since it continually supervises or "monitors" microcomputer operation. The ROM may contain subroutines for frequently performed mathematical operations. Such problem-solving procedures are referred to as **algorithms.** The programming necessary to convert a binary or BCD number representing a peak voltage to its rms equivalent would be an example of an algorithm.

A ROM could also contain **look-up tables,** or data lists. As an example, a look-up table could represent the temperature range of a thermistor, which is an accurate but inherently nonlinear input transducer. Rather than performing a complex and time-consuming algorithm for linearizing the characteristic curve of the thermistor, the output of this transducer could be converted to a binary number by an A/D converter and used as an offset for accessing a look-up table. The output of the A/D converter could be connected to a parallel port of the microcomputer. As part of a simple machine language program, input data could be added to a memory pointer containing the base address of a look-up table in ROM. The data contained at the ROM address accessed by the memory pointer would represent the current value of thermistor temperature. The use of algorithms and look-up tables within microcomputer control systems is covered in the next chapter.

Figure 12–11 is a simplified illustration of how a common-bus microcomputer is organized. (Refer to Figures 12–5 and 12–7 while studying Figure 12–11 to aid in your understanding how the microcomputer's memory interfaces with the microprocessor IC.) Pins 21 through 28 of the 8085 are connected to the upper byte of the microcomputer's address bus; pins 12 through 19 are effectively connected to both the data bus and the lower byte of the address bus. The designations AD_0 through AD_7 indicate that, during a read or write operation of the 8085, the lower byte from the **incrementer/decrementer address latch** is **multiplexed** with the 8-bit data word being sent to or from the microprocessor. (With reference to microcomputers and communication systems, the term *multiplexing* describes the simultaneous transmission of two or more messages on the same circuit or channel.) For the 8085, the primary advantage of multiplexing the lower byte

Figure 12–11
Simplified block diagram of a microcomputer

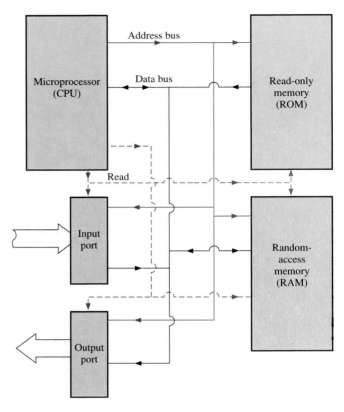

of address information with data is pin conservation. However, as is now demonstrated, multiplexing requires intricate timing for each read and write operation.

Assume that the program counter of the 8085 is pointing to 043B, which is an address in ROM containing the next instruction to be fetched and executed. During T_1, by generating a high pulse at its **Address Latch Enable** (ALE) output (pin 30), the 8085 latches 043B in the address buffer of the ROM. As shown in Figure 12–12, the actual read function occurs during the T_2 and T_3 clock cycles. On the negative transition of T_2, an active-low read request appears at the $\overline{RD}$ output (pin 32) of the 8085. At this time, the contents of the address location pointed to by the address and data/address buffers appear on the data bus of the microcomputer. During the T_3 clock cycle, this op code is latched into the instruction register of the 8085. As indicated in Figure 12–12, the four clock states during which the instruction fetch occurs comprise a **machine cycle.** During subsequent machine cycles, the 8085 executes the instruction fetched from memory. The number of machine cycles required to complete this process depends on the complexity of the instruction being fetched.

Figure 12–13 illustrates the operation of the 8085 microprocessor when the instruction fetched from memory involves a write operation. As an example, assume that 70, the op code for *MOV M, A* is contained at address 402B and the program counter is currently pointing to that address. Assume during one machine cycle, as described for Figure 12–12, op code 70 has been fetched and placed in the instruction register. Thus, during the first clock cycle T_1 in Figure 12–13, since the *MOV M, A* instruction involves storage of data in RAM, the 16-bit address word contained in the H and L register pair is latched into the incrementer/decrementer address latch. Also, during the negative transition of the T_2

Figure 12–12
The 8085 microprocessor instruction fetch

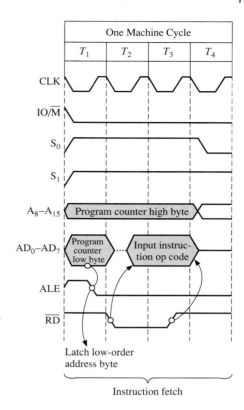

clock cycle, the active-low $\overline{WR}$ output signal appears at pin 31 of the 8085. Assuming the H and L pair form the address word 804C, this location in RAM is enabled to receive data. During the T_3 clock cycle, the data byte contained in the accumulator is stored in RAM location 804C.

To further understand the operation of the memory within a microcomputer, it is necessary to examine the types of ICs that are commonly used in microcomputer memory systems. The embedded microcomputers of industrial control systems typically have limited RAM space. For example, a small microcomputer system (contained on a single printed circuit board) might contain only 8.192 kbytes of RAM, while having a ROM area of 32.768 kbytes. This disproportional relationship of RAM and ROM space is typical for an embedded microcomputer. Ample ROM space is necessary to contain the many subroutines that must be executed by the microprocessor during its control function. However, a relatively small amount of RAM is sufficient for temporary storage of variables that are dealt with during the operation of the embedded microcomputer.

In a modern industrial control system, ROMs exist in a variety of forms. The basic ROM, which could consist of a diode matrix, has one major limitation. Once fabricated, the basic ROM may not be modified. This presents a particular problem in systems that are in the developmental stage, where debugging of subroutines in ROM is likely to be necessary. Also, due to modifications and updates of existing control systems, the programs in ROM may require frequent alterations. To avoid the time delay and cost of having manufacturers replace outdated ROM ICs, microcomputers often contain programmable read-only memories. The advantage of a **programmable read-only memory** (PROM) is that it may be programmed by a relatively small and inexpensive piece of equipment called a **PROM programmer.** Rather than having to order bulk quantities of ROMs

Figure 12–13
The 8085 microprocessor write function

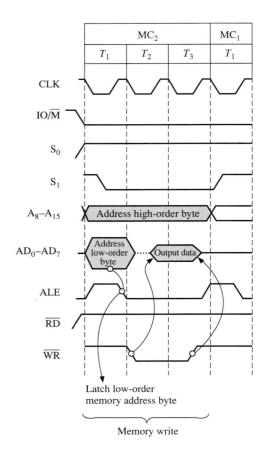

directly from manufacturers during the developmental or prototyping stages of a project, digital designers are able to develop their own PROMs quickly and cheaply. After their monitor programs have been thoroughly debugged, microcomputer developers have the option of ordering bulk quantities of ROMS from another manufacturer or, if their own plant facilities are able to accommodate the task, continue fabricating their own PROMs in-house.

Even greater versatility is provided by the **erasable programmable read-only memory** (EPROM). Like the PROM, this form of memory may be easily programmed by a compact machine. An example of an **EPROM programmer** is shown in Figure 12–14(a). During the developmental stage of a design project, if the program contained within an EPROM must be modified, the obsolete program may be erased simply by exposing the device to ultraviolet light. The EPROM has a small quartz window through which the UV light is allowed to pass during the erasure process. An example of an **EPROM eraser** is shown in Figure 12–14(b). After the EPROM is erased and reprogrammed, the programs in memory are usually protected by placing a label over the quartz window of the device. One obvious advantage of using EPROMs is cost efficiency, since expenditure of parts is reduced. A still more versatile memory device is the **electrically erasable programmable read-only memory** (EEPROM or E^{2}PROM). With this device, erasure is quickly accomplished through a short electrical pulse (as quick as 20 msec) rather than exposure to UV light. Thus the need for an EPROM eraser may be eliminated. EEPROMs are covered further during the discussion of microcontrollers.

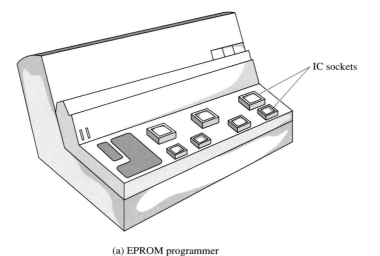

(a) EPROM programmer

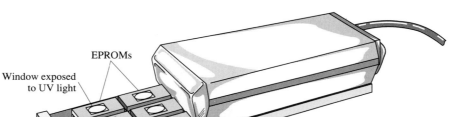

(b) EPROM eraser

Figure 12–14

Figure 12–15 shows the pin assignments and simplified internal structure of either the 8355 ROM or 8755 EPROM, manufactured by Intel Corporation. Both of these tri-state NMOS devices are compatible with the 8085 microprocessor. The 2k $\times$ 8 designation in Figure 12–15(b) indicates these memories contain approximately 2000 8-bit data locations. (In actuality, a 2k memory has 2^{11} or 2048 locations.) In addition to the 2048 ROM locations, the 8355 and 8755 provide two 8-bit ports designated as **port A** and **port B.**

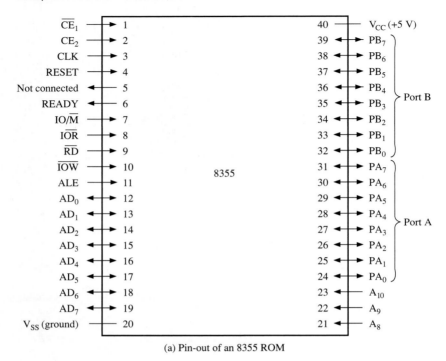

(a) Pin-out of an 8355 ROM

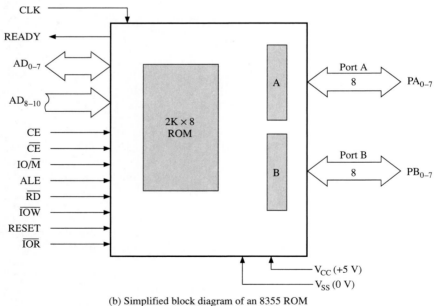

(b) Simplified block diagram of an 8355 ROM

Figure 12–15

As explained later, each of the 8 bit positions of both ports of the 8355 and 8375 may be programmed individually as an input or an output, through use of two data-direction registers. Although the 8355 and 8755 are read-only memories, it is important to realize that, with their I/O capability, these ICs must be able to perform both read and write operations. The following discussion of the control signals involved with these two ICs should clarify how these read and write operations are achieved.

Consider first the conditions that must be met for the 8355 and 8755 to be enabled to operate. Remember that, for a tri-state logic device to be active, it must receive an enabling signal from some external control source. Otherwise, the device remains in its dormant, high-impedance state. For the 8355 and 8755 to be enabled to place data on the data bus, or receive information from the address and data buses of a microcomputer, the active-high **chip enable** line at pin 2 (CE) must be high. At the same time, the active-low chip enable line at pin 1 ($\overline{CE}$) must be at a logic low.

As shown later, a RAM must receive both read and write request signals from a microprocessor. However, since the 8355 and 8755 contain read-only memories, just an active-low read request line at pin 9 ($\overline{RD}$) is required. Since the two ports perform both input and output operations, the IO/$\overline{M}$ input at pin 7 is necessary. This input line receives a signal from an equivalent output of the microprocessor (pin 34 of the 8085). A high at pin 7 of the 8355 or 8755 informs the device that an I/O function is being requested, whereas a low at pin 7 indicates that a memory function (in this case, always a read) is to take place. Since a command to receive data through a port is equivalent to reading data, this action could be initiated by holding the IO/$\overline{M}$ line high while the $\overline{RD}$ input is low. Also, the 8355 and 8755 both have an $\overline{IOR}$ line (pin 8) that may be brought low to initiate an input operation. If the combined IO/$\overline{M}$ and $\overline{RD}$ signals are to be used rather than the single $\overline{IOR}$ input, then $\overline{IOR}$ must be deactivated by tying pin 8 to a logic high. Since an output operation is equivalent to writing to a port, then this action is initiated by a single $\overline{IOW}$ signal (at pin 10).

As explained in the previous section, ports are usually defined as inputs or outputs during the initialization process at the beginning of a program. With the 8355 and 8755, the data-direction registers for ports A and B are addressed as ports 10_2 and 11_2, respectively. Placing a logic high in a bit position within the data-direction register causes the corresponding position in its associated port to function as an output. Placing a logic low in a bit position of the status register causes the corresponding port position to be an input. Assume port A of the 8355 is to function as an output. Immediate data FF_{16} would first be loaded into the accumulator. This action could then be followed by the 2-byte instruction *OUT, 02*. If the $\overline{WR}$ output (pin 31) of the 8085 were connected to the $\overline{IOW}$ input (pin 10) of the 8355, the *OUT, 02* instruction would cause the $\overline{IOW}$ input to the 8355 to go low while 00000010_2 was present at the AD_0 through AD_7 inputs (pins 12 through 19). The logic low present at pin 12 and the high at pin 13 together address the data-direction register for port A. Thus, as accumulator datum 11111111_2 is loaded into the data direction register, all 8 bits of port A are defined as outputs. Thus, a subsequent instruction *OUT, 00* would cause the contents of the 8085's accumulator to be sent to port A of the 8355 or 8755.

Assume port B is to function as an input. As part of a program's initialization process, an *XRA A* instruction followed by *OUT, 11* would send 00000000_2 to the data-direction register of port B, defining that port as an 8-bit input. The execution of an *IN, 01* instruction by the 8085 later in the program would cause a logic high to be present at the IO/$\overline{M}$ input of the 8355 or 8755, while the $\overline{RD}$ input of these devices would be in its active-low state. Thus, as 00000001_2 appears at the AD_0 through AD_7 inputs of the 8355 or 8755, port B performs a read function. The data taken in at port B leaves the memory IC via lines AD_0 through AD_7 and enters the accumulator of the 8085.

Figure 12–16 illustrates the timing relationship of the signals involved in a read function from the 8355 or 8755 to the 8085. Since the address high byte lines of the 8085 are not multiplexed, as represented in the timing diagram, the address high byte may remain on the A_8 through A_{15} output lines throughout the read operation. However, since the lower byte of the address bus and the bidirectional data bus are multiplexed, the lower byte of the address word must get out of the way of the data word being read from memory. Thus, as shown during T_1 and T_2, the lower byte of the address word is placed on the data bus. At the negative transition of the **address latch enable** signal (ALE), the data byte present at the AD_0 through AD_7 inputs of the 8355 or 8755 is latched into the internal address buffer of that IC. After the latching of the lower byte of the address word, while the $\overline{RD}$ line is low, data from the address location being pointed to in ROM are placed on the data bus. Before the positive transition of the $\overline{RD}$ signal, the ROM data byte now present at the AD_0 through AD_7 inputs of the 8085 is clocked into the data/address buffer of that device.

Figure 12–17 shows the pin assignments and simplified internal structure of an 8155, which is a **static RAM** capable of storing 256 bytes of data. Like the 8355 and the 8755 ROM devices, the 8155 is designed for compatibility with the 8085. Along with its basic 256×8 RAM, the 8155 contains two 8-bit ports and a 6-bit port. The operation of these ports, which may be used in I/O routines involving **handshakes,** is more complex than that of the ports of the 8355/8755. An example of an I/O operation with handshakes is covered in detail as part of this section. The 8155 also contains a **timer,** which can be programmed to output either a single active-low strobe pulse or a continuous square wave. As explained in detail later, the duration of the strobe pulse and the frequency of the square wave may be adjusted through programming. Control of port operation and the

Figure 12–16
The 8085 microprocessor read function

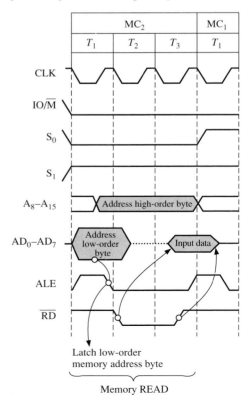

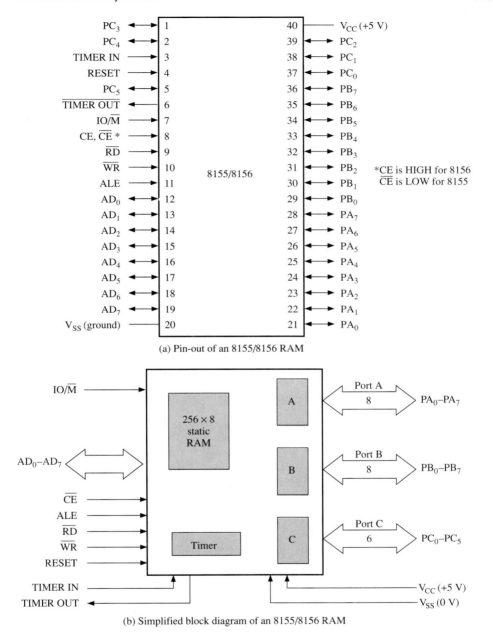

(a) Pin-out of an 8155/8156 RAM

*CE is HIGH for 8156
$\overline{CE}$ is LOW for 8155

(b) Simplified block diagram of an 8155/8156 RAM

Figure 12–17

timer of the 8155 is achieved through addressing the **command register** of that device. The current condition of the command register is read by addressing the **status register** of the 8155.

Since the 8155 is a tri-state device, it may be used within a common-bus configuration. However, unlike the 8355 and 8755, it has only one, active-high chip enable input (CE). An 8156 static RAM is also available. This device is identical to the 8155, except for the fact that its chip enable line is active low ($\overline{CE}$). Since the 8155 must respond to both read and write commands, it has both $\overline{RD}$ and $\overline{WR}$ control inputs, along with the ALE and IO/$\overline{M}$ lines. For the 8085 to accomplish a read function involving the static RAM portion of the 8155, $\overline{RD}$ and IO/$\overline{M}$ must both be low. During a write cycle involving the RAM, $\overline{WR}$ and IO/$\overline{M}$ must both be low. To send data to one of the ports of the 8155, IO/$\overline{M}$ must be high while the $\overline{WR}$ line is low. To receive data from one of the ports, IO/$\overline{M}$ must be high while $\overline{RD}$ is low. Since the 8155 has only 256 memory locations, it needs only eight address lines. For compatibility with the 8085, these address lines are multiplexed with the data bus, thus being designated as AD_0 through AD_7. As with the 8355/8755 read operation represented in Figure 12–16, the negative transition of an ALE input latches the address word in the internal address buffer of the 8155 RAM. While the $\overline{RD}$ line is low, the data contained in the location pointed to by the address word are present on the AD_0 through AD_7 lines and clocked into the data/address buffer of the 8085. With a write function, as just described, the address information is latched into the address buffer of the RAM on the negative transition of the ALE pulse. While the $\overline{WR}$ line is low, the data byte present at the data/address latch of the 8085 is stored at the RAM location pointed to by the 8-bit address word.

The **status/command register** of the 8155 is addressed as a port, being accessed via bits AD_2, AD_1, and AD_0. The command register, which is used to define ports A, B, and C, is accessed when AD_2, AD_1, and AD_0 are all low. Thus, after an immediate data byte in the accumulator is established for defining port and timer operation, the command register may be updated with the command *OUT, 40* (assuming the upper nibble of the port address word must be 0100 to enable the given 8155). The command assignments of the individual bits of the command/status register are summarized in Figure 12–18(a). An *IN, 40* instruction causes the contents of the status register to be written to the accumulator, allowing the 8085 to check the condition of individual bits representing the status of port A, port B, and the timer. The bit definition of the status register is shown in Figure 12–18(b).

As indicated in Figure 12–18(a), the 2^0 and 2^1 bit positions of the command register define the operation of ports A and B. Note that, with the 8155, in contrast to the 8355/8755 method of port definition, the entire port must be defined as an input or output. Thus, if the 2^0 bit is a 0 and the 2^1 bit is a 1, all 8 bits of port A function as an input, while all of port B becomes an output. The 2^2 and 2^3 bit positions of the command register define the operation of port C. The four alternatives for the operation of port C are listed and explained in Table 12–2. With the 2^2 and 2^3 bits both low, the ALT_1 condition exists, allowing all 6 bits of port C to function as a 6-bit input port. With the 2^2 and 2^3 bits both high, the ALT_2 condition exists, allowing all 6 bits of port C to function as a 6-bit output port. The other two alternatives for defining port C involve the operation of this port in I/O with handshakes. They will be explained when handshakes involving ports A and B are introduced.

The 2^4 and 2^5 bit positions of the command register control the interrupt capability of ports A and B. For example, a logic high in the 2^5 bit position would enable the interrupt capability of port B, allowing that port to be used in an I/O with handshaking operation. A logic low in the 2^5 bit position would disable the interrupt capability of port B. The 2^6 and 2^7 bit positions control the operation of the timer within the 8155. When both of these bit positions contain a logic low, the operation of the timer is unchanged. If the 2^6 bit is high while the 2^7 bit is low, the timer is stopped immediately. If the 2^6 bit is low while the 2^7 bit is high, the timer is stopped at the end of its next time out. If the 2^6 bit and 2^7 bit are both high, the timer is started immediately.

Before the I/O operations of the 8155 are introduced, the operation of the timer should be analyzed. Still assuming that the upper nibble of the port address must contain 0100, the timer high byte and low byte are addressed as ports 45 and 44, respectively.

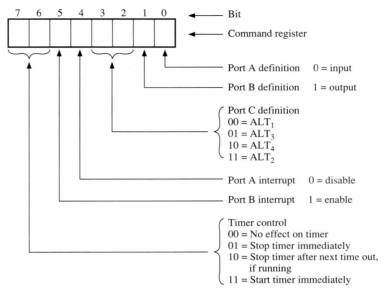

(a) Command register of 8155/8156 RAM

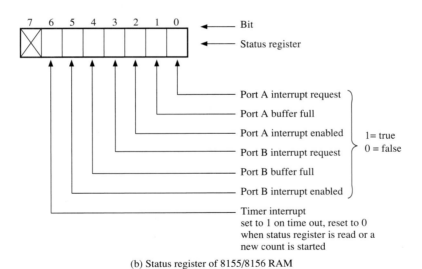

(b) Status register of 8155/8156 RAM

Figure 12–18

The 2 most significant bits of the timer high byte (being sent to port 45) determine the **mode** of operation for the timer. The remaining 6 bits of the timer high byte, along with the 8 bits of the timer low byte, form a variable. In the following explanation, this variable is referred to as N.

All four timer modes, which are illustrated in Figure 12–19, involve the decrementing of the count N to 00000000000000. With each cycle of the Timer In signal (at pin 3 of the 8155), which is typically the same clock signal as that of the microprocessor, the

Table 12–2
Alternatives for operation of port C

Port C bit	ALT1	ALT2	ALT3	ALT4
2^0	Input	Output	Port A interrupt (AINTR)	Port A interrupt (AINTR)
2^1	Input	Output	Port A buffer full (ABF)	Port A buffer full (ABF)
2^2	Input	Output	Port A strobe ($\overline{ASTB}$)	Port A strobe ($\overline{ASTB}$)
2^3	Input	Output	Output	Port B interrupt (BINTR)
2^4	Input	Output	Output	Port B buffer full (BBF)
2^5	Input	Output	Output	Port B strobe ($\overline{BSTB}$)

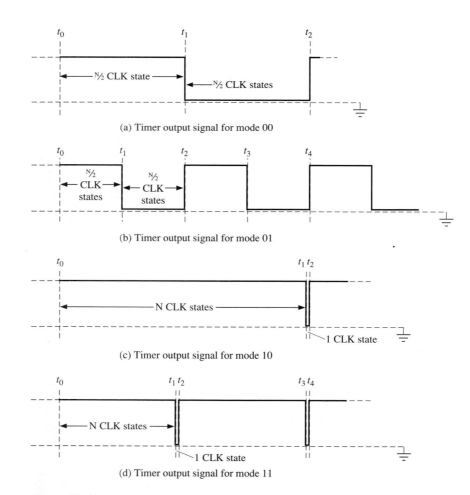

(a) Timer output signal for mode 00

(b) Timer output signal for mode 01

(c) Timer output signal for mode 10

(d) Timer output signal for mode 11

Figure 12–19

count variable becomes $(N - 1)$. In mode 00, with both the timer mode bits low, providing N is an even number, the $\overline{\text{Timer Out}}$ signal is high for a period equal to exactly $N/2$ clock signals then low for an equal number of clock states. If N is an odd number, then the high portion of the output is one clock cycle longer than the low pulse. The output signal produced during **mode 00** can be compared to that of an active-low one-shot, occurring only once with each command to start the timer. In **mode 01,** with 2^{15} low and 2^{14} high, the development of the output pulse is identical to that for mode 00; however, the N variable is automatically reloaded after being decremented to zero. As a result, the $\overline{\text{Timer Out}}$ signal becomes equivalent to that of an astable multivibrator, continuing until the timer is stopped by entering 01 or 10 into the 2^7 and 2^6 positions of the command register. In **mode 10,** with 2^{15} high and 2^{14} low, as the N variable is being decremented, the $\overline{\text{Timer Out}}$ signal remains high until zero is attained. At that instant, the $\overline{\text{Timer Out}}$ signal falls low for one cycle of the Timer In signal. As in mode 00, this output pulse occurs only once, unless the timer is restarted by placing 11 in the 2^7 and 2^6 positions of the command register. In **mode 11,** with 2^{15} and 2^{14} both high, the development of the output pulse is identical to that for mode 10. However, as in mode 01, the output pulse is continuous, occurring until one of the two stop timer codes are entered into the 2^7 and 2^6 positions of the command register.

In simple I/O operations, such as those possible with the 8355/8755 ICs, data bits are either written to an output port or read from an input port with no other communication occurring between the peripheral device and the microcomputer. However, most **real-time** applications of microcomputers, such as those involving operation of industrial machinery, require **handshaking signals.** This explanation concludes with two examples of handshaking, demonstrating the I/O capabilities of the 8155.

In Figure 12–20, the 8085 relies on the 8155 to supervise the operation of an 8703 A/D converter. The A/D conversion process is initiated by a high pulse from the 2^7 output of port B. This signal directs the 8703 to take a sample of the voltage present at its analog input. The 8-bit digital output of the 8703 is connected directly to port A of the 8155, which is defined to function as an 8-bit input. During the A/D conversion process, the data byte at port A is continually changing. At the end of an A/D conversion cycle, when a **valid data** byte is present at the outputs of the A/D converter, an active-low Data Valid pulse from the 8703 signals the 8155 that an input data byte is ready at port A. Note that this signal is connected to the 2^2 bit position of port C of the 8155. According to Table 12–2, in the ALT3 and ALT4 configurations of port C, this bit position serves as the $\overline{\text{ASTB}}$ input. At the termination of the $\overline{\text{ASTB}}$ signal, with valid data from the A/D converter latched in port A, an active-high AINTR signal is generated at the 2^0 bit of port C. This signal is connected to the RST5.5 input of the 8085, causing that microprocessor to initiate an interrupt service routine. This routine places the data present at port A of the 8155 into the accumulator of the 8085. The sequence of pulses involved in the handshake routine is diagrammed in Figure 12–21, and a possible 8085 routine for executing this handshake operation is shown in Program Example 12–8. In this program, it is assumed that a subroutine for processing the data received from the A/D converter is located at address 4050.

As shown next, to provide for the vectored interrupt that occurs when the input data at port A is valid, an unconditional jump instruction must be entered at the vector address of RST5.5:

002C	C3
002D	10
002E	30

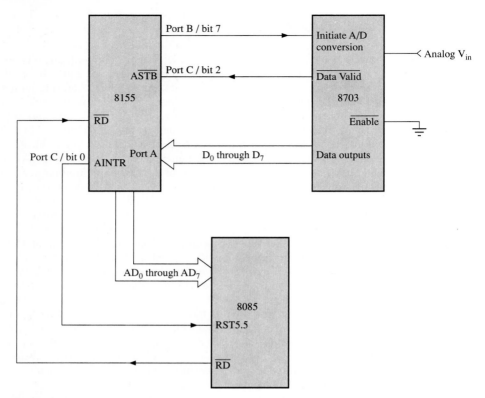

Figure 12–20
A/D conversion involving I/O with handshaking

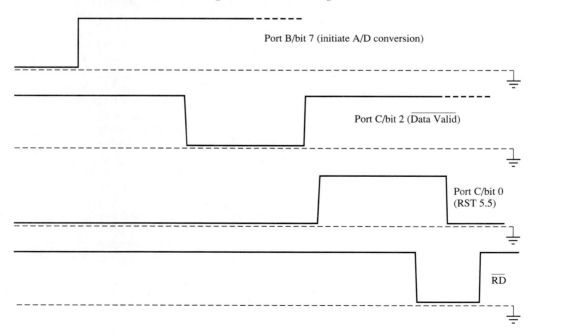

Figure 12–21
Timing sequence for A/D conversion involving I/O with handshaking

Program Example 12-8

3000	31	LXI SP, 5000	Set stack pointer at 5000.
3001	00		
3002	50		
3003	3E	MVI A, FE	Unmask RST5.5.
3004	FE		
3005	30	SIM	
3006	3E	MVI A, 1A	Initialize command register.
3007	1A		
3008	D3	OUT, 40	
3009	40		
300A	FB	EI	Enable interrupt.
300B	3E	MVI A, 80	Initiate the A/D conversion process.
300C	80		
300D	D3	OUT, 42	
300E	42		
300F	76	HLT	.Wait for interrupt.
3010	DB	IN, 41	Input valid data from A/D converter.
3011	41		
3012	CD	CALL, 4050	Goto routine to process input data.
3013	50		
3014	40		
3015	C3	JMP, 3000	Repeat handshake routine.
3016	00		
3017	30		

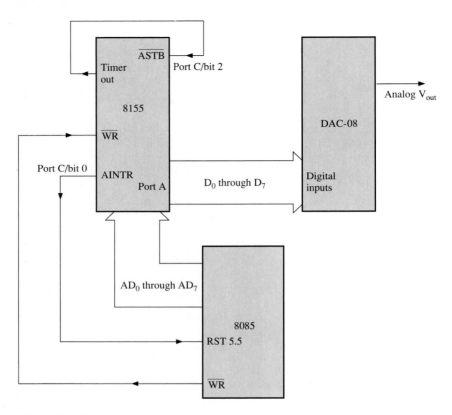

Figure 12–22
D/A conversion involving I/O with handshaking

Program Example 12-9

3000	31	LXI SP, 5000	Set stack pointer at 5000.
3001	00		
3002	50		
3003	3E	MVI A, FE	Unmask RST5.5.
3004	FE		
3005	30	SIM	
3006	3E	MVI A, FF	Define timer low byte.
3007	FF		
3008	D3	OUT, 44	
3009	44		
300A	3E	MVI A, CF	Define timer high byte.
300B	CF		

Program Example 12-9 (*continued*)

300C	D3	OUT, 45	
300D	45		
300E	3E	MVI A, DB	Initialize command register.
300F	DB		
3010	D3	OUT, 40	Start timer.
3011	40		
3012	FB	EI	Enable interrupt.
3013	76	HLT	Wait for interrupt.
3014	CD	CALL, 4050	Goto routine to derive new output data.
3015	50		
3016	40		
3017	D3	OUT, 41	Output data to D/A converter.
3018	41		
3019	31	LXI SP, 5000	Set stack pointer at 5000.
301A	00		
301B	50		
301C	C3	JMP, 3012	Repeat handshake routine.
301D	12		
301D	30		

As shown by the circuit example in Figure 12–22, handshaking can also be used in an output routine. Here, a binary control variable produced by the microcomputer is fed to the DAC-08, which is an 8-bit D/A converter. The D/A conversion process is inherently simpler than A/D conversion, since the microcomputer is not required to produce an output signal initiating the conversion process. Instead, the timer function of the 8155 is used to establish a minimum period between conversions. This period could be as short as 80 μsec, which is the minimum settling time for the analog output of the DAC-08. In applications such as motor control, this period must be longer, to compensate for the rate at which a motor is able to accelerate smoothly.

In Program Example 12–9, after RST5.5 is unmasked, two immediate data bytes representing the minimum settling time for the D/A converter are loaded into the timer high byte and low byte. Since the upper 2 bits of the high byte data are 11, mode 11 is established for operating the timer. Thus, the timer runs continuously, with an active-low trigger pulse occurring at the Timer Out pin of the 8155 each time N is decremented to zero. Since the $\overline{\text{Timer Out}}$ signal serves as the $\overline{\text{ASTB}}$ input, its negative transition initiates a high transition of the AINTR signal, which is connected directly to the RST5.5 input of the 8085. As with the input handshake routine analyzed earlier, the vector address for RST5.5 must be entered as follows:

002C	C3
002D	14
002E2	30

Example 12–5

During the execution of the preceding output handshake routine, assume the clock signal of the microcomputer is also the Timer In signal for the 8155. If the frequency of this signal is 2.85 MHz, what is the D/A conversion rate for the DAC-8 in the preceding example?

Solution:

Step 1: Convert bits 2^0 through 2^{14} of the timer high byte and low byte to a decimal number:

$$\text{0FFF}_{16} = 0011111111111111_2 = 4095_{10}$$

Step 2: Determine the period of the D/A conversion cycle:

$$\text{Period of clock} = \frac{1}{2.85 \text{ MHz}} = 350.877 \text{ nsec}$$

$$350.877 \text{ nsec} \times 4095 = 1.437 \text{ msec}$$

$$\frac{1}{1.437 \text{ msec}} = 696 \text{ conversions per second}$$

SECTION REVIEW 12–2

1. What are the advantages of using EPROMs rather than PROMs during the developmental stages of a microcomputer-based control system?
2. Describe the process used to erase an EPROM.
3. Describe the process used to erase an EEPROM.
4. Define the term *nonvolatile* as it applies to memory devices.
5. Describe the four possible operating modes of the timer within the 8155 RAM. How does the 8155 determine which mode is to be used?
6. Describe a possible application of the timer within the 8155.
7. What signal from the 8085 allows the 8355/8755 to differentiate between a port operation and a read from memory?
8. How are the individual bits within the ports of 8355/8755 defined as inputs or outputs?
9. Describe the four possible operating configurations for port C of the 8155.
10. Assume the clock signal of the 8085 is used as the Timer In signal to the 8155. Also assume the frequency of this signal to be 2.85 MHz. Using 8085 machine language,

write a short program routine that causes the timer to produce a free-running square wave with a frequency of 180 pps, defines port A as an input, defines port B as an output, and enables port C to perform I/O handshaking for ports A and B.

12-3   INTRODUCTION TO MICROCOMPUTER ARCHITECTURE

In this section, the overall structure of a small microcomputer system, typical of those embedded in larger industrial control systems, is introduced. Methods by which RAM, ROM, and the I/O ports of the microcomputer communicate with the microprocessor IC are presented. Also included in this section is a discussion of an alternative form of input/output operation called **memory-mapped I/O.** Methods by which the embedded microcomputer sends data to and receives data using memory-mapped I/O are investigated. Specialized devices used in these operations are also introduced.

An 8-bit, first-generation microprocessor such as the 8085, since it has 16 address lines, is capable of interfacing with 65,536 memory locations. For embedded microcomputers, this much memory is usually not necessary. However, even the smallest microcomputer contains, besides RAM and ROM, several other ICs that must be introduced. One such IC is the **1 out of 8,** or **binary-to-octal decoder,** examples of which include the Intel 8205 and the 74LS138. This form of IC is widely used to decode the memory and ports of microcomputer systems. The pin-out of the 8205 is shown in Figure 12–23, and the truth table for the device is shown in Table 12–3.

The 8205 has three address inputs, A_0, A_1, A_2, and three enable inputs, $\overline{E_1}$, $\overline{E_2}$, and E_3. As indicated by the vincula, $\overline{E_1}$ and $\overline{E_2}$ are active-low, whereas E_3 is active-high. Unless $\overline{E_1}$ and $\overline{E_2}$ are both low at the same time E_3 is at a logic high, all eight of the active-low outputs, $\overline{O_0}$ through $\overline{O_7}$, are held at a logic high, regardless of the logic conditions of the address inputs A_0, A_1, and A_2. In the truth table of Table 12–3, note that these three address inputs are shown forming a binary up-count, equivalent to decimal 0 through 7. With the enable conditions maintained, as shown in the middle column, a corresponding output falls to a logic zero for each of these eight possible input conditions.

Figure 12–24 illustrates how three 8205 ICs can be used to interface between an 8085 and 24.576 kbytes of memory. Ten address lines, A_0 through A_9, are connected directly between the 8085 and each of the twenty-four 1.024k memory ICs, one of which is illustrated in Figure 12–24. This tri-state memory IC is enabled by the $\overline{O_0}$ output line of the first 8205. Referring to the truth table for the 8205, this output line can only be low when the A_0, A_1, A_2, $\overline{E_1}$, and $\overline{E_2}$ inputs to the first 8205 are low simultaneously. (Note that the E_3 input of the first 8205 is tied permanently to a logic high (V_{CC}).) Thus, only when the A_{10} through A_{14} outputs of the 8085 are all at a logic low is the first memory IC enabled. With the low condition of the A_{13} line being present at the E_3 input of the second 8205 and the low condition of the A_{14} line being present at E_3 of the third 8205,

Figure 12–23
Pin-out of 8205 binary-to-octal decoder

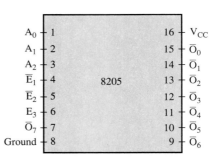

Table 12–3
Truth table for 8205 binary-to-octal decoder

A_2 A_1 A_0	$\overline{E_1}$ $\overline{E_2}$ E_3	$\overline{O_0}$ $\overline{O_1}$ $\overline{O_2}$ $\overline{O_3}$ $\overline{O_4}$ $\overline{O_5}$ $\overline{O_6}$ $\overline{O_7}$
0 0 0	0 0 1	0 1 1 1 1 1 1 1
0 0 1	0 0 1	1 0 1 1 1 1 1 1
0 1 0	0 0 1	1 1 0 1 1 1 1 1
0 1 1	0 0 1	1 1 1 0 1 1 1 1
1 0 0	0 0 1	1 1 1 1 0 1 1 1
1 0 1	0 0 1	1 1 1 1 1 0 1 1
1 1 0	0 0 1	1 1 1 1 1 1 0 1
1 1 1	0 0 1	1 1 1 1 1 1 1 0

all the output lines of these two ICs are held at a logic high. Consequently, the remaining 23 memory ICs are disabled and effectively removed from the common address/data bus of the microcomputer. With the conditions just described, the microprocessor is only able to perform read/write operations with the first memory IC, which is considered to contain memory locations 000000000000000_2 through 0000011111111111_2.

Assume that the address lines of the 8085 toggle from 0000011111111111_2 to 0000100000000000_2. With the A_{10} line now at a logic high, the low condition shifts from the $\overline{O_0}$ output to the $\overline{O_1}$ output, thus enabling the second tri-state memory IC. With the

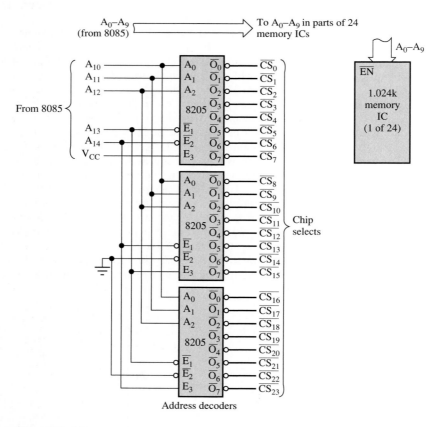

Address decoders

Figure 12–24
Memory interface containing three 8205 binary-to-octal decoders

$\overline{CS}_1$ line now low, the microprocessor is addressing memory locations 000010000000000_2 through 000011111111111_2. At the next toggle point, with the microprocessor addressing 000100000000000_2, the third memory IC is enabled. Assume the address lines of the 8085 have incremented to 001111111111111_2, causing a logic low to be present at the O_7 output of the first 8205. If the address lines now toggle to 010000000000000_2, with the A_{13} line now high, the first 8205 is disabled while the second 8205 becomes enabled. With A_{10}, A_{11}, and A_{12} all low, the $\overline{O}_0$ output of the second 8205 falls low, enabling the ninth memory IC. It should now be obvious how the enabling of the remaining memory ICs must occur, through the continued incrementing of the 8085 address lines. Eventually the third 8205 would become enabled, with the first and second decoders being in the disabled state.

In the previous section, the discussion of I/O operations was limited to those ports contained within RAM and ROM ICs. With this standard form of I/O operation, only the input and output instructions of a microprocessor may be used to address the ports. An alternate form of port operation is referred to as **memory-mapped I/O.** With this form of I/O, the microprocessor does not differentiate between memory and port locations. Instead, memory-mapped I/O allows the input ports of the microcomputer to be addressed through read instructions, while the output ports are addressed through write operations. The primary advantage of memory-mapped I/O is its flexibility, in that all instructions involving the microprocessor's memory can also be used with I/O operations.

Figure 12–25 shows an example of a memory-mapped I/O system involving an Intel 8032 microprocessor, a predecessor of the 8085. This early example of a first-generation microprocessor does not have an I/O/$\overline{M}$ output line and, therefore, does not differentiate between port and memory functions. In fact, the address and data lines of the device, which comprise three 8-bit groups, are designated as ports ($P0_0$ through $P0_7$, $P1_0$ through $P1_7$, and $P2_0$ through $P2_7$). As shown in Figure 12–25, ports $P0_0$ through $P0_7$ form the data bus input/output lines. As with the 8085, these eight data lines are multiplexed with the lower byte of the address bus. Ports $P1_0$ through $P1_7$ comprise a second set of data lines, and ports $P2_0$ through $P2_7$ form the upper byte of the address bus.

The memory mapping for the microcomputer in Figure 12–25 is such that, when the A_{13}, A_{14}, and A_{15} are all at a logic low, the 0 output of the first 74LS138 is low. Note that this line is connected directly to one of the active-low enable lines of the second 74LS138. Since the active-high enable input to the second 74LS138 is permanently tied high via a pull-up resistor, a low on the $\overline{WR}$ line of the 8032 enables the second 74LS138, allowing one of eight output ports to be active. The 74LS373 in Figure 12–25 serves as the address latch for the lower byte of the address word. Although the active-low enabling input of this IC is permanently tied to the ground, its active-high enabling input is connected to the Address Enable (AE) line of the 8032. (The AE output of the 8032 is equivalent to the ALE output of the 8085.) Thus, on the positive transition of the AE signal, the lower byte of the address word, present at the $P0_0$ through $P0_7$ lines of the 8032, is latched into the 74LS373. While the second 74LS138 is enabled, the conditions of the A_0, A_1, and A_2 outputs of the 74LS373 specify the output port to be addressed. The 74LS374, which serves as an output port, is an 8-bit tri-state data latch. The enabling inputs to this device consist of an active-low input $\overline{EN}$, which is permanently tied to circuit ground in this application, and a chip enable signal $\overline{C1}$, which is also active-low. As shown in Figure 12–25, each active-low output of the second 74LS138 is connected to the $\overline{C1}$ input of a 74LS374. When an output port is selected, its $\overline{C1}$ input is brought low. This allows the information present on the data bus of the microcomputer to be latched in the appropriate 74LS374, while the other seven ports remain in the high-impedance state.

Assume, for example, that output port 3 is to receive data from the 8032 microprocessor. For this to occur, address lines A_{13}, A_{14}, and A_{15} must be low, while A_0 is high, A_1 is high, and A_2 is low. A low on the $\overline{WR}$ output line from the 8032 then latches the data

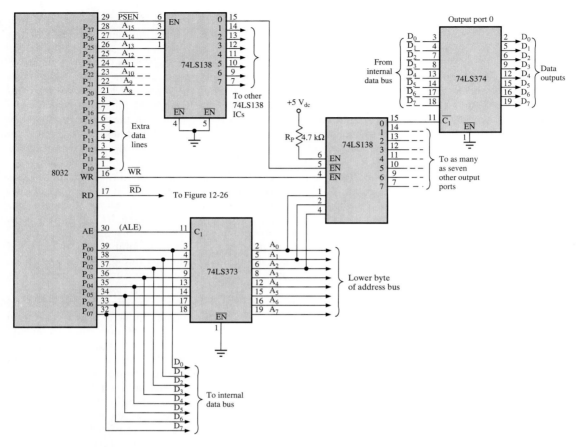

Figure 12–25
Memory-mapped I/O using 8032 microprocessor: output operation

into the 74LS374 serving as output port 3. In many applications of embedded microcomputers, where control signals are sent considerable distances, port outputs are fed to optocouplers. The use of optocouplers in the propagation of digital signals is explained in Section 5–4.

Figure 12–26 illustrates how input ports may be accessed within a memory-mapped I/O scheme. The same 74LS138 that received address lines A_{13}, A_{14}, and A_{15} in Figure 12–25 is shown for clarity in Figure 12–26. The 2^2 output of this device is connected to one of the active-low enable lines of a second 74LS138. Since the active-high enable input of the second 74LS138 is permanently tied high via a pull-up resistor, a low on the $\overline{RD}$ line of the 8032 enables this device, allowing one of eight possible input ports to be active. While this second 74LS138 is enabled, the conditions of the A_0, A_1, and A_2 lines from the 74LS373 in Figure 12–25 specify the input port to be addressed. The 74LS541 is an eight-bit tri-state data latch, which serves as an input port. In this application of the device, both active-low enable inputs (designated $\overline{EN}$) are tied together and connected to one of the active-low outputs of the second 74LS138. Thus, only one of the eight possible input ports is enabled at a given instant, while the other seven are in a high-impedance condition.

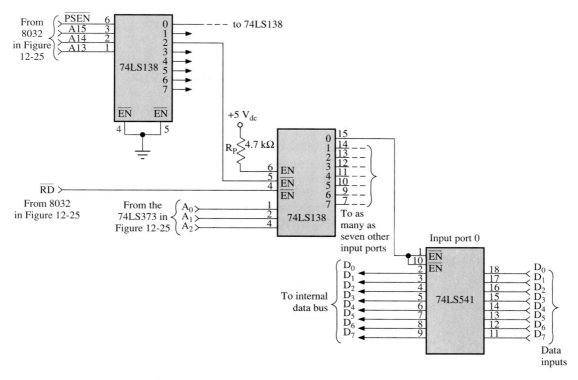

Figure 12–26
Memory-mapped I/O using 8032 microprocessor: input operation

Assume, for example, that input port 7 is to write data to the 8032 microprocessor. For this to occur, address line A_{13} must be low, A_{14} must be high, and A_{15} must be low, causing output line 2 of the first 74LS138 to fall low. Also, address lines A_0, A_1, and A_2 must be high simultaneously. A low on the $\overline{RD}$ output line from the 8032 then enables the 74LS541 serving as input port 7 to place its contents onto the data bus, allowing the 8032 microprocessor to read this data.

SECTION REVIEW 12–3

1. What common type of digital device is represented by the 8205 and 74LS138 ICs?
2. Explain the basic function of an 8205 or a 74LS138 within the memory of a microcomputer.
3. Define memory-mapped I/O. What are the advantages of using this method of accessing ports?
4. Why is it necessary for an earlier first-generation microprocessor such as the 8032 to use memory-mapped I/O?

12–4 INTRODUCTION TO MICROCONTROLLERS

After the appearance of the first generation of microcomputers, development of the microcomputer essentially followed two paths. Along one path, the computer industry is continuing to develop faster and more powerful personal computers, capable of handling data words of 16 and 32 bits. The development of such computers has an immediate impact on industrial process control. These more powerful PCs are assuming increasingly important roles as monitoring and governing devices for industrial systems. A second line of development involves the evolution of ICs that contain *all* the components necessary for a given process control application. Such devices, referred to as **microcontrollers,** are revolutionizing industrial control electronics. Many embedded microcomputers, contained on single circuit boards, can now be replaced by single microcontroller ICs. The need for more sophisticated automotive electronic control systems has been a major incentive in microcontroller development. The arrival of the microcontroller has resulted in the creation of sophisticated engine control modules and antilock braking systems. The need for increasingly sophisticated hand-held test equipment and communication devices has also been a motivating force in microcontroller evolution.

The Intel 8096 is a high-speed **interrupt-driven** microcontroller, designed for rapid response to interrupt service requests from up to 22 different sources, 20 of which may be enabled at one time. A simplified block diagram of the internal circuitry of an 8096 microcontroller is shown in Figure 12–27. Like a microprocessor, the 8096 microcontroller contains a CPU, which is capable of performing arithmetic and logic functions similar to those of the 8085. Representing a design advancement over first-generation microprocessors, the 8096 microcontroller contains its own 8192-byte ROM and 232-byte RAM. The CPU of the 8096 is represented in Figure 12–28. Like the 8085, the 8096 contains a stack pointer and a program counter. However, the CPU of the 8096 contains no accumulator or scratch pad registers! Instead, it uses RAM locations to function effectively as accumulators and work registers. As with the 8085, the upper byte of the PSW contains flag flip-flops for the CPU. Three of these flags, the zero, negative, and carry/not borrow flags, should already be familiar. The lower byte of the PSW comprises the interrupt mask register (abbreviated **INT_MASK).**

A significant feature of the 8096, which facilitates the work required by system design engineers, is an internal 10-bit successive-approximation A/D converter. (Successive-approximation A/D conversion was covered in detail in Section 11–5.) The A/D converter of the 8096 is accessed by way of a specialize input port (port 0), which may serve either as a conventional 8-bit data input port or as an eight-channel analog input. In the second-

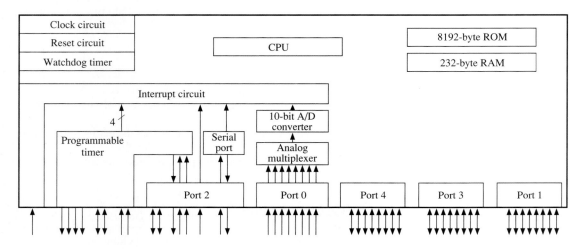

Figure 12–27
Simplified block diagram of 8096 microcontroller

Figure 12–28
Block diagram of Intel 8096 CPU

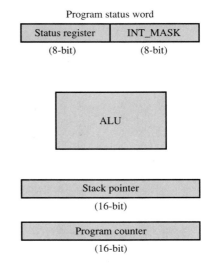

Program status word

Status register	INT_MASK
(8-bit)	(8-bit)

ALU

Stack pointer
(16-bit)

Program counter
(16-bit)

generation (8096BH series) of Intel microcontrollers, internal sample-and-hold circuitry is also provided with the A/D converter, expediting still further the task of the design engineer. As shown in Figure 12–27, the eight analog inputs of port 0 are fed to the internal analog multiplexer, which, in turn, feeds the A/D converter. Along with the eight possible analog inputs at port 0, the 8096BH provides a reference voltage input (V_{REF}) and an analog ground connection (Angnd). Through programming, a sample of one of the eight analog input lines may be fed to the A/D converter and quickly converted to a 10-bit binary number. The internal A/D converter of the 8096BH microcontroller is capable of completing a conversion cycle every 22 μsec, allowing a conversion rate of nearly 45k per second!

Although D/A conversion circuitry is not directly provided by the 8096BH, this device is capable of producing a pulse-width-modulated (PWM) output. Note that, in Figure 12–27, the arrow at the 2^5 bit position of port 2 indicates that this pin is restricted to an output function. This output line does, however, serve a dual function. While normally a conventional digital output, the 2^5 line may also be initialized to produce a PWM signal. This is easily accomplished by setting a bit in an I/O control register of the microcontroller. Control words are then written to an 8-bit **PWM control register.** The number held by the PWM control register is continuously compared to the instantaneous value of a free-running 8-bit **PWM counter.** This counter increments at one-third the frequency of the microcontroller's clock crystal. The PWM output cycle begins when the counter resets to 00_{16}, marking the positive transition of the PWM signal at the 2^5 bit of port 2. This signal remains at a logic high until the instant the number formed by the PWM counter equals the contents of the PWM control register. At this point, the PWM output signal falls low. Thus, the duty cycle of the PWM output is directly proportional to the number contained in the PWM control register. Figure 12–29 illustrates the relationship of PWM control register data and the output duty cycle. The buffered PWM output from the 8096BH could be fed to a dc-to-dc power converter, similar to those introduced in Section 9–5, allowing the microcontroller to regulate precisely an analog voltage level.

Recall that the 8155 RAM may be equipped with a timer function that is initialized by output instructions from the 8085 microprocessor. Also recall that, once initialized and enabled, this timer may run independently of the program being executed by the 8085. The concept of a free-running counter operating simultaneously with the micropro-

Figure 12–29
PWM output available at bit 5 of port 2 of
the 8096 microcontroller

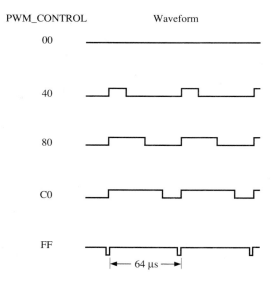

cessor has been refined and incorporated into the design of the 8096 microcontroller. As explained in detail after the interrupt capabilities of the 8096 are introduced, the programmable timer section of this microcontroller contains two 16-bit counters. The clocking frequency of these counters is always one-eighth that of the microcontroller. For example, if the 8096 has a 12-MHz crystal, and the microcontroller operates at one-third that frequency, the clocking frequency of the free running counter is then determined as follows:

$$\frac{12 \text{ MHz}}{3} = 4 \text{ MHz}$$

$$\frac{4 \text{ MHz}}{8} = 500 \text{ kHz}$$

$$T_{\text{CLK}} = \frac{1}{500 \text{ kHZ}} = 2 \text{ } \mu\text{sec}$$

To meet the increasingly complex control requirements of modern industrial machinery, the 8096 microcontroller is equipped with an elaborate interrupt system. As shown in Table 12–4, these interrupts are grouped into nine basic categories.

The maskable interrupts of the 8096 are enabled and disabled in a two-level format that evolved from the interrupt control system of the 8085. For these eight maskable interrupts, an EI instruction enables the interrupt capability and a DI instruction disables all maskable interrupts. Masking and unmasking of each of the first eight forms of interrupt listed in Table 12–14 is accomplished by loading data into an interrupt mask register (abbreviated INT_MASK). (Note that the software interrupt of the 8096, like the TRAP inputs to the 8032 and 8085, is not maskable.) With the 8096, the **I bit,** which represents the enabled or disabled condition for the maskable interrupts, is a flag condition contained in the 2^9 bit position of the PSW. The individual bit assignments of the PSW of the 8096 are shown in Figure 12–30.

As with the 8085, a hardware interrupt to the 8096 automatically vectors to an interrupt service routine, as long as the I bit is set to a logic high and the corresponding bit in the INT_MASK register is also high. As shown in Table 12–4, the vector location for each interrupt consists of two contiguous memory locations. (Only 2 bytes are required for

Table 12–4
Intel 8096 interrupt groups

Interrupt Group (from highest to lowest priority)	Vector Address Locations	Interrupt Mask Register Bit Position
External interrupt	200E and 200F	$0008/2^7$ bit
Serial port	200C and 200D	$0008/2^6$ bit
Software timers	200A and 200B	$0008/2^5$ bit
High-speed input 0	2008 and 2009	$0008/2^4$ bit
High-speed outputs	2006 and 2007	$0008/2^3$ bit
High-speed input data ready	2004 and 2005	$0008/2^2$ bit
A/D conversion complete	2002 and 2003	$0008/2^1$ bit
Timer overflow	2000 and 2001	$0008/2^0$ bit
Software interrupt	2010 and 2011	(none)

each vector location, since the branch instructions for the 8096 consist of 2 bytes rather than 3.)

As seen in Figure 12–27, the programmable timer of the 8096 is directly connected to eight pins of the IC. Four of these lines function as **high-speed outputs** only, these being HSO.0, HSO.1, HSO.2, and HSO.3. Outputs HSO.4 and HSO.5 are available at two bidirectional terminals, which also function as **high-speed input** lines HSI.2 and HSI.3. Two lines, HSI.0 and HSI.1, function as high-speed inputs only. If the **high-speed input data interrupt** is unmasked and the interrupt capability of the 8096 is enabled, a signal at one of the four high-speed input lines initiates an interrupt service routine involving the programmable timer. Due to the presence of a first-in/first-out buffer system (FIFO), the programmable timer of the 8096 is able to track up to eight input events at a time reliably. This FIFO memory system is represented in Figure 12–31.

The FIFO itself contains seven 20-bit locations that hold information about seven pending HSI interrupt requests. The upper 4 bits of the data contained in each of these seven buffer locations identify the source of the interrupt request (HSI.0 through HSI.3). The lower 16 bits represent the time the event occurred. The **holding register,** which comprises 4 bit positions of the **high-speed input status register** (abbreviated HSI_STA-

Figure 12–30
Program status word of 8096 microcontroller

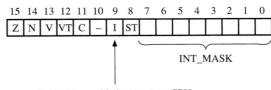

Set I bit to enable interrupts to CPU

Clear I bit to disable all interrupts

Z = Zero flag

N = Negative flag

V = Overflow flag

VT = Overflow trap flag

C = Carry/not borrow flag

ST = Sticky bit flag

Figure 12–31
The 8096 FIFO memory for timing multiple
inputs

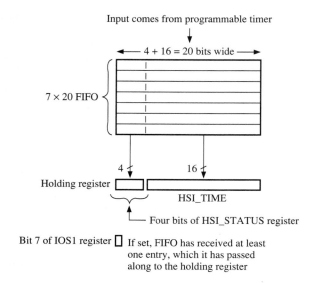

TUS), identifies the source of the oldest interrupt request pending in the FIFO. The 16 HSI_TIME bits represent the time (based on the microcontroller's internal clocking) that interrupt request occurred. When the CPU of the 8096 responds to an interrupt service request involving the HSI lines of the programmable timer, it reads the contents of the holding register. After reading the HSI_TIME bits to determine when the interrupt request took place, the CPU clears the holding register data. If another interrupt request (next oldest with respect to the one currently being serviced) is waiting in the FIFO, its identity and time are loaded into the holding and HSI_TIME registers. The 2^7 bit of input/output status register 1 (IOS1) is then set high, indicating that at least one more interrupt request is waiting in the FIFO. Thus, during an interrupt service routine, prior to returning to the main program, the CPU tests the status of the ISIO 2^7 bit to determine if any additional interrupt service requests are pending.

The programmable timer of the 8096 contains equivalent FIFO memory circuitry for servicing interrupts involving its six high-speed output lines (HSO.0 through HSO.5). As with the input FIFO memory, as many as eight output interrupt service requests may be handled reliably. For each pending output service request, a code for that event and the time for that event are entered into a holding register, which is divided into a **command register** and a **time register.** The contents of the holding register are automatically stored into the output buffer, where they are checked every 2 μsec (assuming the crystal oscillator of the microcontroller operates at 12 MHz).

As shown in Figure 12–32, the programmable timer section of the 8096 consists of a **free-running counter,** a **comparator,** an **output-compare register,** an **output-level bit latch,** and a **D flip-flop.** The free-running counter increments continuously from the instant power is applied to the 8096. Unlike the programmable timer within the 8155 RAM, the free-running counter of the 8096 *cannot* be loaded with data. Instead, during the execution of a program, the instantaneous value of the 16-bit number formed by this counter can be quickly read by the CPU. An offset number can then be added to this instantaneous count, with the result of this addition being placed in the output compare-register. Also, a data bit is written into the output-level bit latch, the output of which is connected to the D input of a D-type flip-flop. At the instant the counter increments to a number equal to that contained in the output-compare register, a positive transition, which serves as the clock pulse for the D-type flip-flop, is initiated by the comparator. This

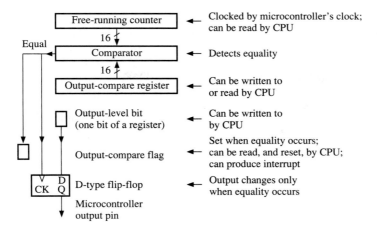

Figure 12–32
The 8096 microcontroller programmable output operation

positive transition causes the data bit contained by the output-level bit latch to be clocked into the Q output of the D-type flip-flop, which connects directly to an output pin of the 8096.

The programmable timer capability of the 8096 may be used in both input and output operations of the microcontroller. In a typical output application, the interrupt capability of the 8096 is first momentarily disabled. Assuming the output line of the programmable timer is to go to a logic high, the contents of the free-running counter is first read. Next, a small offset value is added to this counter number, with the result being placed in the output-compare register. This offset allows sufficient delay time for the ongoing program to set the output level bit to a logic high. As soon as the free-running counter number equals the output-compare register data and the Q output of the D-type flip-flop toggles to a logic high, a number must be added to the current contents of the output-compare register, representing the desired pulse width of the output signal.

Example 12–6

Assume the output pulse width of a signal produced by the programmable timer of the 8096 is to be 25 msec. Also assume the clock state (T) of the free-running counter is 2 μsec. Given these conditions, determine the 16-bit variable that must be added to the contents of the output-compare register.

Solution:

Step 1: Determine the decimal value of the offset variable:

$$N = \frac{PW}{T} = \frac{25 \text{ msec}}{2 \text{ } \mu\text{sec}} = 12.5k$$

Step 2: Determine the hexadecimal value of the offset variable (which become a high byte and low byte of immediate data).

$$12.5k_{10} = 30D4_{16}$$

After the hexadecimal number determined in Example 12–6 is added to the output-compare register data, a logic low is placed in the output-level bit latch. This action allows the output of the programmable timer to fall low when the contents of the free-running counter equals the contents of the output-compare register. Figure 12–33 illustrates the timing relationship between the program in memory and the pulse produced by the programmable timer. Whether or not the interrupt capability of the microcontroller is enabled during the pulse width of the output signal is left to the discretion of the design engineer. This individual must determine if interrupt service routines that might be called during the 25-msec output pulse could introduce a significant error in the pulse width of this signal. If this potential exists, it is better to leave the maskable interrupts in a disabled state.

The programmable timer of the 8096 may be used in a variety of input operations, one of the simplest of which is the measurement of the pulse width of a signal at one of its high-speed input lines. The 8096 has the capability of sensing both a positive-going and a negative-going edge of such a signal. Thus, by counting the number of clock pulses that occur between these two transitions, the 8096 is able to measure the pulse duration. (The degree of accuracy for such a measurement is, of course, limited by the clocking-frequency of the free-running counter within the microcontroller's programmable timer.) By measuring the number of counts occurring between a rising edge and a falling edge, the 8096 is able to determine the duration of an active-high pulse. Conversely, by determining the number of counts occurring between a falling edge and a rising edge, the microcontroller is able to measure the time for an active-low pulse. The components of the programmable timer that allow these input measurement operations are represented in Figure 12–34.

As an example of an input measurement operation, assume the pulse width of an active-high signal is to be measured. To prepare for this process, a code is first entered into the **edge-select control latch,** indicating the 8096 is to respond to a positive-going transition. The **input-capture flag** is then reset to a logic low. At the instant a positive transition occurs at the selected high-speed input line to the 8096, the input-capture flag toggles to a logic high. Also, the number present in the free-running counter at the instant this flag is set is automatically loaded into the **input-capture register.** This setting of the input-capture flag could initiate an interrupt service routine that saves the contents of the input-capture register in a pair of RAM locations.

After the 16-bit variable representing the leading edge of the input pulse is safely stored in memory, a code is entered into the edge-select control latch, causing the microcontroller to respond to a negative-going transition at the selected input line. Also, the input-capture flag is reset a second time. Thus, as a negative transition occurs at the input, the input-capture flag again toggles to a logic high, causing the contents of the free-running timer to be loaded into the input-capture register. With this second setting of the input-capture flag, another interrupt service routine would be initiated, causing the current contents of the input-capture register to be stored in another pair of RAM locations.

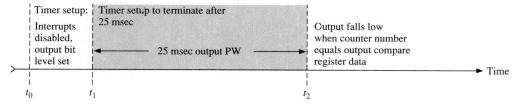

Figure 12–33
Timing relationship between programmable timer output pulse and program in memory

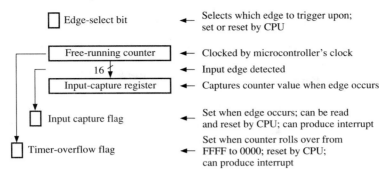

Figure 12–34
Input measurement by programmable timer of 8096

Once these two sample counts, representing the leading and trailing edges of the input pulse, are stored in memory, the CPU of the 8096 has only to subtract the first sampled count from the second. The absolute value of the result of this subtraction represents the pulse width of the input signal. The timing relationship between the program in memory and the input pulse being sampled is shown in Figure 12–35.

Example 12–7

Assume that, as a result of the input sampling procedure just described, on the positive edge of an active-high signal, $32D7_{16}$ is loaded in the input-capture register. On the negative edge of that signal, $17B3_{16}$ is loaded in the input-capture register. How many clock pulses of the free-running counter occurred during the input pulse width? If the free-running counter increments every 2 μsec, what is the duration of this pulse width?

Solution:

Step 1: Expand the two sample count variables to binary:

$$32D7_{16} = 0011\ 0010\ 1101\ 0111_2$$
$$17B3_{16} = 0001\ 0111\ 1011\ 0011_2$$

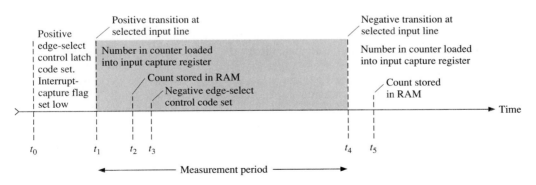

Figure 12–35
Timing relationship between program in memory and sampled input pulse

Step 2: Subtract the first sampled count from the second using two's-complement subtraction:

a. Deriving the two's complement of the subtrahend:

$$2\text{'s comp of } 0011001011010111 = 1100110100101001$$

b. Perform two's-complement subtraction:

$$0001011110110011 + 1100110100101001 = 1110010011011100$$

Step 3: Since the borrow flag of the 8096 would be set during the preceding operation, the absolute value of the result must be determined by deriving its two's-complement value:

$$\text{Two's complement of } 1110010011011100 = 0001101100100100_2 = 1B24_{16}$$

Step 4: Determine the pulse width of the input signal:

a. Convert the absolute value of the difference in count variables to a decimal value:

$$1B24_{16} = 6948_{10}$$

b. Multiply the above result by the clock period of the free-running timer:

$$6948 \times 2 \ \mu sec \cong 13.9 \text{ msec}$$

In microcontroller design, a trade-off must be made between the accuracy of the pulse measurement circuitry just described and the maximum duration of the pulse that may be measured. Perhaps one disadvantage of the Intel 8096 is that its prescaler (or modulo) is fixed. For some microcontrollers, such as the Motorola 68HC11, the prescaler may be adjusted through programming, depending on the degree of accuracy required for a given application. With the 8096, assuming a crystal frequency of 12 MHz, the clocking period of the free-running counter is fixed at 2 μsec. Thus, it would appear that the maximum measurable input pulse width is limited to around 130 msec, which is determined as follows:

$$PW_{max} = T_{CLK} \times 2^{16} = 2 \ \mu sec \times 65{,}536 \cong 131.1 \text{ msec}$$

The measurement of pulse widths of greater duration than 130 msec is possible with the 8096, through use of the **counter-overflow flag.** Each time the free-running counter toggles from FFFF$_{16}$ to 0000$_{16}$, the counter-overflow flag of the 8096 is set. Thus, while being programmed to capture the counts occurring at the leading and trailing edges of an input pulse, the 8096 may also be programmed to monitor the number of times the counter-overflow flag is set during the extended input pulse. One method of tracking the number of overflow points is to increment a memory location (beginning with 00) each time the counter-overflag is set. The CPU may then use this information, along with the input-capture register data, to determine a variable representing pulse width.

The presence of a second, 16-bit counter within the programmable timer section of the 8096 allows the measurement of pulse trains, such as those produced by the optocoupled interrupter module and signal conditioning circuitry of Figure 5–30. Unlike the free-running counter, this second counter may be clocked by an external source. Also, differing from a conventional counter, this second counter increments on both a positive and a negative transition of the clocking source. While this second counter, like its free-running counterpart, may not be directly written to, it does have the capability of being reset at a specified count.

For example, assume a square wave occurs 144 times at point G in Figure 5–30 during a single revolution of a shaft encoder. Point G could serve as the clock input to the second timer within the 8096, being connected to one of two possible port inputs to that device. If this timer is used to monitor the revolutions of the shaft encoder, it could be programmed to reset at count 288. (Remember the counter increments on both rising and falling clock edges.) With each reset condition, a signal could be initiated, indicating the completion of a revolution.

All of the 8096 operations covered thus far could, conceivably, be performed by only the microcontroller. The function of a microcontroller without the aid of peripheral memory and/or port devices is referred to as **single-chip operation.** During single-chip operation of the 8096, the **external address valid input** (EA) must be tied to a logic high (+5 V_{dc}). This action allows the CPU of the microcontroller to address its own internal ROM, reading data and instructions from address 2080 to 3FFF. Also, with the EA line tied high, the CPU vectors to addresses 2000 to 2011 in response to interrupt service requests (see Table 12–4).

Since the 8096, like most other microcontrollers, has limited on-board RAM, it is often necessary for that device to access data and program routines found in an external memory. During this **expanded mode** of operation, the internal ROM of the 8096 may still be accessed, providing the EA line is held high. To access external memory devices, ports 3 and 4 of the 8096, which are 8-bit bidirectional parallel ports, function as the address and data latches for the external address/data bus. This external bus may be configured in one of two ways. The first bussing configuration is similar to that of the 8085. The upper byte of the external bus is accessed through port 4 of the 8096, which functions only as an output latch for the address high byte. Port 3 is multiplexed, serving as an input/output data latch, as well as the output latch for the address low byte. In the second possible address/data bus configuration, both ports 3 and 4 are multiplexed, functioning as 16-bit address and 16-bit data latches. As will be explained momentarily, this allows the 8096 to communicate with memory devices that are configured in an odd/even format.

Initialization of the 8096 for either of the two external bussing configurations just described is accomplished as follows. If the external data bus is to be permanently limited to 8 bits, a logic low must be placed in the 2^1 bit position of the **chip configuration byte** (CCB), which is located within the ROM (address 2018_{16}) of the microcontroller. The CCB is read by the 8096 immediately after every reset condition. If the 2^1 bit of the CCB is set to a logic high, the logic condition of the BUSWIDTH input determines the width of the external data bus. A logic low at the BUSWIDTH pin tells the 8096 that the data bus is only 8 bits wide. A logic high at the BUSWIDTH input indicates the **byte-wide format** (16-bit data bus) is to be used. Making 2^1 of the CCB high and allowing the BUSWIDTH input to control the external bussing configuration provides versatility in memory mapping. For example, assume the A^{15} output of the 8096 (the 2^7 pin of port 4) is tied directly to the BUSWIDTH input. This would allow addresses at or above 8000_{16} to be accessed in a byte-wide format, while addresses at $7FFF_{16}$ or below would be accessed in an 8-bit format.

Figure 12–36 illustrates the possible memory mapping for an 8096 operating in the byte-wide format. Here, both ports 3 and 4 are multiplexed, with the address and the data bus both being 16 bits wide. With the configuration shown here, 16-bit data words are stored in the four static RAM ICs in an odd address/even address format. The Hitachi HM6264LP is a tri-state $8 \times 8.192k$ static RAM. For this IC to be active, one enable pin (26) must be high and the second enable pin (20) low. When the HM6264LP is enabled, data may be written to it via the eight input/output lines (IO_0 through IO_7), providing the WE (write enable) line is at a logic low. During the enabled condition, data may be read from this RAM IC whenever the OE line (output enable) is low.

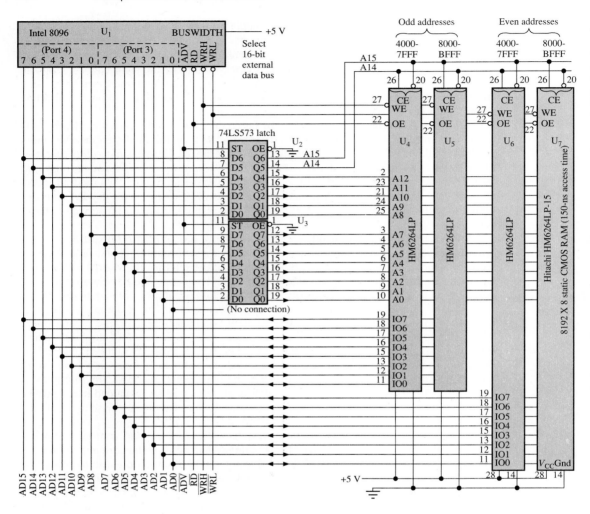

Figure 12–36
Memory mapping for an 8096 microcontroller operating in the expanded mode

The odd/even, addressing system shown in Figure 12–36 requires no address-decoder ICs, such as those discussed in the previous section. The only additional ICs used in this configuration are the two 74LS573 data latches. These low-power Schottky TTL devices together form the 16-bit address latch. With the active-low OE (output enable) lines of both ICs tied to circuit ground, the logic conditions present at the 16 D inputs are latched at the 16 Q outputs whenever the ST (strobe) inputs are low. Note that both strobe inputs of the 74LS573s are connected to the ADV (external address valid) output of the 8096. The function of this output line is essentially the same as that of the ALE (address latch enable) of the 8085. Also note that the two enabling inputs of each of the four RAM ICs are controlled by the two most significant outputs (A14 and A15) from the address latch. For addresses between 4000 and $7FFF_{16}$, with A14 high and A15 low, RAM ICs U4 and U6 are enabled, while RAM ICs U5 and U7 are in a high-impedance state. At and above address 8000_{16}, the enabling condition reverses, with U5 and U7 being enabled and U4 and U6 being held in a high-impedance state.

The method by which the memory locations in RAM are accessed is determined by the type of instructions used within the programming of the 8096. If an STB (store byte) instruction is used, data are stored at the address location that follows the command in the program. If this address has an even number, the data byte being stored is assumed to be the high byte of a 16-bit word. Thus, the **WRH** (external write, upper byte) output of the 8096 goes to its active-low state while **WRL** (external write, lower byte) remains high. Since the WRH line is connected to the WE inputs of U4 and U5, data are stored in the even-address portion of RAM. The program step just described may be followed by an STB command for the odd-numbered address immediately above the location where the high byte was stored. While the 16-bit word strobed into the address latch does not change, the WRL line now goes to an active-low state while WRH is high. The data being stored are now assumed to be the low byte of a 16-bit word. Since the WRH line is connected to the WE inputs of U6 and U7, these low-byte data are sent to a location in the odd-address portion of the memory.

With the expanded memory configuration shown in Figure 12–36, it is possible to store or read the high byte and low byte of a 16-bit word simultaneously, thus saving valuable program time. For this to occur, both the WRH and WRL lines of the 8096 must go low at the same time. This happens automatically during the execution of an ST instruction, which treats the high and low data bytes as a single word. As with the STB command, the high byte is stored in the even-address section of memory, whereas the low byte goes to the odd-address section.

SECTION REVIEW 12–4

1. What industry has been a major force in the evolution of microcontrollers?
2. Describe the structure of the CPU within the 8096. What is contained within the PSW of this device?
3. What form of A/D converter is contained within the 8096 microcontroller?
4. What single output of the 8096 microcontroller would allow it to control precisely the operation of a dc-to-dc power converter?
5. How many interrupt inputs does the 8096 microcontroller have? How many of these interrupts may be enabled simultaneously?
6. What is FIFO memory? What section of the 8096 contains such devices? How do they enhance the ability of the 8096 to handle multiple high-speed interrupts?
7. Identify the components within the programmable timer section of the 8096 that would be used in the measurement of an input pulse width. Explain how this measurement process could be implemented.
8. Explain how the 8096 is initialized for single-chip operation.
9. Explain how the 8096 is initialized to operate in the expanded mode with an 8-bit external data bus and a 16-bit external address bus.
10. Explain how the 8096 must be initialized to operate in the expanded mode using byte-wide data and a 16-bit address bus. Assume the external memory is configured in an odd/even format. With such a memory organization, what is the advantage of using an ST instruction rather than STB when accessing memory?

SUMMARY

With a first-generation microprocessor such as the Intel 8085, a central processing unit, work registers, control circuitry, and memory-interfacing circuitry are contained on a single IC. Microprocessors contained in embedded control microcomputers are often programmed in machine language. The machine language instruction set of the 8085 is divided into the data transfer, arithmetic and logic, branch control, and I/O and machine control groups. While data transfer commands allow information to be exchanged between the microprocessor and memory, as well as between registers within the microprocessor, arithmetic and logic instructions allow computations and Boolean functions to be performed with this information. Branch control instructions include jumps and calls, which may be either conditional or unconditional. Conditional branch instructions involve the carry, zero, parity, and sign flags of the 8085. The software restarts, which use vectored addressing, are also considered branch control commands. The I/O and machine control instructions involve input and output port functions, stack operations, and the hardware interrupts of the 8085.

The two basic forms of memory found within a microcomputer are ROM and RAM. The ROM of an embedded microcomputer, because it contains the monitor program, algorithms, and look-up tables for the system, is likely to be larger than the RAM, which is only needed to store intermediate results during the ongoing control process. The types of ROM devices likely to be found in an industrial control microcomputer include PROMs, EPROMs, and EEPROMs. Erasable nonvolatile memories are advantageous during prototyping of microcomputer systems, since monitor programs and control algorithms are likely to change several times during system development. Special RAM and ROM ICs have been manufactured for use with first-generation microprocessors. The 8355/8755 ROM and the 8155 RAM are designed for compatibility with the 8085. The 8155 is equipped with a timer and three ports, one of which may be configured to provide handshaking signals for the other two. I/O with handshakes is necessary if a microcomputer is to provide real-time control of an industrial process.

With the common-bus organized microcomputer, the external data bus may be considered as an extension of the internal data bus of the microprocessor. This data bus is often multiplexed with the low byte of the address bus. In an embedded microcomputer system, binary-to-octal decoders are frequently used in the enabling of memory and port ICs. In some microcomputer systems, memory-mapped I/O is used instead of conventional I/O operations. With older forms of the first-generation microprocessors, like the Intel 8032, memory-mapped I/O is required, since the device does not distinguish between ports and memory during read/write operations.

While increasingly powerful microprocessors are being created for use in personal computers, more powerful control devices are also being developed for use in embedded control systems. Such devices, called microcontrollers, contain *all* components necessary for a given control function. An example of a state-of-the-art microcontroller is the Intel 8096, which contains, in addition to its CPU, a successive-approximation A/D converter, a pulse width modulator, a programmable timer, and complex interrupt processing circuitry. The interrupt system of the 8096 may operate in conjunction with the programmable timer, allowing for precise measurement of input signals and close timing control of output signals. The 8096 is capable of both single-chip and expanded-mode operation. In the expanded-mode, the 8096 may be configured to process either 8-bit or 16-bit data.

SELF-TEST

1. How does the PSW of the 8085 differ from the PSW of the 8096?
2. What is the primary structural difference between a microprocessor and a microcontroller?

3. True or false?: A nonvolatile memory is one that cannot be erased.
4. What is one advantage of using an EEPROM rather than an EPROM in a microcomputer memory?
5. What is a monitor program? Where is it likely to be contained in a microcomputer memory?
6. True or false?: The ROM within an embedded microcomputer is typically much larger than the RAM.
7. For an 8085, after execution of a *CMP M* instruction, what flag would be set if the data in memory equals the contents of the accumulator? What flag would be set if the data in memory is greater than the contents of the accumulator?
8. Assume for the 8085 that the carry flag is set and the accumulator holds 01011010. After execution of an RAR instruction, what are the contents of the accumulator and carry flag?
9. Assume for the 8085 that the carry flag is reset and the accumulator holds 01101011. After execution of an RLC instruction, what are the contents of the accumulator and the carry flag?
10. As a result of the execution of the following 3-byte instruction by the 8085, what number is placed in the program counter? What numbers are placed on the stack?

2032	CZ
2033	06
2034	41

11. Assume the stack pointer of the 8085 contains 2052. After execution of an RNZ instruction, what would be the contents of the stack pointer?
12. Using 8085 machine language programming and beginning at address 2067, write a short program routine that sets the stack pointer at 503A, unmasks hardware interrupt RST 7.5, then enables the microprocessor's interrupt capability.
13. Write a short routine in 8085 assembly language showing how the current contents of the stack pointer could be stored at memory locations 5069 and 506A *without* destroying the contents of the stack pointer itself.
14. Identify the similarities between the CPUs of the 8085 and the 8096. What is the most significant difference between the two CPU designs?
15. How is it possible for an 8096 operating in the expanded mode to communicate with an external memory, half of which contains 16-bit data words and half of which contains 8-bit data words? What input of the 8096 allows this to occur?
16. True or false?: The DAA command of the 8085 instruction set converts a hexadecimal number to its BCD equivalent.
17. Define the term *microprogramming* as it applies to the operation of a microprocessor.
18. Which interrupt of the 8096 is unmaskable?
19. Identify a possible automotive application of the 8096 microcontroller.
20. Why are microcontrollers better suited than microprocessors for use in portable, hand-held instruments?

13 INTRODUCTION TO PROGRAMMABLE LOGIC CONTROLLERS AND CLOSED-LOOP CONTROL SYSTEMS

CHAPTER OUTLINE

LEARNING OBJECTIVES

On completion of this chapter, the student should be able to:

- Identify the basic components of a programmable logic controller. Explain the advantages of a programmable logic controller over an embedded microcomputer system in implementing an industrial control task.
- Identify and describe the basic components of a process control loop.
- Define discrete control. Explain the advantages and limitations of this form of control.
- Define proportional control. Explain the advantages and disadvantages of this control mode.
- Define integral control. Explain the advantages and disadvantages of this control mode.
- Define derivative control. Explain the advantages and disadvantages of this control mode.
- Identify the three composite control modes. Describe benefits and limitations of each.
- Explain how a programmable logic controller or embedded microcomputer could be used to perform discrete control functions within an industrial process.
- Explain how a programmable logic controller or embedded microcomputer could be used to perform proportional-integral-derivative (PID) control functions within an industrial process.
- Describe various algorithms that could be executed by the CPU of a programmable logic controller or embedded microcomputer during discrete control of an industrial process.

■ Describe various algorithms that might be executed by the CPU of a programmable logic controller or embedded microcomputer during PID control of an industrial process.

INTRODUCTION

The purpose of this chapter is to consolidate ideas already introduced in the previous chapters of this text. Now that many of the tools used in industrial process control have been introduced, methods by which these components may be linked are investigated.

First, the programmable logic controller is investigated. The advantages of this device over an embedded microcomputer in performing industrial control tasks are identified. Next, commonly used industrial control modes are introduced, beginning with relatively simple discrete (on/off) control and ending with the relatively complex composite modes. In this chapter, control circuits introduced earlier in the text are frequently referred to as examples of the control techniques being analyzed. Also, for each control mode being introduced, methods by which a microcomputer or programmable logic controller might fulfill the control requirements of an industrial system are investigated. Methods of connecting the microcomputer to input and output devices are also examined. Finally, standard programming algorithms used to achieve the various forms of closed-loop control are introduced.

13-1   INTRODUCTION TO PROGRAMMABLE LOGIC CONTROLLERS

The idea of a hierarchy of microcomputers performing various tasks within a large industrial system has already been introduced. Although these microcomputers are capable of performing their various control tasks with speed and precision, they do present a few critical problems when used in a typical industrial workplace.

First, these dedicated microcomputers, once placed in a system, are difficult to reprogram. Recall that program routines and look-up tables used by microcomputers on a continuous basis are entered into the nonvolatile memory of the device. In a microprocessor-based system, these routines and data are usually stored in one or more EPROM or EEPROM integrated circuits (ICs) located on the microcomputer circuit board. For a microcontroller-based system, this information could be stored in an EPROM within the microcontroller and/or in a nonvolatile external memory. In either case, modification of these routines and data is a difficult and time-consuming exercise. In most cases, the system must be shut down and the microcomputer circuit board must be removed. An equivalent microcomputer circuit board containing the updated program routines must then be placed into the system. Otherwise, if an updated board is not available, the EPROM must be removed from the original circuit board. The EPROM is then erased and reloaded with updated routines and data. As explained in the previous chapter, this process would require the availability of two specialized pieces of equipment, an EPROM eraser and an EPROM programmer. Once the EPROM is reprogrammed, it must be reinserted into the microcomputer board. Afterward, the microcomputer board must be reinstalled in the control electronics module of the machinery. We also stress that, during the entire process just described, to avoid damage to CMOS logic present on the microcomputer, ESD (electrostatic discharge) precautions must be adhered to strictly.

Also, microcomputers operating in typical industrial environments usually do so under hostile conditions. Because of the high temperatures that are likely to occur in the industrial workplace, microcomputers must often be air conditioned or, at least, cooled by a fan. Due to the presence of dust and hazardous materials, microcomputer circuit boards are often enclosed in bulky cabinets, occupying considerable space within the machinery. Since these cabinets must be ventilated, air filters become necessary in work areas containing excessive dust. Air filters are maintenance-intensive items that require frequent cleaning and replacement. Humidity is also a serious problem in the industrial workplace. Even when contained in cabinets, circuit boards operating in humid areas for prolonged periods may malfunction and, eventually, be destroyed.

Both problems just described, the difficulty of reprogramming and possible damage from a hostile environment, are overcome by a device called a **programmable logic controller** (PLC). Like an embedded microcomputer system, a PLC may report to and receive commands from a higher level computer. However, as the term *programmable* implies, this device may be quickly programmed and reprogrammed independently of the main computer system. Although PLCs exist in a variety of forms, they all have the following distinguishing features:

1. A PLC is easily programmed or reprogrammed, possibly by way of a keyboard or portable keypad. Routines within the memory of some PLCs are easily changed simply by replacing memory modules. Compared to microcomputers, which are usually reprogrammed by software specialists, PLC routines are relatively easy to revise. Thus, the task of reprogramming is often delegated to a machine operator or technician.
2. Unlike an embedded microcomputer, which relies on external power sources for its dc operating voltages, a PLC is equipped with its own power supply. Thus, PLCs may be directly connected to 3-ϕ or single-phase ac power sources. The dc voltages necessary for operating the internal CPU and interfacing hardware of a PLC are automatically provided by the power supply circuitry within the device.
3. To withstand the harsh environment it is likely to encounter within an industrial plant, a PLC is usually contained in a rugged, sealed, container. This allows the device to operate reliably in areas where heat, dust, or humidity could cause an embedded microcomputer to malfunction.
4. Also, to aid in troubleshooting, PLCs are often equipped with their own internal diagnostic systems. LED indicators on the outside of PLC modules may suffice to indicate faults for simpler control applications. In more complex PLC applications, small alphanumeric displays or CRT monitors are often used for indicating machine status and displaying diagnostic messages.

Figure 13–1(a) contains a typical programmable logic controller, while graphic and hand-held programming consoles for such a device are shown in Figure 13–1(b) and (c).

Virtually all of the major components of the PLC have already been covered earlier in this text. Like the microcomputer, a PLC contains a CPU and memory. The CPU of an earlier model of PLC is likely to contain a first-generation (8-bit) microprocessor such as the Motorola Z80 or the Intel 8085. Later model devices could contain Motorola 6800 or Intel 8600 series 16-bit microprocessors. As with the microcomputer, the memory within a PLC consists of both RAM and ROM. The **executive program,** contained within the nonvolatile memory of the PLC, is similar to the monitor program of an embedded microcomputer. Like a monitor program, the executive program continuously governs the operation of the PLC. This program contains routines that control the communication of the PLC with such peripheral devices as keyboards, displays, and printers. Also, the executive program contains those routines that allow the PLC to be programmed easily by an operator or technician. Finally, the executive program contains routines that allow the PLC to run its system diagnostics.

PLCs must be able to control and receive status signals from a variety of actuating devices. For this reason, they are equipped with several forms of input/output module. The simplest form of input/output device contained within the programmable controller

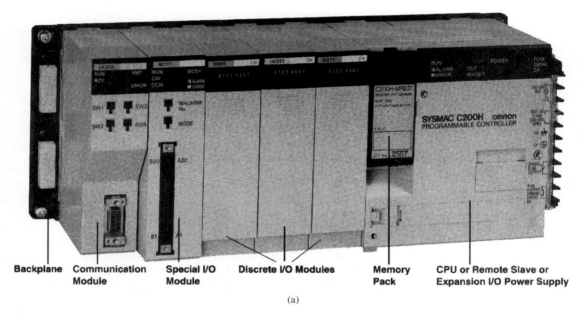

Backplane   Communication   Special I/O   Discrete I/O Modules   Memory   CPU or Remote Slave or
            Module   Module   Pack   Expansion I/O Power Supply

(a)

Figure 13–1
(a) Typical programmable logic control (courtesy of Omron Electronics, Inc.)

is the **discrete I/O module.** This family of I/O modules, which was introduced in Section 11–4, includes ac-to-dc and dc-to-dc input devices, as well as dc-to-dc and dc-to-ac output devices. As the term *discrete* implies, these modules have only two states: totally on or totally off. For example, a dc-to-ac output module, in the form of a zero-point switch, could be used to turn on a single-phase ac motor. Since control is limited to two states, the motor is either turning at its full speed (typically 1800 rpm) or not turning at all.

Advanced forms of PLC may contain **low-level** analog input circuitry, the purpose of which is to amplify low-voltage signals from such devices as thermocouples or load cells. This input signal-conditioning circuitry may contain instrumentation amplifiers, which were introduced in Section 2–4 and implemented in Chapters 3 and 4. Also, active filters and A/D converters may be included in this low-level analog input circuitry.

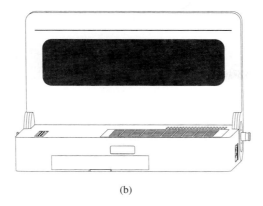

(b)

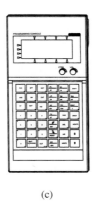

(c)

Figure 13–1 (continued)
(b) Graphic programming console for PLC; (c) Hand-held programming console for PLC

Advanced PLCs might also contain special-purpose I/O modules that are capable of performing control tasks independently of the CPU. Such I/O modules are referred to as **intelligent interfaces.** The use of these devices is referred to as **distributed processing.** Two such intelligent interfaces are the **PID control module** and the **motion control module.** As explained later, PID is the abbreviation for **proportional-integral-derivative.** Although PID is the most complex form of closed-loop control, it is also the most effective and, therefore, frequently implemented in industrial control systems. For this reason, ready-made I/O modules for executing this form of control are now made available in PLCs, expediting the task of system design. The capabilities of this type of I/O module are examined in detail after the five basic control modes are introduced. Due to the wide use of both stepper and brushless dc servomotors in industrial machinery, motion control modules, designed especially for controlling these devices, are often contained in PLCs.

(The internal structure and operation of a servomotor controller is examined in detail in the next chapter.)

Newer PLC models may be equipped with intelligent interfaces specially designed to process data from optical transducers. These **visual input modules,** which are capable of processing data from photodiode arrays, are being used increasingly as recognition devices for industrial robots and automated machines. Due to the high resolution now possible with two-dimensional photodiode arrays, visual input modules are capable of inspecting workpieces being processed by a machine tool. These modules check for such criterion as proper workpiece alignment, gauging, or placement of holes. Photodiode arrays and their associated signal conditioning circuitry were covered in detail in Section 5–7.

Data processing interfaces are intelligent input devices designed to perform specific algorithms involving input data. This input information is usually derived from an analog source such as an input transducer. Although these modules might not contain microprocessors or microcontrollers, they are capable of performing certain fixed mathematical operations at extremely high speeds. For example, the output of a nonlinear transducer such as a thermistor, after being developed by an instrumentation amplifier, is transformed to a binary number by an 8-bit A/D converter. As part of a linearization process, it may be necessary to square this information before sending it to the CPU of the PLC. In automatically performing this mathematical operation with each sample of thermistor data, the data processing interface frees the PLC to perform its tasks with minimal interruption. The data processing interface would interrupt the CPU only when it is ready to send it valid data.

In recent years, significant advances have been made in communication electronics through the exploitation of light as a means of signal transmission. Since light is not affected by the electromagnetic or RF noise that is prevalent at an industrial site, modern PLCs are likely to contain both **fiber-optic output modules** and **fiber-optic input modules.** These modules are usually reserved for transmitting and receiving low-amplitude signals that must travel long distances at an industrial site. The fiber-optic output module converts electrical energy to light, whereas the fiber-optic input module performs the opposite function. Figure 13–2 illustrates the internal circuitry of these types of I/O modules.

PLCs that communicate with such peripheral devices as keyboards and CRT (cathode-ray tube) display monitors must contain an **ASCII interface.** This device is not a true I/O module, since it is not involved in signal conditioning or process control. Rather, it is an input/output port device, necessary for the transmission and reception of alphanumeric information.

Since PLCs were originally designed to replace complex mechanical and electromechanical switching systems, methods of programming PLCs have evolved from relay logic circuits and their associated **control ladder diagrams.** (Electromechanical relays, relay logic, and control ladder diagrams were introduced in Section 1–2.) PLC program routines are designed to poll the input modules on a continual basis, performing algorithms with data acquired from these modules and updating output devices as necessary. One complete cycle through the I/O loop of the PLC is referred to as a **scan.** For most PLCs, the scan time is determined by the length and complexity of the executive program. For other types of PLC, however, the scan time is predetermined by the manufacturer. The two categories of low-level PLC instructions are the **relay type** (sometimes referred to as **basic**) and the **extended** (sometimes referred to as **enhanced**). Eight of the most commonly used basic instructions are identified in Table 13–1. Two relay-type mnemonic symbols used in low-level programming of PLCs, the normally closed (NC) and the normally open (NO) contacts, should already be familiar.

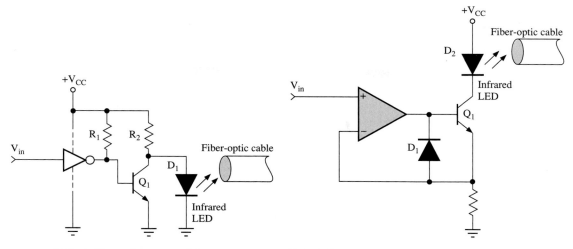

(a) Basic fiber-optic output module
for digital signal transmission

(b) Basic fiber-optic output module for analog signal transmission

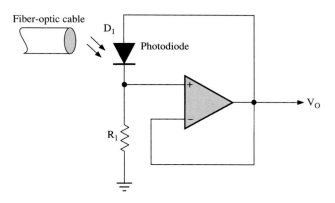

(c) Fiber-optic input module in bootstrap configuration

Figure 13–2

At this point, we stress that, since the PLC replaces complex electromechanical relay logic circuitry, the contact symbols shown in Table 13–1 do *not* represent actual switch contacts. Rather, for a solid-state switching system, placement of one of these contact symbols on a rung of a control ladder indicates that a discrete input signal must be in a specified condition before continuity results across the rung. For example, an NO contact symbol on a rung indicates that a certain input must be in a high (ON) condition for continuity to occur.

The output functions shown in Table 13–1 could represent such load devices as lamps, motors, solenoids, or clutch/brakes. In some PLC instruction sets, the output functions are symbolized by circles rather than parentheses. In either case, however, the output operation is clearly defined by abbreviations placed above the symbols. Examples of such representation of outputs are shown in Table 13–2.

Table 13–1
Eight common PLC instructions

Symbol	Instruction
—| |—	NO (normally open)
—|/|—	NC (normally closed)
—( )—	OUT
—(/)—	OUT NOT
—(L)—	LATCH OUT
—(U)—	UNLATCH OUT
—(TIM)—	TIMER
—(CNT)—	COUNTER

Output functions may be categorized as either positive or negative. The basic **OUT** function in Table 13–1 is a positive function, since continuity through the control ladder rung in which it is placed causes the load it represents to be energized. However, for the **OUT NOT** function, continuity through the rung causes the output load to be de-energized.

As a demonstration of the use of a PLC's basic relay-type instructions, consider the simple ladder diagram in Figure 13–3. On the first rung, an OFF condition for a discrete signal received at an input module causes the PLC to turn on a motor via one of its discrete output modules. For the second rung, an output module holds a red indicator lamp in the ON condition as long as the input signal for the rung is low. However, when the PLC senses the presence of the input signal, it switches off the lamp.

The idea of latching on a relay was introduced in Section 1–2. Once the coil of a relay is latched on by one signal, it does not drop out until a second signal momentarily interrupts the flow of current to the coil. This concept has been incorporated into the instruction set of the PLC. If, on a given rung of a control ladder, continuity is provided to a **LATCH OUT** function, the load represented by that function turns on and remains energized until continuity is provided to the **UNLATCH OUT** function for that load. Figure 13–4 clarifies the operation of the PLC's LATCH OUT and UNLATCH OUT instructions. Here, the arrival of a signal represented by the control contact on rung 1 latches on the load represented as 006. Although the control input on rung 1 disappears,

Table 13–2
Common PLC output functions

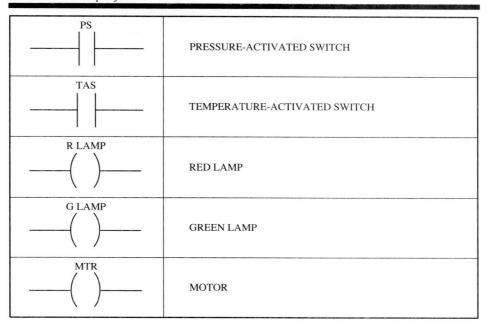

PS ⊣⊢	PRESSURE-ACTIVATED SWITCH
TAS ⊣⊢	TEMPERATURE-ACTIVATED SWITCH
R LAMP ( )	RED LAMP
G LAMP ( )	GREEN LAMP
MTR ( )	MOTOR

the load remains energized until the arrival of the control signal on rung 2. Note that this latching relay action is equivalent to that of the basic set/reset (SR) flip-flop such as a NAND or a NOR latch. The LATCH OUT function is equivalent to SET, and the UNLATCH function is equivalent to RESET.

The notation for latching and unlatching of a load, as shown in Figure 13–4, is cumbersome, since one output function must be represented on two rungs of the control ladder. An alternative method of notation is represented in Figure 13–5. Note the latched output, designated as **KEEP,** is controlled by inputs S and R. The arrival of a set input causes the KEEP function to be latched on, while a reset signal switches the KEEP function to an off state. As shown in Appendix B, the KEEP command is one of the

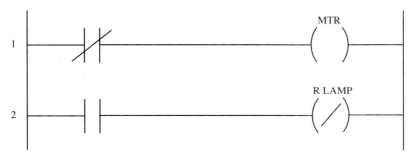

Figure 13–3
Basic PLC ladder diagram

Figure 13–4
Simple example demonstrating latch and unlatch output functions

basic instructions for the Sysmac C200H series PLC (manufactured by Omron Electronics, Inc.).

The **TIMER** function of the PLC evolved from the operation of the time-delay relay. This form of electromechanical relay, which is often pneumatic, produces a delay between the time the coil of the relay energizes and the moment at which the actuation of the contacts occurs. When continuity is provided to a TIMER function, the output it represents is energized after a predetermined wait period. The TIMER function of a PLC is usually governed by the clocking and counting functions within the device's CPU. (The programmable timer functions of microcomputers and microcontrollers were covered in detail in Chapter 12.)

The **COUNTER** function of the PLC must not be confused with the TIMER operation. The COUNTER function may be incremented by an external source. For example, a positive or negative transition from the detector of a light barrier placed across a conveyance path could increment a COUNTER function of a PLC, allowing that device to track the number of objects arriving at a packaging, loading, or palletizing area. The COUNTER function could be made to increment toward or decrement from a preset value that represents the number of items to be placed on a pallet or loaded in a carton.

Simple applications of the TIMER and COUNTER functions are illustrated in Figure 13–6. When the input signal represented by the NC contacts on rung 1 switches to an on state, the contacts open and the TIMER function is activated. The BAS 01 notation below the TIMER output indicates the TIMER function increments by intervals of one second. The PRE 60 notation indicates the TIMER is preprogrammed to count through 60. Thus the counting time becomes one minute. After the one-minute delay, the program

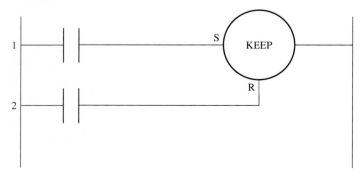

Figure 13–5
Latching and unlatching using a KEEP function

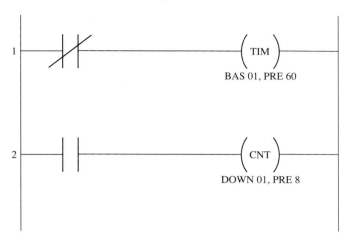

Figure 13–6
Applications of the PLC TIMER and COUNTER functions

sequence moves to the second rung. The DOWN 01 notation below the COUNTER function indicates the counter is decremented each time the signal represented by the NO contact on rung 2 occurs. PRE 8 indicates that the COUNTER is initially preset to the number 8, which could represent the number of items being placed in a carton. Thus, when the counter function decrements to zero, the PLC could initiate the signals necessary for loading and closing the carton.

Eighteen of the most commonly used extended PLC instructions are listed in Table 13–3. The first four instructions are self-explanatory. The arithmetic operations can be performed using constants, data from the PLC, or a combination of the two. The **GET** and **PUT** instructions are based on the read and write operations of the microprocessor. On the control ladder rung containing an extended instruction, the switching contact that initiates execution is referred to as the **control contact.** As shown on rung 1 in Figure 13–7, numeric notation above the extended functions indicates the source and destination addresses for data being moved. In the example shown here, data from memory location 109 is multiplied by data contained in address 214. As indicated above the MULTIPLICATION function, the result of this operation is stored at address 152.

The extended instruction sets of PLCs contain commands that allow these devices to respond to either a positive or negative transition resulting from an arithmetic or logic operation of the CPU. These transitions are represented in the extended instruction set **ONE-SHOT** contacts. An upward-pointing arrow inside the contact symbol indicates its response to a positive transition, whereas a downward-pointing arrow indicates response to a negative-going transition. When an output function is placed on the same rung as a ONE-SHOT contact, closure of that contact causes the output to be active for a period equivalent to one scan. Consider the control ladder in Figure 13–8. When a positive transition occurs as the result of a CPU operation, one-shot 7 sends a high pulse to latching relay 26. Because one-shot outputs remain constant throughout a scan cycle, they are often used to switch on output devices that are to be energized continuously.

Four commands within Table 13–3 are used to isolate or set off portions of the program in the PLC's CPU in much the same way as a subroutine could be removed from the normal flow of events in a microcomputer program. To create such an area or zone within the control ladder diagram, a **MASTER CONTROL RELAY** output function must be placed on the first rung of the zone. The actual routine that is to be isolated then follows, occupying as many subsequent rungs of the ladder as necessary. On the rung

Table 13–3
PLC extended programming symbols

Symbol	Description	Symbol	Description
—(+)—	ADDITION	—(END MCR)—	END MCR INSTRUCTION
—(−)—	SUBTRACTION	—(ZCL)—	ZONE CONTROL LAST INSTRUCTION
—(×)—	MULTIPLICATION	—(END ZCL)—	END ZCL INSTRUCTION
—(÷)—	DIVISION	—(JMP)—	JUMP INSTRUCTION
—| GET |—	GET INSTRUCTION	—(JSB)—	JUMP-TO-SUBROUTINE INSTRUCTION
—(PUT)—	PUT INSTRUCTION	—(RET)—	RETURN-FROM SUBROUTINE INSTRUCTION
—| ↑ |—	OFF-TO-ON TRANSITIONAL ONE-SHOT CONTACT	—| CMP = |—	COMPARE EQUAL
—| ↓ |—	ON-TO-OFF TRANSITIONAL ONE-SHOT CONTACT	—| CMP > |—	COMPARE GREATER THAN
—(MCR)—	MASTER CONTROL RELAY INSTRUCTION	—| CMP < |—	COMPARE LESS THAN

immediately following this routine, an **END MCR INSTRUCTION** is placed. Until the control contact condition on the rung containing the MASTER CONTROL RELAY output function is met, the zone is skipped over during the PLC's scanning process. For this reason, the MASTER CONTROL RELAY output command is often referred to as a **skip** instruction. Note that, in Appendix B, the **JUMP** and **JUMP END** commands of the SYSMAC PLC instruction set are equivalent to the MASTER CONTROL RELAY and END MCR INSTRUCTION.

Figure 13–9 illustrates an application of the PLC's skip instructions. This sequence could represent a small segment within the scan loop of the PLC's executive program.

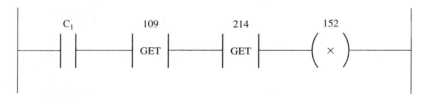

Figure 13–7
PLC routine involving multiplication of data in memory

Figure 13–8
PLC application of ONE-SHOT function

On rung 1 of the control ladder, the control contacts could represent a start signal initiated by the CPU. This signal turns on a green start lamp, indicating that a motor is going to be turned on. On rung 2, as a signal from the detector of a light barrier falls low, the motor is latched on.

For the PLC application in Figure 13–9, it is assumed that the chassis of the motor contains a thermal sensor. Through use of simple signal-conditioning circuitry such as that represented in Figure 3–15, the continuously varying analog signal produced by a sensor such as a thermistor is converted to a discrete signal compatible with the PLC. A logic low from the circuit in Figure 3–15 indicates an over-temperature condition for the motor. Thus, at rung 3, the control contact representing this over-temperature signal is shown as NC. Note that the output function represented on rung 3 is the MCR function. Also note that the END MCR function is placed on rung 6. Unless a motor over-temperature

Figure 13–9

PLC application of the MCR and END MCR functions

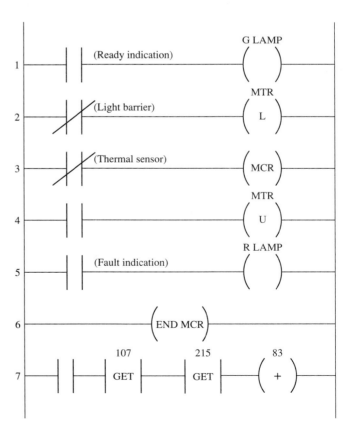

condition exists, the executive program of the PLC skips from rung 2 to rung 7, where arithmetic operations involving data in memory begin to occur. However, if the motor does begin to overheat, the program sequence proceeds to rungs 4 through 6 during the next scan. Through execution of these zone instructions, the motor is turned off and a red fault lamp is turned on.

Many PLC instruction sets contain a **ZONE CONTROL LAST INSTRUCTION** and an **END ZCL INSTRUCTION,** which are similar to the MCR and END MCR commands. For the zone established by the MCR and END MCR instructions, during the time in which the zone is being skipped over, the output functions within the zone are automatically held in the off state. Not until the zone routine is executed are these output functions enabled to turn on. However, when the ZCL and END ZCL commands are used to set off a zone, the output conditions contained within the zone are kept in the last states they held until the point at which the zone routine is executed.

The remaining extended instructions in Table 13–3 should be virtually self-explanatory, since they have counterparts in microprocessor machine language programming. The **JUMP INSTRUCTION** is equivalent to the unconditional jump of a machine language instruction set. The **JUMP-TO-SUBROUTINE INSTRUCTION** can be compared to the **CALL** instruction of the 8085 instruction set. As shown in Figure 13–10, numeric notation is placed over the JMP or JSB instructions, representing the number of the control contact to which the program sequence is to jump. Thus, when the control contact condition for rung 6 is met, the program counter of the PLC's microprocessor is jumped, allowing the program sequence to resume with execution of the subroutine beginning at rung 23. As with the CALL instruction, the subroutine accessed by the JSB command must end with a **RETURN-FROM-SUBROUTINE INSTRUCTION.**

Thus, at the completion of the short subroutine represented in Figure 13–10, program execution resumes at rung 7 of the control ladder, immediately following the rung con-

Figure 13–10
PLC application of JSB and RET instructions

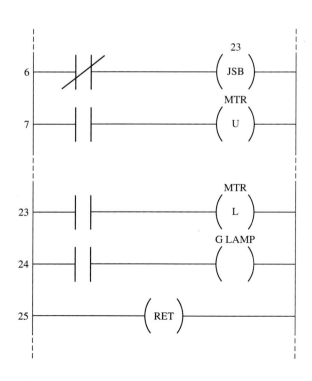

taining the JSB command. Since the machine language instruction set of the 8085 is now familiar, these branch commands and their associated program counter and stack operations should be readily understood. Referring to the SYSMAC PLC instruction set, we should clarify that the JUMP instruction represents a skip rather than a branching action. (Again, JUMP and JUMP END are equivalent to the MASTER CONTROL RELAY and END MCR INSTRUCTION, which are used to establish zones rather than perform branching.)

As with the JMP and JSB commands, the operating principles of the three compare functions in the PLC extended instruction set should already be familiar. As explained in the previous chapter, a compare instruction is based on the subtract operation. Thus, during execution of a PLC's **COMPARE EQUAL** command, equality is detected by testing the zero flag immediately after execution of the internal microprocessor's own machine-level CMP instruction. Recall that, in the operation of a microprocessor, the zero flag is set as the result of a subtract operation only when the subtrahend equals the subtracter. With the **COMPARE GREATER THAN** and **COMPARE LESS THAN** instructions, the carry/borrow flag of the microprocessor is tested. When the subtracter is greater than the subtrahend, the carry/borrow flag remains low. If, however, the subtrahend is greater than the subtracter, the carry/borrow flag is set.

As indicated by the control ladder in Figure 13–11, numeric notation is placed above the CMP=, CMP>, or CMP< command to indicate the source of the variable that becomes the subtracter during the compare operation. Thus, on rung 206, the variable at memory location 114 serves as the subtrahend, being compared to data from location 236. If the two variables are equal, the program sequence jumps to a rung containing control contact 127. If the two variables are not equal, the program sequence continues at rung 207. Here, the same two variables are tested to determine if the variable from location 236 is greater than the data from location 114. If the number at 236 is greater than that present at 114, the program jumps to a subroutine beginning on a rung containing a control contact designated 216. If the number at location 236 is not greater than data at 114, the program sequence continues at rung 208. Here, a third comparison is made, to ascertain if the variable at 114 is greater than the variable at 236. If the number at 114

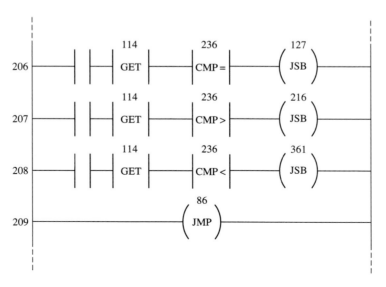

Figure 13–11
PLC application of CMP instructions

is greater, the program jumps to a subroutine beginning at a rung containing a contact designated 361. On completion of this subroutine, or when the number at 114 is not greater than data at 236, an unconditional jump is executed, looping the program sequence back to the rung containing control contact 86.

As shown in Table 13–4, Boolean instructions may be incorporated into the low-level programming of a PLC. Note that several of the mnemonics contained in the Boolean instruction set are identical to those in Table 13–1. Also note that most of these Boolean commands have equivalent instructions in the SYSMAC PLC instruction set.

For complex applications of PLCs, programming directly in low level relay-type or extended instructions might be impractical. Although the control ladder diagram may still be used in these advanced applications, blocks are often used to form high-level commands. Again reviewing the SYSMAC PLC instruction set, it becomes obvious that blocks are used frequently for the symbolic representation of that device's Data Comparison, Data Transfer, Data Conversion, Data Shift, BCD Math, and Binary Math instructions. As seen in Figures 13–12(a) and (b), both the TIMER block and a COUNTER block may have

Table 13–4
PLC Boolean programming mnemonics and symbols

BOOLEAN MNEMONIC	LADDER SYMBOL	BOOLEAN MNEMONIC	LADDER SYMBOL
AND	—| |—	ADD	—(+)—
OR	—| |⌐	SUBTRACT	—(−)—
OUT	—( )—	MULTIPLY	—(×)—
OUT NOT	—(/)—	DIVIDE	—(÷)—
NAND	—|/|—	COMPARE EQUAL	—(CMP =)—
NOR	—|/|⌐	COMPARE GREATER THAN	—(CMP >)—
LOAD	⊢| |—	COMPARE LESS THAN	—(CMP <)—
LOAD NOT	⊢|/|—	JUMP	—(JMP)—
OUT LATCH	—(L)—	JUMP TO SUBROUTINE	—(JSB)—
OUT UNLATCH	—(U)—	MASTER CONTROL RELAY	—(MCR)—
TIMER	—(TIM)—	END MASTER CONTROL RELAY	—(END MCR)—
COUNTER	—(CNT)—		

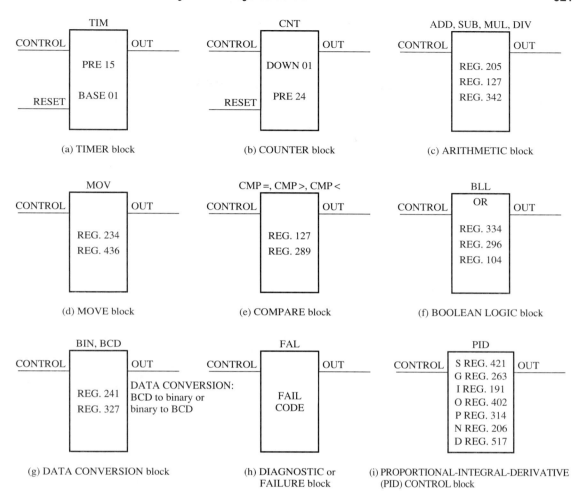

Figure 13–12

two controlling inputs and a single output. For the timer block, the CONTROL input may be a single pulse (represented by a control contact) that initiates the timer. Resetting from another source may be required. However, with many PLCs, reset to a specified initial value occurs automatically as the CONTROL signal arrives. As with the low-level timer representation, PRE represents the initial value (count) to be placed in the TIMER, and BASE represents the counting interval. The high-level CNT function, like its low-level counterpart, relies on an outside control signal to cause it to increment or decrement. Note in Appendix B that the basic instructions of the SYSMAC PLC contain a simple resetable down-count function and a reversible (up-down) counter.

Figure 13–12(c) contains the high-level block representation of an arithmetic operation. While the block for each of these functions has a single input and a single output, it contains at least three spaces in which address locations are entered. For the example shown here, an arithmetic operation takes place involving data contained at address 205 and data at location 127. The result of this operation is then placed at location 342. Referring again to Appendix B, note that fourteen instructions of the SYSMAC PLC's instruction set involve BCD arithmetic operations. In the format just described, source and destination addresses are placed inside these instruction blocks. Since the result of a multiplication or division operation is likely to be a double byte, the destination address for each of these operations is specified as R (low byte) and R+1 (high byte). The output line for an arithmetic operation may be used to indicate an overflow condition for the result register. It may even switch on a warning light, indicating to the programmer that the result of an arithmetic operation is too large for the result register. This programming problem may easily be corrected through a process called **scaling,** which is usually explained in detail in the user's manual of a given PLC. (See the first Special Instruction in Appendix B.)

A versatile PLC is likely to have several forms of data transfer command. The basic **MOVE** instruction shown in Figure 13–12(d) is simply a combination of the extended GET and PUT commands. Here, a data byte at location 234 is moved to location 436. With a **MOVE NOT** instruction, as represented in the Data Transfer instructions of Appendix B, a data word within the source register is complemented before being sent to the destination register.

Many PLCs are also capable of **BLOCK MOVE** operations, an example of which Block Transfer is found in the Data Transfer instructions of the SYSMAC PLC. The block move is a convenient method of transferring several consecutive data words from one stack to consecutive locations within another stack. As shown for the Block Transfer command in Appendix B, the number of data words to be moved is first entered into the instruction block, followed by the initial address of the source block and the base address of the destination. Through further investigation of Appendix B, it becomes evident that the SYSMAC PLC is capable of arithmetic rotating and shifting operations, similar to those described for the accumulator of the 8085 microprocessor. Many PLCs are capable of **BIT TRANSFER** and **DIGIT TRANSFER** operations, examples of which are also found in Appendix B. Through the BIT TRANSFER command, it is possible to move a single bit from a specified position in one register to a specified location in another register. The DIGIT TRANSFER operation is similar to a BIT TRANSFER, except for the fact that it involves a nibble of data rather than a single bit.

As seen in Figure 13–12, the three basic **COMPARE** (CMP) functions and the BOOLEAN LOGIC (BLL) functions can be performed as high-level block operations. Examples of these commands are also contained in Appendix B. With the COMPARE instructions, two address words, representing the sources of the variables being compared, are entered into the block. The output line is high only when the compare condition ($<$, $=$, or $>$) is met. For BOOLEAN LOGIC blocks, as with arithmetic functions, three addresses are required, since the source addresses of the variables and the destination of the result of the operation must be specified.

Since data conversion must be performed frequently in the operation of PLCs, special block instructions are used for this type of operation. As seen in Appendix B, the SYSMAC PLC is capable of performing both BCD-to-binary and binary-to-BCD conversion. It can also perform 4-to-16 bit decoding and 16-to-4 bit encoding. For display purposes, as indicated for the 7-segment Decoder instruction, the SYSMAC PLC also expands 4 data bits into an 8-bit code for controlling a seven-segment display.

As stated earlier in this section, a primary advantage of using a PLC in industrial control systems is the ability of this device to perform its own internal diagnostics. When the **DIAGNOSTIC** (FAL) block shown in Figure 13–12 is energized, it is able to sense a variety of faults within the system being controlled by the PLC. Such faults could include dc power supply failures, tripping circuit breakers, blown fuses, or over-temperature conditions. When one of these faults occurs and is sensed by the PLC, the scan program is interrupted and the DIAGNOSTIC block outputs a failure code corresponding with the type of malfunction being sensed. For some PLCs, this failure code may be limited to a two-character seven-segment display. The technician would then be required to cross-reference the failure code generated by the DIAGNOSTIC function to a list of faults. In more complex, user-friendly systems, the failure code could generate a complete diagnostic message on a display monitor.

Recall that a PLC may be equipped with an intelligent interface module capable of PID control. Such devices relieve the PLC of the task of performing the intensive algorithms required for proportional-integral-derivative (PID) control. However, for PLCs capable of directly performing PID, such as the Sysmac C200H series, control parameters may be entered into a **PROPORTIONAL-INTEGRAL-DERIVATIVE CONTROL** block. As shown in Figure 13–12(i) seven registers can be identified within this block. The S register holds the set point for the process, and the G register holds a variable representing gain. The I register receives input data from the process being monitored, and the O register holds data to be fed back to the process. The three registers designated P, N, and D hold the proportional, integral, and derivative terms that together comprise the output signal. (PID control techniques are introduced in the final section of this chapter.)

As stated earlier, a major advantage of the PLC is that it can be easily programmed by someone other than a software expert. Although use of ladder diagrams and higher level instructional blocks allows the engineer or technician to envision the control task being performed by the PLC, actual programming of the PLC is often done in **computer-type language** (CTL). Many PLCs use CTLs similar to **BASIC,** which is a familiar language to PC users. The term BASIC is deceptive, because BASIC is not at all a ''basic'' language. Rather, it is a high-level, **user-friendly** computer language. (BASIC is an acronym standing for Beginner's All-Purpose Symbolic Instructional Code.) The BASIC commands commonly used in PLC programming are listed in Appendix B. **State language** is another form of CTL that is rapidly gaining popularity among PLC users. More sophisticated than control ladder-based PLC programming methods, state language allows for high-speed monitoring of multiple control processes.

SECTION REVIEW 13–1

1. Describe the problems encountered by a single-PC board microcomputer operating in a hostile industrial environment.
2. Identify the advantages of using a PLC in an industrial machine control application.
3. What type of PLC I/O module is limited to two-state control?
4. Define the term *intelligent interface* as it applies to the operation of a PLC. Identify three examples of an intelligent interface.
5. Define the term *distributed processing*. How is distributed processing beneficial to the operation of a PLC's CPU?
6. What is the advantage of using fiber-optic I/O modules with a PLC?
7. What is the purpose of the ASCII interface module of a PLC?
8. Define the term *scan* as it applies to the operation of a PLC.
9. Describe the functions of the executive program of a PLC.
10. Describe the purpose of the MCR and END MCR instructions in PLC programming. What commands of the SYSMAC PLC instruction set are equivalent to these instructions?
11. What is the difference between an MCR and a ZCL instruction in PLC programming?
12. Describe the difference between a MOVE and a MOVE NOT instruction in PLC programming.
13. Describe the difference between a TIMER and a COUNTER function in PLC programming.

14. What two low-level PLC instructions comprise a high-level MOVE instruction?

15. Identify two forms of computer-type language used in PLC programming.

13–2   INTRODUCTION TO CLOSED-LOOP CONTROL SYSTEMS

The term **process** may be defined as a series of steps, actions, or operations that are followed to produce a desired result. In a manufacturing plant, exact processes must be followed to ensure consistency of an end product. During every industrial process, certain variables such as light, temperature, pressure, flow rate, and humidity must be strictly controlled. For instance, during a drying or baking process, oven temperature must be maintained at a constant level, regardless of fluctuations in the temperature of the environment. As a second, more complex example of a process, consider the system shown in Figure 13–13. Here, during the production of dog food, to make the end product more appealing to the canine palate, syrup is added to the slurry of ingredients (which are best left unmentioned) as it is augured through a tube. To ensure consistency, it is necessary to maintain a uniform mixture of dog food and syrup, regardless of slight deviations in the flow rate of the slurry. Thus, the intake valve must compensate for these changes in flow.

The monitoring and governing of such variables as temperature, pressure, and flow rate during an industrial process are referred to as **process control.** Any process may be represented by equations. A property representing an end product of the process must first be identified. Next, those variables that affect this property are itemized. In a mathematical sense, the given property may be considered as a function of these variables.

$$P = F(V_1, V_2, V_3, V_4, \ldots, V_N, t) \qquad (13\text{–}1)$$

where P represents a given property, V represents a variable affecting that property, and t represents time.

Effective process control is possible once a desired property of a product has been identified and the variables affecting that property are quantified. Consider, for example, the baking industry. Assume that a large baking company decides to distribute a certain type of oatmeal cookie based on a homemade recipe. Once the proper proportions of

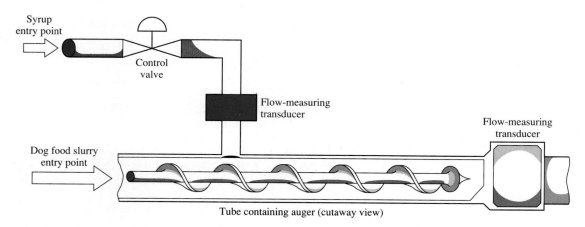

Figure 13–13
Process control example: mixing of ingredients during dog food production

ingredients and optimal baking time and temperature have been established, vast numbers of this type of cookie may be produced at several different sites over a long time; yet all of these cookies look, feel, smell, and taste the same. This is possible if a precise mixing and baking process has been established and effective methods of monitoring and controlling this process have been developed.

The variables that must be controlled during the baking process include oven temperature and baking time. If, during the baking process, the cookie sheets pass through the oven on a conveyer, then the speed of this conveyor becomes a factor that affects the baking time. Thus, oven temperature and conveyor speed become **controlled variables.** An example of an uncontrolled variable would be the temperature of the work area containing the oven. While the temperature of the workplace is certainly governed by an environmental control system, this system is *not* considered as part of the oven temperature control system. The oven temperature must be held at a nearly constant level during the baking process, regardless of deviations in environmental temperature. For instance, to hold oven temperature at a specified point, more current must be provided to the oven's heating elements when environmental temperature is only 65°F than when it is 90°F.

The process of holding a controlled variable at or near a specified value is referred to as **regulation.** This idea should be familiar, since regulatory circuits were already introduced in Chapters 9 and 10. In fact, key concepts of closed-loop control systems have already been introduced for such devices as voltage regulators and dc motor control circuits. Recall for the basic series regulator in Figure 9–42 that a virtually constant output voltage is maintained, despite variations in the output load current. This is possible through the interaction of a resistive voltage divider, an op amp, which functions as an error detector, and a pass transistor, which serves as the final corrective device.

Figure 13–14 contains a block diagram of a **process control loop.** After review of the simple series regulator in Figure 9–42, the correlation between this circuitry and the diagram in Figure 13–14 should become evident. In Figure 9–42, the **process** is the ongoing effort to hold the regulator output at a constant dc level. **Measurement** occurs at point B, where an error signal is produced, reflecting any deviations in output load voltage. The reference voltage in Figure 9–42 is equivalent to a **set point.** The op amp thus performs the combined function of **error detector** and **controller.** The pass transistor may be considered to be the **final control element,** since it directly governs the output level of the regulator.

In an industrial control system, the measurement process is likely to involve one of the transducers examined in Chapters 3 and 4. Assume, for example, that the oven temperature control system uses a thermistor as its measurement device. Recall that, when used in such an application, the thermistor must be placed in some form of resistive network. This allows changes in temperature, which produce corresponding changes in thermistor resistance, to be sensed as changes in voltage. Thus, the control system does not directly measure temperature. Rather, it depends on the temperature transducer to produce changes in voltage analogous to deviations in oven temperature. To predict accurately the operation of the measurement section of a closed-loop control system, the relationship between the **controlled variable** (V_c) and the **measured variable** (V_m) must be clearly defined. For an oven temperature control system, the controlled variable is oven temperature, which may be considered to be the input to the measurement unit. The measured variable, which would be the change in voltage resulting from a change in thermistor resistance, is considered to be the output of the measurement unit. Thus, for the measurement block shown in Figure 13–14, the relationship of the input V_c and the output V_m are expressed mathematically as a **transfer function:**

$$V_m = T(V_c, t) \tag{13–2}$$

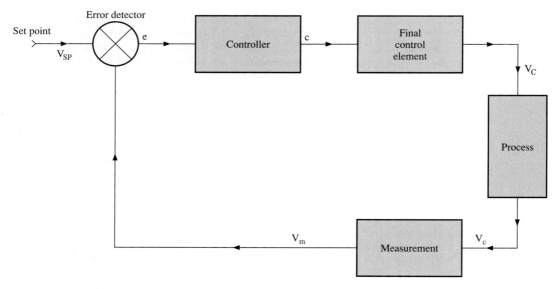

Figure 13–14
Block diagram of a process control loop

where T indicates an operation involving a transfer function. This equation states that the measured variable, an output voltage of the measurement block in Figure 13–14, is a function of the controlled variable. Time (t) is a critical component of the transfer function, since an input transducer, especially a thermistor, requires a significant amount of time to respond fully to a change in a controlled variable. Thus, t in Eq. (13–2) accounts for any propagation delay between a change in the controlled variable and the full response of the measurement block.

In Chapter 3, Figure 3–14(a) contains a passive signal conditioning circuit, in the form of a simple voltage divider. This circuit could function as the measurement block in a control system. Figure 3–14(b) contains the transfer function for this circuit. Although the response curve of the thermistor (as shown in Figure 3–13) is nonlinear, the plot of output voltage versus temperature becomes nearly linear in Figure 3–14(b). As shown in Figure 3–14(a), this near linearity is achieved by placing a thermistor having an NTC at the top of the voltage divider. In many industrial control applications, the response curve produced by the measured variable is nonlinear. As explained later in this chapter, after such nonlinear analog information is converted to digital form, a microcomputer or PLC may perform a linearization algorithm with this information before using it within a process control program.

In an industrial control system, the error detector is often a differential (instrumentation) amplifier. With this type of system, changes in the output of the error amplifier are proportional to changes in the error signal. In systems involving discrete control, the error detector may simply be a comparator that toggles as an error signal passes a set-point level. In either case, the error detector receives the measured variable and a reference voltage that represents the set-point level of the measured variable. Assuming linearity of response for the input transducer, V_m represents V_c, while the reference voltage represents the set-point value of the controlled variable. Thus, the transfer function of the error detector may be stated as:

$$e = (V_m - V_{SP}) \tag{13–3}$$

where e represents the output of the error detector.

The **controller** can be considered to be the "brain" of the closed-loop system. It receives information from the error detector and determines the amplitude of the feedback signal being sent to a corrective device called the **final control element.** In a complex control loop, the controller may be a microcomputer, microcontroller, or PLC with intelligent interface modules. If such devices are used, A/D conversion occurs at the input side of the controller, while D/A conversion occurs at its output. The algorithm executed by the controller depends on the type of response desired for a specific industrial process. In some applications, the controller must react quickly to a disturbance, whereas in other applications, a gradual response may be preferable. Although each of the various control modes has its own transfer function, a general expression can be used to describe the operation of the control block in Figure 13–14. This transfer function states that the amplitude of the signal sent to the final control element as a result of a control calculation (or algorithm) is a function of the amplitude of the error signal and the propagation time of the system.

$$c = C(e, t) \tag{13–4}$$

where c represents the corrective signal sent to the final control element.

The final control element, which might be a motor, solenoid, or heating element, has its own transfer function. As stated in Eq. (13–5), the value of the **controlling variable** (V_C), which could be a valve position, motor speed, or the amplitude of current sent to a heating element, is a function of the output of the controller and the propagation time of the system.

$$V_C = F(c, t) \tag{13–5}$$

The ideal closed-loop control system presents a paradox regarding the regulatory process. Since the ideal control system would allow no deviation of the controlled variable from a set-point value, no regulation could occur! Thus, some error must exist for the controller to be able to initiate a corrective action. In the real-world closed-loop control system, it is impossible to maintain a controlled variable exactly at a set-point value. Therefore the controller is constantly active, attempting to hold the controlled variable close to the set point. For every industrial process, uncontrolled variables always exist. Consequently, the need for process control is always present.

For a closed-loop control system, as represented by Equation 13–3, **error** can be defined as the difference between the actual value of a controlled variable and the set-point or reference value of that variable. Error may be expressed in a variety of ways. One method is to express the error directly in unit values of the controlled variable. For example, assume oven temperature is expressed in degrees Fahrenheit and the set point is 375°F. If the actual oven temperature is 361°F, error is determined simply as follows:

$$\Delta T = 375°F - 361°F = 14°F$$

A problem with this method of error notation is that it does not indicate the severity of the error. In most closed-loop control systems, it is necessary to determine the relative amount of deviation represented by the error. One method of showing this deviation is to express the error as a percentage of the set point. Thus, an error of 14°F would represent a deviation of 3.73% with reference to the set point of 375°F. Being determined as follows:

$$e_{PS} = \frac{14°F}{375°F} \times 100\% = 3.73\%$$

where e_{PS} represents error as a percentage of set point.

If, however, the set-point temperature is 525°F, the 14°F deviation represents an error of only 2.67%:

$$e_{PS} = \frac{14°F}{525°F} \times 100\% = 2.67\%$$

A more meaningful and, therefore, more common method of expressing error is to show it as a percentage of the range of a controlled variable:

$$e_P = \frac{(V - V_{SP})}{V(\max) - V(\min)} \times 100\% \qquad (13-6)$$

where e_P represents the percentage of error, V represents the instantaneous value of the controlled variable, V_{SP} represents the set-point value of the controlled variable, $V(\max)$ represents the upper limit of the range of the controlled variable, and $V(\min)$ represents the lower limit of the controlled variable range.

Example 13–1

Assume an oven temperature control system is able to govern temperature effectively between 235°F and 500°F. Also assume the set-point temperature is 368°F. With respect to the oven control range, what percentage of error is represented by an instantaneous oven temperature of 378°F? What would be the instantaneous value of oven temperature if the percentage of error equals −2.3%?

Solution:
Step 1: Given the instantaneous oven temperature of 378°F:

$$e_P = \frac{(V - V_{SP})}{V(\max) - V(\min)} \times 100\% = \frac{(378°F - 368°F)}{(500°F - 235°F)} \times 100\% \cong 3.8\%$$

Step 2: Given an error of −2.3%, if

$$-2.3\% = \frac{(V - 368°F)}{(500°F - 235°F)} \times 100\%$$

then, through transposition,

$$V = -0.023 \times (500°F - 235°F) + 368°F \cong 361.9°F$$

Two important considerations for a closed-loop control system are its **steady-state response** and **transient response.** The steady-state response of a closed-loop control system is depicted graphically in Figure 13–15. As explained earlier, for the control system to be able to exert control over a given variable, that variable must deviate slightly from its set-point value. With regard to the steady-state response of a closed-loop control system, the slight difference between the desired set-point value of a variable and its actual value is called the **residual error.** For a well-designed closed-loop control system, the residual error is likely to represent only a small percentage of the range of the controlled variable.

In many industrial environments, sudden changes momentarily occur in one or more of the uncontrolled variables, causing the controlled variable to deviate suddenly from

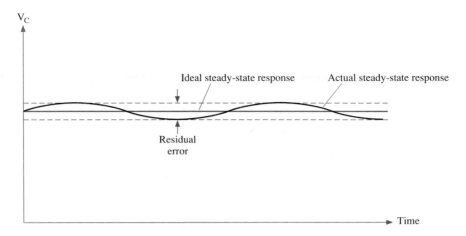

Figure 13–15
Steady-state response of a control system

its set-point value. Such a discontinuous form of error is referred to as a **transient.** As an example of transient response of a control system, consider a condition where, on an extremely cold winter day, the door to the loading dock, which is close to the oatmeal cookie oven, is suddenly swung open and left open for several minutes. Even with the best possible oven temperature control system, a disturbance results, causing the oven temperature to drop temporarily below its steady-state value. The true measure of the effectiveness of a closed-loop control system is how quickly and smoothly it takes corrective action in the event of such a transient disturbance. The transient response of a closed-loop control system may be described as being either **overdamped** or **underdamped.**

Figure 13–16(a) graphically depicts a closed-loop control system exhibiting overdamped response to a transient disturbance. As illustrated here, the controlled variable rapidly drifts to a peak error level. Then, as the control system initiates corrective action, the controlled variable gradually approaches the set point. Eventually, it settles around the residual error level, where it was before the transient disturbance. As shown in Figure 13–16(a), the duration of an overdamped transient response is measured from the instant the controlled variable begins to drift to the point where it returns to its residual error level.

Figure 13–16(b) represents the response of an underdamped closed-loop control system to a transient disturbance. The distinguishing characteristic of underdamped response is **oscillation.** Rather than drifting away from then gradually returning to the residual error point, the controlled variable exhibiting underdamped response fluctuates above and below the set-point level. Initially, the oscillation is abrupt, with the controlled variable attaining a maximum peak error level. Gradually, as the control system reacts to the transient disturbance, the oscillations dampen. The duration of the underdamped response is measured from the point at which the oscillation begins to the point where it dampens to the residual error level. As indicated in Figure 13–16(b), these oscillations are periodic. The period of an oscillation (T) may be measured between two peak error points.

The performance of a closed-loop control system is usually evaluated through determination of its **residual error** and **dynamic error.** The residual error, which has already been defined, is usually determined using an average value of the controlled variable, calculated over an extended time. As explained in the introduction to the previous chapter,

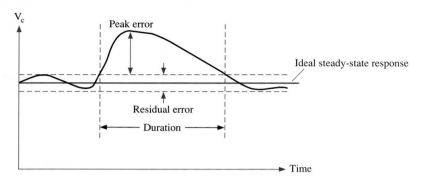

(a) Overdamped response to a transient disturbance

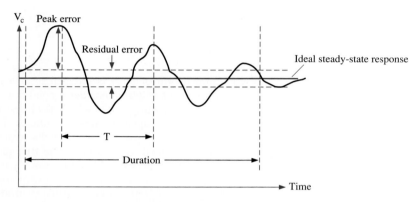

(b) Underdamped response to a transient disturbance

Figure 13–16

such a task could easily be performed by a PC monitoring the performance of an industrial system. After deriving the value of residual error, the computer could compare it to the set point and determine a percentage of error based on the possible range of the controlled variable.

The dynamic error of a closed-loop control system is determined by introducing a transient disturbance of a known value into the process being controlled. For a system exhibiting an overdamped transient response, the peak error and the duration of the response are recorded. For a system with an underdamped transient response, peak error, duration, and oscillation frequency (as a reciprocal of the period) are recorded. The dynamic error for either an overdamped or underdamped closed-loop control system could be accurately determined with a computer.

The primary function of a closed-loop control system is to provide stability during an industrial process. While operating stability is achievable with a well-designed control system, a badly designed or defective system may actually introduce instability into the process. Instability of a process control system is often described as **growth without limit.** As this description implies, a malfunctioning process control system could allow the controlled variable to increase steadily. In systems without safeguards, such uncontrolled growth could produce disastrous results.

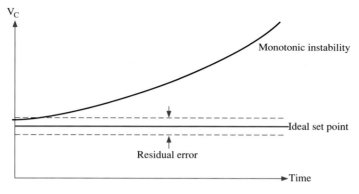

(a) Monotonic instability

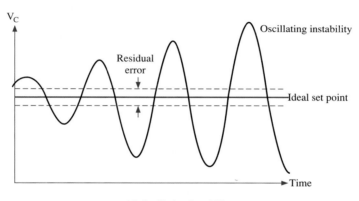

(a) Oscillating instability

Figure 13–17

Monotonic instability usually occurs as a consequence of control system failure. As illustrated in Figure 13–17(a), monotonic instability is characterized by a steady drifting of the controlled variable away from the set point. For example, during an industrial process, the contents of a hopper or vat could be held near a set-point level. However, because of the total failure of a control system and the resultant monotonic instability, the vat or hopper would attempt either to overflow or drain completely. For an open vat, a minimal safeguard against such an occurrence would be a capacitive level-detecting system similar to that shown in Figures 4–22 through 4–28. This system would be equipped with a loud alarm horn that would sound as the vat began to overfill. Also, for a suspended hopper, such as that shown in Figure 4–56, an overweight or underweight condition could be sensed by the tension load cell, resulting in a similar audible alarm signal. In process control applications involving pressurized containers or high-temperature ovens, monotonic instability becomes a direct threat to the safety of personnel. Thus, along with audible alarms, these systems might have automatic shutdown capabilities to prevent fires or explosions.

Oscillating instability, which is represented in Figure 13–17(b), can occur if a control system is inadequately designed for a given process. If the control system is working correctly, it is able to dampen quickly any oscillations that result from a transient disturbance by sending negative, or degenerative, feedback to the process. Since this feedback signal is nearly 180° out of phase with the oscillation, it counteracts the oscillation, gradually bringing the controlled variable closer to the set point. However, at a certain frequency of oscillation, it is possible that enough propagation delay exists to cause the feedback signal to have a positive, or regenerative, effect. Thus, at this frequency, the oscillations are reinforced rather than dampened, gradually increasing in amplitude as shown in Figure 13–17(b). A major advantage of underdamped transient response is that it allows quicker corrective action than overdamped response. However, closed-loop control systems exhibiting underdamped response are more prone to oscillating instability.

SECTION REVIEW 13-2

1. Define the term *process* as it applies to an industrial control system.
2. Define the term *set point* as it applies to the operation of a closed-loop control system.
3. Identify the major functional blocks within a closed-loop control system. Write the transfer function for each block.
4. Define the term *residual error* as it applies to the operation of a closed-loop control system. How is residual error usually measured?
5. Define the term *transient* as it applies to the operation of a closed-loop control system. What are two basic types of transient response exhibited by a closed-loop control system?
6. True or false?: For a closed-loop control system, error is almost always expressed as a percentage of set point.
7. When is a control system likely to exhibit monotonic instability? When is a control system likely to exhibit oscillating instability?
8. A closed-loop oven temperature control system has an error of −1.7%. Assume the set-point temperature is 485°F and is at the middle of the controller's active range. If the actual oven temperature is 483°F, what temperatures define the upper and lower limits of the controller's range?
9. Define growth without limit. Describe some of the safeguards an industrial system could contain to minimize the damage that could result from this phenomenon.
10. Define dynamic error. For a closed-loop control system exhibiting an underdamped response, what three parameters of dynamic error could be measured and recorded using a computer?

13-3   DISCRETE CONTROL SYSTEMS

Discrete, or on/off, control is the oldest and simplest form of industrial control. The simplest discrete control system would involve the control of a single variable. For example, a discrete oven controller, by toggling between an on and off condition, attempts to hold oven temperature at a set-point level. The operation of such a circuit, which is illustrated in Figure 8–15, is covered in Section 8–4. Because of the zero-point switching action of this circuit, the heating element is either fully on or fully off. Ideally, the circuit in Figure 8–15 would keep the heating element fully on while the oven temperature is below the set-point value. This is naturally accomplished through the zero-point switching process. When V_A is higher than V_B, whole cycles of ac current pass to the heating element. However, as the oven temperature attempts to exceed the set point level, V_A falls below V_B, causing the flow of current through the heating element to be interrupted at the 0° point of the subsequent ac source cycle. While V_A remains lower than V_B, load current remains off throughout the 360° of each cycle. Thus, while oven temperature is above the set-point level, the heating element is totally off. Again assuming ideal operation of the control system, as oven temperature falls below the set-point level, the heating element switches fully on at the 0° point of the next source cycle. On the basis of this ideal operation of the oven controller in Figure 8–15, the following mathematical expression may be stated:

$$S_{out} = \begin{bmatrix} ON \ V_m > V_{SP} \\ OFF \ V_m < V_{SP} \end{bmatrix} \qquad (13\text{–}7)$$

where S_{out} represents the condition of the feedback signal to the final control element, V_m represents the measured variable, V_{SP} represents the set-point or reference level of the measured variable, and ON and OFF represent the two possible states of the feedback signal.

Figure 13–18 represents the ideal operation of the discrete oven controller in Figure 8–15. As the oven temperature decreases, due to the NTC of the FR-1M thermistor, the measured variable, equivalent to V_A, attempts to exceed the set-point level. Thus S_{out}, equivalent to the condition of the TRIAC, switches to an ON state. As the oven temperature increases, the measured variable begins to fall below the set-point level, causing S_{out} to switch to an OFF state. It is now evident that, assuming nearly ideal operation for the circuit in Figure 8–15, S_{out} toggles rapidly and continuously between the ON and OFF states.

Although such oscillation is not a problem for the noise-free solid-state controller in Figure 8–15, it would present a serious problem for a discrete control system involving electromechanical relays, motors, or solenoids. Such rapid switching of these components would introduce excessive noise into a system and possibly result in component destruction. Fortunately, any comparator-based control system, such as that in Figure 8–15, has some inherent hysteresis, which tends to limit its rate of oscillation. However, in most discrete control systems, it is necessary to introduce additional hysteresis to dampen the rate of oscillation and provide operating stability.

A control system with adjustable hysteresis has already been introduced in Figure 11–29. Here, through use of a window detector and a D-type flip-flop, the fan is made to turn on as cabinet temperature begins to exceed an upper limit. The fan continues to

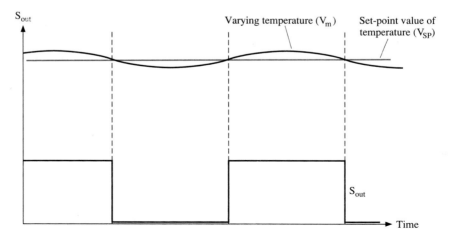

Figure 13–18
Operation of discrete (on/off) oven controller

run until the cabinet temperature falls below a lower limit, which is a few degrees lower than the upper limit. As was explained in Section 11–3 and demonstrated in Example 11–2 of Chapter 11, these threshold temperatures may be established through the adjustment of potentiometers R_3 and R_5.

Thermistor R_2 is strategically placed within the circuit board cabinet such that it can accurately reflect the temperature within the cabinet. Thus, with cabinet temperature as the controlled variable (V_c) the voltage at point A becomes the measured variable (V_m). For the deadband fan control circuit in Figure 11–29, there is no actual set point, representing an optimal cabinet temperature. There is, rather, a maximum allowable temperature for the cabinet. After the circuit boards in the cabinet are energized, they gradually begin to heat, due to the I^2R loss of their components. When the maximum allowable temperature is attained, the fan is switched on, continuing to run until the cabinet cools to the lower threshold temperature. At this point, with the fan switched off, the cabinet again begins to heat. When the maximum safe temperature is exceeded, the fan turns on again. This cycle repeats as long as power is applied to the fan and card cage. Thus, as represented graphically in Figure 13–19(a), this type of control system is naturally oscillating.

In a discrete control system, the difference between the upper and lower switching thresholds is referred to as the **deadband, hysteresis gap,** or **differential gap.** The mathematical expression for a deadband controller is as follows:

$$c = \begin{bmatrix} 0\% \ e < -\varepsilon \\ 100\% \ e > +\varepsilon \end{bmatrix} \quad\quad\quad (13\text{--}8)$$

where c represents the control signal to the final control element, e represents error, $-\varepsilon$ represents the lower limit of the deadband, and $+\varepsilon$ represents the upper limit of the deadband.

Equation (13–8) states that, if the error signal at point A in Figure 11–29 attempts to fall below the level of $-\varepsilon$, the output of the controller (c) brings the final control element to its 0% (totally off) condition. If, however, the controlled variable attempts to exceed the level of $+\varepsilon$, the controller switches on the final control element at its 100% (full-on) condition.

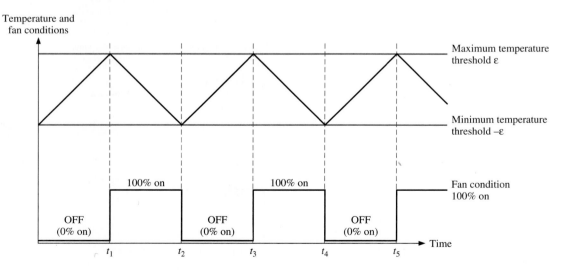

Figure 13–19
Operation of deadband controller

Most microcomputer-based and PLC-based discrete control systems are responsible for monitoring several input variables and making decisions based on the conditions of these variables. As shown in later examples, the design of these systems becomes increasingly complex when the controlled variables are interactive or when these controlled variables must change states at various points within an industrial process. Advanced automation systems often involve such **sequential state control.** In the design of such sequential state control systems, the various states within the given process must be clearly defined. As a simple example of such a discrete system, consider an automated sequential oven controller. For such a system, the baking process could be divided into five definite cycles, or states, including preheat, load, bake, cool-down, and unload. Discrete variables controlled during the baking process, as identified in Table 13–5, include the conditions of the oven door, the heating element, an oven lamp, and a cooling fan.

During the preheat cycle (state 1) of the baking process, both oven doors must remain closed and the green standby lamp is lit. When the oven temperature arrives at the set-point level, the loading process (state 2) is allowed to begin. At this point the front oven door (door 1) opens automatically, allowing the cookies to be conveyed into the oven. The duration of the baking cycle (state 3) is usually a predetermined period, perhaps 15 or 20 minutes. During this time while the red oven lamp is lit, the oven temperature must be held as close as possible to the set point. At the end of the baking cycle, the cool-down period (state 4) begins, with the heating element being switched to the off state. The cool-down time, like the baking cycle, may be a predetermined period. At the end of the cool-down cycle, the unloading cycle (state 5) is initiated. At this time, the exit door (door 2) opens, allowing the cookies to be conveyed out of the oven.

The control sequence outlined in Table 13–5 could easily be implemented through use of a microcomputer, microcontroller, or PLC. The following analysis shows, at a block level, how an 8-bit microcomputer might be used in this application. As shown in Figure 13–20, if the microcomputer must monitor inputs from various sensors within the system and also provide output signals, several I/O lines are required. At the same time, a relatively small amount of RAM is needed. Thus, memory-mapped I/O would be a preferable design concept for the microcomputer block represented in Figure 13–20.

A zero-point switching circuit still serves as the discrete controller for the heating element in Figure 13–20. However, certain design alterations must be made to allow the device to interface with the microcomputer. Discrete I/O modules such as those introduced in Section 11–4 allow the microcomputer to receive signals from input transducers and send discrete control signals to final control elements. As explained in the introduction to Chapter 12, the embedded microcomputer may be connected to the main computer system via a common-bus system. This would allow the operator to enter a code representing the set-point oven temperature at the keyboard of the main computer. The oven control microcomputer reads this information and translates it into a code that is sent to the

Table 13–5
Sequential control of baking process

State	Cycle	Door 1	Door 2	Heating Element	Oven Lamp	Cooling Fan
1	Preheat	Closed	Closed	On	Off	Off
2	Load	Open	Closed	On	Off	Off
3	Bake	Closed	Closed	Hold set point	On	Off
4	Cool-down	Closed	Closed	Off	Off	On
5	Unload	Closed	Open	Off	Off	Off

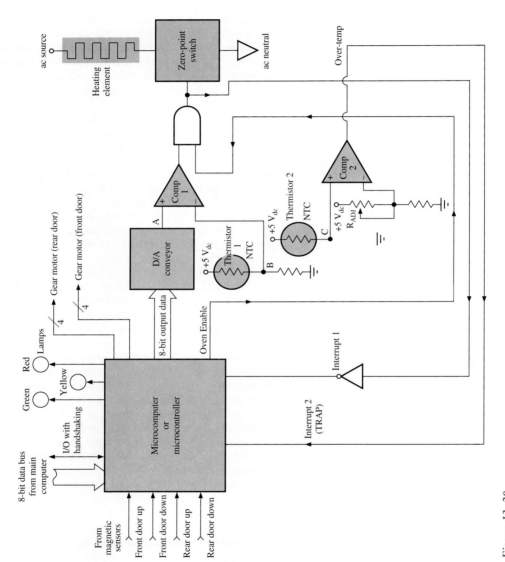

Figure 13–20
Automated sequential oven controller

D/A converter. The higher the value of the 8-bit binary number at the input to the D/A converter, the higher the reference voltage at the noninverting input of the first comparator (point A). If V_A is established at a relatively high level, V_B must attain that level for the output of the first comparator to fall low and turn off the zero-point switch. For this to occur, the oven temperature must have increased to a relatively high value.

If the binary number being sent to the D/A converter is reduced, the set-point voltage at point A is also reduced. Thus V_B attains the level of V_A at a lower value of oven temperature, allowing the zero-point switch to be turned off at that temperature. Part of the design process for the system in Figure 13–20 could be the creation of a look-up table to be placed in the nonvolatile memory of the microcomputer. This table would accurately cross reference each control word arriving from the control PC with an output variable that, when fed to the D/A converter, produces a set-point temperature equal to that represented by the control word. The gain of the D/A converter would be finely adjusted to ensure this correspondence is achieved. This could be accomplished through a one-time adjustment of the D/A converter's gain potentiometer (not shown).

In Figure 13–20, two interrupt lines are shown being fed back to the microcomputer. As part of the initialization process, at the beginning of state 1 in Table 13–4, the hardware interrupt line functioning as Interrupt 1 is unmasked. After the microcomputer outputs the Oven Enable signal to the oven control circuitry, the microcomputer awaits the arrival of a positive transition at its Interrupt 1 input. This interrupt request vectors the microcomputer into state 2. At this point, the interrupt capability of the microprocessor is disabled, preventing the microcomputer from responding to any further transitions at the Interrupt 1 input.

At the beginning of state 2, the microcomputer must initiate signals that raise the entrance door (door 1) of the oven. This may be accomplished using the interface circuitry shown in Figure 13–21. Here, it is assumed that an 8-bit tri-state latch is being used to control the operation of both the entrance and exit doors of the oven. Outputs Q_0 through Q_3 of this latch control door 1, while Q_4 through Q_7 control the operation of door 2.

To understand the control codes shown in Table 13–6, it is necessary to review briefly the operation of dc motors. As explained in Section 10–3, the direction of rotation of a dc motor can be reversed by changing the direction of current flow through the field winding. Figure 13–21 illustrates how such reversal of rotation can be implemented by means of solid-state switching. To raise door 1 of the oven, assume current flow through the motor field winding must be from point A to point B. Analysis of the circuit in Figure 13–21 proves this current flow would result when, as indicated in Table 13–6, the Q_0 and Q_3 outputs of the motor-control port are low. When the oven door is to be closed, the Q_1 and Q_2 outputs of the port are brought low, causing the direction of field current to be reversed. Assuming the Darlington drivers are also connected to the field windings of the door 2 motor as shown in Figure 13–21, then Q_4 through Q_7 of the output port are treated as shown in Table 13–6.

Figure 13–22 illustrates how photosensors and magnetic proximity sensors could be positioned within the automated oven system. (Light barriers were covered in Section 5–5, and magnetic proximity sensors were covered in detail in Section 4–3.) Four proxistors, which function as oven door limit switches, are mounted inside the frames of the oven doors. For each of the two doors, one proxistor senses the door is fully elevated and a second proxistor senses the oven door is fully closed. Also, at both the entrance and exit to the oven, a light barrier is mounted, being positioned nearly flush with the oven conveyer. The light barrier senses when the cookie sheets have either arrived at or cleared the oven doors.

The program routine for state 2, opening the oven door, is initiated by the arrival of a cookie sheet at the oven door. Immediately after the infrared light beam between the emitter and detector of the light barrier is broken by the cookie sheet, the microcomputer

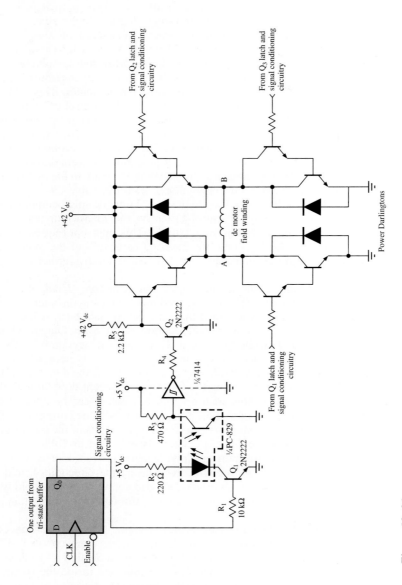

Figure 13–21
Oven door direction control circuitry

Table 13–6
Oven door control codes

Q_7	Q_6	Q_5	Q_4	Q_3	Q_2	Q_1	Q_0	Door Conditions
1	1	1	1	1	1	1	1	Both motors off
1	1	1	1	0	1	1	0	Door 1 open
1	1	1	1	1	0	0	1	Door 1 closed
0	1	1	0	1	1	1	1	Door 2 open
1	0	0	1	1	1	1	1	Door 2 closed

generates the output signals necessary for elevating the oven door. As the cookie sheet enters the oven, the infrared beam is restored. When the microcomputer senses this condition, it generates signals for closing the oven door and initiating the baking cycle.

Figure 13–23 illustrates how the two light barriers (photosensors) and four proximity sensors could be connected to a single 8-bit input port of the microcomputer system. Since the microcomputer module is likely to be located several feet away from the oven, both types of sensor are connected to the port via input modules containing optocouplers.

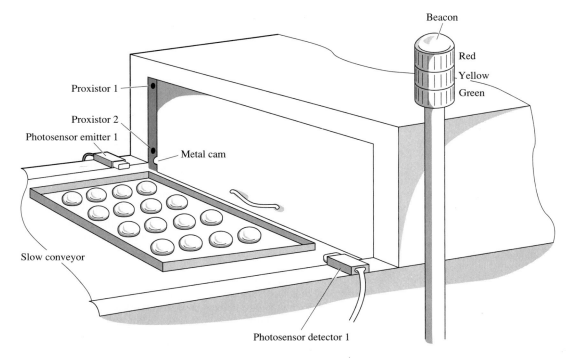

Figure 13–22
Placement of photosensors, proximity sensors, and indicator lamps for automated oven system

Q_7	Q_6	Q_5	Q_4	Q_3	Q_2	Q_1	Q_0
XX	XX	Photo-sensor 2	Photo-sensor 1	Proxistor 4	Proxistor 3	Proxistor 2	Proxistor 1

Figure 13–23
Sensor input port configuration

With the current-sinking outputs of the input modules being connected to Schmitt-triggered inverters, the signals from the proximity sensors are active-high. Also, the breaking of an infrared light beam produces a positive transition at a port input, whereas restoration of the beam produces a negative transition.

Table 13–7 defines the discrete sensor input conditions as they would be read by the microcomputer. In this example, the Q_0 through Q_3 bit positions of the input port are assigned to the four proximity sensors, while Q_4 and Q_5 receive discrete signals from the two light barriers. Positions Q_6 and Q_7 remain unused and are thus always masked out during the microcomputer input operations.

Assuming door 1 is to be opened, the microcomputer could output 11110110 to the motor control port. With door 1 now being elevated, the microcomputer waits for an input signal from a proximity sensor that indicates the door is fully open. An interesting design question now arises. Assume the computer is placed in a wait loop, requiring the arrival of a signal from a proximity sensor to allow it to escape from the loop and continue with the state 2 portion of the program. If the motor experiences an electrical or mechanical failure, the microcomputer is locked in the wait loop, since the signal from the sensor does not arrive! If the microcomputer contains watchdog circuitry, the failure of the microcomputer to reset the watchdog timer could result in the initiation of a general malfunction message, which could be sent to the main computer system.

A better method of monitoring the operation of either door would be to start a timer, either an actual programmable timer or a delay subroutine within the microcomputer monitor program. This timer function could be started immediately after the command to start the motor is made. The period of this timer function must be slightly longer than the time required for opening the oven door. For example, if the oven door normally requires 6 sec to fully open, the period of the timer could be 8 sec. Thus, if the oven door is working normally, the signal from the door 1 open proximity sensor always arrives at the microcomputer sensor input port before the time-out of the timer function. This action causes the microcomputer to deactivate the timer function and continue state 2 operation. However, if the timer is able to count through the allotted period, the microcomputer enters a default routine, in which it declares an oven door malfunction. The microcomputer then shuts down the oven by bringing the Oven Enable signal low and switches on the yellow fault lamp. After a fault message is sent to the main PC via the common data bus, the PC then sends a diagnostic message to its display monitor, possibly stating, "Door 1 of the oven has failed to open."

We must realize that, during the bake cycle (state 3), the microcomputer does not directly control the oven. In fact, during normal operation of the oven-control system, the microcomputer is not part of the control loop. Through the interaction of the first thermistor, comparator 1, and the zero-point switch, oven temperature is held close to the set point. Thus, assuming normal operation, after the oven door is closed, the microcomputer initiates the timer function for the bake cycle and turns on the red lamp, indicating baking is in progress. As with oven temperature, baking time could have been entered as data at the keyboard of the main PC. It is at this juncture that Interrupt 2 (TRAP) becomes significant. Assume, for example, that thermistor 1 fails, causing the voltage at point B

Table 13–7
Definition of input port conditions

Q_5	Q_4	Q_3	Q_2	Q_1	Q_0	Condition Defined
X	X	X	X	0	1	Door 1 open
X	X	X	X	1	0	Door 1 closed
X	X	0	1	X	X	Door 2 open
X	X	1	0	X	X	Door 2 closed
X	1	X	X	X	X	Cookie sheet approaching door 1
X	1	X	X	X	X	Cookie sheet entering oven
1	X	X	X	X	X	Cookie sheet leaving oven
1	X	X	X	X	X	Cookie sheet out of oven

to stay below the set-point level. A growth without limit condition would then ensue, allowing oven temperature to increase steadily. However, assuming thermistor 2 is still operative, a voltage level is eventually attained at point C that causes the TRAP input to toggle to a logic high. This unmaskable interrupt signal causes the microcomputer to shut down the oven, sound an audible alarm, switch on the yellow fault, turn on the cooling fan, and send an over-temperature message to the main PC. As with a door malfunction, a diagnostic message could appear on the display monitor, possibly stating, "Oven over-temperature condition: Emergency shutdown." A crucial adjustment for the over-temperature detection circuitry in Figure 13–20 is the ohmic setting of R_{ADJ}. A one-time adjustment of this potentiometer establishes the maximum allowable oven temperature.

At the time-out of the baking period, the microcomputer switches off the heating element by bringing the Oven Enable signal low. It also turns on the fan, switches off the red lamp, and initializes the timer for the cool-down cycle. As with the baking cycle, a delay algorithm could be used to define the cool-down period. At the end of the cool-down period, during state 4, the microcomputer opens door 2 by means of a procedure identical to that for opening door 1.

The monitor program of the microcomputer system could contain an adjustable delay routine. Such a routine, frequently used with the 8085 microprocessor, is given in Program

Program Example 13–1

Address	Op Code	Mnemonic	Comment
2127	1B	DCX D	Delay variable − 1.
2128	7A	MOV A, D	Or high byte of variable with low byte.
2129	B3	ORA, E	
212A	C2	JNZ, 2127	Loop back until D & E are cleared.
212B	27		
212C	21		
212D	C9	RET	Go back to main program.

Program Example 13-2

Address	Op Code	Mnemonic	Comment
3040	21	LXI H, 7050	Point to motor control output port.
3041	50		
3042	70		
3043	3E	MVI A, F6	Define control word to open door 1.
3044	F6		
3045	01	LXI B, 0E42	Define count as 3650_{10}.
3046	42		
3047	0E		
3048	77	MOV M, A	Start door 1 motor.
3049	21	LXI H, 705A	Point to sensor input port.
304A	5A		
304B	70		
304C	7E	MOV A, M	Read data at sensor input port.
304D	E6	ANI, 0F	Mask out upper nibble.
304E	0F		
304F	FE	CPI, 01	Check status of door open signal.
3050	01		
3051	CA	JZ, 3061	Exit wait loop when door 1 is fully elevated.
3052	61		
3053	30		
3054	11	LXI D, 01FF	Define delay as 2.2 msec.
3055	FF		
3056	01		
3057	CD	CALL, 2127	Go to delay routine.
3058	27		
3059	21		

Program Example 13-2 (*continued*)

305A	0B	DCX B	Count − 1.
305B	C2	JNZ, 304C	Repeat read loop if time-out has not occurred.
305C	4C		
305D	30		
305E	C3	JMP, 6040	Go to routine to declare malfunction.
305F	40		
3060	60		
3061	21	LXI H, 7050	Point to motor control output port.
3062	50		
3063	70		
3064	3E	MVI A, FF	Define control word to stop door 1 motor.
3065	FF		
3066	77	MOV M, A	Stop door 1 motor.
3067	C3	JMP, 5080	Go to state 3 routine.
3068	80		
3069	50		

Example 13–1. For this subroutine, the duration of the delay cycle is controlled by a double byte of data that is placed in the D and E register pair during the main program.

Extended delay periods, lasting from several seconds to several minutes, can be created by using a register counter in conjunction with the delay subroutine. Consider Program Example 13–2, which could be used to fully elevate oven door 1 during state 2 of oven operation. In this program sample, it is assumed that door 1 must be opened within a period of 8 sec. To prevent damage to the door 1 motor, it must be de-energized within a few milliseconds of the arrival of a signal from the door 1 open proximity sensor. Thus, the D and E register pair is loaded with a 16-bit variable that results in a relatively short delay period of approximately 2.2 msec. To provide a time-out period of approximately 8 sec, a count equivalent to 3650_{10} is placed in the B and C register pair. As demonstrated, if the delay subroutine is repeated 3650 times, the desired time-out is produced:

$$2.2 \text{ msec} \times 3650 = 8.03 \text{ sec}$$

In Program Example 13–2, we assume that the delay subroutine resides at memory location 2127. We also assume that the memory map location of the motor control output port is address 7050_{16}, the input port address is $705A_{16}$, and the starting address of the routine for declaring a malfunction is 6040_{16}. The clocking frequency of the 8085 is assumed to be such that loading $01FF_{16}$ into the D and E register produces a delay time of 2.2 msec. Finally, we assume that the state 3 routine, which comprises the actual baking process, begins at address 5080_{16}.

The complete program for sequential state control, contained in the ROM of the microcomputer in Figure 13–20, must contain routines necessary for closing oven door 1, controlling baking time, and opening and closing oven door 2. As demonstrated earlier, all of these routines could use the same delay subroutine. For example, by placing a large 16-bit number in the D and E register pair and a large count in the B and C register pair, a baking time of several minutes can be provided.

SECTION REVIEW 13–3

1. Define the term *discrete* as it applies to the operation of an industrial control system. How does a discrete input signal differ from a continuous-state signal?
2. Define the term *deadband* as it applies to the operation of a discrete control system. Why is some deadband desirable in a discrete control system?
3. What are two other terms for deadband?
4. Write the mathematical expression for a deadband controller.
5. For Program Example 13–2, assume the allowable period for opening an oven door is to be 9.5 sec. If the delay algorithm still has a delay time of 2.2 msec, what hexadecimal numbers must be entered at memory locations 3046 and 3047?
6. For the control system represented in Figure 13–20, explain the advantage of using the microcomputer's trap function to indicate an over-temperature condition.

13–4   CONTINUOUS-STATE CONTROL SYSTEMS

Recall that, with discrete control systems, although the error signal may vary continuously, the input signal sensed by the control circuit is either high or low. With a **continuous-state control system,** however, the error signal sensed by the control system varies in proportion to the deviation between the actual value of the controlled variable and its set-point level. Continuous-state control systems operate in four basic modes: **proportional, integral, derivative,** and **composite.** Composite control systems are created through combinations of the elements of the first three modes of continuous-state control.

With a proportional control system, the control circuitry sends a feedback signal to the final control element. This signal changes in proportion to variations in the error signal. Thus, with proportional control, as an error signal increases by a factor of 7%, the negative feedback signal produced by the control source also increases by a factor of 7%. The concept of proportional control can be expressed mathematically as follows:

$$C_P = (K_P \times e_P) + C_0 \qquad (13\text{–}9)$$

where C_P represents a percentage of change in the output of the control circuitry, K_P represents the **proportional gain** of the control circuitry, e_P represents a percentage of change in the error signal, and C_0 represents a residual output level of the control circuitry.

The proportional control mode may be represented graphically, as shown in Figure 13–24. Here, two different values of proportional gain (K_P) are represented. Since K_{P1} represents a larger gain factor than K_{P2}, its curve is steeper than that of K_{P2}. The effective control range for a proportional control system is referred to as its **proportional band** (PB). The proportional band terminates at the point where the curve for proportional gain intersects the 0% baseline. This baseline represents the lower limit of the range of the

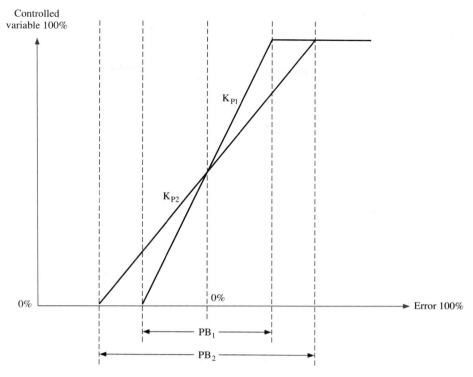

Figure 13–24
Graphic representation of two proportional bands

controlled variable. Beyond the 100% point, which represents the upper limit of the controlled variable range, a saturation condition exists for the controller. Thus any error that could occur beyond this limit is out of the controller's range. Note that, due to its higher gain and resultant steeper curve, the proportional band for K_{P1} is narrower than the PB of K_{P2}.

Example 13–2

Assume that a control system operating in the proportional mode is to govern gas pressure within a range extending from 95 to 147 psig. The set point of the system is to be 121 psig. Assuming K_P equals 3.4 and C_0 is 46%, write the control equation and determine the limits of the proportional band.

Solution:

Step 1: Derive an expression for e_P. Recall that, as demonstrated by Eq. (13–6):

$$e_P = \frac{(V - V_{SP})}{V(max) - V(min)} \times 100\%$$

Substituting:

$$e_P = \frac{(V - 121 \text{ psig})}{(147 \text{ psig} - 95 \text{ psig})} \times 100\% = (V - 121) \times 1.923\%$$

Step 2: Write the control equation:

$$C_P = (K_P \times e_p) + C_0 = (3.4 \times e_p) + 46\%$$

Step 3: Determine the lower limit of the proportional band for e_P:

$$0\% = 3.4 \times e_P + 46\%$$

Thus,

$$e_P = \frac{-46\%}{3.4} = -13.53\%$$

Step 4: Determine the upper limit of the proportional band for e_P.

$$100\% = (3.4 \times e_P) + 46\%$$

Thus,

$$e_P = \frac{54\%}{3.4} = 15.9\%$$

Step 5: Determine the proportional band as a percentage. The proportional band extends from -13.53% to 15.9%, representing a range of 29.43%.

Step 6: Determine the proportional band for the controlled variable. Solving for the value of the controlled variable that represents the lower end of the proportional band:

$$-13.53\% = (V - 121 \text{ psig}) \times 1.923\%$$

$$V(\min) = 121 \text{ psig} - \frac{13.53\%}{1.923\%} \text{ psig} \cong 114 \text{ psig}$$

Solving for the value of controlled variable that represents the upper limit of the proportional band:

$$15.9\% = (V - 121 \text{ psig}) \times 1.923\%$$

$$V(\max) = \frac{15.9\%}{1.923\%} \text{ psig} + 121 \text{ psig} \cong 129.3 \text{ psig}$$

Thus, the proportional band is equivalent to (129.3 psig $-$ 114 psig) or a range of nearly 15.3 psig. Note that, due to the relatively high value of K_P, the proportional band is narrow in comparison to the possible range of the controlled variable.

Figure 13–25 graphically represents the proportional band for the gas pressure control system investigated in Example 13–2. Note that the low end of the curve intersects the 0% point for the controller output at a point equivalent to -13.53% error. The controller output curve saturates at a high end point equal to 15.9% error.

Figure 13–26 illustrates how a PLC could be utilized as a proportional controller within an automated industrial plant. In many industrial sites, small particles of various materials are suspended in liquid, allowing them to be pumped from one point to another through a pipe system. As was explained in Section 4–5, the flow rate of such slurries may be accurately monitored using a venturi or a magnetic flow-sensing transducer. It is also extremely important to monitor and control the consistency of a slurry. In Figure 13–26, this is accomplished by controlling the rate at which diluting water is fed into the slurry flow. If the slurry is allowed to thicken, it could clog the pump and piping system,

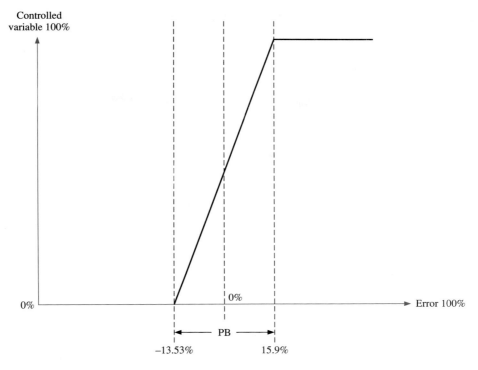

Figure 13–25
Proportional band for gas pressure control system

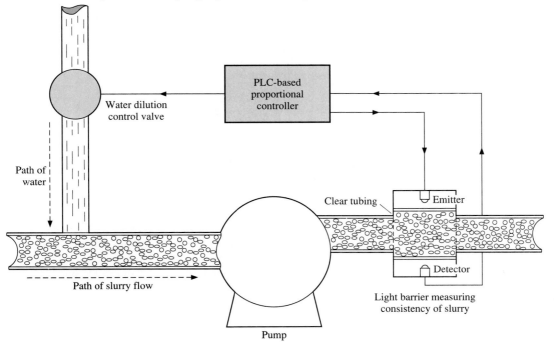

Figure 13–26
PLC-based proportional controller governing the flow of a slurry in an automated industrial plant

delaying production and possibly resulting in costly damage to materials and equipment. Production problems would also result if the slurry is allowed to become too thin. Thus, a system must be established by which an input transducer accurately monitors the consistency of a passing slurry and produces an error signal to be sent to the CPU of the PLC. The PLC then produces an output signal that is proportional to the deviation between the actual consistency of the slurry and its ideal, or set-point consistency. This signal is received by an electronically controlled valve, which functions as the final control element. As the slurry attempts to thicken, the valve opens more, resulting in increased dilution. If the slurry begins to dilute beyond the desired consistency, the valve begins to close, allowing the slurry to thicken slightly.

The consistency of a slurry passing through the system shown in Figure 13–26 would be virtually impossible to monitor directly. (Recall that slurries can easily distort the operation of, or possibly damage, such flow-measuring transducers as turbines and nutating disks.) An ingenious alternative to direct measurement is represented in Figure 13–26. Here, a piece of clear tubing, possibly glass or quartz, is placed in the piping system, immediately downstream from the pump. A light barrier is aligned across this clear tubing. The measurement point is, of course, completely enclosed to avoid interference produced by ambient light. As the slurry attempts to thicken, the number of solid particles suspended in the liquid is increased. Because these particles block a greater portion of the light beam passing through the tubing, a proportional change is produced in the output of the detector. As the slurry thins, less light is blocked within the tube. As a result, a proportional change, now in the opposite direction, is produced at the output of the detector.

The preamplifier circuitry for the detector in Figure 13–26 could be similar to that shown for the *p-i-n* photodiode in Figure 5–18. The output of this signal conditioning circuitry could be connected to a low-level input module of the PLC. This device, which contains A/D conversion circuitry, further amplifies the signal from the detector of the light barrier and converts it to a binary number. This digitized equivalent of the controlled variable (D_V) is periodically sampled by the PLC. A buffer within the CPU also contains a constant (D_{SP}) that represents the set-point value of the controlled variable. During the algorithm for proportional control, the CPU often subtracts this constant from the controlled variable to produce a digitized equivalent of the error signal (D_e). Thus, based on Eq. (13–3),

$$D_e = D_V - D_{SP} \qquad \textbf{(13–10)}$$

This equation could be easily implemented using machine language. For the 8085 microprocessor, if the constant representing the set point (D_{SP}) is treated as immediate data within the program, a SUI command could be used. The result of applying Eq. (13–10) may be positive or negative, depending on whether the input variable (D_V) is above or below its set-point level. The polarity of D_e may be determined through testing the condition of the carry/borrow flag.

Recall from Eq. (13–6) that e_P is a percentage of the range of the controlled variable. Assuming that the low-level input module of the PLC in Figure 13–26 produces an 8-bit output, for the best possible resolution, the gain of this module would be calibrated such that the low end of the range of the controlled variable becomes 00_{16} and the upper limit equals FF_{16}. Where D_{min} represents the digitized equivalent of the low end of the input range and D_{max} represents the high end, a formula for error can be developed from Eq. (13–6). Recall that percentage of error was originally expressed:

$$e_P = \frac{(V - V_{SP})}{V(\text{max}) - V(\text{min})} \times 100\%$$

A formula directly usable in a CPU algorithm may be written as follows:

$$D_{Fe} = \frac{D_V - D_{SP}}{D_{max} - D_{min}} \qquad (13\text{--}11)$$

Here, a binary representation of a fractional error (D_{Fe}) replaces e_P.

If variables are left in their true binary forms while input data bytes are being processed by the CPU, D_{Fe} could be determined through use of a long-division algorithm such as **shift-left-and-subtract.** If variables are first converted to BCD, as could be the case with a SYSMAC PLC in Appendix B, the output of the low-level input module would be converted from binary to BCD. This could be accomplished using the **Binary to BCD** for the SYSMAC PLC. This process, in turn, would allow direct use of the **BCD Divide** instruction.

Recall from Eq. (13–9) that, to derive a percentage of change in controller output, the percentage of error (e_P) is multiplied by the proportional gain (K_P). The result of this operation is then added to a number representing the zero-error output (C_0) of the proportional controller. A complete equation follows for proportional control that can be easily implemented by a microcomputer or PLC:

$$D_{C_p} = (D_{K_p} \times D_{Fe}) + D_{C_0} \qquad (13\text{--}12)$$

where D_{C_p} is a binary number representing the controller output, D_{K_p} is a binary constant representing proportional gain, and D_{C_0} is the binary equivalent of the zero-error output.

For the control system in Figure 13–26, through the process of D/A conversion, changes in the value of D_{C_p} could produce proportional changes in the duty cycle of the analog signal controlling the position of the valve. Figure 1–36 contains a cutaway view of the type of solenoid-operated valve that could be used in Figure 13–26. The duty cycle of the valve could be varied in proportion to the value of D_{C_p}, with D_{C_0} representing the midpoint in duty cycle range.

Offset is a problem inherent to proportional control. Recall that the output of a proportional controller has a zero-error output, which is likely to be near the middle of its proportional band. This zero-error point represents an ideal, or set-point, condition for the controlled variable. The zero-error point value of the controlled variable is derived with the assumption that other variables in the process control loop remain at certain predicted values. The proportional controller is normally able to respond to slight deviations in these variables, holding the controlled variable near the set point. However, it is quite possible for one of these variables to undergo a significant and permanent change. Such a shift in the value of one of these related variables is referred to as a **load change.** To hold the controlled variable near the set point, the proportional controller would need to compensate through a change in its output to the final control element. However, since a purely proportional controller is *not* capable of such a response, a permanent residual error, or offset, results.

One approach to minimizing the offset problem associated with proportional control is the incorporation of the **integral mode** into the control scheme. Because of its ability to correct an offset, returning a controller output to its desired zero error level, the integral control mode is sometimes referred to as the **reset mode.** Integral control, though seldom used in its pure form, could be used in conjunction with proportional control in applications where offset is likely to be severe. (Such composite modes of control are surveyed at the end of this section). Although integral control is almost always used to supplement proportional control, it can best be understood if first examined in its pure form. With purely integral control, the amount of feedback provided by the controller depends on the history of the error. Whereas proportional control is based on the magnitude of the error at a specific instant in time, integral response is based on what the error *has been* through time.

Recall from Section 2–6 that the output amplitude of an integrator is proportional to the time integral of the input signal. A control circuit operating in the integral mode functions on this same principle. In fact, before microcomputers and PLCs assumed active roles as industrial control devices, analog industrial control systems were likely to contain op amp or discrete transistor integrators. For digitally based control systems, however, as explained momentarily, obtaining a history, or time integral, of an error signal is accomplished through use of sample-and-hold gates and A/D conversion circuitry.

The concept of integral control may be expressed mathematically as follows:

$$C_I(t_n) = [K_I \times A_e(t_{n-0})] + C_I(t_0) \tag{13–13}$$

In this equation, C_I represents the controller output as a percentage of its total range; $C_I(t_0)$ represents the controller output at a specified starting time; $C_I(t_n)$ represents the controller output at a given time t_n past t_0; K_I represents the **integral gain** of the controller; and $A_e(t_{n-0})$ represents the **error area,** which is determined for the period between t_0 and t_n.

For an integral mode controller, the net area of the ongoing error, determined as amplitude versus time, must be approximated. A convenient method of accomplishing this task using a sample-and-hold and A/D interface system is **rectangular integration.** As represented in Figure 13–27, with rectangular integration, samples of the instantaneous amplitude of the error signal are taken and converted to binary values. These binary values vary through time, as the amplitude and polarity of the error signal change. However, the value of t, which is equivalent to the control system's sampling rate, remains constant. This sampling rate should be kept as fast as possible, allowing for the greatest possible accuracy for the integration process. Through several successive samples, the microcomputer or PLC is able to approximate the error area, using an algorithm based on Eq. (13–13). The following example serves to illustrate how an integral mode controller implements this equation to calculate the error area and determine the necessary changes to be made in the amplitude of its output signal.

Example 13–3

Assume a microcomputer-based control system operating in the integral mode uses the process of rectangular integration to detect the condition of an error signal. As indicated in Figure 13–28, the control system samples the error signal every 50 msec. At t_0, the output of the controller is assumed to be at 46% of its full range, which represents a set-point condition for the system. Assuming the integral gain of the system to be 2.7, determine the output signal developed by the controller in response to the error samples graphed in Figure 13–28. Plot the controller response directly above these samples of the error signal.

Solution:

Step 1: The error signal sampled between t_0 and t_1 represents a deviation of nearly 3.3% above the set point. Thus, the response of the integral controller is determined as follows:

$$C_I(t_1) = [K_I \times A_e(t_{1-0})] + C_I(t_0) = (2.7 \times 3.3\%) + 46\% = 54.91\%$$

Step 2: The error signal sampled between t_1 and t_2 represents an increase to 4.1%. The output signal is then determined as:

$$C_I(t_2) = [K_I \times A_e(t_{2-0})] + C_I(t_0) = [2.7 \times (3.3\% + 4.1\%)] + 46\% = 65.98\%$$

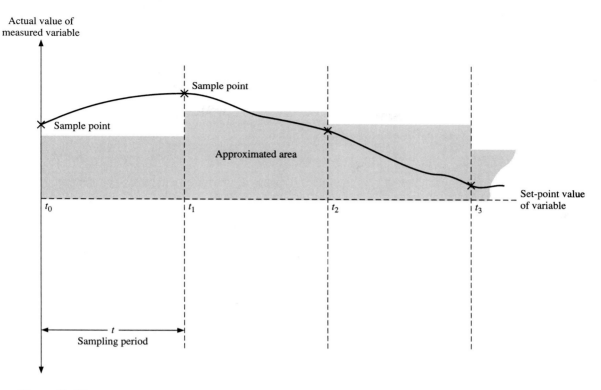

Figure 13–27
Sampling of an analog input using rectangular integration

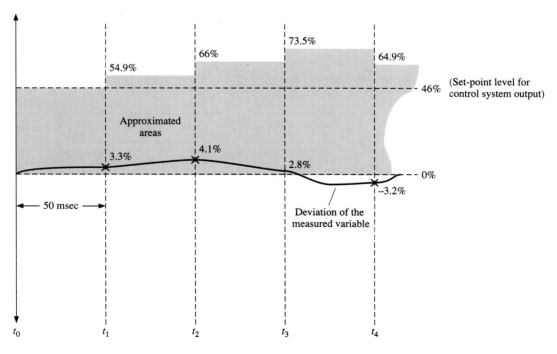

Figure 13–28
Illustration of conditions for Example 13–3

Step 3: The error signal sampled between t_2 and t_3 represents a slight decrease toward 2.8%. Thus, the response of the controller is determined as:

$$C_1(t_3) = [K_I \times A_e(t_{3-0})] + C_1(t_0) = [2.7 \times (7.4\% + 2.8\%)] + 46\% = 73.54\%$$

Step 4: Between t_3 and t_4, the error sample is at a point 3.2% below the zero error level. The controller response is thus determined as:

$$C_1(t_4) = [K_I \times A_e(t_{4-0})] + C_1(t_0) = [2.7 \times (10.2\% - 3.2\%)] + 46\% = 64.9\%$$

Step 5: The plot of the controller response to the error samples is plotted above the deviation of the measured variable in Figure 13–28.

From Eq. (13–13), an algorithm can be developed that allows a microcomputer or PLC to implement integral control. As with the proportional control mode, Eq. (13–11) would first be applied to determine the fraction of error (D_{Fe}). Thus, a sample of the input variable D_V is taken and the value of D_{Fe} is determined. If rectangular integration is being implemented, an error-time approximation of the error area (now designated D_{Ae}) is made. This is accomplished by multiplying a sample of D_{Fe} by a constant (D_t), which represents the sampling time (reciprocal of the sampling rate). The CPU then adds the result of this approximation to the ongoing summation of D_{Ae}, thus updating the value of the error area as follows:

$$D_{Ae} = D_{Ae} + (D_{Fe} \times D_t) \tag{13–14}$$

If the input and output ranges of the controller are the same (i.e., both the input and output binary variables have the same resolution), a simplified approach can be taken in creating the integral control mode algorithm. The digitized samples of the error (now designated D_e) are added directly together. The binary variable called the sum (D_Σ) is then determined as:

$$D_\Sigma = D_\Sigma + D_e \tag{13–15}$$

A simple expression for the integral control mode algorithm would be as follows:

$$D_{CI} = D_{KI} \times D_\Sigma \times D_t \tag{13–16}$$

where D_{CI} is the binary equivalent of a fractional change in the integral component of a controller output and D_{KI} is the binary equivalent of the integral gain constant.

To implement Eq. (13–16), a binary number representing the set-point value of the controller output is placed in a memory location designated to contain D_V. If, initially, D_e is greater than D_V, a positive summation begins to occur. The overall design of the control system must be such that the resulting positive increase in the value of D_Σ produces a change in the value of D_{CI} that counteracts the disturbance that initially caused D_e to be greater than D_V. Eventually, through the corrective action of the controller, D_e becomes less than D_V. Thus, during subsequent summations, since D_e is now negative, D_Σ approaches zero as the controlled variable approaches its set-point value.

In many industrial control applications, it is necessary for a control system to respond rapidly to a sudden change in the value of a controlled variable. A method of control capable of producing such a response is the **derivative mode.** Recall from Section 2–6 that the output of a differentiator is directly proportional to the time derivative of its input signal. This is also the operating principle of a control system using the derivative mode. We stress that the derivative mode of control, like the integral, is never used in its pure form. A purely derivative controller could only respond to an error signal if that signal were in a state of flux. As will soon be explained, derivative control is used in conjunction with the proportional mode or both the proportional and integral modes. However, to understand how derivative control works as part of a composite control scheme, it is first necessary to analyze the derivative control mode in its pure form. The mathematical expression for purely derivative control is as follows:

$$C_D(t_n) = K_D \times \frac{e_P(t_n) - e_P(t_0)}{(t_n - t_0)} \qquad (13\text{--}17)$$

where $C_D(t_n)$ represents a percentage of change in the derivative component of a feedback signal occurring at a specified time t_n; K_D represents the derivative gain; $e_P(t_n)$ represents the percentage of error occurring at t_n; and $e_P(t_0)$ represents the percentage of error occurring at starting time t_0.

From Eq. (13–17), it becomes evident that, for a given change in error input, the amplitude of the derivative feedback is inversely proportional to the time during which the change in error takes place.

For the proportional and integral modes of control, Eq. (13–6) is important, since it allows a percentage of error (e_P) to be determined with reference to the set-point value of the controlled variable (V_{SP}). With the derivative control mode, however, the relationship of the error to the set point is no longer significant. Rather, it is the **rate of change** in error signal that controls the amplitude of derivative feedback. When adapting the derivative mode to an embedded microcomputer or PLC application, we must first establish a fixed rate for sampling the error signal. This, of course, may be the same sampling period set for the proportional and integral components of the control algorithm. This sampling period is equivalent to $(t_n - t_0)$ in Eq. (13–17).

To determine a percentage of change in $[e_P(t_n) - e_P(t_0)]$, we must take a sample of the measured variable at the beginning and end of each sampling period. (The end reading may be equivalent to the initial measurement of the next sampling cycle.) The difference between these two samples of the measured variable must then be considered with respect to the total range of the measured variable. Thus,

$$e_P(t_n) - e_P(t_0) = \frac{[V_m(t_n) - V_m(t_0)]}{[V_m(\max) - V_m(\min)]} \times 100\% \qquad (13\text{--}18)$$

where $V_m(t_n)$ represents the value of the measured variable at t_n, $V_m(t_0)$ represents the value of the measured variable at t_0, and $V_m(\max) - V_m(\min)$ represents the total range of the measured variable.

Example 13–4

Assume that, for a control system using the derivative mode, the sampling time is 750 msec. Also assume that the range of the error signal extends from 1.35 to 8.25 V. As shown in Figure 13–29, samples of the error signal taken at t_0 through t_4 are 5.32, 5.47, 5.38, 5.17, and 5.29 V, respectively. Assume, at t_0, the derivative portion of the feedback signal is at 57% of its total range. If the derivative gain is 3.8, what are the values of C_D (expressed as percentages of total range) at t_1 through t_4? Plot the controller's derivative response directly below the samples of the error signal.
Solution:

Step 1: Solving for $C_D(t_1)$:

$$e_P(t_1) - e_P(t_0) = \frac{[V_m(t_1) - V_m(t_0)]}{[V_m(\max) - V_m(\min)]} \times 100\% = \frac{5.47\ V - 5.32\ V}{8.25\ V - 1.35\ V} \times 100\% = 2.174\%$$

$$C_D(t_1) = K_D \times \frac{e_P(t_1) - e_P(t_0)}{(t_1 - t_0)} = 3.8 \times \frac{2.174\%}{750 \times 10^{-3}} \cong 11\%$$

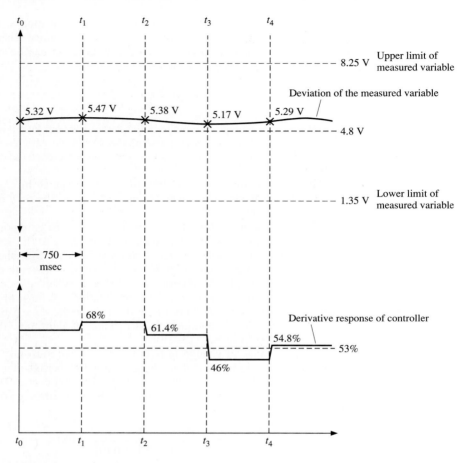

Figure 13–29
Illustration of conditions for Example 13–4

Thus, at t_1, the purely derivative response tends to increase the controller output as follows:

$$C_0(t_1) = 57\% + 11\% = 68\%$$

Step 2: Solving for $C_D(t_2)$:

$$e_P(t_2) - e_P(t_1) = \frac{[V_m(t_2) - V_m(t_1)]}{[V_m(\text{max}) - V_m(\text{min})]} \times 100\% = \frac{5.38 \text{ V} - 5.47 \text{ V}}{8.25 \text{ V} - 1.35 \text{ V}} \times 100\% = -1.3\%$$

At t_2, the purely derivative response tends to decrease the controller output as follows:

$$C_D(t_2) = K_D \times \frac{e_P(t_2) - e_P(t_1)}{(t_2 - t_1)} = 3.8 \times \frac{-1.3\%}{750 \times 10^{-3}} \cong -6.6\%$$

$$C_0(t_2) = 68\% - 6.6\% = 61.4\%$$

Step 3: Solving for $C_D(t_3)$:

$$e_P(t_3) - e_P(t_2) = \frac{[V_m(t_3) - V_m(t_2)]}{[V_m(\text{max}) - V_m(\text{min})]} \times 100\% = \frac{5.17\text{ V} - 5.38\text{ V}}{8.25\text{ V} - 1.35\text{ V}} \times 100\% \cong -3\%$$

$$C_D(t_3) = K_D \times \frac{e_P(t_3) - e_P(t_2)}{(t_3 - t_2)} = 3.8 \times \frac{-3\%}{750 \times 10^{-3}} \cong -15.4\%$$

At t_3, the purely derivative response tends to decrease the controller output as follows:

$$C_0(t_3) = 61.4\% - 15.4\% = 46\%$$

Step 4: Solving for $C_D(t_4)$:

$$e_P(t_4) - e_P(t_3) = \frac{[V_m(t_4) - V_m(t_3)]}{[V_m(\text{max}) - V_m(\text{min})]} \times 100\% = \frac{5.29\text{ V} - 5.17\text{ V}}{8.25\text{ V} - 1.35\text{ V}} \times 100\% = 1.74\%$$

$$C_D(t_4) = K_D \times \frac{e_P(t_4) - e_P(t_3)}{(t_4 - t_3)} = 3.8 \times \frac{1.74\%}{750 \times 10^{-3}} \cong 8.82\%$$

At t_4, the purely derivative response tends to increase the controller output as follows:

$$C_0(t_4) = 46\% + 8.82\% \cong 54.82\%$$

Step 5: The derivative response of the control system analyzed in the above example is plotted below the deviation of the measured variable in Figure 13–29.

Derivative control may be implemented in a microcomputer algorithm as follows:

$$D_{CD} = D_{KD} \times \frac{D_e(t_n) - D_e(t_0)}{D(t_n - t_0)} \tag{13–19}$$

where D_{CD} is a binary variable representing a fractional change in the derivative component of a controller's feedback signal; D_{KD} is a binary constant representing derivative gain; $D_e(t_n)$ is the binary equivalent of the error signal taken at t_n; $D_e(t_0)$ is the binary equivalent of the error signal taken at starting time t_0; and $D(t_n - t_0)$ is the binary constant representing the sampling time of the system. Again we stress that the derivative mode of control is never used by itself. Thus, as demonstrated momentarily, the algorithm represented in Eq. (13–19) must be revised and incorporated into a larger algorithm for one of the composite modes of control.

The **composite control modes** involve the use of proportional control in conjunction with the integral and/or derivative modes. Thus, the composite control modes consist of the **proportional-integral** (PI) mode, **proportional-derivative** (PD) mode, and **proportional-integral-derivative** (PID) mode.

At this point in the chapter, the need for composite control should be apparent, since each of the pure control modes, by itself, has limitations. Recall that purely proportional control, while capable of producing a response relative to the magnitude of an error signal, is susceptible to offset problems. Purely integral response, while capable of eliminating offset, is relatively slow, relying on the gradual compilation of the history of an error. While the purely derivative mode exhibits rapid response to sudden changes in an error, it can only be used in conjunction with other forms of control. The reason for this limitation becomes apparent when considering the derivative response to a constant error. For example, if the value of the controlled variable were to drift 10% above the set point and remain at that level, no derivative response could occur. Unless the error signal is actually in a state of flux, $[e_P(t_n) - e_P(t_0)]$ becomes zero. Thus, $C_D(t_n)$ must also be zero! The mathematical expressions for the three composite modes of control are listed as follows. For the proportional-integral (PI) control mode:

$$C_{P+I} = (K_P \times e_P) + (K_P \times K_I \times A_e) \qquad (13\text{--}20)$$

For the proportional-derivative (PD) control mode:

$$C_{P+D} = (K_P \times e_P) + (K_P \times K_D) \times \frac{e_P(t_n) - e_P(t_0)}{(t_n - t_0)} + C_0 \qquad (13\text{--}21)$$

For the proportional-integral-derivative (PID) control mode:

$$C_{P+I+D} = (K_P \times e_P) + (K_P \times K_I \times A_e) + (K_P \times K_D) \times \frac{e_P(t_n) - e_P(t_0)}{(t_n - t_0)} \qquad (13\text{--}22)$$

Since it is capable of responding to magnitude, history, and rate of flux of an error signal, PID is the most versatile, and therefore most widely used, form of continuous-state control. (Recall from Section 13–2 that the success of PID control has led to the development of specialized PID control modules to be used in conjunction with PLCs.)

In block diagramming of continuous-state control applications of PLCs, the letter K is often used to represent proportional control. The mathematical symbol $\int$, which symbolizes the process of integration, often represents the integral control mode, and mathematical expression d/dt, symbolizing differentiation, represents derivative control. Thus, block representation of proportional control, as well as the three composite control modes, would be as shown in Figure 13–30. Figure 13–30 also shows other important symbols used in block diagramming of control systems. The circle indicates a measurement and/or readout point; a diamond specifies set-point entry. The Greek letter Σ (sigma) represents a summation point, either analog or digital, while the Greek letter Δ (delta) represents a differencing operation.

Figure 13–31 illustrates an actual industrial application of a PLC-based PID control system. In recent years, industrial facilities have become subject to strict rules governing treatment of waste materials and disposal of toxic chemicals. As a result of this movement toward maintaining a cleaner environment, complex control systems have evolved. Such systems monitor the condition of materials that are about to be returned to the environment and control the rate at which these materials are released.

For the control system represented in Figure 13–31, we assume that an industrial complex requiring vast amounts of water is located next to a river. After using this water during a production process, the industrial facility purifies the water and releases it back into the river. Water that has been treated and is ready to be returned to the river is held in an effluent reservoir. From the description given thus far, the control problem might appear quite easy, with the pump being turned on and off as necessary to keep the reservoir water within a minimum and maximum level. If this were the only requirement of the system, simple discrete control would suffice. However, the control process is greatly complicated due to certain environmental considerations, including biological oxygen demand, water color, pH, and temperature. For the system in Figure 13–31, we assume that these parameters may be kept within allowable limits if the ratio of effluent discharge to river flow rate is maintained at an optimal value. Thus, the challenge in the development of the control system in Figure 13–31 is to keep this ratio constant, in spite of changes in river depth and current velocity.

The effluent discharge system in Figure 13–31 demonstrates applications of two forms of transducers introduced earlier in this text. A linear variable differential transformer (LVDT), described in Section 4–2, is used to monitor river depth, while a magnetic flow-

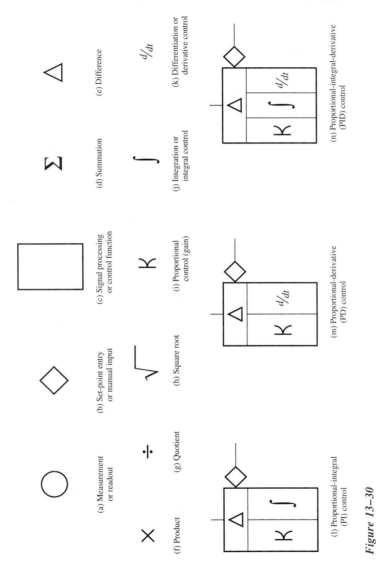

(a) Measurement or readout

(b) Set-point entry or manual input

(c) Signal processing or control function

(d) Summation

(e) Difference

(f) Product

(g) Quotient

(h) Square root

(i) Proportional control (gain)

(j) Integration or integral control

(k) Differentiation or derivative control

(l) Proportional-integral (PI) control

(m) Proportional-derivative (PD) control

(n) Proportional-integral-derivative (PID) control

Figure 13–30
Standard symbols used in control system block diagrams

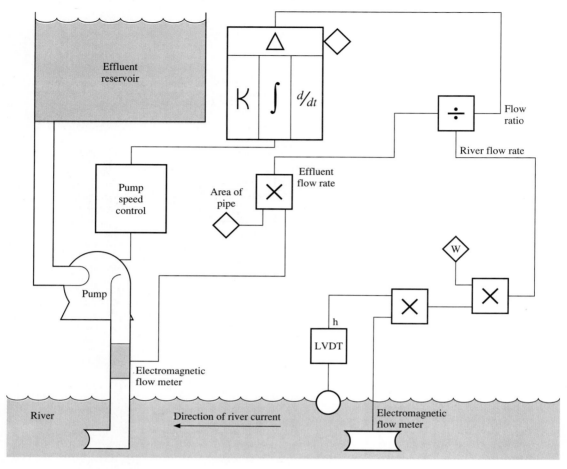

Figure 13–31
Block diagram of PID volumetric flow-control system

sensing transducer, introduced in Section 4–5, is used to measure the velocity of river current. The outputs of these two transducers could feed two of the PLC's low-level analog input modules containing A/D conversion circuitry. The binary variables produced by these modules are used by the PLC's CPU during the computation of the flow rate of the river. The information from the LVDT (D_{DF}) is used in the determination of river depth in feet, while the information from the first magnetic flow-sensing transducer (D_{FM1}) is used to derive the current velocity in feet per minute. The width of the river is already known and assumed to remain nearly constant as depth changes. Thus, within the CPU algorithm for approximation of the **volumetric flow rate** of the river, a binary constant (D_{WF}) represents river width in feet. This algorithm could be implemented as follows:

$$D_{CF1} = D_{WF} \times D_{DF} \times D_{FM1} \tag{13–23}$$

where D_{CF1} represents the volume of water passing the effluent discharge point in cubic feet per minute. Since volumetric flow rate is typically expressed in gallons per minute, the algorithm concludes with a conversion routine. Here, cubic feet per minute are converted to gallons per minute through multiplication by a constant:

$$D_{VFR1} = 7.48 \times D_{CF1} \tag{13-24}$$

where D_{VFR1} represents volumetric flow rate of the river in gallons per minute.

As shown in Figure 13-31, a second magnetic flow-sensing transducer is placed inside the effluent discharge pipe. The purpose of this transducer is to monitor the rate at which water leaves the reservoir and returns to the river. Like the first flow-sensing transducer (placed in the river), this device is connected to one of the PLC's low-level analog input modules having A/D conversion capability. As with measurement of the flow rate of the river water, the binary output of this module does *not* directly represent flow rate. Rather, it could be multiplied by a constant or serve as a memory offset for accessing a look-up table within the PLC's ROM. The binary number obtained from the look-up table or through multiplication becomes a variable representing the flow rate of water through the discharge pipe (in cubic feet per minute). This variable (D_{CF2}) could then be used in an algorithm for determining volumetric flow rate within the discharge pipe.

Volumetric flow rate within the discharge pipe is easily determined as follows. First, assuming the discharge pipe is round, its cross-sectional area is easily determined as:

$$D_{AP} = \pi \times r^2 \tag{13-25}$$

where D_{AP} represents the cross-sectional area of the pipe and r represents the radius of the pipe. Now D_{AP} becomes a constant in the algorithm for determining flow rate through the pipe in feet per minute:

$$D_{CF2} = D_{AP} \times D_{FM2} \tag{13-26}$$

To determine volumetric flow rate through the pipe in gallons per minute (D_{VFR2}), it is necessary to apply the same conversion factor used in Eq. (13-24):

$$D_{VFR2} = 7.48 \times D_{CF2} \tag{13-27}$$

With the volumetric flow rates being determined for both the river and the discharge pipe, the ratio of these two quantities is easily determined as follows:

$$D_R = \frac{D_{VFR1}}{D_{VFR2}} \tag{13-28}$$

where D_R represents the ratio of river volumetric flow rate to discharge volumetric flow rate. Note that this division operation is represented as a functional block in Figure 13-31. The output of this function is fed to the PID block where it is compared to a constant representing the optimal or set-point ratio (D_{SP}). The diamond connected to the PID block represents the use of this set-point constant within the control algorithm. The triangle at the top of the PID block indicates that a differencing operation takes place involving the flow ratio and the set point. Recall that most error calculations are performed in terms of percentage of range [as represented by Eq. (13-6)]. However, within the control algorithm for the system in Figure 13-31, error may be more conveniently expressed as a fractional error with respect to the set point:

$$D_{Fe} = \frac{D_R - D_{SP}}{D_{SP}} \tag{13-29}$$

As the value of D_R approaches the value of D_{SP}, D_{Fe} approaches zero. If, however, the volumetric flow rate of the river (D_{VFR1}) increases or the volumetric flow rate of the pipe (D_{VFR2}) decreases, D_R increases above the set point, causing a positive increase in D_{Fe}. If

D_{VFR1} decreases or D_{VFR2} increases, D_R decreases below the set point, causing a negative increase in D_{Fe}.

Equation (13–30) illustrates how an algorithm for PID control could be developed from Eq. (13–22):

$$D_{CPID} = (D_{KP} \times D_{Fe}) + (D_{KP} \times D_{KI} \times D_{Ae}) + (D_{KP} \times D_{KD}) \times \frac{D_{Fe}(t_n) - D_{Fe}(t_0)}{D(t_n - t_0)} \quad (13\text{–}30)$$

where D_{CPID} is a binary number representing fractional change in the PID control block in Figure 13–31. As shown by the first term in this equation, the PLC exercises proportional control as it multiplies fractional error by the proportional gain. [Note that D_{KP} is a factor within all three terms of Eq. (13–30)]. Thus, the PLC is able to respond in proportion to the deviation between D_R and D_{SP}. If D_{Fe} begins to increase positively, the PLC produces the necessary change in its output signal to increase the pumping action. If D_{Fe} begins to increase negatively, the PLC initiates the necessary change in its output signal to decrease the pumping action.

As implied by the second term of Eq. (13–30), the integral portion of the controller output is a product of proportional gain, integral gain, and error area (D_{Ae}). Here, D_{Ae} represents the running total, or history, of the fractional error (D_{Fe}). This term of the equation is essential for correcting any offset that could be introduced into the control loop. As shown in the third term of Eq. (13–30), the difference between two consecutive samples of fractional error is divided by the sampling rate, with the result of this operation being multiplied by the proportional and derivative gains. The presence of this third term within Eq. (13–30) allows rapid response of the control system to sudden changes in either of the volumetric flow rates.

SECTION REVIEW 13–4

1. Describe a major problem associated with purely proportional control. What other control mode could be used in conjunction with proportional control to help alleviate this problem?
2. Explain why the derivative control mode is never used by itself.
3. Which control mode is capable of responding to the magnitude, history, and rate of change of an error signal?
4. True or false?: For a purely proportional control system, if proportional gain is increased, the proportional band becomes wider.
5. Assume that a control system operating in the purely proportional mode is to govern temperature within a range extending from 85°C to 175°C. The set point of the system is 130°C. Assuming K_P equals 2.7 and C_0 is 48%, write the control equation for the system and determine the limits of the proportional band.
6. For the proportional temperature controller described in Review Problem 5, assume that the four temperature samples shown here are taken at intervals of 2.4 sec:

    t_0: 130°C; t_1: 134°C; t_2: 131°C; t_3: 129°C

    For each of these temperature samples, determine the resultant value of C_P.
7. Assume for a PI temperature controller that proportional gain is 2.4 and integral gain is 2.7. Also assume the sampling time is 185 msec, C_0 equals 48%, and set-point temperature is 124°C. Maximum temperature for the system is 182°C; minimum temperature is 66°C. Given these conditions, assume the following temperature samples are taken at the specified sampling interval of 185 msec:

    t_0: 124°C; t_1: 127°C; t_2: 125°C; t_3: 121°C

    For each of these temperature samples, determine the resultant value of C_{P+I}.

8. Assume the operation of the temperature controller described in Review Problem 7 is enhanced through the addition of derivative control. Assume the derivative gain to be 2.3. For the resultant PID control system, determine the value of C_{P+I+D} for each of the temperature examples given in Review Problem 7.
9. Which control mode is sometimes referred to as reset control? Why is this alternative description appropriate?
10. Describe the process of rectangular integration. What factor within a microcomputer-based control system determines the accuracy of this process?

SUMMARY

Embedded microcomputers contained on exposed printed circuit boards often malfunction or fail completely in hostile industrial environments. This problem may be overcome through use of programmable logic controllers (PLCs). These devices, which contain their own CPUs and power supplies, can be easily programmed by technicians and machine operators. PLCs also contain their own internal diagnostics, allowing for efficient troubleshooting of the circuits they control. PLCs are equipped with a variety of I/O modules designed to interface with industrial control devices. In complex industrial control applications, some PLCs operate in conjunction with intelligent interfaces, two common forms of which are the PID control module and the visual input module. Low-level PLC programming techniques evolved from relay logic circuits and their associated control ladder diagrams. Modern PLCs, however, allow for more ''user-friendly,'' higher level languages, including BASIC and state language.

Within an industrial system, process control involves monitoring and governing of such variables as temperature, pressure, and flow rate. Holding one or more controlled variables as close as possible to a set-point level during an industrial process is referred to as *regulation*. In a closed-loop control system, regulation takes place as a measuring device senses deviations of a measured variable from its set-point value. Error detection circuitry sends an error signal to a controller, which is likely to be a PLC or a microcomputer. The controller, in turn, sends a feedback signal to a final control element in an attempt to return the controlled variable closer to the set point. The merit of a closed-loop control system depends on how smoothly that system responds to sudden changes in a controlled variable. This transient response of a closed-loop control system can be described as being either overdamped or underdamped. While underdamped response is faster than overdamped, a control system exhibiting underdamped response is more prone toward oscillating instability.

Discrete, or on/off, control is the oldest form of industrial control. With discrete control, the final control element is switched totally on or totally off, depending on the condition of a controlled variable. In the design of many discrete systems, for the sake of operating stability, a hysteresis gap or deadband is deliberately introduced. Discrete control becomes increasingly complex when used in sequential state control of automated systems. While the basic control concept of these systems is still on/off, several controlled variables are often involved. Timing within such automated systems is extremely precise, often requiring microcomputer or PLC control.

The pure control modes used in continuous-state control systems include proportional, integral, and derivative. With proportional control, the feedback signal to a final control element varies in proportion to the magnitude of the deviation of a controlled variable from its set-point value. A purely proportional control system is incapable of overcoming a residual error. However, because it responds to the history of an error signal, the integral control mode eventually eliminates an offset problem. For this reason, the integral control mode is often used in conjunction with the proportional, forming a PI control system. The derivative control mode allows a control system to respond according to the rate of change in an error signal. The derivative control mode is either used in conjunction with the proportional mode, forming a PD control system, or combined with the proportional and integral modes, forming a PID control system. PID is the most versatile and, therefore, most common form of continuous-state control.

SELF-TEST

1. Identify the advantages of using a PLC rather than an embedded microcomputer circuit board to control an industrial process.
2. True or false?: PLC programming methods have evolved from relay logic circuits and their associated control ladder diagrams.
3. True or false?: A distinguishing characteristic of an overdamped closed-loop control system is oscillation.
4. Which pure continuous-state control mode is capable of overcoming an offset error?
5. Which pure continuous-state control mode responds to an error signal only when it is changing?
6. True or false?: With the purely derivative control mode, if the instantaneous condition of an error signal is 0%, the derivative feedback must also be 0%.
7. A PID controller is able to respond to
   a. the magnitude of an error signal
   b. the history of an error signal
   c. the rate of change of an error signal
   d. all of the above
8. Describe two methods by which a PLC could initiate PID control.
9. When is it necessary for a PLC to be equipped with an ASCII interface module?
10. Under what conditions is it advantageous for a PLC to be equipped with fiber-optic input and output modules?
11. Define the term *process* as it applies to the operation of an industrial control system.
12. For an industrial control system, error is most likely to be expressed as
    a. a percentage of the proportional band
    b. a percentage of the range of a controlled variable
    c. a percentage of the set point
    d. a percentage of the hysteresis gap
13. Define monotonic instability as it applies to a closed-loop control system. When is monotonic instability likely to occur?
14. How is it possible for oscillating instability within a closed-loop control system to result in growth without limit?
15. Explain the operation of the closed-loop control system in Figure 13–26. What technique is used to measure the consistency of the slurry?
16. What are the three composite forms of continuous-state control? Which of these three control modes is the most versatile?
17. What mathematical operation is represented by d/dt? In a block diagram of a closed-loop control system, what control mode is indicated by this symbol?
18. True or false?: In a block diagram of a closed-loop control system, the Greek letter sigma (Σ) represents either the process of integration or the use of integral control.
19. Describe the operation of the PID control system in Figure 13–31. How is the set point entered into the system? How is fractional error determined? How is the process being governed in Figure 13–31 similar to that represented in Figure 13–13?
20. For the closed-loop control system in Figure 13–31, what does the Greek letter delta (Δ) represent at the top of the PID block?

14 INTRODUCTION TO AUTOMATED MACHINE TECHNOLOGY AND ROBOTICS

CHAPTER OUTLINE

LEARNING OBJECTIVES

On completion of this chapter, the student should be able to:

- Identify the basic components and explain the operation of a servo loop.
- Identify the basic components and explain the operation of a microcomputer-based servocontroller.
- Explain how a microcomputer-based servocontroller could control the operation of a brushless dc servomotor.
- Describe the structure and operation of a mechanical manipulator.
- Define the term *kinematics* as it applies to the operation of an industrial robot.
- Define the term *dynamics* as it applies to the operation of an industrial robot.
- Define the terms *position, orientation,* and *frame* as they apply to the operation of an industrial robot.
- Identify three classifications of robotic actuators.
- Identify two common forms of robotic sensors.

INTRODUCTION

The purpose of this final chapter is to demonstrate how various devices and circuits introduced earlier in this text are combined to form automated manufacturing systems. Two basic classifications of system, fixed automation and mechanical manipulator, are presented. Fixed automation devices, which include various forms of machine tools, are relatively rigid, each being confined to one particular type of task. Mechanical manipulators are the most common form of industrial robot and are, by comparison, more versatile than fixed automation devices. Mechanical manipulators can be programmed to perform such varied tasks as part sorting and spot welding.

Fixed automation and mechanical manipulator systems contain the same basic elements, including controllers, actuators, and sensors. In a modern industrial facility, both automated machine tools and mechanical manipulators contain microcomputer-based controllers. Also, actuators such as servomotors and stepper motors are found in either form of device. Sensors are necessary in machine tools and robots for monitoring the positions of actuators, detecting the presence of workpieces, and measuring such parameters as velocity and force. A detailed study of the structure and operation of control devices, including microcomputers, microcontrollers, and PLCs was presented in Chapter 12 and the first section of Chapter 13. The structure and operation of two important forms of actuating device, the brushless dc servomotor and the stepper motor, were covered in Chapter 10. Also, many of the sensors found in machine tools were introduced earlier in this text. Thus, in this final chapter of the text, automated machine tool and robotic systems can be addressed at a higher, block level, with the assumption that the internal operation of their controlling, actuating, and sensing elements is already understood.

The chapter begins with an introduction to the concept of the servo loop and a demonstration of how it may be implemented in the control of an actuating device. The idea of the servo loop is then developed further when a state-of-the-art, microcomputer-based servocontroller is introduced. After programming methods for this device are surveyed, its application in the precise control of a dc brushless servomotor is investigated in detail. Actual applications of brushless dc servomotors and stepper motors within automated machine tools systems are then investigated.

The text concludes with a brief investigation of industrial robotics. With the understanding that this field, in itself, is the subject of entire texts, no attempt is made to cover this subject in great detail. Rather, the basic components of the mechanical manipulator, sensors used in monitoring manipulator position, and robotic programming methods are surveyed at an introductory level.

14–1   INTRODUCTION TO SERVO SYSTEMS AND MICROCOMPUTER-BASED SERVO AMPLIFIERS

The term **servomotor,** or simply **servo,** is used to describe a position-correcting actuator. The term *servo* is derived from the Latin ''servus,'' meaning slave. Hence, the term *servomotor* implies a ''slave'' motor or a motor that responds directly to a control source. This control source, called a **servo amplifier,** is capable of receiving a relatively weak input signal, representing a desired position for a given load, and amplifying this signal to a power level capable of controlling servomotor speed and/or direction. For a position-control servo system, the input to the servo amplifier might originate from a control synchro, the shaft of which could be connected to a rotary dial. (Control synchros were introduced in Section 10–4.) A Wheatstone bridge containing a displacement potentiometer could also produce the input signal to a servo system. (The resistance thermometer contained in Figure 4–5 is an excellent example of the use of a resistive bridge within such a system.) When connected to a load in one of the mechanical drive configurations introduced in Section 10–2, the servomotor must also be connected to a **servomechanism.** Through mechanical linkage to a position-sensing transducer, the servomechanism provides a negative feedback signal to the servo amplifier. The closed loop formed through the presence of the servo mechanism is often referred to as a **servo loop.** A servo loop is represented in block form in Figure 14–1.

Within a servo system, turning the rotary dial of a control synchro or moving the wiper of a displacement potentiometer produces an error signal, representing the difference

Figure 14–1
Block diagram of a basic servo loop

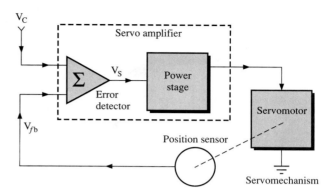

between the desired position of the load attached to the servomotor and its present position. In response to this error signal, the servo amplifier forces the servomotor to move in the direction necessary to place the load in the desired position. We should stress that, with *any* servo system, control of the load is automatic. As the load approaches the desired position, the action of the servomechanism causes the error signal to decrease. In a position-correcting system, as the error signal approaches 0 V, the output of the servo amplifier also approaches 0 V, thus stopping the servomotor.

As represented by the Greek letter Σ (sigma) in Figure 14–1, the error detector, which serves as the input stage of the servo amplifier, is likely to be a summer amplifier. At any instant, the output of the summer amplifier is determined as follows:

$$V_S = (V_C + V_{fb}) \times K \tag{14-1}$$

where V_C is the input taken from a control device, V_{fb} is the feedback signal from the servomechanism, and K is the gain factor of the summer amplifier stage.

Since the servomechanism provides negative feedback to the summing amplifier, a balanced condition is achieved when the absolute value of V_{fb} is equal and opposite to that of V_C. Ideally, with this condition, the output of the summer amplifier is 0 V.

The ability of negative feedback to improve the operating stability of a dc motor has already been demonstrated in Chapter 10. As explained for the self-regulating motor control system in Figure 10–32, an increase in the load on a motor results in an earlier trigger point for the silicon-controlled rectifier (SCR). This action produces more current flow for the motor armature winding, allowing the desired speed of rotation to be maintained. Conversely, when the load on the motor is lightened, the rotational velocity of the motor attempts to increase. However, the resultant decrease in the firing-delay angle causes a reduction in armature current, again allowing the desired speed of rotation to be maintained.

Although the circuit in Figure 10–32 might be adequate for some machine control applications, more sophisticated control is required for motors used as actuators within fixed automation systems. Such motor control systems often operate on the principles of the servo loop. Consider, for example, the circuit in Figure 14–2. Here, a reference voltage, representing a set-point value of motor speed, serves as one input to an op amp summer amplifier. A negative feedback signal from a rotary transducer, such as a dc generator, forms the second input to the summer amplifier. Since this linear transducer is connected to the shaft of the motor, it is capable of producing changes in dc voltage that are proportional to deviations in motor rpm.

The design scheme for the motor control system in Figure 14–2 is as follows. The fixed reference voltage is established through adjustment of potentiometer R_1. The value

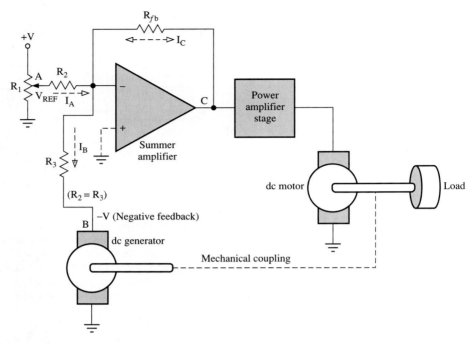

Figure 14–2
Op amp summer amplifier used in dc motor control circuit with negative feedback

of this reference voltage depends on the desired set-point value of motor rpm. If, at the desired rpm, the output of the dc generator (point B) is -3.5 V, the reference voltage at point A is set to $+3.5$ V. Thus, when the motor is rotating at the desired speed, the current leaving point A is equal to and opposite the current entering point B. With this condition, the feedback current for the summing amplifier (flowing through R_{FB}) is nearly 0 A and the output voltage to the power stage is nearly 0 V.

If, due to increased loading, the motor attempts to rotate slower, the negative output of the dc generator begins to decrease toward 0 V. With I_A now greater than I_B, I_C (the algebraic sum of I_A and I_B) flows from the summing point (the virtual ground) toward point C. Thus V_C, which serves as the control input to the power stage, increases in the negative direction. The design of the power stage circuitry is such that, as V_C begins to swing negatively, armature current begins to increase. This increase in armature current produces an increase in armature speed, pulling V_B back in the negative direction. This negative increase in V_B continues until the absolute values of V_A and V_B are again equal. If the motor attempts to rotate faster due to decreased loading, V_B attempts to increase still further in the negative direction. Consequently, with I_B becoming greater than I_A, I_C begins to flow from point C toward the virtual ground point. Thus, as V_C swings positive, the response of the power stage is to decrease the flow of armature current. As a result, V_B begins to decrease. Equilibrium is attained when the absolute values of V_A and V_B are again equal and opposite.

To control a medium-power dc servomotor, the power stage in Figure 14–2 could contain power Darlingtons or MOSFETs. For more powerful 3-ϕ brushless dc servomotors, such as those described in Section 10–5, more sophisticated control is required, as represented by the pulse-width-modulated (PWM) power converter stages in Figures 10–51,

10–52, and 10–53. To expedite the development of fixed automation and robotic systems, manufacturers of servomotors often provide a variety of control modules that have been designed for compatibility with their own line of motors. Along with power converter circuitry such as that represented in Section 10–5, these modules also contain the digital control logic and the interface circuitry necessary for processing signals from such feedback devices as brush-type dc tachometers, brushless tachometers, Hall-effect sensors, synchros, resolvers, and shaft encoders. (These devices were also surveyed in Section 10–5.)

The four **servocontrollers** shown in Figure 14–3 are typical of a generation of microcomputer-based modules capable of exerting precise control over brushless dc servomotors contained in modern automated systems. These servocontrollers, manufactured by IMEC Corporation, are designed for use in conjunction with a line of servomotors produced by Pacific Scientific. Figure 14–4 contains a block diagram of the internal circuitry typical of the servocontrollers contained in Figure 14–3. These servocontrollers could be considered to be dedicated PLCs, due to the following design features. Each device has an internal, microprocessor-based central processing unit (CPU) and switching power supply. It also has its own internal diagnostics. As shown in Figure 14–3, individual light-emitting diodes (LEDs) may be used to indicate the status of a servocontroller. Also, as shown for the third controller, a single seven-segment display can also be provided. Like an advanced form of PLC, the servocontroller is easily programmed in a high-level, user-friendly language, similar in structure to the BASIC commonly used in personal computer (PC) programming. Programming of the servocontroller is introduced in the next section of this chapter.

Since the servocontroller is dedicated entirely to the task of governing the operation of a servomotor, it has no provision for input and output modules. Besides its CPU and dc power supply, it contains only that circuitry necessary for monitoring servomotor speed and position, controlling servomotor velocity, monitoring servomotor current, and monitoring temperature within both the servomotor and its driver stages. A servocontroller can be designed with the assumption that the servomotor is providing sine and cosine feedback signals, generated by a resolver. Thus, as represented in Figure 14–4, the device is equipped with a resolver-to-digital converter (RDC) and the sine wave oscillator required to produce the resolver excitation signal. The signal conditioning circuitry within the RDC allows the servocontroller to determine motor velocity and angular position. (Resolvers were covered in Section 10–4, and the RDC and a possible source of the excitation signal were covered in Section 10–5.) Servomotor velocity is controlled by summer amplifier and power stage circuitry as represented in Figure 14–2. However, in a commercially manufactured servomotor controller, as is soon shown, this analog circuitry is far more complex. The output of the summer amplifier stage drives PWM circuitry that, in turn, controls as many as six power converter stages. These power converter stages are similar to those illustrated in Figures 10–51, 10–52, and 10–53. The servocontroller is designed with the idea that the servomotor being controlled contains a thermistor with a positive temperature coefficient (PTC). As explained later, motor temperature may be monitored by the simple comparator circuitry shown in Figure 14–14.

Figure 14–5 is a simplified functional diagram representing a combination of the detailed schematic diagrams in Figures 14–6, 14–7, 14–8, 14–10, 10–51, and 10–52. Here, the CPU of the servocontroller serves as a starting point in a general overview of system operation. The CPU block in Figure 14–5 represents the complete microcomputer section of the servocontroller, including the microprocessor, ROM, RAM, and I/O interface circuitry. At this point, we assume that the CPU is executing a user program that has been entered into the servocontroller RAM via the keyboard of an external host computer.

Assume that, as part of the user program, the servocontroller is causing the servomotor to increase its speed gradually. For the given direction of rotation, assume this action is to be achieved through gradually increasing the negative dc potential at the **Velocity**

Figure 14–3
Typical industrial servocontrollers (Courtesy of Pacific Scientific)

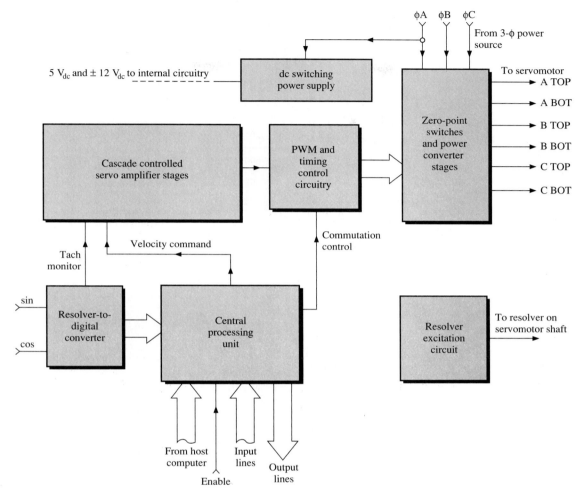

Figure 14–4
Functional block diagram of a commercially manufactured servocontroller

Command (VEL CMD) input. To initiate this process, the CPU sends to the **position loop D/A converter** the necessary sequence of binary numbers to generate the required negative-going ramp for VEL CMD. This ramp signal at VEL CMD produces an error signal that is amplified by the **velocity error amplifier.**

The action described thus far constitutes the operation of a basic servo amplifier. However, there are two overlapping control loops within the summer amplifier circuitry of the SC450. The control configuration represented in Figure 14–5 is often referred to as **cascade control.** The output of the velocity error amplifier is summed with the **IFBMAG** signal, a variable dc voltage that represents the constant component of the three motor phase currents. The output of the **current error amplifier** drives the PWM and power stage circuitry.

The velocity loop in Figure 14–5 is a true servo loop, since it contains an actual servomechanism in the form of a resolver. The velocity feedback path arrives at the servocontroller's CPU by way of the RDC. The RDC allows the CPU to detect both

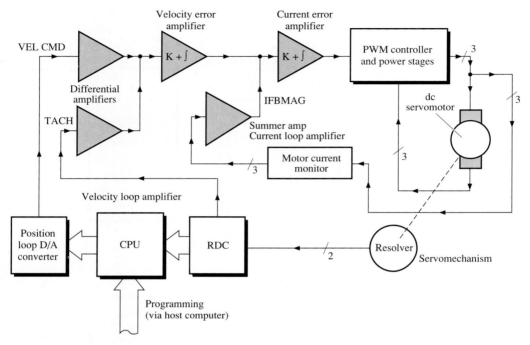

Figure 14–5
Functional diagram of servocontroller operation

servomotor shaft speed and position. Shaft speed information is converted to a variable dc signal called **TACH.** This signal, representing *actual* motor speed, becomes an input to the velocity loop amplifier stage. Also, velocity and shaft position information from the RDC proceed to the CPU via the servocontroller's internal data bus. As the ongoing program routine compares this RDC information with data representing desired speed and shaft position, it formulates command information for the position loop D/A converter. Thus, as actual motor speed approaches desired motor speed, the error signal approaches 0 V. The current loop, which comprises the second closed loop of the summer amplifier stage, completely bypasses the servocontroller's digital circuitry. Current-sensing circuitry, connected directly to the three current-sourcing outputs of the power stage, are able to sense quickly any deviations in motor current. These samples of motor phase current are summed to form the IFBMAG signal.

Note, as indicated in Figure 14–5, that both the velocity and current error amplifiers provide proportional plus integral (PI) control. As explained in subsequent analysis of this control circuitry, the purpose of the PI control scheme is to limit the gain of the ac components of the control signal. This action is necessary to prevent overshoots, oscillations, and resultant rough operation of the servomotor. As explained in the next section of the chapter, proportional and integral gain may also be controlled through programming of the servocontroller. Thus, PI control is achieved in two ways, through operation of the analog control circuitry and through control algorithms executed by the CPU. (Recall that such algorithms were introduced in Section 13–4.)

We now examine in detail the operation of the circuitry just explained at a survey level. As shown in Figure 14–6, the Velocity Command input is a differential signal

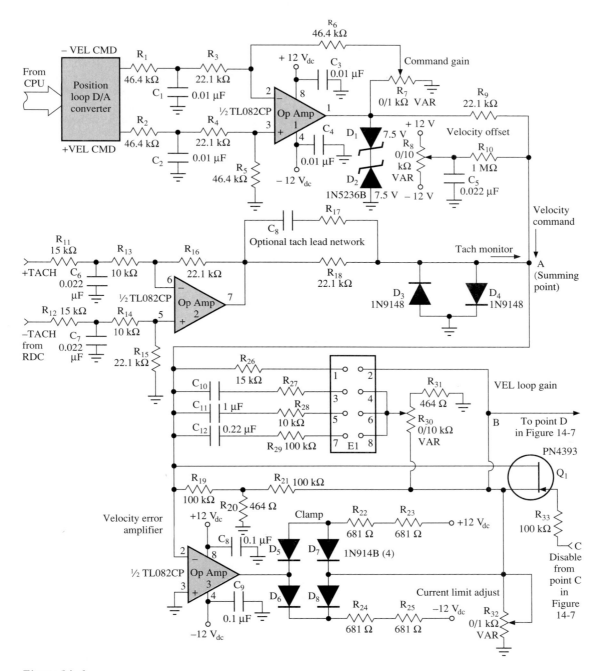

Figure 14–6
Servocontroller cascade control circuitry: velocity loop amplifier

(abbreviated $-$VEL CMD and $+$VEL CMD). The **Tach Monitor** signal is also differential, consisting of signals $+$TACH and $-$TACH. The Velocity Command and Tach Monitor signals are equivalent to the V_C and V_{fb} signals in Figure 14–1, respectively. The VEL CMD signals arrive at a differential amplifier containing op amp 1, the gain of which is controlled through adjustment of the **Command Gain** potentiometer R_7. This adjustment may be considered in terms of rpm/V. Increasing the ohmic value of R_7 increases the voltage gain of the first differential amplifier, resulting in a greater Δrpm for a given ΔV at the input terminals of op amp 1. Diodes D_1 and D_2 form a simple **back-to-back Zener limiter,** preventing the output of Op Amp 1 from exceeding ± 8.2 V. The Tach Monitor signals, differential outputs from the RDC of the servocontroller, serve as the velocity feedback input to the summer amplifier and varies in proportion to servomotor velocity by a factor of 0.56 V/1000 rpm. The Tach Monitor signal, always of opposite polarity to Velocity Command, arrives at the differential amplifier containing op amp 2. After being conditioned by an optional **tach lead network,** consisting of C_8, R_{17}, and R_{18}, the amplified Tach Monitor signal arrives at the summing point (point A) of the velocity error amplifier. Diodes D_3 and D_4 form a **double-shunt limiter,** preventing the voltage at the summing point from exceeding ± 0.7 V. The **Velocity Offset** potentiometer R_8 allows offset adjustment to be performed such that, whenever the Velocity Command and Tach Monitor signals are both at 0 V, the output of the velocity error summer amplifier consisting of op amp 3 is also 0 V.

The velocity error amplifier in Figure 14–6, containing op amp 3, is designed for adaptability to various user applications. E1 is a small terminal block that allows a jumper to be placed in one of four possible positions. Placement of the jumper between terminals 1 and 2 of E1 allows for purely proportional control to take place at this stage of the cascade controller, since the feedback path of the op amp is then purely resistive. By connecting the jumper between terminals 7 and 8 or terminals 5 and 6 of E1, two possible combinations of RC values (0.22 μF and 100 kΩ or 1 μF and 10 kΩ) may be accessed, allowing PI control to be implemented by the velocity error amplifier. This control scheme can be referred to as **VEL loop compensation,** since the RC combination, if properly chosen, counteracts (or compensates for) the effects of the motor's line-to-line inductance. For either of these optional RC combinations, the cutoff, or **lag break,** frequency is determined as follows:

$$f_{CO} = \frac{1}{2\pi \times R \times C} \tag{14–2}$$

Thus, with the jumper placed between terminals 7 and 8:

$$f_{CO} = \frac{1}{2\pi \times 0.22\ \mu F \times 100\ k\Omega} \cong 7.2\ \text{Hz}$$

This result indicates that, while changes in the dc component of the velocity error amplifier output signal are passed easily, any ac noise occurring above the frequency of 7.2 Hz is attenuated.

With the jumper placed between terminals 5 and 6:

$$f_{CO} = \frac{1}{2\pi \times 1\ \mu F \times 10\ k\Omega} \cong 16\ \text{Hz}$$

With this RC combination, although lag break occurs at a higher frequency, the response time of the servocontroller is decreased.

In some servomotor control applications, the user of the servocontroller may require a specific velocity loop compensation that is not achievable with the RC combinations provided by the manufacturer. For this reason, spaces are provided within the servocontroller for insertion of a resistor and capacitor of the user's choice (labeled R_{27} and C_{10}). After installation of these two components, this **VEL loop custom compensation** option can be used by placing the jumper between terminals 3 and 4 of E1. The **VEL loop gain** potentiometer R_{30} allows fine adjustment of the gain of the velocity error amplifier. Lowering the ohmic value of R_{30} tends to reduce the high-frequency component of the control signal (possibly produced by the motor inductance). However, too low an ohmic setting of R_{30} results in an overdamped response within the velocity loop. Too high a setting of R_{30} results in an underdamped condition, which could produce overshoots within the velocity loop and cause rough operation of the motor.

The feedback network of the velocity error amplifier in Figure 14–6 contains a **clamp** circuit, in the form of a diode bridge. The purpose of this circuit is to limit the flow of servomotor current to no more than ±7.4 A during an overload or short circuit condition. Assuming R_{32} is adjusted to its maximum ohmic value of 1 kΩ, the maximum voltage that could be developed at the output of the velocity error amplifier would be around ±4.8 V, being determined as follows:

$$V_O(\text{max}) = (V_{CC} - V_D) \times \frac{R_{32}}{R_{22} + R_{23} + R_{32}} \qquad (14\text{–}3)$$

Thus, with R_{32} adjusted for maximum resistance:

$$V_O(\text{max}) = (12\ V_{dc} - 0.7\ V_{dc}) \times \frac{1\ k\Omega}{681\ \Omega + 681\ \Omega + 1\ k\Omega} = 4.78\ V$$

The current loop amplifier of the servocontroller's cascade control stage is contained in Figure 14–7. The current error amplifier, containing op amp 4, is a summer amplifier that receives the amplified velocity loop signal from point B in Figure 14–6 and a **Current Sense** feedback signal (IFBMAG) from Op Amp 6. The functions of op amps 6 and 7, which are also part of the Current loop amplifier, are explained in detail at a later point in this section. Two 6.8-V Zener diodes, D_9 and D_{10}, are placed in parallel with the RC compensation network for this stage. These components form a back-to-back Zener limiter within the feedback network of op amp 4, keeping the output amplitude of the stage within ±7.5 V. The current loop amplifier has its own compensation network, consisting of R_{44} and C_{11}. These components provide PI control within the system current loop.

Both the Velocity error amplifier in Figure 14–6 and the Current error amplifier in Figure 14–7 may be disabled by means of solid-state switching. The $\overline{\text{Disable}}$ signal, which originates in the digital control circuitry of the servocontroller, first arrives at an inverter formed by op amp 5 in Figure 14–7. (Recall from Sections 2–1 and 2–7 that an op amp, in an open-loop configuration, operates effectively as a comparator.) The digital control signal at the inverting input of op amp 5 is either at a logic high of nearly 12 V or a logic low of 0 V. The presence of a reference level of 3 V_{dc} at the noninverting input of op amp 5 ensures reliable switching of that device. When the $\overline{\text{Disable}}$ input is at a logic high (greater than the reference voltage) the output of op amp 5 falls to a saturation level of slightly less than -12 V. When the $\overline{\text{Disable}}$ input falls to a logic low (less than the reference voltage) the output of op amp 5 switches to nearly $+12$ V. While the $\overline{\text{Disable}}$ signal is in its inactive, logic high condition and the output of Op Amp 5 is at nearly -12 V, the two n-channel junction field-effect transistors (JFETs), Q_1 in Figure 14–6 and Q_5 in Figure 14–7, are in their pinch-off condition, functioning as open switches. Thus, these two JFETs are effectively removed from the circuit, allowing normal operation of both stages of the summer amplifier. However, when the $\overline{\text{Disable}}$ signal switches to its active, logic low condition, the two JFETs switch on, with a V_{DS} of nearly 0 V. Thus,

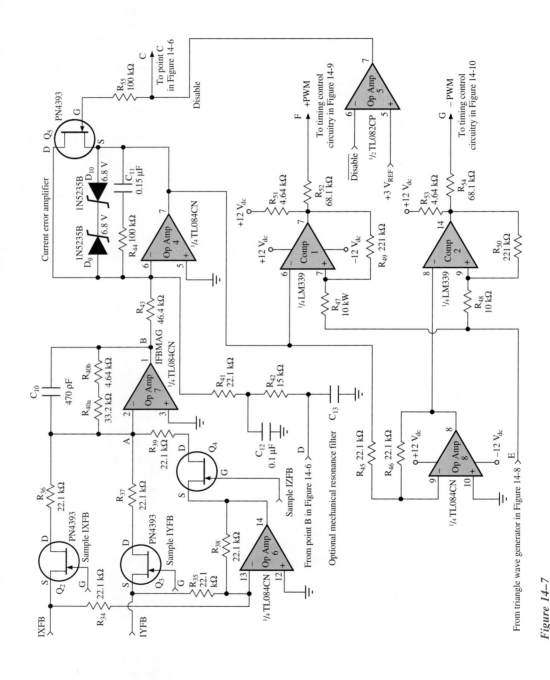

Figure 14-7

Servocontroller cascade control circuitry (continued): current loop amplifier and part of PWM control circuitry since these switches shunt the feedback circuitry of both stages of the cascade controller, the PWM-controlled power stages and the servomotor are held in the off state.

Figure 14–7 also contains a portion of the PWM circuitry for the servocontroller. An op amp and two comparators comprise the first stage of the pulse width modulator. While the output from op amp 4 goes directly to the inverting input of comp 1, this signal is inverted by op amp 8 (which has a voltage gain of -1) before arriving at the inverting input of comp 2. Thus, the signals arriving at the inverting inputs of the two comparators are always equal in amplitude but opposite in polarity. The noninverting inputs of both comparators simultaneously receive a triangle signal from op amp 10, located in Figure 14–8. Op amp 10 and op amp 9 together function as a free-running triangle wave generator, with op amp 9 operating as a Schmitt trigger, and op amp 10 functioning as a Miller integrator.

The operation of op amps 9 and 10 in Figure 14–8 proceeds as follows. Assume that, as shown between t_0 and t_1 in Figure 14–9, the output of op amp 9 is at its positive saturation level of approximately 11 V. With this condition, capacitor C_{16} begins to charge negatively with respect to point C. Since the output of op amp 9 remains near its saturation

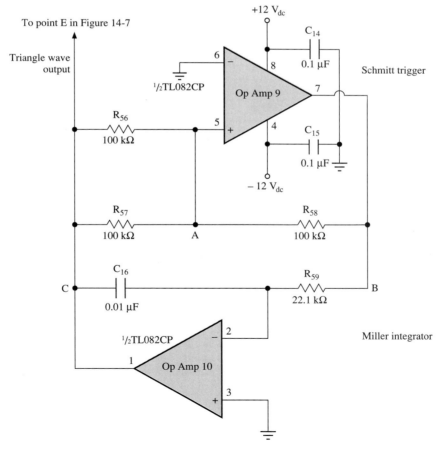

Figure 14–8
Triangle wave oscillator for PWM section of servocontroller

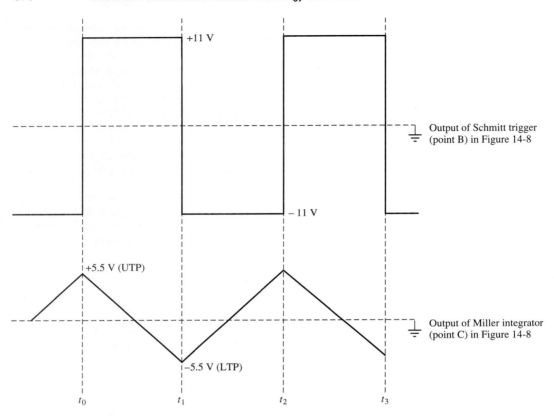

Figure 14–9
Waveforms produced by triangle wave generator in Figure 14–8

level as C_{16} charges, the inverting input of op amp 10 acts as a constant current source, allowing the negative slope of the triangle wave at point C to develop linearly. Capacitor C_{16} continues to charge in the negative direction until the voltage at the noninverting input of op amp 9 (point A) approaches nearly 0 V, which is the reference level established by directly grounding pin 6. As the voltage at point A attempts to fall below 0 V, as seen at t_1 in Figure 14–9, the output of the Schmitt trigger rapidly toggles to its negative saturation level of approximately −11 V. This action causes C_{16} to begin charging positively with respect to point C. Again, the inverting input to op amp 10 functions as a constant current source, allowing the positive slope of the triangle wave to be linear. Capacitor C_{16} continues to charge positively until the voltage at point A approaches 0 V. Then, as shown at t_2 in Figure 14–9, a second cycle of the triangle waveform begins.

To determine the operating frequency of the triangle wave generator contained in Figure 14–8, we must first analyze the operation of the Schmitt trigger. Due to the high impedance of the noninverting input of op amp 9, R_{56} and R_{57} may be considered as being in parallel, forming a 50-kΩ resistance. Again disregarding the high impedance of the noninverting input of op amp 9, the parallel resistance of R_{56} and R_{57} is effectively in series with R_{58}. Having made this simplification, the upper and lower trigger points for the Schmitt trigger can be easily determined. With a reference voltage established at ground potential, assuming the absolute values of $+V_0(\text{sat})$ and $-V_0(\text{sat})$ to be the same, the current flow through R_{58} is calculated using Ohm's law. Assuming $\pm V_0(\text{sat})$ to equal

11 V, $I_{R_{58}}$ is determined as follows:

$$\pm I_{R_{58}} = \pm \frac{V_0(\text{sat})}{R_{58}} = \pm \frac{11\text{ V}}{100\text{ k}\Omega} = \pm 110\text{ }\mu\text{A}$$

Since the sum of the currents flowing through R_{56} and R_{57} virtually equals $I_{R_{58}}$, the upper trigger point (UTP) and lower trigger point (LTP) for the Schmitt trigger are easily determined as follows:

$$\text{UTP} \cong I_{R_{58}} \times \frac{(R_{56} \times R_{57})}{(R_{56} + R_{57})} \cong 110\text{ }\mu\text{A} \times 50\text{ k}\Omega \cong 5.5\text{ V}$$

$$\text{LTP} \cong -I_{R_{58}} \times 50\text{ k}\Omega \cong -5.5\text{ V}$$

These solutions for UTP and LTP, as indicated in Figure 14–9, represent the positive and negative peak amplitudes, respectively, of the triangle wave seen at point C in Figure 14–8. Since the values of UTP and LTP depend on the ratio of R_{58} to the parallel resistance of R_{56} and R_{57}, the operating frequency of the triangle wave generator may be determined as follows:

$$f = \frac{\eta}{4 \times R \times C} \qquad\qquad (14\text{--}4)$$

where η represents the ratio of R_{58} to the parallel resistance of R_{56} and R_{57} in Figure 14–8, R is the integrator resistance, and C is the integrator capacitance. Thus, for the triangle wave generator in Figure 14–8, operating frequency is determined as follows:

$$f = \frac{2}{4 \times 22.1\text{ k}\Omega \times 0.01\text{ }\mu\text{F}} = 2.26\text{ kpps} \cong 2.3\text{ kpps}$$

With the triangle wave from op amp 10 being fed to the noninverting inputs of both comparators in Figure 14–7, two complementary rectangular waveforms, designated +PWM and −PWM, feed the CD4053B IC in Figure 14–10. This CMOS IC is designed to receive up to six input signals from PWM circuitry and three timing control signals from a servocontroller's commutation control logic circuitry. The frequency of the +PWM and −PWM signals remains constant, being equal to that of the triangle wave generator. However, changes in the pulse width of the +PWM signal and the space width of the −PWM signal are both proportional to changes in the differential voltage between the outputs of op amps 4 and 8 in Figure 14–7. Note that the output of comp 1 in Figure 14–7 is connected to the a_X, b_X, and c_X inputs of the CD4053B in Figure 14–10. The output of comp 2 is connected to the a_Y, b_Y, and c_Y inputs of the CD4053B. With this circuit configuration, changes in the duty cycle of the PWMA, PWMB, or PWMC output signals are always proportional to changes in the differential voltage between the outputs of op amps 4 and 8.

The A, B, and C inputs to the CD4053B in Figure 14–10 are commutation signals that control the timing of the PWMA, PWMB, and PWMC outputs, respectively. In some servo systems, Hall-effect sensors form the servomechanism that produces these three square waves. These Hall-effect devices, along with their signal conditioning circuitry, are often permanently installed within the servomotor. (The use of Hall-effect sensors within servomotors was explained briefly in Section 10–5.) Within such systems, the commutation signals produced by the signal conditioning circuitry are sent directly to the commutation control circuitry of the servocontroller. However, within the servocontroller being analyzed, these signals are developed by the CPU as it processes information received from the RDC. The A, B, and C commutation signals are shown in Figure 14–11.

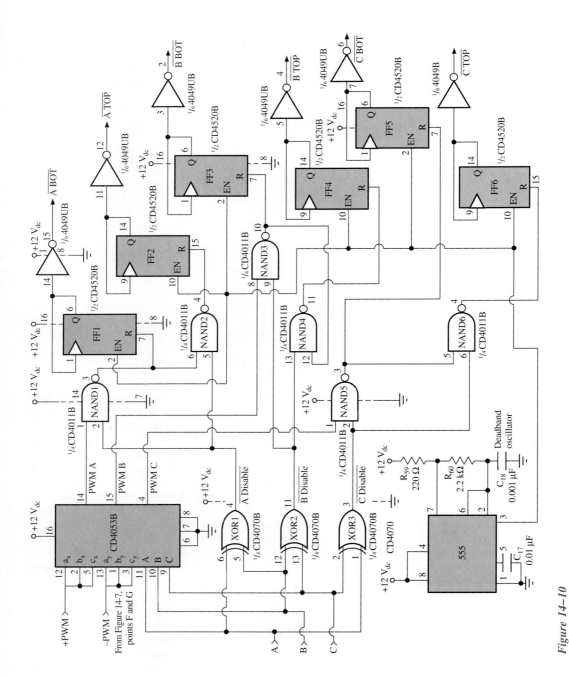

Figure 14–10

Timing control circuitry for PWM-controlled power stages of servocontroller

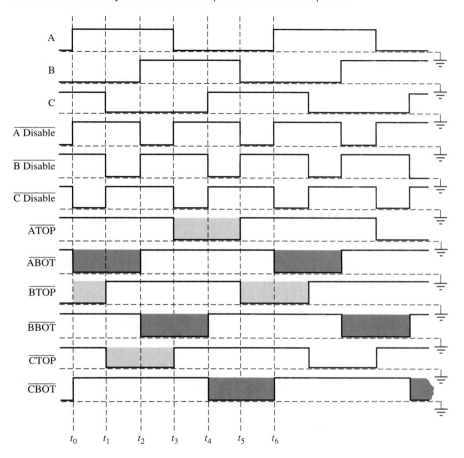

Figure 14–11
Input and output signals for the timing control circuitry in Figure 14–10

One cycle of a commutation signal represents one revolution of the servomotor shaft. Thus, the 120 electrical degree spacing between each of the commutation signals represents 120° of shaft rotation. The frequency of these commutation signals is a function of the servomotor velocity.

Example 14–1

Assume that Hall-effect sensors within a servomotor are producing commutation signals A, B, and C as shown in Figure 14–11. If the servomotor is turning at 1486 rpm, what are the frequency and pulse width of the commutation signals? What is the difference in time between any pair of these commutation signals?

Solution:

Step 1: Solve for the frequency of the commutation signals in pulses per second:

$$f = \frac{1486 \text{ rpm}}{60 \text{ sec}} = 24.77 \text{ pps}$$

Step 2: Solve for the pulse width:

$$T = \frac{1}{24.77 \text{ pps}} = 40.38 \text{ msec}$$

$$PW = 0.5 \times 40.38 \text{ msec} = 20.19 \text{ msec}$$

Step 3: Solve for the difference in time between the commutation signals.

$$\Delta t = \frac{120°}{360°} \times T = \frac{40.38 \text{ msec}}{3} = 13.46 \text{ msec}$$

At this juncture, the commutation logic circuitry of a typical servocontroller is investigated in detail. The commutation logic circuitry in Figure 14–10, represented by the PWM timing control circuitry block in Figure 14–4, is designed to control the six power converter stages represented in Figure 10–51 and detailed in Figure 10–52 of Chapter 10.

Besides serving as the timing inputs to the CD4053B in the PWM circuitry of Figure 14–10, the A, B, and C commutation signals also enter an XOR array. Responding to odd parity, XOR1 produces a logic high only when the A and B commutation signals are in opposite logic states. XOR2 produces a logic high only when the B and C commutation signals are in opposite logic conditions. XOR3 responds to an odd-parity condition for the A and C commutation signals. The **A Disable**, **B Disable**, and **C Disable** input signals, along with the A, B, and C commutation signals, are plotted in Figure 14–11. Note that the three Disable signals occur at twice the frequency of the commutation signals and have a duty cycle of 66.7%.

The purpose of the PWMA, PWMB, PWMC, and three Disable signals is to control the timing of the six active-low output signals of the commutation logic circuitry, consisting of **A TOP, A BOT, B TOP, B BOT, C TOP,** and **C BOT.** In Figure 14–11, the shaded areas of the A BOT, B BOT, and C BOT signals indicate the points at which PWMA, PWMB, and PWMC are enabled to pass to the power converter stages of the servocontroller. Through the action of the first two SR flip-flops and NAND gates 1 and 2 in Figure 14–10, the PWMA signal is channeled to the A BOT output at the proper points in the commutation cycle. The second pair of flip-flops and NAND gates 3 and 4 steer the PWMB signal to the B BOT output. The third pair of flip-flops and NAND gates 5 and 6 steer the PWMC signal to output C BOT.

Referring again to Section 10–5, the A TOP, B TOP, and C TOP signals could control the operation of the **φX, φY,** and **φZ current-source power converter** stages, respectively, in Figure 10–51. The A BOT, B BOT, and C BOT signals could control the operation of the **φX, φY,** and **φZ current-sink power converter** stages. With reference to the detail in Figure 10–52, these signals could arrive at the **COMMUTATION (PWM)** input (point B) of each of the six power converters. As illustrated in Figure 10–55, current flows through two sets of servomotor windings when a current-sourcing power Darlington of one phase and a current-sinking power Darlington of another phase are on simultaneously. Examination of the six output signals in Figure 14–11 yields the switching pattern illustrated in Figure 14–12.

As an example of the operation of the commutation logic circuitry in Figure 14–10, consider the input conditions occurring between t_0 and t_1 in Figure 14–11. With the A commutation and the A Disable signals both at a logic high, PWMA signals arrive at the R (reset) inputs of flip-flops 1 and 2. This causes the Q outputs of flip-flops 1 and 2 to toggle at a rate of around 2.3 kpps, with the pulse width of the signal at the Q output of flip-flop 1 becoming a function of the differential voltage developed between the +PWM

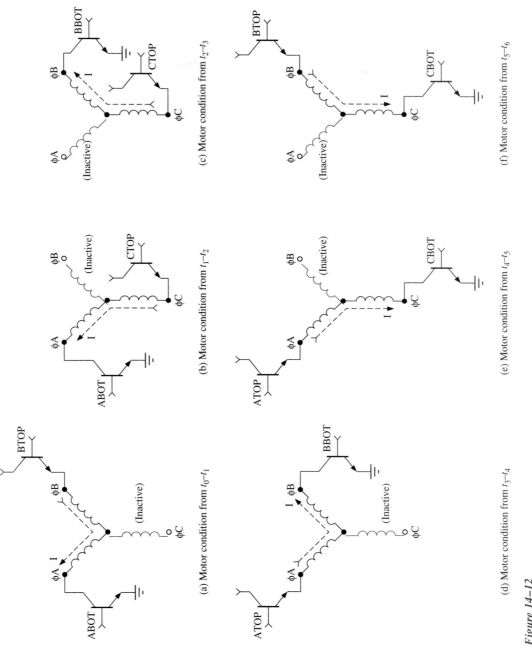

(a) Motor condition from t_0-t_1

(b) Motor condition from t_1-t_2

(c) Motor condition from t_2-t_3

(d) Motor condition from t_3-t_4

(e) Motor condition from t_4-t_5

(f) Motor condition from t_5-t_6

Figure 14–12
Motor conditions corresponding to timing waveforms in Figure 14–11

881

and $-$PWM inputs of the CD4053B. At this point, we should note that the space width of the signal at the Q output of flip-flop 2 also varies as a function of this differential voltage. However, from t_0 through t_3 in Figure 14–11, the $\overline{\text{A TOP}}$ signal has no bearing on circuit operation. We should also note that, through the switching action of NAND gates 1 and 2, the $\overline{\text{A TOP}}$ and $\overline{\text{A BOT}}$ signals are always in opposite logic states. This ensures that the ϕX current-sourcing and ϕX current sinking power Darlingtons represented in Figure 10–51 are never on simultaneously. Although the B Disable signal is high between t_0 and t_1, the PWMB signal remains low, since the B commutation signal is low. Thus, the R input to flip-flop 3 remains high while the R input to flip-flop 4 remains low. Consequently, the Q output of flip-flop 3 remains low while the Q input to flip-flop 4 remains high. Further analysis of the logic conditions occurring between t_0 and t_1 indicates that the R inputs of flip-flops 5 and 6 both remain high, causing their Q outputs to be latched low. As a result of the conditions just described, due to the inversion of the commutation logic's output signals, the space width of the $\overline{\text{A BOT}}$ signal varies as a function of the pulse width of the PWMA signal while the $\overline{\text{B TOP}}$ signal remains at a logic low.

The logic conditions just described for t_0–t_1 result in servomotor current flow as shown in Figure 14–12(a). The constant logic low for the $\overline{\text{B TOP}}$ signal is equivalent to a constant low at point B in Figure 10–52. (Assume that Figure 10–52 represents the ϕY current source power converter.) This condition allows the emitter voltage of the power Darlington at the output of the ϕY current source power converter to increase to its maximum level. However, current entering the collector of the power Darlington of the ϕX current-sink power converter is a function of the space width of the $\overline{\text{A BOT}}$ signal. (Assuming that Figure 10–52 now represents the ϕX current-sink power converter, the $\overline{\text{A BOT}}$ signal appears at point B.) Thus, as the differential voltage between the $+$PWM and $-$PWM inputs of the CD4053B in Figure 14–6 increases, the space width of the $\overline{\text{A BOT}}$ signal also increases. This causes a proportional increase in the flow of base current to the power Darlington of the ϕX current-sink power converter. Since collector current varies as a function of base current, current through the ϕX and ϕY motor windings increases proportionally, tending to increase servomotor velocity. As the differential voltage between the $+$PWM and $-$PWM signals decreases, the space width of the $\overline{\text{A BOT}}$ signal also decreases. A proportional decrease results in the flow of base current to the power Darlington of the ϕX current-sink power converter. This, in turn, causes a proportional decrease in current through the ϕX and ϕY motor windings, tending to reduce servomotor velocity.

Analysis of the commutation logic circuitry conditions occurring between t_1 and t_2 proves that motor current flows as shown in Figure 14–12(b). With the $\overline{\text{C TOP}}$ signal constantly low during this time, the ϕZ current source power converter is enabled to provide current to the ϕX and ϕZ windings. The ϕX current-sink power converter still functions as described for t_0–t_1, with the amplitude of current flowing through the ϕX and ϕZ windings being a function of the $\overline{\text{A BOT}}$ signal. Between t_2 and t_3, with the $\overline{\text{C TOP}}$ signal still low, the ϕZ current-source power converter is still active. However, the ϕY current-sink power converter now controls the flow of current through the ϕZ and ϕY windings, responding to variations in the space width of the $\overline{\text{B BOT}}$ signal. For the sake of brevity, this investigation of the commutation sequence of the servomotor controller ends here. However, as further practice in circuit analysis, the student is encouraged to carry it through the remaining three states.

For the sake of simplicity, the function of the **deadband oscillator** in Figure 14–10 has been deliberately excluded from the analysis of the commutation logic circuitry. The deadband oscillator is an astable multivibrator with an operating frequency of approximately 300 kpps. (The operation of the 555 timer as an astable multivibrator was covered in detail during Section 6–2.) To understand the purpose of the deadband oscillator, it is important to realize that its operating frequency is much greater than that of the PWM circuitry in Figure 14–10. As many as 130 pulses of the deadband oscillator could occur during a single cycle of the PWM circuitry, which is determined as follows:

$$N = \frac{300 \text{ kpps}}{2.3 \text{ kpps}} \cong 130$$

where N represents the number of deadband oscillator cycles occurring during one PWM cycle.

It is also important to note that the output of the deadband oscillator feeds all six EN (enable) inputs of the flip-flops in Figure 14–10 simultaneously. Thus, the oscillator provides a **chop** signal for all six output signals of the commutation logic circuitry. During the 1.67-μsec pulse width of the chop signal, while the flip-flops are enabled, the complement of the R condition of each flip-flop is latched at its Q output. During the 1.67-μsec space width of the chop signal, while the flip-flops are disabled, the Q outputs effectively function as open switches. This chopping action of the deadband oscillator greatly improves the power efficiency of the power stages of the servocontroller.

Now that methods for controlling servomotor commutation and current flow have been introduced, the operation of the **current sense** circuitry in Figure 14–7 can be investigated. One method of sensing changes in the value of servomotor current is shown in Figure 14–13. Within the servomotor controller enclosure, two extremely low value resistances are placed in series with the outputs to the ϕX and ϕY windings. To convert changes in the ϕX and ϕY currents to proportional changes in a voltage, differential outputs are taken across each of these resistors and sent to signal conditioning circuitry within the servocontroller. This signal conditioning circuitry amplifies each of the two current-sensing signals such that they change at a rate of 0.37 V/A.

As shown in Figure 14–7, the amplified outputs of the current sensing circuitry become the IXFB and IYFB inputs to the summer amplifier stage consisting of op amps 6 and 7. Through the designers' clever exploitation of Kirchhoff's current law, the servocontroller is able to monitor the conditions of all three phase currents, although only the ϕX and ϕY currents are being directly measured. Kirchhoff's current law states that the algebraic sum of the currents entering and leaving a node is always 0 A. Thus, assuming the phase windings of the servomotor to be in a wye formation:

$$I_{\phi Z} = -(I_{\phi X} + I_{\phi Y}) \tag{14–5}$$

This mathematical function is always fulfilled by op amp 6 in Figure 14–5, which multiplies the algebraic sum of IXFB and IYFB by a factor of -1. Consider the condition for t_0–t_1 of the commutation cycle, as represented in Figure 14–12(a). Because current leaving the ϕY terminal of the servomotor is equal and opposite to the current entering the ϕX terminal, current flowing through the ϕZ winding must equal 0 A. Thus, the voltage at the IXFB input of the current sense op amps must be equal and opposite to the voltage at the IYFB input. With this condition, the feedback current for op amp 6 (representing IZFB), becomes 0 A:

$$I\text{ZFB} = (I\text{XFB} + I\text{YFB}) \times -1 = 0 \text{ A} \times -1 = 0 \text{ A}$$

For t_1–t_2 of the commutation cycle, as seen in Figure 14–12(b), no current is flowing through the ϕY terminal, causing the IYFB input to also be 0 A. However, with this condition, the current leaving the ϕZ terminal of the servocontroller is equal and opposite to that entering the ϕX terminal. This condition is accurately sensed by op amp 6 as follows:

$$I\text{ZFB} = (I\text{XFB} + 0 \text{ A}) \times -1 = I\text{XFB} \times -1 = -I\text{XFB}$$

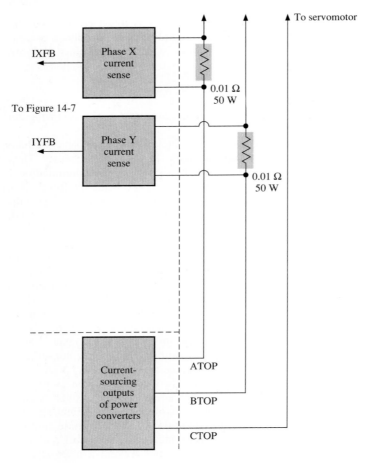

Figure 14–13
Servocontroller current-sensing circuitry

For the sake of brevity, the analysis of the operation of op amp 6 ends here. However, the student is encouraged to study its operation and write the equations for the remaining four points in the commutation cycle.

The three JFET switches shown in Figure 14–7, operating identically to the two Disable JFETs, control the IXFB, IYFB, and IZFB current sense signals, allowing only one of these signals to be enabled at any given instant. These solid-state switches are activated by the A, B, and C commutation signals as seen in Figure 14–11. As a result of this switching action, the IXFB, IYFB, and IZFB motor current feedback signals sequentially appear at the summing point for op amp 7. The cutoff frequency for this summer amplifier stage is about 9 kHz, which is determined as follows:

$$f_{CO} = \frac{1}{2\pi \times (R_{40a} + R_{40b}) \times C_{10}}$$

$$= \frac{1}{2\pi \times (33.2 \text{ k}\Omega + 4.64 \text{ k}\Omega) \times 470 \text{ pF}} \cong 8.95 \text{ kpps}$$

The output current produced by op amp 7, designated IFBMAG, flows to the summing input of op amp 4, where it is algebraically added to the current produced by the velocity error amplifier (op amp 3) in Figure 14–6.

Figure 14–14 shows the temperature-sensing circuitry contained within the servocontroller. Two output terminals of the servocontroller, **MOTOR PTC** and **MOTOR PTC RTN,** allow the comparator within that device to be connected directly to a thermistor contained within the servomotor housing. For proper operation of the comparator circuitry within the servocontroller, this thermistor must exhibit a **positive temperature coefficient** (PTC). As motor temperature increases, possibly as a result of an overload condition, thermistor resistance increases, eventually causing the voltage at the inverting input of the comparator to increase beyond the reference level of 3 V. This condition causes the comparator output to fall low. This active-low signal, **MOTOR OT** (motor over-temperature) goes directly to the servocontroller CPU, which immediately removes power from the servomotor. Meanwhile, the internal diagnostics of that device could generate a fault code on its seven-segment display module, indicating the occurrence of a servomotor over-temperature condition. The servocontroller could also generate an active-low **Fault** output as its internal diagnostic circuitry detects a servomotor malfunction. This message is sent to another device, usually a system control PC. On receipt of this signal, the system control PC could display on its monitor a diagnostic message such as "Servomotor malfunction has occurred." The technician would then observe the seven-segment display on the servocontroller to determine the specific fault.

For the circuit in Figure 14–14, the temperature at which the servomotor must be shut down to prevent damage is referred to as the **trip-out temperature.** The ohmic value of the PTC thermistor that occurs at the trip-out temperature is designated as R_{PTC}. If R_1 is to remain at 100 kΩ as shown, R_{PTC} must be approximately 33 kΩ for the output of the comparator to fall low at the instant servomotor temperature reaches the trip-out point. The manufacturer of the servocontroller usually allows removal and replacement of R_1.

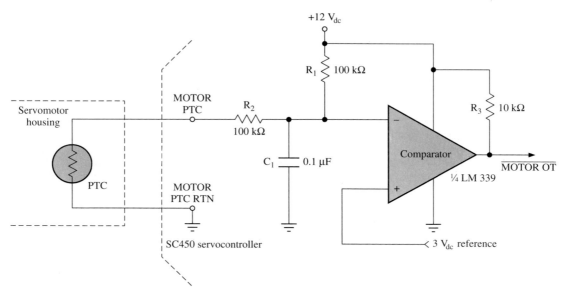

Figure 14–14
Servocontroller over-temperature sensing circuitry

This action could be necessary if the thermistor within the servomotor being controlled by the servocontroller does not have an R_{PTC} of 33 kΩ. For the circuit in Figure 14–14, due to the relatively high ohmic values of the resistors and the relatively low input impedance of the LM339 QUAD comparator, R_1, R_2, and R_{PTC} may not be approached as a simple series voltage divider. Therefore, to expedite the calculation of a replacement value of R_1, the manufacturer recommends the use of the following formula:

$$R_1 = 3 \times (R_{PTC} + 100 \ \Omega) \tag{14–6}$$

Example 14–2

Assume that, for a given servomotor, the thermistor used to detect an over-temperature condition has a resistance of nearly 15.57 kΩ at the trip-out temperature. What standard value of resistor could be used as R_1 in Figure 14–14?

Solution:

$$R_1 = 3 \times (R_{PTC} + 100 \ \Omega) = 3 \times (15.57 \ \text{k}\Omega + 100 \ \Omega) \cong 47 \ \text{k}\Omega$$

SECTION REVIEW 14–1

1. Identify the three major components that comprise a servo loop.
2. What component within a servo system provides negative feedback to the control source?
3. What is the purpose of the clamp circuitry in Figure 14–6?
4. Describe the function and purpose of JFETs Q_1 and Q_5 in Figures 14–6 and 14–7.
5. In Figure 14–7, what is the purpose of the compensation network consisting of R_{44} and C_{11}? What is the lag break frequency established by these components?
6. For the triangle wave generator in Figure 14–8, assume R_{56} and R_{57} both equal 68 kΩ, R_{57} equals 47 kΩ, R_{59} equals 12 kΩ, and C_{16} equals 0.022 μF. Given these values, determine the positive and negative peak amplitudes and the frequency of the triangle wave output at point C.
7. Assume that Hall-effect sensors within a servomotor are producing commutation signals as shown for waveforms A, B, and C in Figure 14–11. If the difference in time between any pair of these commutation signals is 15.6 msec, what is the frequency of these signals? What must be the velocity of the servomotor in rpm?
8. What is the purpose of the Tach Monitor signal in Figure 14–6? What is the source of this signal within the servocontroller? How does the amplitude of the Tach Monitor signal vary in accordance with servomotor rpm?
9. What is the purpose of the IFBMAG signal in Figure 14–7? Explain in detail how this signal is produced. Also explain how the servocontroller is able to detect the amplitude of all three phase currents, although it directly monitors only the current flow through the ϕX and ϕY servomotor windings.
10. For the servomotor temperature-sensing circuitry in Figure 14–14, assume that the user of the servocontroller has replaced R_1 with a 150-kΩ resistor. What must be the value of R_{PTC} for the thermistor within the servomotor if the trip-out temperature is to occur when 3 V is attained at the inverting input to the comparator?

14–2   ## PROGRAMMING AND OPERATION OF
MICROCOMPUTER-BASED SERVOCONTROLLERS

Now that the operation of the analog portion of a commercially manufactured servocontroller has been analyzed, the programming and operation of the digital control section of this device are investigated. As mentioned in the previous section, a servocontroller may be programmed in a manner similar to a modern PLC. A language used to program the types of servocontroller shown in Figure 14–3 is **PacSci Stepper BASIC™,** a high-level, user-friendly language similar in structure to the BASIC commonly used in PC programming. (The PacSci Stepper BASIC™ Version 3.6 Command Summary is contained in Appendix C.) User program routines, called **motion profiles,** are conveniently entered into the servocontroller, either directly from a hand-held terminal or from the keyboard of a host PC. The servocontroller may be programmed to operate in either the **immediate mode** or the **stand-alone mode.** Programming for immediate mode operation allows the servocontroller to be constantly monitored and controlled by another device. Thus, when functioning within a large piece of industrial machinery, where there may be several PLCs and/or embedded microcomputers answering to a system control computer, the servocontroller would operate in the immediate mode. Programming in the stand-alone mode allows for constantly independent operation of the servocontroller. Thus, in a smaller machine tool application, where just one servocontroller and servomotor are required, stand-alone programming is possible. At this point, we should stress that it is not the author's intention to make the student a skilled programmer in Stepper BASIC. Rather, the objective of this section is to provide just enough programming details to give the student a basic insight into the operation of computer-controlled machinery.

Assume a given servocontroller responds to nine active-low inputs. Six of these input lines are dedicated; three others are programmable by the user. The six dedicated inputs, which are explained in detail momentarily, are **Mechanical Home Switch** (input 1), **Jog + (input 5), Jog − (input 6), Start** (input 7), **Stop** (input 8), and **Enable.** The three programmable input lines are inputs 2, 3, and 4. Inputs 1 through 8 (all discrete inputs except Enable) are read as flag conditions **INP1** through **INP8,** respectively. Assume these inputs are sampled by the servocontroller every 1.024 msec. INP5 through INP8 are called **predefined** input flags.

If, as part of the initialization of the servocontroller, the **PREDEF.INP** variable is set to 1, INP5 through INP8 function as Jog +, Jog −, Start, and Stop, respectively. If the PREDEF.INP variable is set to 0, then INP5 through INP8 become user-programmable inputs. For example, assume during the operation of the servocontroller that INP5 through INP8 are always to operate as Jog +, Jog −, Start, and Stop. In the immediate programming of the servocontroller, a *PREDEF.INP = 1* statement would be made. The command **SAVEVAR** would then follow, allowing the servocontroller to retain this information in its nonvolatile memory. If inputs 1 through 8 are to function as programmable inputs, then they may be read individually as flag conditions or as an 8-bit variable. For example, assume that during execution of the user program, a logic low at input 7 indicates that an external device is ready for the servomotor to begin execution of a routine that begins on line 180. As shown here, the servocontroller could be placed in a wait state, requiring the arrival of a logic low at input 7 to resume program execution:

```
10      IF INP7 = 0 GOTO 180
20      GOTO 10
```

If a logic high is present at input 7, then the GOTO command is ignored. If a logic low is present at this input, then, within the 1.024-msec scan time, the program jumps to line 180 and resumes execution at that point.

When all eight discrete inputs of the servocontroller are made programmable, as a result of a *PREDEF.INP = 0* statement, the **INPUTS** variable can be used. This allows all eight discrete input lines to be read as an 8-bit variable. For some servocontrollers, this variable is treated as a **fixed point integer** with a range extending from 0 through 255. In forming this integer, input 8 becomes the most significant bit (MSB), and input 1 becomes the least significant bit (LSB). Consider the following program example:

```
10    IF INPUTS = 234 GOTO 130
20    GOTO 130
```

Here, discrete inputs 1 through 8 must, respectively, form the binary number 11101010 for the program to escape from the wait loop and resume execution at line 130.

The Enable input allows the servomotor to be enabled or disabled from a remote location. A logic low at the Enable input of the servocontroller sets its read-only **ENABLED** flag to a logic high. The monitor program of the servocontroller reads this flag condition as part of its own initialization routine. A logic high for the ENABLED flag is a prerequisite for activation of the servomotor.

The Mechanical Home Switch input of the servocontroller allows that device to know when the servomotor motor has moved a load, possibly a tool or workpiece, into its mechanical home position. Arriving at input 1, this signal could be produced by a mechanical limit switch, the detector circuitry of a light barrier, or possibly a magnetic proximity sensor.

Since these input devices may have either active-high or active-low outputs, Stepper BASIC allows the user to define a **HOME.ACTIVE** flag according to the active state of the input signal. For example, assume that input 1 of the servocontroller is connected to a magnetic proximity sensor with an active-high output. As part of the initialization portion of the user program, the statement *HOME.ACTIVE = 1* could be followed by the command SAVEVAR. The first step defines the Mechanical Home Switch input as having an active-high condition; the second step allows the servocontroller to retain this information permanently in its nonvolatile memory.

Once the HOME.ACTIVE flag condition is defined, the **SEEK.HOME** statement can be placed within the user program to tell the servomotor to place its load in the home position. As this command is being executed, the servocontroller drives the servomotor until it senses an active transition at input 1. Prior to using the SEEK.HOME command, both direction and speed of servomotor rotation, as well as acceleration rate, must be defined. Direction is established by the setting of the **DIR** flag. The statement *DIR = 0* defines clockwise rotation of the servomotor, while *DIR = 1* defines counterclockwise motion. Speed of rotation is defined by the **RUN.SPEED** variable. A typical range of the speed control variable could extend from 0.01 to 10,000 rpm, with a resolution of 0.01 rpm. Stepper BASIC allows this speed information to be entered directly into the program. For example, the statement *RUN.SPEED = 1235* defines motor speed as 1235 rpm.

As the mechanical load approaches its home position, the speed of the servomotor should gradually decrease. This deceleration is necessary to prevent equipment damage that could result from sudden starting and stopping of the servomotor. Stepper BASIC allows both acceleration and deceleration of the servomotor to be defined within the user program. Servomotor acceleration and deceleration are defined in rpm per second. With a typical servocontroller, rates of acceleration and deceleration may be varied from 3 to 10,000,000 rpm/sec, with a resolution of 3 rpm/sec. In Stepper BASIC, both acceleration and deceleration rates may be set by defining the **ACCEL.RATE** variable. For example, consider the program statement *ACCEL.RATE = 150,000*. If this statement precedes a SEEK.HOME statement in the program sequence, the servomotor decelerates at a rate of 150,000 rpm/sec as it moves the load toward its home position.

The following short program sequence illustrates a motion profile for bringing a mechanical load to its home position. Here, the servomotor begins to decelerate from a running speed of 1225 rpm, at a rate of 175,000 rpm/sec, as it brings a mechanical load to its home position.

```
10     RUN.SPEED = 1225
20     ACCEL.RATE = 175,000
30     SEEK.HOME
```

Assume that, when the PREDEF.INP flag has been set to 1, a logic low at the Jog + input of the servocontroller causes the servomotor to start turning counterclockwise at a speed defined by the **JOG.SPEED** variable. A logic low condition at the Jog − input results in clockwise rotation, the speed of which is again defined by the JOG.SPEED variable. Both the Jog + and Jog − functions are disabled while the servocontroller is executing a program. Assume, for example, the statement *JOG.SPEED = 1545* is made as part of the initialization of the servocontroller. When the servocontroller is enabled and no program is running, a logic low at input 5 causes the servomotor to start turning counterclockwise at 1545 rpm. A low at input 6 would initiate clockwise rotation at the same speed.

When the Start input is brought to a logic low, the **RMT.START** (remote start) function of the servocontroller may be activated in one of two ways, providing the PREDEF.INP variable has been set to a logic high. Assume a *RMT.START = 0* statement was made prior to the arrival of a logic low at the Start input. On the falling edge of the Start signal, the servocontroller begins to execute one of four possible GO statements. These GO instructions, which include **GO.ABS, GO.HOME, GO.INCR,** and **GO.VEL,** are explained in detail momentarily. If a *RMT.START = 1* statement was made prior to the negative transition of the Start signal, the servocontroller performs a RUN function, beginning execution of the program in RAM. It is possible for the servocontroller to begin executing the program in memory at the instant power is applied, regardless of the conditions of the PREDEF.INP and INP7 variables. This is accomplished by entering a RMT.START = 2 statement and saving it in the nonvolatile memory of the servocontroller via a SAVEVAR command.

When the Stop input to the servocontroller is brought to a logic low, the program currently being executed by the servocontroller is halted and motor movement is stopped immediately. Thus, the Stop input allows an external control source to interrupt servomotor operation, possibly as part of an emergency shutdown operation.

Assume the given servocontroller has two dedicated discrete outputs, **Fault** (output 8) and **Motor Moving** (output 7), and six discrete output lines (outputs 1 through 6) that can be programmed by the user. For outputs 7 and 8 to assume their predefined tasks of conveying the Fault and Motor Moving signals to an external device, the **PREDEF.OUT** variable must be set to a logic high. This is accomplished through the statement *PREDEF.OUT = 1.* Following this statement with a SAVEVAR command allows outputs 7 and 8 to retain their predefined functions after power is removed from the servocontroller. If the statement *PREDEF.OUT = 0* is made during the initialization of the servocontroller, then outputs 7 and 8 function as programmable outputs, identical to outputs 1 through 6. All eight of these discrete outputs are likely to be open-collector, functioning as current sinks for external loads.

During servomotor operation, the Fault output remains at a logic low, as long as no malfunctions have been detected by the diagnostics within either the servocontroller or the servomotor. A positive transition at the Fault output terminal indicates that a malfunction has been detected by the servocontroller. Stepper BASIC contains a **FAULTCODE** variable. FAULTCODE remains 0 if no faults are detected. However, FAULTCODE toggles to 1 when a malfunction occurs. For some forms of servocontroller, a **PRINT FAULTCODE** statement in the user program allows the monitor of a host PC to display the malfunction detected by the servocontroller diagnostics.

The Motor Moving output switches on whenever the servomotor begins rotating faster than 2 rpm. Thus, output 7 could be connected to an input port line of the system control computer, indicating that the servomotor is operating. As part of a system startup test, the control computer could send a jog + or jog − signal to the servocontroller and wait for a negative transition at output 7. Failure of this signal to appear would indicate a malfunction within either the servomotor or the servocontroller.

Assume the six programmable output lines of the servocontroller are accessed from within a BASIC program. If outputs 7 and 8 have been predefined as Fault and Motor Moving, then outputs 1 through 6 are controlled individually by defining the logic states of variables OUT1 through OUT8. For example, the statement *OUT5 = 1* switches on the open-collector transistor at output 5, bringing that terminal to a logic low. An *OUT5 = 0* statement switches off the open-collector transistor at output 5. If the PREDEF.OUT variable has been set to a logic low, outputs 1 through 8 may be defined as an 8-bit word using the **OUTPUTS** variable.

When all eight discrete outputs of the servocontroller are made programmable, as a result of a *PREDEF.OUT = 0* statement, the OUTPUTS variable may be used. This allows all eight discrete output lines to be used effectively as a single output port. The OUTPUTS variable could be treated as a fixed point integer with a range extending from 0 through 255. As with the INPUTS integer, output 8 becomes the MSB, and output 1 becomes the LSB. Assume, for example, that an *OUTPUTS = 187* statement occurs during a program. As a result, outputs 8 through 1, respectively, form the binary number 10111011.

Besides the Fault and Motor Moving signals, the servocontroller could provide two analog outputs indicating the status of the servo loop. These signals were already discussed during the analysis of the servocontroller's cascade control circuitry in Figures 14–6 and 14–7. The **Motor Current Monitor** output allows an external device, such as a piece of test equipment, to measure a voltage analogous to the IFBMAG signal of Figure 14–7. Representing the average value of current flow to the three phase windings of the servomotor, the Motor Current Monitor may vary between ±4.4 V, at a rate of 0.37 V/A. The **Tach Monitor** output is equivalent to the Tach Monitor signal sent to the velocity error amplifier in Figure 14–6. This signal is produced by the servocontroller's RDC circuitry and varies in proportion to motor speed at a rate of 0.56 V/1000 rpm. Like the Motor Current Monitor output, the Tach Monitor output may be connected to test equipment to check the performance of the servo loop.

In some industrial applications of brushless dc servomotors, such as positioning of a machine tool (or a workpiece being processed by that device), it is necessary to control the exact position of the servomotor shaft. Such precise absolute positioning of the servomotor shaft is possible if the servocontroller contains a high-resolution RDC. The RDC within a typical servocontroller could allow a resolution of 2^{12} (4096) for servomotor shaft position. Thus, the device would be capable of monitoring and controlling servomotor shaft position with an accuracy of 1/4096 of the 360° of shaft rotation. The resolution of the RDC within a servocontroller is usually expressed in terms of the minute (or arcmin), equivalent to 1/60 of a degree. Thus, the resolution is approximately 5.3 minutes, being determined as follows:

$$\text{RDC resolution} = \frac{1}{2^{12}} \times 360° \times 60 \text{ minutes} \cong 5.3 \text{ arcmin} \qquad (14–7)$$

Whereas the resolver attached to the shaft of the servomotor is a purely analog device, producing a continuous-state output, the digital output of the RDC within the servocontroller is a 10-bit binary variable. Thus, as determined earlier, the incremental change (often referred to simply as the **step**) represented by the RDC output can be no more accurate than ±5.3 arcmin. A servocontroller could allow for position control within a range of ±100,000,000 steps, with an accuracy of 15 arcmin. This range may be expressed in terms of servomotor shaft revolutions and is determined as follows:

$$\text{Range} = \frac{100 \times 10^6}{2^{12}} = 24,414 \text{ revolutions} \tag{14–8}$$

Ideally, as the preceding calculation indicates, the servocontroller would be able to control servomotor shaft position within a range of over ±24.4 k revolutions, with an accuracy of 5.3 arcmin. In reality, due to accumulated error, guaranteed accuracy might be only 15 arcmin. However, considering this is an error factor of only one-quarter of a degree, the absolute positioning control capability of the servocontroller is still quite impressive!

We now survey the BASIC statements involved in absolute positioning of a servomotor. These statements involve tracking the position of the servomotor in increments, or steps, of 5.3 arcmin. One of the first tasks to be performed in absolute positioning is the establishment of the **electrical home position,** which is a resolver position considered to represent 0°. Normally, as part of the servocontroller initialization process, the electrical home position is made equivalent to the mechanical home position established by some form of proximity sensor. With the arrival of the mechanical load at the home position, indicated by an active transition at the Mechanical Home Switch input to the servocontroller, the servocontroller CPU establishes a 0 value for a read-only variable called **POSITION.** This variable, which represents the *actual* servomotor shaft position, is updated in increments of 5.3 arcmin within a range of ±134,217,728 steps. The **HMPOS. OFFSET** variable allows an offset to be established between the electrical and mechanical home positions. This offset is defined as an integer with a range extending from 0 to ±4,096,000. This integer represents the number of 5.3-arcmin steps between the 0 position point and the offset point. A positive integer represents a clockwise offset to the 0 position point, while a negative integer defines an offset with a counterclockwise reference.

The **POS.COMMAND** variable represents the most current position command for the servomotor shaft, as opposed to the POSITION variable, which represents the actual shaft position. Ideally the POS.COMMAND and POSITION variables are the same value. However, due to the propagation delay of the servo loop, these two variables can differ significantly. A read-only variable called **POS.ERROR,** which is calculated by the servocontroller monitor program, represents the disparity (in 5.3-arcmin steps) between POS.COMMAND and POSITION. A maximum allowable error (**MAX.POS.ERROR**) may be defined within the user program. The MAX.POS.ERROR is equivalent to the POSITION VERIFY DEADBAND in Appendix C. A **POS.ERROR.GOSUB** (similar to POS.VERIFY.JUMP in Appendix C) command can be used to point to a subroutine that is accessed in the event the POS.ERROR exceeds the defined MAX.POS.ERROR. For example, assume the POS.ERROR.GOSUB = 680 statement is made as part of the initialization of the user program. In the event that POS.ERROR is greater than MAX.POS. ERROR, execution of the user program is jumped to a subroutine beginning at line 680. Depending on the type of program being executed, this subroutine takes the necessary steps to minimize damage or waste that could result from the position error problem.

Two important BASIC programming statements used in absolute positioning of the servomotor shaft are **GO.ABS** and **TARGET.POS.** A GO.ABS statement causes the servomotor to increase to a velocity specified by the most recent RUN.SPEED statement. The rate at which the motor speed increases is defined by the latest ACCEL.RATE statement. The TARGET.POS statement defines the absolute position to which the servomotor shaft is to move, with respect to the electrical home position. The direction of this movement cannot be defined by the user. Rather, direction of servomotor shaft rotation is determined by the servocontroller CPU as it compares the POS.COMMAND variable to the TARGET.POS variable. If the TARGET.POS variable is greater than POS.COMMAND, resultant servomotor rotation is clockwise. If TARGET.POS is less than POS.COMMAND, shaft rotation is counterclockwise.

As a sample application of the program statements just described, consider the following:

```
10    RUN.SPEED = 1275
20    ACCEL.RATE = 175000
30    TARGET.POS = 27623
40    GO.ABS
```

In this example, with the execution of the GO.ABS statement, the servomotor accelerates toward a velocity of 1275 rpm, at a rate of 175,000 rpm/sec. It then turns at a steady velocity of 1275 rpm as it moves toward the target position. As the servomotor approaches the target position, which is 27,623 resolver steps clockwise from the electrical home position, it decelerates at a rate of 175,000 rpm/sec. Again, we stress that the direction of servomotor rotation depends on the current POS.COMMAND value. For example, assume that, prior to the execution of the above routine, that POS.COMMAND equals $-16,527$. The amount and direction of servomotor travel required to attain the target position is determined as follows:

$$\text{Travel} = \text{TARGET.POS} - \text{POS.COMMAND}$$
$$= 27,623 \text{ steps} - -16,527 \text{ steps} = 44,150 \text{ steps} \tag{14–9}$$

This positive result indicates that the servomotor shaft must rotate clockwise 44,150 resolver steps to attain the desired target position. The motion profiles generated by this type of absolute positioning program, as represented in Figure 14–15, are trapezoidal. Between t_0 and t_1 in Figure 14–15(a), as represented by the positive-going slope, the

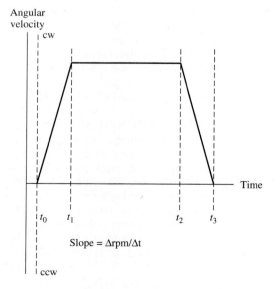

(a) Motion profile for cw rotation of servomotor shaft

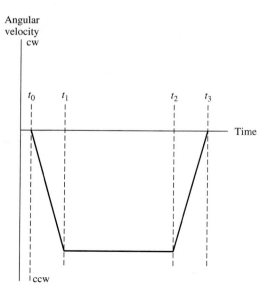

(b) Motion profile for ccw rotation of servomotor shaft

Figure 14–15

servomotor accelerates toward the desired travel speed of 1275 rpm. The slope, Δrpm/Δt, is a function of the ACCEL.RATE variable defined at line 20. Between t_1 and t_2, as represented by the top of the trapezoid, the motor shaft travels at a constant rate of 1275 rpm. Between t_2 and t_3, deceleration occurs, with Δrpm/Δt being the same as for t_0–t_1.

Now assume that, for the same program example, the present value of POS.COMMAND is 49,237. The amount and direction of required servomotor travel is again determined as:

$$\text{Travel} = 27{,}623 \text{ steps} - 49{,}237 \text{ steps} = -21{,}614 \text{ steps}$$

This negative result indicates that the servomotor shaft must rotate counterclockwise 21,614 resolver steps to arrive at the target position. The motion profile representing this shaft movement is seen in Figure 14–15(b).

The correlation between the operation of the servo amplifier within the servocontroller and the motion profiles in Figure 14–15 is interesting. During execution of one of the absolute positioning routines just described, the signal seen at point A in Figure 14–6 would closely resemble one of the motion profiles shown in Figure 14–16. (This signal could be easily captured and displayed by a storage oscilloscope.) For example, to produce the clockwise motion represented in Figure 14–15(a), the CPU of the SC450 outputs a series of increasing binary numbers to its position loop D/A converter. This results in a negative-going slope at point A in Figure 14–6. Ideally, the slope of this ramp, seen between t_0 and t_1 in Figure 14–16, produces the Δrpm/Δt defined by the ACCEL.RATE variable in the motion profile routine. When, through processing the output of the RDC, the CPU senses the servomotor is rotating at the velocity specified by the RUN.SPEED variable, it sustains the voltage at a nearly constant dc level, holding motor speed at the desired rpm. This action is seen between t_1 and t_2 in Figure 14–16. The deceleration slope, as shown between t_2 and t_3, becomes the inverse of the acceleration slope. This is accomplished as the CPU reverses the sequence of control numbers at the input of the position loop D/A converter from that which occurred between t_0 and t_1.

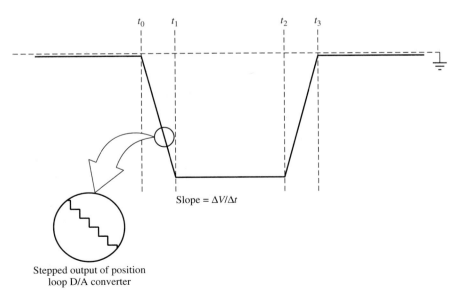

Slope = $\Delta V/\Delta t$

Stepped output of position
loop D/A converter

Figure 14–16
VEL loop gain signal corresponding to motion profile in Figure 14–15

An application of the absolute positioning process just described is illustrated in Figure 14–17. Here, a servocontroller and a medium-power brushless dc servomotor are used to control the lateral position of a machine tool located above a workpiece. For enhanced torque capability, the shaft of the servomotor is connected to a gear reducer. The actuator contains a toothed belt that turns on sprockets contained in two spline boxes. The gear reducer is securely fastened to the left-hand spline box of the actuator, with its splined shaft fitting inside the left-hand sprocket. The machine tool is securely fastened to the actuator. Thus, the tool is able to move toward the right-hand spline box as the motor turns clockwise. Counterclockwise servomotor rotation results in movement of the tool toward the left-hand spline box.

The home position of the tool being moved along the actuator is assumed to be in the exact middle of its lateral range. A magnetic proximity sensor and an exciter cam are used to establish this middle point as the mechanical home position for the servomotor. The exciter cam, resembling a simple thumbtack, is firmly pressed into the toothed belt

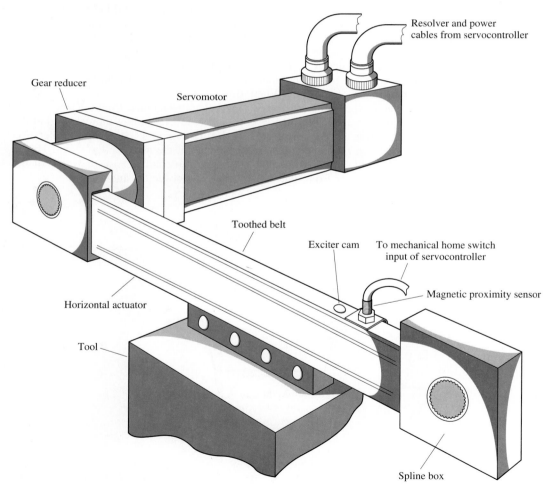

Figure 14–17
Application of a servomotor in a tool positioning system

of the actuator. The proximity sensor is attached to the actuator at a point where, with the tool in its home position, the exciter cam is directly below the sensor tip. The active-high output of the proximity sensor is connected to input 1 (Mechanical Home Switch) of the servocontroller. As explained earlier, by defining the HOME.ACTIVE flag as a logic one, the servocontroller responds to a positive-going transition at input 1.

Example 14–3

For the servocontroller application illustrated in Figure 14–17, assume the gear ratio from the servomotor shaft to the gear reducer is $6:1$. To position the tool in the desired position over the workpiece during a given manufacturing application, assume the tool must be moved toward the left-hand spline box. It has already been determined that, to achieve this motion, the gear reducer must perform four complete rotations plus an additional $167°$. It has also been determined that the constant rotating speed for the gear reducer during this movement must be 245 rpm. Finally, the acceleration/deceleration rate of the servomotor is to be 185,000 rpm/sec. Given these specifications, write a short Stepper BASIC routine that will accomplish the desired move.

Solution:

Step 1: Determine the required servomotor travel in degrees. Since the gear ratio is $6:1$, the servomotor must rotate six times more than the gear reducer. To accomplish this move:

$$\text{Servomotor travel} = 6 \times (4 \times 360° + 167°) = 9642°$$

Step 2: Convert servomotor travel in degrees to resolver steps:

$$\text{Servomotor travel} = 4096 \times \frac{9642°}{360°} = 109{,}705 \text{ resolver steps}$$

Since servomotor travel must be counterclockwise to accomplish the required move, the actual TARGET.POS variable becomes $-109{,}705$.

Step 3: Determine the required motor travel velocity. Due to the gear ratio of $6:1$, the servomotor shaft must rotate six times faster than the gear reducer. Thus,

$$\text{Servomotor speed} = 6 \times 245 \text{ rpm} = 1470 \text{ rpm}$$

Step 4: Write the motion basic routine to accomplish the required move.

```
10      RUN.SPEED = 1470
20      ACCEL.RATE = 185000
30      TARGET.POS = -109705
40      GO.ABS
```

The servocontroller BASIC statements covered thus far have primarily involved the positioning of the servomotor. We now introduce other programming statements that are concerned with the operation of the servo amplifier within the servocontroller. The variables that define these operating parameters include **DACGAIN, FFGAIN, PLGAIN,** and **PLIGAIN.**

The DACGAIN variable defines the gain of the position loop D/A converter as an integer between 0 and 255. (Recall that this D/A converter translates binary control variables from the CPU to the analog $\pm$VEL CMD signals that drive op amp 1 in Figure 14–6.) Typically the DACGAIN variable is set to a midpoint value of around 120. Once defined, through a statement such as *DACGAIN = 122,* this variable may be saved in the nonvolatile memory of the servocontroller through a subsequent SAVEVAR command. The servocontroller FFGAIN variable allows for user definition of the servocontroller's **feed-forward gain factor.** Like DACGAIN, FFGAIN may be defined as an integer between 0 and 255. Typically, FFGAIN is set to a midpoint value of about 120. Setting FFGAIN to a relatively high value tends to reduce the magnitude of the position error. (Recall that position error, read as the POS.ERROR variable, is the difference between the most recent POS.COMMAND reading and the actual servomotor shaft position.) However, too high a setting of the FFGAIN variable is likely to introduce an overshoot problem into the servo loop. Lowering the value of FFGAIN minimizes the possibility of overshoots. However, too low a value of FFGAIN tends to increase the position error. The feed-forward gain factor is established through a programming statement such as *FFGAIN = 127* and saved by a subsequent SAVEVAR command.

    Recall from the previous chapter that the three basic and three composite modes of continuous-state control may be implemented through use of microcomputer algorithms. Within these algorithms, factors such as proportional and integral gain are encoded as integers. As represented in Figure 14–5, if a servocontroller operates on the principles of PI control, both the **position loop proportional gain** and **position loop integral gain** are made definable. Position loop proportional gain is defined through the PLGAIN variable, and position loop integral gain is defined by specifying the PLIGAIN variable. Both of these variables are defined as integers between 0 and 255 and may be saved in the nonvolatile memory through use of the SAVEVAR command.

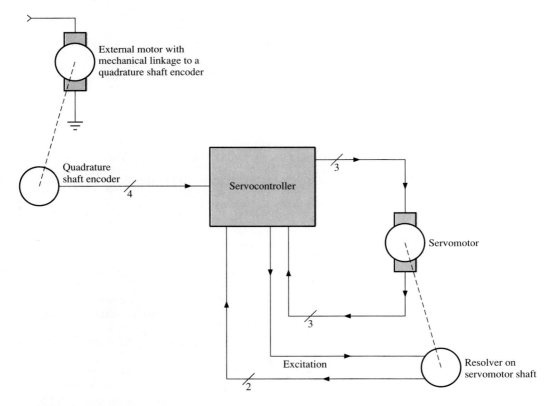

Figure 14–18
Servocontroller configuration for an electronic gearing application

Typical settings of both the PLGAIN and PLIGAIN would be near the midpoint value of 120. Increasing the value of PLGAIN would enhance the ability of the servocontroller to initiate corrective action in proportion to the magnitude of the position error. However, an increase in PLGAIN would also result in a narrower proportional band. Increasing the value of PLIGAIN enhances the ability of the servocontroller to correct a persistent position error.

An important capability of the modern servocontroller is **electronic gearing,** which greatly reduces the need for mechanical linkage within a servo system. When enabled, this feature allows the servocontroller to monitor the signal produced by a shaft encoder within a drive system that is *not* linked mechanically to the servomotor being governed by that servocontroller. This servocontroller processes the external encoder signal and causes the servomotor under its direct control to rotate at a precise ratio of the external encoder speed, thus simulating the action of a mechanical gearhead. This application of the servocontroller's electronic gearing capability is illustrated in Figure 14–18. Here, a shaft encoder is mounted on the shaft of a pulley, which is part of a belt drive system powered by a 3-φ ac induction motor. This servo system configuration allows servomotor velocity to adapt to any changes in belt speed that could result from slipping or changes in load condition.

We assume here that the external encoder feeding the servocontroller is a **quadrature device.** The quadrature shaft encoder has four digital outputs, consisting of A, $\overline{A}$, B, and

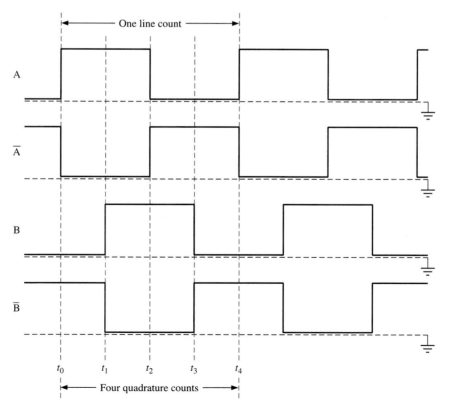

Figure 14–19
Output signals produced by a quadrature shaft encoder

Table 14–1	t_0-t_1	A $\overline{\text{A}}$ B $\overline{\text{B}}$ = 1001
Binary outputs occurring within a line	t_1-t_2	A $\overline{\text{A}}$ B $\overline{\text{B}}$ = 1010
count of a quadrature shaft encoder	t_2-t_3	A $\overline{\text{A}}$ B $\overline{\text{B}}$ = 0110
	t_3-t_4	A $\overline{\text{A}}$ B $\overline{\text{B}}$ = 0101

B. As illustrated in Figure 14–19 and Table 14–1, for every **line count** of the quadrature encoder (from t_0 through t_4), four incremental changes occur in the logic conditions of A, $\overline{\text{A}}$, B, and $\overline{\text{B}}$. Thus, for a quadrature shaft encoder with a line count of 1024, there are 4096 **quadrature counts.** Since the servocontroller receives all four outputs of the quadrature shaft encoder, it is able to track the angular position of the encoder in quadrature counts.

The Stepper BASIC variable used to track the position of the encoder shaft is **ENCDR.POS,** which may be an integer from 0 to $\pm 2,147,483,648$. Once set to 0 by the *ENCDR.POS = 0* statement, the ENCDR.POS variable increments with each quadrature step of the external shaft encoder. The status of the ENCDR.POS variable is read by the CPU of the servocontroller every 1.024 msec.

The electronic gearing capability of the servocontroller is enabled by setting the **GEARING** variable to 1. The statement *GEARING = 0* disables the electronic gearing capability. Regardless of the logic condition of the GEARING variable, the servocontroller automatically scans its external encoder inputs and reads the ENCDR.POS variable every 1.024 msec.

Example 14–4

Assume the A, $\overline{\text{A}}$, B, and $\overline{\text{B}}$ of a quadrature encoder with 256 line counts are connected to a servocontroller. After the programming statement *ENCDR.POS = 0* is made, the ENCDR.POS variable attains a value of 147,932. How many revolutions of shaft encoder travel does this variable represent? How many minutes does each increment of the ENCDR.POS variable represent?

Solution:

Step 1: Determine the number of quadrature steps that occur during each revolution of the external shaft encoder:

$$\text{Quadrature steps} = 4 \times 256 = 1024$$

Step 2: Determine the travel of the shaft encoder between ENCDR.POS = 0 and ENCDR.POS = 147932.

$$\text{Travel} = \frac{147,932}{1024} = 144.465 \text{ rev} = 144 \text{ rev} + 167.4°$$

Step 3: Determine the resolution in minutes for ENCDR.POS:

$$\frac{360°}{1024} \cong 0.3516°$$

$$0.3516° \times 60 \cong = 21.1 \text{ arcmin}$$

There are two methods of implementing the electronic gear ratio capability of the servocontroller: through use of either the **RATIO** or the **PULSE.RATIO** variable. The RATIO variable allows the desired gear ratio to be expressed as the ratio of servomotor shaft movement to encoder shaft movement. When using this variable, it is necessary to inform the CPU of the line count of the external encoder. This is done through definition of the **ENCODER** variable. RATIO may be defined as a floating point integer between $\pm$xx.xxxx; ENCODER can range in value from 200 to 10,000.

Example 14–5

Assume that, during an electronic gearing application, a servocontroller is to monitor the quadrature outputs of a shaft encoder having a line count of 256. The shaft of the servomotor is to turn 27 times for every 39 revolutions of the external encoder. Using the RATIO statement, write a brief BASIC program that establishes this electronic gearing relationship. Also determine the number of ENCDR.POS states that occur within an encoder revolution.

Solution:

Step 1: Determine the electronic gear ratio:

$$N = \frac{\text{Servomotor rev.}}{\text{Encoder rev.}} = \frac{27}{39} = 0.6923$$

Step 2: Write the Motion Basic program:

```
10      ENCODER = 256
20      RATIO = 0.6923
30      GEARING = 1
```

Step 3: Determine the number of ENCDR.POS states occurring within an encoder revolution. For the quadrature encoder, the number of ENCDR.POS states per cycle equals the quadrature count. Thus,

$$\text{ENCDR.POS states per cycle} = 4 \times 256 = 1024$$

The PULSE.RATIO variable can also be used to define an electronic gear ratio. This variable establishes a ratio of motor shaft resolver steps to external encoder resolver steps. As an example of the use of the PULSE.RATIO variable, consider the expression *PULSE.RATIO = 24 : 1*. As a result of this expression, the servomotor shaft moves 24 resolver steps for each quadrature step of an external encoder. The BASIC example that follows is a typical PULSE.RATIO program application. Note that the first statement of the program establishes a starting point for the ENCDR.POS variable. Next, a PULSE: RATIO is defined such that the servomotor travels three resolver steps for every seven quadrature steps of the external shaft encoder. After the electronic gearing is enabled, a limit is defined for the duration of the electronic gearing algorithm. Meanwhile, the program is caught in a loop at step 40 until that limit value of the ENCDR.POS variable is attained. At the point where the servocontroller senses that the ENCDR.POS variable exceeds 15.63×10^6, the routine stops and electronic gearing is disabled.

```
10      ENCDR.POS = 0
20      PULSE.RATIO = 3:7
30      GEARING = 1
40      IF ENCDR.POS < 15630000 THEN 40
50      GEARING = 0
```

SECTION REVIEW 14–2

1. Identify the six predefined inputs of a typical servocontroller. Describe the exact function of each of these inputs. What initial programming steps must be performed to ensure that these inputs assume their predefined functions every time power is applied to the servocontroller?

2. Assume that a proximity sensor with an active-high output is to be connected to the Mechanical Home Switch input of the servocontroller. What initial programming steps should be performed to ensure that the servocontroller responds correctly to the proximity sensor?

3. What are the two predefined outputs of the servocontroller? Describe the exact function of each of these outputs. What initial programming steps must be taken to ensure that these outputs assume their predefined functions?

4. Identify four BASIC variables that directly affect the operation of the servo amplifier section of the servocontroller. Briefly describe the purpose of each of these variables.

5. What problem could result if the FFGAIN variable of the servocontroller is set at too high a value?

6. When the TARGET.POS variable and the GO.ABS statement are used together in a BASIC program, how does the CPU of the servocontroller determine the required direction of servomotor rotation?

7. What two BASIC variables must be defined before the occurrence of a GO.VEL statement in a user program?

8. For the following BASIC program routine, assume the present POS.COMMAND value is −86,527. During execution of this routine, what is the direction of servomotor shaft rotation? How many resolver steps does this motion represent?

```
10      RUN.SPEED = 1458
20      ACCEL.RATE = 197000
30      TARGET.POS = 127952
40      GO.ABS
```

9. True or false?: The motion profile generated by the program routine in Review Problem 8 would be trapezoidal.

10. A quadrature shaft encoder with 512 line steps is connected to a servocontroller as shown in Figure 14–18. How many states of the ENCDR.POS variable would occur during one revolution of the shaft encoder? How many minutes does each increment of the ENCDR.POS variable represent?

11. Using the RATIO variable, write a BASIC routine that causes a servomotor to rotate 19 times for every 16 turns of a 1024 line external quadrature shaft encoder. The electronic gearing capability of the servocontroller should then be disabled after 3,927 revolutions of the external shaft encoder.

12. Using the PULSE.RATIO variable, write a program routine that allows a servomotor to move five resolver steps for every quadrature step of an external shaft encoder. Assume the quadrature count of the shaft encoder to be 2048. This routine should terminate as the ENCDR.POS variable approaches 327,680. (When writing this routine, remember which parameter is represented by the ENCODER variable.)

14–3   INTRODUCTION TO INDUSTRIAL ROBOTICS

The industrial robot emerged as a distinctive automation device during the 1960s. Whereas an automated machine tool is usually confined to one task or one particular type of task, an industrial robot is far more versatile. The industrial robot, typically in the form of a **mechanical manipulator,** can be programmed to perform a wide variety of tasks. The field of robotics enjoys wide popularity, due to the **anthropomorphism** (attribution of human characteristics) associated with robots. Although industrial robots do perform human-like tasks involving gripping, lifting, and moving of parts and tools, present application of mechanical manipulators is limited to relatively simple, repetitive tasks. Compared to the most advanced forms of industrial machine tools, the mechanical manipulator is, as yet, not capable of performing high precision manufacturing procedures. Also, current forms of mechanical manipulators encounter difficulty in the performance of tasks requiring applications of active force.

The study of robotics synthesizes several broad fields of study, including mechanical manipulation, locomotion, computer vision, and artificial intelligence. Control of the operation of mechanical manipulators encompasses the fields of **kinematics** and **dynamics.** Kinematics is the science of motion, without regard to the forces that cause it. Dynamics, complementing the field of kinematics, concerns the study of forces that produce motion. Kinematics involves geometric analysis of the movement of robotic parts. As seen later, kinematics, in itself, becomes quite complex. The dynamics of industrial robotics involves the precise control of actuators necessary for smooth execution of the motions identified through kinematics. Industrial robotics involves applications of a variety of actuators, as well as motion and force sensors. Actuators used in industrial robotics that are already familiar include stepper motors and servomotors. Pneumatic and hydraulic systems are also used in the dynamic control of mechanical manipulators.

Figure 14–20 illustrates a typical mechanical manipulator, which consists of several rigid **links.** Although immobile (except for some degree of rotation), the base of the mechanical manipulator may be considered as the lowest link and, therefore, could be assigned the number 0. The mobile links of the robot, from the base upward, can then be assigned the numbers (1, 2, 3, . . .). The **end effector,** which is placed at the end of the chain of links, could consist of a gripper, welding torch, or electromagnet. **Joints** between the links allow relative motion of one link with respect to another. The term **lower pair** describes a joint where relative motion between a pair of links consists of two surfaces sliding over each other. Figure 14–21 illustrates six basic forms of lower pairs. Within mechanical manipulators, the **revolute** (rotary) joint is the prevalent form of lower pair. **Prismatic** (sliding) joints are also frequently used in industrial robots. Other basic forms of lower pairs include **cylindrical, planar, screw,** and **spherical.**

Figure 14–20
Typical mechanical manipulator

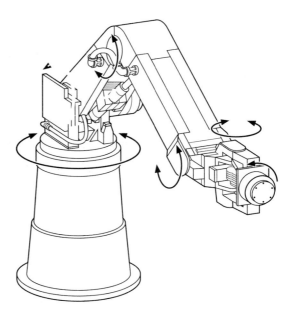

Figure 14–21
Six basic forms of lower pairs

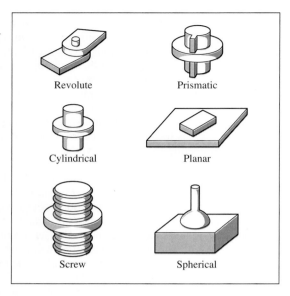

Revolute

Prismatic

Cylindrical

Planar

Screw

Spherical

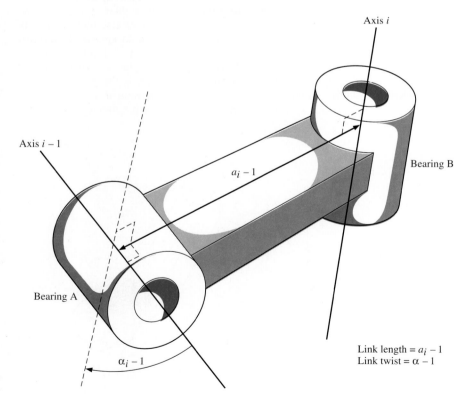

Axis i

Axis $i-1$

a_{i-1}

Bearing B

Bearing A

α_{i-1}

Link length = $a_i - 1$
Link twist = $\alpha - 1$

Figure 14–22
Parameters of a single robotic link: link length and link twist

From a purely kinematic standpoint (disregarding such mechanical concerns as shape, weight, material, and choice of joint bearings), a link is simply a rigid body that defines the relationship between neighboring joint axes. Thus, there are only two significant link parameters, **link length** and **link twist.** As shown in Figure 14–22, link length is measured along a line mutually perpendicular to both joint axes. Link twist is a measure of the relative angular positioning of the two joints at either end of a link with respect to the common perpendicular line.

Again from a kinematic standpoint, when connecting the intermediate links of the robot, two additional parameters become important. As seen in Figure 14–23, **joint offset** is the distance measured along the common axis formed at the junction of two robotic links. The degree of rotation occurring about this common axis is the **joint angle.** At a prismatic joint, joint offset (d_i) becomes variable. At a revolute joint, such as that shown in Figure 14–23, joint angle (θ_i) becomes variable while joint offset is constant.

In kinematics, the operation of mechanical manipulators may be described in terms of **degrees of freedom,** or the number of position variables that must be defined in locating all parts of the link chain. For each of the six basic forms of lower pairs shown in Figure 14–21, there is just one degree of freedom. Thus, since each joint of the mechanical manipulator normally has one degree of freedom, the number of position variables is usually equivalent to the number of joints. Typically, an industrial robot must have five or six degrees of freedom to accomplish its assigned tasks. Thus, the mechanical manipulator must have five or six joints. The joints of a mechanical manipulator are equipped with sensors that allow the robot's control circuitry to monitor accurately the relative positions of the manipulator links. At revolute joints, high-resolution absolute rotary position sensors measure **joint angle.** At prismatic joints, linear position sensors measure **joint offset.**

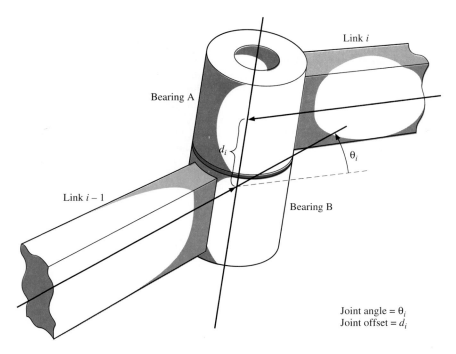

Figure 14–23
Parameters of contiguous robotic links: joint angle and joint offset

The importance of kinematics in industrial robotics becomes evident as the problems of positioning the mechanical manipulator are briefly investigated. The attributes of manipulator placement that must first be defined are **position, orientation,** and **frame.** As illustrated in Figure 14–24, a point in space may be described as the tip of a vector extending in a three-dimensional **coordinate system.** Since, during the kinematic analysis of mechanical manipulators, several coordinated systems must be dealt with simultaneously, these coordinate systems are given letter designations. Thus, for coordinate system A in Figure 14–24, the Z_A axis defines the height of the point P_A within coordinate system A. The Y_A axis defines the lateral position of P_A, while the X_A defines its forward position.

The position of a point in space, in itself, is not sufficient for defining the placement of a robotic part. For the end effector of a mechanical manipulator, it is also important to know the orientation. For example, as shown in Figure 14–25, the focal point of a gripper (the exact point where the fingers converge) is to be placed at a defined point in space. While the exact position of this point may already be specified with respect to the base coordinate system, the orientation, or tilt, of the end effector with respect to this base axis must also be determined. In Figure 14–25, the end effector has its own coordinate system, designated here as B. Whereas the focal point of the gripper in Figure 14–25 is easily determined with respect to coordinate system B, a more challenging problem arises in relating the point in coordinate system B to system A. This process is necessary since the end effector must be tilted into the desired orientation by the robot's actuators. Such complex vector problems, involving **rotation matrices,** are solved through the implementation of robotic programming languages.

The orientation and position of a robotic part together establish its **frame.** The **manipulator position** is a description of the movable **tool frame** (position and orientation of the end effector) with respect to a stationary **base frame** (position and orientation of the base, or link 0). **Forward kinematics** requires the application of static geometry to determine the placement of the end effector, given a complete set of joint angles. For such an approach to kinematics, there is a single solution for the placement of the tool frame. Most practical applications of mechanical manipulators require the process of **inverse kinematics.** Since the desired placement of the tool frame is the known variable, while the joint angles are not specified, the geometric process becomes the reverse of the forward kinematic problem. With inverse kinematics, several possible joint angle solutions might exist for achieving the desired tool frame placement. However, the programming language used to develop a given sequence of robotic movements should arrive at a

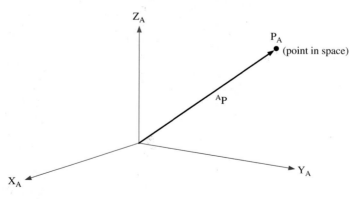

Figure 14–24
Three-dimensional coordinate system

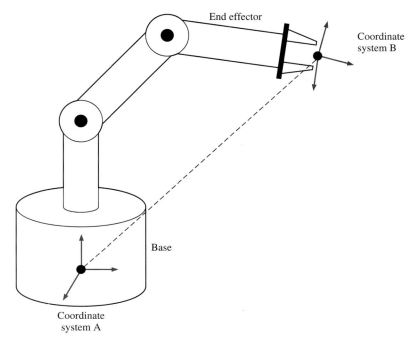

Figure 14–25
Concept of rotation matrices

kinematic solution that allows the smoothest and most efficient combination of joint motions.

In the control of mechanical manipulators, the field of dynamics becomes important in the acceleration and deceleration of the links and end effector. Dynamics involves the direct control of the mechanical manipulator's **joint actuators.** These are the electric, hydraulic, or pneumatic power devices that achieve manipulator motion. In the previous section of this chapter, the concept of generating a precise motion profile from a microcomputer-based servocontroller was introduced. Now imagine a condition where several microcomputer-controlled actuators are involved in moving an end effector at a precise and constant rate. A complex set of torque functions must be applied by the joint actuators to accomplish this task. The motion control problem becomes more challenging still when we consider that the preferred method of manipulator motion is to begin and end all joint movements smoothly and simultaneously!

The computer control of the motion process just described is referred to as **trajectory generation.** During this process, the path of the end effector is defined as a series of intermediate points called **via points.** The smooth passage of the end effector through a set of via points is referred to as a **spline.** As illustrated in Figure 14–26, assume an end effector, consisting of a gripper, is to move downward in a straight trajectory. To implement this action by the preferred method, the computer must calculate a set of joint motions. Once this task is accomplished, the computer sends commands to the dedicated control devices that govern the operation of the joint actuators.

In modern industrial robots, most of the joint actuators are controlled by servo systems. Thus, various position-sensing feedback transducers are utilized within the servomechanisms, providing link and end effector position information to the servo control devices. These controllers, which are likely to be similar in structure and operation to the servocon-

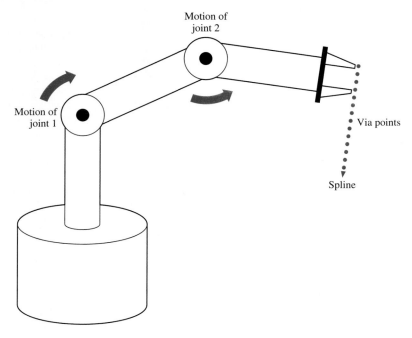

Figure 14–26
Concept of via points and splines

troller examined in the previous section, take corrective action based on the information received from position sensors. Besides position sensors, mechanical manipulators contain force-sensing transducers. These devices are used to measure the force or torque produced by joint actuators. They also serve to detect mechanical overload conditions. As industrial robots emerge from limited, repetitive automation tasks toward higher level **intelligent automation** operations, the need for more sophisticated position and force-sensing transducers increases.

Most of the position-sensing transducers found in mechanical manipulators should already be familiar, because they were examined in detail earlier in this text. Where possible in robotic design, sensors are placed directly on the shafts of actuators, forming **co-located** actuator/sensor pairs. Thus, the prevalent form of robotic position sensor is the **incremental rotary optical encoder.** As seen in Figure 14–19, a 90° lead/lag relationship exists between the A and B outputs of the optical encoder, allowing the direction of shaft rotation to be sensed. Also, the A and B outputs of the device are inverted, forming the complete quadrature output seen in Figure 14–19 and represented in Table 14–1. As shown in Figure 14–19, a single line count is equivalent to four quadrature counts. Depending on the degree of accuracy required in a given application, an optical encoder found in an industrial robot could have a line count as high as 10,000. Assuming the device to have quadrature outputs, actual resolution would be as much as 0.54 arcmin!

Resolvers can also be used as position sensors in robotic systems. Although resolvers can be more reliable than incremental rotary optical encoders, they do present challenges when used in robotic systems. Due to the time required in processing resolver output signals, placement of a resolver directly in a co-located configuration with a rotary actuator might not allow sufficient resolution. Thus, additional gearing may be required just for driving the resolver. In a modern industrial robot, direct measurement of joint actuator rotational speed is seldom performed through use of a velocity-sensing transducer. Although a tachometer or dc generator could directly provide velocity information, the preferred method of monitoring shaft speed is through **numeric differentiation.** This process allows shaft speed to be derived by sampling the outputs of the position sensor signal conditioning circuitry. Recall, for example, that a servocontroller can rely on a resolver for feedback in determining both servomotor shaft position and speed. This is possible since, by sampling RDC outputs at a specified rate and comparing two consecutive readings, the CPU is able to calculate servomotor shaft speed. This form of CPU algorithm can also be applied when sampling the outputs of optical encoder signal conditioning circuitry. Although slight time delay problems may arise when relying on numeric differentiation to measure angular velocity, this technique is, nevertheless, widely used in the control of mechanical manipulators, since it eliminates the need for additional co-located transducers.

The predominant form of force-sensing transducer used in mechanical manipulators is the strain gauge. This device was discussed in detail in Section 4–6. At this juncture, it is advisable to review this information to better understand the operation of a strain gauge as a robotic sensor. The strain gauges used in robotics may be of the conventional metallic foil type or they could be semiconductor devices. Typically, metallic foil strain gauges are bonded to the robotic part that experiences mechanical stress. The orientation of the strain gauge with respect to the lines of applied force depends on whether that force is tensile or compressive. Recall that strain gauges undergo slight, but highly linear, changes in resistance when subjected to changes in applied force. Extensive preliminary signal conditioning, as well as A/D conversion, is required to create digital feedback information from these slight changes in strain gauge resistance.

Strain gauges have three basic robotics applications. At robotic joints, strain gauges are used in the measurement of torques or forces produced by actuators. At a point between the end effector and the last joint of the manipulator, strain gauges are contained in a device called a **wrist sensor.** Two possible forms of wrist sensor are shown in Figures 14–27(a) and (b). Such devices can measure from three to six components of force. As shown in Figure 14–27(c), the fingers at the end of a gripping end effector are likely to contain built-in strain gauges. These devices are necessary for measurement of the tactile force applied by the gripper in material handling operations.

An essential process in controlling industrial robots is the assignment of **body-attached frames** to the manipulator links. The set of kinematic symbols used in this process are referred to as **Denavit-Hartenberg notation.** For convenient reference, these symbols and their respective link parameters are summarized in Table 14–2.

As an introduction to the process of frame assignment, consider the **three-link planar arm** in Figure 14–28. This form of manipulator, which typically serves as the arm extension of a spherical robot, is also referred to as the RRR or 3R. These latter two designations refer to the fact that this type of manipulator consists of three revolute joints. The term *planar* refers to the fact that all three revolute joints originate from a common plane. In Figure 14–28(a), the link lengths of the 3R device are designated as L_1, L_2, and L_3, and the joint angles are designated as θ_1, θ_2, and θ_3. Figure 14–28(b) shows the actual assignment of link frames.

The process of frame assignment for the 3R manipulator begins with the establishment of a stationary reference frame, frame 0, at the base. (Recall that the base of the manipulator is considered as link 0.) At this point, we should stress that, while the entire manipulator is able to rotate on the vertical axis of its spherical base, this degree of freedom does *not* become a consideration in arm kinematics. Thus, in Figure 14–28(b), the base frame is represented by the stationary X_0 and Y_0 axes. (Since the Z axes of all three link frames are parallel and emanate directly out of the page, they are not shown in Figure 14–28(b).)

The establishment of the three movable link frames of the 3R manipulator in Figure 14–28 is as follows. Frame 1 has the same origin as frame 0. Thus, as link 1 moves on

Figure 14–27

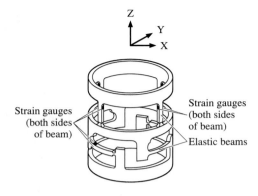

(a) Internal structure of a force-sensing wrist

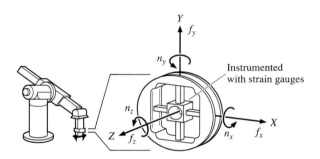

(b) Alternative structure of a force-sensing wrist

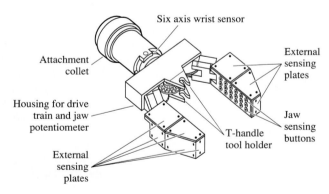

(c) Sensors at end effector of mechanical manipulator

this base axis, it forms joint angle θ_1 with respect to link 0. Note that the X axis of frame 1 (designated X_1) extends along link 1 in the direction of the second revolute joint. Looking back from this second joint toward the base, link length L_1 may be designated as $a_i - 1$ with respect to the axis of joint 2. Frame 2 begins at the axis of joint 2, with its X axis extending along link 2 in the direction of joint 3. As link 2 rotates with respect to link 1, it forms joint angle θ_2. Looking back from the third joint toward the second, L_2 may be considered as $a_i - 1$ with respect to the axis of joint 3. Since link 3 contains the end

Table 14–2
Denavit-Hartenberg notation

Symbol	Parameter
a_i	Link length
α_i	Link twist
d_i	Joint offset
θ	Joint angle

effector, the analysis of the link frames of the 3R manipulator ends here. Analysis of the link frames of the manipulator becomes a separate effort.

The link parameters of the 3R manipulator can be summarized in tabular form. As shown in Table 14–3, the axes of the manipulator are listed in the first column (designated *i*). Column 2 (designated $\alpha_i - 1$) represents the link twists of the manipulator. Since the 3R manipulator has no link twists, this column contains only zeros. The variable $a_i - 1$ represents link length, measured along an X axis from a joint (*i*) to the previous joint. Thus for axis (*i*) 1, $a_i - 1$ equals zero, since frame 1 has the same reference as frame 0. For axis 2, $a_i - 1$ is equivalent to L_1, and $a_i - 1$ for axis 3 becomes L_2. Since there is no joint offset for the 3R manipulator in Figure 14–28, all entries in the d_i column become zero. The last column in Table 14–3 lists the joint angles (θ_i) associated with each of the joint angles.

The robot, shown in Figure 14–29, is typical of most mechanical manipulators found in a modern automated industrial site. This device is a rotary joint manipulator with six degrees of freedom. Because all six joints of this device are revolute, it may be referred to simply as a 6R mechanism. The link frame assignments for this mechanical manipulator are shown in Figure 14–30.

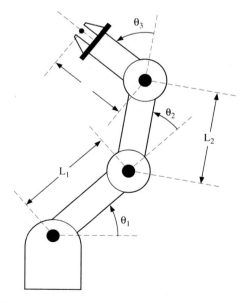

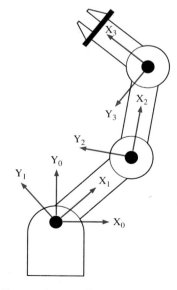

(a) Link lengths and joint angles of a three-link planar arm   (b) Frame assignments for a three-link planar arm

Figure 14–28

Table 14–3
Link parameters for the 3R manipulator

i	$\alpha_i - 1$	$a_i - 1$	d_i	θ_i
1	0	0	0	θ_1
2	0	L_1	0	θ_2
3	0	L_2	0	θ_3

As with the simple 3R mechanism just analyzed, frame 1 in Figure 14–30 is assumed to be coincident with frame 0 (which is not shown). The link parameters of frame 1 with respect to frame 0 are summarized in the first row of Table 14–4. In Figure 14–30(a), it is important to note the difference in orientation between frames 1 and 2. Whereas the frame 1 axis allows rotation of the manipulator around the base, the frame 2 axis allows changes in the tilt of link 2. Note also that frames 1 and 2 are assumed to originate at a common point. These conditions are clearly represented in the second row of Table 14–4. Note that $\alpha_i - 1$ for the second row is $-90°$, indicating that a link twist of $-90°$ exists for the axis of joint 2 with respect to the axis of joint 1. Zero conditions in the $a_i - 1$ and d_i blocks of the second row further indicate that frames 1 and 2 share a common point of origin. Thus, there is no link length or joint offset for joint 2 with respect to joint 1.

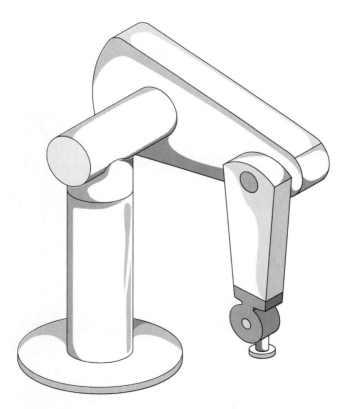

Figure 14–29
Mechanical manipulator with six degrees of freedom

Figure 14–30

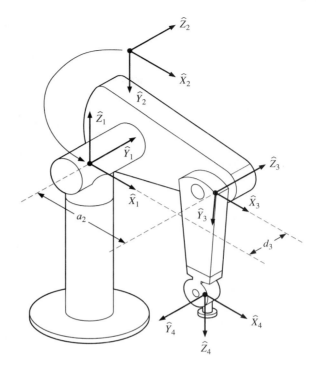

(a) Link frame assignments for mechanical manipulator

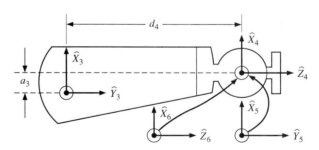

(b) Link frame assignments for the forearm of the mechanical manipulator

Table 14–4
Link parameters for mechanical manipulator with six degrees of freedom

i	$\alpha_i - 1$	$a_i - 1$	d_i	θ_i
1	0	0	0	θ_1
2	$-90°$	0	0	θ_2
3	0	a_2	d_3	θ_3
4	$-90°$	a_3	d_4	θ_4
5	$90°$	0	0	θ_5
6	$-90°$	0	0	θ_6

At this point it would be excellent practice to refer directly to the third row of Table 14–4 and attempt to visualize the relationship of joints 2 and 3 of the robot in Figure 14–29. A zero condition for $\alpha_i - 1$ indicates that no link twist exists from joint 2 to joint 3. Therefore, the axes of these two joints are parallel. The presence of a_2 in the $a_i - 1$ position indicates that joint 3 is separated from joint 2 by the link length of L_2, which is measured along the X axis of frame 2. The presence of d_3 in the d_i block indicates that a joint offset exists from the origin of frame 3 to that of frame 2. Finally, θ_3 is formed as frame 3 rotates with respect to frame 2. Referring back to Figure 14–30(a), the link parameters just described should become evident. Here it is seen that link 2 extends the distance a_2 from the **shoulder joint** (joint 2) to the **elbow joint** (joint 3). It is also evident that the axes of these two joints are aligned in parallel. Finally, since link 3 is offset from link 2 at the elbow joint, a joint offset, d_3, results.

In analyzing the operation of the wrist joint of the mechanical manipulator in Figure 14–29, it is important to realize that frames 4, 5, and 6 have a common point of origin. Frame 4 allows rotation of the entire wrist, whereas frame 5 allows adjustment of the pitch of the end effector. Frame 6 might, at first, appear identical to frame 4. However, frame 6, representing the roll of an end effector, such as a nut driver, is able to rotate freely while frame 4 (at the wrist joint) is stationary. Note that in Figure 14–30(b), due to the $-90°$ link twist between frames 3 and 4, the length of link 3 becomes a_3 while joint offset is the relatively longer distance d_4. The wrist operation just described, with joint axes originating at a common point, is typical for industrial robotic systems. It should now be evident that the mechanical manipulator in Figure 29–30, though not equipped with a gripper, is ideally suited for tasks such as welding and nut driving.

SECTION REVIEW 14–3

1. List the four kinematic link parameters used to describe mechanical manipulator position. Write the Denavit-Hartenberg symbolic notation for each of these parameters.
2. Describe the structure and operation of a three-link planar arm. List two possible abbreviated descriptions for this form of manipulator.
3. For the three-link planar arm, what two link parameters equal zero?
4. Identify the number and types of joints contained within a mechanical manipulator described as RPR.
5. True or false?: The end effector of a mechanical manipulator operates within a coordinate system referred to as the base frame.
6. Define the term *lower pair* as it pertains to the structure of a mechanical manipulator. List six types of lower pair.
7. Identify and describe the most common form of robotic joint.
8. True or false?: The robot in Figure 14–29 may be described as a 6R device.
9. Identify the most common form of robotic position sensor. Describe the most common method of attaching this device to an actuator.
10. Identify the most common form of robotic force sensor. What part of the robot is likely to contain the greatest concentration of these devices?

SUMMARY

Servo systems are often necessary for precise control of automated machinery and industrial robots. In servomotor control systems, the servomechanism often consists of a rotary transducer such as a shaft encoder or a resolver. Although the incremental rotary encoder is the preferred form of co-located position sensor in robotic systems, resolvers also serve as position sensors in many other servomotor applications. The basic servo loop must contain an error-detecting summer amplifier and a power stage for driving the servomotor. In commercially developed servocontrollers, the idea of the basic servo loop evolves into a highly sophisticated design concept. These servocontrollers, which are often microcomputer-based, resemble sophisticated PLCs. Like state-of-the-art PLCs, microcomputer-based servocontrollers are often programmable in a user-friendly, BASIC-type language. To provide extremely precise control of servomotor speed and acceleration, commercially manufactured servocontrollers often contain PWM-controlled power stages.

The most common form of industrial robot is the mechanical manipulator, often consisting of a jointed arm on a rotating base. The maneuverability of an industrial robot is expressed in degrees of freedom, which usually correspond to the number of manipulator joints. The science of robotics involves both *kinematics* and *dynamics*. Kinematics encompasses the study of robotic motion, without regard to the forces that cause this motion. Dynamics involves the study of those complex torques that must be exerted by actuators in the implementation of efficient manipulator motion. Joint actuators found in mechanical manipulators are almost always servo controlled. Thus, to detect both position and force, many sensors are required in industrial robots. Incremental rotary encoders often serve as sensors for revolute joints, whereas strain gauges serve as force detectors.

SELF-TEST

1. The servocontroller represented in Figure 14–5 is capable of
   a. only proportional control
   b. proportional plus integral control
   c. proportional plus integral plus derivative control
   d. proportional plus derivative control
2. Why is frequency compensation necessary in the design of the typical servocontroller? What adverse effects could result if frequency compensation were not implemented?
3. For the velocity error amplifier in Figure 14–6, assume the jumper for terminal block E_1 is connected between terminals 3 and 4. The chosen value of R_{27} is 68 kΩ, while C_{10} equals 0.47 μF. Given these conditions,
   a. the response of the velocity error amplifier is purely proportional
   b. the lag break frequency is 7.2 Hz
   c. the lag break frequency is nearly 5 Hz
   d. the response of the velocity amplifier is proportional plus derivative
4. For the velocity error amplifier in Figure 14–6, assume the jumper for terminal block E_1 is connected between terminals 1 and 2. With this condition,
   a. the response of the velocity error amplifier is proportional plus integral
   b. the lag break frequency is around 16 Hz
   c. both choices a and b are correct
   d. the response of the velocity error amplifier is purely proportional
5. The servo amplifier stages contained in Figures 14–6 and 14–7 constitute
   a. cascade control
   b. feed-forward control
   c. PID control
   d. none of the above
6. True or false?: The current error amplifier in Figure 14–7 performs PID control.

7. For the current-loop amplifier in Figure 14–7, at a given instant in time, assume the IXFB and the IYFB signals are of equal amplitude and opposite polarity. Given these conditions,
    a. I_{R38} equals virtually 0 A
    b. I_{R38} equals $(I_{R34} \times I_{R35})$
    c. I_{R38} equals $(I_{R34} - I_{R35}) \times 2$
    d. I_{R38} may not be determined
8. Briefly explain the proper method for adjusting R_8 in Figure 14–6.
9. True or false?: Increasing the ohmic value of R_7 in Figure 14–6 increases the factor of $\Delta rpm/\Delta V$ for the velocity loop amplifier.
10. Specify *all* logic conditions that must occur in Figure 14–10 to achieve the servomotor condition represented in Figure 14–12(e). (Assume each of the six output stages of the servo amplifier consists of the power converter circuitry in Figure 10–52.)
11. What 555 timer configuration is used to form the deadband oscillator in Figure 14–10? What is the purpose of this circuitry?
12. True or false?: For the deadband oscillator in Figure 14–10, duty cycle is slightly greater than 50%.
13. When a GO.ABS statement follows a TARGET.POS variable in a BASIC program, how does the CPU of the servocontroller determine the required direction of servomotor shaft rotation?
14. Assume the shaft of a servomotor connected to a servocontroller travels the equivalent of 23,434 resolver steps. How many degrees of rotation does this distance represent? How many servomotor revolutions does this distance represent?
15. Describe the operation of an incremental rotary optical encoder. What is the phase relationship of the A and B outputs of this device? How are quadrature outputs created for this device?
16. Assume that an incremental rotary encoder with quadrature outputs is connected to the shaft of a pulley. The resolution of this shaft encoder is assumed to be 1024 line counts. If the input circuitry of the servocontroller senses that the pulley has traveled 13,312 quadrature counts, how many degrees has the pulley rotated? How many revolutions does this motion represent?
17. During an electronic gearing application, how does the CPU of the servocontroller learn the resolution of the quadrature encoder connected to the external drive system?
18. Define via points and splines as they pertain to the dynamics of a mechanical manipulator.
19. Explain the process of numeric differentiation. Identify a disadvantage and an advantage of using this process to determine shaft velocity.
20. Why are resolvers used less frequently than incremental rotary encoders as position sensors at robotic joints?

APPENDIX A: INTEL 8085 INSTRUCTION SET

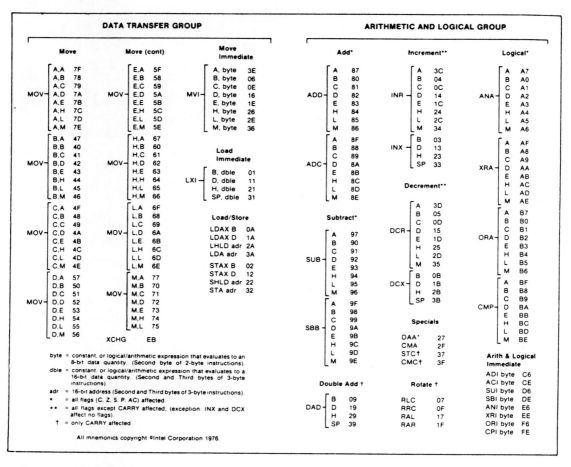

DATA TRANSFER GROUP

Move

MOV		
	A,A	7F
	A,B	78
	A,C	79
	A,D	7A
	A,E	7B
	A,H	7C
	A,L	7D
	A,M	7E

MOV		
	B,A	47
	B,B	40
	B,C	41
	B,D	42
	B,E	43
	B,H	44
	B,L	45
	B,M	46

MOV		
	C,A	4F
	C,B	48
	C,C	49
	C,D	4A
	C,E	4B
	C,H	4C
	C,L	4D
	C,M	4E

MOV		
	D,A	57
	D,B	50
	D,C	51
	D,D	52
	D,E	53
	D,H	54
	D,L	55
	D,M	56

Move (cont)

MOV		
	E,A	5F
	E,B	58
	E,C	59
	E,D	5A
	E,E	5B
	E,H	5C
	E,L	5D
	E,M	5E

MOV		
	H,A	67
	H,B	60
	H,C	61
	H,D	62
	H,E	63
	H,H	64
	H,L	65
	H,M	66

MOV		
	L,A	6F
	L,B	68
	L,C	69
	L,D	6A
	L,E	6B
	L,H	6C
	L,L	6D
	L,M	6E

MOV		
	M,A	77
	M,B	70
	M,C	71
	M,D	72
	M,E	73
	M,H	74
	M,L	75

XCHG   EB

Move Immediate

MVI		
	A, byte	3E
	B, byte	06
	C, byte	0E
	D, byte	16
	E, byte	1E
	H, byte	26
	L, byte	2E
	M, byte	36

Load Immediate

LXI		
	B, dble	01
	D, dble	11
	H, dble	21
	SP, dble	31

Load/Store

LDAX B	0A
LDAX D	1A
LHLD adr	2A
LDA adr	3A
STAX B	02
STAX D	12
SHLD adr	22
STA adr	32

byte = constant, or logical/arithmetic expression that evaluates to an 8-bit data quantity. (Second byte of 2-byte instructions).

dble = constant, or logical/arithmetic expression that evaluates to a 16-bit data quantity. (Second and Third bytes of 3-byte instructions).

adr = 16-bit address (Second and Third bytes of 3-byte instructions).

* = all flags (C, Z, S, P, AC) affected.

** = all flags except CARRY affected; (exception: INX and DCX affect no flags).

† = only CARRY affected.

All mnemonics copyright ©Intel Corporation 1976.

ARITHMETIC AND LOGICAL GROUP

Add*

ADD		
	A	87
	B	80
	C	81
	D	82
	E	83
	H	84
	L	85
	M	86

ADC		
	A	8F
	B	88
	C	89
	D	8A
	E	8B
	H	8C
	L	8D
	M	8E

Subtract*

SUB		
	A	97
	B	90
	C	91
	D	92
	E	93
	H	94
	L	95
	M	96

SBB		
	A	9F
	B	98
	C	99
	D	9A
	E	9B
	H	9C
	L	9D
	M	9E

Double Add †

DAD		
	B	09
	D	19
	H	29
	SP	39

Increment**

INR		
	A	3C
	B	04
	C	0C
	D	14
	E	1C
	H	24
	L	2C
	M	34

INX		
	B	03
	D	13
	H	23
	SP	33

Decrement**

DCR		
	A	3D
	B	05
	C	0D
	D	15
	E	1D
	H	25
	L	2D
	M	35

DCX		
	B	0B
	D	1B
	H	2B
	SP	3B

Specials

DAA*	27
CMA	2F
STC†	37
CMC†	3F

Rotate †

RLC	07
RRC	0F
RAL	17
RAR	1F

Logical*

ANA		
	A	A7
	B	A0
	C	A1
	D	A2
	E	A3
	H	A4
	L	A5
	M	A6

XRA		
	A	AF
	B	A8
	C	A9
	D	AA
	E	AB
	H	AC
	L	AD
	M	AE

ORA		
	A	B7
	B	B0
	C	B1
	D	B2
	E	B3
	H	B4
	L	B5
	M	B6

CMP		
	A	BF
	B	B8
	C	B9
	D	BA
	E	BB
	H	BC
	L	BD
	M	BE

Arith & Logical Immediate

ADI byte	C6
ACI byte	CE
SUI byte	D6
SBI byte	DE
ANI byte	E6
XRI byte	EE
ORI byte	F6
CPI byte	FE

(Courtesy of Intel Corporation)

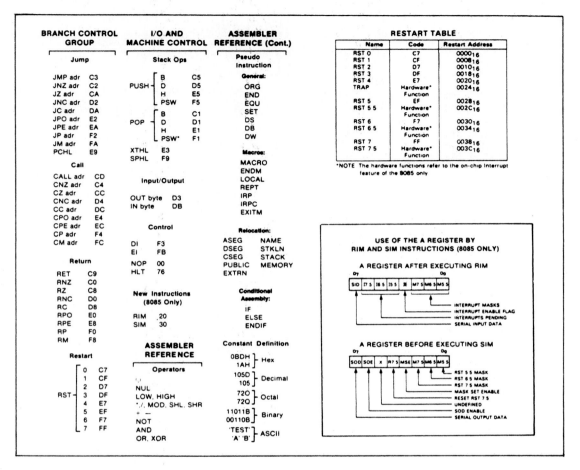

BRANCH CONTROL GROUP

Jump

JMP adr	C3
JNZ adr	C2
JZ adr	CA
JNC adr	D2
JC adr	DA
JPO adr	E2
JPE adr	EA
JP adr	F2
JM adr	FA
PCHL	E9

Call

CALL adr	CD
CNZ adr	C4
CZ adr	CC
CNC adr	D4
CC adr	DC
CPO adr	E4
CPE adr	EC
CP adr	F4
CM adr	FC

Return

RET	C9
RNZ	C0
RZ	C8
RNC	D0
RC	D8
RPO	E0
RPE	E8
RP	F0
RM	F8

Restart

RST	0	C7
	1	CF
	2	D7
	3	DF
	4	E7
	5	EF
	6	F7
	7	FF

I/O AND MACHINE CONTROL

Stack Ops

PUSH	B	C5
	D	D5
	H	E5
	PSW	F5

POP	B	C1
	D	D1
	H	E1
	PSW*	F1

XTHL	E3
SPHL	F9

Input/Output

OUT byte	D3
IN byte	DB

Control

DI	F3
EI	FB
NOP	00
HLT	76

New Instructions (8085 Only)

RIM	20
SIM	30

ASSEMBLER REFERENCE

Operators

( )
NUL
LOW, HIGH
*, /, MOD, SHL, SHR
+ —
NOT
AND
OR, XOR

ASSEMBLER REFERENCE (Cont.)

Pseudo Instruction

General:

ORG
END
EQU
SET
DS
DB
DW

Macros:

MACRO
ENDM
LOCAL
REPT
IRP
IRPC
EXITM

Relocation:

ASEG	NAME
DSEG	STKLN
CSEG	STACK
PUBLIC	MEMORY
EXTRN	

Conditional Assembly:

IF
ELSE
ENDIF

Constant Definition

0BDH	} Hex
1AH	

105D	} Decimal
105	

72O	} Octal
72Q	

11011B	} Binary
00110B	

'TEST'	} ASCII
'A' 'B'	

RESTART TABLE

Name	Code	Restart Address
RST 0	C7	0000_{16}
RST 1	CF	0008_{16}
RST 2	D7	0010_{16}
RST 3	DF	0018_{16}
RST 4	E7	0020_{16}
TRAP	Hardware* Function	0024_{16}
RST 5	EF	0028_{16}
RST 5 5	Hardware* Function	$002C_{16}$
RST 6	F7	0030_{16}
RST 6 5	Hardware* Function	0034_{16}
RST 7	FF	0038_{16}
RST 7 5	Hardware* Function	$003C_{16}$

*NOTE The hardware functions refer to the on-chip interrupt feature of the 8085 only

USE OF THE A REGISTER BY RIM AND SIM INSTRUCTIONS (8085 ONLY)

A REGISTER AFTER EXECUTING RIM

D_7							D_0
SID	I7 5	I6 5	I5 5	IE	M7 5	M6 5	M5 5

INTERRUPT MASKS
INTERRUPT ENABLE FLAG
INTERRUPTS PENDING
SERIAL INPUT DATA

A REGISTER BEFORE EXECUTING SIM

D_7							D_0
SOD	SOE	X	R7 5	MSE	M7 5	M6 5	M5 5

RST 5 5 MASK
RST 6 5 MASK
RST 7 5 MASK
MASK SET ENABLE
RESET RST 7 5
UNDEFINED
SOD ENABLE
SERIAL OUTPUT DATA

Courtesy of Intel Corporation

APPENDIX B: INSTRUCTION SET FOR OMRON SYSMAC C200H/C200HS PROGRAMMABLE CONTROLLERS

Name Mnemonic	Symbol	Function	Operands	CPU
Load LD	B	Used to start instruction line with status of designated bit. Used to define a logic block for use with AND LD and OR LD.	**B:** IR SR HR AR TC LR TR	All models
Load NOT LD NOT	B	Used to start instruction line with inverse of designated bit.	**B:** IR SR HR AR TC LR	All models
AND AND	B	Logically ANDs status of designated bit with execution condition.	**B:** IR SR HR AR TC LR	All models
AND NOT AND NOT	B	Logically ANDs inverse of designated bit with execution condition.	**B:** IR SR HR AR TC LR	All models
OR OR	B	Logically ORs status of designated bit with execution condition.	**B:** IR SR HR AR TC LR	All models
OR NOT OR NOT	B	Logically ORs inverse of designated bit with execution condition.	**B:** IR SR HR AR TC LR	All models
AND Load AND LD		Logically ANDs results of preceding blocks.	None	All models
OR Load OR LD		Logically ORs results of preceding blocks.	None	All models
Output OUT	B	Turns ON B for ON execution condition; turns OFF B for OFF execution condition.	**B:** IR SR HR AR LR TR	All models

Instruction Set

Name Mnemonic	Symbol	Function	Operands		CPU
RESET RESET	─┤ RESET B ├─	Turns B OFF for ON execution condition and remains OFF when execution condition returns to OFF	**B:** IR SR AR HR LR		HS-CPU01 HS-CPU03
SET SET	─┤ RESET B ├─	Turns B OFF for ON execution condition and remains OFF when execution condition returns to OFF	**B:** IR SR AR HR LR		HS-CPU01 HS-CPU03
Output NOT OUT NOT	─⦸ B ⦸─	Turns B OFF for ON execution condition; turns B ON for OFF execution condition (i.e., inverts operation).	**B:** IR SR HR AR LR		All models
No Operation NOP (00)	None	Nothing is executed and program moves to next instruction.	None		All models
End END(01)	──── END(01)	Required at the end of the program.	None		All models
Latching Relay KEEP(11)	S ──┐ ── KEEP(11) ── R ──┘ B	Defines a bit (B) as a latch controlled by set (S) and reset (R) inputs.	**B:** IR HR AR LR		All models
Timer TIM	──── TIM N SV	ON-delay (decrementing) timer operation. Set value: 999.9 s; accuracy: +0/−0.1 s. Same TC bit cannot be assigned to more than one timer/counter. The TC bit is input as a constant.	**N:** TC	**SV:** IR HR AR DM LR #	All models
High-speed Timer TIMH(15)	──── TIMH(15) N SV	A high-speed, ON-delay (decrementing) timer. SV: 0.01 to 99.99 s; accuracy: +0/−0.1 s. Must not be assigned the same TC bit as another timer or counter. The TC bit is input as a constant.	**N:** TC	**SV:** IR SR HR AR DM LR #	All models
Counter CNT	CP ──┐ ── CNT N R ──┘ SV	A decrementing counter. SV: 0 to 9999; CP: count pulse; R: reset input. The TC bit is input as a constant.	**N:** TC	**SV:** IR HR AR DM LR #	All models
Reversible Counter CNTR (12)	II ──┐ DI ── CNTR(12) N SV R ──┘	Increases or decreases PV by one whenever the increment input (II) or decrement input (DI) signals, respectively, go from OFF to ON. SV: 0 to 9999; R: reset input. Must not access the same TC bit as another timer/counter.	**N:** TC	**SV:** IR SR HR AR DM LR #	All models

Instruction Set

Basic Instructions

Name Mnemonic	Symbol	Function	Operands			CPU
Differentiate Up DIFU(13) **Differentiate Down** DIFD(14)	DIFU(13) B DIFD(14) B	DIFU turns ON the designated bit (B) for one scan on the rising edge of the input signal; DIFD turns ON the bit for one scan on the trailing edge.	**B:** IR HR AR LR			All models
Interlock IL(02) **Interlock Clear** ILC(03)	IL(02) ILC(03)	If interlock condition is OFF, all outputs are turned OFF and all timer PVs reset between this IL(02) and the next ILC(03). Other instructions are treated as NOP; counter PVs are maintained.	None			All models
Jump JMP(04) **Jump End** JME(05)	JMP(04) N JME(05) N	If jump condition is OFF, all instructions between JMP(04) and the corresponding JME(05) are ignored. Corresponding JME is one of same number; 01 through 99 only usable once per program (direct jumps); 00 may be used as many times as necessary, but instructions between JMP 00 and JME 00 treated as NOP, increasing scan time compared to other jumps.	**N:** 00 to 99			All models
Totalizing Timer TTIM (87)	TTIM(87) N SV RB	Incrementing timer that increments PV every 0.1 s to time between 0.1s to 999.9 s. Accuracy is +0.0/-0.1 sec. Will time as long as execution condition is ON until it reaches SV or until RB turns ON to reset timer.	**N:** TC	**SV: (word, BCD)** IR AR DM HR LR	**RB:** IR SR AR HR LR	HS-CPU01 HS-CPU03

Name Mnemonic	Symbol	Function	Operands			CPU
Multi-word Compare (@)MCMP(19)	MCMP(19) / S1 / S2 / D	Compares the data within a block of 16 words of 4-digit hexadecimal data (S_1 to S_1+15) with that in another block of 16 words (S_2 to S_2+15) on a word-by-word basis. If the words are not in agreement, the bit corresponding to unmatched words turns ON in the result word, D. Bits corresponding to words that are equal are turned OFF.	**S1:** IR SR HR AR LR TC DM	**S2:** IR SR HR AR LR TC DM	**D:** IR SR HR AR LR TC DM	CPU21 CPU23 CPU31 HS-CPU01 HS-CPU03
Compare (@)CMP(20)	CMP(20) / Cp1 / Cp2	Compares two sets of four-digit hexadecimal data (Cp1 and Cp2) and outputs result to GR, EQ, and LE. Cp1 > Cp2 → GR Cp1 = Cp2 → EQ Cp1 < Cp2 → LE	**Cp1/Cp2:** IR SR HR AR TC DM LR #			All models
Double Compare CMPL(60)	CMPL(60) / S1 / S2	Compares the 8-digit hexadecimal values in words S_1+1 and S_1 with the values in S_2+1 and S_2, and indicates the result using the Greater Than, Less Than, and Equal Flags in the AR area. S_1+1 and S_2+1 are regarded as the most significant data in each pair of words.	**S1/S2** IR SR HR AR LR TC DM			CPU21 CPU23 CPU31 HS-CPU01 HS-CPU03
Block Compare BCMP(68)	BCMP(68) / S / CB / R	Compares 1-word binary value (S) with 16 ranges in comparison table (CB: starting word of comparison block). If value falls within any ranges, corresponding bits of result word (R) will set. The comparison block must all be in the same data area.	**S:** IR SR HR AR TC DM LR #	**CB:** IR SR HR TC DM LR	**R:** IR HR AR TC DM LR	All models HS-CPU01 HS-CPU03
Table Compare (@)TCMP(85)	TCMP(85) / S / TB / R	Compares four-digit hexadecimal value (S) with values in table consisting of 16 words (TB: First word of comparison table). If value equals any values, corresponding bits of result word (R) are set. The entire table must be in the same data area.	**S:** IR SR HR AR TC DM LR #	**TB/R:** IR HR AR TC DM LR		All models

Block Compare diagram:

```
        Lower limit  Upper limit
          CB    |  CB+1   →  1
          CB+2  |  CB+3   →  0
    S ←   CB+4  |  CB+5   →  1
              ⋮
          CB+30 |  CB+31  →  0
    Lower limit ≤ S ≤ Upper limit → 1
```

Table Compare diagram:

```
                          R
    S  →   TB       0
           TB+1     1
              ⋮
           TB+14    0
           TB+15    1
        1:agreement
        0:disagreement
```

Instruction Set

**Data Comparison
Instructions**

Name Mnemonic	Symbol	Function	Operands			CPU
Area Range Compare (@) ZCP (88)	ZCP(88) CD LL UL	Compares a word to a range defined by Lower and Upper Limits and outputs the result to GR, EQ and LE flags.	**CD:** IR SR HR AR TC DM LR #	**LL:** IR SR HR AR TC DM LR #	**UL:** IR SR HR AR TC DM LR #	HS-CPU01 HS-CPU03
Signed binary compare CPS (*)	CPS(−) Cp1 Cp2	Compares two 16-bit (4-digit) signed binary values and outputs the result to GR, EQ and LE flags.	**Cp1:** IR SR HR AR TC DM LR #	**Cp2:** IR SR HR AR TC DM LR #		HS-CPU01 HS-CPU03
Double signed binary compare CPSL (*)	CPSL(−) Cp1 Cp2	Compares two 32-bit (8-DIGIT) signed binary values and outputs the result to GR, EQ and LE flags.	**Cp1:** IR SR AR DM HR TC LR	**Cp2:** IR SR AR DM HR TC LR		HS-CPU01 HS-CPU03
Double area range compare ZCPL (*)	ZCPL(−) CD LL UL	Compares an 8-digit value to a range and outputs the result to GR, EQ and LE flags.	**CD:** IR SR HR AR TC DM LR	**LL:** IR SR HR AR TC DM LR	**UL:** IR SR HR AR TC DM LR	HS-CPU01 HS-CPU03

Instruction Set

Name Mnemonic	Symbol	Function	Operands			CPU
Move (@)MOV(21)	MOV(21) / S / D	Transfers source data (S) (word or four-digit constant) to destination word (D).	**S:** IR SR HR AR TC DM LR #	**D:** IR HR AR DM LR		All models
Move NOT (@)MVN(22)	MVN(22) / S / D	Inverts source data (S) (word or four-digit constant) and then transfers it to destination word (D).	**S:** IR SR HR AR TC DM LR #	**D:** IR HR AR DM LR		All models
Column-to-Word (@)CTW(63)	CTW(63) / S / C / D	Fetches data from the same numbered bit (C) in 16 consecutive words (where S is the address of the first source word), and creates a 4-digit word by consecutively placing the data in the bits of the destination word, D. The bit from word S is placed into bit 00 of D, the bit from word S+1 is placed into bit 01, etc.	**S:** IR SR HR AR LR TC DM	**C:** IR SR HR AR LR TC DM #	**D:** IR SR HR AR LR TC DM	CPU21 CPU23 CPU31
Column-to-Line (@)LINE(63)	LINE(63) / S / C / D					HS-CPU01 HS-CPU03
Word-to-Column (@)WTC(64)	WTC(64) / S / D / C	Places bit data from the source word (S), consecutively into the same numbered bits of the 16 consecutive destination words (where D is the address of the first destination word). Bit 00 from word S is placed into bit C of word D, bit 01 from word S is placed into bit C of word D+1, etc.	**S:** IR SR HR AR LR TC DM	**D:** IR SR HR AR LR TC DM	**C:** IR SR HR AR LR TC DM #	CPU21 CPU23 CPU31
Line-to-Column (@)COLM(64)	COLM(64) / S / D / C					HS-CPU01 HS-CPU03
Block Set (@)BSET(71)	BSET(71) / S / St / E	Copies content of one word or constant (S) to several consecutive words (starting word (St) through ending word (E)). St and E must be in the same data area.	**St/E:** IR HR AR TC DM LR	**S:** IR SR HR AR TC DM LR #		All models

Instruction Set

Data Transfer Instructions

Name Mnemonic	Symbol	Function	Operands			CPU
Transfer Bits (@XFRB(62))	XFER(62) / C / S / D	Copies the status of up to 255 specified source bits to the specified destination bits.	**C:** IR SR AR DM TC HR LR #	**S:** IR SR AR DM HR LR	**D:** IR SR AR DM HR LR	HS-CPU01 HS-CPU03
Block Transfer (@)XFER(70)	XFER(70) / N / S / D	Moves content of several consecutive source words (S: starting source word; N: number of transfer words) to consecutive destination words (D: starting destination word). All source words must be in the same data area, as must all destination words. Transfers can be within one or between two data areas, but the source words and destination word must not overlap. S → D S+1 → D+1 ⋮ ⋮ No. of words S+N−1 D+N−1	**N:** IR SR HR AR TC DM LR #	**S:** IR HR AR TC DM LR	**D:** IR SR HR AR TC DM LR #	All models
Single Word Distribute (@)DIST(80)	DIST(80) / S / DBs / Of	Moves one word of source data (S) to destination word whose address is given by destination base word (DBs) plus offset (Of). S ⌐ Base (DBs) + Offset (OF) (S) → (DBs+Of)	**S:** IR SR HR AR TC DM LR #	**DBs:** IR HR AR TC DM LR	**Of:** IR HR AR TC DM LR #	All models
Data Collect (@)COLL(81)	COLL(81) / SBs / Of / D	Extracts data from source word and writes it to destination word (D). Source word is determined by adding offset (Of) to source base word (SBs). Base (DBs) + Offset (OF) → D (SBs+Of) → (D)	**SBs:** IR SR HR AR TC DM LR	**Of:** IR HR AR TC DM LR #	**D:** IR HR AR TC DM LR	All models
Move Bit (@)MOVB(82)	MOVB(82) / S / Bi / D	Transfers designated bit of source word or constant (S) to designated bit of destination word (D). Rightmost two digits of bit designator (Bi) designate the source bit; leftmost two, the destination bit. S D	**S:** IR SR HR AR DM LR #	**Bi:** IR HR AR TC DM LR #	**D:** IR HR AR DM LR	All models

Data Transfer Instructions

Name Mnemonic	Symbol	Function	Operands			CPU
Move Digit (@)MOVD(83)	MOVD(83) / S / Di / D	Moves hexadecimal content of specified four-bit source digit(s) (S: source word) to specified destination digit(s) (D: destination word) for up to four digits. Source and destination digits specified in Digit Designator (Di) digits (rightmost digit: source digit; next digit to left: number of digits to be moved; next digit: destination digit). 15 00	**S:** IR SR HR AR TC DM LR #	**Di:** IR HR AR TC DM LR #	**D:** IR SR HR AR TC DM LR	All models
Data Exchange (@)XCHG(73)	XCHG(73) / E1 / E2	Exchanges contents of two different words (E1 and E2).	**E1/E2:** IR HR AR TC DM LR			All models

Name Mnemonic	Symbol	Function	Operands			CPU
BCD to Binary (@)BIN(23)	BIN(23) / S / R	Converts four-digit, BCD data in source word (S) into 16-bit binary data, and outputs converted data to result word (R). S (BCD) → R (BIN): $\times 10^0 \to \times 16^0$, $\times 10^1 \to \times 16^1$, $\times 10^2 \to \times 16^2$, $\times 10^3 \to \times 16^3$	**S:** IR SR HR AR TC DM LR	**R:** IR HR AR DM LR		All models
Double BCD to Double Binary (@)BINL(58)	BINL(58) / S / R	Converts BCD value in two source words (S: starting word) into binary and outputs converted data to two result words (R: starting word). All words for any one operand must be in the same data area. S, S+1 → R, R+1	**S:** IR SR HR AR TC DM LR	**R:** IR HR AR DM LR		All models C1000H C2000H
Binary to BCD (@)BCD(24)	BCD(24) / S / R	Converts binary data in source word (S) into BCD, and outputs converted data to result word (R). S (BIN) → R (BCD): $\times 16^0 \to \times 10^0$, $\times 16^1 \to \times 10^1$, $\times 16^2 \to \times 10^2$, $\times 16^3 \to \times 10^3$	**S:** IR SR HR AR DM LR	**R:** IR HR AR DM LR		All models
Double Binary to Double BCD (@)BCDL(59)	BCDL(59) / S / R	Converts binary value in two source words (S: starting word) into eight digits of BCD data, and outputs converted data to two result words (R: starting result word). Both words for any one operand must be in the same data area. S, S+1 → R, R+1	**S:** IR SR HR AR DM LR	**R:** IR HR AR DM LR		All models C1000H C2000H
4 to 16 Decoder (@)MLPX(76)	MLPX(76) / S / Di / R	Converts up to four hexadecimal digits in source word (S) into decimal values from 0 to 15 and turns ON, in result word(s) (R), bit(s) whose position corresponds to converted value. Digits to be converted designated by Di (rightmost digit: indicates the first digit; next digit to left: gives the number of digits minus 1). 15 — 00 S □□□□ 0 to F R	**S:** IR SR HR AR TC DM LR	**Di:** IR HR AR TC DM LR #	**R:** IR HR AR DM LR	All models
8 to 256 Decoder (@)MLPX(76)		Can also convert up to eight hexadecimal digits and turn ON corresponding bits in result words R to R+15.				HS-CPU01 HS-CPU03

Instruction Set

Name Mnemonic	Symbol	Function	Operands			CPU
16 to 4 Encoder (@)DMPX(77)	DMPX(77) S R Di	Determines position of highest ON bit in source word(s) (starting word: S) and turns ON corresponding bit(s) in result word (R). Digits to receive converted value are designated by Di (rightmost digit: indicates the first digit; next digit to left: gives number of words to be converted minus 1).	**S:** IR SR HR AR TC DM LR	**R:** IR HR AR DM LR	**Di:** IR HR AR TC DM LR #	All models
256 to 8 Encoder (@)DMPX(77)		S ┌15────────00┐ R └──────┘ 0 to F Can also determine the position of highest ON bit in one or two groups of 16 words and write the ON bits position (00 to FF) to byte(s) in R.				HS-CPU01 HS-CPU03
7-Segment Decoder (@)SDEC(78)	SDEC(78) S Di D	Converts hexadecimal values from source word (S) to data for seven-segment display. Results placed in consecutive half words starting at the first destination word (D). Di designates digit and destination details (rightmost digit: gives the first digit to be converted; next digit to the left: number of digits to be converted minus 1; next digit: 1 = transfer first digit to left half of first destination word, 0 = transfer to right half). S [┌ │ ┐] D [┌──┐] 0 to F	**S:** IR SR HR AR TC DM LR	**Di:** IR HR AR TC DM LR #	**D:** IR HR AR DM LR	All models
ASCII Code Convert (@)ASC(86)	ASC(86) S Di D	Converts hexadecimal values from source word (S) to eight-bit ASCII code starting at leftmost or rightmost half of starting word (D). Rightmost digit of Di designates first source digit; the next digit to the left, the number of digits; the next digit, the rightmost (1) or leftmost (0) half of the first destination word; and the leftmost digit even (1) or odd (0) parity. S [3 │ 2 │ 1 │ 0] 0 to F D [┌──┐] 8-bit data	**S:** IR SR HR AR TC DM LR	**Di:** IR HR TC DM LR #	**D:** IR HR DM LR	All models
ASCII to Hex (@)HEX(*)	HEX (−) S Di D	Converts ASCII data to hexadecimal data.	**S:** IR SR HR AR TC DM LR	**Di:** IR SR HR AR TC DM LR	**D:** IR SR HR AR TC DM LR	HS-CPU01 HS-CPU03
Hours-to-Seconds (@)HTS(65)	HTS(65) S R —	Converts a time given in hours/minutes/seconds (S and S+1) to an equivalent time in seconds only (R and R+1). S and S+1 must be BCD and within one data area. R and R+1 must also be within one data area.	**S:** IR SR HR AR LR TC DM	**R:** IR SR HR AR LR TC DM	**—:** Not used	CPU21 CPU23 CPU31 HS-CPU01 HS-CPU03

Instruction Set

**Data Conversion
Instructions**

Name Mnemonic	Symbol	Function	Operands			CPU
Seconds-to- Hours (@)STH(66)	STH(66) S R —	Converts a time given in seconds (S and S+1) to an equivalent time in hours/minutes/seconds (R and R+1). S and S+1 must be BCD between 0 and 35,999,999, and within the same data area. R and R+1 must also be within one data area.	**S:** IR SR HR AR LR TC DM	**R:** IR SR HR AR LR TC DM	**—:** Not used	CPU21 CPU23 CPU31 HS-CPU01 HS-CPU03
2's Comple- ment (@)NEG(*)	NEG(−) S R	Converts the four digit hexadecimal content of the source word to its 2's complement and outputs the result to R	**S:** IR SR AR DM HR TC LR #	**R:** IR SR AR DM HR LR		HS-CPU01 HS-CPU03
Double 2's Complement (@)NEGL(*)	NEGL(−) S R —	Converts the eight-digit hexadecimal content of the source word to its 2's complement and outputs the result to R.	**S:** IR SR AR DM HR TC LR	**R:** IR SR AR DM HR LR		HS-CPU01 HS-CPU03
7-Segment Display Output 7SEG(*)	7SEG(−) S O C	Converts 4- or 8- BCD data to 7-segment display format and then outputs the converted data.	**S:** IR SR AR DM HR TC LR	**O:** IR SR AR HR LR	**C:** 000 to 007	HS-CPU01 HS-CPU03

Instruction Set

Name Mnemonic	Symbol	Function	Operands		CPU
Increment (@)INC(38)	INC(38) Wd	Increments four-digit BCD word (Wd) by one, without affecting carry (CY).	**Wd:** IR HR AR DM LR		All models
Decrement (@)DEC(39)	DEC(39) Wd	Decrements four-digit BCD word by 1, without affecting carry (CY).	**Wd:** IR HR AR DM LR		All models
Set Carry (@)STC(40)	STC(40)	Sets carry flag (i.e., turns CY ON).	None		All models
Clear Carry (@)CLC(41)	CLC(41)	CLC clears carry flag (i.e, turns CY OFF).	None		All models
BCD Add (@)ADD(30)	ADD(30) Au Ad R	Adds two four-digit BCD values (Au and Ad) and content of CY, and outputs result to specified result word (R). $\boxed{Au} + \boxed{Ad} + \boxed{CY} \rightarrow \boxed{CY}\ \boxed{R}$	**Au/Ad:** IR SR HR AR TC DM LR #	**R:** IR HR AR DM LR	All models
Double BCD Add (@)ADDL(54)	ADDL(54) Au Ad R	Adds two eight-digit values (2 words each) and content of CY, and outputs result to specified result words. All words for any one operand must be in the same data area. $\boxed{Au+1}\ \boxed{Au} + \boxed{Ad+1}\ \boxed{Ad} + \boxed{CY}$ $\rightarrow \boxed{CY}\ \boxed{R+1}\ \boxed{R}$	**Au/Ad:** IR SR HR AR TC DM LR	**R:** IR HR AR DM LR	All models
BCD Subtract (@)SUB(31)	SUB(31) Mi Su R	Subtracts both four-digit BCD subtrahend (Su) and content of CY from four-digit BCD minuend (Mi) and outputs result to specified result word (R). $\boxed{Mi} - \boxed{Su} - \boxed{CY} \rightarrow \boxed{CY}\ \boxed{R}$	**Mi/Su:** IR SR HR AR TC DM LR #	**R:** IR HR AR DM LR	All models
Double BCD Subtract (@)SUBL(55)	SUBL(55) Mi Su R	Subtracts both eight-digit BCD subtrahend and content of CY from eight-digit BCD minuend and outputs result to specified result words. All words for any one operand must be in the same data area. $\boxed{Mi+1}\ \boxed{Mi} - \boxed{Su+1}\ \boxed{Su} - \boxed{CY}$ $\rightarrow \boxed{CY}\ \boxed{R+1}\ \boxed{R}$	**Mi/Su:** IR SR HR AR TC DM LR	**R:** IR HR AR DM LR	All models
BCD Multiply (@)MUL(32)	MUL(32) Md Mr R	Multiplies four-digit BCD multiplicand (Md) and four-digit BCD multiplier (Mr) and outputs result to specified result words (R and R + 1). R and R + 1 must be in the same data area. $\boxed{Md} \times \boxed{Mr} \rightarrow \boxed{R+1}\ \boxed{R}$	**Md/Mr:** IR SR HR AR TC DM LR #	**R:** IR HR AR DM LR	All models

Instruction Set

Name Mnemonic	Symbol	Function	Operands		CPU
Double BCD Multiply (@)MULL(56)	MULL(56) Md Mr R Md+1 | Md × Mr+1 | Mr → R + 3 | R + 2 | R + 1 | R	Multiplies eight-digit BCD multiplicand and eight-digit BCD multiplier and outputs result to specified result words. All words for any one operand must be in the same data area.	**Md/Mr:** IR SR HR AR TC DM LR	**R:** IR HR AR DM LR	All models
BCD Divide (@)DIV(33)	DIV(33) Dd Dr R Dd ÷ Dr → R + 1 | R	Divides four-digit BCD dividend (Dd) by four-digit BCD divisor (Dr) and outputs result to specified result words. R receives quotient; R + 1 receives remainder. R and R + 1 must be in the same data area.	**Dd/Dr:** IR SR HR AR TC DM LR #	**R:** IR HR AR DM LR	All models
Double BCD Divide (@)DIVL(57)	DIVL(57) Dd Dr R Dd+1 | Dd ÷ Dr+1 | Dr → Quotient R + 1 | R Remainder R + 3 | R + 2	Divides eight-digit BCD dividend by eight-digit BCD divisor and outputs result to specified result words. All words for any one operand must be in the same data area.	**Dd/Dr:** IR SR HR AR TC DM LR	**R:** IR HR AR DM LR	All models
Floating Point Divide (@)FDIV(79)	FDIV(79) Dd Dr R Dd+1 | Dd ÷ Dr+1 | Dr → R + 1 | R	Divides one floating point value by another and outputs floating point result. Rightmost seven digits of each set of two words (eight digits) are used for mantissa, and leftmost digit used for the exponent and its sign.	**Dd/Dr:** IR SR HR AR TC DM LR	**R:** IR HR AR DM LR	All models
Square Root (@)ROOT(72)	ROOT(72) Sq R √ Sq+1 | Sq R	Computes square root of eight-digit BCD value (Sq and Sq + 1) and outputs truncated four-digit integer result to specified result word (R). Sq and Sq + 1 must be in the same data area.	**Sq:** IR SR HR AR TC DM LR	**R:** IR HR AR DM LR	All models

Instruction Set

Name Mnemonic	Symbol	Function	Operands		CPU
Binary Add (@)ADB(50)	ADB(50) Au Ad R	Adds four-digit augend (Au), four-digit addend (Ad), and content of CY and outputs result to specified result word (R). $\boxed{Au} + \boxed{Ad} + \boxed{CY} \rightarrow \boxed{CY}\ \boxed{R}$	**Au/Ad:** IR SR HR AR TC DM LR #	**R:** IR HR AR DM LR	All models
Binary Subtract (@)SBB(51)	SBB(51) Mi Su R	Subtracts four-digit hexadecimal subtrahend (Su) and content of carry from four-digit hexadecimal minuend (Mi) and outputs result to specified result word (R). $\boxed{Mi} - \boxed{Su} - \boxed{CY} \rightarrow \boxed{CY}\ \boxed{R}$	**Mi/Su:** IR SR HR AR TC DM LR #	**R:** IR HR AR DM LR	All models
Binary Multiply (@)MLB(52)	MLB(52) Md Mr R	Multiplies four-digit hexadecimal multiplicand (Md) by four-digit multiplier (Mr) and outputs eight-digit hexadecimal result to specified result words (R and R + 1). R and R + 1 must be in the same data area. $\boxed{Md} \times \boxed{Mr} \rightarrow \boxed{R + 1}\ \boxed{R}$	**Md/Mr:** IR SR HR AR TC DM LR #	**R:** IR HR AR DM LR	All models
Binary Divide (@)DVB(53)	DVB(53) Dd Dr R	Divides four-digit hexadecimal dividend (Dd) by four-digit divisor (Dr) and outputs result to designated result words (R and R + 1). R and R + 1 must be in the same data area. $\boxed{Dd} \div \boxed{Dr} \rightarrow \boxed{R + 1}\ \boxed{R}$	**Dd/Dr:** IR SR HR AR TC DM LR #	**R:** IR HR AR LR	All models
Double Binary ADD (@)ADBL(*)	ADBL(−) Au Ad R	Adds two 8-digit binary valves (normal or signed data) and outputs the result to R and R+1.	**Au:** IR SR AR DM HR LR	**Ad:** IR SR AR D M HR LR **R:** IR SR AR D M HR LR	HS-CPU01 HS-CPU03
Double Binary Subtract (@)SBBL(*)	SBBL(−) Mi Su R	Subtracts an 8-digit binary valves (normal or signed data) from another and outputs the result to R and R+1.	**Mi:** IR SR AR DM HR TC LR	**Su:** **R:** IR IR SR SR AR AR DM D HR M TC HR LR LR	HS-CPU01 HS-CPU03
Signed Binary Divide (@)DBS(*)	DBS(−) Dd Dr R	Divides one 16-bit signed binary valve by another and outputs the 32-bit signed result to R+1 and R	**Dd:** IR SR AR DM HR TC LR #	**Dr:** **R:** IR IR SR SR AR AR D D M M HR HR TC LR LR #	HS-CPU01 HS-CPU03

Instruction Set

Binary Math Instructions

Name Mnemonic	Symbol	Function	Operands			CPU
Double Signed Binary Divide (@)DBSL(*)	DBSL(−) / Dd / Dr / R	Divides one 32-bit signed binary valve by another and outputs the 64-bit signed binary result to R+3 through R	**Dd:** IR SR AR DM HR TC LR #	**Dr:** IR SR AR D M HR TC LR #	**R:** IR SR AR D M HR LR	HS-CPU01 HS-CPU03
Signed Binary Multiply (@)MBS(*)	MBS(−) / Md / Mr / R	Multiplies the signed binary content of two words and outputs the 8-digit signed binary result to R+1 and R	**Md:** IR SR AR DM HR TC LR #	**Mr:** IR SR AR DM HR TC LR #	**R:** IR SR AR D M HR LR	HS-CPU01 HS-CPU03
Double Signed Binary Multiply (@)MBSL(*)	MBSL(-) / Md / Mr / R	Multiplies two 32-bit (8-digit) signed binary values and outputs the 16-digit signed result to R+3 through R.	**Md:** IR SR AR DM HR TC LR	**Mr:** IR SR AR DM HR TC LR	**R:** IR SR AR D M HR LR	HS-CPU01 HS-CPU03

Instruction Set

Name Mnemonic	Symbol	Function	Operands	CPU
Shift Register SFT(10)	I — / P — SFT(10) / St / R — E	Creates a bit shift register from the starting word (St) through the ending word (E). I: input bit; P: shift pulse; R: reset input. St must be less than or equal to E and St and E must be in the same data area. 15 00 15 00 [E] - - - [St] ◄ In	**St/E:** IR HR AR LR	All models
Reversible Shift Register (@)SFTR(84)	SFTR(84) / C / St / E	Shifts data in specified word or series of words to either left or right. Starting (St) and ending words (E) must be specified. Control word (C) contains shift direction, reset input, and data input. St and E must be in the same data area and St must be less than or equal to E. 15 00 15 00 [E] - - - [St] ◄ In [CY] In ► [E] - - - [St] [CY]	**St/E/C:** IR HR AR TC DM LR	All models
Arithmetic Shift Left (@)ASL(25)	ASL(25) / Wd	Shifts each bit in single word (Wd) of data one bit to left, with CY. 15 00 [CY] ◄ [Wd] ◄ 0	**Wd:** IR HR AR DM LR	All models
Arithmetic Shift Right (@)ASR(26)	ASR(26) / Wd	Shifts each bit in single word (Wd) of data one bit to right, with CY. 15 00 0 ► [Wd] ► [CY]	**Wd:** IR HR AR DM LR	All models
Rotate Left (@)ROL(27)	ROL(27) / Wd	Rotates bits in single word (Wd) of data one bit to left, with carry (CY). 15 00 [Wd] ◄ [CY]	**Wd:** IR HR AR DM LR	All models
Rotate Right (@)ROR(28)	ROR(28) / Wd	Rotates bits in single word (Wd) of data one bit to right, with carry (CY). 15 00 [CY] ◄ [Wd]	**Wd:** IR HR AR DM LR	All models
One Digit Shift Left (@)SLD(74)	SLD(74) / St / E	Left shifts data between starting (St) and ending (E) words by one digit (four bits). St and E must be in the same data area. St [ ] ◄ 0 St + 1 [ ] ⋮ E [ ]	**St/E:** IR HR AR DM LR	All models

Instruction Set

Data Shift Instructions

Name Mnemonic	Symbol	Function	Operands		CPU
One Digit Shift Right (@)SRD(75)	SRD(75) E St	Right shifts data between starting (St) and ending (E) words by one digit (four bits). St and E must be in the same data area.	**St/E:** IR HR AR DM LR		All models
Word Shift (@)WSFT(16)	WSFT(16) St E	Left shifts data between starting (St) and ending (E) words in word units, writing zeros into starting word. St must be less than or equal to E and St and E must be in the same data area.	**St/E:** IR HR AR DM LR		All models
Reversible Word Shift (@)RWS(17)	RWS(17) C St E	Creates and controls a reversible asynchronous word shift register between St and E. This register only shifts words when the next word in the register is zero, e.g., if no words in the register contain zero, nothing is shifted. Also, only one word is shifted for each word in the register that contains zero. When the contents of a word are shifted to the next word, the original word's contents are set to zero. In essence, when the register is shifted, each zero word in the register trades places with the next word. The shift direction (i.e. whether the next word is the next higher or the next lower word) is designated in C. C is also used to reset the register. All of any portion of the register can be reset by designating the desired portion with St and E.	**C:** IR HR AR DM LR #	**St/E:** IR HR AR DM LR	CPU21 CPU23 CPU31 HS-CPU01 HS-CPU03

Logic Instructions

Instruction Set

Name Mnemonic	Symbol	Function	Operands		CPU
Complement (@)COM(29)	COM(29) Wd	Inverts bit status of one word (Wd) of data. Wd → W̄d	**Wd:** IR HR AR DM LR		All models
Logical AND (@)ANDW(34)	ANDW(34) I1 I2 R	Logically ANDs two 16-bit input words (I1 and I2) and sets corresponding bit in result word (R) if corresponding bits in input words are both are ON.	**I1/I2:** IR SR HR AR TC DM LR #	**R:** IR HR AR DM LR	All models
Logical OR (@)ORW(35)	ORW(35) I1 I2 R	Logically ORs two 16-bit input words (I1 and I2) and sets corresponding bit in result word (R) if one or both of corresponding bits in input data are ON.	**I1/I2:** IR SR HR AR TC DM LR #	**R:** IR HR AR DM LR	All models
Exclusive OR (@)XORW(36)	XORW(36) I1 I2 R	Exclusively ORs two 16-bit input words (I1 and I2) and sets bit in result (R) word when corresponding bits in input words differ in status.	**I1/I2:** IR SR HR AR TC DM LR #	**R:** IR HR AR DM LR	All models
Exclusive NOR (@)XNRW(37)	XNRW(37) I1 I2 R	Exclusively NORs two 16-bit input words (I1 and I2) and sets bit in result word (R) when corresponding bits in input words are same in status.	**I1/I2:** IR SR HR AR TC DM LR #	**R:** IR HR AR DM LR	All models

Instruction Set

**Subroutine, Program
Step Instructions**

Name Mnemonic	Symbol	Function	Operands			CPU
Subroutine De-fine SBS(91)	──┤ SBS(91) N │	Calls and executes subroutine N.	**N:** 00 to 99			All models
Subroutine Entry SBN(92)	──┤ SBN(92) N │	Marks start of subroutine N.	**N:** 00 to 99			All models
Subroutine Re-turn RET(93)	──┤ RET(93) │	Marks the end of a subroutine and returns control to main program.	None			All models
Macro (@)MCRO(99)	──┤ XNRW(37) I1 I2 R │	Calls and executes a subroutine replacing I/O words but keeping the same subroutine structure.	**N:** 00 to 99	**I1:** IR SR AR DM HR TC LR	**O1:** IR SR AR DM HR LR	HS-CPU01 HS-CPU03

Name Mnemonic	Symbol	Function	Operands	CPU
Step Define STEP(08)	──┤ STEP(08) B │	When used with a control bit (B), defines the start of a new step and resets the previous step. When used without N, defines the end of step execution.	**B:** IR HR AR LR	All models
Step Start SNXT(09)	──┤ SNXT(09) B │	Used with a control bit (B) to indicate the end of the step, reset the step, and start the next step.	**B:** IR HR AR LR	All models

Instruction Set

Name Mnemonic	Symbol	Function	Operands			CPU
Failure Alarm and Reset FAL(06)	── FAL(06) N	FAL is displayed on a programming device. When N is 01 to 99, an error that will not stop the CPU is indicated by outputting N (the FAL number) to the FAL output area. If N is 00, any data in the FAL output area is cleared and any other FAL number recorded in memory replaces it. The same FAL numbers are used for both FAL(06) and FALS(07).	**N:** 00 to 99			All models
Severe Failure Alarm FALS(07)	── FALS(07) N	An error is indicated by outputting N to the FAL output area and the CPU is stopped. The same FAL numbers are used for both FAL(06) and FALS(07).	**N:** 01 to 99			All models
Message Display (@)MSG(46)	── MSG(46) / FM	Displays on the Programming Console, GPC, or FIT 8 words of ASCII code starting from FM. All 8 words must be in the same data area. FM [A][B] [C][D] FM+7 [O][P] [ABCD........OP]	**FM:** IR HR AR TC DM LR #			All models
Bit Counter BCNT(67)	── BCNT(67) / N / SB / R	Counts number of ON bits in one or more words (SB: source beginning word) and outputs result to specified word (R). N: number of words to be counted . All words to be counted must be in the same data area.	**N:** IR SR HR AR TC DM LR	**SB:** IR SR HR AR TC DM LR	**R:** IR HR AR TC DM LR	All models HS-CPU01 HS-CPU03
Interrupt Control (@)FUN(89)	── FUN(89) / CC / N / D	Controls interrupts. CC: control code (defines the process); N: Interrupt Module unit number (004: scheduled interrupt); D: control data.	**CC:** 000 to 002	**N:** 000 to 004	**D:** IR HR AR TC DM LR #	All models HS-CPU01 HS-CPU03
Watchdog Timer Refresh (@)WDT(94)	── WDT(94) T	Sets the maximum and minimum limits for the watchdog timer (normally 0 to 130 ms). New limits: Maximum time = 130 + (100 x T) Minimum time = 130 + (100 x (T−1))	**T:** 0 to 63			All models
I/O Refresh (@)IORF(97)	── IORF(97) / St / E	Refreshes all I/O words between the start (St) and end (E) words. Only I/O words may be designated. Normally these words are refreshed only once per scan, but refreshing words before use in an instruction can increase execution speed. St must be less than or equal to E.	**St/E:** IR			All models
Scan Time (@)SCAN(18)	── SCAN(18) / Mi / — / —	Sets the minimum scan time, Mi, in tenths of milliseconds. The possible setting range is from 0 to 999.0 ms. If the actual scan time is less than the time set using SCAN(18), the CPU will wait until the designated time has elapsed before starting the next scan.	**Mi:** IR SR HR AR LR TC DM #	**—:** Not used		CPU21 CPU23 CPU31 HS-CPU01 HS-CPU03

Name Mnemonic	Symbol	Function	Operands			CPU
Long Message (@)LMSG(47)	LMSG(47) S D —	Outputs a 32-character message to either a Programming Console, or a device connected via the RS-232C interface. The output message must be in ASCII beginning at address S. The destination of the message is designated in D: 000 specifies that the message is to be output to the GPC; 001 specifies the RS-232C interface, starting with the leftmost byte; and 002 specifies the RS-232C interface, starting from the rightmost byte.	**S:** IR HR AR LR TC DM	**D:** #000 #001 #002	**—:** Not used.	CPU21 CPU23 CPU31 HS-CPU01 HS-CPU03
Terminal Mode (@)TERM(48)	TERM(48) — — —	When the execution condition is ON, the Programming Console operation mode is changed to TERMINAL mode. There is no program command available to change the mode back to CONSOLE mode. Pressing the CHNG key on the Programming Console manually toggles between the two modes.	**—:** Not used			CPU21 CPU23 CPU31 HS-CPU01 HS-CPU03
Set System (@)SYS(49)	SYS(49) P — —	SYS(49) must be programmed at program address 00001 with LD AR 1001 at program address 00000. The leftmost 8 bits of P must contain A3. The status of bits 00, 01, 06, and 07 are used to control the 4 operating parameters described below. If bit 00 of P is ON, the battery check will be excluded from system error checks. If bit 01 of P is ON, the PC will enter MONITOR mode at startup, unless a Programming Console connected to the CPU is not set to MONITOR. If bit 06 of P is ON, the Force Status Hold Bit (SR 25211) will be effective at startup. If bit 07 of P is ON, the I/O Status Hold Bit (SR 25212) will be effective at startup.	**P:** #	**—:** Not used		CPU21 CPU23 CPU31
High-density I/O Refresh MPRF(61)	MPRF(61) St E	Refreshes I/O words allocated to the IR area Group 2 High-density I/O Modules with I/O numbers St through E.This will be in addition to the normal I/O refresh performed during the CPU's scan. MPRF(61) can be used to refresh the I/O words allocated to the IR area High-density I/O Modules (IR 030 to IR 049) only. Normally these words are refreshed only once per scan, but refreshing words before use in an instruction can increase execution speed. St must be less than or equal to E.	**St/E:** # (0000 to 0009)			CPU21 CPU23 CPU31 HS-CPU01 HS-CPU03
Trace Memory Sample TRSM(45)	TRSM(45)	Used in the program to mark locations where specified data is to be stored in TRACE memory. Used with LSS only.	Not used			HS-CPU01 HS-CPU03
Average Valve AVG(*)	AVG(−) S N D	Adds the specified number of hexadecimal words and computes the mean value. Rounds off the 4-digit past the decimal point.	**S:** IR SR AR DM HR TC LR	**N:** IR SR AR DM HR TC LR #	**D:** IR SR AR DM HR LR	HS-CPU01 HS-CPU03

Instruction Set

Name Mnemonic	Symbol	Function	Operands			CPU
Digital Switch Input DSW(*)	DSW(−) / IW / OW / R	Inputs 4- or 8- digit BCD data from a digital switch.	**IW:** IR SR AR HR LR	**OW:** IR SR AR HR LR	**R:** IR SR AR DM HR LR	HS-CPU01 HS-CPU03
Frame Check Sum FCS(*)	ECS(−) / C / R1 / D	Checks for errors in data transmitted by a host link command.	**C:** IR SR AR DM HR LR	**R1:** IR SR AR DM HR TC LR	**D:** IR SR AR DM HR LR	HS-CPU01 HS-CPU03
Failure Point Detect FPD(*)	FPD(−) / C / T / D	Finds errors within an instruction block. Used to monitor the time between the execution of FPD(-) and a diagnostic output. If time exceed T a non-fatal error is generated.	**C:** #	**T:** BCD IR SR AR DM HR TC LR #	**D:** IR AR DM HR LR	HS-CPU01 HS-CPU03
Find Maximum (@)MAX(*)	MAX(−) / C / R1 / D	Finds the maximum value in specified data area and outputs that value to another word.	**C:** IR SR AR DM HR LR	**R1:** IR SR AR DM HR TC LR	**D:** IR SR AR DM HR LR	HS-CPU01 HS-CPU03
Find Minimum (@)MIN(*)	MIN(−) / C / R1 / D	Finds the minimum value in specified data area and outputs that value to another word.	**C:** IR SR AR DM HR LR	**R1:** IR SR AR DM HR TC LR	**D:** IR SR AR DM HR LR	HS-CPU01 HS-CPU03
Matrix Input (@)MTR(*)	MTR(−) / IW / OW / D	Inputs data from an 8 input point X 8 output point matrix and records that data in D through D+3.	**IW:** IR SR AR HR LR	**OW:** IR SR AR HR LR	**D:** IR AR DM HR LR	HS-CPU01 HS-CPU03
PID Control (@)PID(*)	PID(−) / S / C / D	PID control is performed according to the operand and PID parameters that are present.	**S:** IR SR AR DM HR LR	**C:** IR SR DM HR LR	**D:** IR SR AR DM HR LR	HS-CPU01 HS-CPU03
Receive (@)RXD(*)	RXD(−) / D / C / N	Receive data via a communications port.	**D:** IR SR AR DM HR TC LR	**C:** IR SR AR DM HR TC LR #	**N:** IR SR AR DM HR TC LR	HS-CPU01 HS-CPU03

Instruction Set

Special Instructions

Name Mnemonic	Symbol	Function	Operands			CPU
Scaling (@)SCL(*)	SCL(–) / S / P1 / R	Performs a scaling conversion on the calculated value.	**S:** IR SR AR DM HR TC LR #	**P1:** IR SR AR DM HR TC LR	**R:** IR SR AR DM HR LR	HS-CPU01 HS-CPU03
Data Search (@)SRCH(*)	SRCH(–) / N / R1 / C	Searches the specified range of memory for the specified data. Outputs the word address(es) of words in the range that contain the data.	**N:** IR SR AR DM HR TC LR #	**R1:** IR SR AR DM HR TC LR	**C:** IR SR AR DM HR LR	HS-CPU01 HS-CPU03
Sum Calculate (@)SUM(*)	SUM(–) / C / R1 / D	Computes the sum of the contents of the words in the specified range of memory.	**N:** IR SR AR DM HR TC LR #	**R1:** IR SR AR DM HR TC LR	**D:** IR SR AR DM HR LR	HS-CPU01 HS-CPU03
Ten Key Input (@)TKY(*)	TKY(–) / IW* / D1 / D2	Inputs 8 digits of BCD data from a 10-key keypad.	**IW:** IR SR AR HR LR	**D1:** IR SR AR DM HR LR	**D1:** IR SR AR DM HR LR	HS-CPU01 HS-CPU03
Transmit (@)TXD(*)	TXD(–) / S / C / N	Sends data via a communications ports.	**S:** IR SR AR DM HR TC LR	**C:** IR SR AR DM HR TC LR #	**N:** IR SR AR DM HR TC LR #	HS-CPU01 HS-CPU03
Expansion DM Read (@)XDMR(*)	XDMR(–) / N / S / D	The contents of the designated number of words of the fixed expansion DM data are read and output to the destination word.	**N:** IR SR AR DM HR TC LR #	**S:** IR SR AR DM HR TC LR #	**D:** IR SR AR DM HR LR	HS-CPU01 HS-CPU03

Special Instructions

Instruction Set

Name Mnemonic	Symbol	Function	Operands			CPU
Value Calculate (@)VCAL(69)	VCAL(69) — C — S — D	Calculates the cosine, or sine of the given degree value, or determines the y-coordinate of the given x value in a previously established line graph. For the sine and cosine conversions, S is entered in BCD as an angle (in the range 0.0 to 90.0 degrees). When calculating the y-coordinate in a graph, S gives the address of the value of the x-coordinate. The calculated data is transferred to the destination word (D). Sine and cosine results are given in BCD. Line graph coordinate calculations (interpolation) can be in BCD or BIN. The data in the control word (C) determines which operation is performed. If C is entered as a constant with a value of 0000 or 0001, the sine or cosine, respectively, of the source data value is calculated. If C is entered as a word designation, it gives the address of the first word of the data table for the line graph. The value of the first two digits gives m−1, where m is the number of data points for which coordinates are given on the line graph. Bits 14 and 15, respectively, specify the output and input data formats (0 indicates BCD, 1 indicates binary).	**C:** IR SR HR AR LR TC DM #	**S:** IR SR HR AR LR TC DM	**D:** IR SR HR AR LR TC DM	CPU21 CPU23 CPU31
Arithmetic Process (@)APR(*)	APR(−) — C — S — D					HS-CPU01 HS-CPU03

Instruction Set

Network Instructions

Name Mnemonic	Symbol	Function	Operands		CPU
SYSMAC NET **SYSMAC LINK** **Write** (@)SEND(90)	SEND(90) S D C	Transfers data from n source words (S is the starting word) to the destination words (D is the first address) in node N of the specified network (in a SYSMAC LINK or SYSMAC NET System). The format of the control words varies depending on the type of system. In both types of systems, the first control word (C) gives the number of words to be transferred. For SYSMAC NET Systems, in word C+1, bit 14 specifies the system (0 for system 1, and 1 for system 0), and the rightmost 7 bits define the network number. The left half of word C+2 specifies the destination port (00: NSB, 01/02: NSU), and the right half specifies the destination node number. If the destination node number is set to 0, data is transmitted to all nodes. For SYSMAC LINK Systems, the right half of C+1 specifies the response monitoring time (default 00: 2 s, FF: monitoring disabled), the next digit to the left gives the maximum number of re-transmissions (0 to 15) that the PC will attempt if no response signal is received. Bit 13 specifies whether a response is needed (0) or not (1), and bit 14 specifies the system number (0 for system 1, and 1 for system 0). The right half of C+2 gives the destination node number. If this is set to 0, the data will be sent to all nodes.	**S:** IR SR HR AR TC DM LR	**D/C:** IR HR AR TC DM LR	CPU31

SYSMAC NET

C	n: no. of words to be transmitted (0 to 1000)		
C+1	0X00	0000	Network no. (0 to 127)
C+2	Destination port no.		Destination node no. (0 to 126)

SYSMAC LINK

C	n: no. of words to be transmitted, 0 to 1000		
C+1	0XX0	Re-trans missions	Response monitor time (0.1 to 25.4 s)
C+2	0000	0000	Destination node no. (0 to 62)

Source node N

S
S+1
S+n−1

→

Destination node

D
D+1
D+n−1

Network Instructions

Name Mnemonic	Symbol	Function	Operands		CPU
SYSMAC NET **SYSMAC LINK** **Read** (@)RECV(98)	RECV(98) S D C	Transfers data from the source words (S is the first word) from node N of the specified network (in a SYSMAC LINK or SYSMAC NET System) to the destination words starting at D. The format of the control words varies depending on the type of system. In both types of systems, the first control word (C) gives the number of words to be transferred. For SYSMAC NET Systems, in the second word (C+1), bit 14 specifies the system (0 for system 1, and 1 for system 0), and the rightmost 7 bits define the network number. The left half of word C+2 specifies the source port (00: NSB, 01/02: NSU), and the right half specifies the source node number. For SYSMAC LINK Systems, the right half of C+1 specifies the response monitoring time (default 00: 2 s, FF: monitoring disabled), the next digit to the left gives the maximum number of re-transmissions (0 to 15) that the PC will attempt if no response signal is received. Bit 13 specifies whether a response is needed (0) or not (1), and bit 14 specifies the system number (0 for system 1, and 1 for system 0). The right half of C+2 gives the source node number.	**S:** IR SR HR AR TC DM LR	**D/C:** IR HR AR TC DM LR	CPU31

SYSMAC NET

C	n: no. of words to be transmitted (0 to 1000)		
C+1	0X00	0000	Network no. (0 to 127)
C+2	Source port no. (NSB: 00, NSU: 01/02)		Source node no. (0 to 126)

SYSMAC LINK

C	n: no. of words to be transmitted, 0 to 1000		
C+1	0XX0	Re-trans missions	Response monitor time (0.1 to 25.4 s)
C+2	0000	0000	Source node no. (0 to 62)

Source node N

S
S+1

→

Destination node

D
D+1

S+n−1

D+n−1

APPENDIX C: PACSCI STEPPER BASIC™ VERSION 3.6 COMMAND SUMMARY

For programmable indexer/drivers
Models 5645, 5445 and 5345

ABS(x)
returns the absolute value of a number.

ACCEL.RATE
= x is the rate the motor accelerates to change speed.

ADDRESS
indicates the multi-drop address of the controller.

AND
performs a logical AND operation.

AUTO
generates program line numbers automatically.

CCW.OT
= x is the counterclockwise software overtravel step limit.

CCW.OT.JUMP
= x is the line number jumped to for counterclockwise overtravel errors.

CCW.OT.ON
= 1 enables counterclockwise overtravel checking.

CHR(x)
Coverts an ASCII code to its equivalent character for display on the terminal.

CINT(x)
converts a value to an integer by rounding off.

CLEAR
clears FLGn, FLTn, and INTn to zero.

CLR.SCAN1; CLR.SCAN2
turns Off Scan 1 and Scan 2.

CONT
causes the program to continue after STOP is encountered.

CONTINUOUS.MOTION
= 1 for continuous motion over multiple moves.

CW.OT
= x is the clockwise overtravel limit.

CW.OT.JUMP
= x is the line number jumped to for clockwise overtravel errors.

CW.OT.ON
= 1 enables clockwise overtravel checking.

DCL.TRACK.ACL
= 1 for deceleration to occur at the same rate as acceleration; = 0 for separate deceleration.

DECEL.RATE
= x is the rate the motor decelerates to change speed.

DELETE
deletes one or more lines from a program.

DIR
= 0 for clockwise rotation (viewed from shaft-end); = 1 for counterclockwise rotation.

ENABLE
= 1 allows power flow to the motor.

ENABLED
= 1 is the read-only indicator that the control is enabled.

ENC.FREQ
= x is the display of the encoder frequency in pulses per second.

ENCDR.POS
= x is the display of external encoder position or a user-entered value to redefine encoder position.

ENCODER
= x specifies the line count of the external encoder for use with electronic gearing.

END
terminates the execution of a program.

FAULTCODE
= 1 is the indicator that a fault has occurred; = 0 indicates no fault or is user entered to clear the code.

FLG1 to FLG8
= x is the user-defined flag (0 or 1) value.

FLT1 to FLT32
= x is the user-defined floating point (decimal) value.

FOR...NEXT
allows a series of statements to be executed in a loop a given number of times.

FREE
displays the number of free bytes of program memory.

GEARING
= 1 enables following the motion of a master control source; = 2 follows clockwise master inputs only; = 3 follows counterclockwise master inputs only.

GO.ABS
commands a motor move to the position specified by TARGET.POS.

GO.FUNC
= x determines the response to Start (Input 7) when RMT.START = 0; = 0 performs GO.VEL; = 1 performs GO.INCR; = 2 performs GO.ABS.

GO.HOME
commands a motor move to the electrical home position.

GO.INCR
commands a motor move an incremental index (INDEX.DIST) from the current position.

GOSUB x...RETURN
branches program execution to a subroutine, executes it, and returns.

GOTO
x causes program execution to jump to line number x.

GO.VEL
moves the motor at constant speed (RUN.SPEED).

HOME.ACTIVE
= 0 for a physical home switch that opens at home (closed otherwise); = 1 if switch closes at home (open otherwise).

HMPOS.OFFSET
= x is the offset distance from the mechanical home position.

IF...THEN
controls program execution based on the truth or falsity of an expression.

INDEX.DIST
= x is the distance the motor moves for each incremental index (GO.INCR).

INKEY( )
returns an ASCII code for the first keystroke in the terminal input buffer.

INP1 to INP16
= 1 is the read-only display for each input that is Off.

IN.POSITION
= 1 is the read-only indicator that the motor is at the commanded position.

INPUT variable
obtains information from the terminal while the program is running.

INPUTS
= x is the BCD read-only display that corresponds to the sum of the Off inputs.

INT1 to INT32
= x is the user-defined integer value.

INT(x)
converts a number into the largest integer less than or equal to the original number.

JOG.SPEED
= x is the speed the motor moves when jogging.

LIST
displays the complete or partial program on the terminal screen.

LOAD

copies saved programs stored in NVRAM into RAM.

LOADVAR

copies stored global variables from NVRAM into RAM.

MAX.DECEL

= x is the rate the motor decelerates in certain conditions.

MIN.SPEED

= x is the minimum start/stop speed for the motor.

MOVING

= 1 is the read-only indicator that the motor is moving.

NEW

clears the program from RAM and sets the value of FLGn, FLTn, and INTn to zero.

OR

performs a logical OR operation.

OT.ERROR

= 0 is the read-only indicator for no overtravel error; = 1 indicates clockwise overtravel error; = 2 indicates counterclockwise overtravel error.

OUTPUTS

= x is the BCD value sum for turning On user- defined outputs.

PACK

speeds up program execution by generating the GOTO table before the program executes.

PAUSE

pauses a program for a time specified by WAIT.TIME.

POS.CHK1; POS.CHK2; POS.CHK3

= x is the commanded position at which outputs 1, 2, and 3 are turned Off or On.

POS.CHK1.OUT; POS.CHK2.OUT; POS.CHK3.OUT

= x specifies and enables or disables an output when POS.CHKn is exceeded.

POS.COMMAND

= x is the display of the commanded position or a user-entered value to redefine motor position.

POS.VERIFY.CORRECTION

= x is the read-only indicator of the number of steps required to compensate for a motor position verification error.

POS.VERIFY.DEADBAND

= x is the maximum step limit allowed before a Position Verification Error triggers.

POS.VERIFY.ERROR

= 1 is the read-only indicator that a position error has occurred.

POS.VERIFY.JUMP

= x is the line number jumped to upon a Position Verification Error.

POS.VERIFY.TIME

= x is the settling time for the encoder reading.

PREDEF.INP10 to PREDEF.INP15

= 1 enables individual predefined inputs 10 to 15; = 0 for discrete input operation.

PREDEF.INP

= x is the BCD value sum for predefining inputs 10 to 15 as a group.

PREDEF.OUT

= 1 enables predefined Output 12 to indicate motor moving; = 0 for discrete Output 12 operation.

PRINT

displays program output on the terminal screen while the program is running.

PWR.ON.ENABLE

= 1 automatically enables the drive on power up.

PWR.ON.OUTPUTS

= x is the BCD value sum for enabling user-defined outputs on power up.

QRY

displays the parameters and current status of the control on the terminal screen.

QRY.PRM

displays the parameters of the control on the terminal screen.

QRY.STAT

displays the current status of the control on the terminal screen.

RATIO

= x is the ratio between the motor shaft and the encoder shaft for use with electronic gearing.

REG.DIST

= x is the distance moved when the Registration input triggers.

REG.ENCPOS

= x is the display of the encoder position when the Registration input triggers.

REG.FLAG

= 1 indicates that a Registration input triggered; = 0 indicates no fault or is user-entered to clear the indicator.

REG.FUNC

= 1 enables a REG.DIST move when the Registration input triggers.

REG.GOSUB

= x is the subroutine line number jumped to when a Registration input triggers.

REM

allows you to put comments in a program. These comments do not affect program execution and are removed when the program is downloaded to the drive.

RENUM

renumbers program lines.

RESET.STACK

clears the software internal stack of subroutine return addresses.

RETURN

returns program execution from a subroutine.

RMT.START

= 0 causes a move based on GO.FUNC when Start (Input 7) triggered; = 1 causes a RUN command to execute when Start triggered; = 2 automatically executes a RUN command on power up. **Note:** For RMT.START = 0 or 1, PREDEF.INP7 must equal 1.

RUN

executes all or part of a program in RAM.

RUN.SPEED

= x is the maximum motor speed.

SAVE

saves program from RAM to NVRAM memory.

SAVEVAR

saves INTn, FLTn, or a group of variables from RAM to NVRAM memory.

SEEK.HOME

moves the motor to the mechanical limit switch.

SET.SCAN1; SET.SCAN2

activates Scan 1 or Scan 2 to respond to the scan trigger.

SETUP

(lockcode 4800) initializes variables to default values. Must be used after a PROM with new software version is installed.

SK1.ENCPOS

records the position of the external encoder when the SCAN condition is satisfied.

SK1.ENCPOS; SK2.ENCPOS

= x is the read-only encoder position displayed upon a Scan 1 or Scan 2 trigger.

SK1.GOSUB; SK2.GOSUB

= x branches program execution to a subroutine upon Scan 1 or Scan 2 trigger.

SK1.JUMP; SK2.JUMP

= x is the line number jumped to upon Scan 1 or Scan 2 trigger.

SK1.OUTPUT; SK2.OUTPUT

= x specifies and enables or disables an output when a Scan 1 or Scan 2 is triggered.

SK1.PCMD

records the value of the position command when the SCAN condition is satisfied.

SK1.TRIGGER; SK2.TRIGGER

= x defines the input condition that triggers a Scan 1 or Scan 2.

/(Slash)

allows log on to a specific drive when using RS-422/485 and enables global command issuance.

STALL.DEADBAND

= x is the range allowed for a difference between commanded and measured steps. When exceeded, a stall error triggers.

STALL.ERROR

= 1 is the read-only indicator that a stall has occurred.

STALL.JUMP

= x is the line number jumped to upon a Stall Error.

STALL.STOP

= 1 to stop the motor when a stall occurs.

STEP.DIR.INPUT

= 0 configures encoder input J6 for quadrature encoder signals; = 1 configures encoder input J6 for step/direction signals.

STEPSIZE

= x is the step size in internal software.

STOP

stops motion and interrupts the program.

STOP.MOTION

stops motion while allowing continued program execution.

TARGET.POS

= x is the target position for GO.ABS.

TIME

= x is a continually running internal timer.

TRON and TROFF

enable and disable tracing of executing program lines.

UPD.MOVE

updates a move in process with new parameters.

VELOCITY

= x is the read-only display of the actual speed of the motor.

VER

displays the current version of the software.

WAIT.TIME

= x is the wait time used with PAUSE.

WHEN

executes an action when a condition is satisfied.

WHEN.ENCPOS

= x is the read-only encoder position display at the time the WHEN statement becomes true.

WHENPCMD

= x is the read-only commanded motor position display at the time the WHEN statement becomes true.

WHILE...WEND

executes a series of statements as long as the expression after the WHILE statement is true.

INDEX